Essentials of Coating, Painting, and Lining for the Oil, Gas, and Petrochemical Industries

Essentials of Coating, Painting, and Lining for the Oil, Gas, and Petrochemical Industries

Alireza Bahadori, Ph.D.

School of Environment, Science, and Engineering
Southern Cross University, Lismore, NSW, Australia

ELSEVIER

AMSTERDAM • BOSTON • HEIDELBERG • LONDON
NEW YORK • OXFORD • PARIS • SAN DIEGO
SAN FRANCISCO • SINGAPORE • SYDNEY • TOKYO

Gulf Professional Publishing is an Imprint of Elsevier

Gulf Professional Publishing is an imprint of Elsevier
225 Wyman Street, Waltham, MA 02451, USA
The Boulevard, Langford Lane, Kidlington, Oxford, OX5 1GB, UK

Notices
Knowledge and best practice in this field are constantly changing. As new research and experience broaden our understanding, changes in research methods, professional practices, or medical treatment may become necessary.

Practitioners and researchers must always rely on their own experience and knowledge in evaluating and using any information, methods, compounds, or experiments described herein. In using such information or methods they should be mindful of their own safety and the safety of others, including parties for whom they have a professional responsibility.

To the fullest extent of the law, neither the Publisher nor the authors, contributors, or editors, assume any liability for any injury and/or damage to persons or property as a matter of products liability, negligence or otherwise, or from any use or operation of any methods, products, instructions, or ideas contained in the material herein.

Library of Congress Cataloging-in-Publication Data
A catalog record for this book is available from the Library of Congress

British Library Cataloguing-in-Publication Data
A catalogue record for this book is available from the British Library

For information on all Gulf Professional Publishing
visit our website at http://store.elsevier.com/

This book has been manufactured using Print On Demand technology. Each copy is produced to order and is limited to black ink. The online version of this book will show color figures where appropriate.

ISBN: 978-0-12-801407-3

*Dedicated to the loving memory of my parents, grandparents,
and everyone who contributed so much to my work over the years*

Contents

About the Author

Alireza Bahadori, Ph.D., is a member of the research staff of the School of Environment, Science, and Engineering at Southern Cross University in Lismore, NSW, Australia. He received his Ph.D. from Curtin University in Perth, Western Australia. During the past 20 years, Dr. Bahadori has held many process and petroleum engineering positions and was involved in several large-scale projects at the National Iranian Oil Co. (NIOC), Petroleum Development Oman (PDO), and Clough AMEC.

He is the author of about 250 articles and 12 books. His books have been published by many major publishers, including Elsevier.

Dr. Bahadori is the recipient of the highly competitive and prestigious Australian government's endeavour international postgraduate research award for his research in the oil and gas areas. He also received a Top-up Award from the State Government of Western Australia through the Western Australia Energy Research Alliance (WA:ERA) in 2009. Dr. Bahadori serves as an editorial board member and reviewer for a large number of journals.

Preface

The surface preparation process not only cleans steel, but also introduces a suitable profile for applying protective coating. The main task of such coating is to prevent or control the external corrosion of buried or submerged steel structures. The coating isolates metal from contact with the surrounding environment.

Since a perfect coating cannot be assured, cathodic protection is used in conjunction with the coating system to provide the first line of defense against corrosion. And since a properly selected and applied coating should provide 99% of the needed protection, it is of utmost importance to know the advantages and disadvantages of available coatings. The right coating material, properly used, will make all other aspects of corrosion control relatively easy. The number of coating systems available requires careful analysis of the many desired properties for an effective pipe coating.

Therefore, the optimum selection and proper application of protective coatings are vital to successful engineering. During extended periods of time, protective coatings deteriorate as a result of contact with moisture, oxygen, chemicals, fluctuating temperatures, abrasion, pressure, and many other factors. Therefore, proper and timely maintenance is required to get the optimum performance from a protective coating.

Meanwhile, selection and application of maintenance coating is more complicated than for initial construction. Climatic conditions, chemical exposure, available time, budget, health and safety, and grade of surface preparation have a significant influence on the planning of optimum design coating. To select the best coating system to fit the environment or oil condition, knowledge of operating and installation conditions is essential. The steel source and job location may limit the coatings available for each project. The choice of a high-quality applicator is the most important (but alas, also most neglected) consideration. Following coating and applicator selection, inspection at the coating, especially on the job site during construction, will go far in ensuring that a high-quality pipe coating system has been installed.

This engineering book covers the minimum requirements for the design and selection of coating systems for the external protection of pipes, storage tanks, and piling systems that will be buried or submerged in water. The contents define the essential elements of surface preparation, selection of coating systems, and repair of coating defects. It is intended to focus on corrosion protection of steel structures of the oil and gas and petrochemical industries, including refineries, chemical and petrochemical plants, gas plants, and oil exploration and production units.

This book covers the minimum requirements for the linings that will be used in various types of equipment (e.g., vessels, storage tanks, and pipelines). It is intended for use in refineries, petrochemical plants, oil and gas plants, and, where applicable, in exploration, production, and new ventures.

Several nonmetallic lining systems are discussed. They are classified into the following main groups:

Thermoplastic materials
Thermosetting materials
Rubbers
Mineral and bitumen materials

This book gives the essential details about the initial construction and maintenance painting of metal surfaces. It also describes the minimum requirements for the surface preparation and painting of piping, plant, equipment, storage tank, building, and other elements that will be exposed to different corrosive environments. Painting schedules, paint systems, and paint color schedules are included in this respect.

Alireza Bahadori
School of Environment, Science, and Engineering
Southern Cross University, Lismore, NSW, Australia
1 November 2014

Acknowledgments

I would like to thank the Elsevier editorial and production teams, Katie Hammon and Kattie Washington of Gulf Professional Publishing for their editorial assistance.

SURFACE PREPARATION FOR COATING, PAINTING, AND LINING

1.1 INTRODUCTION

The term *surface preparation* means the methods of treating the surface of substrate prior to application of coating (painting, coating and lining, etc.). Surface preparation is the essential first-stage treatment of a steel substrate before the application of any coating. It is generally accepted as being the most important factor in the complete success of a corrosion protection system.

The performance of a coating is significantly influenced by its ability to adhere properly to the substrate material. For steel surfaces, residual mill scale is an unsatisfactory base to apply modern, high-performance protective coatings; therefore, it needs to be removed by abrasive blast cleaning. Other surface contaminants on the rolled steel surface, such as oil and grease, are also undesirable and must be removed before the blast cleaning process. Typical contaminants that should be removed during surface preparation include moisture, oil, grease, corrosion products, dirt, and mill scale.

The surface being prepared must achieve a level of cleanliness and roughness suitable for the proposed coating and permit good adhesion of the coating. The money and effort spent on the preparation should be in a reasonable proportion to the purpose and nature of the coating. The contractor performing surface preparation must have the personnel and technical know-how to operate in a technically satisfactory and operationally reliable manner.

The surfaces must be accessible and adequately illuminated. The relevant accident prevention regulations and safety provisions must be observed. All surface preparation work must be properly quality-controlled and inspected. Each subsequent coating may be applied only when the surface to be coated has been prepared in accordance with widely accepted standards.

This chapter gives the minimum requirements for surface preparation of substrates prior to protecting against corrosion, both for initial construction and for maintenance. It includes the minimum requirements for surface preparation of ferrous metals, nonferrous metals and nonmetallic surfaces (e.g., masonry materials and wood). Applicable methods of surface preparation, including degreasing, pickling, manual cleaning, flame cleaning, and blasting, are discussed in this chapter. Recommendations are made regarding the selection of appropriate methods of surface preparation and the coating (including metallic coating and electroplating) to be applied.

The handling of parts or assemblies after cleaning should be kept to a minimum. When handling is necessary, clean gloves or similar protection should be used. Canvas, polyvinyl chloride (PVC), and leather are suitable materials for gloves. Cleaned surfaces should be coated as soon after cleaning as is practical, and before detrimental corrosion or recontamination occurs.

Essentials of Coating, Painting, and Lining. http://dx.doi.org/10.1016/B978-0-12-801407-3.00001-8

There is no single, universal method of cleaning by which all surfaces can be prepared for the application of protective coatings. The cleaning method for any given type of article must be carefully selected and properly carried out.

All materials furnished should be of the specified quality. The entire operation of surface preparation should be performed by experienced workers skilled in the cleaning of surfaces and overseen by qualified supervisors.

1.2 SELECTION OF CLEANING METHOD

The cleaning method should be selected with the following considerations in mind. The choice between blast-cleaning, acid-pickling, flame-cleaning, and manual cleaning is partly determined by the nature of the coating to be applied. It should be understood, however, that coating applied to a properly prepared (e.g., blast-cleaned) surface will always last longer than similar coating applied to flame-cleaned or manually cleaned surfaces.

1.2.1 INITIAL CONDITION OF SURFACE (RUST GRADE)

The initial condition of surfaces for preparation, which, among other factors, determines the choice and mode of execution of the preparation measures and the relevant reference sample to be used, must be determined.

1.2.2 NEW CONSTRUCTION (UNCOATED SURFACES)

Grade of steel, special treatments or methods have an effect on the preparation (e.g., use of cold rolling or deep drawing methods).

Figure 1.1 shows samples of the rust grades of uncoated surfaces.

1.2.3 MAINTENANCE (COATED SURFACE)

Rust level of coated surfaces according to standards such as DIN 53210 and ASTM D 610 (see Figure 1.2);

1. Type of coating (e.g., type of binder and pigment, metal coating), approximate coat thickness, and date when carried out
2. Extent of blistering according to DIN 53209 and ASTM D 714
3. Additional information—e.g., on adhesion, cracking, chemical and other contaminants, and other significant phenomena

1.3 CLEANLINESS OF THE SURFACES
1.3.1 REMOVAL OF CONTAMINANTS/COATS OF MATERIALS OTHER THAN METAL

This stage of the cleaning process includes removal of dirt, dust, soot, ash, concrete, coal slag, sand, moisture, water, acids, alkalis, soap, salts, encrustations, growths, fluxes, oily and greasy contaminants,

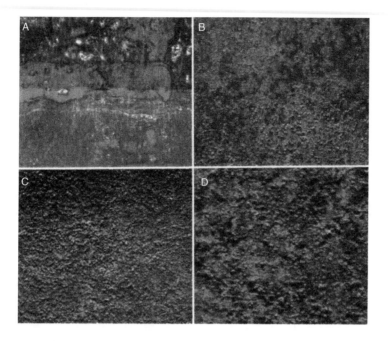

FIGURE 1.1

Examples of rust grades of uncoated surfaces. Rust levels: **(A)** Steel surface covered with firmly adhesive scale and largely free of rust; **(B)** steel surface with the beginning of rust attack; **(C)** steel surface from which scale has been rusted away or can be scraped off, but which exhibits only a few rust pits visible to the eye; **(D)** steel surface from which the scale has been rusted away and exhibits numerous visible rust pits.

and coatings and cementations that are loose, rusted, or unusable as an adhesion surface and corrosion products of metallic coatings. You should use either the cleaning method specified by the standards or one of the methods described in this book.

1.3.2 REMOVAL OF COATED MATERIAL RELATED TO THE METAL (SCALE AND RUST)

Removal of firming adherent scale is only possible with the following methods:

1. Blasting
2. Flame-cleaning
3. Pickling
4. Manual cleaning

With each of these methods, only specific surface conditions can be produced and particular levels of cleanliness achieved. Correspondingly, the appearance of the prepared surface depends not only on the level of cleanliness, but also on the rust removal method used.

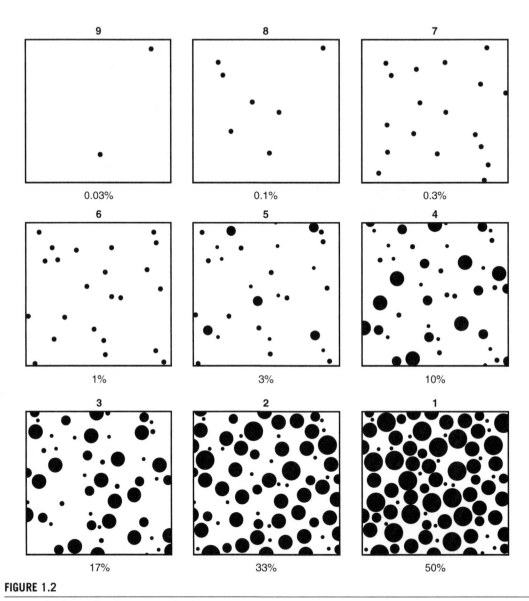

FIGURE 1.2

Rust grade of coated surfaces (example of area percentage).

1.3.3 STANDARD LEVEL OF CLEANLINESS

The standard level of cleanliness for prepared steel surfaces (as listed in Tables 1.1 and 1.2) should apply. Unless otherwise specified, the level of cleanliness of uncoated surfaces should be in accordance

Table 1.1   Standard Levels of Cleanliness for Prepared Steel Surfaces

Standard Level of Cleanliness	Rust Removal Method	Initial Condition of Steel Surface		Reference Sample Photographs	Essential Characteristics of The Prepared Steel Surface	Remarks
		Uncoated	Coated			
Sa 1		B C D	See section 13	B Sa 1 C Sa 2 D Sa 1	Only loose scale, rust and loose coatings are removed	These standard levels of cleanliness apply to blasting
Sa 2		B C D		B Sa 2 C Sa 2 D Sa 2	Virtually all scale, rust and coatings are removed; i.e, only as much firmly adhesive scale rust and coating residues remain (no continuous coats) as to correspond with the overall impression of the reference sample photographs [see explanations)	a. Of uncoated steel surface b. Of coated steel surfaces if the coatings are also removed sufficiently to obtain the required level of cleanliness
Sa 2½	Blasting	A B C D		A Sa 2½ B Sa 2½ C Sa 2½ D Sa 2½	Scale, rust, and coatings are removed to the extent that the residues on the steel surface remain merely as faint shading as a result of coloring of the pores	
Sa 3		A B C D		A Sa 3 B Sa 3 C Sa 3 D Sa 3	Scale, rust and coatings are completely removed (when observed without magnification)	
St 2	Manual or mechanical rust removal	B C D		B St 2 C St 2 D St 2	Loose coatings and loose scale are removed; rust is removed to the extent that the steel surface after subsequent cleaning exhibition a faint luster from the metal	

(Continued)

Table 1.1   Standard Levels of Cleanliness for Prepared Steel Surfaces *(cont.)*

Standard Level of Clean-Liness	Rust Removal Method	Initial Condition of Steel Surface		Reference Sample Photographs	Essential Characteristics of The Prepared Steel Surface	Remarks
		Uncoated	Coated			
St 3		B C D		B St 3 C St 3 D St 3	Loose coatings and loocs scale are removed; rust is removed to the extent that the steel surface after subsequent cleaning exhibits a distinct luster from the metal. In special cases removal of firmly adhesive coatings; eg., by grinding, scraping or by using pickling agents is possible if required, this should be additionally agreed	Normally requires machining
Fl	Flame cleaning according to section 9	A B C D		A Fl B Fl C Fl D Fl	Coatings, scale and rust are removed to the extent that residues on the steel surface remain merely as shades in various hues	Subsequent thorough mechanical brushing is always necessary
Be	Pickling according to section 6	A B C D			Coating residues, scale A and rust are completely removed	Coatings must be removed in a suitable manner before pickling

with blasting and manual cleaning and with the provisions specified in Table 1.1 for flame-cleaning and pickling.

- Blasting
  During blasting, parts that are not to be worked on, parts already coated, and the environment should be protected against the blasting abrasives. With chemically contaminated surfaces, prewashing may be advisable, and with coarse coating of rust on plates, preliminary derusting with impact tools may be advisable.

Table 1.2   Comparison of Standard Levels of Cleanliness According to SIS 055900 with Other Rust Removal Levels or Quality Classes

Standard Level of Cleanliness	Rust Removal Method	SIS 055900	DIN 18364	BS 4232 (only for blasting)	SSPC-VIS
Sa 1	Blasting manual or mechanical rust removal	Sa 1	-	-	Brush off SP 7
Sa 2		Sa 2		Third quality	Commercial SP 6
Sa 2½		Sa 2½	Rust removal level 2	Second quality	Near white SP 10
Sa 3		Sa 3	Rust removal level 3	First quality	White metal SP 5
St 2		St 2	Essentially less than rust removal level 1	-	Hand tool cleaning SP 2
St 3		St 3	Less than rust removal level 1	-	Power tool cleaning SP 3
Fl	Flame cleaning	-	Rust removal level 2	-	Flame tool cleaning SP 4
Be	Pickling	-	Rust removal level 3	-	Pickling SP 8

The standard levels of cleanliness for blasted surfaces are described as follows:

- Sa1 light blast-cleaning
  When viewed without magnification (such as with a magnifying glass), the surface should be free of visible oil, grease, and dirt, and from poorly adhering mill scale rust, paint coatings, and foreign matter (see Table 1.1).
- Sa2 thorough blast-cleaning
  When viewed without magnification, the surface should be free of visible oil, grease, and dirt and of most of the mill scale, rust, paint coatings, and foreign matter. Any residual contamination should be firmly adhered.
  Staining should be limited to no more than 33% of surface area. It may consist of light shadows, slight streaks, or minor discolorations caused by stains of rust, mill scale, or previously applied paint. If the surface is pitted, slight residues of rust and paint may also be left in the bottoms of the pits
- Sa2½ Very thorough blast-cleaning
  When viewed without magnification, the surface should be free of visible oil, grease, and dirt, and of mill scale, rust paint coating, and foreign matter. Any remaining traces of contamination should show only as slight stains in the form of spots or stripes. The reference photographs are ASa2½, CSa2½, and DSa2½ (see Table 1.1). Figure 1.3 shows a blast-cleaned steel surface to the Sa 2½ standard.
- Sa3 white blast-cleaning
  When viewed without magnification, the surface should be free of visible oil, grease, and dirt and of mill scale, rust, paint coating, and foreign matter. It should have a uniform metallic color. The reference photographs are ASa3, BSa3, and CSa3 (see Table 1.1).
  For previously painted surfaces that have been prepared for renewed painting, only photographs with rust grade designations D or C (e.g., DSa2½ or CSa2½) may be used for the visual assessment.

FIGURE 1.3

Blast-cleaned steel surface to the Sa 2½ standard.

- Manual rust removal
  The standard levels of cleanliness of St2 and St3 are used for hand cleaning and power tool cleaning of the surface. The St1 grade is not included here, as it would render the surface unsuitable for painting.
- **St2:** Thorough hand and power tool cleaning
  When viewed without magnification, the surface should be free of visible oil, grease, and dirt, and of poorly adhering mill scale, rust, paint coating, and foreign matter. This grade is accepted only for spot cleaning. The reference photographs are BSt2, CSt2, and DSt2 (see Table 1.1).
- **St3:** Very thorough power tool cleaning
  The treatment is similar to that for St2, but the surface should be treated much more thoroughly, until a metallic sheen arises from the metallic substrate. The reference photographs are BSt3, CSt3, and DSt3 (see Tables 1.1 and 1.2).
- Thermal rust removal and flame cleaning
  Fl: When viewed without magnification, the surface should be free of mill scale, rust, paint, coatings, and ISO foreign matter. Any remaining residue should appear only as a discoloration of the surface. The reference photographs are AFl, BFl, CFl, and DFl (see Table 1.1).
- Chemical rust removal and pickling
  Be: When viewed without magnification, the surface should be free of coating residue, scale, and rust (see Table 1.2).

  The comparison of standard levels of cleanliness according to SIS 055900 with other rust removal levels is shown in Table 1.2.

1.3.4 RUST CONVERTERS, RUST STABILIZERS, AND PENETRATING AGENTS

This section describes the use of so-called rust converters, rust stabilizers, and similar means of chemically converting the corrosion products of the iron into stable iron compounds. This also includes penetrating agents intended to inhibit rust.

1.3.5 INFLUENCE OF ENVIRONMENTAL CONDITIONS ON THE CLEANING AND CLEANLINESS OF SURFACES

The storing of unprotected steel in an urban, industrial, or marine atmosphere (including the practice of "rusting off" scale by weathering) should be avoided because otherwise, a higher level of cleanliness may be necessary owing to the deposit of corrosive substances that will occur. Steel should be prepared and protected in as close to state A or A to B as possible (see Table 1.1).

No surfaces should be prepared for coating during rain or other precipitation. If the possibility of condensation is to be absolutely excluded, the temperature of the surface must be kept above the dew point of the surrounding air. If the work must be done in suboptimal conditions, provision should be made at the planning stage for special measures (e.g., covering, enclosing in a tent, warming the surfaces or drying the air).

Preparatory work in the area of plants subject to the risk of explosion or fire also requires special measures (e.g., low-spark or flame-free methods).

1.3.6 TESTING OF THE CLEANLINESS OF THE PREPARED SURFACES

The surfaces should be tested after subsequent cleaning. For all the standard levels of cleanliness listed in Table 1.2, visual testing is sufficient. The following checks should be made:

- Whether the prepared surfaces exhibit the essential characteristics stated for the appropriate level of cleanliness (as given in Table 1.1) and shown by way of example in the corresponding reference sample photograph
- Whether there is adequate conformity to any agreed reference surfaces
- Whether the reference sample photographs show the interrelationships between the levels of cleanliness

The comparison is made without magnification.

For the Be standard level of cleanliness (see Table 1.2), reference samples are unnecessary. If you have any doubt as to whether the surface is free of oil or grease, be sure to check for this.

1.3.7 DEGREE OF ROUGHNESS (SURFACE PROFILE)

Blast-cleaning produces a roughened surface, and the profile size is important. The surface roughness achieved for each quality of surface finish depends mainly upon the type and grade of abrasive used. Unless otherwise specified by the company, the amplitude of the surface roughness of the steel work should be between 0.1 and 0.03 mm for painting, coating, and lining. Table 1.3 gives the range of maximum and average maximum profile heights of various abrasives to be expected under normal good operation conditions (wheel and nozzle). If excessively high air pressure or wheel speed is used, the profile may be significantly higher.

1.3.8 METHODS OF MEASUREMENT

This section describes some of the suitable methods for measuring surface roughness.

- Sectioning
  In this process, a metallurgical section is prepared and the surface profile examined under a suitable microscope using a micrometer eyepiece.
- Grinding
  The thickness of the blast-cleaned specimen is measured with a flat-ended micrometer. The surface is then ground until the bottoms of the deepest pits are just visible. The thickness is then measured again.
- Direct measurement by a microscope
  The blast-cleaned specimen (or a replica) is viewed through a suitable microscope, first focusing on the peak and then focusing on the lowest adjacent trough, noting the necessary adjustment of focus.
- Profile tracing
  A blast-cleaned specimen is traversed with a diamond or sapphire stylus, and then the displacement of the stylus as it passes over the peaks and roughs is recorded. For instruments and the measurement procedures, see ISO 3274 and ISO 4288.
- Comparator disc
  The comparator disc is a field instrument used to determine the anchor pattern profile depth of a blasted surface. A comparator disc is composed of five sections, each with a different anchor pattern

Table 1.3 Typical Maximum Profiles Produced by Commercial Abrasive Media		
Abrasive	**Typical Profile Maximum (mm)**	**Height Av. Maximum (mm)**
Steel Abrasive:		
Shot S 230	0.074	0.056
Shot S 280	0.089	0.064
Shot S 330	0.096	0.071
Shot S 390	0.117	0.089
Grit G 50	0.056	0.04
Grit G 40	0.086	0.061
Grit G 25	0.117	0.078
Grit G 14	0.165	0.13
Mineral Abrasives:		
Flint Shot (Medium. Fine)	0.089	0.068
Silica Sand (Medium)	0.10	0.074
Boiler Slag (Medium)	0.117	0.078
Boiler Slag (Coarse)	0.152	0.094
Heavy Mineral Sand (Medium. Fine)	0.086	0.066
The profile heights shown for steel abrasives were produced with conditioned abrasives of stabilized operating mixes in recirculating abrasive blast-cleaning machines. The profile heights produced by new abrasives will be appreciably higher, as follows:		
Cast steel shot: Hardness 40–50 Rockwell C.		
Cast steel grit: Hardness 55–60 Rockwell C.		

depth. To use this instrument, place the disc on the blasted surface and visually select the reference section that most closely approaches the roughness.

1.4 TEMPORARY PROTECTION OF PREPARED SURFACES AGAINST CORROSION AND CONTAMINATION

Temporary protection is necessary if the proposed coating (primer or total coating) cannot be applied to the prepared surface before its cleanliness has improved (e.g., by formation of initial rust). The same applies to areas to which the coating is not to be applied.

It is normal to use wash primers, shop primers, phosphatizing adhesive papers, adhesive films, strippable varnishes, and protective materials that can be washed off.

1.5 PREPARATION OF SURFACES PROTECTED BY TEMPORARY COATINGS

Before further coating, all contaminants, as well as any corrosion and weathering products that have been produced in the meantime, should be removed in an appropriate manner. Assembly joints and damaged areas of the primer coating should be derusted again.

1.5.1 PREPARATION OF JOINT AREAS (WELDED, RIVETED, AND BOLTED JOINTS)

The best method of removing residue from welding electrodes and welding or riveting scales is blast-cleaning or manual cleaning.

1.5.2 PREPARATION OF SURFACES OF SHOP PRIMERS, BASE COATING, OR TOP COATINGS

To make sure the coating adheres satisfactorily, it may be necessary to apply some solvent to existing coating or to roughen it either by using sandpaper or steel wool, or by lightly blasting it and then removing the dust. Surfaces of existing coatings (especially primers rich in zinc) must not be burnished or smeared by mechanical brushing or similar methods because it would keep later coatings from adhering satisfactorily.

If an existing shop primer or base coating is not in a condition suitable to provide a base for further coating or is not compatible with such further coating, it should be removed.

1.5.3 PREPARATION OF HOT-DIP GALVANIZED OR HOT-DIP ALUMINUM-COATED SURFACES

Defective areas in metal coating must be prepared and repaired in such a manner that the corrosion-preventing action of the coating is restored and that the adhesion and protective actions of further coating are not impaired. Defective areas repaired by building up with solder or sprayed zinc should be prepared for subsequent coating in the same way as the hot-dip galvanized or hot-dip aluminum coated surfaces. Sprayed zinc-coated and aluminum-coated surfaces should not be cleaned with alkaline detergent.

For the preparation of unweathered hot-dip galvanized or hot-dip aluminum-coated surfaces, during subsequent treatment, transport, or assembly, contamination may occur from such substances as grease, oil, and marking or coding inks. These should be removed by brushing and rinsing with special detergents or solvents. In the case of damage to the metal coating during transport or assembly, the method should be as in standards.

For the preparation of weathered hot-dip galvanized or hot-dip aluminum-coated surfaces, based on the period of weathering and the site where the surfaces are located, a part of surface contamination and various corrosion products of the coating metal or steel may be formed. Soluble or poorly adherent contaminants should be removed, according to the extent and nature of the deposits.

In the case of hot-dip galvanized surfaces (e.g., oxidic compounds and various salts), use dry brushing (brushes with plastic bristles), washing with water with detergent added, or water or steam cleaning. In the case of hot-dip aluminum coated surfaces, use brushing or washing with suitable solvents, cold detergents, or emulsion cleaner. If necessary, it may also be advisable to clean such surfaces by water or steam cleaning or steam cleaning with the addition of a weak phosphoric acid cleaner. For mechanical preparation of severely attacked surfaces, wire brushes, scraper, emery disks, and blasting methods are suitable.

1.6 INSPECTION AND TESTING

The quality control system should meet a minimum set of requirements. If any component is determined not to have been cleaned in accordance with this construction standard, the contractor should reclean the component until the necessary standards have been satisfied.

1.7 QUALITY SYSTEMS

The contractor should set up and maintain such quality assurance and inspection systems as are necessary to ensure that the goods or services supplied comply in all respects with the requirements of this construction standard. The engineers will assess the quality assurance and inspection systems against the recommendations of the applicable parts of ISO 9004, and they should have the right to undertake any surveys that are necessary to verify the quality.

They can and should undertake the inspection or testing of goods or services during any stage of work at which the quality of the finished goods may be affected and to undertake the inspection or testing of raw material or purchased components.

1.8 PROCEDURE QUALIFICATION

Before bulk preparation of components commences, a detailed sequence of operations for the preparation of components should be submitted to monitor compliance with the required standards. The engineers should also specify which cleaned components are to be subjected to the tests for formal approval of the preparation. No preparation should be done until the cleaning procedure has been approved.

1.8.1 PREPARATION PROCEDURE SPECIFICATIONS

The preparation procedure specifications should involve the following, but is not limited to them:

1. Cleaning of components and method of cleaning, including the cleaning of oil and grease
2. The cleaning medium and technique
3. In the case of blast-cleaning, the blast cleaning finish, surface profile, and surface cleaning
4. Dust and abrasive removal
5. Post-drying time and temperature
6. A recleaning technique

1.8.2 PREPARATION PROCEDURE APPROVAL TEST

A batch of 15–20 components should be cleaned in accordance with the approved preparation procedure. Three cleaned components should be selected for preparation procedure approval tests, and they should be subjected to a complete set of tests. The testing should be witnessed, and a full set of records should be submitted for consideration. Cleaning should not be performed until the procedure has been certified as acceptable in writing.

1.9 DEGREASING

Degreasing is used to completely remove oil, grease, dirt, and swarf from the surfaces that are to be protected by painting, coating, and lining. The five following methods, which in turn may consist of different processes, are generally used for the surface cleaning of substrates:

- Hot solvent cleaning
- Cold solvent cleaning
- Emulsifiable solvent cleaning
- Aqueous alkaline and detergent cleaning
- Steam cleaning

The method chosen depends upon the material of the substrate type, shape, and condition of the surface that is to be cleaned. Table 1.4 presents a guide to degreasing methods.

1.10 HOT SOLVENT CLEANING METHOD

Cleaning methods that use hot, chlorinated solvents are used to remove oil, grease, dirt, and swarf from unit parts or simple assemblies. Several methods of processing are available (see Table 1.4), and the choice of the appropriate process depends upon the type and degree of contamination.

1.10.1 HOT SOLVENT CLEANING–NOT WATER RINSABLE

Trichloroethylene (BS 580 type 1) and other chlorinated solvents are used to clean ferrous metals. When parts of aluminum, magnesium, zinc, titanium, or their alloys are degreased in trichloroetylene, the grade of solvent should comply with the requirements of BS 580 type 2.

Trichloroethylene degreasing should not be used on assemblies containing fabric, rubber, or other nonmetallic material unless it is known that no harm will result. Assemblies containing such materials can often be cleaned without damage using 1.1.2-trichoro-1.2.2 trifluoroethane.

Table 1.4   Guide to Degreasing Methods

Degreasing Method	Scope	Sub-method	Cleaning Material	Variation of Cleaning Process [See Appendix A]	Condition of Surface After Treatment	Cleaning Apparatus	Special Points	Safety Precautions
Hot solvent cleaning	All materials and all type of surfaces	Not water rinsable	Trichloroethylene to BS 580 1,1,1, trichloroethane to BS 4487 methylene chloride perchloroethylene to BS 1593 triclorotrifluoroethane	1. Vapor immersion 2. Liquid immersion 3. Jetting 4. Ultrasonic cleaning	Dry	Specially designed apparatus essential continuous or batch operation	Will not remove soaps, sweat or chemical residues. Not suitable for painted articles, or parts containing rubber and certain other non-metallic materials. Ultrasonic cleaning specially suitable for removing fine solid particles	Adequate shop ventilation, and correct operation of plant to avoid excessive inhalation of narcotic vapor. No smoking
		Water rinsable	Emulsifiable blend of cresybic acid (BS 524) and o-dichlorobenzene (BS 2944 grade B)	1. Immersion (bath temp. 60°C)				
Cold solvent cleaning (5.3)	All metals and all types of shapes not heavily contaminated	Not water rinsable	Trichloroethylene to BS 530 1,1,1-trichloroethane to BS 4487 perchloroethylene to BS 1593 trichlorotrifluoro-ethane white spirit to BS 245 white spirit/ solvent naphthas to BS 479 aromatic solvents coal tar solvents	1. Immersion 2. Brushing 3. Or wipings 4. Spraying 5. Ultrasonic system	Dry or from petroleum solvents slightly oily	Tanks or apparatus of special designed. Continuous or batch operation	Complete grease removal not certain. Soaps, sweat or chemical residues not removed	Adequate shop ventilation, no smoking, strict provision against the risk if petroleum solvents are used
		Water rinsable	Dichloromethane based mixture to BS 1994 trichloroethylene based mixture	1. Immersion (liquid) 2. Brushing				

Method	Application	Type	Material	Process	Condition	Apparatus	Properties	Precautions
Emulsifiable solvent cleaning	All metals parts of accessible shape, not heavily contaminated		Hydrocarbon, e.g., white spirit with an emulsifying agent	Immersion (liquid) brushing	Wet and possibly slightly oily	Tanks or apparatus of special design, continuous or batch operation	Will remove sweat and certain chemical residues. Suitable for painted surfaces	Precautions against contact of concentrated material with skin, fire precautions if petroleum solvents are used
				Spraying	Wet and possibly slightly oily	Steam or water jet apparatus	Suitable for large assemblies which cannot be dismantled	
Aqueous alkaline and detergent cleaning	Parts without a highly finished surface and of accessible shape strong alkalis for ferrous metals: milder alkalis for noh- ferrous metals and general purposes cleaning, organic detergent alone (as solution or emulsion) for tin-plate and light duty cleaning	Hot alkaline	Mixture of sodium hydroxide and sodium metasilicate pentahydrate	1. Immersion 2. Electrolytic cleaning 3. Ultrasonic cleaning	Wet	Tanks or apparatus of special design, continuous or batch operation	Will remove sweat and certain chemical residues milder alkalis Suitable for certain types of painted surface not suitable for composite items containing rubber, leather, fabric or wood. Ultrasonic cleaning specially suitable for removing fine solid particles	Strong alkalis require protective measures eyeshields and rubber gloves to prevent possible damage to eyes and skin during handling
		Mild alkaline	Mixture of metasilicate pentahydrate and sodium carbonate					
		Detergent	Detergent	Ultrasonic immersion	Wet and possibly slightly oily	Ultrasonic apparatus		
Steam cleaning			Steam alone or mixture of steam with detergent or alkaline material	Steam cleaning-	Wet	Steam injector apparatus	Suitable for large assemblies which cannot be dismantled	

Trichloroethylene is not flammable, but flames can cause decomposition of solvent vapor and the production of harmful acidic gases. Therefore, degreasing equipment should be kept away from flame. Contact of solvent with the hand should be avoided as well, as the solvent will remove the natural grease from the skin.

The hot solvent-not water rinsable method is applied in immersion (vapor and liquid), jetting, and ultrasonic processes. Heavily contaminated thin sheet material may not degrease satisfactorily with one treatment in vapor, but spraying with clean condensed solvent after vapor treatment or immersion in boiling liquor is preferred in such cases.

During hot solvent processing, the articles should be placed on hooks or racks or in suitable containers. After cleaning and drying, articles should not be handled with bare hands. Clean gloves or similar protection should be worn and handling kept to a minimum.

After the process is completed, the articles will be hot and dry. Materials of a complex construction, including crevices and capillaries, may require further drying at 100–120°C.

The methods used for chemical checking of solvent are described in BS-1133 and BS 580, and standard guidelines of emission control in the vapor process are described in ASTM D 3640-80.

Today, most cleaning is based on aqueous (water-based) technology. Washing, rinsing, and dry cleaning has become the norm. In the past, chlorinated and fluorinated solvents were the cleaning "benchmark" for many years. Virtually every manufacturing facility cleaned materials using a vapor degreaser with a solvent other than water (since water does not work in vapor degreasers). This all changed with the Montreal Protocol, which went into effect in 1989. The purpose of this protocol was to ban or drastically reduce pollution by limiting the emission of solvents that linger for a long time when released into the Earth's atmosphere, including trichlorotrifluoroethane (Freon TF) and other fluorinated and chlorinated hydrocarbon solvents. The protocol and other efforts also sought to reduce the use of solvents that have been either proven or suspected to be cancer-causing agents. In an interesting paradox, it seems that the more likely a solvent is to be stable in the atmosphere (and thereby create an environmental problem), the less likely it is to cause cancer, and vice versa.

Open-top vapor degreasers are batch-loaded boiling degreasers that clean via the condensation of hot solvent vapor on colder metal parts. Vapor degreasing uses halogenated solvents (such as perchloroethylene, trichloroethylene, and 1,1,1-trichloroethane) because they are not flammable and their vapors are much heavier than air.

A typical vapor degreaser (Figure 1.4) is a sump containing a heater that boils solvent to generate vapors. The height of these pure vapors is controlled by condenser coils, a water jacket encircling the device, or both. Solvent and moisture condensed on the coils are directed to a water separator, where the heavier solvent is drawn off the bottom and is returned to the vapor degreaser.

A device called a *freeboard* extends above the top of the vapor zone to minimize vapor escape. Parts to be cleaned are immersed in the vapor zone, and condensation continues until they are heated to the vapor temperature. Residual liquid solvent on the parts rapidly evaporates as they are slowly removed from the vapor zone. Lip-mounted exhaust systems carry solvent vapors away from operating personnel.

The cleaning action is often increased by spraying the parts with solvent below the vapor level or by immersing them in a liquid solvent bath. Nearly all vapor degreasers are equipped with a water separator that allows the solvent to flow back into the degreaser. Figure 1.5. shows an open-top vapor degreaser.

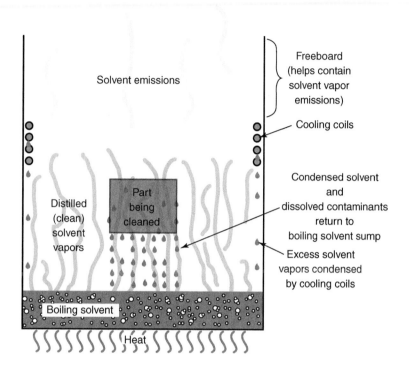

FIGURE 1.4

A typical vapor degreaser. These degreasers were common until the time of the Montreal Protocol. Cleaning was accomplished as condensing solvent vapors dissolved any existing contaminants.

Emission rates are usually estimated from solvent consumption data for the degreasing operation in question. Solvents are often purchased specifically for use in degreasing and are not used in any other plant operations. In these cases, purchase records provide the necessary information on emission rates, and an emission factor of 1,000 kg of volatile organic emissions per milligram (2,000 lb/ton) of solvent purchased can be applied based on the assumption that all solvent purchased will eventually be emitted. When information on solvent consumption is not available, emission rates can be estimated if the number and type of degreasing units are known. The factors for degreasing operations (shown in Table 1.5) are based on the number of degreasers and emissions produced nationwide and may be considerably in error when applied to a particular unit.

As a first approximation, this efficiency can be applied without regard for the specific solvent being used. However, efficiencies are generally higher for more volatile solvents. These solvents also result in higher emission rates than those computed from the average factors listed in Table 1.5.

1.10.2 HOT SOLVENT CLEANING–WATER RINSABLE

This method is normally used for paint removal or for assisting in the removal of carbonaceous deposits. It is based on an emulsifiable blend of cresylic acid and o-dichlorobenzene. It is normally used

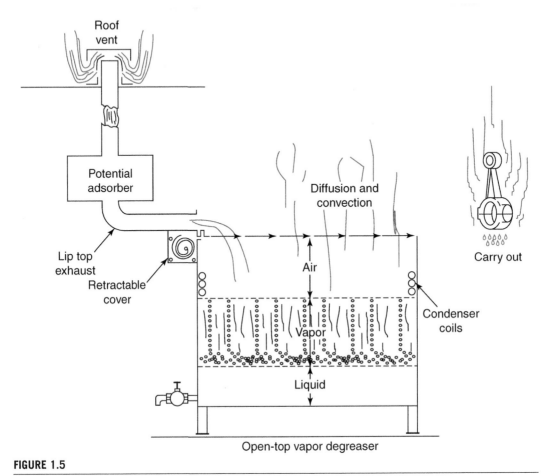

FIGURE 1.5

Open-top vapor degreaser.

with a water seal in a bath (immersion) operated at temperatures up to 60°C. After processing, articles should be rinsed with clear water and then dried.

1.10.3 COLD SOLVENT CLEANING

These cleaning methods are related to the use of cold petroleum solvents, as well as halogenated hydrocarbons, for the removal of oil, grease, dirt, and swarf from unit parts or simple assemblies having easily accessible surfaces. These solvents may also be applied to in situ cleaning of large units, assemblies, or machinery that cannot be accommodated in degreasing equipment.

Several methods of processing are available. The choice of the appropriate process depends upon the type and degree of contamination and the size and shape of the parts. These solvents should not be used for assemblies containing fabrics, rubber, or other nonmetallic materials unless it is known that no harm will result.

Table 1.5   Solvent Loss Emission Factors for Degreasing Operations

Type of Decreasing	Activity Measure	Uncontrolled Organic Emission Factor[a]	
All[b]	Solvent consumed	1,000 kg/Mg	2,000 lb/ton
Cold cleaner	Units in operation	0.30 Mg/yr/unit	0.33 tons/yr/unit
	Surface area and duty cycle[d]	0.165 Mg/yr/unit	0.18 tons/yr/unit
Entire unit[c]		0.075 Mg/yr/unit	0.08 tons/yr/unit
Waste solvent loss		0.06 Mg/yr/unit	0.07 tons/yr/unit
Solvent carryout		0.4 kg/hr/m^2	0.08 Ib/hr/ft^2
Bath and spray evaporation			
Entire unit			
Open top vapor	Units in operation	9.5 Mg/yr/unit	10.5 ton/yr/unit
	Surface area and duty cycle[e]	0.7 kg/hr/m^2	0.15 1b/hr/ft^2
Entire unit			
Entire unit			
Conveyorized, vapor Entire unit	Units in operation	24 Mg/yr/unit	26 tons/yr/unit
Conveyorized, nonboiling Entire unit	Units in operation	47 Mg/yr/unit	52 tons/yr/unit

[a]*100% nonmethane volatile organic compound.*
[b]*Solvent consumption data will provide much more accurate emission estimates than any of the other factors presented.*
[c]*Emissions generally would be higher for manufacturing units and lower for maintenance units.*
[d]*For trichloroethane degreaser.*
[e]*For trichloroethane degreaser; does not include waste solvent losses.*

Cold cleaners are batch-loaded, nonboiling solvent degreasers, which usually provide the simplest and least expensive method of metal cleaning. Maintenance cold cleaners are smaller, more numerous, and generally use petroleum solvents as mineral spirits (petroleum distillates and Stoddard solvents). Manufacturing cold cleaners use a wide variety of solvents, which perform more specialized and higher-quality cleaning with about twice the average emission rate of maintenance cold cleaners. Some cold cleaners can serve both purposes.

Cold cleaner operations include spraying, brushing, flushing, and immersion. In a typical maintenance cleaner (Figure 1.6), dirty parts are cleaned manually by spraying them and then soaking them in a tank. After cleaning, parts are either suspended over the tank to drain or are placed on an external rack that routes the drained solvent back into the cleaner.

1.10.4 COLD SOLVENT–NOT WATER RINSABLE

The choice of solvent in this method is largely governed by consideration of the toxicity, volatility, and flammability of solvents, as well as the environmental conditions. Some solvents, especially coal tar solvents and aromatics, will also dissolve paint, but they are more toxic and have a low flash point.

Benzol (benzene) is the most toxic of these substances, so it should not be used. Xylol, toluol, and high-flash naphtha may be used when their concentration in the air does not exceed safe breathing levels (see Table 1.6). Petroleum solvents are flammable, so suitable fire precautions should be taken.

Immediately after cleaning, the articles should be dried with compressed air or, if their size permits, in a vented oven at 65°C. Articles too large for an oven may be dried by wiping. The cold solvent, not water rinsable method operates by immersion brushing or wiping, spraying, and ultrasonic processes.

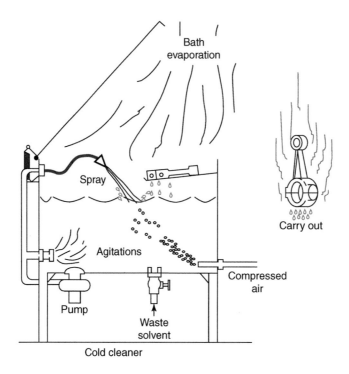

FIGURE 1.6

Degreaser emission points for a Cold cleaner.

1.10.5 COLD SOLVENT–WATER RINSABLE

Dichloromethane-based mixtures are mainly used to remove paint and carbonaceous deposits, but they also act as effective cleaners. It should contain a minimum of 70% (wt/wt) of dichloromethane, a thickening and emulsifying agent. Corrosion inhibitors may also be added during applications to close tolerance parts or where slight etching of metal surface is undesirable. In immersion operation also contain cresylic acid for increased efficiency and are usually used under a 100-mm water seal.

Trichloroethylene-based mixtures can be used on metal surfaces adjacent to certain plastics that would be affected by dichloromethane-based mixtures. A typical composition is as follows:

Trichloroethylene to BS 580, Type 1	70 parts by weight
Toluene to BS 805	6 parts by weight
Industrial methylated spirits to BS 3591	10 parts by weight
Cresylic acid to BS 524, Grade D	4 parts by weight
2-Ethoxyethanol to BS 2713	4 parts by weight
Paraffin wax	Remainder to a total of
A cellulose ether	6 parts by weight but
Emulsifier	paraffin wax should not
Corrosion inhibitors	exceed 2 parts by weight

Table 1.6 Threshold Limit Values (TLVs) for Solvents				
	Adopted Values TWA.TLV[1]		Adopted Values STEL. TLV[2]	
Substance	**ppm**	**mg/cu m**	**ppm**	**mg/cu m**
Acetone benzene (benzol)-skin butylcellosolve-skin	750 10 25	1780 30 120	1000 25 75	2375 75 360
Carbon tetrachloride -skin cyclohexane epichloro-hydrin-skin	5 300 2	30 1050 10	20 375 5	125 1300 20
Ethyl acetate ethanol (ethyl alcohol) ethylene dichloride (1, 2-dichloroethane)	400 1000 10	1400 1900 40	N/A N/A 15	N/A N/A 60
Ethylenediamine furfuryl alcohol-skin	10 10	25 40	N/A 15	N/A 60
Methanol (methyl alcohol) methylene chloride (dichloromethane)	200 100	260 350	250 500	310 1740
Naphtha, coal tar[3] naphtha, petroleum[3] perchloroethylene-skin	N/A N/A 50	N/A N/A 335	N/A N/A 200	N/A N/A 1340
Isopropyl alcohol-skin standard solvent toluene	400 100 100	980 525 375	500 200 150	1225 1050 560
Trichloroethylene turpentine xylene (xylol)	50 100 100	270 560 435	200 150 150	1080 840 655

[1]TWA.TLV stands for "threshold limit value–time-weighted average," the time-weighted average concentration for a normal 8 hour workday or a 40 hour work week.
[2]STEL.TLV stands for "threshold limit value–short-term exposure limit," the maximum concentration to which workers can be exposed for a period of up to 15 min.
[3]In general, the aromatic hydrocarbon content will determine what TLV applies.

This method is operated by immersion and brushing.

After cleaning, articles should be rinsed with clear water and then dried.

Note: Standard guidelines for emission control in cold solvent cleaning operation are described in ASTM D 3640.

1.10.6 EMULSIFIABLE SOLVENT CLEANING

These cleaning methods relate to the removal of oil, grease, and dirt from parts or simple assemblies by immersing them in a cleaning mixture in which grease and dirt are loosened, but not detached, and then washing away the solid contaminants with water. This method has the advantage that no heating is required, and a succession of dirty articles can be cleaned without rapidly fouling the cleaning medium itself.

This method gives a lesser degree of cleanness than that obtained by hot solvent or aqueous alkaline cleaning methods, but it may be employed if a residual trace of the cleaning medium may be tolerated by the subsequent process (e.g., phosphating) or if the size of the parts makes in situ cleaning essential.

Emulsifiable solvent cleaning generally can be used on any metallic parts, but it should not be used on assemblies containing fabrics, rubber, or other organic material unless it has been reliably ascertained that no harm will result.

Emulsion cleaners are broadly classified into three groups on the basis of stability; stable, unstable, and diphase. Stable emulsion cleaners are applied by immersion or spraying. These cleaners should not be used to clean heavily contaminated surfaces.

Unstable emulsion cleaners perform more efficiently in removing heavy shop soils such as oil-based rust and many lubricants used in stamping and extruding. This method is applied by immersion or spraying.

Diphase emulsion cleaners are utilized to remove the most difficult hydrocarbon soils, such as lapping compounds, buffing compounds, and oxidized oils. They are capable of providing a greater degree of cleanness than can be obtained with stable or unstable emulsions. Diphase cleaners are most commonly used in immersion processes.

Typical cycles for the immersion and spraying processes are shown in Table 1.7. Typical compositions and operating temperatures of different types of emulsion cleaners are shown in Table 1.8.

Table 1.7   Typical Cycles for Immersion and Spray Emulsion Cleaning

| Process Sequence | Cycle Time, Minute | | | |
| | Easy Cleaning[a] | | Difficult Cleaning[b] | |
	Immersion[c]	Spray[d]	Immersion[e]	Spray[f]
Clean[g]	2–4	½–1	4–10	1–2½
Rinse[h]	¼–½	¼–½	¼–½	¼–½
Rinse[i]	¼–1	¼–1	¼–1	¼–1
Air dry[j]	½–2	¼–2	½–2	¼–2

[a]*Removing cutting oils and chips from machined surfaces. Shop dirt and oil from sheet metals, and drawing compounds from automotive trim.*
[b]*Removing embedded buffing compounds impregnated carbonized oils from cast iron unit parts and quenching oil from heat treated forgings.*
[c]*Concentration of cleaner (15–60 g/l)*
[d]*Concentration (3.75–15 g/l)*
[e]*Concentration (30–90 g/l)*
[f]*Concentration (7.5–15 g/l)*
[g]*(60–71°C)*
[h]*Unheated rinse*
[i]*(54–71°C)*
[j]*(21–71°C)*

	Composition, Parts by Volume		
Component	**Stable**	**Unstable**	**Diphase**
Petroleum solvent[a]	250–300	350–400	250–300
Soaps[b]	10–15	15–25	None
Petroleum (sulfonates)[c]	10–15	None	1–5
Nonionic surface-active			
Agents[d]	5–10	None	1–5
Glycols, glycol ethers[e]	1–5	1–5	1–5
Aromatics[f]	5–10	25–50	5–10
Water[g]	5–10	None	None
Typical compositions and operating temperature range[h]			
	27–66°C	27–66°C	27–60°C

Table 1.8   Typical Compositions and Operating Temperatures for Emulsion Concentrates

[a]Two commonly used hydrocarbon solvents are deodorized kerosene and mineral seal oil.
[b]Most commonly used soaps are based on rosin or other short-chain fatty acids, saponified with organic amines or potassium hydroxide.
[c]Low-molecular-weight petroleum sulfonates (known as mahogany sulfonates) are used for good emulsification and some rust protection. High-molecular-weight sulfonates (with or without alkaline-earth sulfonates) offer good rust inhibition and fair emulsification.
[d]Increased content generally improves stability in hard water but increases cost.
[e]Glycols and glycol ethers are generally used in amounts necessary to act as couplers in the stable and unstable types of emulsions. These agents are frequently used with diphase and detergent cleaners to provide special cosolvency of unique or unusual types of soils.
[f]Aromatic solvents are frequently used to provide cosolvency for special or unique soils; sometimes they also serve to inhibit odor-causing or souring bacteria.
[g]Water, fatty acids, or both are used to adjust the clarity and stability of emulsion concentrate, particularly those of the stable and unstable types.
[h]The maximum safe temperature depends on the flash point of hydrocarbon (petroleum) solvent used as the major component.

1.11 AQUEOUS ALKALINE AND DETERGENT CLEANING

These cleaning methods relate to the use of cleaning solutions containing alkalis, organic materials, or both for the removal of oil, grease, dirt, and swarf from unit parts or simple assemblies having easily accessible surfaces.

Several methods of processing are available; which process to use depends upon the nature of the metal to be cleaned, the type and degree of contamination, and the composition, size, and shape of the articles. After cleaning the parts, rinse and dry them.

Solutions containing alkalis should be used with caution on articles with highly finished surfaces because they may dull the surface. Porous parts and assemblies that would trap liquid should not be cleaned with such solutions, owing to the difficulty of rinsing away all traces of the solution and of drying the articles afterward.

Alkalines have advantages over other solvents for the removal of certain types of contaminants, including soaps and salts. The cleaning action is based on the saponifying and emulsifying effects of aqueous alkalis, often reinforced by sequestrating, complexing, and surface-active agents. The ingredients usually can include sodium hydroxide, sodium carbonate, sodium metasilicate, trisodium phosphate (TSP), sodium pyrophosphate, sodium borate, complexing agents [such as ethylenediamine tetraacetic acid (EDTA) gluconate, heptonate, polyphosphate, and cyanide] and organic surfactants. Application normally is done by immersion, electrolytical, ultrasonic immersion, or steam mixture jetting, with subsequent water rinsing (see Table 1.9).

Table 1.9   Typical Alkaline Cleaner Formulations for Various Metals

	Aluminum		Copper			Cu Plate	Iron and Steel			Magnesium		Zinc		
	A	B	A	B	C	C	A	B	C	A	B	A	B	C
Builders	Composition of Cleaner (% by Weight)													
Sodium hydroxide, ground	N/A	N/A	20	15	15	55	20	20	55	20	20	N/A	15	15
Sodium carbonate, dense	N/A	N/A	18	N/A	N/A	8	18	29	8.5	18	29	N/A	N/A	N/A
Sodium bicarbonate	21	24	N/A	34	34	N/A	N/A	N/A	N/A	N/A	N/A	N/A	35	34
Sodium tripolyphosphate	30	30	N/A	N/A	10	N/A	20	20	10	20	20	90	10	10
Tetrasodium pyrophosphate	N/A	N/A	20	10	N/A	10	N/A	N/A	1/4	N/A	N/A	N/A	N/A	N/A
Sodium metasilicate, anhydrous	45	45	30	40	40	25	30	30	25	30	30	N/A	40	40
Surface active (wetting) agents	N/A	N/A	N/A	N/A	N/A	N/A	N/A	N/A	N/A	N/A	N/A	N/A	N/A	-
Sodium resinate	N/A	N/A	5	N/A	N/A	N/A	5	N/A	N/A	5	N/A	5	N/A	-
Alkyl aryl sodium sulfonate	3	3	5	N/A	N/A	N/A	5	N/A	1	5	N/A	5	N/A	-
Alkyl aryl polyether alcohol	N/A	N/A	2	N/A	N/A	1	2	N/A	N/A	2	N/A	N/A	N/A	-
Nonionics high in ethylene oxide	1	1	N/A	1	1	1	N/A	1	0.5	N/A	1	N/A	N/A	1
Operating temperature of solution (°C)	71	71	82	76	71	82	93	76	82	93	76	82	76	82
Concentration of cleaner (g/l)	30	7.5	60	7.5	60	60	60	7.5	60	60	7.5	30	7.5	45

All detergent and alkaline cleaning should be followed by adequate draining, but the drainage time should not be so long as to allow the cleaning solution to dry on the articles. Rinsing with water should follow draining.

In conveyorized jetting and spraying machines, the rinsing is usually done with water jets. But sometimes the articles will be transferred from the cleaning tank to a rinse tank with cold running water.

Rinsing should be thorough, and unless the cleaning solution has contained little or no alkali, it is preferable to arrange two rinse tanks in cascade with the water flowing the opposite way from the articles. This much reduces the amount of water needed to give effective rinsing.

If the articles are not to pass to other aqueous processes, a final immersion for 30 to 60 s in hot water (80–95°C) will facilitate drying. Immediately after final rinsing, the parts should be dried unless further cleaning (i.e., removal of rust and miscellaneous residues) is to be undertaken.

As soon as the parts are finally clean and dried, the appropriate protective coating should be applied without delay; otherwise, corrosion may occur very quickly. It is usually desirable to allow the article to cool to 5°C above room temperature before applying the coating. Cooling to room temperature may cause microscopic condensation on bare metal surfaces.

In immersion tanks, the heating can be arranged such that convection currents assist the circulation of the cleaning medium around the articles; preferably, agitating may be provided by compressed air or with an air impeller or circulating pump.

A wide variety of proprietary cleaning mixtures are available for cleaning ferrous and nonferrous metals. Solutions containing caustic alkalis should be used only when no metals other than iron and steel, copper, nickel, chromium, and titanium are present. A typical strong alkaline cleaner for these metals may consist of a mixture of sodium hydroxide (caustic soda), sodium metasilicate (or other silicates of a higher soda/silica ratio), TSP (or other phosphate), and sodium carbonate (soda ash).

Aluminum, lead, zinc, and tin, including galvanized surfaces and tin-plate, are liable to be damaged by any alkaline cleaners that are not specifically formulated for use with such metals. Mild alkaline cleaners made up of sodium metasilicate (or sodium silicates of a lower soda/silica ratio), sodium phosphate, and a suitable proportion of surfactants may be used; sodium carbonate is permissible if silicate is present in such a proportion as to keep the soda/silica ratio below 1:2. A proportion of sodium sulfite also may be included as a means of avoiding feathering of tin-plate by mild alkaline cleaners.

The action of both strong and mild alkaline mixtures used in simple immersion cleaning is much improved by the inclusion of up to 5% of an organic surfactant, commonly a sulfated fatty alcohol or a fatty alcohol/ethylene oxide condensate. In electro-cleaning tanks or jetting systems, such ingredients usually cause foaming and should be omitted or included only in very small proportion. Table 1.9 lists typical alkaline cleaner formulations for various metals.

Where hard water only is available, this should be softened. The addition of a small proportion of sodium hexametaphosphate is useful; usually 1 g/l will be adequate (see also Table 1.10). Solutions of organic surfactants alone are usually harmless to any metal, but as they have a less vigorous action, they require a longer period of contact than alkaline cleaners, so they are unsuitable for dealing with very greasy surfaces. Such solutions are an effective means of cleaning tin-plate without damage, but not all surfactants are suitable for this purpose, and a selection should therefore be made in consultation with the manufacturer.

Aqueous cleaning solutions should be devised in accordance with the manufacturer's recommendations. if any general a concentration of 30–60 g/l of water of an alkaline mixture is necessary for immersion cleaning, and of 5–30 g/l water for jet cleaning, an operating temperature of 80–95°C is

Table 1.10   Minimum Phosphate Additions Required for Softening Water

Hardness of Water PPM CaCO₃	Minimum Addition for Softening 100 g/l	
	Tetrasodium Pyrophosphate	Sodium Tripolyphosphate
	Water at 25°C	
75	520	115
150	810	230
300	1020	450
	Water at 60°C	
75	250	70
150	420	150
300	620	260

Note: Actual (not theoretical) minimum amounts required for water softening only.

generally recommended. If surfactants are used alone, then concentrations of the order of 1–10 g/l of water are normally used at room temperature; however, if jet cleaning is employed, the concentration should be low to avoid excessive foaming.

A alternative method of applying similar active ingredients is to dissolve an organic emulsifying agent in a solvent such as kerosene or white spirit and use it in the form of a weak emulsion in water, to which a small proportion of a mild alkali such as sodium metasilicate may be added. Such an emulsion is generally used in a spraying machine at an elevated temperature.

1.12 STEAM CLEANING

This cleaning method relates to the use of a jet of high-pressure steam for the in situ cleaning of large unit parts, assemblies, and machinery that cannot be accommodated in a cleaning apparatus. Steam cleaning should not be used where delicate mechanisms are found. The cleaning may be carried out with pure steam or with an aqueous detergent solution/steam mixture. After cleaning, the parts may have traces of the original contamination and of water or cleaning solution on their surfaces.

This method is not applicable to the cleaning of interior surfaces of assemblies that cannot be drained readily. The steam and hot water tend to remove the oils, greases, and soaps by thinning them with heat, emulsifying them, and diluting them with water. When used to remove old paint, the steam cooks the vehicle so that the paint loses its strength and bonding to the metal. It can then be easily removed by further washing. When detergent is used, its higher affinity for the metal also causes the oil, grease, and paint to loosen, thereby increasing the rate of cleaning.

The mechanical action of the steam usually permits the use of a lower concentration of detergent than is required for immersion cleaning (Figure 1.7). Alkali cleaners used in steam cleaning will attack aluminum and zinc alloys, unless specifically inhibited against such action.

FIGURE 1.7

Steam cleaning.

They should be used selectively over painted surfaces to ensure that the paint is not damaged if removal is not desired. The equipment required is a pressure-jet steam cleaner. On completion of detergent solution/steam cleaning, straight steam should immediately be directed over all cleaned surfaces so as to wash away deposits from the solution, especially if an alkaline solution has been used. The raised temperature of the surfaces subjected to high-pressure steam helps the drying, but any parts that retain moisture should be dried immediately by blowing with compressed air.

Strong alkalis attack the eyes and the skin; goggles and protective clothing should be worn when carrying out alkaline solution/steam cleaning with strong alkalis, particularly in confined places. Due care should be taken to avoid steam burns.

1.13 TESTING FOR LACK OF GREASE
1.13.1 WATER BREAKS TEST

A commonly used test for the removal of grease and oil is to inspect for water breaks. This is best done after pickling or activation, and involves visual observation after a final rinse in clear, cool water. Some experience is necessary to judge whether there is a break in the film of water. A continuous sheet of water on a part usually indicates a clean surface (although certain precious metal surfaces, such as gold, may exhibit water breaks even if they are clean). A specific drainage time (about 30 s) should be used before observation.

1.13.2 GREASE RED TESTING

On a grease-free horizontal surface, the drop quickly spreads out and a circular outline remains. On a grease-free vertical surface, the run-off path is short and an oval outline remains. On a horizontal surface that is not free of grease, the drop remains at its original size, and after evaporation, there is a sharply serrated outline. On a vertical surface that is not free of grease, a long run-off path forms. The tested areas must be carefully cleaned before coating.

1.13.3 FLUORESCENCE TESTING

Various oils and greases exhibit fluorescent effects when irradiated with ultraviolet light (must be screened from incident daylight). If the type of contamination and its fluorescent properties are known, this method can be used for testing cleaned surfaces.

1.14 NOTES ON DEGREASING
1.14.1 SELECTION OF SOLVENT

Strong alkalis are commonly used for cleaning steel, but not on substances that they can damage (e.g., aluminum, tin, zinc, and brass). Mild alkaline cleaner will not attack metals. Articles that are porous or have fine capillary spaces should not be cleaned in alkaline solutions.

For articles that are partially painted or varnished, trichloroethylene and alkali solutions are not generally suitable. Certain emulsifiable solvent cleaners or petroleum solvents may be used.

Only unit parts or very simple assemblies should be cleaned in aqueous cleaners. More complex parts and assemblies should not be cleaned owing to the risk of trapping any cleaning solution that would not be removed by subsequent rinsing and drying.

1.14.2 REUSE OF CLEANING SOLUTION

Cleaning agents are weakened and contaminated by material and soil being removed from surfaces as they are cleaned. It may be impractical or uneconomical to discard solution after a single use, even in precision-cleaning operations. When solution is reused, care must be taken to prevent the accumulation of sludge at the bottom of cleaning tanks. Periodic cleaning of vats and degreasing tanks, periodic decanting, agitation of solution, and similar provisions are essential to maintain the effectiveness of solutions. Care must be taken to prevent water contamination of trichloroethylene and other halogenated solvents, both in storage and in use. Redistillation and filtering of solvents are necessary before reuse.

Make up is often required to maintain the concentrations and pH of cleaning solutions at effective levels. Do not overuse chemical cleaners, particularly vapor-degreasing solvents. If light films or oily residues remain on the metal surfaces after the use of such solvents, additional scrubbing with hot water and detergent, followed by repeated rinsing with large quantities of hot water, may be necessary.

1.14.3 PROTECTION OF CLEANED SURFACES

Walking on cleaned surfaces should be avoided. If personnel must do this, they should wear clean shoe covers each time they enter, or protective material should be laid on the areas being walked on.

Workers handling critical items whose surfaces have been cleaned should never wear clean lint-free cotton, nylon or Dacron cloth or polyethylene film gloves. Rubber or plastic gloves are suitable during precleaning operation or cleaning of noncritical surfaces.

Keep the openings of hollow items (pipe, tubing, valves, tanks, pump, pressure vessels, etc.) capped or sealed at all times except if they must be open to do work on them. They should be closed off using polyethylene, nylon, tetrafluoroethylene (TFE)/fluoro carbon plastic, or stainless steel caps, plugs, or seals. The reuse of caps, plugs, or packaging materials should be avoided unless they have been cleaned. Do not remove the wrapping and seal from incoming materials and components until they are ready to be used or installed imminently.

1.14.4 HAZARDS OF SOLVENT

All solvents are potentially hazardous, and their concentration in air being breathed by workers needs to be kept low enough for safety. When these substances are used in closed spaces (pit or vessel) where the safe concentration is exceeded, personnel should wear gas masks. Smoking should be prohibited near degreasing equipment.

1.15 PICKLING

Pickling is a method of preparing metal surfaces by chemical reaction, electrolysis, or both. In pickling, rust and scales are removed by chemical reactions with mineral acids and with certain alkaline materials. Various acids used in commercial pickling include sulfuric, hydrochloric or muriatic, nitric, hydrofluoric, and phosphoric acid, and mixtures of these can be used as well.

Pickling is considered a desirable method of removing rust and mill scale from structural shapes, beams, and plates in workshops when the cost of such removal is felt to be justified. Properly accomplished, this method produces a surface that will promote long paint life with most coatings.

Steelworks prepared by pickling should be primed as soon as they are dry and warm, with the specified primer at a specified dry film thickness (which should be a minimum of 50 microns).

1.15.1 PICKLING METHODS

The six main pickling methods, which in turn may consist of different processes, are as follows:

- Sulfuric or hydrochloric acid pickling
- Phosphoric acid pickling
- Footner pickling
- Sulfuric acid-sodium dichromate pickling
- Alkaline derusting
- Electrochemical treatment

The pickling method to be used in any particular scenario depends on the nature of the rust and scale to be removed such as the material that the articles are made of, the degree of desire pickled surface, and the availability and cost of pickling material and equipment.

1.15.2 HYDROGEN EMBRITTLEMENT

When metal dissolves in acid during pickling, a definite volume of hydrogen is produced. Atomic hydrogen, absorbed or dissolved in steel, apparently affects its flexibility and ductility.

The bubbles of molecular hydrogen that form at the metal surface from conjugation of atomic hydrogen are extremely light. This characteristic is known as *hydrogen embrittlement* or *acid embrittlement*. Blisters on sheet or plate during pickling and galvanizing are caused by the same phenomenon. Hydrogen bubbles rise rapidly through a poorly inhibited bath. As they reach the surface of liquid, they break violently and produce fumes that can affect the health of workers and rapidly corrode metalwork and masonry in the pickling room. Inhibitors minimize acid fumes by reducing the hydrogen that causes them.

1.16 CLEANING AND PREPARATION OF METAL PRIOR TO PICKLING

Heavy deposits of oil, grease, soil, drawing compounds, and foreign matter other than rust, scale, or oxide must be removed by any appropriate cleaning method. Small quantities of such foreign matter may be removed in the pickling tanks, provided that no detrimental residue remains on the surface.

Heavy deposits of rust, rust scale, and all paints by suitable mechanical method(s) such as rust deposits that can be removed in a reasonable amount of time may be removed in the pickling tanks. Paints and other types of marking. They can normally be removed mechanically or with solvents.

1.17 THE PICKLING PROCESS

1.17.1 SULPHURIC OR HYDROCHLORIC (MURIATIC) ACID PICKLING

Pickling with these acids is of particular value in descaling, but it should be used only for parts that can be easily and thoroughly washed free of acid. Therefore, it may be unsuitable for parts with a complicated shape, particularly those that have narrow channels or blind holes that cannot be properly washed out or porous surface layers. The following types of components should not be treated with these acids:

- Parts built up by riveting, spot welding, or similar methods.
- Cast-iron parts, owing to the possibility of occlusion of pickling acid in porous surface layers.
- Ferrous articles with associated nonferrous or nonmetallic parts should not be pickled because of the risk of attack or of electrolytic effects, and because acid may be trapped at the join.
- High-strength constructional alloy and heat-treated alloy steels should not be pickled by sulfuric acid, which can be used for low-carbon structural steels. Some higher-carbon and alloy steels burn in acid very easily, making surface smut more of a problem. One method to help solve this is to add rock salt to the sulfuric acid bath. Some specifications call for the bath to contain 1.5% wt. sodium chloride.

1.17.2 OPERATIONS

For the operation of sulfuric or hydrochloric acid pickling, various types of acid-resistant tanks are available [e.g., lead (for sulfuric acid), glass, or glazed earthenware; wood, steel, or concrete lined with rubber or other acid-resisting material] may also be used. Several efficient inhibitors to reduce acid attacking on the base metal are available commercially, and one of these should be employed; it is necessary to adhere strictly to the manufacturer's instructions regarding suitability, concentration, and method of addition.

In hydrochloric acid pickling, the concentration of acid may vary between 1% and 50% of concentrated hydrochloric acid by volume, according to the nature and amount of scale or rust and the time available for pickling. Higher concentrations of acid remove rust and scale more rapidly, but they may attack the steel more severely. Hydrochloric acid pickling works reasonably well without external heating. Often, the heat of reaction between the acid and the scale is sufficient to keep the bath at 30–40°C, and then rapid pickling takes place.

In sulfuric acid pickling, the concentration may be between 5% and 20% of sulfuric acid by volume, and the bath should be heated (e.g., by steam coils) to a temperature of about 60–85°C. It is very uneconomical to pickle in any solution besides hot sulfuric acid because of the relative slowness of the process.

Note: To avoid accidents from overheating, always add the acid slowly to cold water while stirring continuously. Do not add the water to the acid.

Pickling may be accelerated to some extent by mechanical agitation of the parts or of the solution, and sometimes by lightly scrubbing off deposits that have loosened in the acid bath. Alternatively, the pickling action may be facilitated by the use of ultrasonic vibration. Completion of pickling is best judged by periodic inspection.

The acid content of the bath should be checked frequently; add acid as needed to maintain the correct strength. The pickling should be stopped when the iron content has risen to such an extent that the process is seriously slowed. The limiting iron (Fe) contents are of the order of 35 g/l for hydrochloric acid pickling and 16 g/l for sulfuric acid pickling.

1.17.3 PHOSPHORIC ACID PICKLING

Phosphoric acid pickling poses less danger of corrosion from residue or during drying, and therefore it is preferable to the other mineral acids. Nevertheless, with certain exceptions, it is necessary to wash the articles thoroughly after pickling, particularly if the parts are of a complicated shape, contain narrow channels or blind holes, or were made by such methods as riveting or spot-welding.

Light rust can be removed by immersion in cold phosphoric acid or proprietary liquids based on phosphoric acid and substantially free from other mineral acids, diluted for use according to the manufacturer's instructions. The optimum strength of acid is approximately 25% by volume. Generally, the procedure used is dip application.

In the dipping process, immerse the part to be cleaned in the rust-removing solution, if necessary assisting the derusting action by brushing with a steel-wool pad. Alternatively, warm the solution to 60°C to speed up rust removal. The immersion should not be longer than is required for complete derusting; normally, up to an hour (or 15 min at the higher temperature) should suffice. A lead-lined tank is recommended for the rust-removing solution. Rinse well in clean, cold water, and finally in clean, hot water. Dry as quickly as possible.

If composite articles are to be treated, care should be taken to avoid excessive attack on nonferrous metals. Generally, the solution should not be used for leaf springs or springs under stress. Locally hardened or hardened and tempered steel and spring steels should be given further treatment for 30 min in boiling water.

Heavy scale can be removed by phosphoric acid only at higher temperatures; e.g., 85°C for 25% v/v acid. It is not generally necessary to use an inhibitor in the bath. The bath should be discarded when the concentration of iron (Fe) reaches 20 g/l (otherwise, powdery deposits may be formed on the metal surfaces).

Economy in the use of phosphoric acid may be affected by the use of the Footner process for descaling steel plate and other forms of structural steel prior to the application of a protective coating. This process consists of pickling in 5–10% v/v sulfuric acid at 60–65°C for 12–15 min, or until all scale and rust is removed. The bath should contain an inhibitor. Further sulfuric acid should be added when the pickling time increases appreciably. The bath should be discarded when the accumulation of sediment and other detrius and the concentration of iron in the solution interfere with the pickling and result in deposits on the surface of the article.

This occurs when the specific gravity reaches about 1.18-1.20 or there is 1.6% of iron (Fe) (16 g/l) in the solution. After the article is lifted from the acid bath, it should be allowed to drain for 15–30 s before immersion in the water bath.

Wash the item in warm water (60–65°C) by immersing twice before the final bath. There should be a very small flow of water through the water-wash bath to prevent the total acidity as determined by titration against phenolphthalein, from exceeding 0.1 g H_2SO_4 per 100 cm^3. The necessary flow of water can be established after a short time.

Immersing for 3 to 5 min in 2% phosphoric acid solution maintained at a minimum temperature of 85°C. When the iron (Fe) content exceeds 5 g/l proportion, the bath should be discarded and then restored by adding clean water and phosphoric acid.

The lower phosphoric concentration and shorter immersion tends to produce thinner and less porous phosphate coating. This type of coating is an excellent base for most types of paint.

When removed from the hot phosphoric acid bath in the Footner process, the plates dry rapidly and carry a protective, dull gray phosphate film. No subsequent washing is required. Protective coatings should be applied immediately after the pickled surfaces are dry and still warm.

The process is also applicable to lighter material than structural steel, but drying in an oven may be necessary for light materials that do not carry sufficient heat from the bath to air-dry.

1.17.4 SULFURIC ACID-SODIUM DICHROMATED PICKLING

This method is often used in shipyards. Pickling is performed by 5% (by volume) sulfuric acid at 75–80°C, with sufficient inhibitor added to minimize attacks on the base metal, until all rust and scale is removed, followed by a two minute rinse in hot water at 75–80°C.

Next, immerse the pickled and rinsed steel for at least 2 min in a hot, inhibitive solution maintained above 85°C and containing about 0.75% sodium dichromate and about 0.5% orthophosphoric acid.

1.17.5 ALKALINE PICKLING

Rust and scale may be removed and the surface prepared for electroplating or painting of nonferrous metals by one of the following alkaline pickling methods:

- Removal of light rust by immersing in a solution based on caustic soda and a chelating agent such as sodium heptonate or gluconate.
- Removal of heavy rust

Electrolytic treatment in an alkaline solution consisting essentially of sodium hydroxide in water, In that case, the concentration being adequate to ensure good electrical conductivity. The solution may contain substances that assist in the removal of the rust, such as sodium cyanide (solution a) and ethylenediamine tetracetic acid. Typical compositions are:

a. Sodium hydroxide	200 g/l to 300 g/l
Sodium cyanide	20 g/l to 30 g/l
Anionic surface active agent	1.0 g/l to 1.5 g/l
b. Sodium hydroxide	100 g/l to 200 g/l
EDTA	100 g/l to 150 g/l
Nonionic wetting agent	1.0 g/l to 1.5 g/l

The solutions may be operated from room temperature to 60°C, but it is recommended that solution (a) not be exposed to temperatures above 40°C to prevent rapid decomposition of the cyanide. The parts should be treated anodically or cathodically or with periodic reversal of the current densities of the order of 250–500 A/m². After derusting, parts should be washed thoroughly in running water, with special attention paid to any crevices.

Molten caustic alkali descaling processes, which are based on molten caustic alkalis containing additives such as sodium hydride, have less of a tendency to induce hydrogen embrittlement in steel than pickling, but their use on high-strength steel is not allowed. These processes are particularly suitable for castings and for parts in heat-resisting steel stress relieved after fabrication. Molten hydride is, however, liable to embrittle titanium alloys by hydrogen absorption.

1.18 ELECTROCHEMICAL PICKLING

Rust and scale can be removed by the electrolytic methods described in the next sections.

1.18.1 CATHODIC TREATMENT IN ACID SOLUTION

Removal of rust and scale may be accelerated using this method compared with ordinary pickling, acid is economized and attack on the metal is reduced, but hydrogen embrittlement may be serious. Inhibitors such as tin or lead salts should be used. This process is useful for descaling irregular-shaped objects and those with difficult recesses. The work is made cathodic at 645 amp/m² in hot 10% v/v sulfuric acid containing a small amount of lead or tin salts. Tin or lead plates on the descaled areas and the action is diverted to other areas. The plated tin or lead is usually removed by electrolytic alkali cleaning. Some organic inhibitors can substantially reduce metal loss during cathodic pickling.

1.18.2 CATHODIC TREATMENT IN ALKALINE SOLUTION

Hydrdogen embrittlement is less than with the method discussed in Section 1.18.1, but derusting is usually slower than in acid solutions.

1.18.3 ANODIC TREATMENT

Anodic treatment may be carried out in either an acid or alkaline solution. Passivating conditions are established and oxygen, not hydrogen, is produced at the surface.

Hydrogen embrittlement is usually avoided with the anodic process, where there is a slight risk, especially with highly stressed parts of hydrogen being formed in the acid process when the current is switched off and while the work is being removed from the bath.

Examples of this process include the following:

- An anodic process for parts made entirely of steel, based on sulfuric acid solution (sp.gr : 1.22) used at a temperature not exceeding 25°C in a lead tank with lead cathodes, with a high anodic current directly maintained on the steel surfaces (not less than 1075 amp/m^2)
- An anodic alkaline process for parts made entirely of steel, based on a solution of caustic soda containing cyanide or an organic complexing agent

1.19 TREATMENT OF METAL AFTER PICKLING

1.19.1 COLD RINSING

When metal is removed from the pickle bath, a thin film of pickling acid and salts resulting from the reaction of acid with metal clings to it. The acid and salts, with the exception of some produced from phosphoric acid, actually stimulate rust formation and must be completely removed before they dry. An ample supply of clean water must be available for rinsing, which may be accomplished by any convenient means. Steel, wood, or concrete tanks provided with a skimming trough to take care of an ample overflow of water are generally used, although water also can be applied liberally with a hose.

Pickled work should be rinsed promptly, particularly if the acid is hot. If the film dries, it is difficult to rinse away residues, which can cause trouble in many of the operations described next.

1.19.2 FINAL (HOT) RINSING—NEUTRALIZING

When pickling acid and iron salts are removed or diluted, metal must be suitably treated in preparation for any operations performed later. The treatment prevents steel from rusting and prepares it for painting. Weak alkali solutions, such as 1.9–3.7 g/l of sodium carbonate or TSP, are used in a boiling rinse following a cold rinse, as described earlier in this chapter.

The alkaline surface does not rust rapidly, but if it is to be stored indefinitely or exposed to weather, it should be oiled. Alkali cleaning solutions (Neutralizing) are suitable for the application of oil, but not for the application of paint. Also, there are other treatments that can be used to prevent rusting, described next.

1.19.3 PREPARING METAL FOR PAINTING

Most paints do not adhere well and blister in a humid atmosphere if applied to an alkaline or neutral surface. For best painting results, the surface pH should be slightly acid. The best results occur when the surface has a pH between 3 and 5. There are exceptions when using special paints, such as inorganic zincs, which normally are applied to neutral surfaces. In pickling processes for inorganic zinc applications, no further treatment is normally used after the hot-water rinse.

For paints other than those containing inorganic zincs, it is important that the proper pH be maintained. Phophoric or chromic acids, or mixtures of the two, produce the best results. Muriatic or sulfuric acids should not be used because their residues stimulate the formation of rust under paint.

It is desirable to further clean and treat pickled and rinsed steel in a phosphoric acid solution prior to painting. Good results can be obtained by adding approximately 0.25% by weight of concentrated phosphoric acid to the hot-rinse bath in a steel tank, and maintaining this rinse at a pH of 3 to 5 by continuing to add acid as needed.

The cleanliness of the boiling rinse is important because this is where a satisfactorily cleaned surface can be spoiled for painting. For best results, the bath should be discarded and the tank cleaned daily. This is not practical for large-scale structural pickling operations, however, and good painting results can be obtained by merely maintaining a water rinse temperature at 60°C or higher and painting promptly while steel is warm and dry.

1.20 APPEARANCE OF PICKLED SURFACES

Surfaces should be pickled to a degree suitable for the specified painting system. Uniformity of color may be affected by the grade, original surface condition, and configuration of the material being cleaned, as well as by discolorations from mill or fabrication marks and the shadowing from pickling patterns. Visual standards of surface preparation agreed upon by the contracting parties should be used to further define the surface.

1.21 TEST METHODS—DETERMINING THE ACID AND IRON CONTENT OF PICKLE BATHS

Methods regularly used in the laboratory should be used to titrate pickle baths for both acid and iron. It is recommended to install such apparatuses near the pickle or measuring tanks and to have titrations made at regular intervals, usually by the pickle foreperson, who ensures that acid is added in measured and recorded quantities to maintain the proper strength.

1.21.1 EQUIPMENT REQUIRED

Below is the list if required equipment:
- 1 5-cm^3 pipette
- 1 1-cm^3 pipette
- 2 25-cm^3 burettes
- 1 5-cm^3 measuring cylinder
- 1 glass indicator bottle
- 1 burette stand
- 2 250-cm^3 glass beakers
- 2 stirring rods

1.21.2 REAGENTS REQUIRED

- 1.0 normal sodium hydroxide solution
- 0.1 normal potassium permanganate solution
- Methyl orange—1 g/l of H_2O
- C. P sulfuric acid, concentrated (*C. P.* stands for "chemically pure," without impurities detectable by analysis)

1.21.3 DETERMINING ACID AND IRON CONTENT IN SULFURIC AND HYDROCHLORIC ACID BATHS

Measurement of acid content is performed as follows:

1. Measure a 5-cm^3 sample of the pickling solution with a 5-cm^3 pipette and transfer it to a clean 250-cm^3 beaker.
2. Add about 100 cm^3 (half a beaker) of clean, fresh tap or city water and two or three drops of indicator solution (methyl orange).
3. Fill a burette exactly to the zero mark with 1.0 normal sodium hydroxide.
4. Stir the test sample constantly with a stirring rod and slowly add 1.0 normal sodium hydroxide until the exact moment when the red color changes to yellow.
5. Record the reading taken on the graduated burette. This is the number of cubic centimeters of 1.0 normal sodium hydroxide used.

The number of cubic centimeters of 1.0 normal sodium hydroxide used, multiplied by the appropriate factors shown in Table 1.11, gives the desired quantity of 66° or 60° Be sulfuric acid or 20° or 18° Be muriatic acid.

Measurement of iron content is performed as follows:

1. Measure a 1-cm^3 sample of the pickling solution with the 1-cm^3 pipette and transfer it to a clean 250-cm^3 beaker.
2. Add about 100 cm^3 (half a beaker) of fresh, clean water. Measure 5 cm^3 of concentrated sulfuric by means of the 5-cm^3 measuring cylinder and pour it slowly into the beaker, stirring constantly.
3. Fill a burette exactly to the zero mark with 0.1 normal potassium permanganate.

Table 1.11   Conversion Factors for Sulfuric and Hydrochloric (Muriatic) Acid Concentrations

	66°Be Sulfuric	60°Be Sulfuric
Percent by volume	0.573	0.74
Grams per 100 ml	1.053	1.263
Pounds per gal	8.771	10.525
	20°Be Muriatic	18°Be Muriatic
Percent by volume	1.999	2.288
Grams per 100 ml	2.319	2.612
Pounds per gal	19.353	21.796

* BE (Baume hydrometer scale): A calibration for liquids that is reducible to specific gravity by the following formulas: for liquids heavier than water, specific gravity = 145 ÷ (145 - n) (at 60°C) ; for liquids lighter than water, specific gravity = 140 ÷ (130 + n) (at 60°C). (n) is the reading on the Baume scale, in degrees Baume.

4. Stir the test sample continuously with a stirring rod and slowly add the permanganate solution until the color changes to a faint pink that lasts for at least 15 s.
5. Record the reading taken on the graduated burette. This is the number of cubic centimeters of 0.1 normal potassium permanganate used.

The number of cm3 of 0.1 normal potassium permanganate used, multiplied by 0.5580, equals the number of grams of iron (Fe) per 100 cm^3 of pickling solution.

Measurement of iron content in the phosphoric acid bath is done as follows:

1. Take 1 cm^3 of phosphoric acid bath sample, measured accurately with a pipette, and add to a 125-cm^3 Erlenmeyer flask.
2. Add 1 cm^3 of 50% C. P. sulfuric acid and about 25 cm^3 of distilled water.
3. Add 0.18 normal potassium permanganate from a titration burette, while stirring, to solution in a 125-cm^3 Erlenmeyer flask until the solution first turns a pink color of permanganate.
4. Record the number of cubic centimeters of permanganate solution used.

Each cubic centimeter of permanganate solution used is equivalent to 0.96 g of iron per 100 cm^3 of phosphoric acid bath.

If 3.0 cm^3 were required to obtain the pink color, then 3.0 × 0.96 g/100 cm^3 = 2.88 g/100 cm^3 iron in pickle bath. A titration requiring 12 cm^3 of permanganate solution would equal an iron concentration of 12 g/100 cm^3.

1.21.4 **RECORDS**

If there are facilities to analyze the bath, pickling can be efficient. Keeping complete records is essential. A simple procedure for recording the strength and temperature of the pickle bath, its iron content, when and how much acid is added, the number of tons pickled, and other elements of the process is to plot the data at regular intervals on a chart or graph.

These records show the amount of acid consumed per ton. Other pertinent data can be calculated. The chart or graph indicates whether the bath has been discarded with too much acid or before enough iron has been dissolved. Such records can show the effect of different pickling procedures over various periods of time.

Other records help to determine cost accounting and compare one practice with another. An example is comparing the effect of an inhibitor throughout the life of a pickle bath. Data and calculation sheets provide for calculation of the cost per ton. When these figures are compared for two or more pickling practices, many advantages can be seen.

1.22 **PRECAUTIONS**

Do not stack pickled steel surfaces touching one another until they are completely dry and apply paint before visible rusting occurs. In addition, water should never be added to strong acids. Even when properly adding concentrated sulfuric acid to water, enough heat could be generated to boil and blow the acid around, which could cause damage. Further, personnel who are handling acids should use proper protection, including face shields, rubber gloves, and rubber protective clothing.

1.23 MANUAL CLEANING (HAND AND POWER TOOL CLEANING)

Manual cleaning is the method of preparing the surface of metals using hand or powered tools, such as wire brushes, chipping hammers, chisels, scrapers, and vibratory needle guns. It is the least satisfactory method of preparation, especially for steelwork exposed to severe or moderate conditions. It does not provide a satisfactory base for many coatings. But the size of the areas to be treated or other circumstances may preclude the use of more effective methods.

Hand cleaning is acceptable only for spot cleaning, and power tool cleaning is acceptable where blast cleaning is impractical or not economical. The case should be specified by the company. The surface produced should be in accordance with SIS 055900 St3.

1.23.1 SURFACE PREPARATION BEFORE AND AFTER MANUAL CLEANING

Before manual cleaning, remove visible oil, grease, soluble welding residues, and salts by degreasing methods and remove flux residues and loose mill scale. After manual cleaning and prior to painting, reclean the surface if it does not conform to these specifications.

After manual cleaning, and prior to painting, remove dirt, dust, or similar contaminants from the surface. Acceptable methods include brushing, blowing with clean, dry air, and vacuum cleaning. This type of mechanically cleaned surface should be treated with primer on the same day.

1.23.2 HAND TOOL CLEANING

Hand tool cleaning is one of the oldest processes used to prepare surfaces prior to painting. It does not remove tight mill scale and all traces of rust, so it is only acceptable for extremely small areas (i.e., spot cleaning). The tools used for hand cleaning include wire brushes, nonwoven abrasive pads, scrapers, chisels, knives, chipping hammers, and, in some instances, conventional coated abrasives. Specially shaped knives are sometimes necessary as well.

Wire brushes may be of any practical shape and size. Two general types are an oblong brush with a long handle, and the Block type of brush. The bristles are made of spring wire. Brushes should be discarded when they are no longer effective because of badly bent or worn bristles. Nonwoven abrasives come in simple pad form, or they can be applied to a backup holder with a handle. They can be cut to fit various applicators.

Scrapers are designed in a number of convenient ways. They should be made of tool steel, tempered, and kept sharp to be effective. Some scrapers are made by sharpening the ends of 3.5–5-cm-wide flat files or rasps and fastening them to a handle. The handle may be up to 1.5 m long to increase the area that can be reached. Other chipping and scraping tools are made from old files or rasps that are sharpened at both ends. The file is bent at right angles at a point several centimeters from one end.

Hand-chipping hammers are advisable in maintenance work where rust scale has formed. A chipping hammer is about 10–15 cm long, and there are two wedge-shaped faces at either end of the head, one face perpendicular to the line of the handle and the other at right angles to the first face. Auxiliary equipment includes dust brushes, brooms, various sizes of putty knives, and conventional painters, scrapers, coated abrasives, and safety equipment such as goggles and dust respirators.

Hand cleaning of painted surfaces should remove all loose, nonadherent paint in addition to any rust or scale. If the paint to be removed is thick, the edges of the old paint should be feathered to improve the quality of the cleaning job. After cleaning, the surface is brushed, swept, dusted, and blown off with compressed air to remove all loose matter.

1.24 **POWER TOOL CLEANING**

Power tool cleaning provides a better foundation for the priming of paint than hand tool cleaning, especially when surfaces are covered with heavy rust, scale, or other firmly adherent deposits. Generally, power tool cleaning is suitable only for small areas because it is relatively slow. Power tools do not leave as much residue or produce as much dust as abrasive blasting, and they are frequently used where blasting dust could damage sensitive surroundings. However, they may polish or damage the surface if they are used at too high a speed or kept in one spot for too long.

Power tool cleaning should overlap by a minimum of 25 mm into any adjacent coated surface. It should produce a surface of Grade st3 in accordance with SIS 055900.

Power tools used for surface cleaning fall into three basic families:

- Impact cleaning tools
- Rotary cleaning tools
- Rotary impact cleaning tools

Tools in each family have unique characteristics that make them adaptable to different cleaning operations and requirements. The type of power tool to be used depends upon whether an acceptable profile exists on the surface to be cleaned.

Rotary cleaning tools produce a profile of approximately 10–15 μm, whereas impact and rotary impact tools produce a profile of 25 μm or more. The profile depends on the abrasive embedded in the rotary flaps or the diameters of the needles in the needle guns.

1.24.1 **IMPACT CLEANING TOOLS**

Impact cleaning tools are characterized by chipping and scaling hammers. With these tools, a chisel is struck by an internal piston and strikes the work surface.

Chisels can be adapted for scraping and chipping. This type of tool is useful when heavy deposits of rust scale, mill scale, thick old paint, weld flux, slag, and other brittle products must be removed from metal. They have different shapes and are made of various materials.

Cleaning a surface with scaling and chipping hammers is a comparatively slow process. However, when considerable rust scale or heavy paint formation must be removed, it may be the best and most economical method.

Other impact power cleaning tools include needle gun scalers and piston scalers. Needle scalers are more effective on brittle and loose surface contaminations. Piston scalers can be used for cleaning large surface areas.

Impact cleaning tools may be used to remove tight mill scale and surface rusting, but they are not the most practical or economical tools because they gauge metal, which must be smoothed to do a thorough job. Tools must be sharp or they may drive rust and scale into the surface.

1.24.2 **ROTARY CLEANING TOOLS**

Rotary power tools do most hand cleaning jobs rapidly. There are three basic types of cleaning media for these tools:

- Nonwoven abrasives
- Wire brushes
- Coated abrasives

These media can be used on two basic types of tools:

- Straight or in-line machines
- Vertical or right-angle machines

The straight or in-line machine style is used with radial wire brushes, coated abrasive flap wheels, and nonwoven abrasive wheels. The vertical machine style is suited for cup wire brushes, coated abrasive disks, nonwoven abrasive disks, cup wheels, and wheels. The type of machine varies with job conditions. It is advisable to have both types on hand because generally both are used on field jobs. Rotary cleaning machines may be operated by pneumatic or electric motor.

These machines should be compatible with the size and speed rating of the cleaning media and should produce enough power to perform the operation efficiently. Most air-powered machines contain governors to limit the free operating speed. Governors respond to tool load resulting from thrust applied to the work surface and supply more air to the motor, increasing power output and maintaining its rated speed while under load. Electrically driven machines operate at a fixed speed.

Nonwoven abrasives come in cup wheel, radial (wheel) and disk form. Nonwoven abrasive wheels are recommended where base metal should not be removed, but wire brushes are not aggressive enough. These wheels wear at a controlled rate. Fresh working abrasive provides a constant rate of surface cleaning with minimal loading. Nonwoven abrasives are also recommended for removal of light mill scale and coatings because susceptibility to loading is lower compared to coated abrasive.

Wire brushes come in radial (wheel) and cup form. They can be composed of differently shaped and sized wire bristles that may be crimped (crinkled) or knotted. The style and type of bristle or nonwoven abrasive composition employed should be based on trials.

In power wire brushing, it is possible to cut through some mill scale by using the toe of a very stiff brush and bearing down hard. It is impractical to remove tight mill scale by power wire brushing. Generally, removal of only loose mill scale and rust is required. Too high a speed should not be used with rotary wire brushes, and the brush should not be kept on one spot for too long. Detrimental burnishing of the surface may occur. Under such circumstances, the surface is smooth and develops a polished, glossy appearance that provides a poor anchor for paint. It is clear that doing too much surface work is detrimental. Rotary wire brushes are particularly notorious for spreading oil and grease over the surface. Oily or greasy surfaces must be cleaned with solvent before power brushing.

Coated abrasives are used in disk and flap wheel form. Such abrasives are used for metal removal such as weld grinding. Tight mill scale cannot be removed with such media, but loose scale can be. Disks and flap wheels are used to remove loose mill scale, old paint, and other substances similar to the way wire brushes are used.

1.24.3 **ROTARY IMPACT TOOLS**

Rotary impact tools operate on the same basic principle as other impact tools, through cutting or chipping action. But these tools use a centrifugal principle where cutters or hammers are rotated at high speed and thrown against the surface.

Rotary chipping tools use three major types of media: cutter bundles (or stars), rotary hammers, and heavy-duty rotary flaps. All three can be used on pneumatic or electric-powered tools.

1.25 **MANUAL CLEANING OF NONFERROUS METALS**

Adhesion of paint to nonferrous metals may be helped by abrading the surface with fine emery cloth or abrasive paper, and white spirit; this method may be used for small areas prepared on site. Care should be taken not to abrade through thin films of pure aluminum on composite ("clad") sheets; steel wool or hard abrasives should never be used on these surfaces.

Unless metal-sprayed coatings receive a protective coating, corrosion may develop fairly rapidly, especially in damp or chemically charged atmospheres. Although they are not necessarily detrimental to the metal coating, they may affect paint adhesion and should be removed before painting. Scrubbing with clean water with stiff bristle or nylon (but not wire) brushes, followed by rinsing with clean water is usually effective at removing these substances.

Removal of the layer of corrosion that may form on aluminum after several years of exposure can be difficult, and abrasion may be necessary (e.g., with stainless steel wire wool or nylon pads), using water or white spirit as a lubricant. Mild steel, brass, or copper wire should not be used, as broken strands may become embedded in the surface and cause corrosion.

If nonferrous metal coatings on iron and steel are damaged or become eroded over long exposure, rusting of the exposed base metal is likely to occur. Rusted areas should be cleaned by wire-brushing or abrasion, with care being taken to avoid damaging adjacent sound zinc or aluminum.

1.26 **PRECAUTIONS**

Safety practices include the following considerations. All people in the area, including the users of the tools, should wear eye protection to guard against flying particles. Ear protection should be considered as well when impact tools are used. Particular care should be taken when using several tools simultaneously in close proximity.

Hand tools should be selected for the necessary purpose and properly maintained. Hammers should be properly heat-treated and striking faces maintained to eliminate "mushrooming" and flying fragments. All sharp-edged tools deserve respect and proper consideration.

Electrical tools should be run in dry environments. They should be grounded or double-insulated. Power cords should be kept in good repair.

Impact tools should be operated only when the scaling tool is in position and in contact with the work piece. Tools should not be used if ejection of an accessory might endanger personnel.

Gloves and leather aprons are additional safeguards to avoid injury from loose wires. In addition, recommended guards should always be used.

Nonwoven abrasive wheels should be operated in the proper direction of rotation. The wheel or disk should be put on the tool and tightened securely while the tool is disconnected from the power supply. Guards should be used. Protective clothing should be considered. With pneumatic tools, maintaining the proper air pressure is vital. The correct revolutions per minute (rpm) level should be checked with a tachometer for all tools before use.

Respirators should be used if contaminants in the breathing zone exceed applicable threshold limits. This is of particular importance when cleaning paints containing lead, chromate, or coal tar.

If fire or explosive hazards are present, proper precautions should be taken before any work is done. Since the cleaning operations can produce sparks, care must be exercised when cleaning in the area of combustibles and volatile vapors. When such conditions cannot be avoided, only nonsparking tools should be used. If the structure previously contained flammable materials, it should be purged of them. If the structure being cleaned is near flammable material or fumes, nonsparking tools should be used.

1.27 BLAST CLEANING

Blast cleaning is a method of surface preparation in which abrasive particles are directed at high velocity against a metal surface. They may be carried by compressed air or high-pressure water, or thrown by centrifugal force from an impeller wheel.

Blasting operations should never be allowed in the vicinity of painting work, near wet paint, anywhere that blast abrasive, grit or fallout could impinge on a freshly painted surface, or on any uncovered primed surface. Blast cleaning should not be conducted on surfaces that will be wet after blasting and before coating, when the surface temperature is less than 3°C above the dew point, when the relative humidity of the air is greater than 85%, or when the ambient temperature is below 3°C. Blast cleaning is permitted only during the daytime.

Where rectification has been necessary on a blast-cleaned surface, the particular area should be reblasted to remove all rust and slag, and to provide adequate paint adhesion. Blast cleaning should overlap by a minimum of 25 mm into any adjacent coated areas. Any steel work not primed or wetted by rain or moisture should be reblasted prior to being painted if rust develops.

Steel may be blast-cleaned either before or after fabrication, and sometimes it may be necessary to do this at both times. Where steel is cleaned before fabrication, it should be protected with a suitable blast primer to avoid rusting. During fabrication, the blast-primer will inevitably be destroyed or damaged in places (e.g., by welding). Such areas should be cleaned and reprimed as soon as possible. Where steel is cleaned after fabrication, the first coat of the full protective system can be applied, but it may still be necessary to apply a blast-primer first.

The roughness of a prepared surface results from primary roughness already present in the initial state and which is exposed by the mechanical preparation methods, particularly by blasting. The roughness parameters (such as the peak-to-valley height) are factors used for determining the minimum coat thickness necessary for satisfactory embedment and covering. The surface roughness of steel work should be within 0.1 mm to 0.03 ± 0.005 mm for painting, coating, and lining.

Unless otherwise specified by the company, the level of cleanliness for prepared steel surfaces should be Sa3 for zinc silicate and zinc rich epoxy primers, and Sa2½ for other primers and organic coatings and for metal coatings, claddings, and linings.

A qualified, well-trained operator should be employed for the blast-cleaning job.

When sheets of less than 4 mm thickness are blasted, deformation may occur. This should be avoided by using the following:

1. Low air pressure, small grain sizes, and a blasting abrasive of low bulk density
2. A low angle of blasting with sharp-edged grain
3. A short duration of blasting

Note that the first two measures will result in less efficient blasting. With adhesive scale or fairly thick rusting, it may be necessary to carry out preparatory work by some other methods (e.g., manual cleaning or pickling, before cleaning). All blasted steel surfaces should be primed before visible rerusting occurs or within 4 h, whichever is sooner.

1.28 CHOICE OF METHOD OF BLAST CLEANING

The blast-cleaning method will be determined mainly by the factors discussed in the next sections.

1.28.1 SHAPE AND SIZE OF STEELWORK

Centrifugal methods are economical to use for plates and simple sections; they can also be used for large prefabricated sections (e.g., bridge sections), but only in specially designed plants. Misses, gaps in the coating, discovered by inspection can be cleaned with air blast techniques. For large throughput of shaped items such as pipes, both pressure and vacuum air-blasting techniques can be used in continuous and automatic plants.

1.28.2 EFFECT OF THE STAGE AT WHICH IS CARRIED OUT

For blast cleaning on site, pressure or vacuum air-blasting methods have to be used as on large fabricated sections. However, it may be impracticable to use centrifugal methods.

1.28.3 THROUGHPUT

Centrifugal plants are economical to use for a high throughput, but even with a low throughput, this method may still be preferable to large-scale air blasting.

1.28.4 ENVIRONMENTAL CONDITIONS

Despite its relatively high cost, vacuum air blasting may be necessary to avoid contaminating the immediate area with abrasive. It should be ensured that the blast-cleaning process does not affect adjacent materials.

1.28.5 TYPES OF SURFACE DEPOSIT TO BE REMOVED

Water-blasting methods using abrasives are particularly suitable for removing entrapped salts in rust and for abrading hand-painted surfaces (e.g., two-pack epoxies) before recoating. Various methods are listed in Table 1.12, with notes on their advantages and disadvantages.

Table 1.12   Method of Blast Cleaning

Methods	Advantages
Dry Methods Using Compressed Air or Centrifugal Force	
Centrifugal blasting	High production rate, lowest cost, no moisture problems, can be coupled to automatic application of primer, dust problems contained
Air blasting (pressure and suction type)	Simple to operate, very flexible, and mobile to use in both indoor cabinets and special rooms or on site, low capital and maintenance costs
Air blasting (vacuum type)	No dust problems, no special protective clothing needed for operators, fairly low capital costs
Water blasting (hydroblasting, based on projecting water at very high pressure)	Simple to operate, very flexible and mobile to use, suitable for removing soluble contaminants, at very high pressure can remove mill scale, no dry dust hazard
Water blasting (based on projecting water at high pressure and entraining abrasive into the water stream)	Simple to operate, very flexible and mobile to use, suitable for removing all firmly held contaminants as well as soluble contaminants
Water blasting (based on injecting low-pressure water into a compressed air stream that is carrying an abrasive) Water blasting using steam cleaning	Simple to operate, very flexible and mobile to use, suitable for removing all firmly held contaminants as well as soluble contaminants

Disadvantages:

High capital cost, high maintenance cost, lack of flexibility (i.e., not suitable for recessed areas).

High cost of compressed air, low efficiency, liable to moisture entrainment from the compressed air, manually operated and a variable profile can result, operator requires protective clothing, serious dust problems.

Can be very slow and therefore expensive to use on awkward profiles and girder sections. Where flat-plate or gun-head automation is possible, it may be considered, but liable to moisture entrainment from the compressed air.

1.29 ABRASIVES

The grades of metallic and nonmetallic abrasive are selected with the goal of having surface roughness with a maximum amplitude of 0.1 mm, and the grades of nonmetallic abrasive are selected with the goal of having surface roughness with a maximum amplitude of 0.18 mm. It is essential to avoid using contaminated abrasives, as the following three types of contamination may occur:

1. Dry dust and detritus from the surface and the smaller fines from the breakdown of abrasives. They can be removed by automatic and recirculatory plants. Abrasives should not be reused unless they have been cleaned via this process.

2. Water, either on the surface, in the compressed air, or from conditions of very high humidity, forms agglomerates of dust and abrasive particles, which will inhibit cleaning. In this case, the abrasve should be dry before cleaning.
3. Oil and grease on the surface or from the equipment preclude the reuse of abrasives. Such oil and grease should be removed before blast cleaning.

The choice of abrasive will be determined mainly by economic considerations.

1.30 BEFORE AND AFTER BLAST CLEANING
1.30.1 BEFORE BLAST CLEANING

The steel to be blast-cleaned by dry methods should be dry, and the operating conditions should be such that condensation does not occur during the work. When compressed air is used, the surface should be dry and free of oil. Weld defects such as pin holes and discontinuities should be rectified, and weld undercutting should be filled or dressed. Excessive weld spatter should be dressed off (only light spatter will be adequately removed by blast cleaning). Welding slag should be cleaned off. Laminations, laps, and shelling should be dressed off completely. Sharp edges should be smoothed off. Burrs and grease or oil contamination should be removed.

1.30.2 AFTER BLAST CLEANING

The following steps should be done after blast cleaning:

- Residual shot, grit, and dust should be completely removed, preferably by vacuum cleaning, but otherwise by oil- and water-free air blasting or fiber brushing.
- Care should be taken not to contaminate blast-cleaned surfaces prior to painting.
- Blast-cleaned surfaces may be protected for short periods by a thin coat of pretreatment or prefabrication primer. It is imperative that such primers should be applied as a continuous coating in an even manner to achieve a minimum film thickness of 20 μm. Such primers do not replace the full thickness of permanent primer.
- The prepared blast-cleaned surface should be completely primed the same day, before any visible rusting or deterioration of the surface occurs. No blasted surface should stand overnight before coating. If such surfaces are not primed in accordance with these specifications, they should be reblasted.
- Care should be taken not to contaminate blast-cleaned surfaces prior to painting.

1.31 TESTING THE CLEANLINESS OF THE BLASTED SURFACE
1.31.1 VISUAL OR FIELD TESTING

The rust grades for steel surfaces and preparation grade prior to protective coating are useful for assessing cleanliness.

Rubbing a surface with clean, lint-free, white cotton cloth, commercial paper, or filter paper moistened (but not saturated) with high-purity solvent may be used for evaluating the cleanliness of surfaces not accessible for direct visual inspection. Wipe tests of small diameter tubing are made by blowing a clean white felt plug, slightly larger in diameter than the inside diameter of the tube, through the tube with clean, dry, filtered, and compressed air. Cleanness in wipe tests is evaluated by the type of contamination rubbed off on the swab or plug. The presence of sludge on the cloth is evidence of contamination. If the harmful nature of the contamination is not known, a sample of the sludge may be transferred to a clean quartz microscope slide for infrared analysis. Sometimes the test is repeated with a black cloth to disclose contaminants that would be invisible on a white cloth.

Formation of rust stains may be accelerated by periodically wetting the surface with preferably distilled or deionized water or clean, fresh, potable tap water. The wet-dry cycles should be such that the sample remains dry for a total of 8 h in a 24-h test period. After completion of this test, the surface should show no evidence of rust stains or other corrosion products.

When blast-cleaning heavily rusted and rust-pitted steel, a surface appearance that suggests a standard of cleanness equivalent to white metal or Swedish standard Sa2½ or Sa3 can be obtained, which, after an hour or two and sometimes after standing overnight, develops rust-spotting at points corresponding to the rust-pitting. The cause is the presence of rust-producing soluble salts of iron, which are practically colorless and are located at the lowest point of the rust pits. In the presence of moisture, they hydrolyze to form iron oxides and acids; the acids dissolve more iron to form iron salts, thus eventually producing large volumes of rust, which will break the adhesion bond of protective coatings if applied before the salts are removed. In such circumstances, protective coatings are not allowed until the cause of the rust-spotting has been eliminated.

Using potassium ferricyanide test papers, any remaining soluble iron-salt contaminants can be detected as follows:

1. Spray a fine mist of water droplets on a small area of the blast-cleaned surface using a hand spray (such as a scent-spray type of bottle).
2. Allow the water droplets to evaporate, and at the moment when they have disappeared but the surface is just perceptibly wet, apply a small piece of test paper and press with the thumb for 2 to 5 s.

If soluble salts remain, these will be drawn by capillary action into the test paper and will react with the potassium ferricyanide to give a characteristic Prussian blue complex, as blue dots appear on the paper corresponding to the contaminated pits on the blast-cleaned steel.

1.31.2 INSTRUMENTAL OR LABORATORY TESTING

In electrical testing, the contact resistance between the blasted steel surface to be tested and a probe (sprung contact in the form of a sphere of 1-mm diameter) is measured with an ohmmeter not requiring a mains electrical supply and which provides the highest possible accuracy of reading in the range 0–1 Ω. The position of the measurement points is selected arbitrarily, and the number of such points should be agreed by standards. The arithmetic mean is calculated from the individual measured values. Reproducible measured values can be obtained only if the inherent short circuit impedence of the measuring circuit is eliminated. The information provided by this method will be improved if, instead of reading the measured values and calculating the mean value, a determination is made, using an electric counter, of the number of points of application of the probe at which the contact resistance is so small (i.e., the surface is so clean) that a current flows in a particular measurement circuit.

The measured results are affected by the blasting abrasive and the blasting method (type and number of impacts) and also by the initial state of the blasted surface.

In a high-humidity test, subject the surface to a 95–100% humidity at (38–46°C in a suitable humidity cabinet for 24–26 h. After completion of this test, the surface shows no evidence of rust stains or other corrosion products.

Dry the cleaned surface after finish-cleaning at 49°C for 20 min. The presence of stains or water spots on the dried surfaces indicates the presence of residual soil and incomplete cleaning. The test is rapid but not very sensitive.

- The copper sulfate test for passivated surfaces is performed as follows:

Swab the component for a few minutes with an aqueous solution containing the following:
- Copper sulfate ($CuSO_4 \cdot 5H_2O$) 4 g/l (approx.)
- Sulfuric acid ($d = 1.84$) 1.5 ml/l (approx.)

After drying, the presence of any areas of deposited copper will confirm that the surface is unsatisfactory and should be repassivated before being submitted for reapproval.

An alternative test may be carried out on a component, if suitable, or on a test piece with a flat smooth surface made from material of similar composition to the parts being processed. It is performed as follows:

Within 30 min of processing, place one spot of the following solution on a flat passivated surface and allow it to remain for 3 min:

- Palladium chloride, 0.5 g
- Hydrochloric acid ($d = 1.16$), 20 cm^3
- Water, 98 cm^3

Wash the spot with cold running water. Do not swab. If passivation has been effected, no trace or only a slight trace of a dark deposit will be evident on the tested area.

In the refractometric method, grades of blast cleaning are selected in relation to the conditions of use and the protective system chosen. Specifications for blast cleaning include pictorial standards (SIS 055900) or sketches (BS 4232), but these can be used for site control only if the lighting is good and the inspectors are highly experienced. Steel coupons blast-cleaned to an agreed standard are better, but they should be carefully stored and handled to prevent deterioration.

For convenience, it is best to use instruments that enable figures to be recorded. A light reflectometer can be used to assess the effectiveness of removal of mill scale, absence of rust, and general approach to "white metal" condition. This instrument measures the amount of blue light reflected from the surface compared with that from a standard gray tile. Repeatedly blasting the same area of steel causes an increase in reflection until a steady value (Rmax) is reached that corresponds to white metal (dependent on the abrasive and the steel), as shown in Figure 1.8. The acceptance level should be fixed as a percentage of Rmax for Sa2½, the average of all readings on the work should be at least 90% of Rmax, and not more than 1 reading in 10 should be below 80% of Rmax. When using this control method, the value of Rmax and the acceptance levels should be defined by the company at the start of the work.

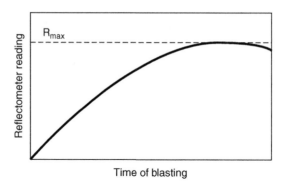

FIGURE 1.8

Assessment of steel cleaning.

1.32 PRECAUTIONS
1.32.1 DRY BLASTING

If fire or explosion hazards are present, proper precautions should be taken before any work is done. If the structure previously contained flammable materials, it should be purged of dangerous concentrations of these substances.

Nozzle blast operators exposed to blast dust should wear a suitable helmet. Filter-type air respirators should be worn by all who are exposed to blast dust, and safety goggles should be worn by all persons near any blasting operation. Adequate protection for personnel from flying particles should also be provided in any blasting operation. The blast nozzle should be properly grounded through use of a hose with antistatic lining. The blasting pot will be equipped with ground wire.

1.32.2 WATER BLASTING

Hydroblast cleaning uses high pressure, so extreme caution should be exercised with the equipment. Instruction and training of operators about correct use and equipment operation is essential. If the surfaces are not metal, they can be damaged with high-pressure water and should be protected from those effects.

The operator of a hydroblast unit must have sound, safe footing. Extra caution should be taken on rigid scaffolding. Swinging stages and bosun chairs are not normally recommended for use with hydroblasting. The operator should wear a rain suit, face shield, hearing protection, and gloves. Everyone within 17 meters of the work area should be warned of hazards associated with hydroblast cleaning.

The blast gun should have automatic control to release pressure when the operator releases the trigger. A dump valve on the gun serves this purpose.

1.33 FLAME CLEANING

Surface preparation by flame cleaning is designated by the letters *fl*. In flame cleaning at a high temperature, oxyacetylene flame is passed over the surface to be cleaned. The effect of the heat is to remove scale and rust, partly by differential expansion and partly by evolution of steam from moisture in the rust.

Prior to flame cleaning, any heavy layers of rust should be removed by manual cleaning. This method does not remove all rust and scale and is in no way a substitute for blast cleaning. The coating applied to a blast-cleaned surface will always last longer than similar coating applied to a flame-cleaned surface. Reasonably portable flame-cleaning equipment can be used in some locations where blast cleaning is not practical. Flame cleaning is not allowed for new work, but it may be used to prepare a surface for maintenance painting, particularly in damp weather.

Care is needed to ensure that surface temperatures are not allowed to become excessive. In the case of structural steelwork, the upper limit is 200°C. However, thinner sections may have to be cleaned at a lower temperature. Uncoated surfaces thinner than 0.5 mm should not be flame-cleaned. Care should be taken to consider the effect of heat transference to other parts of the structure, particularly to any materials touching the other side of the steel.

Execution by injector-mixing chamber burners and oxyacetylene flame (surplus oxygen), which is passed one or more times over the surface with cooling after each passage is an option. Ensure that the rate of flame movement is sufficiently slow to prevent moisture from being deposited. Selection of the correct ratio of gases is essential to give best results.

Flame cleaning should not be used near high-strength. Friction grip bolts or cold-worked high-tensile steel, the existence of pipes and cables should be noted. Flame cleaning of high-strength friction grip bolted joints is totally prohibited.

Flame cleaning should not be used to remove thick coating of tar or bitumen, and it is not recommended for stripping old paint. As a rough guide, its use should be considered only when the breakdown on the old paint has reached grade Re 6 (roughly 20% of rust by area) of the European scale or beyond. It should be noted that flame cleaning does not provide a suitable surface for many coatings, such as zinc silicate, epoxide, and polyurethane two-pack coating.

After flame cleaning, the surface should be cleaned by power tool wire-brushing. Hand wire-brushing does not achieve a satisfactory surface for painting.

Apply priming paint to surfaces that are still warm (but not hot) from the flame-cleaning process. It may be desirable to limit the surface temperature to a maximum of 40°C before paint is applied.

Flame cleaning is usually the quickest and most economical method for the removal of paint from general surfaces of woodwork. It may also be used occasionally to cleaning ship decks before repainting.

1.34 SURFACE PREPARATION OF IRON AND STEEL
1.34.1 CAST IRON

Prior to phosphating and painting, cast-iron parts should be cleaned with one or more of the following methods: degreasing, pickling, hand and tool cleaning, blasting, anodic etching and chemical smoothing. If inhibited solutions are used for acid pickling, absorbed inhibitor must be removed before phosphating or application of painting.

Preparation for hot dip galvanizing should be carried out using one of the following procedures:

- Abrasive blasting with chilled iron or steel, to remove molding sand and iron scale. After blasting, pickling may be used, and where necessary, a flux coating should be applied.
- Hydrofluoric acid (30% HF) diluted to 10% by volume aqueous solution may be used, followed by hosing with water to remove any gelatinous products. After hosing, pickling is used, and where necessary, a flux should be applied.

Prior to metal spraying, the surface should be degreased and then roughened by compressed air or centrifugal blasting with a suitable abrasive, for protection against corrosion and high-temperature oxidation. Immediately before spraying, it should be free of grease, scale, rust, moisture, or other foreign matter. It should be comparable in roughness to a reference surface produced in accordance with standard and should provide an adequate key for the subsequent sprayed metal coating.

1.34.2 STEELS, NONCORROSION-RESISTING

Prior to phosphating and coating, the parts should be cleaned by one or more of the following methods:

- Degreasing in an organic solvent.
- Degreasing in an aqueous alkaline solution with applied anodic direct current of desired. Neither cathodic nor alternating current treatment should be used.
- Alkaline de-rusting.
- Blast cleaning with grit abrasive (It is preferred for high tensile steels and some alloy steels).
- Anodic etching.
- Acid pickling is an inhibited acid solution, is, followed with rinsing prior to phosphating.

Preparation for hot dip galvanizing of steel (other than casings) should be carried out using one of the following procedures:

- The degreasing procedures followed by acid pickling. Where necessary, a flux coating of dried iron salts, zinc ammonium chloride, or a proprietary flux should be applied.
- Abrasive blasting with chilled iron or steel grit to remove welding slag or paint. After blasting, the pickling methods may be used, and where necessary, a flux should be applied.

Preparation of steel casings should be carried out using one of the following procedures:

- Abrasive blasting with chilled iron or steel grit, to remove molding sand and iron scale. After blasting, and where necessary, a flux coating should be applied.
- Hydrofluoric acid (30% HF) diluted to 2% to 10% by volume aqueous solution may be used, followed by hosing with water to remove any gelatinous products.

Prior to metal spraying, the surface should be degreased and then roughened by compressed air or centrifugal blasting with a suitable abrasive for protection against corrosion and high-temperature oxidation. Immediately before spraying, it should be free of grease, scale, rust, moisture, and other foreign matter. It should be comparable in roughness with a reference surface produced in accordance with standard, and should provide an adequate key for the subsequent sprayed metal coating.

For sheet steel up to and including 1.6 mm thick, some relaxation on the degree of roughness may be agreed between the company and the contractor in order to minimize distortion during surface preparation. Distortion can be reduced by blast cleaning on both sides.

Prior to electroplating, the part should be cleaned using one of the following procedures:

- Degreasing with cold solvent, hot solvent, emulsifiable solvent, detergent cleaning, or steam cleaning with or without alkaline or solvent emulsion.

- Intermediate alkaline cleaning removes solvent residues and residual soil which has been softened by degreasing the operation data as described in the next sections of this chapter.
- Rinsing. Double rinses are desirable to reduce the concentration of cleaner in the rinse. Warm rinsing (in 60°C water) is desirable following alkaline immersion cleaning. Agitation of the rinse water is desirable, and, in the case of cold-water rinses, a spray upon leaving the tank is beneficial. The time of rinsing depends in part upon the shape of the part, but it should be no less than 10 s.
- Anodic etching
- Acid pickling, the part should be rinsed in cold water after pickling
- Activation

If electroplating is to be done in alkaline solutions, such as cadmium, copper, tin, or zinc, no further treatment should be necessary. For nearly neutral or acid electroplating processes, however, parts should be immersed for 5 to 15 s, at room temperature in a solution containing 50–100 ml of sulfuric acid. Activated parts should be rinsed in cold water.

1.35 CHEMICAL SMOOTHING PROCESS FOR STEEL

The parts should be treated as follows:

1. Degrease the parts with solvents or alkaline.
2. Dip them in sulfuric acid
3. Rinse thoroughly in cold water.
4. Immerse them in an aqueous solution containing oxalic acid ($C_2O_4H_2 \cdot 2H_2O$), 25 g/l; hydrogen peroxide, 13 g/l; and sulfuric acid, 0.053 cm^3/l

The solution should be used at room temperature.

The treatment may last a few minutes or several hours, depending on the application. The rate of metal dissolution for mild steel is approximately 10 m/h. The process is ineffective on corrosion-resisting steel and on low-alloy steel containing more than about 1% chromium.

Hydrogen peroxide decomposes quite rapidly so amounts equal to the original quantity need to be added every 20 min during use. The complete solution should be replaced when the rate of dissolution falls to an unacceptable level.

1.36 ANODIC ETCHING FOR FERROUS METALS

Etching of ferrous metals should be carried out using one of the methods discussed in this section.

1.36.1 METHOD A

Method A is suitable for all parts, so it is the preferred method. The parts should be immersed in an aqueous solution composed of sulfuric acid ($d = 1.84$), 300 cm^3/l. The solution strength is maintained by periodic additions of sulfuric acid to keep the relative density $d > 1.30$.

The voltage should be set so that the initial current density is not less than 1,000 A/m^2 (4–8 volts) and is preferably twice that amount. The temperature should preferably not exceed 25°C, but for removal of scale, it can go up to 70°C. An inhibitor that is stable in the solution may be added if desired to reduce the formation of smut and absorption of hydrogen by the steel, but wetting agents should be avoided, as they promote the absorption of hydrogen.

Removal of scale can be made easier by soaking the item in an alkaline solution of sodium gluconate at up to the boiling point prior to the procedure. The current density required to remove scale can be reduced by interrupting the current for a few seconds from time to time. If current interruption is to be used on steels of tensile strength 1,400 N/mm^2 and stronger, a safe procedure should be established to the satisfaction of the purchaser.

1.36.2 **METHOD B**

With Method B, the parts should be immersed in an aqueous solution ($d = 1.74$) containing sulfuric acid ($d = 1.84$), 750 cm^3/l with or without a small addition of chromic acid. The solution strength is maintained by periodic additions of sulfuric acid to keep the relative density $d > 1.70$.

An initial current density of at least 1,000 A/m^2 (4–12 V) should be applied. The temperature should not exceed 25°C.

1.36.3 **METHOD C**

For Method C, an alternative aqueous solution ($d = 1.22$) for use on stainless steel prior to a nickel strike should contain sulfuric acid ($d = 1.84$), 200 cm^3/l. The solution strength is maintained by periodical additions of sulfuric acid to keep the relative density $d > 1.20$.

A current density of 2,000 A/m^2 to 2,500 A/m^2 (approximately 6 volts) should be applied. The temperature should not exceed 20°C. The immersion time should range from 1 to 3 min.

The following points should be noted about anodic etching:

- Contamination of the solutions with chloride should be avoided.
- The essential requirement of anodic etching of steel is to render the steel passive, a condition that will be indicated by a sharp increase in voltage between the part and the cathode, a corresponding drop in current, and the onset of gas evolution from the part.
- Passivity should be maintained until the surface acquires a light gray color, free of dark smut.
- The parts may be withdrawn for examination, and if necessary, replaced in the bath for further treatment.
- Anodic etching may not be effective and may cause pitting on the inner surfaces of tubular parts unless auxiliary cathodes are employed. If you want to clean only the external surfaces, the part should be plugged to prevent ingress of the electrolyte.

1.37 **CLEANING OF SHOP-PRIMED STEEL SURFACES PRIOR TO OVERCOATING**

Any primed surface that has been exposed for more than a few days will become contaminated and should be cleaned with fresh water and allowed to dry before overcoating. Prefabrication primer or blast primer exposed to the atmosphere longer than 6 months should be removed by blast cleaning before applying the specified coating system. Prefabrication primer or blast primer less than 6 months old and still in good condition should be cleaned thoroughly with clean, fresh water before applying the paint system and should be removed only if they are not compatible with the subsequently specified paint system.

If it is necessary to protect austenitic stainless steel surfaces during transport and storage at the site, a proofed material such as plastic sheeting should be used unless the surfaces are provided with a prefabrication primer or blast primer. Shop-primed surfaces should be cleaned thoroughly with clean, fresh water before applying the subsequent layers.

The cleaning and patch painting of damage spots and weld areas should occur in addition to the complete specified coating system. If previously coated pipework is to be cut and welded on site, all coating must first be removed from the area of welded joint.

Primed and coated surfaces that have been exposed to marine environments including shipment, will be contaminated with salt and should be lightly wire-brushed and then washed with fresh water before overcoating.

Although zinc-rich primers are very effective at preventing rusting, extended exposure develops a surface contaminated of zinc-corrosion products, which can impair the adhesion of subsequent coats. Zinc-rich primers, both organic and inorganic, that have been exposed long enough to develop white surface staining should be prepared for overcoating by one of the following methods:

- Light blast cleaning and dust removal
- Wire brushing, followed by water washing with water
- Scrubbing with fresh water, using bristle brushes

Damaged surfaces should be repaired with a 0.10-m overlap.

1.38 SURFACE PREPARATION OF STAINLESS STEEL

For preparation of stainless steel prior to metal spraying and painting, parts should be cleaned by one or more of the methods described in the next sections, depending on the requirements of the coating to be applied and the surface condition of the substrate.

1.38.1 DEGREASING OF STAINLESS STEEL

Prior to descaling, degreasing of stainless steel should be applied according to standard. A degreasing material (e.g., alkaline, emulsion, hot, cold, or vapor solvents, synthetic detergent, steam, and cleaning agents] can be used.

For effectiveness of preparation, the cleaned surfaces should be inspected with tests of freedom from grease and tests of cleanliness for blasted surfaces.

1.38.2 CLEANING OF WELDS AND WELD-JOINT AREAS

The joint area and surrounding metal several centimeters from the joint preparation, on both faces of the weld, should be cleaned immediately before starting to weld. Cleaning may be accomplished by brushing with a clean, stainless steel brush, scrubbing with a clean, lint-free cloth moistened with solvent, or both. When the joint cools after welding, remove all accessible weld spatter, welding flux, scale, arc strikes, and other materials by grinding. According to the application, some scale or heat tempering may be permissible on the nonprocess side of the weld, but it should be removed

Table 1.13 **Cleaning with Chemical Solutions**

		Purpose—General Cleaning		
Alloys	**Condition**	**Solution Volume (%)**	**Temperature (°C)**	**Time (min)**
200, 300, and 400 series (except free-machining alloys) precipitation hardening alloys	Fully annealed only	Citric acid, 1 weight % plus $NaNO_3$, 1 weight %	21	60
	Same	Ammonium citrate 5–10 weight %	49–71	10–60
Assemblies of stainless and carbon steel (for example, heat exchanger with stainless steel tubes and carbon steel shell)	Sensitized	Inhibited solution of hydroxyacetic acid, 2 weight % and formic acid, 1 weight %	93	6 h
Same	Same	Inhibited ammonia-neutralized solution of EDTA (ethylene-diamene-tetraacetic acid) followed by hot-water rinse and dip in solution of 10 ppm ammonium 100 ppm hydrazine	121	6 h

from the process side if possible. If chemical cleaning of the process side of the weld is deemed necessary, all precautions must be observed. Austenitic stainless steel in a sensitized condition should not be descaled with a nitric-hydrofluoric acid solution. Welds should be cleaned as described in Table 1.13.

1.38.3 ACID PICKLING OF STAINLESS STEEL

Suggested solution contact times and solution temperatures for descaling of various types of stainless steel are given in Table 1.14.

For removing light scale from an austenitic stainless steel, use a solution with the following specifications:

- Ferric sulfate liquor (40% by weight) ($d = 1.5$) 200–300 cm^3/l
- Hydrofluoric acid (40% by weight) 50–75 cm^3/l
- Temperature, 60–70°C
- Time, 2–30 min

Overpickling must be avoided, and continuous exposure to the pickling solution for more than 30 minutes is not recommended. The item should be drained and rinsed after 30 min and examined to check the effectiveness of the treatment. Most pickling solutions will loosen weld and heat treating scale, but they may not remove them completely.

Table 1.14 Acid Pickling of Stainless Steel				
Alloy[1]	**Condition**	**Solution, Volume (%)[2]**	**Temperature (°C)**	**Time (min)**
200, 300 and 400 series; precipitation-hardening and maraging alloys (except free-machining alloys)	Fully annealed only	H_2SO_4, 8–11% [3]	66–82	5–45 Max[4]
200 and 300 series; 400 series containing Cr 16% or more; precipitation- hardening alloys (except free-machining alloys)	Fully annealed only	HNO_3 15–25% PLUS HF, 1–8% (5,6)	21–60	5–30[4]
All free-machining alloys and 400 series containing less than Cr 16%	Fully annealed only	HNO_3 10–15% plus HF, ½–1½ % [5,6]	Up to 140, with caution	5–30[4]

[1]This table is also applicable to the cast grades equivalent to the families of wrought materials listed.
[2]The solution is prepared from reagents with the following characteristics: weight %: H_2SO_4, 98; HNO_3, 67; HF, 70.
[3]Tight scale may be removed by a dip in this solution for a few minutes, followed by water rinse and nitric-hydrofluoric acid treatment as noted.
[4]The minimum contact time necessary to obtain the desired surface should be used in order to prevent overpickling. Tests should be made to establish correct procedures for specific applications.
[5]For reasons of convenience and handling safety, commercial formulations containing fluoride salts may be found useful instead of HF for preparing nitric-hydrofluoric acid solutions.
[6]After pickling and water rinsing, an aqueous caustic permanganate solution containing NaOH, 10 weight %, and $KMnO_4$, 4 weight %, (71 to 82°C) 5 to 60 min, may be used as a final dip for removal of smut, followed by thorough water rinsing and drying.

After chemical descaling, the surface must be thoroughly rinsed to remove residual chemicals, and a neutralization step is sometimes necessary before final rinsing. To minimize staining, surfaces must not be permitted to dry between successive steps of the acid descaling and rinsing procedure, and thorough drying should follow the final water rinse.

1.38.4 MANUAL CLEANING OF STAINLESS STEEL

Manual cleaning methods include power brushing and sanding with coated abrasive tools; grinding and chipping are used for the descaling of stainless steel surfaces. Grinding is usually the most effective means of removing localized scale, such as that which results from welding.

Particular care must be taken to avoid damage by manual methods when descaling thin sections, polished surfaces, and close-tolerance parts. After manual descaling, surfaces should be cleaned by scrubbing with hot water and fiber brushes, followed by rinsing with clean, hot water.

Grinding wheels should not contain iron, iron oxide, zinc, or other undesirable materials that may cause contamination of the metal surface. Grinding wheels, sanding material, and wire brushes previously used on other metals should not be used on stainless steel. Wire brushes should be made of a stainless steel that is equal in corrosion resistance to the material being worked on.

1.38.5 ABRASIVE BLASTING OF STAINLESS STEEL

Clean, previously unused glass beads, iron-free silica, or alumina sand are recommended for abrasive blasting. Steel shot or grit is not recommended because of the possibility of embedding iron particles.

After blasting, surfaces should be cleaned by scrubbing with hot water and fiber brushes, followed by rinsing with clean, hot water.

1.38.6 FINAL CLEANING

If proper care has been taken in earlier fabrication and cleaning, a final cleaning may be performed, consisting of little more than scrubbing with hot water or hot water and detergent such as TSP using fiber brushes.

Detergent washing must be followed by a hot-water rinse to remove residual chemicals. Spot cleaning to remove localized contamination may be accomplished by wiping with a clean, solvent-moistened cloth.

1.38.7 PASSIVATION OF STAINLESS STEEL

When stainless steel parts are to be used for applications where corrosion resistance is a prime factor to achieve satisfactory performance and service requirements or where product contamination must be avoided, passivation followed by thorough rinsing several times with hot water and drying thoroughly after the final water rinse is recommended. This treatment may be essential if iron or iron oxide remains embedded in the surface, such as from abrasive blasting or pickling.

A suitable process consists of immersion for 10–30 min at 20–50°C in a solution made up of the following ingredients:

- Nitric acid, 200 to 300 cm^3/l
- Sodium dichromate ($Na_2Cr_2O_7$, $2H_2O$), 25 g/l (which may be omitted when passivating austenitic stainless steels)

This treatment should be followed by rinsing in water and, in the case of ferritic and martensitic steels, by immersion for approximately 30 min in a solution containing 50 g/l of sodium dichromate at approximately 65°C and final rinsing in water.

The term *passivation* is used to refer to a chemically inactive surface condition of stainless steel.

1.39 PREPARATION OF STAINLESS STEEL FOR ELECTROPLATING

Parts should be cleaned by one or more of the following methods.

1.39.1 DEGREASING

This can be done with applied anodic direct current if desired. Neither cathodic nor alternating current treatments should be used.

1.39.2 BLAST CLEANING

This method, especially blasting with grit, leaves the surface very readily corrosible, and further processing should be given without delay.

1.39.3 **ANODIC ETCHING**

Anodic etching should be done in a sulfuric acid solution, subject to close control. Additionally, parts should be connected and the current switched on before the parts are immersed in the solution, and after treatment, the parts should be withdrawn rapidly before the current is switched off and then washed immediately. The sulfuric acid should at no time contain more than 11 ppm oxidizable material calculated as sulfur dioxide. (This precaution is necessary because reduced sulfur, phosphorus, and arsenic compounds in the acid promote hydrogen absorption during washing.)

1.39.4 **ACID PICKLING**

Acid pickling is done in an inhibited acid solution, and then by baking prior to further treatment at not less than 190°C for not less than 4 h.

1.39.5 **ELECTROLYTIC DERUSTING**

Electrolytic derusting is performed using an alkaline electrolyte and anodic current. Cleaning should normally be followed by a nickel strike before plating with the metal(s) required.

1.39.6 **NICKEL STRIKE**

To do a nickel strike, one of two treatments should be applied. In the first, parts should preferably be etched anodically for not more than 2 min and then treated cathodically for 5 min at approximately 1,500 A/m^2 to 2,000 A/m^2 in an aqueous solution containing nickel sulfate ($NiSO_4 \cdot 6H_2O$), at 225 g/l; and sulfuric acid ($d = 1.84$) 27 ml/l. Maintain the temperature of the electrolyte at 35°C to 40°C. Insoluble anodes, e.g., lead, should normally be used, the use of nickel anodes or a proportion thereof is also permissible.

In the second treatment, the parts should be made anodic in an aqueous solution of nickel chloride ($NiCl2 \cdot 6H2O$), at 250 g/l; and hydrochloric acid ($d = 1.16$), at 100 ml/l for not more than 2 min, and the current then reversed, so that they are cathodic, for about 5 min. Nickel electrodes to BS 558 should be used. The solution should be maintained at room temperature and a current density of about 300 A/m^2 employed.

When current reversal is not feasible, the short anodic treatment may be replaced by immersion in the solution without current flow for 15 min. The work then being made cathodic for about 5 min. Separate tanks may be used for the anodic (or immersion) and the cathodic treatments.

After any of these treatments have been performed, rinse the parts and transfer them to the final plating bath.

1.40 **SURFACE PREPARATION OF NONFERROUS METALS**

Surface preparation of nonferrous metals should be as they are described in Tables 1.15 through 1.30.

Table 1.15 Surface Preparation of Aluminum and Its Alloys

Surface Preparation for Protective Coatings	Permissible Methods						
	14-A Degreasing	14-B Etching	14-C Pickling	14-D Chemical Polishing	14-E Electro Polishing	14-F Zincate Treatment	14-G Blasting
1.Preparation Prior to Anodizing (Apply Process 14-A and Then One or More of 14-B to 14-E)		See Table 1.16	See Table 1.17.	See Table 1.18.	See Table 1.18		
2. Preparation Prior to Chemical Conversion Coatings (Apply Process 14-A and Then One or More of 14-B to 14-E)		See Table 1.16.	See Table 1.17.	See Table 1.18	See Table 1.18		
3. Preparation Prior to Metal Spraying (Apply Processes 14-A and 14-G)							
4. Preparation of Unanodized Parts for Painting (Apply Processes 14-A and 14-B)		See Table 1.16 (note 1)					
5. Preparation for Electroplating (Apply Process 14-A, Either 14.B Or 14-C, and Then 14-F)		See Table 1.19.	See Table 19			See Table 1.19	

Note: *Prior to the application of primer other than pretreatment primer, etching treatment should be used.*

Table 1.16 Etching Treatment of Aluminum and Its Alloys

Processes	Cleaning Solution	Operating Conditions	
		TEMPERATURE ($^\circ$C)	APPROX. TIME (MIN)
1.Bath (Immersion) Etching Treatment (see notes 1, 2, and 3)	Sulfuric acid (d = 1.84) 150 Cm^3/L chromic acid (Cro^3) 50 g/l demineralized water	50–65	25
2. Brush or Spray Etching Treatment (see notes 4 and 5)	Phosphoric acid in aqueous alcohols in which is suspended kaolin and a green pigment		
3. Etching and Desmutting Treatment (see notes 6,7)	Sodium hydroxide 25–50 g/l Sodium heptonate or sodium sluconate 0.75–1.0 g/l	60–65	0.25–1

Notes:
1. Bath treatment is unsuitable for alloys containing 6% or more copper or aluminum parts in contact with steel or copper alloys.
2. The etched part should be washed thoroughly and rapidly in water (not exceeding 65°C) and dried. The final rinse should be in demineralized or distilled water.
3. Contamination of the solution with chloride, copper, or iron may cause pitting of the metal. The chloride content should not exceed the equivalent of 0.2 g/l. Sodium chloride and the copper or iron contents should not exceed 1 g/l each.
4. The free acidity should be equivalent of 6% to 6.5% by weight of phosphoric acid.
5. Where parts are unsuitable for bathing, brushing or spraying should be performed.
6. This process is used prior to anodizing if a polished finish is not required.
7. The smut from alloying metals in aluminum is left on the surface, and this is removed by dipping in nitric acid 30% to 50% by volume (d = 1.42). For high silicon-containing alloys, 10% hydrofluoric acid (40 wt% HF) is added to the nitric acid desmutting solution.

Table 1.17 Acid Pickling of Aluminum and Its Alloys

Cleaning Solution	Operating Temperature	Notes
Sulfuric acid ($d = 1.84$), 90–120 cm³/l Sodium fluoride (NaF) 7.5–15 g/l	Room temperature	See note 2
Sulfuric acid ($d = 1.84$) 100 cm³/l Potassium fluoride (KF) 40 g/l	Room temperature	See note 2
Sulfuric acid ($d = 1.84$) 100 cm³/l Hydrofluoric acid (40 wt% HF)(1) 15 cm³/l	Room temperature	See note 2
Orthophosphoric acid ($d = 1.50$) 200 cm³/l Hydrofluoric acid (40 wt% HF)(1 m) 7.5 cm³/l	Room temperature	See note 2
Sulfuric acid ($d = 1.84$) 100 cm³/l O-toluidine acid stable wetting agent, 10 cm³/l	90–98°C	

Notes:
1. %wt = grams of HF in 100 g of water.
2. After acid pickling, the part should be rinsed in cold water and transferred to a cold aqueous solution containing approximately 500 cm³/l nitric acid (d = 1.42) for approximately 1 min and then thoroughly washed in clean water at a temperature not exceeding 50°C.

Table 1.18 Polishing of Aluminum and Its Alloys

Processes	Cleaning Solution	Operating Conditions	
		Temperature (°C)	Immersion Time (min)
1. Chemical polishing treatment (see notes 1, 2, and 3)	Phosphoric acid ($d = 1.75$) 75 cm³/l Sulfuric acid ($d = 1.84$) 20 cm³/l Nitric acid ($d = 1.42$) 5 cm³/l	95–105	1–2
2. Electropolishing treatment (see note 4)	Phosphoric acid (H_3PO_4) 400–800 g/l Sulfuric acid ($d = 1.84$) 100–200 g/l CHROMIC ACID (CrO_3) 40–100 g/l	70–80	2–5

Notes:
1. During the operation, nitric acid is lost and should be maintained in the range of 5–10% by volume.
2. This process gives rise to the emission of nitrous fumes, which must be removed by adequate exhaust equipment. This solution is strongly acidic, and appropriate precautions should be taken to protect the operator.
3. After chemical polishing, the work rinsed and further cleaned by immersion in a 30% by volume solution of nitric acid (d = 1.42) or 10 g/l chromic acid (CrO₃) followed by further rinsing.
4. The current density of this process is approximately 200 A/m², and DC voltage is 12–15 volts.

Table 1.19 Pickling and Zincate Treatment of Aluminum Prior to Electroplating

Processes	Cleaning Solution	Operating Conditions	
		Temperature (°C)	Immersion Time (min)
1. Acid pickling (see notes 1, 2, and 3)	Hydrofluoric acid (40 %wt HF) 100 cm³/l	Max. 40	1
	Nitric acid (d = 1.42) 100 cm³/l		
2. Zincate treatment (see note 4)	Zinc oxide (ZnO) 100 g/l Sodium hydroxide (NaOH) 540 g/l	Room temperature	3

Notes:
1. *This solution emits toxic fumes, so efficient fume extraction should be provided.*
2. *Alternatively, etch in a chromic acid/sulfuric acid solution (see Table 1.16). This etching solution has been found to be less satisfactory than acid pickling (process 1 of this table), but its less vigorous attack on the material provides a smoother cleaned surface.*
3. *Rinse thoroughly and proceed immediately to the zincate treatment.*
4. *Prior to zincate treatment, immerse parts for 1 min in an aqueous solution containing approximately 500 cm³/l nitric acid and then rinse thoroughly.*

Table 1.20 Surface Preparation of Zinc Plate and Zinc-Coated Parts

Surface Preparation for Protective oatings	Permissible Methods			
	19-A Hand tool	19-B Degreasing	19-C Electrolytic cleaning	19-D Acid pickling
Removal of corrosion product from zinc-coated parts	See note 1	See note 2		
Preparation for coating (see note 6)	See note 3			
Preparation for electroplating (see note 4)			With alkaline cleaner	See note 5

Notes:
1. *Scrubbing with clean water and stiff bristle on nylon brushes.*
2. *Degreasing with white sprit.*
3. *When parts are to be treated with wash primer, they should be degreased. When other primers (chromate and phosphate treatment) are to be used, degreasing will not be required.*
4. *After processes 19-C and 19-D, rinse the part thoroughly in cold, running water.*
5. *Dip in very dilute sulfuric acid (e.g., 0.5% by volume).*
6. *The method of preparation of zinc for coating is also applicable to cadmium.*

Table 1.21   Surface Preparation of Copper and Its Alloys

Surface Preparation for Protective Coatings	Permissible Methods				
	20-A Degreasing	20-B Blasting	20-C Etching	20-D Pickling	20-E Electropolishing
Preparation for coating (apply process 20-A and then one or more of processes 20-C to 20-E)		Blasting with nonmetallic abrasive	See Table 1.22	See Table 1.22	See Table 1.22
Preparation for electroplating (apply process 20-A and then one or more of process)		Water blasting with nonmetallic abrasive	See Table 1.22, (notes 1 and 2)	See Table 1.22	See Table 1.22

Notes:
1. Etching treatment is suitable for beryllium copper parts and should be followed by rinsing in water and immersion in a solution containing 100 cm³/l of sulfuric acid when plating in an acid electrolyte is to be the next treatment.
2. Soft soldered parts (other than beryllium copper parts) should be dipped in a solution containing approximately 100 ml/l fluorobaric acid (40% wt HBF_4) and rinsed thoroughly in water and copper flashed in an alkaline copper electrolyte at a PH of 10–12 and a temperature of 60 ± 10°C.

Table 1.22   Treatment of Copper and Its Alloys

Process	Cleaning Solution	Operating Conditions	
		Temperature	Time
1. Etching treatment (note 1)	Sodium dichromate ($Na_2Cr_2O_7$, $2H_2O$) 200 g/l Sulfuric acid ($d = 1.84$) 40 cm3/l	Room temperature	Up to 2 min
2. Scale dip pickling	Sulfuric acid ($d = 1.84$) 100 cm3/l	25°C–50°C	N/A
3. Bright dip pickling (note 1)	Sulfuric acid ($d = 1.84$) 500 cm3/l	Room temperature	N/A
4. Chemical smoothing (note 2)	Hydrogen peroxide 33–50 g/l Sulfuric acid ($d = 1.84$) 103 cm3/l	Room temperature	15–60
5. Nitric acid pickling	Nitric acid ($d = 1.42$) (15–20%) by volume	Room temperature	2–5 min
6. Electropolishing (note 3)	Orthophosphoric acid ($d = 1.75$) 70% wt Aliphatic alcohol Water	20–25°C	N/A

Notes:
1. Efficient fume extraction is essential, as the process results in the release of highly toxic fumes.
2. After applying chemical smoothing, the part should be immersed for 20 s in sulfuric acid.
3. Current density of 200–500 A/m² and copper cathodes should be used.

Table 1.23   Surface Preparation of Nickel and Its Alloys

Surface Preparation for Protective Coatings	Permissible Methods						
	22-A Degreasing	22-B Blasting	22-C Alkali Pickling	22-D Acid Pickling	22-E Chemical Polishing	22-F Electropolishing	22-G Strike & Etching Treatment
1. Preparation for coating (apply one or more of processes 22-A to 22-F)				See note 2	See Table 1.24	See Table 1.24	--
2. Preparation for electroplating (note 3)			See note 1	See note 2	See Table 1.24	See Table 1.24	See Table 1.24 (note 4)

Notes:
1. *Descaling in molten caustic alkali.*
2. *The acid pickling solution is based on nitric/hydrofluoric acid, with or without the addition of ferric sulfate.*
3. *Parts should be treated as in 22-G (see Table 1.24, note 4) preceded by one or more of processes 22-A to 22-F.*
4. *After etching treatment and nickel strike, rinse the parts thoroughly and transfer to the final plating bath.*

Table 1.24   Treatment of Nickel and Its Alloys

Processes	Cleaning Solution		Operating Conditions	
			Temperature	Time (min)
Chemical polishing	Glacial acetic acid Nitric acid ($d = 1.42$) Phosphoric acid ($d = 1.75$) Sulfuric acid ($d = 1.84$)	50% by volume 30% by volume 10% by volume 10% by volume	N/A	N/A
Electropolishing (note 1)	Sulfuric acid ($d = 1.84$) water	70% by volume	Room temperature	N/A
Etching treatment	Ferric chloride ($FeCl_3$) or ferric chloride ($FeCl_3 \cdot 6H_2O$) hydro chloric acid ($d = 1.16$)	150 to 200 g/l 250 to 330 g/l 150 to 200 cm3/l	Room temperature	Up to 1 min
Nickel strike (notes 2,3)	Nickel chloride ($NiCl_2 \cdot 6H_2O$) hydrochloric acid ($d = 1.16$)	250 to 400 g/l 100 cm3/l	Room temperature	Up to 2 min

Notes:
1. *A current density of 250 A/m² is employed with the work anodic to the lead lining of the tank. Great care should be taken in making up this solution, and the sulfuric acid must be added slowly to the water and mixed thoroughly. The solution becomes hot during mixing and should be cooled to room temperature before use.*
2. *This process is not required prior to chromium plating.*
3. *A cathode current density of approximately 1,500 A/m² should be maintained for not more than 2 min.*

Table 1.25   Surface Preparation of Titanium and Its Alloys

Surface Preparation For Protective Coatings	Permissible Methods				
	Degreasing	Blasting	Alkali Pickling	Acid Pickling	Etching Treatment
1. Preparation for Coating	See notes 1 and 2	N/A	N/A	See Table 1.25	N/A
2. Preparation for Electroplating	N/A	N/A	N/A	See Table 1.26	See Table 1.26

Notes:
1. The parts of the Ti-Al-Sn alloy made by cold forming or welding should not be exposed to hot chlorinated hydrocarbon solvents prior to stress relief.
2. The time of immersion in hot trichloroethylene liquid or vapor should not exceed 30 min for any one degreasing operation.

Table 1.26   Treatment of Titanium and Its Alloys

Process	Cleaning Solution	Operating Conditions	
		Temperature (°C)	Time (min)
Acid pickling (see note 1)	Hydrofluoric acid 50 cm3/lit (40% wt% HF) Nitric acid ($d = 1.42$) 200 cm^3/l	Up to 65	N/A
Acid pickling (see note 1)	Hydrofluoric acid 120 cm^3/l (40% wt HF) Nitric acid ($d = 1.42$) 400 cm^3/l	Room temperature	N/A
Etching treatment (See notes 2 and 3)	Hydrochloric acid ($d = 1.18$)	30	10–120

Notes:
1. This solution must be handled with care. Efficient fume extraction is essential.
2. If the solution is maintained at 90°C to 110°C, an immersion time of approximately 5 min is adequate.
3. After etching, transfer rapidly to a rinsing solution containing 50 g/l Rochelle salt (sodium potassium tartrate) at room temperature and agitate the solution thoroughly.

Table 1.27   Surface Preparation of Magnesium Alloys

Surface Preparation for Protective Coatings	Permissible Methods				
	Degreasing	Blasting	Fluoride Anodizing	Removal of Fluoride Film	Activation
Preparation for Coating	N/A	N/A	See Table 1.28	See Table 1.28	N/A
Preparation for Electroplating	Cathodic degreasing (see Table 1.28)	N/A	N/A	N/A	See Table 1.28

Table 1.28   Treatments of Magnesium Alloys

Processes	Cleaning Solution	Operating Conditions	
		Temperature (°C)	Time (min)
Fluoride anodizing (see notes 1 and 2)	Ammonium bifluoride (NH_4HF_2) 150–250 g/l Water	Up to 30	10–15
Remove of fluoride film (see notes 3 and 4)	a) Chromic acid (CrO_3) 100–150 g/l b) Caustic soda 50 g/l c) Hydrofluoric acid 150–200 g/l	Boiling solution Boiling solution Room temperature	Up to 15 10 5
Cathodic degreasing (see note 5)	Sodium hydroxide 30 g/l (approx.) Sodium cyanide 30 g/l (approx.) Sodium carbonate 15 g/l (approx.)	Room temperature	1
Activating (see note 5)	Phosphoric acid ($d = 1.75$) 130–150 cm^3/l Potassium fluoride 60–80 g/l	Room temperature	30 S

Notes:
1. The bath should be lined with hard rubber or suitable plastic material resistant to acid fluoride solution.
2. Alternative current is applied, and the voltage raised to 90–120 volts.
3. The fluoride film should be removed by process A (chromic acid) followed by process B (caustic soda) or process C (hydrofluoric acid).
4. This process is not necessary if the subsequent chromate treatment is carried out in an acid bath capable of removing fluoride film.
5. After this process, the part should be rinsed.

Table 1.29   Surface Preparation of Tin and Its Alloys

Surface Preparation for Protective Coatings	Permissible Methods		
	Solvent Degreasing	Alkali Degreasing	Acid Pickling
Preparation for painting	See sections 5.2 and 5.3	N/A	N/A
Preparation for electro plating	See sections 5.2 and 5.3	See notes 1.3	See notes 2, 3

Notes:
1. Clean either by immersion or electrolytically, in a mild alkaline cleaner.
2. Dip for approximately 2 min in a solution containing 10% by volume of hydrochloric acid (d = 1.18). For tin alloys containing lead, approximately 100 cm^3/l fluoroboric acid (40% wt HBF4) should be substituted.
3. After this process, rinse the part thoroughly in cold, running water.

Table 1.30   Surface Preparation of Chromium and Lead Parts

Surface Preparation For Protective Coatings	Chromium Coating
Preparation of chromium for painting	See note
Preparation of lead for painting	N/A

Note:
Electroplated coatings of chromium should be immersed in a chromic acid solution (10 g/l). The solution is used at room temperature with an immersion time of approximately 5 min.

1.41 SURFACE PREPARATION OF METALLIC SURFACES FOR MAINTENANCE

All painted or metallic coated structures, eventually need maintenance treatment, no matter how well they are protected.

1.41.1 CHOICE OF MAINTENANCE METHOD

The decision whether to patch-paint or recoat completely is largely an economic one. It will be influenced by the accessibility of the structure. If much scaffolding is needed, and therefore the cost of access is a high proportion of the total cost, it may be economical to recoat the whole structure on site. This is certainly the case when an excessive amount of patch-painting of small areas (say, more than 10%) would be involved.

If maintenance is undertaken at the right time, it should not be necessary to strip all the old coating before repainting the structure. The ideal situation, in which none of the old paint needs more than a thorough washing-down, is rare. Some parts of the structure will need repainting before the paint itself has neared the end of its life. Where the majority of the structure covering is in good condition and only a small part of it is substandard, it may be better to simply repair the defective areas by spot painting.

1.41.2 CHOICE OF PROCEDURES

Unless the standard specifies otherwise, surface preparation should be carried out to a satisfactory level for the material using the suitable method.

1.41.3 SPOT REPAIR (PATCH-PAINT)

This option would be favored under the following circumstances:

- Repairs are hidden or in a low-visibility area and thus unimportant to the aesthetics of the structure.
- Agency maintenance crews are available for this type of work.
- Structures are small, not requiring extensive scaffolding or hard-to-access areas.
- Corrosion and degradation are limited to isolated areas and relatively small sections amounting to less than 10% of the total area.
- The decision has been made to upgrade small isolated areas, such as bearing areas, crevices, or areas subject to leakage or condensation or chemical splash.

It is important that the procedure used to do spot surface preparation be appropriate to provide a good surface for painting and that it does not adversely affect the other areas. Spot-cleaning procedures include abrasive blast cleaning, manual cleaning, and pressurized water blasting.

The damaged areas will be spot-blasted to remove all the rust and paint to achieve Sa2 or greater. The remaining areas of intact paint will then receive Sa1 to remove loose paint or other surface contaminants. Great care must be exercised when spot blasting to avoid overblast damage to the adjacent intact paints. Spot-blasting is not recommended for areas less than about 0.1 m^2 or for more than 5–10% of the surface area.

Manual cleaning (hand and power tool cleaning) are more suitable for small areas. Pressurized water-jet cleaning is capable of removing loose rust and mill scale under normal conditions.

After spot cleaning, the repaired areas are to be coated with a suitable primer. In this particular case, the remaining paint is not overcoated.

1.41.4 SPOT REPAIR PLUS FULL TOPCOAT

This technique is similar to those described in previous sections, but with the addition of a full finish coat over the entire surface, including spot-repaired areas and the intact paint areas.

This approach is favored under the following circumstances:

- The existing paint is in relatively good condition and still resilient and does not have excessive film build (e.g, not above 0.38 mm).
- Structure access is relatively difficult.
- Intact surface can be readily cleaned with pressurized air, water, hand tool, or power tool cleaning or solvent or detergent wiping.

1.41.5 COMPLETE REPAINT

When the paint condition is in poor condition, a decision to repaint the structure in its entirety is usually made. This involves removing the old coating and all of the corrosion product before applying the primer, intermediate coats, and topcoat finish. Painting a subunit of a structure may be a variation of this alternative when funds are short or when the remaining portions of the structure are in good condition.

Surface preparation procedures for complete repainting include dry abrasive blast cleaning, wet abrasive blast cleaning, vacuum blasting, and pressurized water jetting. Normally, this latter method requires sand or abrasive injection to effect the complete removal of existing rust scale and paint and to add a profile. An important consideration in blast cleaning is the disposal of the abrasives and paint residue, particularly if paint contains lead or other potentially hazardous materials. The surface should be degreased prior to blast cleaning. Flame cleaning is also capable of removing old paint. It should be used in damp weather when blast cleaning is not practicable.

1.41.6 GENERAL NOTES FOR GUIDANCE

All removable parts impeding access to the work should be removed before work commences. Treatment of organic growth on painted surfaces should be according to standards and must be applied prior to blast cleaning if growths of algae impede access to the work.

Washing down before repainting is always desirable, especially in marine environments, because it may remove chlorides from the surface, but it is unlikely to remove sulfates. Consideration should be given to problems of drying after washing.

Some paints become very hard when fully cured, and special preparation, such as light blasting or the use of emery (with coated abrasive tools), may be required to obtain good undercoat adhesion.

Thick films such as those obtained with pitches or bitumens often develop deep cracks. These materials can be overpainted, but the cracking cannot be eliminated without complete removal of the film.

The maintenance system chosen depends to some extent on the original process and the standard of surface preparation which can be achieved. Some two-pack materials are not always suitable for maintenance, and trials and compatibility tests should be carried out before using them.

The standard of protection required varies with atmospheric pollution and for different parts of the structure. For instance, areas subject to condensation and restricted air movement require better protection than fully exposed steelwork. Sometimes the film becomes embrittled and loses its adhesion. In these cases, complete stripping is essential.

Coatings on old structures frequently develop numerous lengthy cracks and fissures. It will save time and produce a better result to caulk them after applying the primary coat.

Alkaline and solvent-type paint remover may be used to remove oil-based paint from metallic or nonmetallic surfaces when the nature of the substrate precludes removal by other method of preparation.

Paint remover attacks aluminum and zinc, and their residues are difficult to remove, especially from porous surfaces.

Particular attention should be paid to the cleaning of crevices and other places where dust and dirt have coated. Debris tends to build up at ground level.

Crevices frequently occur that are positioned or dimensioned such that standard scrapers, brushes, and other tools will be ineffective. In such cases, it is essential that special tools should be devised to ensure adequate cleaning and painting.

1.42 SURFACE PREPARATION OF IMPERFECT METALLIC SURFACES

Surface imperfections such as edges, projections, crevices, pits, weld prosity, and laminations can cause premature failure when the service is severe. The timing of such surface repair work may occur before, during, or after preliminary surface preparation has begun.

1.42.1 WELD SPATTER

Weld spatter should be removed prior to blast cleaning. Most weld spatter, except that which is very tightly adherent, can be readily removed using a chipping hammer, spud bar, or scraper. Tightly adhering weld spatter may require removal by grinding.

1.42.2 POROSITY

Areas of unacceptable porosity should be filled with suitable filler material or closed over with a needle gun or peening hammer prior painting.

1.42.3 SHARP EDGES

Sharp edges, such as those normally occurring on rolled structural members or plates, as well as those resulting from flame cutting, welding, grinding, and especially shearing, may be removed by any suitable method (e.g., grinding, mechanical sanding, or filling). Care should be taken to ensure that during the removing operations, new sharp edges are not created.

1.42.4 PITS

Deep corrosion pits, gouges, clamp marks, or other surface discontinuities may require grinding prior to painting. The surface will require filling.

1.42.5 LAMINATIONS AND SLIVERS

Rolling discontinuities (laps) may have sharp protruding edges and deep penetrating crevices, and such defects should be eliminated prior to painting. Various methods (e.g., scraping and grinding) can be used to eliminate minor slivers. All sharp fins, projections, and edges should be removed.

1.42.6 CREVICES

Areas of poor design for corrosion protection, such as tack or spot-welded connections, back-to-back angles, crevices, etc., may require special attention. Where possible, such deficiencies should be corrected by structural or design modification. Where this is not possible, particular consideration should be devoted to minimize the effect of such deficiencies.

1.43 SURFACE PREPARATION OF CONCRETE

Concrete is a durable material that does not usually require painting for protection, except to prevent further penetration of water and salts after repair of deteriorated concrete. Sometimes, new concrete needs to be painted or coated if the thickness of concrete is insufficient to provide protection.

Concrete should be permitted to age at least 28 days under good conditions prior to applying a coating system. Paintable curing compounds may be used to permit coating in 7 days.

Air and water blasting and hand and power tool cleaning are the most effective methods of surface preparation.

There should be no evidence of laitance (which is a very fine, light powder that floats to the surface when concrete is cast) on the concrete surfaces before coating. Laitance can be removed by the use of light sand blasting or light wet blasting. All soft or loosely bonded surfaces should be cleaned down to a hard substrate, preferably by abrasive blasting.

Mechanical cleaning, in most cases, is preferred to etching because concrete surfaces should be brought to a humidity of not more than 4% wt before paint is applied. The cleaned surfaces should be free of oil, grease, loosely adhering concrete, laitance, and other contamination.

1.43.1 TYPES OF EXPOSURES

Three types of exposure are discussed in this section:

- Architectural—Cementitious surfaces coated primarily for aesthetic purposes or weathering resistance.
- Light maintenance—Cementitious surfaces subject to occasional chemical spillage or moderate fumes or for ease of cleaning, or to minimize damage due to freezing and thawing.
- Heavy-duty maintenance or immersion—Cementitious surfaces to be protected from continuous immersion, continuous or frequent spillage, heavy chemical fumes, severe abrasion, or physical

abuse. Examples of such use include tank linings, pump bases, waste water sewers, tank bases, pumps, floors and building foundations.

1.43.2 TYPES OF CEMENTITIOUS SURFACES

There are four types of cementitious surfaces:

- Poured concrete or precast slab—These surfaces have two problems. First, a weak surface layer (i.e., laitance) causes poor adhesion of coating unless it is removed. Second, various sizes of air bubbles ranging from minute up to 50 mm or larger, present at surface.
- Concrete block walls—These walls have an irregular surface and do not have a weak surface or air pockets.
- Gunite surfaces (shotcrete)—This type of surface results when concrete is sprayed onto a vertical surface.
- Concrete floors—These surfaces have a weak surface layer, without surface air bubbles. Concrete floors must be cleaned and neutralized to pH 7–8 and washed before painting.

1.44 TYPES OF PREPARATION

Surface preparation of concrete surfaces include the following methods as appropriate with reference to service condition standards as defined by the company (see Table 1.31).

1.44.1 AIR BLASTING

This method is intended to remove debris, dust, and loosely adherent laitance from concrete walls and ceilings. Air blasting should consist of cleaning the surface with a compressed air stream at 5.5–7 bar through a blasting nozzle held approximately 0.6 m from the surface. Vacuum cleaning may is required to remove redeposited dust.

Table 1.31   Surface Preparation of Concrete

Type of Surface	Service	Water Blasting	Sand Blasting	Tool Cleaning
Poured concrete or precast slab	Light maintenance	3	4	2
	Heavy-duty maintenance	NR[1]	1	2[2]
	Architectural	2	NN	2
Concrete block	Chemical or architectural	2	NN	1
Gunited surface	Chemical or architectural	2	3	1
Floors	Chemical or architectural	NR	1	2[3,4]

Note: The numbers 1 to 4 in the table indicate the preferred order of surface preparation methods, from best to worst.
NR = Not recommended
NN = Not Needed
[1]Not recommended for dense concrete, as it cannot open voids, but it may be useful in some instances.
[2]Hand impact tools will do an adequate job but their use is not recommended, as they are too slow.
[3]May be used if concrete is hard and irregular. For instance, broom finishing can be used, consisting of sweeping the surface with a clean industrial stiff-bristled broom or similar device.
[4]Power scarification may also be used on smooth floors.

1.44.2 WATER BLASTING

This method should consist of cleaning the surface with a jet of high-pressure water or steam with 240–308 bar, at a sufficient level to remove heavy deposits of grease and oil. When detergent or other emulsifying agents are mixed with water or steam, after cleaning and before the surface dries, the surface should be flushed thoroughly with portable water. Surfaces cleaned with detergent or nonsolvent emulsifying agents should be tested for pH in accordance with test method and may be tested for moisture content in accordance with standard test method prior to applying coating.

1.44.3 HAND OR POWER TOOL PREPARATION

The use of hand impact, hand and power grinding, and hand wire-brushing using a detergent (following by rinsing) removes loose and powdery weak concrete, as well as oil, grease on the surface.

1.44.4 TREATMENT OF ORGANIC GROWTHS

Concrete surfaces on which organic growths are present should be treated according to standards.

1.44.5 STOPPING AND FILLING

Large voids and air holes should be filled with masonry cement or epoxy resin mortars. Minor surface defects should be fixed with interior or exterior grades of water-mixed filler or with masonry cement.

Alternatively, in oil-based paint systems, oil-based stoppers and fillers may be used after priming. Application of cement paint or "bagging" with a cement/sand slurry will reduce surface roughness and fill minor imperfections.

1.44.6 MAINTENANCE OF CONCRETE SURFACES

When repainting with the same type of existing paint system, and the old paint is in good condition (i.e., if loose paint, curled edges, and blisters are not present). The painted surface should be cleaned with a hand wire-brush, using detergent solution or solvent, followed by rinsing.

If the old paint is not in good condition, it should be removed according to standards.

When repainting with a type of paint other than the existing paint, the old paint should be removed completely, using air blasting with fine sand abrasive, water blasting, or paint strippers. Any cracks and holes in the concrete should be filled in the same manner as for new concrete.

1.45 TEST METHODS
1.45.1 STANDARD TEST METHOD FOR pH MEASUREMENT OF CHEMICALLY CLEANED CONCRETE SURFACES

Residual chemicals that are not removed by rinsing with water may adversely affect the performance and adhesion of coatings applied to prepared concrete surfaces. This test method is designed to determine that residual chemicals have been removed by measuring the acidity or alkalinity of the final rinsed surface. Unless otherwise specified by the company, tests should be conducted in accordance with the following procedure.

In this process, test paper with a minimum range from 1 to 11 pH units is used. It has a capability of measuring in increments of 0.5 pH units. Potable water is used to rinse chemically cleaned or etched concrete surfaces. Wet the concrete surface after the final water rinse and before the rinse water has completely drained off the surface.

Tear off a strip of test paper, wet with water, and after the color develops, compare with color chart to determine the pH reading. The pH of the water used for rinsing should be determined to establish acceptance criteria. Readings should be taken at the beginning and end of the final rinse cycle.

At least two surface pH readings should be taken every 50 m^2 or portion thereof. Readings should be taken at randomly selected locations immediately following the final rinse and before all the rinse water has drained off the surface.

The pH readings following the final rinse should not be more than 1.0 points lower or 2.0 points higher than the pH of the rinse water unless otherwise specified by the company. (A point is equivalent to 0.5 pH units.)

The presence of laitance may be detected by scraping the surface with a putty knife. If a loose powdery material is observed, excessive laitance is present. Adhesion could be adversely affected by this laitance, so it should be removed.

1.45.2 STANDARD TEST METHOD FOR INDICATING MOISTURE IN CONCRETE USING PLASTIC SHEETS

Capillary moisture in the concrete may be detrimental to the performance of certain coating systems that cannot tolerate moisture on or within the surface boundary. This test method is used prior to the application of coatings on concrete.

The materials used are as follows:

- Transparent polyethylene sheet, commercially available, approximately 0.1 mm thick
- Adhesive tape, 50 mm wide

This test should be conducted when the surface temperature and ambient conditions are within the established parameters for application of the coating system. Avoid direct sunlight, direct heat, or damage to the plastic sheet, as such treatment affects the reliability of the results.

To perform the test, tape a 457×457-mm segment of plastic sheet tightly to the concrete surface, making sure that all edges are sealed. Allow the plastic sheet to remain in place for at least 16 h. When the proper amount of time has elapsed, remove the plastic sheet and visually inspect the underside of the sheet and the concrete surface of the patch for moisture.

- Sampling

Floors one test area per 46 m^2 or portion thereof, of surface areas unless otherwise specified by the company.

Walls and ceilings—One test area per 46 m^2 or portion thereof, of surface area unless otherwise specified by the company.

The recommended practice is to conduct one test for each 3 m of vertical rise at all elevations starting within 300 mm of the floor.

Report the presence or absence of moisture.

1.46 SURFACE PREPARATION OF PRECAST CONCRETE BLOCKS

1.46.1 TYPES OF BLOCKS

The main types of blocks are as follows:

- Aerated concrete blocks are usually made from mixtures of cement and siliceous materials, such as sand or pulverized fuel ash or a mixture of these, together with an aerating agent and water.
- Dense and lightweight blocks are made from cement and dense or lightweight aggregates, molded and compacted by vibration or pressure.

1.46.2 CLEANING

Brushing down with stiff (not wire) brushes to remove loose material is usually required for cleaning.

1.46.3 STOPPING AND FILLING

Cracks, holes, and damaged areas should be fixed with cement mortar, masonry cement, or, in dry interior conditions, water-mixed fillers. The overall filling of the surface of blockwork by conventional methods is not recommended. On finer-surfaced blocks, a cement/sand slurry or cement paint according to BS 4764 scrubbed into the surface will smooth out the texture and fill small holes. Thick-textured coatings are also useful for this purpose.

1.47 INSPECTION AND REJECTION

1.47.1 SURFACE INSPECTION

The prepared surfaces should be inspected by qualified personnel.

1.47.2 ACCESS OF INSPECTOR

The inspector should have access to the construction site and to the parts of all plants that must conform to the necessary standards.

1.47.3 FACILITIES FOR INSPECTOR

The contractor should furnish the inspector reasonable facilities and space, for the inspection, testing, and collection of all desired information regarding the character or equipment and materials used, application, method(s) of surface preparation, progress and manner or the work, and the results obtained.

1.47.4 INSPECTION GUIDE

The inspection guide (displayed in Table 1.32) shows many of the types of defect that may be found during or after the various preparation operations. The suggested actions in the guide should be performed by the contractor. The aim should be not only to remedy defects already evident, but to prevent their recurrence.

Table 1.32   Inspection Guide

Work Stage and Code	Potential Detects	How Determined	Likely Cause	Suggested Action	Notes
A. Raw Steel					
1	Inaccessibility for treatment of surfaces	Visual	Design precludes access for specified treatment	Inform engineer that the work cannot be performed as specified. Provide a record of work	N/A
2	Oil or grease	Visual	Cranes, jigs, machinery, trucks, etc.	Scrub with emulsion cleaner. Rinse with fresh water	BS 77 73; not applicable for galvanizing
3	Excessive corrosion with pitting	Visual	Old stock or severe storage condition	Allow for longer or improved surface preparation; then check for removal of soluble corrosion products	Where appropriate, refer to steel inspector
4	Welding discontinuities residues and spatter	Visual	Poor welding practice	Refer to engineer for rewelding as necessary; remove residues and spatter	Refer to engineer
5	Sharp edges, rough welds, burrs, and undercut	Visual	Poor fabrication	Radius edges; fettle welds and burrs as necessary. Grind out lamination and shelling. Undercut (where the extent of treatment depends on type of coating and severity of exposure)	N/A
6	Distortion, dishing, buckling, and undercut	Visual	Poor fabrication	Refer to steel inspector	
B. Blast-Cleaned Steel					
1	Oil or grease, dirt, dust, detritus, and water	Visual or adhesive tape or filter paper or reflectometer	Poor working conditions	Treat oil or grease as for A2 specification. Shield working area from contamination. Improve vacuum cleaning	
			Contaminated grit	Discard grit and reblast with fresh supply. If greasy, treat as for A2	
			Contaminated compressed air		
2	Surface too rough relative to coating thickness	Visual	Excessively corroded raw steel	Longer blasting time and extra coats; refer to steel inspector	A much rougher surface is acceptable for sprayed metal coatings than with paint coatings

(Continued)

Table 1.32   Inspection Guide (cont.)

Work Stage and Code	Potential Detects	How Determined	Likely Cause	Suggested Action	Notes
		By instrumental comparison of blast profile with standard	Incorrect grit size, angle of blast, or both	Extra costs to be agreed. Change grit for subsequent work, improved angle of blast, or both	
3	Residual mill scale	Visual and/or "Surclean" comparison with standard	Inadequate blasting	Reblast to the required standard. Heavy mill scale may require initial treatment with abrasive with larger pieces of grit, or a switch to a mixture of grit and shot	"Surclean" may have limitations in the determination of percentage of mill scale
4	Residual rust	Visual	Inadequate blasting	Reblast to the required standard (heavily corroded steel may require abrasive with smaller grit), and then check for removal of soluble corrosion products	N/A
5	General darkening or rusting	Visual and/ or sur clean, comparison with standard	Too lengthy a delay between cleaning and coating, conditions too humid, or metal too cold	Reblast to the required standard and cost; quickly improve conditions	N/A
6	Lamination/ shelling	Visual	Poor rolling of steel at mill	Refer to steel inspector/ engineer for possible quality concerns regarding the steel	N/A
C. Pickled Steel					
1	Rough or pitted surface	Visual	Incorrect concentration of acid and/or inhibitor. Too much time in the pickling tank. Corroded surfaces before pickling	Correct acid and inhibitor concentrations and temperatures of the pickling bath	N/A
2	Dark patches	Visual	Residual mill scale or rust	Repickle	N/A

Table 1.32   Inspection Guide *(cont.)*

Work Stage and Code	Potential Detects	How Determined	Likely Cause	Suggested Action	Notes
3	Dark patches with sticky surface	Test with PH paper; test with chloride paper	Residual acids; residual chlorides	Rerinse or scrub small areas with clean water	N/A
4	Powdery or crystalline deposits	Visual or PH-paper testing	Poor rinsing,	Rerinse	N/A
5	Carbon deposits or smut	Visual or with white cloth	Contamination of pickling baths	Brush off and treat again if severe filter reagents and replace rinse	N/A
6	Aside surface	PH paper	Poor rinsing or over acidity of final process	Rerinse or treat again after PH adjustments of the tanks concerned	N/A
7	Too heavy phosphates Deposit	Laboratory check	Too long or too strong a process	Special retreatment necessary	N/A
8	Cold surface	Contact thermometer	Too long a delay before priming	Adjust programmed to ensure that the advantage of coating or a warm surface is maintained	N/A
D. Manually Cleaned Steel (includes use of handheld power tools)					
1	Rust, slag, and loose scale	Visual comparison with standard, written instructions, or both	Inadequate work, tools, or both	Further work with less worn or more suitable tools, if necessary	N/A
2	Burns, sharp cuts, peakes of steel	Visual and touch inspection	Improper use of power tools	Remove the defect by the most appropriate means and control the methods of working with power tools	When chipping, avoid gouging and production of sharp edges
3	Burnished surface	Visual	Too vigorous brushing (manual or power)	Mark area for emulsion cleansing before painting and control more closely the wire brushing; use of a pretreatment primer would also foster subsequent adhesion	N/A

(Continued)

Table 1.32   Inspection Guide *(cont.)*

Work Stage and Code	Potential Detects	How Determined	Likely Cause	Suggested Action	Notes
E. Environment During Application					
1	Too cold	Air and contact thermometers	Poor air conditioning in works or extremely inclement weather	Improve temperature to acceptable level, but little or no license should be allowed with the specified limit. Open-flame heaters should not be used	N/A
2	Too hot	Contact thermometer	Heated surfaces	Reduce temperature below 35°C or to whatever higher temperature is allowed with special paints	If necessary, rearrange program to avoid overheating

It is emphasized that the inspection guide should be used in conjunction with the relevant specifications and the inspection schedule. Where defects arise from processes not covered in these standards, agreement between the parties concerned is necessary before remedial action can take place.

Under the various "work stages" in Table 1.32 are suggestions for inspection instruments and equipment. In addition, the inspector should have available for general inspection work items such as liquid-sample containers, specimen bags, torches, mirrors, magnifying glasses, sharp knives, marking chalk, and any other appropriate standards, material data sheets, specifications, and other materials that are needed for the work in question.

1.47.5 QUALIFIED INSPECTORS

The extent and time of any inspection by the user or his duly appointed representative should be a part of the job specifications.

A qualified inspector should have the following prerequisites:

- Complete knowledge of the job specifications and their requisites.
- A practical knowledge of all phases of coating application work, including (1) preapplication surface finish requirements, such as grinding of welds, sharp edges, etc.; (2) surface preparation; (3) coating application techniques and workmanship; (4) coating materials; (5) continuity, thickness and cure tests, and tolerances of standards; and (6) equipment and tools used in all phases of coating application work.
- Adequate experience and training in the inspection of coating applications and the instruments used for inspection and evaluation of coating applications.

The specification should so stipulate if final acceptance of the work is to be made by a duly appointed representative of the user rather than the user. In such a case, it is recommended that that representative be a qualified inspector.

1.48 APPLICATION OF DEGREASING METHODS

The cleaning methods described in previous sections are applied by several processes as follows:

1.48.1 IMMERSION CLEANING PROCESSES

To remove simple films of oil and grease, articles may be subjected to the vapor immersion process, in which the parts are immersed in a bath of solvent vapour. The vapor condenses on the cold surfaces of the articles, and the condensate dissolves the oil and grease, taking it to the base of the tank. To ensure maximum condensation, the articles should be as near as possible to room temperature (and at any rate, should not be above it) at the time of immersion; they should be passed through or suspended in the solvent vapor until no further condensation occurs, after which point no further degreasing will take place.

Light metal articles that reach the vapor temperature rapidly and articles with a very heavy film of grease may need a second immersion after cooling. Alternatively, before removal from the vapor, a stream of liquid solvent may be applied over the surface, and this also removes loose dirt deposits.

This process used for cleaning various types of work by the commonly used degreasing systems are described in Table 1.33 and indicated schematically in Figure 1.9.

The vapor immersion equipment consists of one or more vessels or compartments with the means of heating the liquid solvent contained in the lower part by heating coils and a condensing zone near the top, provided by water-cooled coils, to control the vapor level.

Table 1.33   Typical Procedures Used in Vapor Immersion Processes with Trichlorethylene Cleaning Solven

System	Step 1	Step 2	Step 3
For cleaning flat parts with light soils and little contamination			
Vapor only	Vapor; 87°C; 1 min	None	None
For cleaning parts with adhering particles			
Vapor-spray-vapor	Vapor; 87°C; 15–30 s	Spray 60–71°C 15–30 s	Vapor 87°C 30–45 s
For cleaning parts with relatively little contamination			
Warm liquid-vapor[1]	Warm liquid 60–71°C 30–45 s	Vapor 87°C 30–45 s	None
For cleaning heavily soiled parts[2]			
Boiling liquid-warm liquid-vapor[1]	Boiling liquid 87–90°C 30–45 s	Warm liquid 60–71°C 15–30 s	Vapor 87°C 30–45 s

[1]*Work may be held in vapor zone until condensation ceases before being placed in liquid.*
[2]*Some parts may require a solvent dip before step 1.*

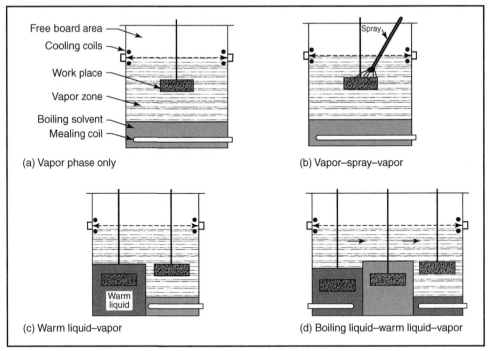

FIGURE 1.9

Principal systems of vapor immersion degreasing process.

1.48.2 LIQUID IMMERSION PROCESS

Loosely bond contamination that is too great for vapor immersion, such as swarf and road dirt, can be removed by immersing the parts in vigorously boiling solvent or hot or cold solvent. It is particularly suitable for hand cleaning in a small tank.

Brushing or scrubbing will aid and accelerate cleaning. Articles with cavities that may hold the solvent should be immersed so that the holes are filled and then removed at an angle that will ensure that they are emptied.

Repeated dipping and agitation may be necessary to flush out solid material such as swarf. After the first cleaning process, the articles may be immersed in a second tank of clean solvent (agitated if possible) for at least 1 min to remove the film of contaminated solvent from the first tank; care should be taken to carry over as little solvent as possible from one tank into the other. After cleaning, all excess solvent should be drained from the articles.

1.49 SPRAY OR JET CLEANING PROCESS

This process may be used to remove oil, grease, and light dirt or swarf contamination from unit parts or very simple assemblies that are capable of being rigidly suspended in a draining position while passing through the spraying zone. It also can be used for small parts that may be held in a cleaning zone (solvent, detergent, or steam) and turned over if necessary while a handheld jet or spray is directed on them.

1.49.1 SOLVENT/DETERGENT CLEANING

Articles with obstinate dirt deposits that cannot be removed by immersion may require jetting at high pressure with hot or cold solvent, emulsifiable solvent, and detergent solution. This process is mainly used in specially designed machines that may be operated by hand or mechanically; in the latter, it is essential for the articles to be suspended, or placed in a basket, in such a position that jets or sprays can reach all surfaces and it may need to be cooled. Proprietory equipment is also available for spray or jet cleaning, in which small articles are placed in a cabinet with a transparent hood, and then solvent is directed on them by a nozzle held in a gloved hand.

Usually, jet or spray cleaning is carried out in a specially designed machine filled with rows of fixed jets and featuring a conveyor belt to carry the articles continuously through the cleaning, rinsing, and drying stages.

1.49.2 STEAM CLEANING

The equipment required is a pressure-jet steam cleaner. A separate solution tank or drum may be required to prepare the cleaning compound. One type of steam cleaner stores the concentrated cleaning solution and mixes it with water at a constant rate to produce a uniform cleaning solution through a heating unit in which it is partially vaporized and put under pressure.

The hot solution and steam are forced through the nozzles onto the surfaces to be cleaned. The same equipment can be used for cleaning with dry steam or with cold water under high pressure. This type of steam cleaner may be either portable or stationary.

Another type of portable pressure jet steam cleaner, sometimes called a *hydro-steam unit,* requires an outside steam source. The cleaning solution is mixed and stored in a container or tank that is not part of the steam cleaner. No water is mixed with the solution in the steam cleaner, so the solution is made up at a lower concentration than that used for the pressure-jet type of cleaner. The solution and steam are mixed in the cleaner and discharged through the nozzle of a steam-cleaning gun. The same equipment can be used for cleaning with dry steam.

The cleaning guns may be furnished with interchangeable nozzles. Round nozzles are used for most cleaning, and flat nozzles are used for flat surfaces.

1.50 BRUSHING OR WIPING PROCESS

This process is intended for the removal of oil, grease, and light contamination from the bare metal areas of assemblies with painted surfaces or nonmetallic inserts that might be damaged by a general application of solvent. It may also be used for the in situ cleaning of articles that are too large for immersion tanks or spray cleaning systems.

Solvent should be applied to the contaminated areas with a clean brush or a cloth. The application of clean solvent with scrubbing or wiping should be repeated until all the contamination has been removed. Care should be taken to apply the solvent to the contaminated metal areas only.

If large amounts of water cannot be used, wiping with wet cloths may suffice if it is done thoroughly. This method is used for removal of contamination from metal surfaces with a cold solvent or emulsifiable solvent.

1.51 ELECTROLYTIC CLEANING PROCESS

This is applicable to solutions that involve electrolytes, such as alkaline cleaners and acid pickles. In this process, the articles to be cleaned are attached to suitable fixtures and immersed in a solution kept at room temperature. A low-voltage current is then passed through the articles in the solution, releasing gas bubbles onto the surfaces being cleaned. In forming and escaping to the surface of the solution, these bubbles exert a "throwing off" and scrubbing action that aids cleaning.

Electro-cleaning in an alkaline solution is more effective than simple immersion, especially for the removal of solids. The electrical equipment for alkaline electro-cleaning should furnish a direct current with a density of 270–540 amp/m^2 to the surface being cleaned. The voltage will normally be between 4 and 8 volts, but it depends upon the electrical resistance of the cleaning solution, the articles, and the suspension racks.

If one tank with current-reversing switches or two tanks are provided, articles for which cathodic cleaning is permissible should be cleaned as the cathode for 1–5 min, and then as the anode for 15–30 s. If only one tank without current-reversing switches is available, the articles should be cleaned for 1–5 min as the anode only. The solution can be either hot or cold depending upon the particular application, but if the articles have had no previous cleaning, the solution is usually heated.

Alkaline electro-cleaning (cathodic or anodic) should not be used for components made entirely or partly of nonferrous metal unless it has been reliably ascertained that the process will not be harmful. This applies particularly to aluminum, magnesium, and zinc, as well as alloys consisting principally of one of these metals. It also applies to electroplated coatings, which may suffer blistering.

The risk of hydrogen embrittlement should be kept in mind when electro-cleaning treatment may be cathodic or anodic. Thus, cathodic alkaline electro-cleaning should not be used on hardened steels (particularly spring steels) when under stress, owing to the danger of cracking. However, hnodic alkaline electro-cleaning is safe for use on these steels. Hydrogen embrittlement through cathodic treatment is not prevented by a short subsequent anodic treatment.

1.52 ULTRASONIC CLEANING PROCESS

Ultrasonic energy can be used in conjunction with several types of cleaners, but it is most commonly applied to chlorinated hydrocarbon solvents, water alone, and water with surfactants. Ultrasonic cleaning, however, is more expensive than other methods because of the higher initial cost of equipment and higher maintenance cost. Consequently, the use of this process is largely restricted to applications in which other methods have proved inadequate.

Typical situations in which ultrasonic methods have proved advantageous are as follows:

- Removal of tightly adhering or embedded particles from a solid surface.
- Removal of fine particles from powder metallurgy parts.
- Cleaning of small precision parts, such as those for cameras, watches, or microscope components.
- Cleaning of parts made of precious metals.
- Cleaning of parts with complex configurations, when extreme cleanness is required. Examples include precision parts used in fuel injection or control, which require the ultimate in cleanness; even a few minute particles of soil may impair the functioning of such components.
- Cleaning of parts for hermetically sealed units. For example, one manufacturer utilizes ultrasonic cleaning for hermetically sealed refrigerator parts. In this application, parts are placed in aluminum racks and cleaned ultrasonically (400 kc/s) in trichlorethylene.

Despite the high cost of ultrasonic cleaning, it has proved economical for applications that otherwise would require cleaning by hand.

1.52.1 SIZE OF PARTS

The size of the part being cleaned is a limitation, although no definite limits have been established. The commercial use of ultrasonic cleaning has been limited principally to small parts, such as those indicated in the abovementioned examples. The process is used only as a final cleaner, after most of the soil is removed by another method.

Ultrasonic transducers which convert electrical energy into ultrasonic vibrations, come in two basic types: electrostrictive (barium titanate) and magnetostrictive. The latter is capable of handling larger power inputs. Barium titanate transducers generally are operated over a range of 30 to 40 kc/s; magnetostrictive transducers usually operate at about 20 kc/s, but they may operate at frequencies up to about 50 kc/s.

Cleaning efficiency in the liquid phase of a vapour-degreasing cycle can be considerably augmented by the use of ultrasonic energy. However, ultrasonic cleaning is expensive, so it is seldom used in a degreasing cycle unless other modifications have failed to attain the desired cleanness or the parts being cleaned are too small or too intricate to receive the maximum benefit from conventional degreasing cycles.

The inside walls of hypodermic needles can be thoroughly cleaned by ultrasonic degreasing. Other examples of parts cleaned by ultrasonics because they failed to respond to conventional degreasing methods are small ball bearing and shaft assemblies, printed circuit boards (for instance, removal of soldering flux), intricate telephone relays, plug-type valve inserts (contaminated with lapping compounds), and strands of cable (for instance, removal of oil and other manufacturing contaminants trapped between the strands). Note that because of the increased fire risk, the use of ultrasonic techniques with flammable solvents should be avoided.

1.53 APPLICATION PROCESS OF BLASTING
1.53.1 GUIDE TO SPECIFICATIONS OF ABRASIVES

The selection of a suitable abrasive for a specific application is influenced by the type of surface contamination to be removed, size and shape of the piece being cleaned, surface finish, type and efficiency of cleaning equipment, and required production rate.

1.53.2 METALLIC ABRASIVES

Metallic abrasives used for blast cleaning include the following:

- Cast steel shot and grit
- Chilled iron shot and grit
- Malleable iron shot and grit
- Cut steel wire shot

Cast steel shot and grit and chilled iron shot and grit are commonly used for the blasting of steel, but the other types are not recommended. The types of metallic abrasives recommended for some typical application are shown in Table 1.34.

Table 1.34   Types of Steel Abrasives Most Commonly Used for Various Structural Steel Blast Cleaning

	Abrasive Type			Hardness (RC)[2]	
	Shot	**Grit**	**Size Range**[1]	**40 to 50**	**55 to 60**
New Steel	X		S170 to S390	X	
Fabricated new steel	X		S170 to S390	X	
		X	G50 to G25	X	X
Heat-treated steel		X	G50 to G25		X
Heavy steel plate	X		S230 to S390	X	
Corroded steel		X	G50 to G25		X
Weld scale	X		S170 to S280	X	
Brush blast	X		S170 to S280	X	
Repair work		X	G50 to G40	X	
Maintenance		X	G80 to G18	X	X

[1]*The term* size range *refers to the working mix (operating mix) for recirculating abrasive blast systems.*
[2]*RC, which stands for "Rockwell," is the unit of hardness for the measurement of size and hardness of metallic abrasive, as described in BS 2451-1963.*

Shot and grit should be designated by the letters *S* and *G* and graded by the shot or grit number according to SAE specifications (see Tables 1.35 and 1.36).

1.53.3 NONMETALLIC ABRASIVE

Nonmetallic abrasives used for blast cleaning may be classified as follows:

- Naturally occurring
- By-product
- Manufactured abrasive

A set of physical data on nonmetal abrasives is summarized in Table 1.37.

Natural nonmetallic abrasives include silica sand, which consists of sand and flint with a medium size range of 0.85–0.425 mm. These types are most commonly used and are effective for new steel and for maintenance in noncritical areas. Another is nonsilica sand or heavy mineral, which has a medium size of 0.212–0.150 mm. It is tough and dense and generally finer than silica sand. Heavy mineral sand can be reused many times, and it is effective for blast cleaning of new steel. Garnet is a natural, high-cost, tough, and angular abrasive that is suitable for cleaning steel parts and casting in a closed system that permits recycling abrasive materials.

Other natural nonmetal abrasives include zircon, a high-cost, tough, and angular abrasive suitable for removal of fine scale, leaving a smooth and mat finish; and novaculite, a very pure, siliceous rock that is well suited to clean precision tools and castings.

By-products include slag abrasives, which are available in a size range of 2.36–0.150 mm, have a sharply angular shape and low (less than 1%) free silica content. Slag abrasives are suitable for cleaning new and painted surfaces and produced the least dust in operation. The breakdown rating of slag is %29–%51, and the resulting surface profile on steel are 0.083–0.092 mm. (The surface profile was measured using electrometer gage.

Other by-products are agricultural shells, such as walnut shells, peach pits, and corn cob. These products have a size range of 2.0–0.150 mm and are excellent for removing paint, fine scale, oil, and grease without altering the metal substrate.

Manufactured abrasives are 10–15 times more expensive than by-product slags, 30 to 40 times more expensive than sand, and generally adaptable to recycling as many as 20 times. These abrasives are suitable for special uses (i.e., silicon carbides for etching, aluminum oxides for blasting of stainless steel, and glass beads for peening and removing oxide film on plastic molds).

1.54 ABRASIVE BREAKDOWN

Forces that develop the cleaning capability of metallic abrasives also tend to reduce the size of abrasive particles and to cause their eventual breakdown to dust. The greater the particle breakdown, the poorer the cleaning rate. From the standpoint of relative metallic abrasive consumption, the following guidelines exist:

- Chilled cast iron abrasives have a breakdown rate that is as much as one-third more than full, hard (65 plus Rc), and untempered steel grit.
- Malleable iron abrasives have a breakdown rate of 50% to 100% more than steel abrasive in the 40–50 Rc hardness range.
- Steel grit breaks down slightly faster than steel shot of the same size and hardness range. However, the smaller the grit involved, the greater the difference in breakdown compared to shot.
- Similarly, increasing the hardness of steel shot or grit in a given operation increases the breakdown rate. Obviously, however, the more rapid breakdown of a harder steel abrasive becomes academic if it has been determined that a lower hardness will not do the job.

1.54.1 ABRASIVE BREAKDOWN TEST AND RATING FORMULA

The abrasive breakdown test procedure is as follows:

1. A sieve analysis as outlined in ASTM D451-63 is run on a representative split sample containing approximately 200 g of test abrasive, and data is recorded.
2. Breakdown test equipment is hooked to an air supply capable of maintaining 6.54 bar of dry air at the blast machine when the machine is operating. A sample of the abrasive being tested is run through the machine, and the flow valve is adjusted to give free, unchoked abrasive flow. (The flow is free if a bluish-white hue is visible through the path of air and abrasive as it comes from the nozzle.) The equipment (both drum and blaster) is then cleaned so that no dust or abrasive remains.

Table 1.35   SAE Shot Size Specifications with Suggested Removal Sizes, and Cast Shot Specifications for Shot Peening or Blast Cleaning

Screen Opening Sizes and Screen Numbers with Maximum and Minimum Cumulative Percentages Allowed on Corresponding Screens SEA Shot Number

NBS Screen Number	Standard (mm)[3]	Screen Size (in)	S1320	S1110	S930	S790	S660	S550	S490[4]	S390	S330	S290	S230	S170	S110	S70
4	4.75	0.187	All pass	N/A	N/A	N/A	N/A	N/A	N/A	N/A	N/A	N/A	N/A	N/A	N/A	N/A
5	4	0.157	N/A	All pass	N/A	N/A	N/A	N/A	N/A	N/A	N/A	N/A	N/A	N/A	N/A	N/A
6	3.35	0.132	90% min	N/A	All pass	N/A	N/A	N/A	N/A	N/A	N/A	N/A	N/A	N/A	N/A	N/A
7	2.8	0.111	97% min	90% min	N/A	All pass	N/A	N/A	N/A	N/A	N/A	N/A	N/A	N/A	N/A	N/A
8	2.36	0.0937	N/A	97% min	90% min	N/A	All pass	N/A	N/A	N/A	N/A	N/A	N/A	N/A	N/A	N/A
10	2	0.0787	N/A	N/A	97% min	85% min	N/A	All pass	N/A	N/A	N/A	N/A	N/A	N/A	N/A	N/A
12	1.7	0.0661	N/A	N/A	N/A	97% min	85% min	N/A	All pass	N/A	N/A	N/A	N/A	N/A	N/A	N/A
14	1.4	0.055	N/A	N/A	N/A	N/A	97% min	85% min	5% max	All pass	N/A	N/A	N/A	N/A	N/A	N/A
16	1.18	0.0469	N/A	N/A	N/A	N/A	N/A	97% min	85% min	5% max	All pass	N/A	N/A	N/A	N/A	N/A
18	1	0.0394	N/A	N/A	N/A	N/A	N/A	N/A	96% min	85% min	5% max	All pass	All pass	N/A	N/A	N/A
20	0.85	0.0331	N/A	N/A	N/A	N/A	N/A	N/A	N/A	96% min	85% min	5% max	10% max	All pass	N/A	N/A
25	0.71	0.0278	N/A	N/A	N/A	N/A	N/A	N/A	N/A	N/A	96% min	85% min	N/A	10% max	All pass	

	Opening (mm)	Opening (in)												
30	0.6	0.0234	N/A	N/A	N/A	N/A	N/A	N/A	N/A	96% min	85% min	N/A	10% max	All Pass
35	0.5	0.0197	N/A	N/A	N/A	N/A	N/A	N/A	N/A	N/A	97% min	85% min	N/A	10% max
40	0.425	0.0165	N/A	N/A	N/A	N/A	N/A	N/A	N/A	N/A	N/A	97% min	80% min	N/A
45	0.355	0.0139	N/A	N/A	N/A	N/A	N/A	N/A	N/A	N/A	N/A	N/A	90% min	80% min
50	0.3	0.0117	N/A	N/A	N/A	N/A	N/A	N/A	N/A	N/A	N/A	N/A	N/A	90% min
80	0.18	0.007	N/A	N/A	N/A	N/A	N/A	N/A	N/A	N/A	N/A	N/A	N/A	N/A
120	0.125	0.0049	N/A	N/A	N/A	N/A	N/A	N/A	N/A	N/A	N/A	N/A	N/A	N/A
200	0.075	0.0029	N/A	N/A	N/A	N/A	N/A	N/A	N/A	N/A	N/A	N/A	N/A	N/A
Suggested Removal Size for Cleaning Structural Steel[2]			**0.0232**	N/A	**0.0165**	**0.0165**	**0.0138**	**0.0117**	**0.0117**	**0.0082**	0.007	0.0059	0.0049	**0.0029**

Note: Courtesy of the Society of Automotive Engineers (SAE).

The shot number is roughly the nominal size of the shot pellets in ten-thousands of an inch amounts are percentages by weight of the total sample. For example, "90% min" means that at least 90% by weight is retained, and at most 10% pass.

[2]See discussion of the work mix (SSPC volume 1).

[3]Corresponds to ISO recommendations.

[4]This is the coarsest size in common use for blast cleaning structural steel for painting.

Table 1.36   SAE Cast Grit Size Specifications for Blast Cleaning

NBS Screen No.	Standard (mm)[1]	Screen Size (in)	Screen Opening Sizes and Screen Numbers with Minimum Cumulative Percentages Allowed on Corresponding Screens SEA Grit Number											
			G10	G12	G14	G16	G18[2]	G28	G40	G50	GSO	G120	G200	G228
5	4.00	(0.157)	N/A	N/A	N/A	N/A	N/A	N/A	N/A	N/A	N/A	N/A	N/A	N/A
6	3.35	(0.132)	N/A	N/A	N/A	N/A	N/A	N/A	N/A	N/A	N/A	N/A	N/A	N/A
7	2.80	(0.111)	All	N/A	N/A	N/A	N/A	N/A	N/A	N/A	N/A	N/A	N/A	N/A
8	2.36	(0.0937)	Pass	All	N/A	N/A	N/A	N/A	N/A	N/A	N/A	N/A	N/A	N/A
10	2.00	(0.0787)	N/A	Pass	All	N/A	N/A	N/A	N/A	N/A	N/A	N/A	N/A	N/A
12	1.70	(0.0661)	80%	N/A	Pass	All	N/A	N/A	N/A	N/A	N/A	N/A	N/A	N/A
14	1.40	(0.0555)	90%	80%	N/A	Pass	All	N/A	N/A	N/A	N/A	N/A	N/A	N/A
16	1.18	(0.0469)	N/A	90%	80%	N/A	Pass	All	N/A	N/A	N/A	N/A	N/A	N/A
18	1.00	(0.0394)	N/A	N/A	90%	75%	N/A	Pass	All	N/A	N/A	N/A	N/A	N/A
20	0.850	(0.0331)	N/A	N/A	N/A	85%	75%	N/A	Pass	N/A	N/A	N/A	N/A	N/A
30	0.710	(0.0278)	N/A	N/A	N/A	N/A	N/A	N/A	N/A	All	N/A	N/A	N/A	N/A
35	0.600	(0.0234)	N/A	N/A	N/A	N/A	85%	70%	N/A	Pass	N/A	N/A	N/A	N/A
35	0.500	(0.0197)	N/A	N/A	N/A	N/A	N/A	N/A	N/A	N/A	N/A	N/A	N/A	N/A
40	0.425	(0.0165)	N/A	N/A	N/A	N/A	N/A	N/A	N/A	N/A	All	N/A	N/A	N/A
45	0.355	(0.0139)	N/A	N/A	N/A	N/A	N/A	80%	70%	N/A	Pass	N/A	N/A	N/A
50	0.300	(0.0117)	N/A	N/A	N/A	N/A	N/A	N/A	N/A	N/A	N/A	All	N/A	N/A
80	0.180	(0.007)	N/A	N/A	N/A	N/A	N/A	N/A	80%	65%	N/A	Pass	All	N/A
120	0.125	(0.0049)	N/A	N/A	N/A	N/A	N/A	N/A	N/A	75%	65%	N/A	Pass	All Pass
200	0.075	(0.0029)	N/A	N/A	N/A	N/A	N/A	N/A	N/A	N/A	75%	60%	N/A	N/A
325	0.045	(0.0017)		N/A		N/A	N/A	N/A	N/A	N/A	N/A	70%	55%	20%
			N/A	N/A	N/A	N/A	N/A	N/A	N/A	N/A	N/A	N/A	65%	

Note: Courtesy of the Society of Automotive Engineers.
The grit number is roughly the nominal size of the shot pellets in ten-thousands of an inch. Amounts are percentages by weight of the total sample. For example, "90%: means that at least 90 percent by weight is retained and at most 10% pass.
[1]Corresponds to ISO recommendations.
[2]This is the coarsest size in common use for blast cleaning structural steel for painting.

3. A total of 10 lbs. of test abrasive having a sieve analysis as determined in step 1 is introduced into the blaster, and all of it is expelled at predetermined settings into the 208-l drum fitted with a dust bag. After blasting, the cone-shaped bottom of the drum is opened, and all accumulated abrasive and dust are collected and transferred to the original weighing container and then reweighed. The difference in weight is recorded.

Table 1.37   Physical Data on Nonmetallic Abrasives

Abrasives	Hardness (Mohs Scale)	Shape	Specific Gravity	Bulk Density	Color	Free Silica (wt %)	Degree of Dusting	Reuse
Naturally Occurring Abrasives (kg/l)								
Silica	5	Rounded	2–3	1.6	White	90+	High	**Poor**
Heavy mineral	5–7	Rounded	3–4	2	Variable	<5	Med	**Good**
Flint	6.5–7	Angular	2–3	1.3	Grey-white	90+	Med	Good
Garnet	7–8	Angular	4	2.3	Pink	None	Med	Good
Zircon	7.5	Cubic	4.5	2.9	White	None	Low	Good
Novaculite	4	Angular	2.5	1.6	White	90+	Low	Good
By-Product Abrasives								
Boiler	7	Angular	2.8	1.3–1.4	Black	None	High	Poor
Copper	8	Angular	3.3	1.6–1.9	Black	None	Low	Good
Nickel	8	Angular	2.7	1.3	Green	None	High	Poor
Walnut shells	3	Cubic	1.3	0.7	Brown	None	Low	Poor
Peach pits	3	Cubic	1.3	0.7	Brown	None	Low	Poor
Manufactured Abrasives								
Silicon carbide	9	Angular	3.2	1.7	Black	None	Low	Good
Aluminum oxide	8	Blocky	4	1.9	Brown	None	Low	Good

4. A split sample of approximately 200 g is taken from the spent abrasive for sieve analysis.

   The breakdown rate is calculated from the sieve analysis, as follows:

$$\text{Breakdown rate} = \frac{\%\,\text{spent abrasive retained} \times \text{average sieve opening}}{\%\,\text{as} - \text{received abrasive retained} \times \text{average sieve opening}}$$

For the screens, the expression becomes:

$$*\text{Rc} = \text{Rockwell} - \text{unit of hardness}$$

Breakdown factors range from 1.0, for an abrasive showing no reduction from original size after blasting, to approximately zero, for large grains that are reduced to dust. Most mineral abrasives will have a rating of approximately 0.6.

1.55 GENERAL NOTES ON USE OF ABRASIVES

The following general points relating to the performance of abrasive particles may be helpful:

- The smaller the size, the finer the surface finish.
- The larger the size, the greater the impact.
- The harder the abrasive, the faster the cleaning action.

Table 1.38 presents a guide for the selection of a suitable abrasive for various specific applications.

When abrasives that have become contaminated by dislodged fragments of coating or by grease, oil, and the like are reused, there is a risk that the adhesion and effectiveness of the coating may be impaired. Blasting abrasives that have become contaminated during use by corrosive materials (e.g., chlorides, sulfates, and the like) must not be reused. Furnace slags and copper slags should not be used in tank linings or other applications where abrasive residues embedded in the steel surface may have a detrimental effect on coating performance.

Surface preparation for coatings less than 0.25 mm thick requires grit or sand, sized not larger than 1.18 mm (16-mesh) for sand or G-40 for grit. The abrasive size for coatings thicker than 0.2 mm is not as critical, so long as the profile is deep enough. A new thin film coating of 0.036–0.05 mm will require a shouldower anchor pattern, which is achieved with a finer abrasive to 0.25 mm (60 mesh) size.

When sheets that are less than 4 mm thick are blasted, deformation may occur. This should be avoided by controlling the following:

Low air pressure, small grain sizes, and the use of a blasting abrasive of low bulk density
Blasting at a low angle, with a sharp-edged grain and a short duration

With adhesive scale or fairly thick rusting, it may be necessary to carry out preparatory work by other methods (e.g., grinding or pickling) before blasting.

1.56 BLASTING PROCESS AND EQUIPMENT
1.56.1 AIR-BLAST CLEANING

Proper surface preparation by air-blasting provides a foundation for the paint, resulting in a clean surface, uniform etching, and a long, economical coating life. In abrasive air-blast cleaning, surface preparation can be achieved on parts or weldments that are not uniform in size or shape.

Air-blast equipment contains and channels abrasive into a compressed air stream through conveying hoses and nozzles to the piece. In effect, the part being cleaned is eroded by a mass of abrasive particles until a firm, clean surface results. Abrasive blast cleaning with a compressed air source, air hose, abrasive blast machine, abrasive hose, and nozzle imparts a velocity to the abrasive particles that turns them into a working force.

Various abrasives are used in this process, but the most widely used abrasive is silica sand that has been processed for a blasting abrasive. Respiratory protection must be given to the operator and workers in the blast-cleaning area because of spent abrasive and the contamination being removed from the surface. Selection of the abrasive in this process becomes a major factor in cleaning speed, surface etching, and coating adhesion. The trend is toward a finer size of abrasive because of increased cleaning

Table 1.38   Abrasives as Used in Abrasive Blast Cleaning

Recommended Service	Silica Abrasives	Slag Sand Abrasives	Slag Shot	Flint Abrasives	Natural Mineral Abrasives	Synthetic Abrasives	Special Abrasives	Vegetable Abrasives	Glass Abrasives	Chilled Iron Grit	Chilled Iron Shot	Annealed Abrasives	Steel Grit	Steel Shot
General blast cleaning where abrasives can be recycled and reused economically	X	X	X	X	X	Y		X		Y	X	X		
General blast cleaning where abrasives cannot be economically reclaimed	Y	X	X	X										
Pre-metallizing blasting	X	X	X	Y	X	X				Y	X			X
Blasting where metal tolerances cannot be changed							X	Y		Y				
Blasting in rooms and cabinets	X	X	X	X	X	Y	X	X	X		X	X		
Blasting where the elimination of food contamination or nonmagnetic abrasives are required								Y						
Blasting to obtain a high luster on aluminum, brass, and other metals					X		X	X	Y					
Liquid Hone – Hydro Hone – Wet hone blasting	X					Y	Y	Y	Y					
Centrifugal wheel blasting												X	Y	Y

Note: X = most commonly used abrasives; Y = preferred abrasives

speed on new or lightly rusted steel; a coarser size of abrasive is used for more corroded steel or harder-to-clean surfaces. It is important to maintain a proper size of abrasive for air-blast cleaning.

The air-blast cleaning process should be performed with one of the following pieces of equipment as defined by the company:

- Pressure-type blast equipment
- Suction blast equipment
- Vacuum blast equipment

1.56.2 PRESSURE-TYPE BLAST EQUIPMENT

In a pressure-type abrasive blast system, the abrasive machine is under the same pressure as the entire system (i.e., the compressor, air lines, abrasive blast machine, abrasive blast hose, and nozzle). This cleaning method is the most productive. The efficiency is largely dependent on actual nozzle pressure, which should be in the range of 6.3–7 bars. The pressure blast machine, or "pot," varies in size, but it must be under pressure for an even flow of abrasives. Velocity of the abrasive in the pressure method is greater than the abrasive velocity found in suction blast equipment.

1.56.3 SUCTION BLAST EQUIPMENT

This equipment utilizes the suction-jet method of obtaining abrasive from the abrasive tank that is not under pressure. The jet of air blasts the abrasive against the surface after sucking abrasive from the container. The cleaning speed is approximately one-third slower than that of pressure blast cleaning with similar-size air jets. Its use should be limited to touch-up or spot cleaning jobs, where high-speed cleaning is not a factor.

1.56.4 VACUUM BLAST EQUIPMENT

In the vacuum-blast cleaning method, air and abrasive are captured in a rubber-hooded enclosure. They are drawn by suction back to the blast unit, where reusable abrasive is separated from blast-cleaned surface contaminants, recycled, and reused.

This is considered a dust-free abrasive blast cleaning because it shields the blast surface area from flying particles and dust. It will not disturb adjacent machinery and workers. The cleaning speed is limited because the surface is not visible to the operator.

There are two methods of vacuum-blast cleaning. In the suction type, the abrasive is siphoned from the container to the blast head. The pressure-type machine delivers sand under pressure through a blast hose to the surface. The pressure method provides greater production. The process is limited to the use of a reusable abrasive, such as metallic, steel shot or steel grit, aluminum oxide, or garnet. In some cases, where moisture is a problem due to high humidity, a mixture of steel grit and aluminum oxide or garnet is recommended because it keeps the metal abrasive from lumping or congealing due to moisture.

1.56.5 RECOMMENDATIONS ON AIR BLASTING

The components of air blast equipment inclue the air supply, air hose and couplings, abrasive blast machines, abrasive blast hose and couplings, nozzles, operator equipment, air-fed hoods and control valves, and oil and moisture separators. A manager of an abrasive air blast operation should maintain a checklist of each component to ensure peak performance.

The compressed air supply for blast cleaning should be adequate in pressure and volume to meet the work requirements. It also should be sufficiently free of oil and water contamination to ensure that the cleaning process is not impaired. Oil and moisture separators require solvent cleaning to remove oil and routine replacement of filters.

There are many types of blast nozzles. Ceramic and cast iron are nozzles with a short life, and carbides are nozzles with a long life. The shape of nozzles can provide a great advantage: Venturi style nozzles (with a large throat that converges to the orifice and then diverges to the outlet) propel abrasive particles rapidly through the nozzle, making the cleaning rate faster than that of a straight bore nozzle of the same length.

The nozzle size in Table 1.39 indicates air consumption (l/s) at 7 bars without abrasive going through the nozzle. When choosing the compressor size, the next larger compressor available for the nozzle should be used. It is also wise to consider other air requirements from the compressor, such as for an air-fed hood (at 9.5 l/s) and air-driven ventilating equipment (at approximately 56.7 l/s). A separate air source for air-fed hoods may be required unless a carbon monoxide detector is installed in the air system. Insufficient air supply results in excess abrasive and slower cleaning rates.

The recommended size of the air supply hose should be three or four times the nozzle orifice; for lines over 30 m, four times should be the minimum. The recommended size of the blast hose is also three to four times the nozzle size, except near the nozzle end. A typical air supply hose will be 30 m of 31.75 mm and blast hose will be 3 m of 25.4 mm for a 9.37-mm nozzle.

A hose is normally constructed of a 6.25-mm-thick rubber tube, with carbon black compounding for the dissipation of static electricity generated by an abrasive flow through the tube. Dissipating static electricity prevents buildup and transmitting a shock to the operator.

The tube is covered by 2- or 4-ply wrapping to provide strength for pressure requirements. Normal working pressure should not exceed 8.6 bars. The normal pressure drop of an air-blast hose with a 9.37-mm nozzle is 0.35 bar per 15.2 m. Therefore, it is important for the hose to be as short as possible.

A word of caution: too large an abrasive hose 25.4 mm on small nozzles 4.7 mm may result in an uneven abrasive flow.

Table 1.39 Air Consumption		
Nozzle Orifice [mm (in)]	**L/s Required 7 Bars (100 psi)**	**Abrasive Consumption (kg/h)**
4.69 (3/16")	28	117
6.25 (1/4")	50	220
7.81 (5/16")	76	365
9.37 (3/8")	110	518
10.93 (7/16")	148	712
12.5 (1/2")	194	910
15.62 (5/8")	274	1133
18.75 (3/4")	396	1428

As with any production job, efficiency results in good production rates and lower unit costs. This is especially true in abrasive air blast operations, where a small drop in pressure rapidly increases consumption of abrasive and decreases the cleaning rate.

In a 2-min blast test, differences in nozzle pressure were compared. At 4.1 bars, the rate of cleaning is the rate at 7 bars and abrasive usage is more than double.

Special consideration should be given to interior blast cleaning. A wide variation in production rates exists for interior cleaning [as much as 2:1 (half the exterior cleaning rate)] because of visibility problems, ventilating problems, and inaccessibility. These variables can be minimized with good lighting, ventilation, and scaffolding techniques.

1.57 CENTRIFUGAL BLASTING

In centrifugal or airless blast cleaning, the spinning of large paddle wheels creates a force that throws the abrasive at the surface. The essential components of all centrifugal-blast cleaning systems are blast wheels, a blast enclosure, an abrasive recovery and recycling system, and a dust collector.

Among the most common applications of centrifugal blast cleaning is the surface preparation of structural steel for coating. The major advantages of centrifugal-blast cleaning over air-blast cleaning are savings in time, labor, energy, and abrasive consumption. A further advantage is the automation of cleaning operation, which provides superior, more uniform cleaning of steel, which is a more acceptable result.

Centrifugal blast cleaning machines are divided into two groups: fixed place (which includes table type, continuous-flow and blasting-tumbling) for shop blasting, and portable machines for on-site blasting. The choice of which of these to use depends on the size and shape of article, shop layout, and capability and degree of the cleanliness desired.

The cleaning rate that can be achieved with portable units is many times greater than that produced by air-blast cleaning. Touch-up cleaning by air blast (or various types of powered hand tools) is required around narrow peripheral areas and protuberances. Otherwise, the operation is environmentally clean and economical. Because it is essentially an automated process, it provides greater consistency and uniformity of cleaning than air-blasting does.

1.57.1 TABLE-TYPE MACHINE

The table-type machine is a form of cabinet machine that contains a power-driven rotating worktable; within the cabinet, the blast stream is confined to approximately half the table area. The work is positioned on the slowly rotating table, and the abrasive particles are propelled by an overhead centrifugal whee6., When the doors are closed, blast cleaning continues for a predetermined time cycle. Some machines of this type are designed with one or more openings in the cabinet.

These openings are shielded by curtains and permit manual adjustment of the parts during the blast cycle, as well as continuous loading and unloading. The table-type machine is used in a batch system and handles single pieces.

1.57.2 CONTINUOUS-FLOW MACHINES

Continuous-flow machines equipped with proper supporting and conveying devices are used for continuous blast cleaning of steel strips, coils, and wire. These machines also are used to clean castings and forgings at a high production rate, using skew rolls, monorails, and other continuous work-handling mechanisms.

In operation, items are loaded outside the blast cabinet and then conveyed into it through a curtained vestibule, which is designed with 90-degree turns to prevent the escape of flying abrasive particles. The conveyor indexes each item to the center of each blast station and rotates it for complete blast coverage.

A machine of this type incorporates abrasive-recycling facilities and an exhaust system for removing dust and fines. Continuous-flow machines can clean the external surfaces of plate and structural members prior to fabrication or a wide variety of fabricated sections, including massive girders and trusses, for the construction of highways, power plants, and industrial buildings.

1.57.3 BLASTING-TUMBLING MACHINES

Blasting-tumbling machines consist of an enclosed, endless conveyor, a mechanical blast-propelling device, and an abrasive-recycling system. They simultaneously tumble and blast the work and are made in various sizes to accommodate workloads from 28 to 2,800 l. The work usually is loaded into the conveyor by means of a skip-bucket loader. As the conveyor moves, it gently tumbles the work and exposes all work-piece surfaces to each abrasive blast. At the end of the cleaning cycle, the conveyor is reversed and the work automatically discharged from the machine.

Blasting-tumbling machines are used for cleaning unmachined castings, forgings, and weldments whose size, shape, and material allow them to be tumbled without damage. This equipment is not employed for cleaning parts after machining, however, because tumbling would damage the machined surfaces. Blasting-tumbling machines remove dry contaminants, such as sand, rust, scale, and welding flux, and provide surface preparation for enameling, rubber bonding, electroplating, and etching prior to tinning.

1.57.4 PORTABLE EQUIPMENT

When parts to be cleaned are too large to be placed in blasting machines, portable equipment can be brought to the work-piece. For structural steel cleaning, this method can presently be used on ship decks, ship hull bottoms and sides, storage tank exteriors (both floater and cone tops and shells), and the wet side of tank bottoms. It can be used to remove heavy, anti-skid coating from aircraft carrier decks. These machines are used during construction and also for maintenance painting. In principle, portable machines use the basic components required for stationary installation; i.e., the blast wheel, abrasive recovery and recirculation, system ventilation, dust removal and collection, and a "work conveyor".

These machines incorporate one or two blast wheels and clean a swath that is approximately 0.7–1.9 m wide. Smaller units employ a single blast wheel and clean a swath that is 0.5 m wide.

In many applications, where small amounts of dust are generated during blasting or where minor dust effluent from the ventilation/collector system is permissible, blast-cleaning units are completely self-contained except for the power supply. For applications where great amounts of blast residues are generated and/or where effluent dust cannot be tolerated, the total system includes a supplementary, large-capacity dust collector.

1.57.5 SOME CONSIDERATIONS ABOUT CENTRIFUGAL BLASTING

Centrifugal-blast cleaning machines incorporate one or more wheel units, positioned so the abrasive blast will reach the entire surface. Generally, the abrasive from each wheel is thrown out in a fanlike pattern covering an area about 70–100 mm wide and 0.9 m long. The number of wheels needed is determined by the size, complexity, and shape of the surface.

Centrifugal blasting machines may contain air-wash separator systems to remove contaminants from returned abrasives. Centrifugal blast wheels are available in several sizes and are equipped with drive motors of up to (100 hp) 75 kW for high-production applications.

Generally, motor sizes of 11–45 kW (15–60 hp) with wheels ranging from 38 to 50 mm inches in diameter are used for structural steel cleaning. Under average conditions, velocities of abrasive propelled by airless wheels are about 71 m/s for special applications. The velocity can be decreased to 91–96 m/s.

1.57.6 ABRASIVES

The abrasive used for structural steel cleaning consists of tiny particles of alloy steel, generally ranging in diameter from 0.12 to 1 mm. These particles, unlike sand, resist fracturing despite repeated impacts at high velocity. Much greater wear results from the use of nonmetallic abrasive such as sand, aluminum oxide, and silicon carbide.

1.57.7 VENTILATION

It is essential to provide sufficient ventilation to ensure that air pressure within that blast enclosure is lower than the ambient pressure that allows dust generated by the blast cleaning to be drawn into the dust collector and prevents it from escaping the blast enclosure into adjacent work areas.

1.58 WATER BLASTING (HYDROBLASTING)

Water blast cleaning may take place under high or low pressure, hot or cold, and with or without an abrasive or detergent to prepare metal prior to painting. Water blast cleaning is not meant to replace abrasive blast cleaning. Water alone cannot etch a metal surface.

Injection of dry cleaning abrasive at the nozzle achieves a surface etch. Water blasting has wide acceptance in situations where dry abrasive blast cleaning dusts and contamination present a hazard to personnel and machinery (see also standard, RP-01-72 surface preparation of steel by water blasting).

In maintenance painting, where job specifications require only the removal of all loose paint scale and flaky rust and a thoroughly washed surface, hydroblasting is more economical than hand or power-tool cleaning. Hydroblast cleaning may also be preferred where there are restrictions on dry abrasive blast cleaning.

Abrasive water blasting can be used to clean irregular shapes, back-to-back angles, corroded valves, marine vessels with seawater corrosion, heat exchangers, boilers, flaking tars, clogged piping, rubber molds, plant filter screens, and latency from concrete surfaces. It is not used to shop cleaned new weldments, and its use is not recommended prior to the application of inorganic zinc primer due to rust forming between the drying period and the coating application.

Water alone, at low (up to 138 bars) or medium pressure (130–700 bars), cannot etch a metal surface. It is especially well suited for removing oil and grease accumulation, but high-pressure (up to 130 bars) water jets, used to clean pipeline coating prior to the maintenance or inspection of line for corrosion. Injection of dry abrasive via a nozzle achieves a surface etch.

Components

The basic water-blast unit consists of an engine-driven pump, inlet water filter, pressure gauge, hydraulic hose, gun, and nozzle combination. Abrasive is injected into the system after water is pressurized by means of a suction head to prevent pump damage. It is usually injected at the blast gun before the nozzle.

Any type of abrasive that is commonly used with abrasive blast cleaning can be used in water blast cleaning. Because the abrasive is normally not dried, screened, and recycled, less expensive abrasives are commonly used, with sand being the most common. Table 1.40 describes various abrasives used in water blasting and lists typical applications of these substances.

The use of a very small amount of abrasive results in minimal action, while too much abrasive causes the formation of a paste that could not be properly circulated. A range of 20% to 35% abrasive (by volume) is satisfactory for most applications.

Table 1.40   Characteristics and Typical Applications of Abrasives Used in Water Blasting

Abrasive	Size mm (Mesh)	Characteristics and Applications
Silica	0.425 to 0.180 (40 to 80)	Fast-cutting used for deburring steel and cast iron, removing oxides from steel close tolerances can not be held
Silica	0.180 (80)	Fast-cutting used for deburring steel and cast iron, roughening surfaces for plastic bonding or rough plating. has peening action. tolerances cannot be held
Quatrz (ground)	0.180 (80)	Very fast-cutting. used for removing heavy burrs, light or medium scale, excessive rust. can be used on nickel alloy steels. tolerances cannot be held
Novaculite	0.150 (100)	Fast-cutting used for cleaning carbon from piston and valve heads; deburring brass, bronze and copper. can be used on crankshafts. tolerances cannot be held
Quartz (ground)	0.150-0.106 (100, 140)	Fast-cutting used for blending-in preliminary grind lines on steel, brass and die castings, removing meduim-hard carbon deposits; blasting radii of 0.012 to 0.024 mm
Silica	0.106 (140)	Used for removing small burrs from steel, copper, aluminum and die castings; rough cleaning of dies and tools; removing metal. tolerances cannot be held
Novaculite	0 045 (325)	Slow-cutting used in first stage for cleaning master rods and glass, and in second stage for cleaning aluminum pistons, crankshafts, impellers, valves. holds tolerances to 0.006 mm
Aluminume oxide	0 038 (400)	Fast-cutting. used on stainless steel and on zinc and aluminum die castings. excellent for oil-contaminated surfaces
Novaculite	0.010 (1250)	Used in second stage for cleaning crankshafts, impellers, rods, pistons, valves, gears and bearings. also for polishing metals, tools, dies and die castings. tolerances can be held
Novaculite	0.0025 (5000)	Used for obtaining extra-fine surfaces on parts
Bright shot (glass beads)	0.85 TO 0.038 (20 TO 400)	Used for removing light scale or discoloration after heat treating, removing light oxide from jet-engine and electronic components. produce peening effect

Table 1.41   Water Blast Cleaning Rates

Surface Conditions	Water Only = W Sand Injection = Si	0–138 Bars; 1.1 M³/h	207–414 Bars; 1.3-2.3 M³/h	640 Bars; 2.3 M³/h
Easy to clean, dusty settlement, flaky, flat	W	14	32.5	46.5
Light oil or grease	SI	18.6	41.9	60.5
Average rusty surface	W	7	18.6	23.3
Angles and piping	SI	9.3	21	32.5
Heavily corroded surface	W	1.9	7	11.6
Rust scale, irregular shape	SI	2.3	9.3	16.3

Note: Hydroblast surface comparable to Sa2 condition. Abrasive cleaned surface comparable to Sa2½ condition. All rates are given in m2/h.

When abrasive is injected into the water stream, a secondary washing procedure must follow to remove spent slurry. This procedure includes a rust inhibitor, which must be compatible with the coating system.

The speed of cleaning depends on the highest manageable working pressure and volume of water. Depending on surface conditions, hydroblasting compares favorably with dry or wet abrasive blasting. Table 1.41 presents a guide to cleaning rates.

Flash rusting is a light rust layer that is of concern when using water blast cleaning. To avoid this, rust inhibitors such as sodium or potassium dichromate or phosphate are often used during or after water blasting. These inhibitors may retard rusting for up to seven days. This is particularly useful when working in tanks. The entire surface can be cleaned prior to coating.

One inhibitive treatment after water blast cleaning is rinsing with water containing 0.32% sodium nitrate and 1.28% by weight secondary ammonium phosphate.

1.59 DRYING PARTS AFTER AQUEOUS CLEANING AND BEFORE COATING

All surfaces should be dried before final protection against corrosion is applied. There are five methods that can be used for the removal of moisture.

1.59.1 DRYING BY HEATING

There are several procedures for drying by heating. Heating in an oven at a controlled temperatures of 120–170°C will effectively dry articles that have been cleaned in an aqueous solution followed by rinsing in water. The air in the oven should be circulated to accelerate drying and be continuously replaced by means of a fan to prevent its becoming saturated with moisture.

Another method is drying by means of blown warm air, which is accomplished by means of a fan moving air through or across a heating element. The source may be fixed or portable, and a suitable flexible nozzle or extension may be used as necessary.

Drying by means of hot water is usually employed as a continuation of an aqueous cleaning operation. The articles are immersed in clean water at 80–95°C until they have acquired the same temperature, and then they are exposed to a clean, dry atmosphere after draining off excess water. Should water be trapped in pockets, it must be blown out with clean dry air as soon as possible after taking the article out of the hot water.

Drying by infrared radiation is achieved by conveying the articles through a tunnel fitted with infrared elements. This method gives very flexible control of the drying time and heat by means of adjusting the number of elements used and their spacing and distance from the articles, combined with the length of the tunnel and the speed of the conveyor.

1.59.2 CENTRIFUGING

Centrifuging is recommended only for the drying of small metal parts that have been cleaned or immersed in an aqueous solution prior to treating with corrosion protectives. The typical equipment used in this procedure comprises an enclosed perforated drum that is approximately 48 cm in diameter and 27 cm deep, with a capacity of approximately 14 l. It is belt-driven from an external motor and has an electrically heated air intake to raise the temperature. The drum should not be filled to more than three-quarters capacity.

1.59.3 DRYING WITH HOT SOLVENT

Water can be quickly removed from any free-draining metal surface by immersing articles in a boiling chlorinated solvent containing a suitable additive. This procedure is similar to (though distinct from) trichloroethylene liquor degreasing and calls for specially designed equipment containing a drying compartment and one or more rinsing compartments.

Articles to be dried in hot solvent should first go through at least two cold running water rinses, but the use of soft water for these is not essential. They are then dipped into the boiling solution containing the additive for 1 min or less to remove the water from the metal. In one process for which perchloroethylene is the recommended solvent, the removal is achieved by evaporating the water in the form of an azeotrope; in another using trichloroethylene, the primary action is to displace the water from the metal surface by a film of the solution. The articles are then dipped for 1 min or less in clean boiling solvent in the rinsing compartment.

Some articles that are not free-draining can be dried by the hot solvent process if they are loaded in containers that can be tilted or rotated; more than one solvent rinse can be used if necessary. This method is particularly applicable where articles have a bright surface (e.g., after acid pickling) that would be dulled by hot-air drying or would show stairs left by the evaporation of hard water. It is normally used after solvent cleaning methods (section 5), except where emulsifiable solvent cleaning is used without rust removing methods to follow.

1.59.4 USE OF WATER-DISPLACING FLUIDS

These materials are volatile solvents which, by the addition of special substances, are enabled to displace water in the cold from metal surfaces. The articles to be dried are immersed in the liquid without agitation, and the water displaced falls to the bottom of the tank and can be drained; the design of a dip-tank for automatic drainage of the water is shown in Figure 1.10.

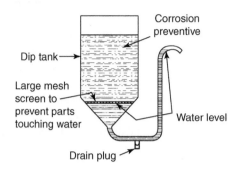

FIGURE 1.10

Possible design of the dip-tank for water-displacing fluid.

1.59.5 BLOWING WITH COMPRESSED AIR

Compressed air should be dry and free from oil and dirt. Moisture traps should be placed at the lowest point in the air delivery pipe and as close as practicable to the jet. Air filters either combined with or separate from the moisture traps should be fitted. Traps should be drained frequently during operation and filters should be cleaned regularly. Air receiver tanks usually have a drain valve or blow-off cock at the bottom; this should be open until the compressor is used again. On reusing the compressor, air should be allowed to blow through the drain valve for a minute to blow out any condensed moisture in the receiver end and the valve closed.

The air jet should then be tested for moisture at the delivery end by permitting the air to blow on a polished metal part at room temperature, and observing any condensation. The pressure of air need not be above 6.17 bars.

1.60 SURFACE PREPARATION OF PLASTER, BRICK, STONE, AND WOOD

Brick, concrete, plaster, stone, or similar surfaces to be painted should be allowed to dry out completely. Any efflorescence should be removed by wiping, first with a dry coarse cloth and then with a damp cloth, and then the surface should be left alone for 48 h to ensure that no further efflorescence occurs. Plastered interior walls must be allowed to dry thoroughly, usually for 30 days, before painting. Ventilation while drying is essential, and in cold damp weather, heat must be used. Damaged places should be repaired and sanded smooth.

1.60.1 PREPARATION OF PLASTER

Plasters in general use for internal work comprise the following:

- Calcium sulfate (gypsum) types (see BS 1191-Part 1), used neatly.
- Lightweight plasters (see BS 1191 Part 2) initially hold more water than other types and may take longer to dry out, particularly in winter. Premixed, lightweight cement plaster under coatings may be used in some circumstances.
- Thin wall plasters are based on organic binders, and as they are used in this layer, they dry out rapidly.

- Cement, cement/sand, or cement/sand/lime plasters may be used where strong, hard, or moisture-resistant surfaces are required. Cement plaster and premixed, lightweight cement plaster are strongly alkaline and, until they substantially dried out, they are likely to attack oil-based paints.
- Nonhydraulic lime plasters, made with high purity lime and clean sand, are free of soluble salts and caustic alkali. It should be assumed that all lime plasters present a risk of alkaline attack until they are dried out.

1.60.2 CLEANING

Dirt and loose surface deposits can usually be removed by dry brushing. Plaster nibs and splashes should be scraped off, with care taken to avoid damaging the surface. Mold growth may occur on plastered surfaces if drying out has been prolonged, especially in poor ventilation.

1.60.3 STOPPING AND FILLING

Cracks, holes, and surface imperfections should be stopped and filled with plaster, water-mixed filler, or, in cement plasters, with masonry cement before first or priming coats are applied. Alternatively, in oil-based paint systems, oil-based stoppers and fillers may be used after priming.

1.61 PREPARATION OF BRICK AND STONE

1.61.1 TYPES

Bricks and stones, which are generally used for external and internal works, comprise the following:

- Clay bricks consisting of common bricks, facing bricks, and engineering bricks (see also BS 3921). Most clay bricks contain soluble salts, and these, combined with water and alkalis in cement mortars, may promote efflorescence.
- Calcium silicate (sand lime or flintlime) bricks (see also BS 187). The surface of calcium silicate bricks is usually smoother than that of clay bricks. Paint adhesion is generally satisfactory.

1.61.2 CONCRETE BRICKS

These are strongly alkaline, and oil-based paints are likely to be attacked if they are applied to brick work before it has substantially dried out.

1.61.3 STONE

The many varieties of natural stone differ considerably in hardness and porosity (e.g., from virtually nonporous granite or marble to porous limestone or sand stone).

1.61.4 CLEANING

New surfaces (e.g., fair-faced brick work) normally require little more than brushing down with stiff (not wire) brushes to remove loose material. Old, unpainted surfaces require more vigorous treatment, including washing down, to remove accumulated dirt. If washing is necessary, time should be allowed for drying out before painting. Organic growths may be present on old external surfaces and should be treated as described in Appendix E.

1.61.5 FILLING AND STOPPING

On old, unpainted surfaces, repairs and repointing should be carried out well in advance of painting to facilitate drying out. Minor surface defects should be treated with interior or exterior grade water-mixed filler or with masonry cement. Alternatively, in oil-based paint systems, oil-based stoppers or fillers may be used after priming.

1.62 PREPARATION OF EXTERNAL RENDERING

1.62.1 TYPES

The main types of external rendering are as described in the following sections. The rendering in general use are cement-based and may incorporate lime.

Note: If repairs to old renderings are necessary, they should be used with cement-based mixtures (see also BS 5262).

1.62.2 STUCCO

Stucco consists of lime/sand that is often applied to the external rendering and usually painted. It is often found in older buildings.

1.62.3 CLEANING

New surfaces normally require little more than brushing down with stiff (not wire) brushes to remove loose material. Efflorescence should be treated. Old unpainted surfaces, especially if rough or textured, require more rigorous treatment, including washing down, to remove accumulated dirt before painting. If washing is necessary, time should be allowed for drying out. Organic growths may be present on old surfaces. These should be treated as described in Appendix E.

1.62.4 STOPPING AND FILLING

Minor cracks, holes, and surface defects should be fixed with exterior grade water-mixed filler or masonry cement before the primer or first coat of paint is applied. Alternatively, in oil-based paint systems, oil-based stoppers or fillers may be used after priming.

1.63 PREPARATION OF WOOD

Surface preparation for new or maintained surfaces depend on wood specification and the type of coating to be applied. The preparation procedures may consist of cleaning with chemicals and flame, sanding with abrasive paper, and preservative treatment.

1.63.1 CLEANING

Dirt and surface deposits, exuded resin, and soluble salts arising from preservative treatment should be removed. Light contamination should be removed by wiping with a clean cloth and white spirit or household detergent, followed by washing and rinsing.

Low-pressure water cleaning equipment may be used when soiling is heavy and the area to be cleaned is extensive. Flame cleaning is usually the quickest and most economical method to use for the removal of paint from general surfaces of woodwork.

1.63.2 SANDING

The main purpose of sanding is to smooth the surface, but it also helps to reduce paint penetration by closing the wood vessels, thus producing a thicker, more uniform film than with unsanded wood. Sanding may also remove the degraded outer layer of wood that has been exposed to weather. Wood may be sanded mechanically or manually; the latter, using abrasive paper, is more common for site work. It is essential to use a grade of abrasive paper appropriate to the surface; excessively coarse grades will damage the wood fibers, damaging the appearance and possibly the performance of the paint system. Care should also be taken not to damage molding and arris edges.

1.63.3 PRESERVATIVE TREATMENT

The following sections indicate treatments likely to be suitable for components that will subsequently be painted. Treatments for external wood working building and not in ground contact consist of the following:

- Organic solvent preservatives; e.g., copper naphthenate (BS 5056) and pentachlorophenol
- (BS 5707), tributyl tin oxide, zinc naphthenates, applied by immersion or vacuum processes
- Treatment with disodium octaborate, which to be carried out on freshly filled timber and cannot be applied to seasoned timber

Treatments for fencing and gates in or out of ground contact (see also BS 5589).

1.63.4 STOPPING AND FILLING

Oil-based stoppers and fillers are generally preferred for wood, especially for external wood, but should be used only on primed surfaces. Otherwise, oil will be absorbed, leaving the material underbound; the same requirement applies to linseed-oil glazing putty. Water-mixed (emulsion) stoppers and fillers may be used on primed or unprimed wood; if used over primer, the stopped or filled areas may require repriming to prevent absorption of the binder from subsequent coats.

With most natural finishes, putty glazing is not suitable. If filling nail holes is necessary on exterior work, an oil-based hard stopper or linseed oil putty should be used after application of the first coat of finish, with the stopper or putty tinted to match the color of the treated wood. For interior work, water-soluble fillers may be used before application of the first coat of finish.

1.64 TYPES AND APPLICATION OF FILLERS AND STOPPERS FOR NONMETALLIC SURFACES

1.64.1 MATERIALS

Fillers are the paste (more fluid consistency than stopper) and used for filling and leveling shouldow depressions, open grain, surface roughnesses, and fine cracks. Stoppers are stiff pastes used for stopping up screw holes, wide cracks, open joints, and similar imperfection.

Oil-based stoppers are preferred for woodwork, especially externally, and may be used on other substrates in oil-based paint systems. Cost considerations and the toxic hazards with lead pigments generally preclude the use of the traditional "hard stopping" made up by the painter from paste white lead, whiting, and gold size. Proprietary oil-based, lead-free stoppers are available.

Linseed oil putty is widely used as a stopper, but it is slow to harden and tends to shrink; it is improved by the addition of gold size or oil-based undercoat. Limited experience with emulsion-based fillers (D.6.1.2) indicates that some of these substances may be acceptable as stoppers for exterior woodwork. For stopping deep holes in woodwork and repairing damage (e.g., to moldings and arris), two-pack polyester or polyurethane fillers, as used for car body repairs, can give satisfactory results.

Stoppers and fillers in general use are usually water-mixed and come in three types. Powder fillers are supplied in powder form for mixing with water and are usually based on water-soluble cellulose and gypsum or white Portland cement. Those containing gypsum are suitable for general use as stoppers and fillers in dry interior situations. They are not usually suitable for use for external work, especially on wood, but the manufacturer's recommendations in this respect should be observed. Portland cement types are suitable for external use on masonry surfaces. They should not be used on woodwork.

Emulsion-based fillers are usually based on vinyl or acrylic resin emulsions and are supplied in paste form, ready for use. Materials of this type vary considerably in characteristics and usage. The manufacturer's recommendations, especially regarding suitability for exterior use, should be observed; some emulsion-based fillers may be acceptable alternatives to oil-based stoppers for external woodwork.

Surface fillers are formulated specifically for use as fillers in the traditional sense, and are preferred to the general-purpose types when a high standard of finish is required, especially with oil-based paint systems. They are supplied ready for use as smooth, finely ground, creamy pastes, usually applied with a wide filling knife or spatula. The types in general use are emulsion-based, but oil-based formulations are also available.

1.64.2 APPLICATION

Stopping and filling should be carried out as early in the coating system as the type of material used and the nature of the substrate will allow. Most materials used for stopping and filling are absorbent and, if applied late in the system (e.g., immediately beneath finishing coats), may cause sinkage and variations in gloss, sheen, or color.

1.64.3 STOPPING

The materials used for stopping are usually fairly stiff in consistency and are generally applied with a putty knife or stopping knife. The stopper should be pressed firmly into cavities in order to drive out air that may prevent proper levelling. It should be knifed flush with the surrounding surface and not allowed to spread beyond the cavity. Deep cavities may require stopping in two stages, allowing an interval for the first application to harden.

When oil-based stoppers are used, surfaces should be primed and the primer allowed to dry before stopping. This prevents absorption of the binder from the stopper, which may cause it to shrink and fall out. Priming is not usually required prior to the use of emulsion-based and powder fillers.

1.64.4 FILLING

Depending upon their type and consistency, fillers may be applied with a wide, flexible filling knife or spreader or by brush. On broad surfaces, filling knives or spreaders are generally used, and brushes are used for moldings.

Oil-based fillers are usually applied after priming, but some may be suitable for direct application to unprimed surfaces; the manufacturer's recommendations in this respect should be followed. Emulsion-based and powder fillers may be applied to unprimed surfaces, but on absorbent surfaces, priming may be necessary to facilitate uniform spreading.

Skilled application of fillers by knife or spreader should leave a surface that requires little subsequent treatment, but some rubbing down may be required for final leveling. With oil-based fillers, waterproof abrasive paper may be used, with water as a lubricant. Waterproof abrasive paper should be used "wet" and is more effective for abrading painted surfaces. It should also be used when rubbing down surfaces painted with lead-based paint. Emulsion-based and powder fillers should be rubbed down dry. After rubbing down, surfaces should be rinsed or carefully dusted to remove residues of filler. If the surfaces have been wetted, they should be allowed to dry before painting.

1.65 TREATMENT OF ORGANIC GROWTHS

The information in this section is based upon that given in BRE Digest No. 139 (Building Reasearch Establishment (BRE) Digest London-HMSO), to which reference is made for a more comprehensive treatment.

1.65.1 EXTERNAL SURFACE

Lichens and mosses are often found on roofs and external walls, especially in rural areas. Their appearance is often regarded as pleasing, but if the surfaces are to be painted, the growths have to be removed. Algae grows in most districts when water runs freely over a surface; it causes unsightly green or black stains and have to be removed before painting. Molds and algal growths resembling dirt deposits may occur outdoors on paints and wood stains; in the early stages, they can be washed off, but molds may penetrate and damage the existing coatings and affect new ones.

Where growths are present on surfaces that are to be painted, preparatory treatment should include application of a toxic wash. BRE Digest No. 139 lists a number of nonproprietary and proprietary materials, of which the main categories are described next.

a. Sodium Hypochlorite

Household bleach and some proprietary fungicides are examples of sodium hypochlorite products. They are effective against many organic growths and will bleach the color of darker ones. They may change the color of paints and some building materials, and there is no active residue. Use as a solution of 1 part bleach to 4 parts water or, with proprietary materials, as directed. Rinse with cold water after use.

b. Sodium Borate

Use sodium borate as a 5% solution in water. It is generally safe to handle. Wash off after use on non-porous surfaces.

c. Zinc or Magnesium Fluorosilicate

These substances were formerly described as "silicofluoride." Use them as a 4% solution in water. They have a long-term effect.

d. Formalin (40% Formaldehyde)

Use formalin as a 5% solution in water. It has a choking smell, so good ventilation is essential. It has no bleaching action and evaporates, leaving no active residue behind.

e. Sodium Orthophenyl-Phenate

This substance is employed in proprietary materials, mainly fungicides. Use as 2% solution in water or as directed. Have some residual effect and may be left on for painting or papering.

f. Quaternary Ammonium Compounds

Benzalkonium chloride and proprietary materials are examples of these substances. Effective against algae and lichens. Use as a 1% solution in water or as directed. It presents very little health hazard, and it is nonstaining, but it should be washed off before emulsion paints are applied.

g. Amino Compounds

These are proprietary materials effective against algae, lichens, and molds. Use as a 1% to 5% solution in water. It leaves an active residual film, nonstaining and need not be washed off.

The efficiency of toxic washes varies according to circumstances. Sodium hypochlorite, in the form of household bleach, is readily available and is effective to treat a wide range of growths, but it has no residual activity.

When using toxic washes, precautions should be taken to protect the eyes and skin against splashes. Toxic washes take a few days to take effect and, in wet weather, they may be washed away before they have time to act. They should preferably be applied during a dry period of weather. Their action is hastened if thick surface growths are partly removed or torn with a wire brush before brushing the wash well. The wash, which should be repeated if necessary, kills the growth, but the dead organism takes some time to disappear. The dead matter should be removed by scraping and wire brushing, and this is often more effective when the growth is dry. However, note that it is important that asbestos-cement be not scraped or wire brushed except when wet. After removal of the dead growth, a further application of toxic wash (with types having a residual effect) can delay the reestablishment of the growth.

Blast cleaning and high-pressure water cleaning are effective methods of removing heavy growths from external surfaces. These methods are preferred when circumstances permit to warrant their use. Subsequent treatment with a toxic wash is advisable.

1.65.2 **INTERNAL SURFACE**

Mold (mildew) is the form of organic growth most likely to affect internal surfaces. The conditions and environments favorable humidity exceeds 70% to its development and measures to prevent or control it.

Preparatory treatment of painted or papered surfaces will depend upon the type of decoration and the extent to which it is affected. Surface coverings should always be stripped if mold is evident on or beneath them. Removal of paint is advisable if it is severely affected and the growth is deep-seated. In disposing of stripped paper or paint, care should be taken to avoid contaminating unaffected surfaces. After removal of surface covering or paint, the surface should be treated with a toxic wash and kept under observation for a week or so. A further wash can be applied if growth is renewed. When growth has ceased and the surface is dry, it can be redecorated.

Where infection is slight and the coating is of a type that will withstand application of a toxic wash, it may be sufficient to clean a surface without removing the coating and then apply the wash, allow to dry, and repaint.

ENGINEERING AND TECHNICAL GUIDELINES FOR PAINTING

2

This chapter covers corrosion protection of steel structures of oil, gas, and petrochemical industries, mainly for refineries, chemical and petrochemical plants, gas plants, aboveground facilities of gas transmission and distribution systems, marine structures, and, where applicable, production and new ventures in exploration. Most of the discussion is about liquid paints (usually applied via either brush or spray) and metallic coating materials that are commonly used for corrosion protection in atmospheric or immersion service. The rate of base metal corrosion where such coatings are used should not exceed approximately 1.3 mm (50 mils) per year. For corrosion rates above this, both in atmospheric and immersion service, or where a failure might happen, these coatings should not be used, and corrosion-protective measures should include the use of more corrosion-resistant alloys, cladding, and special coatings and linings. In addition, internal protection of pipes for water supply and external protection of underground structures are covered. Definitions, types of corrosion environments, and standards of surface preparation are also specified herein.

It should be noted that 10 different paint groups, including 54 paint systems are indicated for a number of applications (paint schedule). Therefore, the user should decide which paint system is to be used at the design stage or at the start of a project.

The basic principles of corrosion prevention by paints and the characteristics of some paint systems are discussed here. In addition, the text introduces typical painting systems for storage tanks, refinery, fresh water vessels, and ships.

The potential life of a protective system is unlikely to be realized unless

- The correct choice of system is made
- The materials used in the system can be supplied when required and with the properties attributed to them when making the choice
- The materials are applied in conditions and with widely accepted standards
- The handling, transportation, and storage (over which the main contractor has minimal control) of all materials and coated components results in no damage to the integrity of the materials or coating that cannot be completely restored
- The erection procedures cause no damage to the coatings that cannot be completely restored
- Such restoration of damaged areas results in a protection at least as good as that of the undamaged areas

There are many variables (both natural and otherwise) that can influence the fulfillment of all these conditions for success. It follows that no two projects can be exactly alike, and this is one reason why a standard specification should always be included in the set of contract documents.

Essentials of Coating, Painting, and Lining. http://dx.doi.org/10.1016/B978-0-12-801407-3.00002-X

2.1 CORROSIVE ENVIRONMENTS

2.1.1 RURAL ENVIRONMENT

A rural environment is defined as one in which interior and exterior atmospheric exposure is virtually unpolluted by smoke and sulfur gases, and which is sufficiently inland to be unaffected by salt contaminations or the high humidity of coastal areas.

2.1.2 INDUSTRIAL ENVIRONMENT

This category includes atmospheric exposures that include urban communities, manufacturing centers, and industrial plants (but would not include heavy industrial environments such as coke plants, which fall into the category of chemical environments). The atmosphere has a considerable amount of gas containing sulfur and industrial fumes which increase the rate of corrosion and adversely affect the paint life. This type of environment is divided into three different categories as follows:

- Mild—Consists of normal indoor and outdoor weathering where light concentration of chemical fumes and light humidity conditions exist
- Moderate—Consists of industrial environment where moderately aggressive chemicals, acids, or caustic fumes exist, unpolluted coastal atmospheres, areas where moderate humidity is found, and areas where there is condensation, splash, spray, spillage, and frequent immersion of fresh water in intermediate temperatures up to 80°C
- Severe—Consists of coastal polluted atmospheres; where manufacturing centers and industrial plants are; atmospheres that contain a considerable amount of gas containing sulfur and industrial fumes; areas of high humidity; and areas where there is condensation, splash, spray, spillage, and frequent immersion of salt water and mild corrosive products

2.2 WATER IMMERSION ENVIRONMENTS

Water immersion environments fall into the following categories:

- Nonsaline water—means permanent immersion in fresh and potable water
- Seawater—means permanent immersion in salt water and other saline and estuary waters

2.3 CHEMICAL ENVIRONMENTS

Chemical environments (which include heavy industrial environments) are exposures in which strong concentrations of highly corrosive gases, fumes, or chemicals (either in solutions or as solids or liquids) touch the surface. The severity may vary tremendously from mild concentrations in yard areas to complete immersion in chemicals. This type of environment is categorized into three different areas as follows:

- Mild—Chemical exposure to acidic, neutral, and alkaline environments (pH 2 to 12)
- Moderate—Chemical exposure to mild chemical solvents and intermittent contact with aliphatic hydrocarbons (mineral, spirits, lower alcohols, glycols, etc.)

- Severe—Severe chemical exposure to oxidizing chemicals, strong solvents, extreme pH, or contamination of these with high temperature; substrates in contact with severe chemical environments need a special lining for protection.

2.4 MARINE ENVIRONMENTS

Marine environments consist of three different corrosive environments, described in the next sections.

2.4.1 ATMOSPHERIC ZONES

The atmospheric zone is the zone of the platform that extends upward from the splash zone and is exposed to sun, wind, spray, and rain.

2.4.2 SPLASH ZONES

Splash zones include fixed or floating offshore or onshore structures such as platforms, which are alternately in and out of the water because of the influence of tides, winds, and seas. Excluded from this zone category are surfaces that get wet only during major storms.

2.4.3 SUBMERGED ZONES

Submerged zones extend downward from splash zones. They include the position of a platform below the mudline.

2.5 UNDERGROUND ENVIRONMENTS

Underground environments include all buried surfaces in direct contact with soil (inland, onshore, subsea). This category encompasses earth, sand, and rock. For protection of underground structures, proper coating should be used.

2.6 ADHESION OF PAINT TO THE SUBSTRATE

The primer is the critical element in most coating systems because it is most responsible for preserving the metallic state of the substrate. In addition, it must anchor the total system to the steel, which it accomplishes in one of two ways, depending upon the nature of the primer vehicle. Most coatings adhere to metal via purely physical attraction (e.g., hydrogen bonds) that develop when two surfaces are brought close together. The other method of adhesion is chemical. Paint vehicles with polar groups (-OH, -COOH, etc.) have good wetting characteristics and show excellent physical adhesion (epoxies, oil paints, alkyds, etc.). Much stronger chemically bonded adhesion is possible when the primer can actually react with the metal, as in the case of wash primer pretreatment and phosphate conversion coating.

For adhesion to take place, the coating and substrate must not be separated from one another. Any contaminant on the steel will increase the separation and decrease paint film adhesion. Moreover, reactive sites on steel where adhesion can occur are masked not only by contamination, but also by chemically bound species, which may themselves satisfy sites on the steel that would otherwise be available for reaction with the paint vehicle. Thorough surface preparation removes such contamination and exposes many more reactive sites, thereby dramatically increasing the amount of surface area where adhesion can occur.

2.7 FACILITIES FOR THE APPLICATION OF COATINGS

Surface preparation is normally done by the contractor applying coating. Blast cleaning is the most effective method to use, but if that is not available, choose a paint that is compatible with the method of surface preparation to be used. The advantages of some chemical-resistant paints are lost if they are applied to inadequately prepared surfaces. When planning the work, factors to be considered include the following:

- The sequence of operations (e.g., blast cleaning before fabrication is normally cheaper than blast cleaning after fabrication).
- The application time (e.g., length of the drying or curing time for coating).
- Method of application (e.g., airless spray or brush).
- The possible advantages or disadvantages of applying the final coat(s) on site. Some coatings (e.g., galvanizing) cannot be applied on site.

2.8 REQUIREMENTS

The general technique of surface preparation and the prepared surface before the application of the primer coat should be in accordance with the applicable grade as specified in Table 2.1. Unless otherwise stated, the surface should be prepared in accordance with the applicable grade as specified in Swedish standard SIS 055900 or SSPC, BS 4232, or DIN 18364 standards. The painting system in general should be specified in accordance with Table 2.1.

Color schemes should comply with the color schedule in current use at the particular center or the paint color schedule. Paint systems are generally specified by the dry film thickness of the coat(s) and total dry film thickness of the primer, intermediate, and topcoat, rather than by the number of coats.

Unless otherwise stated in the schedule, the total dry film thickness of paint systems should be at least 150 μm. All paints and paint materials used should comply with the specifications given in standards for paint materials, and they should be obtained only from approved manufacturers.

All materials should be supplied in the manufacturer's original containers, durably and legibly marked with a description of the contents. This should include the specification number, the color reference number, the method of application for which it is intended, the batch number, date of manufacture, and the manufacturer's name, initials, or recognized trademark. No intermixing of different brands or types of paints will be permitted. The storage and preparation of paints and other coating materials should be in accordance with the manufacturer's instructions.

The products of only a single paint manufacturer should be used for each complete paint system. The use of different paint manufacturers' products for successive coats on a single surface or piece of equipment is not permitted.

Table 2.1 Paint Schedule

Service	Typical Paint Life Time to First Maintenance (years)	Applicable Paint System									
		Rural Environments	Industrial Environments			Chemical Environments		Marine and Water Environments			
			Mild	Moderate	Sever	Mild	Moderate	Marine Atmosphere	Splash Zone	Sea and Saline Water	Nonsaline Water
Structural steelwork, columns, vessels, heat transfer equipment, steel stacks, etc.	Long (10–20 years)	2A, 4C	2A, 5B	2E, 3B 4B 5C	2B, 6G, 9F, 5D	3E, 5D 6G, 7I	7D, 8A	3E, 6I, 8D, 9F, 7K	N/A	N/A	N/A
Bolting, floor grating, ladders, stair-tread hand-railing, electrical, fixtures, fencing, etc. Note: Hot dip galvanizing group is the preferred protection method	Medium (5–10 years)	1E, 4B	1F, 4D, 5C	1F, 4B, 6F, 9B 5B	1F, 4B, 5C, 9E	3B, 6F, 5C, 7B	5D, 7C	3D, 6F, 8C, 7H	N/A	N/A	N/A
	Short (less than 5 years)	4A	1D, 4B	6B, 4C	1F, 6B, 5B	6E, 5B	7B	7B, 8B, 9D	N/A	N/A	N/A
Structural steelwork, fabricated steelwork, Columns, vessels, heat transfer equipment, steel stacks, and other external surfaces	Long (10–20 years)	7A, 2B	4C, 7B	3B, 7C, 6E	3D, 7D, 6G	3E, 7D, 6G	7I, 6G	3E, 6I, 7G	N/A	N/A	N/A
	Medium (5–10 years)	1E, 6C	1F, 6D	4B, 6E, 3A	3B, 9B, 7C, 6D, 5C	3C, 7C	7H	3D, 6F, 7F	N/A	N/A	N/A
	Short (less than 5 years)	1C	6A, 1B, 7A	6B, 4A, 7A	1F, 6B	3B, 6E	7B	7E, 8B	N/A	N/A	N/A
For surface temperatures up to 74°C	Long (10–20 years)	2A, 4B	2B, 4C	2B, 4B, 5D	2B, 7C, 7F	2C, 7D	7D	2C	N/A	N/A	N/A
	Medium (5–10 years)	4A	4B, 7A	4B, 7B	4B, 7B	2B, 7C	7C	2B	N/A	N/A	N/A
	Short (less than 5 years)	4A	4A, 5A	4A, 7A	4B, 7A	4B, 7A	7B	2A	N/A	N/A	N/A

(Continued)

Table 2.1 Paint Schedule (*cont.*)

Application / Surface	Durability										
For surface temperatures from 74°C to 120°C	Long (10–20 years)	2A	2B, 4C	2B, 4B, 5D	2B, 7C	2C, 7D	7D	2C	N/A	N/A	N/A
	Medium (5–10 years)	4B	4B, 7A	4B, 7B	4B, 7B	2B, 7C	7C	2B	N/A	N/A	N/A
	Short (less than 5 years)	4A	4A	4A, 7A	4A, 7A	4B, 7A	7B	2A	N/A	N/A	N/A
Heat exchangers (internal) up to 120°C	Long (10–20 years)	7I	N/A	N/A	N/A	N/A	N/A	N/A	N/A	N/A	N/A
	Medium (5–10 years)	7D	N/A	N/A	N/A	N/A	N/A	N/A	N/A	N/A	N/A
Water boxes of coolers and condensers Fresh-water cooling; seawater cooling	Long (10–20 years)	N/A	N/A	N/A	N/A	N/A	N/A	N/A	N/A	N/A	3D, 7I
Galvanized surfaces		6H, 7A	6H, 7A	6I, 7B	6I, 7C	6I, 7D	6I, 7D	N/A	6I, 7D	6I, 7C	N/A
Storage tanks (internal) Note 1: For fresh water only paint systems, 7I or 10A should be used	Long (10–20 years)	N/A	N/A	N/A	N/A	3E, 7L	7I	N/A	3E, 6G, 7I, 8D, 9F	N/A	3D, 4C, 5D, 6G, 7D, 7I, 8C, 10A, 10B
Note 2: For large potable water, only paint system 10A should be used for small potable water tanks	Medium (5–10 years)	N/A	N/A	N/A	N/A	3D, 7K	7H	N/A	N/A	3C, 6G, 7H, 8C, 9E, 7D	3C, 4B, 5C, 6G, 7B, 7K, 10A, 10B

Group 8 of Table 2.2 is preferred (see BS 5493, note "n")

Note 3: For potable water, see BS 3416 and AWWA-C-210

Marine and Offshore Structures Jetty Structures Immersion Portion (including 1 m below the sea bed level/mud level)	N/A	N/A	N/A	N/A	N/A	N/A	N/A	N/A	N/A	7L	N/A
Superstructure Zone	N/A	N/A	N/A	N/A	N/A	N/A	N/A	3E, 7F	N/A	N/A	N/A
Single-Point Mooring Buoys External Painting of Buoy Body Pipe Sections Guides Pipes, Frame Work and Platform in The Center Well Under the Water Line up to 500 mm Above. Loaded Draft Level and Deck of Buoy, Turntable Derrick Winch Buoy, Light Fitting, Piping on the Turntable, Central and Above-Board Swivels Buoy Body Pipe Section and Guide Pipes in the Center Well from 500 mm Above Loaded Draft Level	N/A	N/A	N/A	N/A	N/A	N/A	N/A	N/A	7F	7J	7D
Internal Painting	N/A	N/A	N/A	N/A	N/A	N/A	N/A	N/A	N/A	N/A	3E 7I
Water-Tight Compartments Including the First Stiffening Ring	N/A	N/A	N/A	N/A	N/A	N/A	N/A	N/A	N/A	N/A	3E 7I

(Continued)

Table 2.1 Paint Schedule (*cont.*)

Rest of Compartment	N/A	N/A	N/A	N/A	N/A	N/A	N/A	N/A	N/A	N/A	10A, 10B
Compartment in the Way of Chain Stoppers	N/A	N/A	N/A	N/A	N/A	N/A	N/A	N/A	N/A	N/A	10A, 10B
Central Pipe, Turntable Piping and Overboard Pipe Swivels	N/A	N/A	N/A	N/A	N/A	N/A	N/A	N/A	N/A	N/A	3E, 7D, 7I
Plant Building Exposed Steelworks	N/A	1E, 4A	1F, 4B, 7A	1F, 4B, 7B	7D	7D	7D	N/A	N/A	N/A	N/A
Built-in Steelwork	N/A	1E	1F	4A	4A	5B	7I	N/A	N/A	N/A	N/A
Galvanized or Aluminum Windows and Doors Etc	N/A	2A	2A	2A	N/A	7D	7D	N/A	N/A	N/A	N/A
Concrete	N/A	1B	1B	1B	7A	7B	7C	7I	7J	7J	7I
Swimming Pool	N/A	N/A	N/A	N/A	N/A	N/A	N/A	N/A	N/A	N/A	6F
Woodwork	N/A	1A	1A	1B	1B	1C	1D	1D	7J	7J	N/A

2.9 SURFACE PREPARATION

Paint life depends primarily on surface preparation, which should remove enough foreign bodies to allow the type of primer used to wet the surface thoroughly and develop adequate adhesion. For surface preparation, the following are some general requirements for surfaces to be painted:

- Steel surfaces should be free of rust, mill scale, salts, oil, grease, moisture, and other substances.
- The use of so-called rust converters, rust stabilizers, and similar means for chemically converting the corrosion products of the iron into stable iron compounds is not permissible for steel structures. This also applies to penetrating agents intended to inhibit rust.
- After the surface preparation of the substrate, grit, dust, and other substances should be removed and a layer of primer applied before any detrimental corrosion or recontamination occurs.
- The priming paint is a good "wetting" type, such as blast primer, and is normally applied by brush.
- Fabrication should preferably be complete before surface preparation begins.
- If hot-dip galvanized steel, stainless steel, and nonferrous metal surfaces are to be painted, suitable pretreatment in the form of a light blast cleaning with a suitable abrasive such as aluminum oxide should be given to ensure proper adhesion of the subsequently applied paint system.
- Woodwork should be treated with a shop-applied wood preservative to prevent conditions under which organisms are likely to damage it. This preservative treatment should not have any adverse effect on the subsequent painting. Wood to be painted or varnished should be made free of contamination. If necessary, a stopping putty (paste filler) should be used for the stopping of holes for unevenness in the woodwork.
- Concrete surfaces to be painted should be clean, dry, and structurally sound, have adequate strength, be free of laitance, and have some roughness to ensure proper paint adhesion. This surface preparation should be obtained by blast cleaning, wire brushing with power tools, or etching with very dilute hydrochloric acid followed by rinsing with excess water. Mechanical cleaning is preferred to etching with hydrochloric acid and excess water because concrete surfaces should be dried until they reach a humidity of not more than 4% wt. before paint is applied.

2.9.1 SELECTION OF CLEANING METHOD(S)

The cleaning method should be selected with reference to the following considerations. The choice between blast cleaning, acid pickling, flame cleaning, and manual cleaning is partly determined by the nature of the paint to be applied. It should be appreciated, however, that paint applied to a properly prepared (e.g., blast-cleaned) surface always lasts longer than similar paint applied to flame-cleaned or manually cleaned surfaces.

2.9.2 STANDARDS OF SURFACE PREPARATION

2.9.2.1 Blast Cleaning

The following grades of surface finish are defined in accordance with SIS 055900:

- SA 3, blast cleaning to pure metal—Mill scale, rust, and foreign matter should be removed completely. Finally, the surface should be cleaned with a vacuum cleaner, clean and dry compressed air, or a clean brush. It should then have a uniform metallic color.

- SA 2½, very thorough blast cleaning—Mill scale, rust, and foreign matter should be removed to the extent that the only traces remaining are slight stains in the form of spots or stripes. Finally, the surface should be cleaned with a vacuum cleaner, clean and dry compressed air, or a clean brush.
- SA 2, thorough blast cleaning—Almost all mill scale, rust, and foreign matter should be removed, and the surface should be cleaned with a vacuum cleaner, clean and dry compressed air, or a clean brush. It should then be grayish in color.
- SA 1, light blast cleaning: Loose mill scale, rust and foreign matter should be removed.

The grade of surface finish compares with some other internationally recognized standards, as shown in Table 2.2.

2.9.3 MANUAL CLEANING

Manual cleaning using mechanical cleaning tools to remove mill scale and rust is the least satisfactory method of surface preparation. It should be used only where blast cleaning is impractical or not economical. The manual cleaning methods should not be used for the preparation of steel, where high-quality long-lasting systems are to be used.

Table 2.2 Standard Levels of Cleanliness In Different Standards

Type of Surface Preparation	SIS 055900 Sweden	SSPC & Nace USA	BS 4232 UK	DIN 1S364 Germany	Remarks
Manual cleaning	St 2 Thorough scraping, brushing, grinding etc.	SP 2 N/A Hand-tool cleaning	N/A	N/A	ST2 or SP2 and Hand-tool cleaning is recommended only for spot cleaning
	St 3 very thorough scraping, brushing, grinding etc.	SP 3 N/A Power-tool cleaning	N/A	N/A	N/A
Blast cleaning	Sa 3 pure metal	SP 5 TMO1-70-NO1 White Metal	First quality	N/A	N/A
	Sa 2 1/2 very thorough	SP 10 TMO1-70-NO2 Near White	Second quality	2	N/A
	Sa 2 thorough	SP 6 TMO1-70-NO3 Commercial	Third quality	3	N/A
	Sa 1 light	SP 7 TMO1-70-N04 Brush-off	N/A	N/A	N/A
Chemical cleaning	Acid pickling	SP 3	N/A	N/A	N/A
	N/A	SP 1 Solvent cleaning	BS 5493 CP 3012	N/A	N/A

2.9.4 **CHEMICAL CLEANING**

Various chemical cleaning methods are discussed in this section.

Acid pickling should be carried out in the workshop. The surface produced should be in accordance with standards. The process normally comprises the following:

1. Immersion in a bath of warm dilute sulfuric acid or phosphoric acid, to remove both mill scale and rust.
2. Washing in a bath of warm water to remove all traces of sulfuric acid, a step that may be omitted when phosphoric acid pickling is used.
3. Immersion in a bath of hot dilute phosphoric acid to provide a passivated surface suitable for painting. The contents of the baths should be discarded when the concentration is below what is indicated specifically for the bath, or if the accumulation of sediment stains the work.

Solvent cleaning should be used prior to the application of paint and in conjunction with the surface preparation methods specified above for the removal of rust, mill scale, or paint. Petroleum solvents, such as kerosene, mineral spirits, and chlorinated solvents such as trichloroethylene or 1.1.1 trichloro-ethene, can be used to dissolve and remove soil. Chlorinated solvents are also effective for removing heavy oils, greases, and waxes.

2.9.5 **SURFACE PREPARATION FOR MAINTENANCE**

Initially, all areas of loosened paint and scale and all points of rusting should be located and treated in a manner that is essentially similar to the preparation detailed in the previous section. If the areas being prepared are sufficiently numerous as to make exact location and definition difficult, then the entire surface should be prepared by sand blasting, chipping, scraping and power wire brushing until the standards detailed in the previous section are met. All prepared areas should then be degreased where necessary by solvent cleaning.

2.10 **PAINT APPLICATION**

Paint application should be in accordance with standards. This section details some general requirements for painting.

All surfaces should receive an appropriate paint system as specified in Table 2.1, with the following exceptions:

- Any equipment furnished completely painted by the manufacturer, unless it is specially required to match a color scheme or to repair damage to the paint film
- Hot-dip galvanized steel, weathering steel, stainless steel and nonferrous metals, monel, brass, copper, and aluminum jacketing, unless it is specially required
- Nonmetallic surfaces
- Name plates, code stampings, and push buttons
- Surfaces to be fireproofed
- Concrete, brickwork, tile, glass, and plastic, unless specially required
- Machined surfaces

- Insulation weatherproofing material or sheeting
- Rubber hoses, belts, flexible braided connectors, stainless steel tubing and fittings, gauges, valve stems, and motor shafts
- Any surface particularly indicated as not to be painted

The painting requirements often depend on the surface temperature of the substrate involved. The surface temperature should be raised to the maximum operating temperature.

2.10.1 PAINT APPLICATION REQUIREMENTS

All paint should be stirred thoroughly to give uniform consistency before use. The grade and quantity of thinners should be to the specifications of the paint supplier; excessive thinning will cause the work to be rejected. Paint supplied as more than one component should be thoroughly mixed in the proportions given by the supplier and applied within the specified time limit after mixing. Coatings containing heavy or metallic pigments that have a tendency to settle should be kept in suspension by a mechanical agitator or stirrer.

No paint should be used in which the vehicle has set hard and which cannot readily be reincorporated by correct mixing. Similarly, no paint should be used that has jellified or thickened to such an extent that too much thinner is required to achieve a brushing consistency. No paint should be used which has livered, gelled, or otherwise deteriorated during storage. In addition, paint should not be applied in the following conditions:

- When the temperature of the surfaces is less than 3°C above the dew point of the surrounding air, the relative humidity is higher than 80%, or both
- When the temperature is below 4°C
- When there is the likelihood of an unfavorable change in weather conditions within 2 h after coating
- When there is a deposition of moisture in the form of rain, condensation, or frost, on the surface; this is likely to occur when the relative humidity is over 80% and the temperature is below 15°C
- When the surface temperature is more than 35°C; each layer of paint should be allowed to dry for a period of time within the limits prescribed by the paint manufacturer before the next layer is applied

Subsequent layers of the paint system should have differences in tint or color. Particular attention should be paid to the painting of corners, edges, welds, etc., especially with respect to the specified minimum dry-film thickness (MDFT). During both application and drying, adequate ventilation and lighting should be provided if the work area is enclosed.

All steel constructions or plates should be provided with a priming or coating system to protect the steel surfaces during the transport, storage, construction, and joining stages (e.g., welding) of the project. All surfaces inaccessible after assembly should be fully painted before assembly.

2.10.2 STANDARD OF PAINT APPLICATION

In all methods of application, the aim is to produce a uniform coating of the specified thickness, which is free of pinholes, missed areas, runs, sags or curtains, and wrinkling or other blemishes, which may impair durability. The following methods of paint application are applied:

- Brushing—The film produced by brushing should be free of brush marks, consistent with good practice for the type of paint used.
- Spray application—This should give a full wet coating with a minimum of spray patterns (i.e., the orange-peel effect; see ISO 12944). Each successive pass should overlap the previous pass to give a uniformly even film. This applies equally to air and airless spraying.
- Metallic coatings—Four methods for applying metallic coatings are in general use:
  - Hot-dip galvanizing—For structures, fittings, and claddings
  - Sherardizing—For fittings, fasteners, and small items
  - Electroplating—For fittings, fasteners, and small items
  - Metal-spraying—For structures and fittings (including fasteners when done after fabrication)

2.11 **PRIMING AND PAINTING**

Prepared surfaces should be primed generally within 4 h or before visible rerusting occurs. Paints and other coating materials should be applied in accordance with the manufacturer's instructions and good industrial practice. The prepared surface should be as specified in widely accepted standards.

Unless otherwise stated, depending on the type of paint, the surface preparation criteria detailed in Table 2.3 should apply.

The degree of surface roughness or peak-to-valley height required after blast cleaning generally depends on the type of paint but should not exceed the following values:

- 80 μm for structural steelwork
- 80 μm for tank lining
- 50 μm for drum lining

The minimum peak-to-valley height should be 30 μm.

Unless otherwise specified, the priming coat or coats on steel should have a total dry film thickness of not less than 50 μm. When primer is being applied in two coats, a shade contrast between the coats is recommended.

In order to minimize contamination between successive coats of paint, overcoating of the preceding coat should be done as soon as permitted by the particular specifications, and not delayed beyond the period specified. When delays are unavoidable, the painted surface should be thoroughly cleaned and dried to the satisfaction of the engineer before overcoating may take place.

Table 2.3 Surface Preparation Criteria

Paint Type	Required Surface Preparation in Accordance with SIS 055900
Zinc Silicate	GRADE SA 3
Epoxy	GRADE SA $2^1/_2$
Oil-based (Alkyd)	GRADE SA $2^1/_2$
Red lead	GRADE SA $2^1/_2$
Others	GRADE SA $2^1/_2$

The total dry film thickness of the paint systems (i.e., primer plus intermediate and topcoat) should be in accordance with the local standards. When painting larger areas, full use should be made of paints formulated for high build and applied by airless spray, thus reducing the number of coats for the required minimum thickness.

Any defect or damage that may occur should be repaired before the application of further coats; if necessary, the particular surfaces should be made paint free.

Areas that are to be overcoated should be thoroughly cleaned, free of grease, oil, and other foreign matter and should be dry. The surfaces then should be prepared to the standard as originally specified for each paint system (for large damaged areas) or prepared to the highest possible standard using mechanically operated tools (for small local damaged spots up to 1 m).

Subsequently, additional compatible coats should be applied until they meet the specifications. These additional coats should blend with the final coating on adjoining areas.

2.11.1 STEEL PREPARED BY BLAST CLEANING

Blast-cleaned steel prepared to SIS standards Sa3 or Sa 2 1/2 will frequently be coated either in the shop or on site with a thin coat of a prefabrication primer. For this purpose, a blast primer should be used.

For further priming over the above primer, or for direct application to blast-cleaned steel, the priming paints should be in accordance with the paint schedule (Table 2.1).

It is essential that the total film thickness of the priming coats (the blast primer plus second primer) should meet the prescribed minimum thickness of 75 μm in all areas. Figure 2.1 shows a corrosion pit on the outside wall of a pipeline at a coating defect, before and after abrasive blasting.

2.11.2 STEEL PREPARED BY ACID PICKLING

Steelwork prepared by acid pickling should be treated with 2% phosphoric acid solution. This leaves a thin phosphate coating on a warm steel surface, to which the paint should be applied immediately. This method is not generally used outside the pipe industry, but large plates for storage tanks have been pickled in this way. Generally, pickling is done by specialist firms.

Zinc-rich primers, both organic and inorganic, are not suitable for application to phosphate pickled surfaces. Zinc chromate primers are also not recommended for use on pickled surfaces, and their use would generally be confined to aluminum surfaces.

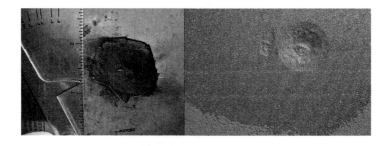

FIGURE 2.1

A corrosion pit on the outside wall of a pipeline at a coating defect, before and after abrasive blasting.

Regardless of the type of primer used on pickled steel, it is essential that the total film thickness of the priming coats meet the prescribed minimum film thickness of 50 μm, and this would normally be achieved by the site application of a further coat of priming in accordance with the paint schedule (Table 2.1). Figure 2.2 shows the difference between a pickled surface and a hot-rolled surface.

2.11.3 STEEL PREPARED BY WIRE BRUSHING

Priming paints for steelwork prepared by wire brushing must have good wetting properties and must be applied by brush to ensure a high standard of adhesion to the steel. The red lead alkyd priming paint, when applied as a two-coat system to an MDFT of 50 μm, is recommended for wire-brushed surfaces.

2.11.4 SHOP PREPARATION, PRIMING, AND PAINTING

Full protection applied in the shop immediately after fabrication normally results in a longer life for the protective system. However, damage during transportation and erection may subsequently require widespread repair or touching up of coatings, so the structural steelwork, surface pipework, towers,

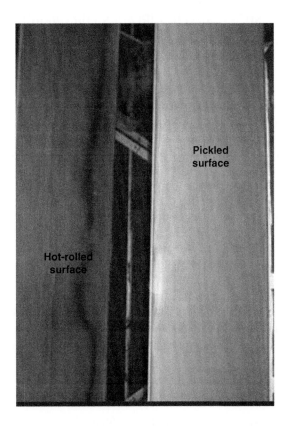

FIGURE 2.2

A pickled surface as opposed to a hot-rolled surface

vessels, heat-exchanger shells, and similar containers that will not be lagged should be treated in the shop as follows:

- Mild and moderate industrial environments, not more than three months—Blast cleaning, second-quality, SIS 055900 Grade Sa 2½. Prefabrication primer or blast primer, with a dry film thickness of at least 20 μm. For moderate industrial environments, the dry film thickness of the primer should be 50 μm.
- Mild and moderate industrial environments, more than three months—Blast cleaning, second-quality, SIS 055900 Grade Sa 2½. Prefabrication primer or blast primer overall, 20 μm dry film thickness, followed by full primer coat(s) 75 μm, and one intermediate coat overall but kept back 100 mm from welding edges. For moderate industrial environments, the dry film thickness of the shop primer should be 50 μm minimum.
- Severe industrial environments; e.g., when stored in severe corrosive environments such as deck cargo shipments—Blast cleaning, second-quality, SIS 055900 Grade Sa 2½. Full primer coats at a minimum of 75 μm overall, plus an intermediate coat and topcoat, to a minimum total dry film thickness of 190 μm.

2.11.5 PREPARING OF SHOP-PRIMED SURFACES FOR OVERCOATING

Any primed surface that has been exposed for more than a few days will become contaminated and should be cleaned with fresh water and allowed to dry before overcoating. Prefabrication primer or blast primer exposed longer than six months to the atmosphere should be removed by blast cleaning before applying the specified paint system. Prefabrication primer or blast primer less than six months old and still in good condition should be cleaned thoroughly with fresh water before applying the paint and should be removed only if they are not compatible with the subsequently specified paint system.

If it is necessary to protect austenitic stainless steel surfaces during transport and storage on site, a proofed material (e.g., plastic sheeting) should be used unless the surfaces are provided with a prefabrication primer or blast primer.

Shop-primed surfaces should be cleaned thoroughly with fresh water before applying the subsequent layers. The cleaning and patch painting of damaged spots and of weld areas should be in addition to the complete specified paint system. If previously painted pipework is to be cut and welded on site, all paint must first be removed from the area of the welded joint.

Primed and painted surfaces that have been exposed to marine environments, including shipment, will be contaminated with salt and should be lightly wire-brushed and then washed with fresh water before overcoating.

Although zinc-rich primers are very effective at preventing rusting, extended exposure develops a surface contaminated with zinc corrosion, which can impair the adhesion of subsequent coats. Zinc-rich primers, both organic and inorganic, that have been exposed long enough to develop white surface staining should be prepared for overcoating by one of the following methods:

- Light blast cleaning and dust removal
- Wire brushing, followed by water washing
- Scrubbing with fresh water, using bristle brushes

Damaged surfaces should be repaired in accordance with the paint schedule (Table 2.1), with 10 cm overlap.

2.12 **TREATMENT OF SPECIFIC SURFACES**

2.12.1 **EDGES**

It is imperative that all sharp edges be coated to the same film thickness as the adjacent steelwork to prevent premature breakdown from this area. Corners, crevices, bolt heads, and rivet heads require similar attention. If there is any doubt that these areas have received an adequate film thickness, engineers may direct that an additional strip coat of paint be applied to ensure the full film thickness.

2.12.2 **FAYING SURFACES OF FRICTION-GRIP JOINTS**

The term *faying surface of a friction-grip joint* refers to a surface that, when in contact with another surface, transmits a load across the interface by friction. The faying surfaces of friction-grip bolted joints (see BS 3294 and BS 4604) require special attention. If left bare, all points where moisture could gain access should be effectively sealed. The alternative is to protect the faying surfaces, but in this case, the effect of the protective schemes on the slip factor has to be closely investigated, and the joints' behavior under static, dynamic, and sustained loads should be considered. If adequate test results are not available, they should be obtained.

Consideration should also be given to the possibility that pre-tension could be lost due to the behavior of protective coatings on fasteners and in friction grip joints. Sprayed aluminum or zinc, hot-dip galvanizing paint of the zinc silicate type, or special paint with abrasive addition should be considered.

2.12.3 **FASTENERS**

Fasteners that are exposed after assembly, such as steel pipe and cable-hangers, are zinc- or aluminum-coated, or blast-cleaned and primed before being welding on (if not blast-cleaned with the structure). Fixing nuts and bolts must be galvanized (see BS 729), sherardized (see BS 4921), or electroplated (see BS 3382 and BS 5493, Table 4). An adequate thickness of zinc should be applied, and when the zinc coating on fasteners (applied by galvanizing, electroplating, or sherardizing) is too thin for the life requirement, further coatings should be applied on surfaces exposed after assembly as follows:

- Zinc-dust paints—To a total thickness suggested by Figure 2.3 for appropriate environment and life to first maintenance
- Other coatings—To a thickness that will offer an equivalent protection to that given to the main structure

2.12.4 **WELDS**

As-rolled steel may be blast-cleaned and protected with a blast primer before fabrication and welding (see BS 5493). This prevents the serious development of rust, which would be difficult to remove. The use of steel that has rusted heavily during storage should be avoided for the same reason. When welding metal-coated or zinc-dust-painted steel, it is necessary to remove the coating near the weld area, or mask off the weld area before coating. Most painted steel can be cut and welded satisfactorily, provided that the coating thickness is less than 25 μm, but welds that

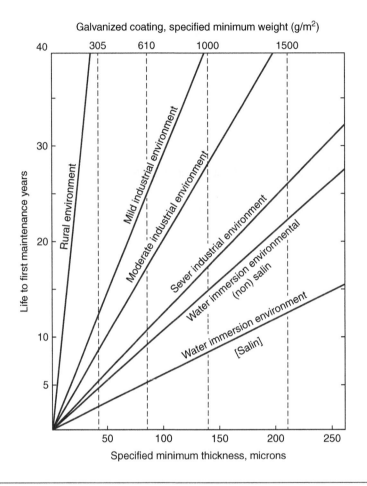

FIGURE 2.3

Typical lives of zinc coatings in selected environments.

are likely to be heavily stressed should be examined by engineers for porosity. After welding, scale- and heat-damaged coating should be removed by local blast-cleaning and the area renovated by repainting the original coating. Galvanized or metal-sprayed surfaces should be made good by doing any of the following:

- Metal-spraying on site
- Application of zinc-rich paints to reinstate the original dry film thickness
- Application of low-melting-point zinc alloys heated by torch until it reaches a pasty condition (these contain fluxes that should be removed)

To avoid the need for early maintenance of site welds on painted structures, they should be blast-cleaned before protection.

2.12.5 SPRAYED METAL PLUS SEALER SYSTEM

A sealer (see BS 5493, Table 4C, Part 2) that fills the metal pores and smooths the sprayed surface, improve the appearance and life of a sprayed-metal coating. It also simplifies maintenance, which then requires only the renewal of the sealer.

Sealers should be applied immediately after spraying the metal coating. Pretreatment primer may be applied before sealing (see BS 5493, Table 4F, Part 4). Metal spraying and sealing are performed by specialist contractors who are equipped to apply the full protective system in shop or on site.

2.12.6 METAL PLUS PAINT SYSTEM

Metal-coated steel is painted only in any of the following circumstances:

- The environment is very acidic or very alkaline (i.e., when the pH value is outside the range of 5–12 for zinc or 4–9 for aluminum)
- The metal is subject to direct attack by specific chemicals
- The required decorative finish can be obtained only by painting
- When additional abrasion resistance is required

The thickness of the paint is related to the type of environment, generally one or two coats of paint (dry film thickness, 100 μm) may be sufficient except in abnormally aggressive environments (like acid environment). Sealed sprayed metal is usually preferable to painted sprayed metal. Appropriate paints usually have a longer life on metal coating than on bare steel, and rusting and pitting of the steel is reduced or prevented.

2.12.7 ZINC COATINGS PLUS PAINT

The paints used in this procedure should be compatible with the sprayed or galvanized surfaces. Most paints, other than those containing drying oils, are suitable for application to zinc-coated steel that has been pretreated based on standards.

Newly galvanized surfaces will not normally be painted unless exposed to environments where the progressive loss of zinc will ultimately cause the corrosion of steel to occur within a 3–4-year period. In practice, this would include galvanized steel subjected to industrial severe corrosive environment and marine atmospheres, as well as painting to improve appearance. Where new galvanized steel is to be painted, it should be thoroughly degreased by wiping the surfaces down with white spirits and primed with a brush-applied coating of blast primer. Subsequent finishing with alkyd paints would then be permissible.

Where new galvanized steel is likely to be subjected to a particularly acid environment (i.e., a moderate chemical environment), including contact with acids or close proximity to acid gases and then cleaning to remove grease and salts (with clean water if necessary), the cleaned galvanized surface should initially be pretreated by a brush-applied coating of a two-pack blast primer, followed by at least two coats of epoxy paint in accordance with standards.

Weathered galvanizing that has been subjected to a period of at least 12 months in its permanent location should be thoroughly cleaned to remove all salts with clean water, dried, and then primed with a brush-applied coating of a two-pack blast-primer. Subsequent finishing with alkyd paints would then be permissible.

Certain fixings for galvanized sheeting, such as zinc-coated strapping, can normally be obtained only with a zinc coating that is thinner than 75 μm. Such fixings should be prepared and primed after erection as described above, followed by an intermediate coat and topcoat. Certain zinc-coated steel surfaces, including zinc metal spray, electroplated zinc, and sherardizing, can also be treated as discussed above.

For storage or transport of chemicals, specialist advice should be sought. The effect of the coating on the chemical should be considered, as well as the protection of the steel.

When subject to splashes of acid or alkaline chemicals, painting recommendations are similar to those given in groups 4, 5, and 7 of Table 2.2, except that oil-type paints (groups 1 and 2) should not be used and zinc silicates are not recommended for acid conditions. Metallic zinc is generally suitable for chemicals when the pH value is in the range of 5–12; metallic aluminum is generally suitable when the pH value is in the range of 4–9. Only specially formulated two-pack chemical-resistant paints and silicates are suitable for solvents and petroleum products. In splash conditions, coal tar epoxy 7J (for more than 10 years), 7I (for 5 to 10 years), 7H (for less than 5 years), and 5D (for more than 10 years) may be suitable.

2.12.8 STAINLESS STEEL PLUS PAINT

Stainless steel is not normally required to be painted, but where insulation is to be applied to stainless steel equipment, piping, or such items to be stored in the open air for long periods or are to be shipped as deck cargo, protective coating should be specified by the company. Zinc-containing paints are not allowed for this purpose. For potential fire situations, where hot-dip galvanizing or zinc coating is present, austenitic stainless steel equipment should be specially protected against the possibility of zinc embrittlement, which may result in rapid-fire escalation. Such equipment should be located in a shielded position, which will reduce the risk of molten zinc falling onto the steel. Any bare stainless steel parts (e.g., flanges) that are within reach of zinc should be protected with a painted steel shield. Where adequate shielding is impractical, hot-dip galvanized, or zinc-coated components should not be used in close proximity to (and particularly above) the stainless steel concerned. Stainless steel plates in storage should be stacked on their edges.

2.13 HIGH-TEMPERATURE SURFACES

Resistance to heat is influenced mainly by the nature of the temperature cycle, the maximum service temperature, and its duration. Furthermore, the behavior of the coating varies considerably depending on whether the surface remains dry (even when cold). When warm, the presence of hot gases will have specific effects.

Only general recommendations are given in standards, and the treatment for bare hot surfaces will be subject to a particular specification for each project, which will be specified in the contract documents. For temperatures up to 200°C, sealed sprayed aluminum or sealed, sprayed zinc may be considered for long or even very long life to first maintenance, depending on the circumstances. A special silicone alkyd over a zinc silicate primer system may be considered for medium periods of time. Where silicones cannot be tolerated, a silicone-free aluminum paint may be specified;

advice should be sought from the paint suppliers. The maintenance period is related to the operating temperature.

For temperatures up to about 550°C, aluminum (175 μm nominal thickness) is suitable as sprayed. Arc-sprayed aluminum should preferably be specified where there may be cyclical temperature fluctuations. The sprayed aluminum may also be silicone-sealed and, for temperatures typically about 250°C, silicone-sealed, sprayed aluminum can have a very long life (20 or more years). Zinc silicate systems are also recommended (such as system 5B); in moderate industrial environments, they can last for up to 10 years before maintenance is needed.

The zinc silicate/aluminum silicone treatment is to be preferred for severe industrial environments. For temperatures up to 900°C, aluminum (175 μm nominal thickness) applied by an electric arc under controlled conditions may be considered for some purposes. BS 2569, part 2, gives some alternatives.

For components to be used at service temperatures up to 1,000°C, BS 2569, part 2, specifies a nickel-chromium alloy. With sulfurous gases present, the nickel-chromium is followed by aluminum, and in each case, there is a subsequent heat treatment.

The life to first maintenance of coatings recommended for temperatures up to 550°C, up to 900°C, and up to 1,000°C will depend on the exact combinations of conditions in service but will usually be less than 10 years. However, the sprayed aluminum coating may last longer if the maximum temperature and other conditions are not too severe.

Notes:

- Oil-based aluminum paint should not be applied to surfaces above 100°C where flammable vapor or explosive dust may be present.
- Heat-resisting silicone aluminum paint will become sufficiently difficult to handle at ambient temperatures; however, in general, maximum performance is obtained only after the paint has been exposed for at least 1 h to 200°C.
- If, for the detection of hot spots on the outside surface of cold wall reactors (i.e., reactors with an internal refractory lining), temperature-indicating paints are specified, the surface preparation and paint application should be in accordance with the manufacturer's instructions. Only products from approved manufacturers should be used.

2.14 REFRIGERATED SURFACES

For refrigerated surfaces (as low as −30°C), the low temperature reduces the corrosion rates but facilitates condensation conditions. Where water is present, an effective barrier layer is required on the steel.

The general recommendations for treatment of such surfaces are as follows:

- Sealed or unsealed sprayed-metal coatings and bare galvanizing are suitable.
- Typical coating systems include zinc silicate (system 5B) or zinc-rich epoxy paint (system 4A) for 5–10 years; chlorinated rubber paint system (system 6C) for less than 5 years; epoxy paint (system 7B) for 5–10 years; and system 7A for less than 5 years. For temperatures below −30°C, specialist advice should be sought.

2.14.1 CONCRETE SURFACES

If concrete surfaces are to be protected, the first paint layer should be applied by brush after filling any surface imperfections with an appropriate putty.

2.14.2 ENCASEMENT IN CONCRETE

Steel requires no protection when fully encased in alkaline (uncarbonated) concrete. Protection by sprayed zinc, by galvanizing, or by zinc-rich coatings is beneficial in the zone that may become carbonated (see BS 5493, in the "Protection of Steel by Cement and Allied Products" and "Concrete" sections). Where the coated steel enters the concrete, a bituminous coating may subsequently be applied at any interface where water may tend to remain (see Figure 2.4).

2.14.3 UNDER INSULATION SURFACES

Although a specific coating system cannot be recommended, the requirements that must be met by the coating can be specified. However, the company should specify the primer system depending on circumstances.

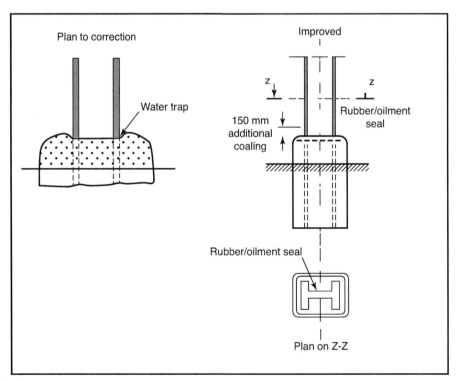

FIGURE 2.4

Protection of a stanchion at ground level.

In the insulation industry, a primer is sometimes used, particularly under spray-on foam insulation materials. It should be understood that the purpose of the primer is to present a clean surface for the insulation to bond to. Primer paints are not intended to and will not prevent corrosion by hot water. The key to specifying a paint system is to remember that it will be exposed to hot-water vapor—a very severe environment for paints.

The use of inorganic zinc paint as a primer, coupled with the hot water, will not cause accelerated corrosion of the steel. Zinc and the silicate binder are both dissolved by hot water. Even if conditions favor the zinc becoming cathodic to steel, the protective coating binder would dissolve and the coating would break down.

The advisability of using inorganic zinc under other protective coatings is less clear. Some tests seem to indicate good performance, but others show that inhibited primers with topcoats achieve maximum performance in hot-water systems. Therefore, the main criterion for a protective coating under insulation is that it resist hot water and water vapor. These severe conditions also require high-quality surface preparation.

2.14.4 MARINE STRUCTURES

Structures exposed to marine conditions, including offshore equipment, jetties, wharfs, dolphins, buoys, pipework, manifolds, hose-handling gear, lifting equipment and machinery, deck, landing stages, and separators for oil/water and oil/gas, require a high grade of protection and should be coated as specified in Table 2.1. All immersed and partially immersed marine structures, including piling, submarine pipelines, harbor craft and buoys, oil/water separators, and cooling water systems, should be provided with cathodic protection where necessary in addition to protective coatings.

2.14.5 STEEL PIPING AND VALVES

Unless otherwise stated, only shop-fabricated, aboveground steel piping with a diameter equal to or greater than 100 mm should be painted in the shop. Steel piping of smaller diameters should be painted on site. All aboveground valves should be painted in the shop. The paint systems used should be in accordance with Table 2.1.

2.15 PAINT SCHEDULE

Table 2.1 indicates by system numbers, the painting schemes to be used for new steel structures, at site, which will be installed in different environments and also for maintenance working where applicable (see NACE Standard RP-01-84). The systems classified by environment, where applicable, by typical lifetime to first maintenance, by surface temperature of the structure and by MDFT, of the paint system indicate the options available to the specified surfaces. The applicable paint systems for severe chemical environment are not included in Table 2.1.

For some of these applications, several systems offer acceptable protection, so the selection has to be guided by other characteristics. These include availability, convenience of application, ease of maintenance, and economic use for the specific structures and situations involved. It may be necessary to assess the life of each part of a structure separately. For each assessment (whether or not more than one is required), the following points should be taken into account:

- Required life of structure
- Decorative aspects (the decorative life of a coating is not usually as long as the protective life)
- Irreversible deterioration if scheduled maintenance is delayed
- Difficulty or ease of access for maintenance
- Technical and engineering problems in maintenance
- Minimum acceptable period between maintenances
- Total maintenance costs, including shutdown of plant, closure of roads, and access costs

In terms of the paint life (i.e., time to first maintenance), the paint systems are classified into three categories as follows:

- Long—Typically 10–20 years
- Medium—Typically 5–10 years
- Short—Typically less than 5 years

Mechanical damage to coating during handling, transport, and erection is not considered in Table 2.1. The recommended treatments listed for longer lives will always protect for shorter periods and are frequently economical for these shorter lives as well. The recommendations indicate the minimum requirements to ensure protection.

Surface preparation, priming, and coating should be carried out before assembly/erection. At least a prefabrication primer or blast primer should be applied to carbon steel and intermediate alloy steel. Shop-primed surfaces should be cleaned thoroughly with clean, fresh water before applying the subsequent layers.

The cleaning and patch painting of damaged spots and of weld areas should be done in addition to the application of complete specified paint systems. Wherever possible, all defects should be repaired with the paint systems as originally specified.

2.16 PAINT COLOR SCHEDULE

Included in Table 2.4 are colors for appearance and identification of buildings, steelworks, pipeworks, instruments, tanks, plant equipment, machineries, and safety and fire equipment, as well as safety colors and safety signs to be adopted by the oil, gas, and petrochemical industries. This specification is basically selected from the British Standard 381 C and specifies surface (opaque) colors for painting.

Tables 2.5 and 2.6 specify the standard colors for painting gas cylinders in use at all plants and medical centers throughout the oil, gas, and petrochemical industries, with specific reference to BS 1319, BS 349, and BS 381C.

2.17 CORROSION PREVENTION BY PAINT COATINGS

Carefully formulated and selected paint systems prevent corrosion. Normally, paint is applied in two or three coats; apart from a single coat of an inorganic zinc silicate, a one-coat paint system is rarely used and would require excessive care to ensure freedom from pinholes and minor discontinuities and

Table 2.4 Paint Color Schedule for Buildings, Structures, Pipeworks, Tanks, and Safety and Fire Equipment

Item	Color

1. Building
  a. External
  b. Internal
  c. Doors and windows
	As specified on drawings

2. Steelwork
  a. Structural steelwork
  b. Stairways including stair treads
  c. Platforms:
    Upper ladder
  d. Hand rails
    Mid-rail
    Kick plates
  e. Fire escapes and ladders including safety
    loops
  f. Pipes, columns, supports in walkways
	As primed or specified on drawings or black
	As primed
	Light orange, to BS 381C. No. 557 black
	Light orange, to BS381C. No. 557
	Signal red, to BS 381C. No. 537
	Diagonal white and canary yellow
	(BS 381C. No. 309) bands of 15 cm

3. Pipework
3.1 Inside plant Boundaries
  a. Product lines
  b. Steam lines
  c. Water lines
  d. Fire mains
  e. Foam lines
  f. Dangerous, poisonous chemicals and gases
  g. Natural gas lines
  h. Sour gas lines
  i. Crude oil lines
  j. Glycol lines
  k. Valves, flange to flange on gas lines
    On oil lines
  l. Air supplies
  m. All pumps and valves:
    Smaller than 7 cm (3″)
    Larger than 7 cm (3″)
  n. Line contents:
  o. Identification:
	White
	Aluminum
	Aluminum
	Signal Red, to BS 381C. No. 537
	Signal Red, to BS 381C. No. 537
	Canary Yellow, to BS 381C. No. 309
	
	Dark yellow middle
	Middle brown, to BS 381C. No. 411
	Black
	White and canary yellow, to BS 381C. No. 309
	Bands
	White and canary yellow, to BS 381C. No. 309
	Bands
	Arctic blue, to BS 381C. No. 112
	Canary yellow, to BS 381C. No. 309
	Canary yellow bands
	To be written in black on aluminum
	Background with arrow for flow direction
	To be bands 10 cm. (4″) wide at valves and
	Flanges on aluminum background

4. Instruments a. Panels, control consoles and instruments	Brilliant green, to BS 381C. No. 221

(Continued)

Table 2.4 Paint Color Schedule for Buildings, Structures, Pipeworks, Tanks, and Safety and Fire Equipment *(cont.)*

Item	Color
5. Tanks	White
a. Avgas, mogas, naphtha, pentane	Light gray, to BS 381C. No. 631
b. Kerosene	Light straw, to BS 381C. No. 384
c. Gas Oil	Sea green, to BS 381C. No. 217
d. Diesel	Black
e. Fuel Oil	Middle brown, to BS 381C. No. 411
f. Crude Oil	Aluminum
g. Water Tanks	Black
h. i. Identification numbers on tanks other than black	White
ii. On black tanks	
6. Heaters and heat exchanger	Black
a. Heater casing and stack	White
b. Heat exchangers	
7. Mobile plant	Canary yellow and black diagonal bands
a. Back and front shields	Canary yellow, to BS 381C. No. 309
b. Forks of fork lifts sling hooks	Canary yellow, to BS 381C. No. 309
c. Guards, safety parts on rotating machinery belts, coupling, pumps and motors	
8. Machinery	Light gray (battleship), to BS 381C. No. 631
a. Pumps	Light gray (battleship), to BS 381C. No. 631
b. Base plates	White high gloss
c. Outdoor and indoor elec. Motors	Light orange, to BS 381C. No. 557
d. Moving parts including couplings, flanges, areas covered by guards	Arctic blue, to BS 381C. No. 112 Brilliant green, to BS 381C. No. 221
e. Cable boxes	Indoor: manufacturers standard
f. Machine tools	Outdoor: galvanized or zinc
g. All other electrical items as fuse boxes, junction boxes	Spray finish or aluminum
h. Lighting towers, lighting fittings etc.	Light gray, to BS 381C. No. 631
i. Major switch gear, transformers and control panels	
9. Miscellaneous	All sumps containing cooling water and or effluent valves to have enameled diagram plates attached to guardrail. Plates to have white background with lettering in black
a. Cooling water sump	
b. Identification of equipment	All items of equipment and plant to have identification name and W. I. N. Painted on in white letter over black background
W. I.N. = Works identification number	

Table 2.4 Paint Color Schedule for Buildings, Structures, Pipeworks, Tanks, and Safety and Fire Equipment *(cont.)*

Item	Color

10. Safety Color Scheme
   a. Fire fighting equipment including fire pumps. Hose stations, fire monitors, supporting towers and fire hydrants CO_2 bottles in substations
   b. Safety showers and other safety installations
   c. Dangerous obstructions
   d. Relief and safety valves
   e. Dangerous or exposed part of machinery
   f. Danger points of electrical installation
   g. First aid equipment
   h. Breathing apparatus masks and other safety equipment
	Signal red, to BS 381C. No. 537
	Brilliant green, to BS 381C. No. 221
	Alternative bands of black and canary yellow canary yellow, to BS 381C. No. 309
	Canary yellow, to BS 381C. No. 309
	Arctic blue, to BS 381C. No. 112
	White lettering brilliant green, to BS 381C. No. 221
	brilliant green, to BS 381C. No. 221

11. Notices on Signboards
   a. Plants using caustic soda or acids
   b. Plants with high concentration of hydrogen sulphide
	Black letters on yellow background
	Red letters on yellow background

12. Stenciling
   a. Contractors shall put stencil marking on tanks, vessels, equipment numbers in English / Farsi
   b. Storage tanks shall have digits stenciled in black minimum of 1.5 M. High on at least two sides	

Table 2.5 Paint Color Schedule for Some Industrial Gas Containers

Name of Gas	Chemical Formula	Body and Band Color	Color Number To BS 381C.
Acetylene	C_2H_2	Maroon	541
Air	N/A	Light gray	631
Ammonia	NH_3	Black	N/A
		Signal red band near valve fittings, canary yellow band between red band and ground color	537 and 309
Argon	Ar	Peacock blue	103
Carbon dioxide	CO_2	Black	N/A
Carbon monoxide Chlorine	CO Cl_2	Signal red canary yellow band near valve fittings canary yellow	537 309 309

(Continued)

Table 2.5   Paint Color Schedule for Some Industrial Gas Containers *(cont.)*

Name of Gas	Chemical Formula	Body and Band Color	Color Number To BS 381C.
Coal Gas	N/A	Signal Red	537
Ethyl chloride	C_2H_5CL	Light gray signal red band near valve fittings	631 537
Ethylene	C_2H_4	Dark violet signal red band near valve fittings	796 537
Ethylene oxide	C_2H_4O	Dark violet signal red band near valve fitting canary yellow band between red band and ground color	796 537 309
Helium	He	Middle brown	411
Hydrogen	H_2	Signal Red	537
Hydrogen cyanide	HCN	Peacock blue canary yellow band near valve fitting	103 309
Methane	CH_4	Signal Red	537
Methyl bromide	CH_3Br	Peacock blue black band near valve fittings	103
Methyl chloride	CH_3CI	Light brown, with green, signal red, band near valve fitting	225 537
Neon	Ne	Middle brown black band near valve fitting	411
Nitrogen	N_2	Light gray black band near valve fittings	631
Oxygen	O_2	Black	N/A
Phosgene	$COCL_2$	Black, peacock blue band near valve fittings. Canary yellow band between peacock blue band and ground color	103
Propane (commercial)	N/A	Signal red	537
Sulphur dioxide	so_2	Light brown with green. Canary yellow band near valve fittings	225 309

Note: For cylinders with a valve-protecting extension, the color of the band near the valve fitting should be painted on the protecting extension.

to obtain an even thickness. The sequence of coats is (in order of application) primer, intermediate coat (undercoat), and topcoat (finish). The primer is of the utmost importance to the paint system, as it is the tie-coat between the metal and subsequent coats. Primers should have the following characteristics:

Table 2.6   Paint Color Schedule for Medical Gas Cylinders

Name of gas	Symbol	Valve End Color	Body Color
Oxygen	O_2	White	Black
Nitrous oxide	N_2O	Arctic blue	Arctic blue
Cyclopropane	C_3H_6	Light orange	Light orange
Carbon dioxide	CO_2	Light gray	Light gray
Ethylene	C_2H_4	Violet	Violet
Helium	He	Light brown	Light brown
Nitrogen	N_2	Black	Light gray
Oxygen and helium mixtures	O_2+He	White and middle brown	Black
Air (medical)	Air	White and black	Gray
Oxygen and nitrous oxide mixture	O_2+N_2O	White and arctic blue	Arctic blue

- Adhere well to the steel
- Be anticorrosive
- Form a suitable base for intermediate coat

The adhesion to the steel is primarily a function of the resin binder in the primer. The anticorrosive nature is largely encouraged by the pigments. The formation of a good base for the topcoat is partly a function of the binder, but the ratio between the quantities of pigment and binder also plays an important part.

The choice of the type of binder for the primer is also governed by the choice of topcoat, which in turn is determined by the necessary level of resistance to the environment. An all-purpose primer that can be covered with all types of topcoats would simplify the painting task, but this ideal is not yet available.

For a paint coating to prevent corrosion, it is evident from the previous reactions that this can be achieved by three methods, which can be used either alone or in combination:

- Barriers to restrict the access of moisture, oxygen and salts
- Electrical method
- Chemical inhibition

These methods are explored in the next sections.

2.18 BARRIERS

As all coatings are permeable to both water and oxygen to some extent, it is impossible to completely keep them from reaching the steel surface. However, some coatings are more impermeable than others in certain situations. Catalyzed epoxies, coal-tar epoxies, solution vinyl, and chlorinated rubber paint systems are used in water-immersed environments. These are typically known as *self-primer materials*,

Table 2.7 Site Treatment of Previously Metal-Coated Steelwork

Initial Condition	Present Conditions	Surface Preparation	Replacement of Metal Where Required	Paint Treatment Over	
				Sprayed Metal	Galvanizing
Bare metal coating	Area of corrosion and/or some rusting of substrate	If metal is to be replaced, blast-clean the surface	Spray metal to required specifications	Not normally necessary if overcoating is required	N/A
		If metal is not to be replaced, clean corroded areas by the best means available	N/A	Build up cleaned areas with suitable paint system and, preferably, apply a chemical-resistant finish overall	Build up cleaned areas with suitable paint system and, preferably, apply a chemical-resistant finish overall
	Area with some white corrosion products	If decoration is required, wash to remove salts using a stiff brush if necessary; remove loose material with a nonmetallic brush	N/A	Apply sealing coat and chemical-resistant finish for maximum life	Apply suitable surface pretreatment followed by, preferably, a chemical-resistant finish
		If decoration is not required, no action is necessary	N/A		
	Areas in sound condition	If decoration is required, wash to remove salts using a nonmetallic brush	N/A	Apply suitable paint, which should be chemical-resistant for maximum life	Apply suitable surface pretreatment followed by suitable paint, which should be chemical-resistant
		If decoration is not required, no action is necessary	N/A		

Table 2.7	Site Treatment of Previously Metal-Coated Steelwork *(cont.)*				
				Paint Treatment Over	
Initial Condition	**Present Conditions**	**Surface Preparation**	**Replacement of Metal Where Required**	**Sprayed Metal**	**Galvanizing**
Sealed or painted metal coating	Area of corrosion or some rusting of substrate	If metal is to be replaced, blast-clean the surface	Spray metal to the required specifications	Consider one or two coats overall, preferably chemical-resistant	Consider one or two coats overall, preferably chemical-resistant
	N/A	If metal is not to be replaced, remove corrosion by the best method available	N/A	Build up cleaned areas with suitable paint, applying one or two coats overall, preferably chemical-resistant	Build up cleaned areas with suitable paint, applying one or two coats overall, preferably chemical-resistant
	Areas with some degradation of paint, disspation of sealer, or loss of adhesion	Remove loose material with a nonmetallic brush	N/A	Apply further coats of paint or sealer, preferably chemical-resistant	Apply further coats of paint or sealer, preferably chemical-resistant
	Areas in sound condition	If decoration is required, dust the surfaces	N/A	Apply further coats of paint or sealer, prefer-ably chemical-resistant	
	N/A	If decoration is not required, no action is necessary	N/A	N/A	
It should be confirmed that the apparent rusting emanates from the substrate.					

which means that the intermediate coat or topcoat is used as the first coat on the steel and an anticorrosive primer is not required—that is, they prevent corrosion only by a barrier effect.

These highly impermeable coatings relay on excellent adhesion to the surface and the maintenance of film integrity as water diffuses through the coating. For this reason, it is essential that such coatings do not contain any water-soluble components. Otherwise, blistering with eventual rupturing and exposure of bare steel will occur due to osmosis, which is the passage of water through a semipermeable membrane (the paint film) when different concentrations of salts exist on each side of the paint coating.

Equally damaging is the presence of salt from sea air or seawater osmosis. It is essential that steel be washed with fresh water to remove these or other soluble salts before any surface painting or preparation (including blast cleaning), as they can be blasted into the steel surface and become trapped beneath the paint coating.

2.19 ELECTRICAL METHODS

The electrical method of preventing corrosion involves minimizing the flow of corrosion current so that, if negligible current flows, negligible corrosion results. Thus, in a corrosion:

$$I = E/R$$

where I is the corrosion current, E is the polarized potential difference between local anodes and cathodes, and R is the total electrolyte resistance, which must be much greater than the metal conductor resistance.

Thus, the higher the value of R is, the less the value of I is, so by making the electrolytic path of the current of high resistance, the movement of ions is impeded. Paint resins with the highest electrical resistance are catalyzed epoxies, phenolics, vinyls and chlorinated rubbers, each with values in the order of 10 ohm/cm^2.

The addition of coal tar further adds to the resistance, and the addition of extended pigments in the first coat such as talc, china clay, mica and iron oxide also assists in increasing the resistance. As mentioned above, the removal of soluble materials on the surface is necessary because their presence will short circuit the resistance of the paint film so that the value of I increases, possibly to the stage where rusting will occur.

The film thickness is variable and, as the diffusion of water through a paint film is inversely correlated to the film thickness, it will take longer for moisture to diffuse through a thick film than for a thin film. Similarly, the thicker the coating, the higher the electrical resistance, and, for immersed or buried surfaces, the dry film thickness of paints needs to be between 250 and 500 μm. An alternative method of electrically preventing corrosion of iron is to employ a metal that is more anodic (i.e., less noble) than iron, such as zinc. Thus, zinc-rich paint coatings will protect the steel; as the iron is no longer the anode of the electrical circuit, the zinc metal becomes the anode, and the iron will not corrode.

2.20 CHEMICAL INHIBITION

The term *chemical inhibition* refers to the use of special pigments in the primer coat. These anticorrosive pigments function in varying ways, but in broad terms, the following can be stated:

- Blanketing either the anode or cathode areas
- Interfering with either the anode or cathode reaction, hence preventing the formation of ferrous hydroxide
- In the case of zinc chromate, by oxidizing the ferrous corrosion product to ferric oxide which plugs the anodic areas and increases the electrical resistance

2.21 CHARACTERISTICS OF PROTECTIVE SYSTEMS

Paint systems usually consist of primer, intermediate coats or undercoats, and a topcoat or finish. Each component normally contains pigment (solid particles) suspended in a solution of binder (resin solution). The choice of pigment and ratio of pigment to binder depends on the function of the paint; e.g., more binder will be used in a penetrating primer for sprayed metal and more pigment in a high-build intermediate coat.

The binder more clearly defines the essential characteristics of the coating (see Appendix B of BS 5493). Drying-oil-type paints (like alkyd paints) dry in the presence of atmospheric oxygen; the action is catalytically promoted by metallic soaps. One-pack, chemical-resistant coating (such as chlorinated rubber and vinyl paints) usually dry by the evaporation of solvent, but moisture-cured polyurethanes are also in this group. Two-pack, chemical-resistant coatings (epoxy paints) form by chemical reaction. The two components have to be mixed just before use.

Pigments may inhibit corrosion, reinforce the dry film, provide color, absorb or reflect ultraviolet radiation, or any combination, thus improving the durability and stability of the coating.

High-build formulations permit much greater film thicknesses per coat. They are usually applied by airless spray, but can be applied by roller or brush.

All paints within a system should have compatibility between the coats and with the metal substrate (i.e., there should be adequate adhesion to the substrate and between coats over the operating temperature range, and there should be no under-softening to cause the lifting, wrinkling, or bleeding through of stains). For this and other reasons, it is generally advisable to obtain all the components of a paint system from the same source; otherwise, assurance of such compatibility should be obtained. If cathodic protection is applied to the structure, the paint system should be compatible with it.

Solvent modification of paint composition is frequently necessary to allow for the characteristics of different matters of application. Two-pack, chemical-resistant paints withstand reasonable handling and can be readily stacked when fully cured, but most other paints are relatively easily removed down to the primer level. One-pack, chemical-resistant paints tend to stick on stacking, but drying-oil-containing paints can be stacked with care.

Touching up on site is easy for most paints, but initial abrasion may be needed to provide adequate adhesion of touch-up treatments on two-pack, chemical-resistant paints.

2.21.1 PRIMERS

Primers usually consist of binders, inhibitive pigments, and solvents.

The primers commonly used are as follows:

- Prefabrication primer, which is used for steel before fabrication
- Blast primers, which are used for steel either before or after fabrication
- Drying oil primers
- One-pack, chemical-resistant primers
- Two-pack, chemical-resistant primers
- Wash primer

2.21.2 ZINC-RICH PAINTS

Metallic zinc-rich paints (commonly called *zinc-rich coatings*) may be organic or inorganic. With 90% or more of zinc dust content (which may contain up to 4% zinc oxide) in the dry film, the coating will afford cathodic protection but will be slightly permeable. The formation of zinc salts gradually renders the coating impermeable, and it will be a barrier coating. If damage to the coating exposes the steel, the zinc will again become cathodically sacrificial to prevent rust spreading. Suitable sealer coats improve the appearance of zinc-rich paint coatings. The advice of the coating supplier should be sought regarding the type of sealer to be used, especially if the surface is exposed between applications. The various uses of zinc-rich paints as blast primer, fabrication primer, or main coating are recommended. Members and assemblies coated with zinc-rich paints may be handled or stacked as soon as the coating is dry, but exposure of freshly applied zinc silicate paints to moisture within a stack can result in deleterious changes.

2.21.3 DRYING-OIL-TYPE PAINTS

Drying-oil-type paints cover a wide range of material, ranging from largely obsolescent simple oil paints, which were slow-drying but tolerant of less than perfect surface preparation, to phenolic varnishes and epoxy ester paints, which dry well even at low temperatures. Recoating usually presents no problem, but chemical resistance is poor to moderate, and weather resistance is moderate to good.

Silicone alkyds are more expensive than other drying-oil-type paints, but they keep clean more easily and retain color and gloss better than most other coatings.

2.21.4 ONE-PACK, CHEMICAL-RESISTANT PAINTS

One-pack, chemical-resistant paints dry in any ventilated conditions, but where coatings are built up, thickly retained solvent may keep films soft and prone to damage for days or even weeks. Recoating is easy for all but heavily contaminated surfaces because the films remain soluble in suitable hydrocarbon solvents. Where the highest chemical resistance is not required, a system of anticorrosive drying oil type primer, carefully selected for compatibility with a chemical-resistant finishing system, allows some relaxation of the steel preparation standards.

2.22 TWO-PACK, CHEMICAL-RESISTANT PAINTS

Two-pack, chemical-resistant paints are resistant to acids, alkalis, oils, and solvents, but they should not be used unless the highest quality of surface preparation and application can be ensured. Cured films are hard and solvent-resistant, so that intercoat adhesion may be doubtful, particularly where surface contamination may occur. Coal tar epoxy and urethane/tar coatings are cheaper and may be easier to apply, but they are restricted to darker colors and have lower solvent resistance. Hot-applied, solvent-free epoxies have a particular usefulness for tank linings where flammable solvents could be a hazard.

Where the greater part of a two-pack, chemical-resistant system is to be applied in the shop and where travel or erection damage may have to be touched up on site, it may be advantageous to use a system that incorporates a chlorinated rubber travel coat consisting of a two-pack primer and intermediate coat overcoated with a one-pack, chemical-resistant finish or travel coat (tie coat) and site finish. A travel coat (tie coat) is a paint with a binder of chlorinated rubber modified with alkyd in a ratio of 2 (min) to 1, which will readily accept a further chlorinated-rubber coat after erection.

2.23 BITUMINOUS COATINGS

Bituminous coatings are low-cost coatings whose protective properties depend on film thickness. There is a wide range of materials based on either mineral bitumen or coal-tar fractions applied as unheated solutions, hot solutions, or hot melts; bituminous emulsions are not used very often to protect steel. Specially developed materials, based on powdered coal dispersed in pitch, are widely used for the protection of underground pipes. Blast cleaning before coating gives the best performance.

Bituminous coatings have good resistance to dilute acids and alkalis, salt solutions, and water, but are not resistant to vegetable oils, hydrocarbons, and other solvents. They may become brittle in cold weather and soften in hot weather. Bitumen-coated articles should not be stacked. Bitumen solutions and emulsions are readily applied by brush or spray and are often used as priming coats for the heavy-duty materials, which can be applied hot or cold at the works or on site. The specifier should consider inhibitive oleo-resin-based primers for heavy-duty bitumen, provided that sufficient drying time (several weeks) can be allowed to pass before overcoating.

2.23.1 COAL-TAR ENAMELS

Coal tar enamels have high resistance to moisture and good adhesion to steel, so they are very suitable for structures that are immersed in water (especially foul water) or buried in the ground. The appropriate water supply authority should be consulted before coal-tar-based material is used in conjunction with potable water. Coal tar enamels are less readily softened by hydrocarbon oils.

Prolonged exposure to weather and sunlight causes surface chalking because of oxidation and loss of plasticizing components, so coal-tar enamels and bitumens should never be specified for such conditions (unless they are overcoated with asphaltic material in solution or emulsion form), nor should they be used in very hot conditions (such as may arise in a pipeline downstream of a compressor).

The coatings may be reinforced with glass fiber or asbestos wrapping, especially for the protection of pipelines. Wrappings made from vegetable fibers such as cotton or hessian are liable to microbiological attack.

2.23.2 ASPHALTIC COATINGS

Asphaltic coatings have much better resistance than coal-tar enamels to sunlight, weather, and exposure to the direct heat of the sun. Resistance to breakdown under sunlight can be improved with flake aluminum. Asphaltic coatings are recommended for buried or submerged conditions and they are best used with inhibitive primers.

2.23.3 APPLICATION OF COAL-TAR AND ASPHALT ENAMELS

The materials are heated as needed in boilers near the application site. For vertical surfaces, the material is daubed on with a stiff brush, covering small rectangular areas with short strokes and overlapping to form a continuous coating. In weld areas, the brushstrokes should be in the direction of the weld; a second coat should then be applied in the opposite direction. For horizontal surfaces, the material can be poured on and then trowelled out and if unevenness occurs where a smooth surface is required, it may be permissible to play a blow-lamp onto the surface and finish by troweling. Considerable skill is required in all these operations.

In general, only bituminous material should be used for overcoating bituminous material. It is, however, possible to overcoat with some emulsion paints or cement paints, and these may be desirable to reduce surface heating under sunlight.

2.24 CHARACTERISTICS OF METALLIC COATINGS
2.24.1 ZINC COATINGS

Four methods for applying zinc coatings (other than zinc-rich paints) are in general use:

- Hot-dip galvanizing—For structures, fittings, and claddings
- Sherardizing—Mainly for fittings, fasteners, and small items
- Electroplating—Mainly for fittings, fasteners, and small items
- Metal spraying—For structures and fittings (including fasteners when done after fabrication)

The desirable weight or thickness of bare zinc for use in different environments can be derived from Figure 2.3. The metal corrodes at a predictable and uniform rate, which increases as sulfur dioxide pollution of the environment increases.

Areas of discontinuity or insufficient thickness of a metallic zinc coating, however caused, may be rectified at any stage by the application of sprayed zinc, special zinc-alloy solder sticks, or zinc-rich paints.

Cleaned steel is immersed in a bath of molten zinc; a partial alloying action result in a metallurgically bonded coating. As soon as the steel is cool (after withdrawal from the bath), it may be stacked and transported or it can be primed for overcoating.

The size of structural assembly that can be galvanized is limited by the size of the largest bath in each galvanizer's works. The standard does not refer to the galvanizing of steel thinner than 5 mm. Thinner components have thinner galvanized coatings (see BS 729), and coatings on continuously galvanized sheets are specified in BS 2989 (with further reference in DD 24). The coating weight specified for sheets is the total weight on both sides of the metal; for components, the rate of coating on one side only is specified.

For sections not less than 5 mm thick, refer to BS 729, where a minimum specified weight of zinc (610 g/m^2) is equivalent to a thickness of 86 μm (shown as "85 μm minimum" in the tables) on each face. The thickness of coating varies with the thickness of steel, surface preparation, and conditions of immersion. It may be increased to 140 μm ($1,000 \text{ g/m}^2$) if either the steel is grit-blasted before coating or the steel contains silicon (typically more than 0.3%).

The coating thickness may be increased to 210 μm ($1,500 \text{ g/m}^2$) by using silicon (typically more than 0.3%) or silicon-killed steel; consultation with the steel supplier and galvanizer is essential if these thick coatings are required. Brown staining may occur early in the life of steel containing silicon. This is a surface phenomenon and does not affect the protective value of the coating. The different lives of some galvanized coatings can be estimated and the appropriate life requirement to first maintenance may be assessed (see Figure 2.3).

The sherardizing process is used mainly for small parts and fasteners, particularly for threaded work where only small changes of dimension are acceptable. After suitable surface preparation, the items are tumbled in hot zinc dust. The thickness of the coating varies with the processing conditions; two grades (15 μm and 30 μm) are specified in BS 4921.

Zinc-plating of small parts by the electrolytic deposition of zinc from zinc-salt solutions is done only by specialist firms. It is rarely economical to apply electroplate thicker than 25 μm. Cadmium plating is an alternative process used for special purposes. BS 1706 specifies coating techniques for threaded parts.

2.24.2 SPRAYED-METAL COATINGS

The metals commonly used for spraying structural steel are zinc and aluminum. The technique of spraying metals is applicable to structures and fittings that are either in the shop or on site. An atomized stream of molten metal is projected from a special gun (fed by either wire or powder) onto a surface prepared in accordance with BS 2569. There is no size limit, and the process is especially economical when the area/weight ratio is low.

All grades of steel can be sprayed. The steel surface remains cool, and there is no distortion, nor is there any effect on the metallurgical properties of the steel. Coating thicknesses less than 100 μm are not usually specified unless the sprayed metal is to be sealed or painted immediately. For most atmospheric environments, there is no advantage in spraying aluminum to a thickness greater than 150 μm.

2.25 CHARACTERISTICS OF SOME OTHER PROTECTIVE SYSTEMS
2.25.1 POWDER COATINGS

Coatings formed from pigmented resins, applied as dry powders and fused by heat, have been developed for the protection of lightweight steel components. Some types of powder coatings also find applications on pipes and hollow sections (e.g., lighting columns, where simple shapes facilitate coating and heat curing). They are unlikely to be economical for use on heavy sections because of the high temperatures required for fusing or curing.

Powder coatings are of two main types:

- Thermoplastic powders, based on such substances as polyethylene, polypropylene, vinyl copolymers, and nylon-II, which fuse to form films without any chemical change
- Thermosetting powders, based on such substances as epoxy, polyester, acrylic, and polyurethane resins, which cure to chemically cross-linked films after being fused by heating

Thermoplastic powders are frequently applied by a fluidized-bed technique in components that have been preheated so that the powder will stick readily. Coatings can be built up to a thickness of 200–300 μm for one operation and are highly protective where a complete wraparound can be achieved. Such powders are usually applied by electrostatic spraying to a film thickness of usually between 50 and 100 μm. Epoxy powders are easily applied and form tough coatings.

Polyester and polyurethane powders afford better resistance to weather. Selected polyurethane and nylon powders are particularly useful where impact and abrasion resistance is important, as on certain types of fasteners. Unless anticorrosive pretreatments or solvent-borne primers are first used, powder coatings often have indifferent adhesion, so protection may fail rapidly once the coating has been broken.

2.25.2 GREASE PAINTS

Coatings based on various types of greases have two uses:

- As permanent noncuring coatings for application to the insides of box sections
- As temporary protection (see BS 1133) for components in store or for machined surfaces before assembly

Thick grease films form effective barriers to moisture, but inhibitors are added to increase effectiveness. Grease-based coatings should be used in conjunction with tapes or other wrappings on components that are to be stacked or are liable to rough handling.

2.25.3 WRAPPING TAPES AND SLEEVES

Wrapping with adhesive tape protects ferrous metals, particularly pipelines, joints, valves, and other fittings, by excluding the environment from the substrate. For further protection against accidental damage and to promote adhesion of the wrapping tape, it is desirable to clean any rust products thoroughly from the substrate and to prepare the surface with an inhibitive primer before taping.

Buried pipelines are often wrapped at works, with bitumen or coal tar reinforced by glass fiber, and only the joints require wrapping at site. When applying wrapping tape, an overlap of at least half the width of the tape is recommended and for a coating pipe with a diameter of up to 300 mm, it is good practice to use tape of a width that approximately matches the diameter of the pipe. Application by hand works satisfactorily for small jobs, but for large installations, such as long pipelines, fully automatic or semiautomatic methods are used. The necessary skills are obtaining a consistent tension throughout the operation, uniform bonding, and the avoidance of air pockets. Three types of wrapping are commonly available, described next.

Petroleum-jelly tapes consist of fabric of natural or synthetic fiber or glass cloth impregnated with a mixture of petroleum jelly and neutral mineral filler. They should be used in conjunction with a petroleum-jelly primer. The coating is permanently plastic; it is suitable for application to irregular profiles and should be smoothed by hand, taking care to avoid any air pockets. When used aboveground, these tapes should be protected by a bituminous-tape overwrap in situations where they may be subject to damage by abrasion. They are also suitable as insulation to avoid bimetallic contacts.

The most readily available synthetic tapes are polyvinyl chloride (PVC) and polyethylene tapes. These polymer strips, usually 125–250 μm thick with a fabric core, are coated on one side with a compact adhesive, normally with a synthetic rubber base. They are usually available in a range of colors if pipe identification is required.

Synthetic resin or plastic tapes are suitable as insulation to avoid bimetallic contacts, particularly in damp or dirty conditions. The best level of protection is obtained if the steel is first cleaned and coated with a conventional rust-inhibitive primer. For exterior exposure, black polyethylene tape is preferred to PVC because its surface degrades much less upon exposure to sunlight and weather.

Coal tar and bitumen tapes are used mainly for buried pipelines. They have a high resistance to moisture and good adhesion to steel. The fabric reinforcement is usually made from glass fiber. First, the steel should be cleaned and give a coating of coal tar or bitumen primer. According to the temperature expected in service so can the low properties of the tape be varied. For some high temperatures, and especially for heavy-duty requirements, the grade of coal tar or bitumen used is such that the tape must be heated to sufficiently to obtain the best seal.

In the two-pack taping method, a woven tape is impregnated after wrapping with a two-pack solventless composition, normally a polyester or two-pack epoxy. The technique used is similar to that used in the preparation of glass fiber molding, except that the resin-impregnated glass cloth is intended to adhere to the metal substrate. This method is used especially for shafting exposed to marine conditions and for surfaces that may be subject to cavitation corrosion.

The surface to be protected should be thoroughly cleaned to a bright metal, then primed with a two-pack epoxy (or similar). It should then be wrapped with glass cloth in sheet or tape form and the cloth impregnated with a two-pack polyester or two-pack epoxy composition. It is good practice to apply a thin coating of the impregnating resin before applying the reinforcing substrate; then several layers are built up and the final surface is troweled smooth. The coating will then set to a hard, glasslike, and tough protective coating that is impermeable to water.

Polyethylene and similar plastic may be used to make protective sleeves on pipes and are sometimes shrunk onto the metal. For spun iron pipes and castings, the application of a nonadherent but snug-fitting polythene sleeve gives good protection. The sleeving is applied at the time of laying the pipe, and joints are taped with adhesive strip.

2.25.4 PROTECTION OF STEEL BY CEMENT AND ALLIED PRODUCTS

Cement-mortar linings are widely used for the internal protection of water mains. Special formulations and coating procedures are used. They have limited impact resistance but may be repaired on site by fresh applications. Conversely, steel structures (and steel reinforcement) may be in contact with or embedded into concrete. The BS codes CP 110 and CP 114 to CP 117, relating to concrete and steel/concrete structures, contain relevant information. Exposed steel should first be covered with a suitable bitumen coating that is resistant to water penetration while the plaster or cement is curing, and then it may be covered with gypsum plaster and magnesium oxychloride cements.

2.25.5 CATHODIC PROTECTION

The degree of protection afforded by a cathodic system may be enhanced by paint coatings, but not all paint systems are compatible with cathodic protection systems and specialist advice should be sought. Tables 2.8–2.11 provide more useful information.

Table 2.8   Typical Organic Zinc-Rich Paints (Prefabrication Primer)

Number	Binder	Main Pigment	Volume Solids (Nominal %)	Main Pigment in Total Pigment (Weight % min)	Dry-Film Thickness (μm Per Coat Minimum Advised)
1	Two-pack-epoxy	Zinc dust	35	95	50

Notes: Quality covered by maximum of 75 μm recommended by spraying for each layer.
An initial prefabrication primer may be only 25 μm

Table 2.9   Typical Blast Primers

Number	Binder	Main Pigment	Volume Solids (Nominal %)	Main Pigment in Total Pigment (Weight % Min)	Dry-Film Thickness (μm Per Coat Minimum Advised)	Additional Information
1	Two-pack epoxy	Zinc phosphate	25	40	20	See BS 4652 Essential to avoid
2		Zinc dust	30	95	40	settlement of pigment also
3	Two-pack polyvinyl butyral	Zinc tetroxy-chromate	10	85	15	has uses other than as a blast primer
4	Two-pack polyvinyl butyral/phenolic	Zinc tetroxy-chromate	10	85	15	Suspect with cathodic protec-
5	One-pack polyvinyl butyral/phenolic	Zinc phosphate	22	20	20	tion if there is any discontinuity covering system
6		Zinc chromate	22	20	20	

Table 2.10   Typical Drying Oil Primers

Number	Binder	Main Pigment	Volume Solids (Nominal %)	Main Pigment in Total Pigment (Weight % min)	Dry-Film Thickness (μm Per Coat Minimum Advised)
1	Blend of raw and process drying oils	Red lead	75	98	40

Table 2.11 Typical Resistance Chart for Some Paints

Generic Type	Cure Mechanism	Acid	Oxidizing Acid	Alkali	Salt	Solvent	Weather	Max. Temp. (Dry Heat)
Chlorinated rubber	Solvent evap.	VG	VG	G	VG	P	G	66°C
Epoxy (polyamide)	Chem. crosslinking	E	G	E	E	VG	G	121°C
Silicon (alum.)	Solvent/heat	P	P	F	G	F	G	538°C
Vinyl	Solvent evap.	E	E	VG	E	P	VG	121°C
Zinc-rich (inorganic)	Hydrolysis	E*	E*	E*	E	E	E	400°C*
Zinc-rich (organic)	Chem. crosslinking	VG*	VG*	VG*	VG	VG	VG	149°C**

Rating Scale:
  (E) Excellent—No effect; best choice if performance and appearance retention are desired.
  (VG) Very good—No effect on performance; very little degradation of appearance.
  (G) Good—Little effect on performance, some degradation of appearance.
  (F) Fair—Performance and appearance affected by exposure.
  (P) Poor—Not suitable; coating attacked.
*Results indicate zinc-rich coating performance when topcoat is applied. Use of these coatings without a topcoat in chemical environments is not recommended.
**Limited by topcoats in the system.

2.26 INTERNAL PAINTING OF ABOVEGROUND STEEL TANKS, LPG SPHERES, AND VESSELS

2.26.1 HORIZONTAL TANKS/VESSELS

When a tank or vessel rests on a saddle of concrete or brickwork, the bearing surface should be heavily coated with petroleum jelly tape. Shop treatment should be given to both sides of the plates of horizontal tanks/vessels that are to be used to store specific products.

If treatment is not applied in the shop, the surfaces should be blast-cleaned on site to grade Sa 2½. The outer surface should then be painted as indicated in the paint schedule (Table 2.1) and the inner surface treated as indicated below:

- Petroleum products, except for food grade hydrocarbons—The shop coat on horizontal tanks/vessels for petroleum products need not be made good and further painting is not necessary.
- Food grade hydrocarbons—Painting is not required if a prefabrication or shop primer has been applied; rather, it should be removed by blast cleaning.
- Benzene, toluene, and xylene tanks for products containing more than 60% aromatics—Painting is not required if an oil-based red lead shop primer has been used; painting should be removed by blast-cleaning.

- Chemical and solvents such as alcohols, ketones, and ethers—Horizontal tanks are used for storing a wide range of chemicals (particularly processing chemicals) at manufacturing plants. Special tank lining materials are often required.

2.26.2 SPHERES AND CYLINDRICAL VESSELS FOR LIQUEFIED PETROLEUM GAS (LPG) UNDER PRESSURE

Shop treatment should be given to both sides of the plates of spheres and cylindrical vessels for liquefied petroleum gas (LPG) under pressure. No further painting is required on the inside, but the outside should be painted as indicated in the paint schedule (Table 2.1).

2.26.3 VERTICAL TANKS

For external painting, see the paint schedule (Table 2.1); for internal painting, see Tables 2.12 through 2.23.

Table 2.12 Internal Painting of Crude Oil Tanks			
Tank Parts and Conditions			**Paint Systems**
Bottom plates	Below 60°C and not severely corrosive		Shop-primed only (no treatment at site)
	Below 60°C when severely corrosive		7L
	Above 60°C		7L
Shell	Bottom course		As for bottom plates
	Top course 1,500 mm		As for rest of shell
	Fixed roof tanks	a. Generally, except b	7L
		b. Sour crude	1C
	Floating roof tanks		As roof external
	Rest of shell		Shop-primed*
Roofs	Fixed roof sheets/trusses	a. Generally, except b	
		b. Sour crude	7L
	Underside floating roof		As for rest of shell
*By approval of an authorized representative			

Table 2.13   Internal Painting: Aviation Gasoline Grades and Their Component Tanks and Naphtha Tanks

Tank Parts		Paint Systems
Bottom plates		Sand blast Sa$_2$ or shop prime
Shell plates	Bottom course (1,000 mm)	As for bottom plate
	Top course (1,500 mm)	As for bottom plate
	Rest of shell	Shop-primed only or sand blustedatsite
Roofs	Underside of floating roofs	Shop-primed only or sand blustedatsite

Table 2.14   Internal Painting: Motor Gasoline and Kerosine

Tank Parts		Paint Systems
Bottom plates		7B,7E
Shell plates	Bottom course (1,000 mm)	7B,7E
	TOP course (1,000 mm)	As exterior
	Rest of shell	Blast-cleaned (Sa$_2$)
Roofs	Sheets/trusses/compartment intervals	Blast-cleaned (Sa$_2$)

Table 2.15   Internal Painting: Wet-Treated (Merox) Kerosenes and Kerosene-Type Jet Fuel Tanks

Tank Parts		Paint Systems
Bottom plates		7B
Shell plates	Bottom course	As for bottom plates
	Rest of shell	As for bottom plates
Roofs	Sheets/trusses	

Table 2.16   Internal Painting: Gas Oil and Distillate Diesels and Their Components

Tank Parts		Paint Systems
Bottom plates		7A
Shell plates	Bottom course	7A
	Rest of shell	Shop-primed or sand blasted to Sa$_2$
Roofs	Sheets/trusses	7A

Table 2.17   Internal Painting: Tanks for Food Grade Hydrocarbons

Tank Parts		Paint Systems
Bottom plates		Blast-cleaned to Sa_2 (no painting)
Shell plates	Bottom course	Blast-cleaned to Sa_2 (no painting)
	Rest of shell	Blast-cleaned to Sa_2 (no painting)
Roofs	Sheets/trusses	1A

Table 2.18   Internal Painting: Benzene, Toluene, and Xylene Tanks and Tanks Containing More Than 60% Aromatics

Tank Parts		Paint Systems
Bottom plates		Blast-cleaned to Sa_2
Shell plates	Bottom course	Blast-cleaned to Sa_2
	Rest of shell	Blast-cleaned to Sa_2
Roofs	Sheets/trusses	1A

Table 2.19   Internal Painting of Teepol Tanks

Tank Parts		Paint Systems
Bottom plates		7B
Shell plates	Bottom course	As for bottom plates
	Rest of shell	As for bottom plates
Roofs	Sheets/trusses	As for bottom plates

Table 2.20   Internal Painting: Industrial Water Tanks

Tank Parts		Paint Systems
Bottom plates		10A or 10B or 7K
Shell plates	Bottom course	As for bottom plates
	Rest of shell	As for bottom plates
Roofs	Sheets/trusses	10A 10B

Table 2.21   Internal Painting: Ballast Water, Slop, and Demineralized Water Tanks (at Temperatures Below 60°C)

Tank Parts		Paint Systems
Bottom plates		7K
Shell plates	Bottom course	As for bottom plates
	Rest of shell	As for bottom plates
Roofs	Sheets/trusses	10A or 10B or 7K

Table 2.22   Internal Painting: Drinking Water Tanks

Tank Parts		Paint Systems
Bottom plates		71 (8C for small tanks)
Shell plates	Bottom course	As for bottom plates
	Rest of shell	As for bottom plates
Roofs	Sheets/trusses	As for bottom plates

Table 2.23   Internal Painting: Refrigerated Storage Tanks

Tank Parts	Paint Systems
Outer shell plate Roof plates Suspended deck Steel lining shell and bottom (both sides) Tank internal Pump shafts Inner tank shell and bottoms both sides; fine Grained C. MN steel	4C
9 % N1 steel	No painting

2.27 TYPICAL REFINERY PAINTING SYSTEMS SCHEDULE FOR STEEL SURFACES RECOMMENDED SCHEME

Table 2.24   Typical Refinery Painting Systems Schedule for

Surface	Paint System	Comments
Buried piping	System 10 B, 1 glass and 1 felt wrap and finish coat of whitewash	N/A
Buried structures other than piping	7L	N/A
Docks (above water) and mooring buoys	7C	If black coating is acceptable, the alternative has a lower cost and is just as effective
Docks (below water) and sheet piling, before driving 7L In addition to the coating, cathodic coating should be considered for maximum protection	7L	In addition to the coating, cathodic coating should be considered for maximum protection
Fence fabric, chain link	8C	N/A
(Uninsulated) exchangers, vessels, heaters, stacks, aboveground piping, and other surfaces to 93°C in mild and moderate industrial environments	5C	N/A
In seacoast and severe industrial environments	5D	(1) With some items, galvanizing may be considered instead of coating (2) Compatibility of topcoat with a zinc-rich primer must be determined before application
93–260°C	5D	(1) These are not perfect materials, but are probably the best available recommendations for hot surfaces
204–371°C	5D	(2) Zinc-rich inorganics do a good job of corrosion protection, but where color is important, some coating suppliers may be able to furnish a suitable silicone topcoat

Table 2.24 Typical Refinery Painting Systems Schedule for (cont.)

Surface	Paint System	Comments
Insulated surfaces under insulation; if metal surface operating temperature will be under 180°C	7K	N/A
Prepainted items (compressors, pumps, motors, etc.)	None normally required	Number of coats depends on the amount of hiding required
Structural steel in mild and moderate industrial environments	8A	N/A
In seacoast and severe industrial environments	8D	If galvanized or shop-coated with zinc-rich inorganic material, clean and blast welds and touching up with compatible zinc-rich coating after erection in field
Walkways, handrails, ladders, line supports, nuts, bolts, and miscellaneous hardware	8D	It is recommended that micarta blocks (or similarly effective materials) be cemented and sealed under pipelines where they rest on supports
Brine or wastewater,	7L	N/A
clean water, or condensate	7D	If potable water, coating should be nontoxic
Crude bottom and up to 45 cm on shell	7L	N/A
Crude, lower, and middle shell plates, and roof	No coating	N/A
Crude, top ring (floating roof only)	7B	Use alternative only if crude, not sour
Tanks, exterior above grade, shell, and cone roof in mild and moderate industrial environments	1D	For light products, a chalking white finish is preferred to minimize evaporation losses
In seacoast and severe industrial environments	7F	Where pickup of dirt in the atmosphere is a problem on light-colored finishes, overcoating with a soil-retardant solution should be considered
Floating roof	5D	For alternative, if post-cured zinc-rich inorganic material is used, make certain that the curing agent is removed before topcoating
Below-grade shell	7L	In addition to coating, cathodic protection should be considered for maximum protection
Bottom	Apply cathodic protection	Set tanks on sand, pulverized limestone, or concrete pad slightly above grade where possible

Table 2.25   Typical Refinery Painting Systems Schedule for Steel Surfaces Other Than the Recommended Scheme

Surface	Surface Preparation	Paint System	Finish	Comments
Aluminum	None normally required			Do not use lead base primer on aluminum
Concrete interior walls	Clean	One or two coats of white latex block filler (until voids are filled)	One or two coats of two-package polyester, or catalystcured epoxy (an alkyd, vinyl latex, or acrylic latex may be substituted where washability is not important	Number of coats depends on hiding required. For previously painted walls, check with the block filler manufacturer on possibility of adhesion problems
Concrete exterior walls	Clean	One or two coats of white latex block filler (until voids are filled)	One or two coats of two-package polyester (and alkyd, exterior vinyl latex, or exterior acrylic latex may be substituted).	
Concrete floors	Muriatic acid etch, wash with detergent, and allow to dry	One coat of catalyst-cured urethane	Two coats of catalyst-cured urethane	(1) Preferred system cannot be applied over conventional alkyd or oleoresinous paints. (2) Consideration should also be given to tile, other flooring materials, and tinted concrete
Tanks, brine, and wastewater	Clean and dry	Coal tarepoxy (400 μm)*	N/A	Sandblasting or acid-etching may be required
Copper-	None normally required			N/A
galvanized,	No coating normally required for several years			N/A
plastered walls, offices, halls, etc.	Clean	Two or three coats latex (acrylic or vinyl)	N/A	N/A

Table 2.25   Typical Refinery Painting Systems Schedule for Steel Surfaces Other Than the Recommended Scheme *(cont.)*

Surface	Surface Preparation	Paint System	Finish	Comments
Washrooms, etc.	Clean	One coat emulsion-type primer-sealer	Two coats, two-package polyester or catalyst-cured epoxy	(1) With catalyst-cured epoxy, wall sealer may not be required depending on the manufacturer (2) Topcoat should be mildew-resistant in humid areas
Wood (general)	Clean and dry	One coat of wood primer	Two coats of alkyd or oleoresinous	Topcoats should be mildew-resistant for humid areas
Wood (outside walls)	Clean and dry	One coat of wood primer for latex	Two coats of latex house paint	N/A
Wood (floors)	Sanded, clean and dry	Three coats ure-thane floor varnish		N/A
Insulation coverings (canvas)	Clean	Two coats of emulsion-type, fire-retardant insulation sealer		N/A
Bitumen mastics	N/A			N/A
Urethane foam	In accordance with foam manufacturer's recommendations			Elastomeric coating most useful

Note: *Film thicknesses shown are to be measured dry. The color of the finish coat should be selected as per the paint color schedule. All coatings are to be applied strictly in accordance with the manufacturer's recommendations.*
*To be applied in the number of coats required to achieve this film thickness, but in no case should this be fewer than two coats.

Table 2.26   Typical Painting Systems for Freshwater Marine Vessels

Surface to be Painted	Typical Paint Systems
Towboat and barge hull exteriors	3C, 6G, 7E, 7L
Towboat and barge decks and covers	3B, 6F, 7E
Towboat superstructures and interiors	3B, 6E, 7C, 7E
Coal and acid-carrying barge decks and hoppers	3C, 6G, 7D
Barge rake interiors	1D
Barge innerbottoms and wings	1D

Table 2.27   Typical Ship-bottom Painting System

Number	Paint System	Anti-Fouling	Repainting Procedure	Application Equipment
1	10A or 10B	Rosin base, cuprous oxide toxic 75 μm MDFT	Freshwater spot-blast or power tool clean bad areas clean bad areas	Spray recommended; may be rolled
2	3C	Vinyl-rosin base 100 microns. MDFT toxic usually cuprous oxide	Freshwater wash, spot-blast bad areas	Spray recommended; small areas can be rolled or brushed
3	7B	Vinyl anti-fouling 100 microns MDFT	Freshwater wash. Spot-blast bad areas, step back antifouling in way of repair	Epoxy airless spray recommended; anti-fouling spray small areas can be rolled
4	6E	Chlorinated rubber anti-fouling, 100 Microns MDFT	Freshwater wash, spot-blast bad areas	Airless spray recommended; small areas can be rolled
5	7L	Vinyl anti-fouling, 100 microns. MDFT	Freshwater wash. Spot-blast bad areas, step back anti-fouling in way or repair	Airless spray recommended for epoxy spray anti-fouling small areas can be rolled

Table 2.28   Typical Boot-top and Topside Painting System

Number	Paint System	Repainting Procedure	Application Equipment
1	7E or 6B	Wash and remove contaminants—abrasive blast or power tool clean damaged or failed areas. Touch up using same system as applied during construction	Airless spray preferred-air spray can be used also roller and brush for small areas
2	7C or 7F	Wash and remove contaminants—abrasive blast or power tool clean damaged or failed areas. Touch up using same system as applied during construction	Airless spray preferred-air spray can be used also roller and brush for small areas
3	3C	Wash and remove contaminants—abrasive blast or power tool clean damaged or failed areas. Touch up using same system as applied during construction	Airless spray preferred-air spray can be used also roller and brush for small areas

APPLICATION METHODS OF PAINT AND PREPARATION FOR USE

3

This chapter provides the requirements for the initial construction and maintenance painting of surfaces, the painting of ferrous metals and nonferrous metals which will be exposed to different corrosive environments, and the painting of nonmetallic surfaces (e.g., plaster concrete, brick, stone, block, and wood). It is focused on corrosion protection of structures in oil, gas, and petrochemical industries, mainly refineries, chemical and petrochemical plants, gas plants, aboveground facilities of gas transmission and distribution systems, marine and offshore facilities and ships, buildings, and, where applicable, in exploration, production, and new ventures.

The chapter focuses on the liquid applied (usually by brushing or spraying) for paint and coating materials, which are commonly used for corrosion protection in atmospheric or immersion services. Definitions and test methods for quality control are also specified.

3.1 INTRODUCTION

All surfaces should receive an appropriate paint system, with the following exceptions:

- Any equipment furnished completely painted by the manufacturer, unless it is specially required to match a color scheme or to repair damage to the paint film
- Hot-dip galvanized steel, weathering steel, stainless steel and nonferrous metals, monel, brass, copper, and aluminum jacketing, unless it is specially required by designer
- Nonmetallic surfaces
- Nameplates, code stampings, and push buttons
- Surfaces to be fireproofed
- Concrete, brickwork, tile, glass and plastics, unless specially required
- Machined surfaces
- Insulation, weatherproofing material, or sheeting
- Rubber, hoses, belts, flexible braided connectors, stainless steel tubing and fittings, gauges, valves, and motor shafts
- Any surface that is indicated as not to be painted

Paint systems are generally specified by the dry film thickness (DFT) of coats and the total DFT of primer, intermediate, and top coat, rather than by the number of coats. Unless otherwise specified by the contract, the total DFT of the paint system should be at least 100 μm. All paints and paint materials used should comply with the specification given in relevant standards. Material standards should be obtained only from approved manufacturers. All materials should be supplied in the manufacturer's original containers, durably and legibly marked according to relevant standards.

Paint life depends primarily on surface preparation, which should be done in accordance with Chapter 1 of this book. Fabrication should preferably be complete before surface preparation begins.

All painting should be carried out in full conformity with this standard. Particular attention should be paid to instructions on storage, mixing, thinning, pot life, application conditions, application technique, and recommended time intervals between coats. Coatings should not be applied to wet or damp surfaces.

No paint should be used when it cannot readily be reincorporated by correct mixing. Similarly, no paint should be used that has jellified or thickened to such an extent that too much thinner is required to achieve brushing consistency.

Except in special cases, paint should not be applied under the following conditions:

- When the temperature of the surface is less than 3°C above the dew point of the surrounding air, or the relative humidity is higher than 80% (see Table 3.1 for dew point determination).
- When the temperature is below 4°C.
- When the surface temperature is higher than 35°C.
- When there is the likelihood of an unfavorable change in weather conditions within 2 h after coating.
- When there is a deposition of moisture in the form of rain, condensation, frost, and other moisture on the surface. This is likely to occur when the relative humidity is over 80% and the temperature is below 15°C.

Table 3.1 Dew Point Determination

Dew Points (°C) at various relative humidities

Air temp.	30%	40%	50%	60%	70%	80%	90%	100%
−1	—	—	—	—	−6.5	−4	−2	−1
4	—	−6.5	−4	−2	0.5	1.5	3.5	4.5
10	−6.5	−3.5	0.5	2	3.5	5.5	8.5	10
15.5	0	2	4	8	10	11.5	14	15.5
21	3	6.5	10	13	15	18	19.5	21
26.6	7	12	15.5	19	21	23.5	25	26.5
32	13	16.5	20.5	24	25.5	28.5	30.5	32
38	18	22	25.5	29	31	33.5	36	38

Note: *It is essential to ensure that no condensation occurs on blasted steel or between coats during painting.*
Air at a given temperature can only contain a certain maximum amount of water vapor. This proportion is lower at lower temperatures.
The dew point is the temperature of a given air-water vapor mixture at which condensation starts, since at that temperature, its maximum water content (saturation) is reached.
In practice, a safety margin must be kept, whereby the substrate temperature is at least 3°C above dew point.

Each layer of paint should be allowed to dry for a period of time specified by the paint manufacturer before the next layer is applied. Subsequent layers of a paint system should have a difference in tint.

Particular attention should be paid to the painting of corners, edges, welds, and other surfaces, especially with respect to the specified minimum dry-film thickness. During both application and drying, adequate ventilation should be provided if the work area is enclosed. All steel constructions or plates should be provided with a priming or coating system to protect the steel surfaces during the transport, storage, construction, and joining stages (e.g., welding) of the project. All surfaces that will be inaccessible after assembly should be fully painted before assembly.

It should be the contractor's responsibility to coordinate work so that items are primed and painted with compatible coating, as specified in the contract, the standards, or both. Coatings should be applied by conventional or airless spray in exact accordance with this standard or the manufacturer's instructions.

Two-pack paints should be carefully mixed in strict accordance with standards, manufacturer's instructions, or both. The pot life of such paints should be carefully noted, and any mixed paint that has exceeded its pot life must be discarded, regardless of its apparent condition.

The application should leave no sags, runs, marks, or other defects. The drying and application time between coats should adhere to the coating manufacturer's recommendations, with temperature and humidity conditions taken into account, and they should generally be kept to the minimum amounts to prevent contamination between coats. If contamination occurs between coats, this must be completely removed, generally by washing per the paint manufacturer's recommendation or with suitable detergent and then rinsing with clean water. The paint surface should be dry before application of overcoating.

The greatest precautions should be taken with the spraying of inorganic zinc primers to ensure proper cohesion and adhesion, with care taken not to exceed the maximum film thickness. Any and all holes and surface imperfections should be cleaned and filled in an approved manner before painting.

The number of coats should be the minimum needed to achieve the specified film thickness. The DFT should not exceed the maximum specified by the company.

All equipment should be maintained in good working order. Equipment should be thoroughly cleaned daily. Worn parts should be replaced. Effective oil and water separators should be used and serviced regularly.

All damage to paintwork incurred at any stage of the work, including site welding operations, should be reprepared to the original standard and recoated with the specified priming and finish coats to restore the film thickness. In all such instances, preparation should extend 25 mm into the sound paintwork, and a further 25 mm of sound paintwork should be lightly blasted to etch the surface. Repainting should then cover the prepared surface and the etched paintwork.

When painting insulating flanges with paints containing metallic pigments, insulating materials will be covered with protective stripping to prevent breaking or electrically shorting the insulating barrier. Unless otherwise specified, the minimum allowable time before application of intermediate or finish coats should be 3 h. The maximum allowable time between application of intermediate and finish coats should be as recommended by the paint manufacturer, but often not less than 8 h.

Any surfaces to be coated should be rendered dust-free prior to the application of the prime coat. This should be accomplished by blowing the surface with clean, dry air or by using an industrial vacuum cleaner.

3.2 PAINT MATERIALS

3.2.1 SELECTION OF PAINT MATERIALS

Before painting is permitted, the paint system and the application process should be defined and qualified in accordance with the painting procedure approval specified in the appropriate section. The paint materials supplied should be certified by the manufacturer in accordance with the relevant standards.

3.2.2 IDENTIFICATION OF PAINT MATERIALS

The contractor should ensure that all materials supplied for painting operations are clearly marked according to the applicable standards for paints.

3.2.3 STORAGE OF PAINT MATERIALS

All paint materials consigned to the coating site should be properly stored in accordance with the manufacturer's instructions at all times to prevent damage and deterioration prior to use. Materials should be used in the order in which they are delivered.

3.3 PAINT APPLICATION COLOR CODING

The manufacturer and product color codes identify the shade of color to be applied. Unless otherwise specified on drawings, all equipment should be color-coded using the color code. All painted color-coding bands should be 50 mm wide.

Painted color-coding bands will be placed on both sides of every valve and flange along the line where space permits. For small facilities where line space is limited between valves and flanges, the lines should be adequately marked with painted color-coded bands so that the bands should be visible from each valve and flange (see Figures 3.1, 3.2, 3.3, and 3.4).

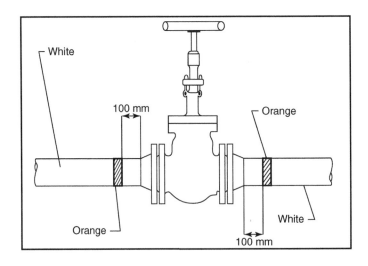

FIGURE 3.1

Painted color-coding bands for valves and flanges in sour hydrocarbon liquid pipeline.

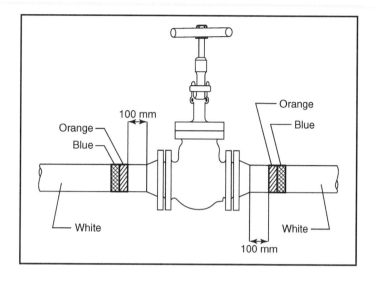

FIGURE 3.2

Painted color-coding bands for valves and flanges in sour produced water pipeline.

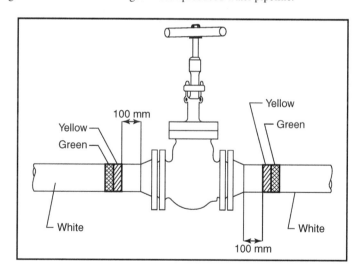

FIGURE 3.3

Painted color-coding bands for valves and flanges in sweet hydrocarbon liquid pipelines,

3.4 COMPONENTS OR WORK PIECES
3.4.1 COMPONENT IDENTIFICATION

All identification markings, whether internal or external to the component, should be carefully recorded before surface preparations begin. When applicable, identification plates should be carefully removed and, after the painting has been accepted, replaced using an adhesive compatible with the paint. Special care should be taken to ensure that the original data is reaffixed to the correct component.

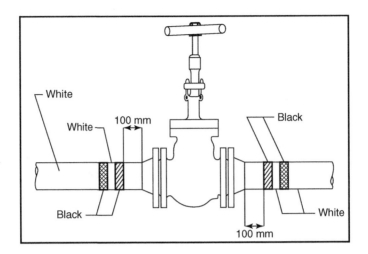

FIGURE 3.4

Painted color-coding bands for valves and flanges in liquid sulfur pipeline.

3.4.2 PROTECTION OF WELD AND PREPARATIONS

Weld end preparations should be protected from mechanical damage during handling, storage, surface preparation, and the coating processes. The methods used should also ensure that no damage occurs to the internal surface of the component.

Weld end preparations should be protected from paint during the paint application process by a method based on standards. The paint should not be applied within 150 mm ±15 mm of the ends of the component before welding.

3.4.3 SURFACE PREPARATION

Unless otherwise specified, the method of surface preparation should be specified by the contractor as part of the painting procedure qualifications and should take into account the requirements specified in Chapter 1, plus any special requirements specified by the company. All surfaces should be inspected immediately after preparation for compliance with the requirements specified in Chapter 1. Any components or parts found to have defects that exceed the levels permitted in the relevant component or part specification should be set aside for examination.

3.4.4 PAINTING PROCEDURE TESTS

The painting application process should comply with the procedure established in the relevant coating procedure qualifications. Any changes in coating materials, component dimensions, or the coating process may necessitate a new coating procedure test at the discretion of the company. Additionally, approved procedure tests should be confirmed at intervals of not more than 1 year for each type of paint used by the contractor and for each size of component as stipulated by the company (see Table 3.2).

Table 3.2 Minimum Quality Control Requirements
Requirements
a) Check cleanliness of components immediately prior to cleaning
b) Monitor size, shape, and cleanliness of the blast-cleaning material and process
c) Check visually, in good light, the surface of the components for metal defects, dust, and entrapped grit
d) Check component surface blast profile
e) Check for residual contamination of component surfaces
f) Check temperature control of the component surface by an agreed method
g) Check the weather conditions
h) Check the paint thickness
i) Check the cure of paint
j) Check the paint adhesion
k) Supervision to ensure the adequate and proper repair of all defects
l) Check on paint color and appearance; e.g., uniformity and flow
m) Check for damage to weld and preparations

3.5 INSPECTION AND TESTING

The quality control system should include as a minimum the requirements listed in Table 3.2. If, in the opinion of the company after examination or test, any component has not been cleaned or painted in accordance with this construction standard, the contractor should be required to remove the paints that are considered defective or inadequate and to reclean and recoat the component to the satisfaction of the company.

3.6 QUALITY SYSTEMS

It is necessary to set up and maintain such quality assurance and inspection systems as required to ensure that the goods or services supplied comply in all respects with the requirements of this chapter. The company will assess such systems against the recommendations of the applicable parts of ISO 9004 and should have the right to undertake such surveys as necessary to ensure that the quality assurance and inspection systems are satisfactory. Also, the company should have the right to undertake the inspection or testing of the goods or services during any stage of work at which the quality of the finished goods may be affected and to undertake inspection or testing or raw material or purchased components

3.7 PROCEDURE QUALIFICATIONS

Before bulk painting, component requirements to be met and a detailed sequence of operations to follow on the painting of components should be submitted for checking compliance with standards.

3.7.1 PAINTING PROCEDURE SPECIFICATIONS

The painting procedure specifications should incorporate the following items, among others:

- The paint systems to be used with the appropriate data sheets
- Cleaning of the component and method of cleaning

- Cleaning medium and technique
- Blast-cleaning finish, surface profile, and type of abrasive and surface cleaning in the case of blast cleaning
- Dust removal
- Painting method (brushing or spraying)
- Preheat time and temperature, if any
- Powder spray, if any, including the use of recycled material
- Curing and quenching time and temperature
- Post-curing time and temperature
- Surface treatment medium and technique
- Repair technique
- Painting stripping technique

3.7.2 COATING PROCEDURE APPROVAL TESTS

A batch of 15–20 components (representing a normal production run) or, in the case of painting structure, three to five different sections with the minimum of 2 m² for each section should be painted in accordance with the approved painting procedure. No painted component should be dispatched or no painting performed until the procedure has been certified in writing as acceptable.

3.8 METAL SURFACE TREATMENT (CONVERSION COATINGS)

Metal surface treatment or prepaint treatment required by the job will be applied by chemical means, whereby the chemically active metal surface, subject to oxidation, will be converted to one that is less active and more resistant to corrosion. Conversion coating serves as a substrate for the subsequent bonding of metal to paint or other organic finishes. The corrosion resistance of the painted article can be measured by humidity (ASTM D-822), salt-spray (ASTM B-117), and other standardized corrosion tests.

Metal treatment does not replace anti-corrosion primers, which should be applied as soon as the conversion painting is dry. Wash primer, phosphate coating, and chromate coating are the types of conversion coatings which are applied over ferrous and nonferrous metals.

Generally, wash primer uses for treatment of ferrous and nonferrous metals, phosphate treatment uses for steel and zinc-coated steel and chromate treatment is used only for nonferrous metallic surfaces.

3.9 PREPARATION OF PAINT BEFORE USE
3.9.1 STORAGE AND APPEARANCE

Paint should be stored in a well-ventilated room, free of excessive heat or direct sunlight and maintained at a temperature between 4°C and 27°C. Open-air storage should be avoided particularly of heavy paints, such as primers and undercoats.

The maximum storage time for paints should be as recommended by relevant standards. Paints should not be stored in open containers, even for a short time. All containers of paint should remain unopened until required for use. Those containers that have been previously opened should be used first. The label information should be legible and should be checked at the time of use.

The settlement of heavy paints such as red lead oxide primers, enamel undercoats, and wood priming paints should be lessened by rolling the drums in which they are stored every six weeks. Touring the drums on their ends is not allowed. The normal finishing paints and drum paints do not require rolling during the storage period.

If paint has thickened to such an extent that more than 5% by volume (10% by volume for priming paints) of the correct thinners is required to bring it to brushing consistency. Paint that has levered, gelled, or otherwise deteriorated during storage should not be used; however, thixotropic materials that can be stirred to attain normal consistency may be used. The oldest paint of each kind should be used first.

The paint temperature may be excessively high or low depending on storage or shipping conditions. If so, warm or cool the paint to a temperature of 10–32°C before mixing and use.

3.9.2 PREPARATION OF PAINT FOR APPLICATION

Paints should preferably be mixed by powered mixers and shakers. Only small quantities are suitable for hand mixing, and then should be mixed only by an efficient method such as boxing (i.e., the process of mixing paint by pouring from one container to another. The maximum container size should be 20 l). Avoid shaking partly full cans of latex paint, as it causes foaming.

The paint should be mixed in a manner that will ensure the breakup of all lumps, complete dispersion of pigment, and a uniform composition. The lumpy or stiff paste should be broken up with a mechanical agitator, or in some cases, with a wide strong paddle made of wood or iron.

If a skin has formed in the container, the skin should be cut loose from the sides of the container, removed, and discarded. If the volume of the skins is more than 2% of the remaining paint, the paint should not be used.

Mixing in open containers should be done in a well-ventilated area, away from sparks or flames. Paint should not be mixed or kept in suspension by means of an airstream bubbling under the paint surface.

Dry pigments that are separately packaged should be mixed into paints in such a manner that they are uniformly blended and all particles of the dry powder are wetted by the vehicle. Pastes should be made into paint in such a manner that they should be uniformly blended and all lumps and particles broken up to form a homogeneous paint.

Tinting pastes or colors should be wetted with a small amount of thinner, vehicle, or paint and thoroughly mixed. Next, the thinned mixture should be strained. Finally, it should be added to the large container of paint and mixed until the color is uniform.

Paint that does not have a limited pot life (time interval) or does not deteriorate on standing may be mixed at any time before using, but if settling has occurred, it must be remixed immediately before using.

Paint should not remain in spray pots, painter's buckets, or other containers overnight. Rather, it should be stored in a covered container and remixed before use.

Catalysts, curing agents, or hardeners that are separately packaged should be added to the base paint only after the latter has been thoroughly mixed. The proper volume of the catalyst then should be poured slowly into the required volume of base with constant agitation. Do not pour off the liquid that has separated from the pigment and then add the catalyst to the settled pigment to aid mixing. The mixture should be used within the pot life specified by the manufacturer. (For example, the pot life limit for some chemically cured paints is more than 20 min and less than 8 h after mixing.) Therefore, only enough paint should be catalyzed for prompt use. Most mixed, catalyzed paints cannot be stored, so unused portions of these should be discarded at the end of each working day at the contractor's expense. When specified, special continuous mixing equipment should be used according to the manufacturer's directions.

Drum paints should be rolled on their side for a few minutes before the container is opened. The entire paint contents should be poured into an empty clean drum or can, ensuring that no heavy paste remains in the original container. If paste remains, some of the paint should be poured back and the mixture again stirred thoroughly.

All pigmented paint should be strained after mixing, except when application equipment is provided with strainers. Strainers should be a type that removes only skins and undesirable matter, not the pigment. Cheesecloth of a fine metal gauze, approximately 0.15 mm (80 mesh), is recommended as the strainer.

When mixing two-component paints, check and remix each component individually. Then blend the two components at a low speed until the mixture is completely uniform in color. Often, the two components are supplied in different colors so that a good mix can be readily determined. Do not mix more than a few liters at a time since the exotherm caused by the mixture may be at such a high level as to make the paint solidify in the container. Hand-mixing of paints should be permitted only for containers up to 5 l. All larger containers should be mixed by mechanical agitators and brought to a uniform consistency. If pigment separation can readily occur, such as with heavy or metallic pigments, there should be continuous mixing during application to prevent this.

Do not thin the paint unless recommended by the supplier or needed for spray application or in case of paint thickening. If thickening of paint prevents proper application by brush, not more than 5% by volume of the correct thinner may be added; for oil-based primers containing red lead, up to 10% by volume is acceptable. For enamels that are applied by spraying, special enamel thinners should be used. For drum paints and priming paints, white spirit (mineral turpentine) should be used as a thinner. Emulsion paints normally require thinning up to 12% by volume of clean fresh water. The addition of more water than is necessary to obtain a satisfactory brushing consistency is not allowed. When thinning the paint, be sure that it is well mixed before adding the thinner. Then thinner should be added slowly to the paint during the mixing process.

If the paint is cold, do not add thinner to make application easier. Instead, bring the paint to a temperature of 10–32°C. Paint heaters can be used to reduce viscosity for spray application, thus avoiding the addition of thinners. Do not apply warm paint to cold steel. The results will be best if both the paint and the surface are similar in temperature.

3.10 APPLICATION METHODS OF PAINT

The accepted methods of applying coating on site are brushing and spraying; applying paint by paint pad and paint glove is also permissible in same cases. The choice of method is usually determined by the nature of the work, the type of material to be applied, or both. The manufacturer's recommendations regarding the suitability of coatings for application by particular methods should also be considered.

Whatever method is employed, operators should be skilled and experienced in the techniques of application, as well as in the care and maintenance of tools and equipment and, where relevant, in the setup and adjustment of equipment to obtain optimum results.

This section describes the general characteristics of the methods referred to above and the types of tools and equipment employed. On the assumption that operators will be skilled and experienced, techniques of application are not described beyond the basic principles. With all methods of application, the aim is to produce a uniform coating of the film thickness specified, free of pinholes, missed area, runs, sags or curtains, and wrinkling or other blemishes that may negatively affect durability.

3.10.1 BRUSH APPLICATION

Brush application may be used under the following circumstances:

- When an area cannot be properly coated by spray application for any reason
- For repairs to localized, damaged paint
- When the manufacturer considers the coating material suitable for brush application
- For applying the initial coat of paint (primer) to corners, crevices, or other irregular surfaces prior to spray application

The brush should not be dipped more than one-third of the bristle length into the paint to avoid overloading the bristles and filling the heel with paint. The brush should be held at an angle of about 75° to the work.

The paint should be spread to hide the surface and provide a uniform coating. Work from dry to wet surface. Excessive pressure should not be applied to the brush.

When the surface has been completely covered with paint, the wet area should be brushed crosswise to ensure uniformity and finally brushed lightly to smooth out brush marks and laps. On large areas, this final light brushing should be in a vertical direction. The film produced by brushing should be free of brush marks, consistent with recommended practice for the type of paint used.

Brushes should be made from a high-quality hog bristle for solvent-based paints and synthetic filament (nylon and polyester) for water-thinned paints and caustic material such as cement paint. Avoid brushes with filament that are not flagged. Round or oval brushes are suitable for irregular surfaces and wide, flat brushes are suitable for large flat areas. The construction and dimensions of brushes should comply with requirements of BS 2992.

3.10.2 SPRAY APPLICATION

Spray application is preferred, especially for application to a large area. There are several types of spray equipment: air, airless, hot, and electrostatic spray. The spray equipment should be kept sufficiently clean so that dirt, dried paint, and other foreign materials are not deposited on the paint film. Any solvents left in the equipment should be completely removed before using. Paint should be applied in a uniform layer, with overlapping at the edges of the spray pattern. During application, the gun should be held perpendicular to the surface and at a distance that will ensure that a wet layer of paint is deposited on the surface. The trigger of the gun should be released at the end of each stroke.

All runs and sags should be brushed out immediately, or else the coating should be removed and the surface repainted. Before spraying each coat, all corners, edges, welds, cracks, crevices, blind areas of all rivets and bolts, nuts, and all other inaccessible areas should be prepainted by brush, dauber, or sheepskin to ensure that these areas have at least the minimum specified dry-film thickness.

Paint should be suitable for the particular spray application method used. Particular care should be observed with respect to the type of thinner, amount of thinner, paint temperature, and operating techniques in order to avoid deposition of paint that is too viscous, dry, or thin. In some cases, the paint may have to be reformulated to suit the application method.

Caution must be exercised so that hot coatings are not applied to cold surfaces and vice versa. Possible other limitations to spray application include the following:

- Possible hazards to health or safety should be recognized and avoided.
- Spray application on exterior work in windy weather could cause difficulties.
- The conventional primers for building surfaces should not be applied by spraying, except for zinc-rich epoxy primers.

3.10.3 AIR SPRAYING

The original method of spray application was by air atomization. In this process, a compressor supplies air under pressure via an air hose to a spray gun, which atomizes the paint to produce a fine spray that is projected onto the surfaces.

According to the design of the equipment, the working air pressure should be between 2 and 4.7 bars. The air caps, nozzles, and needles should be those recommended by the manufacturers of the material being sprayed and the equipment being used.

Traps or separators should be provided to remove oil and condensed water from the air. The traps or separators must be of adequate size and drained periodically during operations. The air from the spray gun impinging against a clean surface should show no condensed water or oil.

The pressure on the material in the pot and of the air at the gun should be adjusted for optimum spraying effectiveness. It should be adjusted when necessary for changes in elevation of the gun above the pot. The atomizing air pressure at the gun should be high enough to properly atomize the paint, but not so high as to cause excessive fogging of paint, excessive evaporation of solvent, or loss by over-spraying.

If compressor units for air spray equipment are powered by a gas or diesel engine, they should be located outside the building. If this is not practical, exhaust fumes should be conveyed directly to the open air.

For normal use, gravity-fed or suction-fed guns fitted with a paint cup of required size should be used. If the work requires the continuous use of more than 0.5 l of paint, the installation of the pressure pot having a capacity up to 4.5 l may be provided.

This pot should have a filter and water trap inserted into the line. Where several guns are being employed, the air line couplings should be interchangeable.

Spraying should be done in such a manner as to produce a full, uniform film of paint without runs, rays, uneven surfaces, pin holes, to the standards of the company.

At the end of each day or upon completion of the job, the gun should be cleaned thoroughly by spraying the thinner through it and forcing it into a container by holding cloth over the air cap. The air cap should be removed and the fluid tip washed with thinner. The cap should be immersed in thinner and the holes cleaned with a matchstick, not a nail or wire. The spray gun should never be immersed in the thinner. The gun should be lubricated at regular intervals.

3.10.4 **LOW-VOLUME, LOW-PRESSURE (LVLP) SPRAY**

Low-volume, low-pressure (LVLP) spray guns (Figure 3.5) operate at a lower pressure, but they use a low volume of air compared to conventional and high-volume, low-pressure (HVLP) equipment. This is a further effort at increasing the transfer efficiency (amount of coating that ends up on the target surface) of spray guns while decreasing the amount of compressed air consumption. Figure 3.6 shows different types of nozzles and sprays, and Figure 3.7 show details of the construction of an air spray gun.

FIGURE 3.5

LVLP spray.

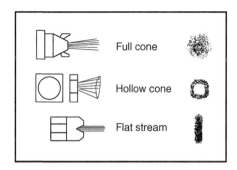

FIGURE 3.6

Types of nozzles and sprays.

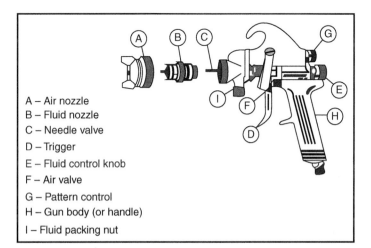

A – Air nozzle
B – Fluid nozzle
C – Needle valve
D – Trigger
E – Fluid control knob
F – Air valve
G – Pattern control
H – Gun body (or handle)
I – Fluid packing nut

FIGURE 3.7

Construction of an air spray gun.

3.10.5 AIRLESS SPRAY APPLICATION

Airless spray relies on hydraulic pressure rather than air atomization to produce the desired spray. An air compressor electric motor or gas engine is used to operate a pump to produce pressures of 71–430 bars. Paint is delivered to the spray gun at this pressure through a single hose. Within the gun, a single paint stream is divided into separate streams that are forced through a very small orifice, resulting in atomization of paint without the use of air. This results in more rapid coverage with less overspray (Figure 3.8).

Caps with capacities ranging from approximately 0.25 l/min to 5 l/min are available, and care should be taken to select the correct cap for the particular application. For best results, specially formulated paints are necessary.

Heavier coatings are usually used, and because of the lower degree of control given by the airless spray gun and the high paint flow rate, greater resistance to sagging and tearing is required from the paint. For good results, the gun is held at right angles to the work and about 300 mm away, and the operator should start at the bottom and work upward. The speed of the operating strokes should be much faster than for normal spraying. The trigger movement must be abrupt and the spray started and stopped just after starting the work and just before completion. Successive passes of the gun should overlap only slightly since the spray pattern is of uniform thickness throughout its width.

Paint must never be allowed to dry in the gun. Cleaning instructions must be strictly followed.

Airless spray usually is faster, cleaner, more economical, and easier to use than conventional air spray and applies thicker film. Most types of coatings can be applied, with the possible exceptions of those containing coarse aggregates or fibers, such as some masonry paints (e.g., cement paints).

Because of the very high pressures involved, caution should be exercised in the handling of airless spray equipment. In particular, the spray gun should never be pointed toward any part of the body while the equipment is in operation.

Fluid tips should be of the proper orifice size and fan angle, and the fluid control gun construction should be as recommended by the manufacturer of the material being sprayed and the equipment being used. The fluid tips should be of the safety type, with shields to prevent penetration of the skin by the high-pressure stream of paint.

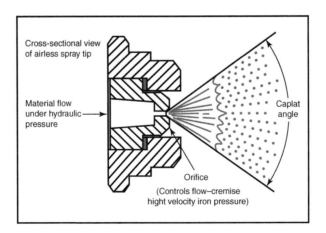

FIGURE 3.8

Airless spray action.

The air pressure to the paint pump should be adjusted so that the paint pressure to the gun gives optimum spraying effectiveness. This pressure should be sufficiently high to properly atomize the paint, but not considerably higher than those necessary to properly atomize the paint.

Spraying equipment should be kept clean and utilize proper filters in the high-pressure line so that dirt, dry paint, and other foreign materials are not deposited on the paint film. Any solvents left in the equipment should be completely removed before paint is applied. Because of very high-pressure, paint must be thoroughly screened to prevent clogging of the nozzles.

The trigger of the gun should be held fully open during all spraying to ensure the proper application of paint. Airless paint spray equipment should always be provided with an electric ground wire in the high-pressure line between the gun and the pumping equipment. Further, the pumping equipment should be suitably grounded to avoid the buildup of any electrostatic charge on the gun. The manufacturer's instructions on the proper use of the equipment are to be followed.

3.10.6 **HOT SPRAY**

In this method, the material is passed through a heater unit before delivery to the spray gun. Heating enables the solvent content of the material to be reduced, so allowing a thicker film to be applied. Coatings require to be specially formulated for hot spray application; water-thinned coatings are not suitable. The hot-spray method can be used in conjunction with either air spray or airless spray equipment, but is rarely used for site application.

Avoid applying heated paint to cold steel. Heated spray units have a number of advantages over unheated units:

- Application is faster, and it's also dry
- Lower pressure (under 69 bars or 1,000 psi) and less power and air required
- Increased thickness per coat if solids are increased

3.10.7 **ELECTROSTATIC SPRAY**

Electrostatic spraying should not be used for on-site work. This method is ideal for painting wire fence, angles, channels, cables, and pipes. It involves imparting to the material an electrical charge (up to 60,000 volts) of opposite potential to the surface to be painted, causing the charged particles to be attracted to the surface. This reduces wastage of material and also creates a wraparound effect so that components such as pipes and railing may be coated all around from one direction.

The electrostatic spray technique uses expensive equipment and has a slower operation than other spray applications. In addition, it is unsuitable for large steel structures.

3.10.8 **PAINT PADS AND GLOVES**

Paint pads and gloves should be used only for coating pipes, railings, and surfaces that are inaccessible to other methods. It is impossible to obtain a high standard of finish by this method.

3.11 **APPLICATION PROCEDURES FOR GENERIC GROUPS OF PAINT**

The materials covered in this section are to be applied as specified. In case of conflict with any other part of this standard, these special provisions should prevail. Minimum DFTs are indicated, but thicker coatings should be applied when specified by the company, recommended by the manufacturer's instructions, or both. Materials that are not specifically covered in this standard should be applied in accordance with the directions of the manufacturer.

3.12 VINYL AND CHLORINATED RUBBER PAINT

Vinyl and chlorinated rubber finish paint should be applied by spraying, with application by brush limited to small areas and touch-ups. Primers may be brushed or sprayed. These paints should be thinned as recommended by the manufacturer. They should be applied with a coverage that will result in a DFT according to the standards. When vinyl or chlorinated rubber paints are applied by brush, paints should be applied to the surface with a minimum of brushing so that there is little or no lifting or softening of the undercoats.

3.12.1 BITUMINOUS COATINGS

The term *bituminous paint (thin film)* refers to low-consistency solutions of coal tar or asphalt.

The term *cold-applied bituminous coating (medium film)* refers to high-consistency filled solutions of coal tar or asphalt. They should be applied by brushing or spraying. If spray applied, special, heavy-duty, pump-type spray equipment should be used. This material should be stirred without thinning until it attains the proper consistency for application. It should be applied at a coverage that will result in the DFT specified by the company, or if not specified, the DFT as recommended by the manufacturer.

The term *cold-applied bituminous coating (thick film)* refers to very high-consistency filled solutions of coal tar or asphalt. They should be applied by brushing or spraying. If spray is applied, special, heavy-duty, pump-type spray equipment should be used. These materials must be stirred without thinning until they attain the proper consistency for application. They should be applied with a coverage that will result in the DFT specified by the company or, if not specified, the DFT as recommended by the manufacturer. The expected range of DFT for the cold applied bituminous coating (thick film) is from 375–450 μm, per coat, and the necessary number of coats should be applied to provide a minimum DFT of 625 μm unless another thickness is specified.

The term *cold-applied bituminous mastic (extra-thick film)* refers to very thickly applied filled solutions of coal tar or asphalt applied by brushing, troweling, or spraying. If spray applied, special, heavy-duty, pump-type spray equipment should be used. Thinning should not be necessary, and it should not supplant adequate stirring. They should be applied with a coverage that will result in the DFT specified by the company, or if not specified, the DFT as recommended by the manufacturer. The expected range of DFT for the cold-applied bituminous mastic (extra-thick film) is about 1,000–1,700 μm per coat, and it is preferable that it be applied in two coats.

3.12.2 CHEMICALLY CURED COATINGS

Two-pack, chemically cured coatings (such as catalyzed epoxies and coal-tar epoxies) should be stored, mixed, thinned, applied, and cured in accordance with the manufacturer's instructions. Also, any special precautions by the manufacturer should be followed. For example, the time between coats for a coal-tar epoxy should not exceed that indicated by the manufacturer; otherwise, it may become necessary to roughen the previous coat to obtain proper intercoat adhesion. Chemically cured coatings should not be applied when the surface, paint, or air is below 13°C. Low temperatures greatly reduce the curing rate of chemically cured coatings.

A reaction time should be required after introducing and mixing the catalyst with the pigmented component. Allow a reaction time of 1 h unless otherwise specified by the manufacturer. Epoxy coatings may be thinned, but they should not be reduced by adding more than 0.50 l epoxy thinner to each 3.7 l of epoxy enamel.

Precautions should be taken to protect surfaces other than that being covered from splatter, drip, or overlap. Each coat should be flowed on (i.e., do not spread the paint out thin) in a deep wet film with a thickness of 150 µm. Epoxy coatings may be applied over base coats after a minimum cure time of 8 h.

3.13 ZINC-RICH PAINT
3.13.1 INORGANIC, ZINC-RICH PAINT

Inorganic, zinc-rich paint should be applied by airless spray. If the zinc powder is packaged separately, mix with the vehicle just before use. They should be applied at a coverage rate recommended by the standards. Unless otherwise specified by the standards, the DFT should not be less than 50 µm. Prior to top coating, a barrier or tie coat may be required for overcoating with certain generic coatings. The manufacturer's recommendations should be followed.

Complete curing of zinc-rich primer is necessary before topcoat is applied. Overspray of the zinc-rich primer will result in improper adhesion of the topcoat. Overspray should be removed with a stiff bristle brush or wire screen. Popping will be eliminated by scraping the painted surfaces with soft sandpaper before overcoating.

3.13.2 ORGANIC, ZINC-RICH PAINT

The abovementioned provisions may also apply to organic zinc-rich paints, except that they may be applied by brushing as well.

3.14 URETHANE COATINGS

Single-component (one-pack), moisture-cured urethane coatings that meet ASTM D-16, "Definition of Terms Relating to Paint, Varnish, Lacquer, and Related Products," type II urethane coating may be applied by brushing, conventional spraying, and airless spraying.

Special care should be taken to ensure that all spray equipment is moisture-free. Since these coatings cure by reacting with moisture in the air, it should be noted that application on days when the humidity is low will result in slow curing. The manufacturer's directions should be followed concerning thinning and application parameters. One-pack, moisture-cured urethane coatings should be mixed mechanically prior to application. This should be done slowly so as not to create a vortex and introduce moisture into the coating that could reduce the pot life. One-pack, moisture-cured urethane coatings are extremely susceptible to moisture contamination, so they should not be applied unless temperatures both during and up to 2 h after application will be at least 3°C above the dew point.

Two-component, polyisocyanate polyol-cured urethane coatings may be applied by brushing, conventional spraying, or airless spraying. Special care should be taken to ensure that all spray equipment is moisture-free. The manufacturer's directions should be followed concerning thinning and application parameters. During the mixing operation, the catalyst should be poured slowly into the base component and then both components should be mixed mechanically. Mixing should be done slowly so as not to create a vortex and introduce moisture into the coating that could reduce the pot life. These urethane coatings are extremely susceptible to moisture contamination, so it should not be applied unless temperatures both during application and up to 3 h after application will be at least 3°C above the dew point.

3.15 LATEX PAINTS FOR STEEL

Latex paints may be applied by spray, preferably, or brush if spraying is not suitable. Cross-brushing or cross-spraying is highly desirable. Application by spray tends to provide the best leveling. Conventional or airless spray can be used with most latex coatings. Since one-coat systems have very limited protective properties, multiple-coat systems should always be applied. For structural steel, the preferred system is two coats of primer and one coat of topcoat, for a total thickness of 190 μm.

The atmospheric conditions at the time that the latex paint, especially the primer, is applied are extremely important. A latex primer should not be applied at a temperature below 10°C or above 49°C. The best conditions for storing latex paints are at temperatures between 4°C and 27°C. Latex paints should never be allowed to attain temperatures over 49°C or be subjected to repeated freezing and thawing.

3.16 PAINTING OF FERROUS METALS

Paints should be applied in accordance with this standard and good industrial practice. Manufacturer instructions should also be considered. The paint film should not be exposed to moisture and contamination before it has dried. Priming and painting under controlled conditions in the shop are preferred. The degree of cleanliness of the surface will be determined by the requirements with respect to the paint system.

The number of coats applied after priming and the total DFT of paint will be determined by the requirement with respect to the paint system and conditions in terms of standards. By no means should the total DFT of paint be less than 125 μm for moderate exterior environments, and severe environments need an even greater film thickness.

3.16.1 PRIMING OF FERROUS METALS

Prepared surfaces should be primed generally within 4 h, or before visible rerusting occurs.

Blast-cleaned surfaces may be protected for short periods by thin coat of a pretreatment primer. Such a primer should be applied as a continuous coating in an even manner to achieve a minimum film thickness of 20 μm. This type of primer does not replace the full thickness of permanent primers.

The priming coat or coats on steel should be as specified by standards, but in no case should they have a total DFT below 50 μm. When applied in two coats, a shade contrast between coats is recommended.

In order to minimize contamination between successive coats of paint, overcoating of the preceding coat should be done within the period of time recommended by the manufacturer and should not delayed beyond the period specified. When delays are unavoidable, the painted surface should be thoroughly cleaned and dried to the satisfaction of the company before overcoating may take place. The primer is applied by spraying, except when another method of application is preferred by the company or required by the job or material.

Primed steelwork, especially if it has been exposed for a lengthy period, should be examined carefully before further coats of paint are applied. If the primer has been deteriorated (e.g., perished, eroded, or poorly adhering) or damaged so that corrosion can develop, the affected areas should be reprepared and primed. If there is evidence of widespread corrosion beneath the primer, it should be removed and the surface again prepared and primed. Removal of salt deposits by washing from surfaces primed with zinc-rich primers is especially important, as the corrosion products formed by reaction between the salts and the zinc can affect the performance of subsequent coats.

With a single coat of primer, it is difficult to obtain films of uniform thickness and free of pinholes (the points at which corrosion can start). In all but mild interior environments, application of two coats of primer is suitable. If application of two coats cannot extend to the entire surface, a second coat should be applied to vulnerable points (e.g., along external angles and to bolts and rivet heads).

When a factory-applied prefabrication primer has been used and a paint system of conventional type is to be applied, the second coat can be a drying-oil chromate or zinc phosphate type. Surface primed with red-lead primer should not be exposed to the weather for more than 1 month.

3.16.2 PRIMING OF STEEL PREPARED BY BLASTING

Blast-cleaned steel prepared to SIS grade Sa 3 or Sa 2½ should frequently be coated either in the shop or on site with a pretreatment or prefabrication primer or with original primer. For prefabrication and blast primers.

For further priming over prefabrication primers or for direct application to blast-cleaned steel, the priming paint should be in accordance with relevant standards for paints. Unless otherwise specified, it is essential that the total film thickness of the priming coats (blast or prefabrication primer plus second primer) should meet the prescribed minimum thickness of 75 μm in all areas.

3.16.3 PRIMING OF STEEL PREPARED BY PICKLING

Steelwork prepared by acid pickling should be treated by a phosphating process. This method is not generally used outside the pipe industry, but large plates for storage tanks have been treated in this way.

Zinc-rich primers, both organic and inorganic, are not suitable for application to phosphate-treated surfaces. Zinc chromate primers are not recommended; their use would generally be limited to aluminum surfaces.

Regardless of the type of primer used on pickled steel, it is essential that the total film thickness of the priming coats meets the specified film thickness (but not less than 50 μm). This would normally be achieved by the site application of a further coat of primer on prefabrication primer.

3.16.4 PRIMING OF STEEL PREPARED BY WIRE BRUSHING (TOOL CLEANING)

Priming paints for steelwork prepared by wire brushing must have good wetting properties and must be applied by brush to ensure a high standard of adhesion to the prepared steel surface. Red lead alkyd priming paint, when applied as a two-coat system to a minimum DFT of 50 μm, is recommended for wire-brushed surfaces.

3.16.5 SHOP PAINTING OF STEEL

Full protection applied in the shop immediately after fabrication normally results in longer life of the protective system. However, damage during transportation and erection may subsequently require widespread repair or touch-up of coating, so the structural steelwork, surface pipework, towers, vessels, heat-exchanger shells, and similar containers that will not be lagged can be treated in the shop.

The shop treatments will be determined by requirements with respect to the paint system and conditions such as type of environment transportation, economic, etc. The handling and storage of shop-treated items should be such that damage to the treated surfaces is prevented. Damage resulting from handling in the shop following painting, such as during storage or loading, is to be repaired as a part of the painting operation. If the shop coat is damaged in fabrication, it should be repaired before leaving the shop.

Contact surfaces should be painted or not, as specified in the procurement documents. If the surface is painted, at least the first coat should be applied in the shop with subsequent coats being applied in the field while the surfaces are still accessible, unless otherwise specified.

If the paint specified is harmful to the welding operator or is detrimental to the welding operation or the finished welds, the steel should not be painted within 100 mm of the areas to be welded.

3.16.6 SHOP PAINTING IN COLD CLIMATES

The paint shop should be enclosed and heated to keep the temperature at a minimum of 4°C. If practical, the temperature should be kept at 18–21°C.

Note: Temperature and humidity have considerable effects on the quality of the paint job. Most cases of paint failure due to mill scale lifting occur on steel that was fabricated, cleaned, and painted during the winter.

3.17 FIELD AND TOUCH-UP PAINTING OF STEEL

Previously applied shop coatings must be dry and free of dirt, oil, and other contaminates. The manufacturer's instructions should be followed if special surface preparation procedures are required before application of the field coats. All shop-primed items that have deteriorated as a result of transshipment to the extent that either crumbling or white staining of the coating is evident should receive a superficial sweep blast cleaning sufficient to remove the degradation and to reprepare exposed degraded metal substrate and dust.

Shop-coated steel items should preferably be field-painted after they are erected. They may be field-painted on the ground beforehand, provided that such painting is touched up in any damaged spots afterward with the same number of coats and kinds of paint. However, the last complete coat of paint should be applied after the structure is erected. The first field coat of paint should be applied within a reasonable period after the shop coats, and in any event, before the weathering (and required touch-up) of the shop coat becomes excessive.

When the type of paint for field coats is not specified, it should be determined that the paint to be used is compatible with the shop-applied coats. Paint used in the first field coat over shop-painted surfaces should not cause wrinkling, lifting, or other damage to the underlying paint.

Contact surfaces should be painted or not, as specified in the procurement documents or required by the job. Surfaces (other than contact surfaces) of fabricated assemblies that are accessible before erection but not afterward should receive all field coats of paint before erection. All cracks and crevices should be filled with paint if it can be done practically.

The final coat of paint on steel structures should not be applied until all concrete work is finished. In addition to the cleaning specified in standards, all cement or concrete spatter and drippings should be removed before any application of paint. If any paint is damaged, the damaged surface should be cleaned and repainted before the final coat is applied.

Wet paint should be protected against damage from dust or other detrimental foreign matter as much as is practical. Steel stored pending erection should be kept free of contact with the ground and positioned such as to minimize water-holding pockets, soiling, contamination, and deterioration of the paint film. Such steel should be cleaned and repainted or touched up with the specified paint whenever necessary to maintain the integrity of the film.

All field welds and all areas within 100 mm of the welds should be cleaned before painting, using surface preparation methods at least as effective as those specified for the structure itself. All welds should either be blast-cleaned, thoroughly power wire-brushed, chemically scrubbed, or water-scrubbed of all detrimental welding deposits as required.

3.18 MAINTENANCE PAINTING OF STEEL

Surface preparation for maintenance work should be as specified in Chapter 1. Paint that curls or lifts after application of the spot or priming paint should be removed, and the area should be repainted.

On structures that are known to have been originally pretreated with basic zinc chromate wash primer or other methods, the cleaned areas should be similarly pretreated before applying the prime coat of paint unless otherwise specified.

All prepared surfaces should be primed before any deterioration of the preparation occurs or within 4 h, whichever is less. Where patch priming is being carried out, this should extend 50 mm to the adjacent sound paintwork. The minimum DFT of individual coatings, the total DFT of the complete paint system, and wet film thickness (WFT; especially where existing paint surfaces are overcoated) should be determined at the discretion of the company with reference to standards.

On repair work, epoxy coatings and inorganic zinc coatings should be applied only to newly blasted surfaces. If pinholes are present, they should be treated as follows, depending on their extent:

- If pinholes are few and local, the areas should be rubbed down and an additional coat or coats should be applied by brush.
- If the areas are extensive, the area should be made paint-free and be repainted at the contractor's expense.
- The word *pinhole* is synonymous with *holiday* and *pore* (see the section "Pinhole and Holiday Detection," later in the chapter, for more information).

3.18.1 **PAINTING OF SPECIFIC SURFACES**

Unless otherwise specified by the company, the following practice should be executed regarding painting of contact surfaces:

- The areas of steel surfaces to be encased or embedded in concrete should not be painted.
- Steel to be completely enclosed in brick or other masonry should be given at least one coat of shop paint.
- The areas of steel surfaces that are to be in contact with wood should be painted.
- Surfaces that are to be in contact only after field erection should be painted, except where the paint interferes with assembly or where indicated otherwise.
- Steel surfaces that are not in direct bonded contact, but inaccessible after assembly, should receive the full specified paint system before assembly.
- Bearing-type joints may be painted. Contact surfaces of members to be joined by high-strength bolts in friction-type joints are a special case. Unless specifically authorized to the contrary, they should be left unpainted and free of oil, grease, and coatings.

3.18.2 **EDGES**

All sharp edges should be coated to the same film thickness as the adjacent steelwork to prevent premature breakdown from this area. Corners, services, bolt heads, and rivet heads require similar attention. Where there is any doubt that these areas have received adequate film thickness, the engineer may ask that an additional strip coat of paint be applied to ensure full film thickness at no additional cost.

3.18.3 **WELDS**

Rolled steel may be blast-cleaned and protected with blast primer before fabrication and welding. This prevents the serious development of rust, which would be difficult to remove after fabrication. The use of steel that has rusted heavily during storage should be avoided for the same reason. When welding metal-coated or zinc-dust-painted steel, it is necessary to remove the coating near the weld area, or mask off the weld area before coating.

Most painted steel can be cut and welded in a satisfactory fashion, provided that the coating thickness is less than 25 μm. After welding, scale and heat-damaged coating should be removed by local blast cleaning and the area renovated by repainting the original coating.

3.19 **PAINTING OF STAINLESS STEEL**

Stainless steel does not normally need to be painted, but where insulation is to used with stainless steel equipment, piping, or items that are to be stored in the open air for long periods or shipped as deck cargo, protective coating should be applied as specified by the company. Paints that contain zinc are not allowed for this purpose.

For potential fire situations, where hot-dip galvanizing or zinc coating is present, austenitic stainless steel equipment should be specially protected against the possibility of zinc embrittlement failure, which may result in rapid fire escalation. Such equipment should be located in a shielded position, which will reduce the risk of molten zinc falling onto it. Any bare stainless steel parts (e.g., flanges) that are within reach of zinc should be protected with a painted steel shield.

Where adequate shielding is impractical, hot-dip galvanized or zinc-coated components should not be used in close proximity to the stainless steel concerned (especially above it). Stainless-steel plates should be stacked on edge while in storage.

The company should indicate whether austenitic stainless steel surfaces below 50°C or above 200°C are to be painted. When painting of stainless steel surface is needed, the paint should be applied on surfaces prepared in accordance with standards. The paint system and DFT of the paint should be as determined by the company, the job requirements, or both.

3.20 PAINTING OF NONFERROUS METALS, INCLUDING METAL-COATED SURFACES

Nonferrous metals are more resistant to corrosion than iron and steel. For this reason, they are often used as alternative materials.

In most situations, painting is not necessary except for appearance, but paint should be applied in some environments, such as acid or marine conditions. Zinc, aluminum, and some other nonferrous metals should be pretreated to improve paint adhesion.

3.20.1 PAINTING OF ALUMINUM AND ITS ALLOYS

Unless when the type of primer specified by expert, similar primers are used for aluminum and its alloys. Primers containing zinc or other chromates, but not lead or graphite pigments, are suitable. The chromate pigment should constitute about 20% by mass of the dried paint film, but factory-applied red oxide/chromate primers with about 5% chromate can be satisfactory if the alloy is resistant to corrosion and the conditions of exposure are not severe.

Pretreatment primers, especially of the two-pack wash primer type, are particularly suitable for aluminum and its alloys. It assists adhesion on smooth surfaces, such as sheets, extruded sections, and aluminized steel. Finishing paints for aluminum should not contain lead or graphite pigment, which stimulates corrosion, as the priming coat may not prevent them from touching the metal.

3.20.2 PAINTING OF ZINC AND ZINC-COATED STEEL

After a zinc-coated surface is prepared, it should be treated with chemicals, followed by the paint system specified by the paint schedule. Lead-containing primers should not be used on zinc and aluminum.

Table 3.3 shows a site treatment procedure for previously metal-coated steelwork as a guideline. The case and the paint system should be defined by the company.

3.20.3 PAINTING OF COPPER AND ITS ALLOYS

Copper alloys (such as brass and bronze) are rarely painted except for appearance enhancement. Adhesion of paint may be assisted by surface preparation and application of pretreatment primer. Direct application of alkyd gloss finish after preparation and treatment is suitable. The finishing system for copper is similar to that of iron and steel.

Table 3.3 Site Treatment and Painting of Previously Metal-Coated Steelwork

Initial Conditions	Present Conditions	Surface Preparation	Replacement of Metal Where Required	Paint Treatment Over	
				Sprayed Metal	Galvanizing
–	Areas of not normally necessary corrosion and/or some rusting of substrate	If metal is to be replaced, blast clean it	Sprayed metal to appropriate	Not normally necessary	N/A
		If metal is not to be replaced, clean the corroded areas by the best means available	Not applicable	Build up cleaned areas with a suitable paint system and apply chemical-resistant finish overall	Build up cleaned areas with a suitable paint system and apply chemical-resistant finish overall
–	Areas with some white corrosion products	If decoration is required, wash to remove salts, using a stiff brush if necessary Remove loose material with a nonmetallic brush	Not applicable	Apply sealing coat and chemical-resistant finish for maximum life	Apply suitable surface pretreatment followed by a chemical-resistant finish
		If decoration is required, no action is necessary	Not applicable		
–	Areas in sound condition	If decoration is required, wash to remove salts, using a nonmetallic brush	Not applicable	Apply suitable paint, which should be chemical-resistant for maximum life	Apply suitable surface pretreatment, followed by paint, which should be chemical-resistant
		If decoration is not required, no action is necessary	Not applicable		

(Continued)

Table 3.3 Site Treatment and Painting of Previously Metal-Coated Steelwork (*cont.*)

Initial Conditions	Present Conditions	Surface Preparation	Replacement of Metal Where Required	Paint Treatment Over	
				Sprayed Metal	Galvanizing
Sealed or painted metal coating	Areas of corrosion or some rusting of substrate	If metal is to be replaced, blast clean it	Sprayed metal to appropriate	Consider one or two coats overall, preferably chemical-resistant	Consider one or two coats overall, preferably chemical-resistant
		If metal is not to be replaced, remove corrosion by the best method available	Not applicable	Build up cleaned areas with suitable paint; apply one or two coats overall, preferably chemical-resistant	Build up cleaned areas with suitable paint; apply one or two coats overall, preferably chemical-resistant
	Areas with some degradation of paint, dissipation of sealer, or loss of adhesion of either paint or sealer	Remove loose material with a nonmetallic brush	Not applicable	Apply further coats of paint or sealer, preferably chemical-resistant	Apply further coats of paint or sealer, preferably chemical-resistant
	Areas in sound condition	If decoration is required, dust the surface	Not applicable	Apply further coats of paint or sealer, preferably chemical-resistant	Apply further coats of paint or sealer, preferably chemical-resistant
		If decoration is not required, no action is necessary	Not applicable		

3.20.4 PAINTING OF LEAD

Surface preparation and treatment with phosphating solution are satisfactory methods. Many conventional metal primers also can work, provided that they do not contain graphite. The finishing system for lead is similar to that of iron and steel.

3.20.5 PAINTING OF CHROMIUM, NICKEL, TIN, AND CADMIUM

New chromium and nickel coatings rarely require painting, but it may be necessary if they become corroded. Corrosion should be removed before pretreatment primer is applied to the surface. Tin plate presents few difficulties in painting; most paints will adhere satisfactorily after surface preparation.

Cadmium should not be weathered prior to painting. Phosphate treatment or abrasion followed by pretreatment primer will provide a key for subsequent coats.

3.21 PAINTING OF PLASTER, CONCRETE, BRICK, BLOCK AND STONE

The paint materials and systems, finish type, primer, finish systems, and colors should be as specified by the company. The finished materials should be manufactured from the highest-quality materials and should meet the minimum requirements of the Institute of Standard Specifications.

3.22 TREATMENT OF STAINS

Brown stains with no appreciable surface deposit sometimes appear on emulsion paints but not normally on oil-based paints. They are usually derived from substrates, notably certain types of brick, plaster, hollow clay pot, or cinder block, containing soluble salt or coloring material, or from sands containing organic matter that reacts with alkali. If it is suspected that this type of staining is likely to occur, a coat of alkali-resisting primer will usually prevent it. This protection may also be applied to stained emulsion paint to prevent staining of succeeding coats.

3.23 PLASTER

Priming is required when oil-based paint systems are applied to plaster; however, it is not usually necessary with emulsion paints. When emulsion paints are applied to plaster of high or variable porosity, differential absorption can cause difficulties in application or variations in color or sheen, which may persist through several coats. A well-thinned first coat of emulsion paint, sometimes referred to as a "sealing" or "mist" coat, often solves this problem, but it is likely to have relatively poor opacity. If such a coat is required, it should be regarded as an additional coat in the system. Where this proves inadequate, it is usually necessary to apply a coat of alkali-resisting primer or primer sealer, but this should be done only if the substrate is substantially dry. The primer or primer-sealer may require thinning to ensure that it does not provide a glossy surface to which emulsion paint may not adhere properly.

3.24 **CONCRETE**

If concrete surfaces need to be protected, the first paint layer should be applied by brush after any surface imperfections have been filled with an appropriate putty. Priming is required when oil-based systems are applied, but it is not usually necessary with emulsion paints.

3.25 **BRICK AND STONE**

Priming is necessary with oil-based systems. Because mortar joints are likely to be alkaline, an alkali-resisting primer should be used.

Priming is not usually necessary with emulsion paint to accommodate variations in surface porosity and facilitate application. Primers of first coats may require thinning in accordance with the manufacturer's instructions.

3.26 **PRECAST CONCRETE BLOCKS AND EXTERNAL RENDERING**

The priming and finishing of precast concrete blocks and external rendering are done similar to that for brick and stone.

3.27 **METHOD OF MEASURING MOISTURE CONTENT**

The selection of paint systems for nonmetallic surfaces is generally related to the dryness or humidity percentage of the substrate. The following methods are used to measure the moisture content of building surfaces.

3.27.1 **WEIGHING**

Weighing the moisture lost during oven-drying of samples obtained by methods such as drilling is the most accurate way to measure moisture content. This method may require access to laboratory facilities, although a calcium carbide meter can be used with drilled samples for on-site determination of the moisture content of walls, hence avoiding the need for oven-drying. However, any method of test involving drilling may be impracticable for general use.

3.27.2 **HYGROMETER**

The equilibrium humidity produced in an airspace that is in contact with the substrate can be measured by using an accurate hygrometer. The space may be formed by a sealed and insulated box in which the hygrometer is mounted; another option involves having a sheet of polyethylene taped to the substrate with the hygrometer inside (though it is less reliable). In either case, however, several hours should be allowed for equilibrium to be reached.

3.27.3 ELECTRICAL MOISTURE METERS

Two types of electrical moisture meters in general use are conductivity meters and capacitance meters. Conductivity meters measure the electrical resistance between two steel probes forced into the substrate; the higher the moisture content, the lower the resistance to the flow of current. With hard, dense substrates, it may be necessary to drill holes in order to obtain readings. Resistance is reduced by the presence of soluble salts, so the readings may be higher than the amount of moisture actually present. Capacitance meters have two flat electrodes that are pressed against the surface, avoiding damage. They register the amount of moisture present only in the upper 1–2 mm of the substrate and are inaccurate on rough surfaces; as with the conductivity types, soluble salts may affect the accuracy of readings of capacitance meters.

Electrical moisture meters, although less accurate than the other devices described in this section, are easy and convenient to use. They enable a number of readings to be taken quickly. Their use is preferable to relying on surface appearance or rule-of-thumb methods.

If a wall is believed to be damp but meter readings at a shallow depth indicate low moisture content, the area should be covered with a sheet of polyethylene and rechecked 24 h later. If there is adjacent woodwork, check its moisture content; as there is less likelihood of soluble salts being present in wood, readings will be more reliable, especially if they are significantly lower for the wood.

3.28 PAINTING OF WOOD

The natural finishes are put into two groups: varnishes and stains. Like paints, natural finishes may not prevent the entry of moisture. Accordingly, their use, even of wood coatings with preservative or water-repellent properties, does not obviate the need for preservative treatment.

Exterior wood coatings (e.g., paint, varnish, and stain) are conveniently grouped into two major types: water-borne (emulsion) and solvent-borne (oil-based). Water-borne coating should not be used over linseed oil putty.

Note the following:

- Varnishes do not completely obscure the grain of wood. They give a film with little or no ability to hide the grain. They give one of the most attractive finishes to wood, but when used outdoors, they can flake off, primarily due to embrittlement of the film by solar radiation combined with photodegradation of the wood surface.
- Stains are defined as solutions that color wood by penetrating it without hiding the grain or leaving any perceptible surface film.
- Water repellency is the ability of a coating to resist wetting by water; i.e., any water on the surface forms discrete droplets and does not wet the material. Water repellency can be increased by the addition of certain compounds (e.g., silicones, waxes).

The finished materials should be manufactured from the highest-quality materials as defined by standards and should meet the requirements standard as a minimum.

3.29 **PAINT SYSTEMS (OPAQUE COATING) FOR WOOD**

Paint systems of varying types and ranges from matt through semi-glass to glass are used extensively for external and internal woodwork, and offer good resistance to the weather including solar "radiation". Glass finishes are suitable for external use and in severe internal environments, and they should be applied in four coats, including primer. Mid-sheen and matte finishes should be used in moderate mild internal environments and should be applied in three coats, including primer.

Moisture can be very damaging to wood, both through its effects on dimensions (i.e., it shrinks wood) and by providing conditions under which microorganisms can attack the surface. The moisture content at the time of painting should not exceed about 18%. The moisture content should be measured. If timber has received preservative treatment, time should be allowed for drying out or evaporation of solvent before priming.

If site priming is necessary, it should be carried out immediately after delivery of the joinery, provided that its moisture content is at a satisfactory level. Joinery is usually factory-primed. Primed joinery, if not fixed or erected immediately, should be properly stored. Both transparent and pigmented primers (BS 5082 and BS 5358) are suitable for paint systems.

If primed woodwork has been exposed for a lengthy period, the condition of the primer should be checked before continuing application of the paint system. It is important that areas of defective or poorly adhering primer be cleaned and the exposed areas reprimed. If the primer is firmly adhering but is chalking or powdery, it should be rubbed down with abrasive paper and a further coat applied.

The color of the coating will influence the surface temperature when exposed to sunlight, with black and other dark colors reaching much higher temperatures than white or other pale colors.

3.30 **NATURAL FINISH SYSTEM (TRANSPARENT COATING) FOR WOOD**

Natural finishes for wood are those which, unlike paint, do not completely obscure the grain of wood, although they may modify its color to varying degrees. The use of varnishes for exterior woodwork is best avoided. In special cases when exterior woodwork is to be varnished, the wood should first be treated with a preservative that is effective against blue-stain. Blue-stain fungi often contribute to the darkening of weathered wood. With adequate maintenance, the fungicide in exterior wood stains will usually inhibit the development of blue-stain.

With most natural finishes, putty glazing is not suitable. Varnishes are usually applied to unprimed wood. The stains can be used with both exterior and interior surfaces and should contain water-repellent preservative and fungicides for exterior use.

Many stains, especially low-solids, are suitable for use on plywood. For initial treatment of new external wood, a minimum of two and preferably three coats of exterior wood stain should be applied.

Decorative wood stains have no protective or preservative properties and serve essentially to impart color to wood prior to the application of varnish or other clear finish. Transparent and pigmented primer can be used for staining.

3.31 DRYING AND HANDLING

3.31.1 DRYING OF PAINTED SURFACES

No coat of paint should be applied until the preceding coat has been dried. The paint should be considered dry for recoating in two circumstances: (1) when another coat can be applied without the development of any film irregularities such as lifting or loss of adhesion of undercoats; and (2) when the drying time of the applied coat does not exceed the maximum specified for it as a first coat. The minimum drying time between coats should comply with the manufacturer's instructions.

The maximum practical amount of time should be allowed for the paint to dry before recoating. Some paints may dry excessively hard for good adhesion of subsequent coats; these should be recoated within the time period specified by the manufacturer. If not recoated within the specified time, then the previously applied coatings should be roughened prior to recoating.

No paint should be force-dried under conditions that will cause checking, wrinkling, blistering, formation of pores, or damage to the protective properties of the paint. No drier should be added to paint on the job unless specifically called for by the manufacturer. Paint should be protected from rain, condensation, contamination, snow, and freezing until it is dry to the fullest extent practical. No paint should be subjected to immersion before it is thoroughly dried or cured.

3.31.2 HANDLING OF PAINTED SURFACES

Painted surfaces should not be handled, loaded for shipment, or shipped until the paint has been dried except as necessary in turning for painting or stacking for drying. Paint that is damaged during handling should be scraped off and touched up with the same number of coats and kinds of paints as were previously applied to the surface.

3.31.3 CLEANUP

All paint application tools and equipment must be carefully cleaned. Dried paint in the equipment will ruin it. To do this, remove as much paint as possible. With solvent paints, clean thoroughly with a compatible solvent. Use a detergent solution with latex paint. Clean two or three times with fresh solvent (or warm mild detergent), and then wipe clean and dry. Well-cleaned tools and equipment will last longer and always stay in good condition.

Be sure to clean brushes down to the heel since paint tends to dry in this less-visible area, and that can make the bristles shorter and less flexible. After washing, twirl brushes to remove excess water and comb them to straighten the bristles. Finally, wrap brushes in paper or place in a brush keeper and lay flat until dry. Never allow a brush to rest on its bristles, which can cause permanent damage.

Using a large container, wash the paint mitt used for solvent paints in three types of solvent or warm mild detergent, depending on the type of paint. The solvent-cleaned mitts should then be washed in mild detergent. Rinse mitts in clear warm water, then hang up to dry. Place clean solvent (or detergent) in pots and pass it through hoses and spray guns. Be sure to remove the tip from airless spray guns and wash separately. Never immerse the gun in solvent because this can ruin the packing.

Clean with three changes of solvent (or detergent), and then dry. When cleaning after spraying water-based paint, be sure to finish rinsing with a water-miscible solvent, such as alcohol. Otherwise, some parts of the spray equipment may rust. Make sure that all hoses are flushed thoroughly. Completely nonrusting spray equipment should be used with water-based paints to prevent rusting. Try to leave the solvent in the system when possible to avoid build-up of paint in the hose.

3.32 INSPECTION PROCEDURE
3.32.1 INSPECTION GUIDE

Finished paintwork should have the correct shade, degree of gloss, and evenness, and it should be free of tackiness after drying/ curing and from cracks, holidays, runs, sage, wrinkles, patchiness, brush marks, or other defects that may be deleterious to the quality of the coating. Prior to final acceptance of the paintwork, an inspection should be made. The contractor and the company should both be represented, and they should sign an agreed inspection report (such as what is shown in Table 3.4).

Table 3.4 Inspection Guide

Work Stage and Code	Potential Defects	How Determined	Likely Cause	Suggestions for Action	Notes
Environment during application					
1	Too cold (Below 4°C)	Air and contact thermometers	Pool air conditioning in works or externally inclement weather	Improve temperature to acceptable level, but little or no license shall be allowed with limit specified. Open-flame heater shall not be used	
2	Too hot	Contact thermometer	Heated surfaces	Reduce temperature below 35°C or to whatever higher temperature is allowed with special paints	If necessary, re-arrange program to avoid hot conditions
3	Too damp	Hydrometer of visual	Poor air conditioning in works or externally inclement weather	Improve air conditioning to acceptable level	Try to Avoid Depositions of Dew Betas Paint is 'set'
4	Too wet	Visual and or Moisture	A) Rain. Sleet or Snow B) Condensation	Protect operations with suitable sheeting, but avoid causing condensation provide suitable ventilation	Where cleaning and painting in adjacent areas additional Covers are required to prevent contamination of the paint
5	Too dark	Visual	Insufficient lighting and/or diet on glass cladding	Improve lighting: clean glass	Where appropriate confirm with luxmeter
6	Too dusty	Visual	(A)Too much wind (B)Poor dust extraction	Shield work or delay il till wind abates Protect work from blasting dust improve extraction	—

(Continued)

Table 3.4 Inspection Guide *(cont.)*

Work Stage and Code	Potential Defects	How Determined	Likely Cause	Suggestions for Action	Notes
7	Air too foul	Smell and/or instruments	Poor ventilation	Improve ventilation to point where threshold limit values (TLVs) of solvents, etc are not exceeded and or provide fresh air supply to operators	—
Paint mixing and storage*					
1	Wrong weight of container paint	Spring balance	Poor filling of containers or wrong paint	Reject all full under weight containers. A check on other containers with the same batch number will probably indicate if material is at fault	
2	Wrong specific gravity	SG cup and balance (see ASTM D 19G3-S5)	(A) Poor mixing of paint	Reconstitute if possible	
			(B) Thinning or over-thinnine	Reject all paintwork where such paint has been used and either remove by stripping or. if agreed by the company apply an extra coat of the correct paint	
3	Thin consistency	Visual and'or flow cup (See ASTM D 1200-SS)	(A) Defective or wrong paint	} □□□□□□ As for 1 find 2 above, as applicable	
			(B) Poor mixing		
			(C) Thinning or over-thinning		
4	Thick consistency	Visual and'or flow cup	(A) Defective of wrong paint	} □□□□□□ As for 1 and 2 above, as applicable	
			(B) Poor mixing		
			(C) Thinning not as specified	Control the correct use of thinners, adjust viscosity	
			(D) Outside the pot life	Discard all paints affected	

Table 3.4 Inspection Guide *(cont.)*

Work Stage and Code	Potential Defects	How Determined	Likely Cause	Suggestions for Action	Notes
5	Wrong color in container	Visual comparison with wet standard	(A) Wrong paint supplied	Obtain supplies of correct paint	
			(B) Poor color matching or variation in raw materials	Set aside until manufacturer certifies supply as satisfactory	
			(C) Not thoroughly mixed	Re-mix and check as for 2 or 3	
6	Wrong color and other characteristics when thoroughly mixed	Visual, instrumental	Wrong proportions of multi-pack materials	Discard unless proportions used are known and can be corrected	
7	Paint contaminated with moisture	Visual	Dirty paint store, inadequately heated and/or ventilated	Clean up paint store, provide adequate heating and ventilation	
8	Deterioration of paint in tins, setting. separation, gelling. etc.	Visual	(A) Old stocks outside the stipulated storage period	Refer stocks to manufacturers for clearance before permitting farther use	
			(B) Use of stocks out delivery order, leading to cause (A) above		
			(C) Storage conditions too hat or too cold		
Priming					
1	Over-thick areas, sagging, curtaining	Visual and/or instrumental comparsion	Failure to bush out property or too heavy spray application	Scrape off excess, wipe off while still wet, or otherwise remove. Re-prime to give the correct thickness	Particular attention is required to ensure that each primer is not applied too thickly. Loss of adhesion to subsequent coats may occur

(Continued)

Table 3.4 Inspection Guide *(cont.)*

Work Stage and Code	Potential Defects	How Determined	Likely Cause	Suggestions for Action	Notes
2	Thin area, grinning through	Visual and/or instrumental	Not enough paint	Further application	
3	Dry Spray	Visual or Touch	Incorrect spraying technique or high winds especially with zinc epoxide and zinc silicate paints	If noticed at the time of application, brush surface with hard bristle brush and re-apply. If noticed at a later stage, re-blast and re-prime	
4	Corrosion products	Visual	Too long storage in primed state	(A) If rusting or rust spotting, re-blast and re-prime (B) If zinc corrosion products show on a metallic zinc primer, especially in a sheltered situation, scrub with fresh water and dry	
5	color change or soft paint near welds	Visual or touch	Non-removal wielding flux, residues have saponified the paint	Remove all the affected paints, by most appropriate method, to bare metal. scrub affected areas with water and mild detergent , rinse thoroughly and allow to dry. Re-prepare surfaces as specified and prime immediately	
Painting					
1	Sagging, curtaining	Visual	Over-application of paint	Remove paint by scraping or other effective method and re-paint	
2	Area of low thickness, grinning through	Visual, Instrumental	Under-application of paint, poor spray pattern	Apply extra coat or coats, as appropriate, wthin the recommended limits of the recoating period	

Table 3.4 Inspection Guide *(cont.)*

Work Stage and Code	Potential Defects	How Determined	Likely Cause	Suggestions for Action	Notes
3	"Orange-peel" effect	Visual	Poor spraying techniques	Correct all adjustment to spraying technique	May be difficult to avoid with heavy compositions
4	Lifting, Wrinkling etc.		(A) Incompatibility of solvents with the state of the previous coat	Reduce all affected paint by most suitable means to the firm sound substrate and replace all coats as necessary	
			(B) wrong interval between coats		
			(C) Drying conditions too fast		
			(D) Surface contaminated		
			(E) Over-thick coat		
5	Poor Inter-coat Adhesion	Visual	(A) Surface contaminated, see 'Notes' column	Remove and abrade surface, recoat	There are several other causes but these are commonest in practice, e.g. with oil, grease, water powdery deposits
			(B) For epoxy or urethane types, too long interval between coats		
			(C) Incompatibility of the painting system, see 'Notes' column		E.G. previous coating insufficiently cured to withstand solvents
6	Loss of gloss, rough surface	Visual or touch	(A) Airborne dust and dirt, and overspray	Rub down surface with suitable abrasive paper before further coating. if the affected coat was the final coat was the final coat it shall be repeated	

(Continued)

Table 3.4 Inspection Guide *(cont.)*

Work Stage and Code	Potential Defects	How Determined	Likely Cause	Suggestions for Action	Notes
			(B) Condensation during drying	Refer to paint manufacturer before proceeding with further coats, if the affected coat was the final coat, re-apply	
7	Slow drying	Touch	(A) Unsuitable, ambient conditions, or too heavy coating	Defer further coats until point is completely dry	There are several other causes, but these are the commons in practice
			(B) Mixing error for two-pack material	Remove the coating and re-apply using correct two-pack mix	
8	Pinholing	Visual	(A) Contamination or spray airlines, see 'Notes' column	⎫ ☐☐☐☐☐ Remove and re-paint affected area	Particularly with silicones, oils or water
			(B) Mixing error for two-pack material	⎭	
9	Cissing	Visual	Contamination of surfaces, mainly with oils and greases	Remove and re-paint affected area	Particularly with airless spray
10	No color difference between paints for successive coats	Visual	Wrong paints or specification	Supply in stock may be tinted to give intercoat color contrast but consult paint manufacturer	
Handling and transport					
1	Contamination of cleaned surface by dirt or sweat	Visual	Bare hands or dirty lifting tackle on cleaned surface	Re-cleanse area with clean water	

Table 3.4 Inspection Guide *(cont.)*

Work Stage and Code	Potential Defects	How Determined	Likely Cause	Suggestions for Action	Notes
2	Easily damaged coating	Visual or touch	Insufficient drying period before hardening	Repair damage to coating, allow longer drying and hardening times	
3	Damage by lifting tackle and other handling gear	Visual	Lifting point not included in design, no purpose-made gear provided	Repair damage, agree improved methods of handling	
4	Chafing in transit	Visual	No special support packings or lashings provided on wagons, poor storage	Repair damage, no further loading without adequate measures to prevent damage	
5	Components adhering together	Visual	(A) No special packaging provided	Repair damage or return for recoating. agree packaging methods	
			(B) Stacked before paint is thoroughly dry		This happens especially with chlorinated rubber coatings

*Whenever there is anything unusual about the paint other than the potential. Defects listed in paint mixing and storage (1 to 4) refer to the paint manufacturer.

3.32.2 **TEST METHODS**

Wet film thickness (WFT) measurement is used to aid painters and inspectors in determining how much material to apply to achieve the specified DFT. WFTs on steel and most other metallic substrates are considered guideline thicknesses, with the DFT being the thickness of record. However, when coating concrete or nonmetallic substrates, WFT is often the accepted value because DFT can be determined only by destructive means.

The WFT gauge is generally a standard notch gauge (Figure 3.9), although circular dial gauges (such as interchemical thickness gauges) are also used. The instrument is pressed firmly into the wet film perpendicular to the substrate and withdrawn.

The WFT/DFT ratio is based on the percent solids by volume of the specific material being applied. The solids by volume of the coating material is information readily available from the manufacturer and is commonly included in their product data sheets:

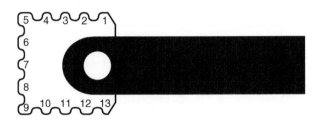

FIGURE 3.9

A WFT gauge measures the coating thickness during application by progressively deeper steps (marked in mils).

For unthinned coating materials, the following equation applies:

$$\text{WFT} = \frac{\text{Desired Dry Film Thickness}}{\%\,\text{Solids by Volume}}$$

For thinned coating materials, the following equation applies:

$$\text{WFT} = \frac{\dfrac{\text{Desired Dry Film Thickness}}{\%\,\text{Solids by Volume}}}{(100\% + \%\,\text{thinner added})}$$

Assume that a material contains 78% solids by volume and is to be applied in one coat to a DFT of 200 μm (0.2 mm). Without thinner added, the required WFT is determined as follows:

$$\text{WFT} = \frac{200}{0.78} = 256\ \mu\text{m}\ (0.26\,\text{mm})$$

If the coating in the same example is thinned by 20%, the new required wet film is calculated as follows:

$$\text{WFT} = \frac{200}{\dfrac{0.78}{1.2}} = \frac{200}{(0.56)} = 308\ \mu\text{m}\,(0.308\,\text{mm})$$

Thus, without thinning, a 256-μm wet film is required to obtain a 200-μm dry film. After thinning, however, the solids by volume value drops from 78% to 65% and the required WFT increases nearly 50 μm.

3.32.3 DFT MEASUREMENT

DFT measurement instruments consist of magnetic gauges that measure the thickness of a dry film of a nonmagnetic coating applied to a magnetic substrate. Magnetic stages are grouped into two types:

- Type 1—Pull-off-gauges (such as Mikrotest, Inspector, Tinsley Thickness gauges).
- Type 2—Fixed-probe gauges (such as Elcometer thickness gauge, Minitector, General Electric type B, thickness gauge, verimeter, permascope, and Dermitron).

For calibration and measurement procedures, see SSPC PA2 and BS 3900 part C5.

3.32.4 **PINHOLE AND HOLIDAY DETECTION**

After all the coats of paint have been applied, the inspector should verify that the appropriate cleanup is done and that any abrasions, nicks, or scrapes are repaired as required.

pinhole, or spark testing is to be used to find the nicks, scrapes, and pinholes in the coating film, particularly if the coating is intended for immersion service. Holiday testing should be required after application of either the next-to-last or the last coat of paint. Usually, such testing is done before final curing of the coating has occurred so that any repair material applied will successfully bond to the underlying coat.

Pinhole and holiday detectors come in three types: low-voltage wet sponge, DC high-voltage, and AC electrostatic types. When testing conductive lining applied over steel substrate (i.e., conductive rubber lining), the AC electrostatic type should be used.

If the continuity of the coating (e.g., for tank linings) is checked with a high-voltage spark test, the pinhole detection device should be set as high as is practicable, with a minimum of 5 watts/ζm of average coating thickness. The test apparatus should be a low-pulse DC detector of a type approved by the company.

To detect holidays such as pinholes and voids in thin film paints and coatings from 0.0254 to 0.254 mm (1 to 10 mls) in thickness, a low-voltage holiday detector should be used. This method may be considered a nondestructive test because it incorporates an applied voltage of less than 100 volts DC.

It is effective on films up to 0.508 mm (20 mils) and is not satisfactory for the thickness over 0.508 mm (see ASTM G 62-87). The voltage between the electrode (sponge) and the metal surface upon which the coating lies should not exceed 100 volts DC and is according to the manufacturer's instructions.

3.32.5 **FIELD ADHESION TESTING**

Usually, there is a need to test the adhesion of the coatings after application. There are different types of adhesion testing methods used from the simple penknife to more elaborate testing units. The use of a penknife generally requires a subjective evaluation of the coating adhesion based on some previous experience. Generally, one cuts through the coating and probes at it with a knife blade, trying to lift it from the surface to ascertain whether the adhesion is adequate.

A modified version of this type of testing is the cross-cut test, which consists of cutting an X or a number of small squares or diamonds through the coating done to the substrate. Then tape is rubbed vigorously onto the scribes and removed firmly and quickly. The cross-hatch pattern is evaluated according to the percentage of squares that are delaminated or remain intact. The X and cross-cut tape adhesion tests are described in ASTM D-3359, in the section "Measuring Adhesion by Tape Test."

There are also instruments available for testing the tensile adhesion strength of coatings. They apply a value to the adhesion strength in g/m^2, eliminating some of the subjectivity of the abovementioned tests. An instrument for tensile testing is the pull-off adhesion tester (Figure 3.10), which consists of the test unit itself and aluminum test dollies or lugs, each with a surface contact area of 12.7 mm^2. The dollies are cemented to the coating surface using an adhesive. After the adhesive has cured, the coating around the periphery of the dolly is cut down to the substrate. The claw of the test instrument is placed under the lip of the dolly and the unit tightened by hand, using as uniform a turning speed and motion as possible.

FIGURE 3.10

Pull-off adhesion testers

The test unit applies a pulling force onto the head of the dolly, ultimately breaking contact with the surface. The point of the break is read from the scale on the instrument in g/m^2. Both the numerical value and the type of break are important when using this instrument. For example, there is a significant difference in the test results if one finds a clean break to the substrate or between coats, compared to finding a cohesive break within a coat. Many times, one may experience a failure of the adhesive. If this occurs, it establishes that the coating tensile adhesion strength is at least as good as the pressure that broke the adhesive.

It is generally recommended that two-component epoxy adhesives be used rather than single-component, fast-drying, cyanoacrylate types. When testing zinc-rich coating, for example, it has been found that the thin cyanoacrylates have a tendency to penetrate and bind the zinc particles together, resulting in a much greater tensile pull than should be expected. In other cases, the adhesive appears to soften and cause premature failure of the coating systems.

3.32.6 EVALUATING CURING

When a coating is to be used in immersion service, the applied coating film must be allowed to dry-cure for a given length of time prior to being placed into service. This dry-cure time is generally shown on the manufacturer's product information. Alternatively, forced-heat curing may be used to reduce the time between curing and servicing.

Determining the cure of coatings is generally difficult. ASTM D1640 outlines one method, but there are no universally reliable field tests for such purposes. Solvent rub tests can be used, as well as sandpaper tests. When most coatings are suitably cured, rubbing them with sandpaper will produce a fine dust. If the sandpaper gums up, depending upon the coating, it may not be cured properly. Certain phenol-containing coatings may discolor upon heating, and the cure of phenolic tank lining coatings is often determined by comparison of their color with color reference coupons supplied by the coating manufacturer.

Because a coating is dry or hard does not necessarily mean that it is cured. In fact, for most coatings, hardness is not synonymous with curing. The only coating types for which this is true are solvent-deposited coatings such as chlorinated rubbers and vinyls. Even then, residual retained solvents (and moisture in water emulsion coatings) may take a long time to escape from the paint film under certain atmospheric conditions of temperature and humidity. Final attainment of film properties will be acquired only upon satisfactory loss of these entrapped solvents. In some cases, this evaporation process may take two or three weeks, or even longer.

3.33 SAFETY

This section describes typical hazards to health and safety that may be encountered in the painting process and makes general recommendations for dealing with them. Potential hazards include the following:

- Fire and explosion.
- Toxic hazard.
- Health hazard.

Information in noise levels, e.g., from compressors and mechanical tools, and code of practice for reducing the exposure of employed persons to noise.

Hazard may also be created by the use of unsuitable or defective scaffolding.

3.34 PREVENTION OF FIRE AND EXPLOSION

Flammable, volatile solvents in paints constitute a major hazard with regard to fire and explosions wherever flame or spark exposure is possible. No painting should be done within 45 m of steel welding or torch cutting.

When painting is required in a confined area, all flame sources (such as pilots and lights) should be extinguished and no smoking permitted. Painting should cease whenever this condition may be reached, or when solvent vapor concentration reaches hazard levels (see Table 3.5).

Vapor exhaust equipment should be used to maintain minimum levels of solvent concentration. The superintendents should acquaint themselves and their forepersons with the fire hazards inherent in the job and on the job site. The use and storage of flammable materials are to be only in restricted areas, which are to be well marked with appropriate signs.

Table 3.5 Threshold Limit Values (TLV) for Solvents

| Substance | Adopted Values TWA-TLV[1] | | Adopted Values STEL-TLV[2] | |
	ppm	mg/m	ppm	mg/m
Acetone	750	1,780	1,000	2,375
Benzene (benzol) — skin	10	30	25	75
Butylcellosolve — skin	25	120	75	360
Carbon tetrachloride — skin	5	30	20	125
Cyclohexane	300	1,050	375	1,300
Epichlorohydin — skin	2	10	5	20
Ethyl Acetate Ethanol (ethyl alcohol)	400	1,400	N/A	N/A
ethylene dichloride	1,000	1,900	N/A	N/A
(1. 2-dichloroethane)	10	40	15	60
Ethylenediamine	10	25	N/A	N/A
Furfuryl alcohol — skin	10	40	15	60
Methanol (methyl alcohol) — skin	200	260	250	310
Methylene chloride	100	350	500	1,740
(dicliloromethane)				
Naphtha, coal tar[3]	N/A	N/A	N/A	N/A
Naphtha, petroleum[3]	N/A	N/A	N/A	N/A
Perchloroethylene — skin	50	335	200	1,340
Isopropyl alcohol — skin	400	980	500	1,225
Stoddard solvent	100	525	200	1,050
Toluene	100	375	150	560
Trichloroethylene Turpentine	50	270	200	1,080
Xylene (xylol)	100	560	150	840
	100	435	150	655

[1]*Threshold limit value-time weighted average (TWA-TLV): The time-weighted average concentration for a normal 8-h workday or a 40-h workweek.*
[2]*Short-term exposure limit-threshold limit value (STEL-TLV): The maximum concentration to which workers can be exposed for up to 15 min.*
[3]*In general, the aromatic hydrocarbon content will determine what TLV applies.*

Fire extinguishers and fire hoses should be placed at locations that are agreed upon or designated by the safety manager. Provide adequate ventilation in all working areas to prevent a buildup of explosive concentrations of solvent vapor. Check regularly in confined areas or closed spaces to be sure that vapor concentrations are below explosive limits. Do not use metal ladders in confined areas or within 3 m of exposed electric wiring.

Clean up before, during, and immediately after painting operations.

Threshold limit value (TLV) is the amount of a solvent, expressed as parts per million (ppm) to air, that an operator may exposed to during an 8-h working day with no ill effects.

3.35 PREVENTION OF TOXIC HAZARDS

Several components used in organic coating materials are toxic. Most solvents are toxic to some degree, depending upon exposure. The degree of toxicity in substances can be measured by checking TLV (see Table 3.5).

Some pigments are toxic. The most common of these contain lead compound and chromate. Precautions should be taken when applying or removing paints containing these pigments. Lead and chromate paints should not be used on surfaces that may be licked or chewed by animals.

Some paint additives, such as the mercurial compounds used to impart fungicidal properties, may be toxic if ingested. A few binders are toxic to some degree if exposure is excessive. Typical of these are epoxies, acrylics (but not latex), polyurethane, and polyesters.

If permissible exposure limits, as determined by an industrial hygienist, are exceeded, respiratory protection becomes necessary.

3.36 PREVENTION OF HEALTH HAZARDS

Several actions can help prevent health hazards when handling these substances. Identify and seal all toxic and dermatitic material when not in use. Adequately ventilate all painting areas. Air used to fulfill the requirements of coating will require monitoring prior to, during, and after use to ensure proper quality. Effluent treatment is required for the removal of fumes, vapors, and particulates when monitoring analyses indicate that levels exceed the acceptable discharge limits.

Wear goggles and proper respirators when spraying or performing any operation where an abnormal vapor or dust is formed. Eye protection should be worn during paint spraying or when painting overhead. Cleaning the hands with paint solvents or thinners may cause serious chronic skin complaints. Wear appropriate gloves and clothing when handling dermatitic material, and change and clean work clothing daily. Avoid touching any part of the body when handling dermatitic materials. Wash hands, face, and arms thoroughly before eating and at the end of the day, and try to take a shower at or near the job site.

All alkaline cleaners should be handled with care. Rubber gloves and face or eye shields should be worn when these materials are added to cleaning tanks. Should these materials touch the skin, it should be flushed with water as soon as possible. These precautions also apply to the handling of acids used in phosphating and chromating.

Liquid acid should never be drawn from a carboy by using air pressure to force it out, even when using so-called air pressure reducers. There is always the danger that the carboy will break and spray or splash acid onto the operator. This also holds true for drums.

3.37 PAINT SCHEDULES FOR NON-METALLIC SURFACES

Tables 3.6–3.14 provide details of paint schedules for non-metalic surfaces.

Table 3.6	Paint Systems for Plaster, Concrete, Brick, Block, and Stone (Internal)			
Substrate Condition	**Finish Types**	**Primer**	**Finish System**	**Typical Life To First Maintenance[2]**
Dry R.H. below 75%	Alkyd gloss, mid-sheen, or matt. (A5/1), (A5/2, A5/3)	Alkali resisting primer (A3/1) or, plaster only, water-thinned primer (A3/2)	(a) Gloss finish. one coat oil-based (A4/1) or emulsion (A4/2) undercoat; one coat alkyd gloss finish (A5/1) (b) Mid-sheen, finish, two coats alkvd mid-sheen, finish (A5/2) (c) Matle-finish. two coat; alkyd matt finish (A5/3)	Five years or more Five years or more Up to five years
	Emulsion paint (A5/4, A5/5)	Primer not usually required A well-thinned first coat of emulsion paint may be required on surfaces of high or variable porosity	d) Matle or mid-sheen, finish. two or three coats general purpose emulsion paint, matle or mid-sheen, (A5/4) (e) Matle, high-opacity finish. two coats 'contract' emulsion paint A5/5 one coat, spray-applied, may suffice in some situations	Five years or more Up to five years
	Multi-color (A5/13)	Primer or basecoat as recommended by manufacturer	(f) Usually one coat multi-color finish (A5/13) spray-applied but refer to manufacturer's instructions	10 years or more
	Textured (A5/13)	Primer not usually required but refer to manufacturer's instructions	(g) 'Plastic' texture paint (A5/14). normally one coat but may require overpainting (h) Emulsion-based masonry paint, heavy-texture (A5/9) normally one coat but refer to manufacturer's instructions	Indefinite in environments in which normally used but likely to require periodic overpainting to maintain appearance
	Cement paint (A5/12) (not on gypsum plaster)	Primer not required	(i) One or two coats cement paint (A5/n)	Five years or more in situations for which cement paint is generally used
Drying some damp patches R.H. 75% to 90%	Emulsion paint (A5/4. A5/5)	As for dry substrates	As (d) and (e) above	Generally as for similar systems on dry as substrates but some risk of failure at higher moisture levels
	Multi-color finishes (A5/13) possible but consult manufacturer	As for dry substrates	As (f) above	
	Textured paints (A5/14) possible but consult manufacturer	As for dry substrates	As (g) and (h) above, if overcoating is nccessary, emulsion paint shall be used	
	Cement paints (A5/12) (not on gypsum plaster)	Primer not required	As (i) above	As for dry substrates

Table 3.6 Paint Systems for Plaster, Concrete, Brick, Block, and Stone (Internal) *(cont.)*

Substrate Condition	Finish Types	Primer	Finish System	Typical Life To First Maintenance[2]
Damp obvious damp patches R.H. 90% to 100%	Emulsion paint (A5/4, A5/5) possible	Primer not recommended	As (d) ami (e) above 'contract' types (A5/5) are usually more permeable than general purpose (A5/4) types and less prone to failure on damp substrates	High risk of early failure
	Cement paint (A5/12) (not on gypsum plaster)	Primer not required	As (i) above	As for dry substrates
Wet moisture visible on surface R.H. 100%	Cement paint (A5/12) (not on gypsum plaster)	Primer not required	As (i) above	Generally as for dry substrates but some risk of failure

[1] R.H. *refers to the relative humidity in equilibrium with the surface.*
[2] *Life expectancies shown assume application to dry, sound substrates, qualified as indicated for other substrate conditions, and based on performance in moderate internal environments.*

Table 3.7   Paint Systems for Renderings, Concrete, Brick, Block, and Stone (External)

Substrate Condition	Finish Type	Primer	Finish System	Typical Life To First Maintenance[2]
Dry R.H. below75%[1]	Alkyd gloss (A 5/1)	Alkali-resisting primer	(a) One coat oil-based undercoat (A4/1); One or two coats alkyd gloss finish (A5/1)	Three years to five years or more
	Emulsion paint general purpose (A.5/4) if suitable for external use	Primer not usually required	(b) Two coats general purpose emulsion paint (A5/4)	Up to five years
	Masonry paints, solvent-thinned A5/10 or A5/11	Alkali-resisting primer (A3/l) or as recommended by manufacturer	(c) Smooth or fine-textured (A5/10) types, solvent-thinned two coats	Five years or more
			(d) Thick, textured (A5/11) types, solvent-thinned, usually one or two coats applied by spray, often by specialist applications	10 years or more
	Masonry paints, emulsion-based (A5/8 or A5/9)	Primer not usually required with A5/8 types	(e) Smooth or fine-textured (A5/8) types. emulsion-based, two coats	Five years or more
		Primer not usually required with A5/9 types but refer to manufacturer's instructions	(f) Heavy-textured (A5/9) types, emulsion-based, usually one coat applied by roller	10 years or more
	Cement paint (A5/12)	Primer not required	(g) Two coats cement paint (A5/12)	Up to five years

(Continued)

Table 3.7 Paint Systems for Renderings, Concrete, Brick, Block, and Stone (External) *(cont.)*

Substrate Condition	Finish Type	Primer	Finish System	Typical Life To First Maintenance[2]
Drying some damp patches may be visible R.H. 75% to 90%	Emulsion paint. general purpose (A5/4)	As for dry substrates	As (b) above	Potentially, as for dry substrates but some risk of earlier failure at higher moisture levels
	Masonry paints, emulsion-based (A5/8 or A5/9)	As for dry substrates	As (e) or (f) above	
	Possibly solvent-thinned masonry paints (A5/10, A5/11) but refer to manufacturer's recommendations	As for dry substrates	As (c) or (d) above	
Damp obvious patches R.H. 90% to 100 %	Cement paint (A5/12)	Primer not required	Two coats cement paint (A5/12)	As for dry substrates
	Possibly emulsion-based masonry paints (A5/8, A5/9) but refer to manufacturer's recommendations	Primer not recommended	As (e) or (f) above	Potentially as for dry substrates but high risk of earlier failure
Wet moisture visible on surface. R.H. 100%	Cement paint(A5/12)	Primer not required	2 coats cement paint (A5/12)	As for dry substrates but some risk of earlier failure

[1]*R.H. refers to the relative humidity in equilibrium with the surface.*
[2]*Life expectancies shown assume application to dry, sound substrates, qualified as indicated for other substrate conditions, and are based on performance in moderate external conditions.*

Table 3.8 Miscellaneous Primers for Building

Ref.	Description	General Composition	Characteristics And Usage
A 3/1	Alkali-resisting	Typically, alkali-resistant drying-oil resin type binder, lightly pigmented	Although described as "alkali-resisting", primers of this type are intended for use on substantially dry, possibly alkali-containing, surfaces mainly beneath drying-oil type finishes; they will not necessarily prevent attack by alkalis on subsequent coats if the structure is very damp. Color: typically low-opacity white or off-white
A 3/2	Water-thinned primer or primer-undercoat	Emulsion-type binder (typically based on an acrylic polymer) may be identical in composition to A4/2	These primers are for use on dry plaster and similar surfaces; they may also be suitable for priming wood and building boards and as undercoats; see A4/2. Color: typically white or off-white

Table 3.8 Miscellaneous Primers for Building *(cont.)*

Ref.	Description	General Composition	Characteristics And Usage
A 3/3	Primer-sealer, other descriptions include "stabilizing primer or solution" "penetrating primer" or "masonry sealer"	Variable TUT typically drying-oil / resin type binder, lightly pigmented or semi-transparent with a "marker" pigment to assist even application	The essential function of a primer-sealer is to "bind down" powdery or friable residues of previous coatings which can not be removed completely, to much reliance should not be placed on the ability of this type of material to penetrate unsound coatings of substantial thickness, and as much as possible of the old material shall be removed, primer sealers may also be used to reduce absorption on surfaces of high or uneven porosity, "masonry sealers" are usually formulated specifically for use on exterior surfaces and may not be suitable for interior work: with other types, the reverse may apply, reference should be made to the manufacturer's recommendations Color: variable
A3/4	Universal primer	Typically, drying-oil / resin type binder with white or light-colored pigments, usually including lust-inhibitive types; e.g., zinc phosphate	Primers of this type are convenient for small scale maintenance work involving patch-priming of a variety of substrates, e.g., wood, metal and dry plaster; for new and large-scale work, primers formulated for specific substrates are generally be preferred Color: typically white or light gray

Table 3.9   Undercoats for Building

Ref.	Description	General Composition	Characteristics And Usage
A 4/1	Oil-based undercoat	Typically, drying-oil resin type binder, pigmented with titanium dioxide and / or colored pigments	This type of undercoat provides a matte or low-sheen surface for subsequent application of drying-oil type finishes, especially gloss finishes; in normal conditions, overnight drying is usually required before application of further coats. Most undercoats are suitable for interior and exterior use Color: wide range
A 4/2	Water-thinned undercoat or primer-undercoat	Emulsion-type binder (typically based on an acrylic polymer), pigmented as A4/1 may be identical in composition to A3/2	These undercoats are quicker-drying than A4/1 and, in normal conditions, may permit same-day recoating. General usage as for A4/1 as primers they may be suitable for priming dry plaster and similar surfaces, building boards, and wood Color: may be limited to white and pale tints, especially with "primer-undercoat" types

Table 3.10   Pigmented Finishes for Building

Ref.	Description	General Composition	Characteristics And Usage
A5/1	Alkyd gloss finish	Typically, drying-oil alkyd resin binder, pigmented with titanium dioxide and or light-fast, colored pigments, some types contain) small amounts of polyurethane or other resins to increase hardness or extend durability	These are high-gloss protective and decorative finishes suitable for interior and exterior use on most building surfaces; they have good durability and wearing properties in most conditions except those of direct chemical attack or very high humidity, drying and recoating times are variable according to composition but, usually, overnight drying between coats is required Color: wide range available
A 5/2	Alkyd "mid-sheen" finish, other descriptions include "eggshell", "satin" and "silk"	Generally as for A5/1 but adjusted to provide a lower level of a gloss	The durability and protective properties of this type of finish are of a lower order than those of 5/1, and exterior use, especially on wood and metal, is not usually recommended. Internally, this finish is suitable for use on most building surfaces. Drying and recoating characteristics arc usually similar to those of A5/1 Color : wide range available
A5/3	Alkyd matte or flat finish	Generally as for A5/1 adjusted to provide a matte finish	Essentially, this is a decorative finish for interior use only and is not generally recommended for "hard-wear" situations or in humid conditions. Drying and recoating characteristics are usually similar to those of A5/l. Color : range may be limited
A 5/4	Emulsion paint, general purpose matte or "mid-sheen". other descriptions include "plastic emulsion". "latex" and "vinyl", the latter term being used increasingly as a generic description for emulsion paints rather than an indication of com-position, descriptive terms for sheen levels may be described at for A5/2. note: "glossy" emulsion paints are available, but experience with them is limited and they are not in widespread general use at the present time, reference should be made to the manufacturers for information regarding suitability for specific applications	Emulsion-type binder, typically based on vinyl or acrylic polymers or combinations of these, with titanium dioxide and/or colored pigments	Emulsion paints of this type are essentially decorative rather than protective in function, although their wear-resistance and washability may approach those of A2/2 finishes. They have advantages over the latter in case of application to large areas, speed of drying (same-day recoating is usually possible), absence of solvent odor and ease of clean-up. They are widely used for ceilings and walls, mainly internally although some may be suitable for use on exterior walls. Choice of sheen level may depend largely on location and aesthetic considerations but, in general terms washability and resistance to soiling increase proportionately to the degree of sheen. Emulsion paints are more permeable to moisture than drying-oil type finishes, and some may be unsuitable for use in situations; e.g., in some kitchens and bathrooms, where humidity is high for long periods, color: a wide range is available

Ref.	Description	General Composition	Characteristics And Usage
Table 3.10 Pigmented Finishes for Building *(cont.)*			
A5/5	Contract "Emulsion paint"	Composition, typically, is similar to that of A5/4 but with higher pigment content	Higher pigment gives increased opacity, compared with 5/4 type, but at the expense of some reduction in wash ability and resistance to soiling. Finish is usually matle. Emulsion paints of this type are frequently used for new interior walls and ceilings where, apart from economy in use, their permeability allows drying-out of contained moisture Color: usually limited to white and pale tints
A 5/6	Aluminum paint, general purpose	Typically, drying-oil / alkyd resin binder pigmented with flake aluminum	The laminar nature of the pigment imparts excellent moisture resistance to aluminum paints, and then leflective and heat-resistant properties are useful in many situations. They are suitable for use on most building surfaces, but then widest application is on structural steelwork, storage vessels, and heated metal surfaces
A 5/7	Micaceous iron oxide paint, general purpose	Typically, drying-oil or drying-oil / resin type binder pigmented with micaceous iron oxide. Small amounts of other pigments may be incorporated to modify the natural color (dark gray) of the main pigment	Because of the laminar nature of the pigment and the substantial thickness of the film, micaceous iron oxide paints provide good protection and are widely used extensively on structural steel, bridges, harbor installations and electricity transmission towers, on weathering, the coating develops a metallic "sparkle", the dark color of the pigment restricts the range of colors available, but overcoating with alkyd gloss finish may be possible
A 5/8	Masonry paint, emulsion-type, smooth or fine-textured	Emulsion- type binder, typically based on acrylic, vinyl and other polymers formulated to have the degree of flexibility required for exterior used and pigmented with titanium dioxide and/ or lime-resistant colored pigments. Some types are smooth, others contain fibre, sand or other agent to give a fine texture, most contain an additive to inhibit mould and algal growth	These are essentially "decorative" coatings but have the weather resistance and durability required in coatings for exterior walls, They are not generally suitable for application to wood and metal, formulations containing fibre sand, etc., usually provide thicker films than the smooth types. Manufacturers should be consulted regarding the use on interior surfaces of masonry paints containing additives to inhibit mould and signal growth because of the possible toxic hazard Color: a wide range is available

(Continued)

Table 3.10 Pigmented Finishes for Building *(cont.)*

Ref.	Description	General Composition	Characteristics And Usage
A 5/9	Masonry paint, emulsion-based heavy-textured, some types may be described as "organic renderings"	Compositions may vary considerably but consist generally of a heavy-bodied emulsion-type binder, reinforced with coarse "extenders" and, in some cases, fibrous material and aggregates, pigmentation is similar to that of A5/8 and mould algal inhibitions may be incorporated	Coatings of this type have been used in Europe for external wall surfaces for many years but have become available in Britain only comparatively recently, they provide thick (1 mm to to 2 mm), weather-resistant coatings, and experience in Europe indicates that they are capable of durability in excess of 10 years, their heavy texture is derived partly from their composition specially-designed rollers, in addition to their use externally, coatings of this type may be suitable for internal walls where a hard-wearing textured finish is required Color: a wide range is available
A 5/10	Masonry paint, solvent-thinned, smooth or fine-textured, the latter type may be described as "stone" paints	Composition varies considerably, older types may be based on A0 drying-oil / resin type binder, others on modified or synthetic rubber film-formers, "stone" paints contain fine sand, stone or mica, pigmentation is generally as for other masonry paints	The appearance of this type of coating is similar to that of the A5/8 types, and film thickness is of the same order, permeability is usually lower, and they may therefore provide greater resistance to moisture penetration but should be applied to dry substrates, brush application will generally be shower than with A5/8 types, Color: may be limited
A 5/11	Masonry paints, solvent-thinned, thick, textured	Composition variable but typically employs a drying-oil/alkyd resin binder, may be described as "polyester" (= alkyd). Usually contains coarse 'extender' and/or fibrous material. Pigmentation as for other masonry paints	These provide coatings of substantial thickness (0.6 mm to 1.0 mm), of relatively low permeability, and therefore have good resistance to moisture penetration, coatings of this type are usually applied by spray or roller, often by specialist applications, material and application costs are likely to be high, Color: may be limited
A 5/12	Cement paint to BS 4764	Based on white Portland cement with titanium dioxide or colored pigments, and additives to assist application and increase water-repellency. Supplied in powder form and mixed with water immediately before application	Low-cost coatings giving a rough finish and used mainly for exterior wall surfaces, although they may be used inside except on gypsum plaster: not recommended for application over other types of paint, on lengthy outside exposure, pastel shades may tend to lighten and darker colors to become patchy, Color: range is fairly limited

Table 3.10 Pigmented Finishes for Building *(cont.)*

Ref.	Description	General Composition	Characteristics And Usage
A 5/13	Multi-color finish	In one type, droplets of pigmented nitrocellulose or vinyl resin solution are dispersed in an aqueous medium so that the coating is water-thinnable. The pigmented droplets remain discrete in the dry film, so providing the multi-color effect. In another type, a pigmented emulsion-type basecoat is followed, when dry, by spray-applied 'spatter' coats of similar material. Finally, an emulsion-type clear glazecoat may be applied, in a third type, a pigmented emulsion-type basecoat is followed, when dry, by a clear emulsion glaze in which are suspended multi-colored flakes of solid material	This is quick-dryling, hard-wearing finish for interior wall surfaces, often used in circulating areas, cloakrooms and similar "hard-wear" locations, application is usually by spray. but materials of the third type may be suitable for brush or roller application, Color: a wide range of multi-color effects is available
A 5/14	"Plastic" texture paint	Typically, based on gypsum and supplied as dry powder for mixing with water before use, but ready-for-use emulsion-based types are available, the term "plastic" indicates that the material can be "worked" after application to provide relief texture effects	This is essentially a "decorative" coating for interior use only. it may be used as a substitute for plaster skimming on plaster board ceilings or to disquise rough or cracked walls and ceilings, texture is achieved either directly by spraying or by brush or trowel application and subsequent combing, stippling or other treatment, some types are sell-colored and require no further treatment; others may require over-painting, e.g., with emulsion paint

Table 3.11   Paint Systems for Wood, Internal, and External (Excluding Hardwood Surfaces Not Usually Painted)

Application	Requirements For Preservative Treatment[1]	Primers And References In Table A.7	Finish Systems And Product References In Tables A.4 and A.5	Typical Life To First Maintenance[2]
Window Joinery, softwood, internal and external	Essential	Low-lead, oil-based (A7/1). Aluminum A7/2 (preferred for resinous woods, e.g. douglas fir. and possibly for timber treated with metallic naphthenate preservatives). water-thinned A7/3 (compatibility with water-repellent organic solvent Preservatives should be clicked)	External gloss One Coat undercoat (A4/1 or A4/2) Two coats alkyd gloss finish (A5/1). (or two undercoats and two coat finish)	Three years to five years
			Internal gloss One Coat undercoat (A4/1 or A4/2) One Coat alkyd gloss finish	Five years or more
			Internal mid-sheen two coats alkyd mid-sheen finish (A5/2)	Up to five years
			Internal matle two coats alkyd matle finish (A5/3)	As mid-sheen finish

(Continued)

Table 3.11 Paint Systems for Wood, Internal, and External (Excluding Hardwood Surfaces Not Usually Painted) *(cont.)*

Application	Requirements For Preservative Treatment[1]	Primers And References In Table A.7	Finish Systems And Product References In Tables A.4 and A.5	Typical Life To First Maintenance[2]
External silts, hardwood	Optional but necessary it excessive sapwood present	Low-lead, oil-based (A7/l) aluminium (A7/2) filling necessary with open-grain timber	As for external gloss above	Generally as for window joinery but depending on species and nature of timber
Doors and frames internal softwood plywood	Optional None	As for window joinery As for window joinery filling recommended on open-grain veneers	As for window joinery	As for window joinery
Doors and frames, external softwood plywood	Desirable Optional; required if it contains non-durable species	As for window joinery low-lead, oil-based (A7/1) aluminium (A7/2) filling recommended on open-grained veneers or if checking occurred	As for window joinery	As for window joinery
Skirtings, softwood	Desirable	As for window jointly	As for window joinery	
Cladding, barge-boards, fascias and soffits softwood plywood	Required by building regulations for some species (notwestern red cedar) Optional: required if it contains non-durable species	As for window joinery Low-lead oil-based (A 7/11) aluminium (A7/2) paper overlay desirable and may allow primer to be omitted	Textured coatings finish types A5/9. A5/10 and A5/11 may be suitable for use on timber cladding consult manufacturers	As for window joinery As for window joinery Up to 10 years
Gates and fences softwood hardwood	Essential	Low-lead, oil-based (A7/1) alumimnn (A7/2)	Gloss finish as for window joinery	Up to five years, depending on design and degree of exposure
	Desirable and may be essential in some circumstances: see BS 5589	Note: unless painting is necessary for appearance, consideration shall be given to preservative treatment of gates and fences initially and for subsequent maintenance (See Note[1])		

[1]*Preservative treatments are summarized in the standards; also refer to BS 5589.*
[2]*Life expectancies shown assume application to dry, sound woodwork which, if necessary, has received appropriate preservative treatment and is in moderate environments.*

Table 3.12   Primers for Wood

Ref.	Description	General Composition	Characteristics And Usage
A7/1	Low-lead, oil-based primer see BS 5358	Drying-oil/resin type binder with "low lead" pigmentation	Primers of this type have a lead content of less than 1% (as determined by the method described in BS 4310) but are equivalent in performance to the traditional "pink" (white lead/red lead) wood primers. They are suitable for general use on wood not highly resinous and not treated with metallic naphthenate preservatives: also for fiber boards and wood chipboards not fire-retardant treated Color: typically white or pink
A 7/2	Aluminum wood primer to BS4756	Drying-oil / resin type binder with aluminum, pigment	Alternative to A7/1. and more suitable for woods which are resinous or have been treated with metallic naphthenate wood preservatives or creosote, may also be used as primers for fiber boards and wood chipboards (not fire-retardant treated) and as 'sealers' for surfaces that have been coated with bituminous materials Color: aluminum
A 7/3	Water-thinned primer see to BS 5082	Emulsion-type binder (typically based on acrylic polymer) with lead-free pigmentation	These primers dry more rapidly than A7/1 and A7/2 types, usually allowing same-day recoating if required. They do not contain flammable solvents and permit tools and equipment to be cleaned with water. Their durability without top-coats on exterior exposure is equivalent to that of A7/1 and A7/12 types but, as they are more permeable, they may be less effective in excluding raise grain than oil-based primers They may also be used as primers for fibre building boards and woodchip boards not fire-retardant treated, color: typically white, light gray or pink

Table 3.13   Natural Finish Systems for Wood, External And Internal

Application	Product Type And Reference In Table A.9	System	Typical Life To First Maintenance[1]
External window joinery doors and frames			
Hardwood	Varnish Not recommended Exterior wood stain Low solids (A 9/7) High solids (A9/8), Opaque (A9/9)	Two to three Coats	Variable according to product type but unlikely lo exceed three years on full exposure
	Varnish Exterior grade, fuel gloss (A9/2)	Four Coats	Unlikely to exceed three years
	Exterior wood stain As for softwood	Two to three Coats	As for softwood
Plywood e.g., door panels	Varnish Not recommended	N/A	N/A
	Exterior wood stain Low solids (A9/A)	Two to three Coats	As for softwood, salt-staining possible

(Continued)

Table 3.13 Natural Finish Systems for Wood, External And Internal (*cont.*)

Application	Product Type And Reference In Table A.9	System	Typical Life To First Maintenance[1]
External boarding cladding, bargeboards, soffits fascias			
Softwood	Varnish Not recommended	N/A	N/A
	Exterior wood stain As far window joinery Also Madison formula (A9/6)	Two to three Coats	As for window joinery
Hardwood	Varnish Exterior grade, full gloss (A9/1)	Four Coats	As for window joinery
	Exterior wood stain As for window joinery As Madison formula (A9/6)	Two to three Coats	As for window joinery
Plywood	Varnish Not recommended	N/A	N/A
	Exterior wood stain As for window joinery Also Madison formula (A9/6)	Two to three	As for window joinery. salt-staining possible
Gate, fences, handrails			
Softwood	Varnish Not recommended	N/A	N/A
	Exterior wood stain Law solids (A9/A)	Two to three Coats	As for window joinery
Hardwood	Varnish Exterior grade, full gloss (A9/1)	Four Coats	As for window joinery
	Exterior wood stain Low solids (AS/7)	Two to Three	As for window joinery
Internal general joinery surfaces, linings and fitment			
	Varnish Interior grade full gloss (A.9/2) Or mid-sheen (A 9/3) Polyurethane, two-pack or Moisture-curing (A 9/4) for Exceptional abrasion resistance	Two to three Coats	Variable according to type and service conditions but typically up to five years in average wear environments
Softwood, hardwood plywood[2]	Wood stain Some exterior wood stains, e.g., A9/7 and A.5/8, may be Suitable for interior use but Refer to manufacturer's Recommendations	One to two Coats	Variable according to type and service conditions, may give lifetime service in some situations

* Over decorative wood stain (49/5) if required.
[1]*Life expectancies shown assume application to dry, sound timber which, if necessary, has received preservative treatment.*
[2]*When a high standard of finish is required on internal hardwood surfaces, special wood finishes, such as French polish or lacquer, are generally used.*

	Description	**General Composition**	**Characteristics And Usage**
A 9/1	Varnish, exterior grade, full gloss	Typically, drying-oil, phenolic or alkyd resin	This provides a tough, flexible, water-resistant coating, used principally as a clear protective finish for exterior hardwood
A 9/2	Varnish, interior grade, full gloss	Typically, drying-oil/alkyd, urethane or urethane/alkyd resin	Harder than A9/1 type and is more suitable for use on interior hardwood joinery, some types may be sufficiently abrasion-resistant to be suitable for use on hardwood floors, counter tops and similar hard wear locations
A 9/3	Varnish, eggshell, satin or matle finish	Composition generally as for A 9/2 but adjusted to provide a lower level of gloss	Generally as for A9/2 but is likely to be less suitable for use in hard wear locations
A 9/4	Varnish, polyurethane, two-pack or moisture-curing one-pack	Two-pack types are supplied as separate base and activator. which are mixed before use to initiate chemical curing, with one-pack moisture-curing types, the reaction is initiated by absorption of moisture from the atmosphere or from the surface to which the material is applied	These coatings provide extremely hard, strong films with exceptional resistance to abrasion, the stresses set up within the film may lead to peeling and flaking especially on exterior woodwork whose surface through long exposure without protection, has become degraded, in general, the use of this type of coating is best confined to interior woodwork where exceptional resistance to abrasion and possible chemical attack is required
A 9/5	Decorative wood stain	Drying-oil, spirit or water media, with colored pigments or dyes	Used essentially to modify or enhance the appearance of wood without obscuring its grain and is usually overcoated with clear finishes
A 9/6	Exterior wood stain, semi-transparent, Madison formula	Drying-oil binder (boiled linseed oil), paraffin wax, fungicide and pigment	Water-repellent penetrating; stain that imports an oiled appearance to the wood, suitable for brush application, film remains soft and slightly tacky so will retain dirt and is not suitable for situations where it is likely to be abraded or come in contact with clothing, wax component may cause difficulty if over-painting is subsequently required
A 9/7	Exterior wood stain, semi-transparent, low solids	Resin solution of low viscosity with fungicide and pigment	Water-repellent penetrating stain suitable for brush application, imports little or no sheen to surface of wood, because of very low film thickness. offers little resistance to passage of water vapor and, in consequence, moisture content of wood may fluctuate considerably. Stains of this type can be used on interior woodwork but before doing so, it should be as certain from the manufacturer that the fungicide contained does not constitute a health hazard

Table 3.14   Natural Finishes for Wood

Table 3.14 Natural Finishes for Wood *(cont.)*

	Description	**General Composition**	**Characteristics And Usage**
A 9/8	Exterior wood stain, semi-transparent, high solids	Generally as for A 9/1 but higher resin content	Because of its higher resin content, this type of stain will normally import a noticeable sheen to the surface, it is less penetrative than A 9/7 and offers greater resistance to water vapor movement, so fluctuations in the moisture content of the wood are less pronounced
A 9/9	Exterior wood stain, opaque	May be a solvent-thinned resin solution or an emulsion-type, with pigment and fungicide	This type may be regarded as intermediate between a stain and a paint, it has low gloss, but the texture of the wood remains evident because of the differences in penetration within the growth rings Some opaque stains may be used over weathered but sound paintwork

3.38 CONVERSION COATING (METAL SURFACE TREATMENTS)

Wash primer, phosphate coating, and chromate coating are conversion coatings that are applied over ferrous and nonferrous metals.

3.38.1 WASH PRIMER

Basic zinc chromate-vinyl butyral wash primers are a pretreatment medium for metals; they react with the metal and form a protective vinyl film that contains an inhibitive pigment to help prevent rusting and improve the adhesion of paint.

Wash primer should be mixed with the diluent in quantities that will be applied within 6–8 h after mixing. Primer that cannot be used within a maximum of 8 h after mixing with diluent should be discarded.

Apply the wash primer by spraying or brushing. Spraying is generally the preferred method, but brushing may be desirable when the primer is applied to rough or poorly prepared steel. Roller coating should not be used.

Wash primer should be applied to dried film thicknesses of 10–13 μm dry, or approximately 100–130 μm wet. Note that at these thicknesses, which should not be exceeded, the base metal will show through the coating, as evidenced by uneven coloring. This is the normal appearance; do not attempt to hide the base metal completely with the primer.

When spot treating, cover only spots free of old paint. A slight overlap of existing paint is generally not harmful, provided that adherence of the wash primer to the old paint is satisfactory and the old paint is not lifted. The next coat of paint may be applied as soon as the wash primer is dry, usually from a half-hour to 4 h later, except when otherwise authorized by the inspector.

Wash primer should be applied over clean, dry metal; however, a slightly damp surface may be painted over, provided that adequate normal butyl alcohol is used in the thinner. If the surface is excessively wet, the vinyl butyral resin will be ejected from the solution and form a gel, or the dried film will turn white, become brittle, and lack adhesion to the metal.

Wash primer is not intended for use as a shop coat for metal, and it should be recoated with the prime coat of paint before exposure, preferably within 24 h. It is especially effective when applied to galvanized steel or aluminum, but is not effective if applied over another primer or a phosphated or chromated surface because the reaction will be impeded and adhesion destroyed. It must be applied directly over a bare metal substrate.

The wash primer should not be white in spots; when dry, it should be tested for adhesion to the substrate by scraping it away with a knife. Wash primer permits many types of conventional paint to be used over steel and galvanizing surfaces, especially an emulsion cleaner water blasting is used for preparation of surfaces. It is also used to provide adhesion of paint to stainless steel and aluminum and should be required by some vinyl paint system.

Keep wash primer away from heat, sparks, and open flame during storage, mixing, and application. Provide sufficient ventilation to maintain vapor concentration at less than 25% of the lower explosive limit.

3.39 PHOSPHATE COATING

3.39.1 USES

Phosphate coating is the treatment of ferrous metal with a dilute solution of phosphoric acid and other chemicals, whereby the surface of the metal, reacting chemically with the phosphoric acid, is converted to an integral, mildly protective layer of insoluble crystalline phosphate. Phosphate coatings transform the metal surface into a new surface with nonmetallic and nonconductive properties. They are used under paint, plastic coating, metallic coating, wax, and rust-preventive oil, for protection of the surface against corrosion and for improving adhesion. Phosphate coatings are generally used for ferrous metals and applied by immersion and spraying methods.

Two principal types of phosphate coatings are in general use: zinc and iron. The other types of phosphating include manganese phosphating, solvent phosphating, and phosphating in low temperatures; they are used for special purposes. Phosphate coating is applied in a thickness of 2.5–50 μm. Coating weight (in milligrams per square decimeter of coated area), rather than coating thickness has been adopted as the basis for expressing the amount of coating deposited. The types of phosphate coating are as follows:

- Type 1.A—Heavy weight; coatings that consist essentially of manganese and/or iron phosphate and have coating weights of not less than 7.5 g/m^2 of the treated surface.
- Type 1.B—Heavy weight; coatings that consist essentially of zinc phosphate and have coating weights of at least 7.5 g/m^2 of treated surface.
- Type 2—Medium weight; coatings that consist essentially of zinc and/or other phosphates and have coating weights within the range of 4.5–7.5 g/m^2 of the treated surface.
- Type 3—Light weight; coatings that consist essentially of zinc or other metal phosphates and have coating weights of 0.2–4.5 g/m^2.

Iron phosphating has a lower initial capital investment and fewer processing stages than zinc phosphating, but zinc phosphating has a greater coating weight than iron phosphating and permits the application of heavier paint finishes with a potentially longer life expectancy.

3.39.2 ZINC PHOSPHATE COATING

Zinc phosphate coating is formed by crystallization on the surface by a chemical reaction. Zinc phosphating should be applied by spray or immersion, and it can be used for any phosphating application.

Spray coating on steel surfaces ranges in weight from $1–11$ g/m^2, and immersion coating from $1.6–43$ g/m^2. Zinc phosphate coating of $1.6–4.5$ g/m^2 is used for steel prior to painting, and coating of $1.6–43$ g/m^2 is used for zinc and zinc that was plated prior to painting. The type of phosphate coating and the coating weight of zinc phosphate coating for specific jobs should be approved by the company.

3.39.3 ZINC PHOSPHATING PROCESS

The zinc phosphate system should be applied in six stages:

1. Degreasing with weakly alkaline cleaner (concentration 1%, temperature 50–55°C)
2. Rinsing with water by immersion or spraying
3. Zinc phosphating
4. Rinsing with water
5. Performing a passivating rinse
6. Drying

Note that usually only one rinse is required. If the water supply is so high in mineral content that residue remains on the parts after rinsing, a rinse in deionized water should be required. An intermediate stage may be interspersed between the second and third stages for the purpose of improving the crystal size of the zinc phosphate by using a colloidal titanium salt.

Zinc phosphating component-zinc phosphating should be applied to a surface by either immersion or spraying. A fine uniform crystal is desirable when gloss is desired for the paint film (coarse crystals promote dullness). When the coating is applied to provide lubricity, a coarse crystal may be preferable. In this method, the surface is treated with a chemical solution prepared by diluting a propriety concentrate to the 2–4% level. Immersion baths are more concentrated than spray baths.

The essential components of a phosphating bath are zinc salt, neutralized phosphoric acid, nitric acid, an oxidant, and an accelerator (sodium nitrite).

With a passivating rinse, phosphated parts should be given a chromic acid rinse following the post-phosphating water rinse. The purpose of the chromic acid rinse is to neutralize any phosphating acid that may remain on the parts and to leave a light chromate coating, which improves the corrosion resistance of the part. The phosphated parts should be immersed in or sprayed with a chromic acid solution with a concentration of 15–31 ml of acid per 100 l of water.

When paint sensitive to chromic acid is to be applied subsequently, phosphated parts should be rinsed with deoinized or distilled water to remove excess chromic acid.

3.39.4 IRON PHOSPHATE COATING

Most iron phosphate coatings are applied by spraying. The iron phosphate system is applied by a four-stage process:

1. Cleaning and phosphating
2. Rinsing in hot water (70–80°C) by immersion or spraying

3. Performing a passivating rinse.

4. Drying

The components of the phosphating solution are salt of phosphoric acid, ammonium or potassium dihydrogen phosphate (70–80 wt%- solid content), activator (sodium nitrate 8–10 wt% solid content) and cleaning agent (nonionic surfactant 10 wt% solid content). The concentration (volume percent) of these materials normally range from 0.5–2% iron phosphating solution. Iron phosphate coatings are generally of a very fine structure which is amorphous (noncrystalline) in appearance.

Because these coatings are used primarily as bases for paint or to assist in the bonding of metal to a nonmetallic surface (fabric, wood and others), the fine structure is desirable. To maintain the required specifications for iron phosphate coating, chemical control of processes is necessary.

3.40 OTHER TYPES OF PHOSPHATE COATING
3.40.1 MANGANESE PHOSPHATE COATING

Manganese phosphating is applied only by immersion, requiring times ranging from 5–30 min. Because manganese phosphate crystal is softer, and therefore breaks down more readily than fine phosphate crystal, these coatings should not be used prior to painting. They should be used as an oil base.

3.40.2 SOLVENT PHOSPHATE COATING

Solvent phosphate coating uses trichloroethylene (or methylen chloride) as the base for cleaning, phosphating, and subsequent finishing (Table 3.15). The method involves four stages and requires special equipment, as follows:

1. Vapor degreasing in a boiling bath of trichloroethylene at 86.5°C
2. Phosphating by either spraying or immersion in the organic acid phosphates
3. Nonaqueous rinse; the phosphated surface, while wet, is typically immersed or flooded with a chromium containing rinse composition (solvent base)

Table 3.15 Phosphating Specifications

Phosphating Method	PH	Temperature°C	Phosphate Coating Weight	
			g/m²	g/ft²
Zinc phosphating 1) Immersion 2) Process	1.8–2.4	67–77	6–43	0.16–4.6
Spray process	3–3.4	38–64	1–11	0.12–1.2
Iron phosphating	3–6	50–67	0.2–0.9	0.02–0.92
Manganese phosphatiug (immersion process)	1.8–2.4	90–99	5–32	0.67–3.4

4. Coating with an organic finish; solvent phosphating produce a water resistant coating and is suitable for water-based finishes

Note that light to medium zinc phosphate coating of (1.6–4.5 g/m²) and light iron phosphate coating (0.2–0.9 g/m²) are used for paint bases.

3.41 CHEMICAL CONTROL OF PHOSPHATING PROCESSES

To achieve a satisfactory phosphate coating on steel surfaces, phosphating solutions must be chemically controlled within limits. Even the mineral content of plain water rinses need to be controlled to avoid leaving a residue on parts. The specific limits vary depending on the phosphating concentrate used. Solutions should be tested on a regular schedule; unless otherwise specified by the company, the frequency of tests is determined by the workload of the phosphating line. General methods of controlling zinc, Iron and manganese phosphating solutions are described later in this chapter.

3.41.1 ZINC PHOSPHATING SOLUTIONS

When zinc phosphating solutions become unbalanced, the results are poor coatings, excessive sludge buildup, and insufficient coating weights. Three chemical tests are usually made on a zinc phosphating solution to determine its suitability for coating. These are tests of total acid value, accelerator content, and iron concentration.

3.41.2 TOTAL ACID VALUE

Zinc phosphate solutions have a total acid value established that should be maintained for satisfactory performance. To determine the total acid value, a 10-ml sample of the solution is titrated with 0.1 N of sodium hydroxide (1 ml of sodium hydroxide used = 1 point), using phenolphthalein as an indicator. One commonly used solution is controlled at 25–27 points. The end point is reached when the solution changes from colorless to pink.

3.41.3 ACCELERATOR TEST

Sodium nitrite is used as an accelerator in some zinc phosphate solutions. It is usually controlled at 1 point. Before the test for sodium nitrite is run, the phosphate solution should be checked for the absence of iron. This is done by dipping a strip of iron-test paper in the phosphate solution. If the paper does not change color, no iron is present in the solution. If the paper changes to pink, however, iron is present, and small amounts of sodium nitrite should be added until an iron-test paper shows no change.

The sodium nitrite test is made using a 25-ml sample of the phosphate solution. Between 10 and 20 drops of 50% sulphuric acid are added carefully to the solution, and the solution is then titrated with 0.042 N potassium permanganate. The end point is reached when the solution turns from colorless to pink (1 ml = 1 point).

3.41.4 IRON CONCENTRATION

Because iron is constantly being dissolved from parts being zinc-phosphated, the concentration of iron may build up until the efficiency of the solution is impaired. Zinc phosphating solutions usually operate best when the iron concentration is maintained between 3 and 4 points. Production experience with a particular solution will indicate whether the iron content can be expanded without affecting the quality of the coating.

To determine the iron content, a 10-ml sample of the solution is first acidulated with a sufficient amount of a 50% mixture of sulfuric and phosphoric acid to ensure a low pH while titrating (2 or 3 drops may be suffipotassium permanganate until a permanent pink color is obtained (1 ml = 1 point). This titration is used for immersion zinc phosphating solution. Spray zinc phosphating usually does not involve a buildup of iron in the solution because of the oxidizers that are present.

3.41.5 IRON PHOSPHATING SOLUTIONS

If recommended chemical limits are not maintained in iron phosphating solutions, the results will be low coating weights, powdery coatings, or incomplete coatings. To maintain required balance in iron phosphating solutions, titration checks should made to determine total acid value and acid-consumed value.

3.41.6 TOTAL ACID VALUE

Total acid value is determined by titration of a 10-ml sample of the phosphating solution with 0.1 N of sodium hydroxide, using thymolphthalein as an indicator. The end point is reached when the solution changes from colorless to blue. The number of milliliters of the 0.1 N sodium hydroxide is the total acid value (in points) of the phosphating solution. A normal concentration would be 10 points.

3.41.7 ACID-CONSUMED VALUE

Acid-consumed value is determined by titration of a 10-ml sample of the phosphating solution with 0.1 N sulfuric acid, using bromocresol green as an indicator. The end point is reached when the solution changes from blue to green. A normal range for the acid-consumed value for a solution with a 10-point total acid value would be 0.3–0.7 ml of 0.1N sulfuric acid.

3.41.8 MANGANESE PHOSPHATING SOLUTIONS

The manganese phosphating solutions used to produce wear-resistant and corrosion-protective coatings should be maintained in balance by controlling the total acid, free acid, acid ratio, and iron concentration values. Since the phosphate solutions are acid, these values are determined by titration methods using a standard basic solution. The frequency of control checks on manganese phosphating solutions depends a great deal on both the amount of work being processed through the tank and the volume of the solution. Normally, however, one to two checks per shift would be sufficient.

3.41.9 TOTAL ACID VALUE

Total acid value is determined by titration of a 2-ml sample of the phosphating solution with 0.1 N sodium hydroxide, using phenolphetalein as an indicator. The end point is reached when the solution changes from colorless to pink. The number of milliliters of 0.1 N sodium hydroxide multiplied by 5 (because a 2-ml sample is used rather than a 10-ml sample), equals the point value of the total acid. A normal total acid value would be 60 points (12 ml × 5).

3.41.10 FREE ACID VALUE

Free acid value is determined by titration of a 2-ml sample of the phosphating solution with 0.1 N sodium hydroxide, using methyl orange xylene cyanole as an indicator. The end point is reached when the solution color changes from purple to greenish gray. The number of milliliters, multiplied by 5, is the free acid value in points. A normal free acid value for a 60-point total acid value would be 9.5–11.0 ml or points.

3.41.11 ACID RATIO

In order to obtain satisfactory coatings, the ratio of total acid to free acid content of a manganese phosphating solution should be maintained within certain limits. For a solution with a 60–70-point total acid value, this ratio should be between 5.5 to 1 and 6.5 to 1. Low-ratio solutions produce incomplete coatings, poorly adherent coatings, or coatings with a reddish cast. High-ratio solutions also result in poor coatings.

3.41.12 IRON CONCENTRATION

Since iron is continually dissolved from the parts going into the phosphating bath, the concentration of ferrous iron in the bath gradually builds up. Some manganese phosphate coating problems that can be traced to high iron concentrations are light gray instead of dark gray to black coating, powdery coatings, and incomplete coatings in a normal time cycle.

The concentration limits of iron will depend on the type, hardness, and surface condition of the steel being treated. A normal manganese phosphating bath will operate satisfactorily with an iron concentration ranging from 0.2–0.4%.

Production experience will indicate whether the iron concentration limits can be expanded without affecting the quality of the coating. To determine iron concentration, a 10-ml sample of the phosphating solution is used. To this sample, 1 ml of 50% sulphuric acid is added. The solution is then titrated with 0.18 N potassium permanganate. The end point is reached when the solution changes from colorless to pink; 1 ml of the 0.18 N potassium permanganate is equivalent to 0.1% iron.

3.41.13 IRON REMOVAL

If iron removal becomes necessary, the ferrous iron in the solution is oxidized with hydrogen peroxide, which causes the iron to precipitate and also liberates free acid in the bath. Since the free acid in the bath will increase and thus lower the acid ratio, it is necessary to neutralize the liberated free acid by adding manganese carbonate.

The approximate amount of hydrogen peroxide needed to lower the concentration of iron by 0.1% and neutralize the liberated free acid is 1.25 ml per 1 l of manganese carbonate. Iron removal may be unnecessary if the area of steel being processed and the volume of the phosphate bath limit the amount of iron buildup.

3.42 REPAIR OF PHOSPHATE COATINGS

Small parts that did not accept a satisfactory phosphate coating can easily be stripped, cleaned, and rephosphated. Large parts with a faulty coating or with a coating that was damaged in processing are less easily handled, and repair of the phosphated surface may therefore be preferable to stripping and rephosphating.

The simplest method is to blast the phosphate film until all defective coating is removed and clean bare metal is exposed. A proprietary phosphating solution compounded for this type of application is brushed or wiped on the area to be rephosphated, and it is allowed to remain in place for a prescribed length of time (usually measured in seconds); the surplus solution is then removed by thoroughly rinsing with water and wiping it dry with clean rags.

These wash-off or wipe-off solutions usually consist of phosphoric acid, butyl celosolve, and a suitable wetting agent, plus 50–70% water. If the volume of repairs is considerable, a portable steam spray unit may be used, which will spray hot phosphating solution, water, and chromic acid rinsing solution through a hose and nozzle.

3.43 INSPECTION METHODS

Most phosphate coating quality control methods are based on visual inspection. For zinc and manganese phosphate, the coating must be continuous, adhere well to the surface, and be of uniform crystalline texture that is suitable for the intended use. Color ranges from gray to black. Loose smut or white powder (due to dried phosphate solution), blotchiness, excessive coarseness, and poor adhesion can cause rejection. Crystal size may be observed by using micrographs at magnifications of 10x to as much as 250x, depending on the type of coating.

Iron phosphate coatings have no apparent crystalline texture; rather, they appear to be amorphous, with color varying from blue to brown. Loose or patchy coatings can cause rejection.

3.43.1 TEST FOR THE PRESENCE OF A PHOSPHATE COATING

When testing for phosphate coating, reagent needs to be prepared by dissolving 8 g of ammonium molybdate in 80 ml of distilled water, and then adding 12 ml concentrated hydrochloric (d = 1.14), 20 g ammonium chloride, and 10 ml of saturated potassium persulfate solution. The reagent should be fresh.

Add one drop of the reagent to the test surface. The appearance of blue within 30 s indicates the presence of a phosphate coating. A plain untreated surface of the same basis metal should be used as a control.

3.43.2 DETERMINATION OF COATING WEIGHT

The sample should be weighed on a balance with a precision of 0.1 mg and immersed in a fresh solution of concentrated hydrochloric acid (d = 1.14) containing 20 g/l of antimony trioxide at room temperature. When the coating has dissolved (a period of 5 min is normally sufficient), the sample should be removed from the solution and washed for 30 s in running water; any nonadherent matter should be rubbed off with a wet swab. The sample should then be thoroughly dried and reweighed and the weight of the phosphate coating calculated in grams per square meter. A solution containing 50 g/l of chromic acid, operated at approximately 75°C, may be used as an alternative to the previously specified test procedure.

Coating voids or spots not covered may be checked by the following process:

1. Use a clean, dry phosphated specimen.
2. Soak a piece of filter paper, 40–50 cm^2 in area, in a solution containing 7.5 g potassium ferricyanide and 20 g sodium chloride per liter. Allow excess solution to drain off.
3. Apply wet filter paper to the phosphate sample for 5 min. Remove and observe blue spots, which indicate noncoated areas.

The method of rating may vary with different processes and requirements; one general method is as follows:

- Excellent—Zero to three fine spots, up to 1 mm
- Good—No more than 10 fine spots
- Satisfactory—No more than 20 fine spots or up to 3 large spots

3.44 CHROMATE TREATMENT
3.44.1 COMPONENT

Chromate conversion coating is employed to impart brightness and improve the corrosion resistance of bare metal and as a substrate to provide improved paint adherence on nonferrous surface. It can be used on aluminum, zinc, tin plate, copper, cadmium, magnesium, and zinc-plated surfaces. These coatings are achieved by immersion and electrochemical processes.

There are several chromate treatments, using hexavalent chromium (VI) and trivalent chromium (III), as well as resinous coating containing chromate. Table 3.16 shows some recommended compositions of chromate solutions.

Chromate treatments are also used as a sealing, an anticorrosive or a coating for iron or steel pretreated with a phosphate solution, and as an insulating or anticorrosive coating for electrolytic iron plate. Chromate coating will not prevent the growth of metallic filaments, commonly known as "whiskers." Chromic acid or chromium salts, such as sodium or potassium chromate or dichromate, together with reducing agents and certain anions that act as activators, are used in chromating solutions.

Table 3.16 Recommended Compositions of Chromate Solutions		
Phosphate coating and type of sealant or organic coating	**Recommended compositions in term of CrO$_3$ trivalent Cr**	
	Minimum	**Maximum**
Coatings with subsequent demineralized water rinse to be finished with paint, varnish, or lacquer	0.2 g/l	1.0 g/l
Coatings without subsequent demineralizer water rinse to be finished with paint, varnish, or lacquer	0.1 g/l	0.5 g/l
Coatings to be finished with oil. grease, or wax	0.1 g/l	2.5 g/l

3.44.2 **CHEMICAL CONTROLLING OF CHROMATING PROCESS**

The concentration of active ingredients (primarily hexavalent chromium) in the bath should be maintained within ±10% of the initial makeup value. Close control of pH of the bath is a useful aid in maintaining uniform results.

3.44.3 **CHROMIC ACID SOLUTIONS**

The concentration of chromic acid solutions may be determined by two (or sometimes three) titration checks for total acid value, free acid value, and when necessary, chromate concentration.

3.44.4 **TOTAL ACID VALUE**

Total acid value is determined by titrating a 25-ml sample of the chromic acid solution with 0.1 N sodium hydroxide, using a phenolphtalein indicator. The end point is reached when the color changes from amber to a reddish shade that lasts at least 15 s. Each ml of 0.1 N NaOH required equals 1 point total acid.

3.44.5 **FREE ACID VALUE**

Free acid value is determined by titrating a 25-ml sample of chromic acid solution with 0.1 N sodium hydroxide, using bromocresol green as an indicator. The end point is reached when the color changes from yellow to green. Each ml of 0.1 N NaOH required equals 1 point free acid. The concentration of free acid in chromic acid solutions is usually 0.2–0.8 ml.

3.44.6 **CHROMATE CONCENTRATION**

Chromate concentration may be determined by placing a 25-ml sample of the solution into a 250-ml beaker, adding 25 ml of a 50% sulfuric acid solution and two drops of orthophenanthroline ferrous complex indicator, and titrating with a 0.1 N ferrous sulfate solution. Each ml of 0.1 N ferrous sulfate solution of the amount required to change to solution from blue to a reddish-brown color is 1 point of chromate concentration.

3.44.7 **INSPECTION METHODS**

Colored chromate coatings are preferred for maximum corrosion resistance and the preferred paint base color ranges between yellow and brown. The appearance of a visible colored film having good continuity and adhesion (i.e., thoroughly dried coating cannot be wiped off by gently wiping with clean white cloth), without visible flows or defects in the basis metal and coating may, under many conditions, be a sufficient test of quality.

Another test method is used on colorless chromate coating used for decorative purposes and with low electrical resistance. A drop of chemical solution that, when in contact with the basis metal, results in a change in color of the spot test solution, may be used to determine the presence of a protective colorless film when compared to a similar drop of solution on a clean but untreated surface.

Solution A

Solution Make up Reagent grade ferric nitrate—Fe $(NO_3)3. 9 H_2O$, 2 g
Reagent grade
Hydrochloric acid—HCl (36.5–38.0 percent), 20 ml
Distilled or deionized water, 473 ml

Solution B

Reagent grade
Potassium ferricyanide—K3 Fe(CN)6, 2 g
Distilled or deionized water, 473 ml

When stored separately, the solutions are stable. When they are mixed, storage life is limited to a maximum of 1 week. The mixed solution (equal volumes of solutions A and B) should be kept in a brown bottle.

An untreated surface from which the surface oxide has been removed by acid or alkali should be used for comparison. The treated surface should be allowed to dry thoroughly before testing. A drop of test solution is applied by an eyedropper to both the treated and untreated surfaces. The solution will turn blue-green very quickly on an untreated surface, while the time for it to turn blue-green on the treated surface will vary with the thickness of the chromate film. Aged coatings dried at elevated temperatures (that is, dehydrated coating) develop a colored spot more quickly than fresher coatings.

Still another test method is used only with coatings that are free of secondary supplementary coatings, such as oil-, water-, or solvent-based polymers or wax. Determine the presence of a colorless (clear) coating by placing a drop of lead acetate testing solution on the surface. Allow the drop of testing solution to remain on the surface for 5 s, and then remove it by blotting gently, taking care not to disturb any deposit that may have formed. A dark deposit or black stain indicates the absence of a coating. Prepare the test solution by dissolving 50 g of lead acetate trihydrate [Pb $(C_2H_3O_2)2, 3 H_2O$] in 1 l of distilled or deionized water. The pH of the solution should be between 5.5 and 6.8. Any white precipitate formed during the initial preparation of the solution may be dissolved by small additions of acetic acid, provided that the pH is not below 5.5. Upon formation of a white precipitate, the solution should be discarded.

For comparative purposes, treat an untreated surface similarly. On an untreated surface, a black spot forms almost immediately.

To determine the chromatic coating weight for either a colored or colorless coating, perform the following procedure:

1. Use a 71-by-71 mm test panel, and wipe it with solvent to remove oil, grease, and other surface soil.
2. Weigh the panel on an analytical balance to an accuracy of ±0.2 mg.
3. Immerse it for 2 min in molten salt bath of the following composition:
   - Reagent grade sodium nitrate $(NaNO_2)$
   - Temperature ranging from 326–354°C
4. Remove the panel from the salt bath and rinse thoroughly in cold water.
5. Immerse it for 30 s in 1 part (by volume) of concentrated nitric acid (reagent grade) and 1 part (by volume) water at room temperature.

6. Rinse thoroughly in cold water and blow dry.
7. Reweigh it on analytical balance.
8. Repeat steps 1–7. The loss in weight on the second stripping should normally be less than 0.6 mg on a panel of this size.

This process is satisfactory for coatings dried or baked at any temperature up to normal paint baking temperatures.

For test methods for adhesion resistance, salt spray resistance, and abrasion resistance, see ASTM B201, B449, and B117.

3.45 SURFACE TREATMENT OF FERROUS METALS

For surface treatment of ferrous metals, phosphating or applying of wash primer should be performed. These include treatments to form oxide and carbide surface layers of improved corrosion resistance, processes for handling rust, for improving subsequent coating, and for special situations in processing steel sheets and the manufacture of steel products.

Cast iron should be phosphated with zinc and manganese phosphate coating to protect against corrosion in outdoor storage for 1 year. The stainless steel and alloy steels should not be treated by phosphating. Ferrous metal surfaces should be prepared prior to surface treatment and painting.

3.46 TREATMENT OF NONFERROUS SURFACES

3.46.1 ZINC AND ZINC-PLATED SURFACES

Conversion coating improves the paint-holding capacity of zinc surfaces after this preparation and prevents the formation of "wet storage" staining and paint bonding to zinc-coated steel. Three types of conversion coating can be applied to zinc and zinc-coated surfaces: phosphate coating, chromate coating, and wash primer.

3.46.2 PHOSPHATING TREATMENTS OF ZINC AND ZINC-PLATED SURFACES

The phosphating treatment of zinc and zinc-coated steel convert the surface to a nonreactive zinc phosphate. It consists of six stages:

1. Cleaning with mild alkaline-type cleaner
2. Rinsing with cold water
3. Rinsing with hot water
4. Zinc phosphating (applied by brushing, spraying, or immersion)
5. Rinsing in warm water
6. Rinsing in chromic acid

A phosphating formulation that can be recommended by standards has the following ingredients:

- 66 g zinc oxide.
- 66 g nickel carbonate.
- 170 cm^3 phosphoric acid (75% wt).
- 2 cm^3 lactic acid.
- Enough water to make up 1 l of solution.
- Nickel salt, which serves as an accelerator. (Nitrites, copper, and cobalt salts can be used for the same purpose.)
- An additional accelerator, when there is aluminum in quantities of 0.1–0.5% in hot-dip zinc coating. Ammonium salts of fluorine or other soluble metal fluorides are used for this purpose.

Phosphate coating is used as a base for painting zinc and zinc-plated surfaces, in weights ranging from 1.6 to 43 g/m^2 of the surface.

3.46.3 CHROMATE TREATMENT OF ZINC AND ZINC-PLATED SURFACES

Chromate conversion coatings are applied to zinc and galvanized steel for wet storage stain control. The steps involved in a typical chromating process include the following:

1. Degreasing, pickling, and rinsing to remove oil, grease, oxides, and heavy metallic impurities
2. Chromate treatment by immersion or an electrochemical process
3. Applying dye (when desired) and rinsing
4. Drying

Care must be taken so that overheating does not occur (i.e., temperatures above 65°C), as this would reduce the corrosion resistance of the film owing to its dehydration. Careful handling of the freshly coated parts is critical because the films are relatively soft and gelatinous in nature, and may be marred easily until completely dried.

3.46.4 WASH PRIMER FOR ZINC AND ZINC-PLATED

While the range of paints that will adhere well to such films is somewhat less than for phosphating, a system based upon wash primer should be used. Their major advantage is that no rinsing is required. The DFT should not exceed 12.5 μm.

The wash primer should not be applied to a wet surface, nor to a previously chromated or phosphatized surface, because the reaction will be impeded and adhesion destroyed. It should be primed and topcoated promptly, since the inherent resistance of wash primer films to moisture is low.

3.46.5 ALUMINUM AND ALUMINUM ALLOYS

The surface treatment of aluminum involves processes for improved weathercoating for short-term protection. Unanodized aluminum surfaces should be treated by chromate treatment after preparation and prior to painting.

3.46.6 CHROMATE TREATMENT OF ALUMINUM SURFACES

Aluminum surfaces should be chromate-treated when they are to be used in marine and humid environments. The process of chromating an aluminum surface consists of the following stages:

1. Alkaline cleaning and rinsing
2. Deoxidizing or desmutting with mineral acid solution (incorporating nitric or hydrofluoric acid)
3. Rinsing with clean fresh water spray
4. Treating with hexavalent chromium, applied by immersion or spraying
5. Final rinsing with water (at a maximum temperature of 70°C); deionized water is recommended
6. Drying, during which the chromated surfaces must not reach a temperature greater than 70°C

Chromate conversion coating should be applied at weights from 0.1–1.07 g/m^2 for use as a paint base. The weight of coating should be approved by the company.

ENGINEERING AND TECHNICAL FUNDAMENTALS OF LINING

4

This chapter covers the minimum requirements for lining that will be used in equipment (e.g., vessels, storage tanks, and pipelines) intended for use in refineries, petrochemical plants, oil and gas plant and, where applicable, in exploration, production, and new ventures. The lining (both bonded and loose) is applied to equipment fabricated in metal or concrete.

There are several nonmetallic lining systems, which are classified in the following main groups:

- Thermoplastic material
- Thermosetting material
- Rubbers
- Mineral and bitumen material

The requirements for design and fabrication of the equipment and the state of preparation necessary for the surfaces to be lined are specified in this chapter. Details about the metallic lining (cladding) of pressure vessels and pipelines includes a hardness comparison chart for plastics and rubbers and the procedure for testing the effect of lining materials on the quality of drinking water.

4.1 PURPOSE OF LINING

Preventing corrosion by using a lining is desirable for several reasons:

- To extend the life of equipment
- To prevent shutdowns
- To prevent accidents resulting from corrosion failures
- To avoid product contamination

The potential life of a protective system is unlikely to be fully realized unless the correct choice of system is made; the materials used in the system can be supplied when required and with the properties attributed to them when making the choice; the materials are applied in appropriate conditions and standards of workmanship; the handling, transportation, and storage of all materials and lined components results in no irreparable damage to the integrity of the materials or coating; the erection procedures cause no damage to the lining that cannot be completely restored; and such restoration of damaged areas results in a protection at least as good as that of the undamaged areas.

If the lining is applied after the equipment has been installed, due consideration should be given to proper surface preparation and quality control after application. If it is applied before the equipment is installed, due consideration should be given to lining the field joints.

Many variables (both natural and otherwise) can influence the successful fulfillment of all these conditions. It follows that no two projects are exactly similar.

Essentials of Coating, Painting, and Lining. http://dx.doi.org/10.1016/B978-0-12-801407-3.00004-3

4.2 LINING DESIGN

4.2.1 INFORMATION TO BE SUPPLIED BY THE PURCHASER

The purchaser should state either precise details of the lining requirements or the service conditions for which the lining is required so that the lining contractor can make recommendations. In the latter case, the purchaser should provide the following relevant information:

- Design and fabrication details of the equipment to be lined, including drawings if possible
- Details of gasket materials
- Details of the contents of the vessels or equipment, including trace materials
- The design temperature
- The design pressure
- The details of any solids to be handled
- Any mechanical properties required of the lining, including suitability for machining
- The methods of heating, cooling, or both
- Cleaning methods, such as water washing, solvent washing, boiling, or steaming.
- The location where the lining work will be done
- Site conditions that may affect the work, such as availability of services
- Preparation of surface

4.2.2 INFORMATION TO BE SUPPLIED BY THE LINING DESIGNER

The contractor should provide the following relevant information:

- Any special requirements for the materials and system used in surface preparation
- The type and specification of material to be used for the lining
- The minimum thickness of the lining and, where applicable, the maximum thickness
- The provision of test plates
- The types of solvents, with their flash points and threshold limit values; when the lining is to be applied outside the applicator's works and the primers and adhesive to be applied contain solvents
- The specified sheet or rubber lining
- The minimum and maximum allowable temperatures and allowable humidity, which are required for the correct application of the lining
- The minimum temperature required during the vulcanization of the lining (for rubber linings) and the vulcanization schedule, where applicable
- Inspection techniques and procedures to be employed, acceptance levels, and stages of inspection
- Handling, transport, storage, and installation procedure
- Any special information needed or recognized with regard to inorganic linings

4.3 COMPLIANCE

4.3.1 FABRICATOR

The fabricator of metal equipment to be lined should comply with the design and fabrication details of the equipment to be lined and provision of drawings. It also is responsible for the inspection and testing of the equipment as fabricated; the equipment after preparation of surfaces, welds and edges in the case of metals, and edges alone in the case of concrete; and handling, transport, storage and installation procedures.

4.3.2 APPLICATOR

The applicator should comply with the following standards:

- Selection of lining
- Method of lining
- Inspection and testing of the equipment after blast cleaning, the lining material, lining after fitting, and lining after welding
- Method of continuity testing for lined equipment
- Rectification of faults in fully cured linings
- Handling, transport, storage, and installation procedure

4.4 DESIGN, FABRICATION, AND SURFACE FINISHING OF EQUIPMENT TO BE LINED

The basic design, fabrication, and testing for mechanical reliability of the equipment to be lined should be carried out to the appropriate standards. Linings will affect heat transfer and tolerances required for joints; both features need to be considered at the design stage of the equipment.

4.4.1 DESIGN OF METAL EQUIPMENT TO BE LINED WITH ORGANIC MATERIAL

Equipment to be lined with organic material (such as thermoplastics, thermosets, rubber, bitumen, and coal tar) should be sufficiently rigid that there is no possibility of deformation, which would result in damage to the lining during transportation, installation, and operation. When stiffeners are required, they should normally be applied to the unlined side of the equipment. In the case of hard rubber and ebonite linings, this requirement is particularly important.

The arrangements for the lifting of equipment should be determined at the design stage. The design of all equipment should allow for access during the preparation of the surface and application of the lining, and for venting of fumes evolved during the operation.

In completely enclosed vessels, there should be at least one manhole that, after lining, complies with BS 470, and one additional branch with a bore of not less than 77 mm to permit adequate circulation of air. It is recommended that, where practicable, the minimum diameter of a manhole should be 600 mm. Riveted constructions should not be used. Bolted joints should be permitted only if they can be dismantled for lining. When calculating clearances, allowance should be made for the thickness of the lining. Surfaces to be lined should be of a smooth contour.

Discontinuities, crevices, and sharp projections should be avoided. Fittings that have to be installed after the completion of the lining process should be designed to be lined or fabricated from materials that will not be affected by the conditions of the process. Normally, all connections to lined parts of equipment should be flanged. If for any reason screwed connections are required,

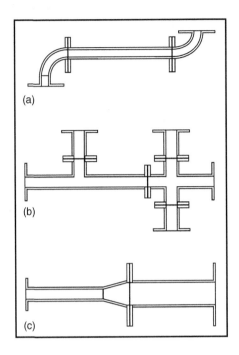

FIGURE 4.1

Pipework details; the details shown in (a), (b), and (c) are permissible.

then these parts should be fabricated from corrosion-resistant materials. All nozzles and outlets should be as short as possible and be both straight and flanged. Flange faces should have a plain surface to allow the lining to continue over each face.

Branches and outlets, except for plain ended pipes, should be as short as possible and flanged so that the lining may go over the flange face to prevent the ingress of process fluid between the lining and the substrate. Acceptable flange systems are shown in Figure 4.1. Plain-ended pipes should be designed so that the lining may be turned over the ends and the outside of the pipe.

Where pads are required, they should be designed as shown in Figure 4.1(d). Drilled and tapped holes in pads for studs should not penetrate the shell of the equipment. Heating coils, immersion heaters, and sparger pipes should be installed after the completion of the lining process and should be located so that no part is closer than 100 mm from a lined surface.

In the case of nozzles through which heating coals and other elements enter equipment, a smaller clearance is permitted, provided that the temperature of the pipe through the nozzle does not exceed 80°C. In no case should this clearance be less than 25 mm. Fluids introduced through sparger pipes or dip pipes should not impinge directly onto the lining. External heating of equipment should not be permitted without full consultation with the applicator of the lining. The design of the equipment

Table 4.1 Maximum Lengths Between Flanges	
Nominal Pipe Size (mm)	**Length Between Flanges (mm)**
25	2,000
32	2,500
40	3,000
50	3,500
65	4,000
80–600	6,000

should be such that there is easy hand access to areas where thermoplastic sheets need to be welded, and the design of pipework should allow ready access to welds and bends for weld and surface preparation and permit the fitting of extruded liners (see Figure 4.1).Pipe systems made from straight lengths with separate standard (1.5 D) bends and tees meet this requirement. Bends should not be greater than 90°.

Where pipe systems include nonstandard fittings, their suitability for lining should be established. The maximum dimensions of pipes and fittings that can be rubber-lined should be determined during the exchange of data. Typical maximum lengths of straight pipe that can be rubber lined are listed in Table 4.1.

Cast-iron pipework and fittings manufactured to the BS 2035 standard are of suitable dimensions and should be permitted for lining. Pipes with diameters less than 25 mm should not be lined with rubber, nor should screwed fittings. Pipes of less than 450 mm in diameter should be of seamless construction only.

Depending on the fabrication method, where flat-face or raised-face flanges are used, they should be welded square to the pipe or fitting according to normal manufacturing tolerances.

Flanges that are attached to a pipe by fillet welding (e.g., slip-on flanges) should be suitably vented to allow any air trapped between fillet welds to escape into the atmosphere during lining and vulcanizing. This may be accomplished by using a vent hole in the flange so long as it meets standard piping design requirements.

For pipe spools longer than 400 mm, venting is achieved by drilling two vent holes of diameters ranging from 2.5 to 3.5 mm into the pipe wall, perpendicular to each other, at a distance of 150 mm from each flange. Hence, four holes should be put into each pipe spool. For pipe spools with a length 400 mm, one vent hole of at least 2.5 mm diameter at each flange is acceptable. Hence, two holes should be put into each pipe spool. The metallic equipment to be lined should be capable of withstanding heat up to a specific processing temperature (or drying) of the polymer (e.g., 450°C for PTFE lining).

Unless otherwise specified, surface treatment of metal components to be lined should be as shown in Table 4.2.

Table 4.2 Fabrication Requirements for Specified Parts

Component Part	Fabrication Requirements
Outside corners	Round to a minimum radius of 3 mm
Sharp inside corners	Fill to a smooth radius with an uncured rubber fillet strip over the primed surface
Welds	Welds should be continuous; grind to a smooth and uniform surface.
Porous areas, cavities, and pockets	Fill in with weld metal

No welding is permitted on the interior or exterior surfaces of equipment during or after application of the lining. Hydrostatic tests on equipment should be performed prior to application of the lining.

The design of equipment should allow adequate access and venting. All branches should be flanged, and the lining should be taken over the flange face to prevent ingress of the process liquid behind the lining.

Sharp changes of contour in the surface to be lined should be avoided. Such changes should be finished by grinding with mortar such that the internal radius of the lining is not less than 3 mm. Vent holes to prevent air trapped in welded joints may sometimes be necessary (see Figure 4.2).

If necessary, connections for stairs, supports, and other elements should be made before the rubber lining is applied, since welding is not allowed on lined equipment and piping. Heating elements should be situated at least 100 mm from the rubber lining to avoid overheating.

Impingement of steam on the rubber surface should be avoided when using steam injection heating. When the equipment is subject to approval by authorities, the lining should be carried out after hydrotesting.

Pipes of less than 600 mm in diameter should not be lined with thermoset materials such as epoxy. Pipes of less than 686 mm in diameter should be joined in a manner that eliminates the need for entering the pipe to complete or repair the coal-tar or bitumen lining at the joints. Pipes that are 686 mm or larger in diameter may be joined by any suitable method, including welding.

The types of pipes to which standards are applicable include both welded and seamless pipes of nonalloy steel used for the conveyance of fluids. The types of fittings are mainly bends, tees, reducers, and collars.

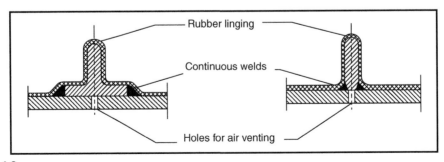

FIGURE 4.2

Air vent holes, to be drilled at regular distances; the diameter of the holes depends on the dimensions of the vessel, but is generally 5 mm.

4.5 DESIGN OF METAL EQUIPMENT TO BE LINED WITH INORGANIC MATERIAL

For lining with inorganic materials such as glass, ceramic, brick, cement-mortar, and refractory, the following consideration should also be carried out:

4.5.1 DESIGN OF METAL EQUIPMENT TO BE LINED WITH GLASS

Because glass is very strong in compression but weak in tension, it is formulated to have less thermal expansion than the base metal to which it is applied. Therefore, during cooling, very high compressive stresses are maintained in the glass coating. Sharp edges will be not lined well due to the surface tension, but even edges that are ground with a radius are not suitable if the radius is not generous enough. Glass will tend to shear off of radii that are too sharp. This requirement for curved surfaces and rounded edges requires the use of swaged openings, Vanstone flanges, and flanges on covers and clamp-top vessels.

Jacket aprons and sealers must be so designed that no undue stresses are introduced into the liner shell during welding of the jacket. Excessive heat applied to the liner shell will cause the steel to expand, cracking the glass by stress tension. The heating (firing) procedure used when glassing steel must be carefully controlled. For this reason, the design of the equipment must be such that there are no sudden changes in metal thicknesses, such as with very heavy flanges. The base metal must be thick enough to withstand the operating pressures and to support the vessel from sagging during firing at the elevated temperatures encountered in glassing. The unit must be designed such that any stresses or combinations of stresses occurring during the actual operation of the vessel do not exceed the yield point of the base metal.

Sharp edges on castings should be avoided because neither the wet- nor dry-process coatings will cover sharp edges adequately. Inside and outside corners should be rounded to uniform thickness and generous radii provided for fillets and outside corners.

Special styling techniques should be used for designing appendages, internal passages, and lug-fastening faces so as not to emplace a mass of metal near an otherwise-uniform enameling surface. These design considerations should include a thorough review of the available mold-making techniques in conjunction with the pattern designer.

4.5.2 DESIGN OF METAL EQUIPMENT TO BE LINED WITH CERAMIC

All areas of high stress that will result in movement of the steel structure, such as oil canning or vibration, should be externally supported. Anchorage is required to secure the lining to the shell and to distribute shrinkage and cracking evenly over the whole area.

Anchors, such as v-type or long horn studs, should be installed on the proper center lines in a diamond or staged pattern with the lines randomly oriented. For lining up to 50 mm thick, anchorage may be provided by mild steel, welded-wire fabric. Anchorage should be used in both single- and dual-layer systems.

4.5.3 DESIGN OF METAL EQUIPMENT TO BE LINED WITH BRICK

It is important to have smooth, continuous steel surfaces when membrane lining is used. The grades of steel generally used fall within the requirements of ASTM specification A-283 or A-285.

Tanks of cylindrical design are used frequently as opposed to a rectangular shape for reasons of economy, pressure rating, and stability of the brick linings. The thickness of steel in shell plates for cylindrical vessels should be as follows:

When the temperature differential between brick lining and steel shell is negligible, or when expansion joints are provided in the brick lining, it is suggested that steel tanks be designed in accordance with the usual rules for accommodating forces imposed by the internal pressure, the weight of the vessel, lining, and contents, and by wind, shock, or other loads that may occur. If operating pressure is to be 1 bar (15 psig) or less, consideration should be given to designing the vessel to permit hydrostatic testing at twice the operating pressure. If vessels are to operate at pressure above 1 bar, it is suggested they be designed to meet requirements of the ASME Code. The exception to this is that a 0.6-mm minimum thickness is recommended for tanks with diameters over 1.25 m.

When a temperature differential between the brick lining and steel shell will exist and expansion joints are not provided in the brick lining, it is suggested that the thickness of steel should be determined by the following equation:

$$t_s = \frac{t_b E_b}{S_s}\left(a_b \Delta T_a - a_s \Delta T_s - \frac{S_s}{E_s}\right) + \frac{PD}{2S_s} \tag{4.1}$$

where *ts* is the thickness of the shell (in.); *tb* is the thickness of the brick lining (in.); *Es* is the modulus of elasticity for steel (psi; use 30×10^6 psi); *Eb* is the modulus of elasticity in compression for the brick lining (psi; may vary from 2×10^6 to 7×10^6 psi); *Ss* is the allowable stress for the shell (psi; use 13,750 psi); *Sb* is the allowable compressive stress for the brick lining (psi; may range from 4,000 to 10,000 psi); *as* is the coefficient of expansion for steel (inches/°F); 6×10^{-6}; *ab* is the coefficient of expansion for brick (inches/°F); 3×10^{-6} to 6×10^{-6}; ΔTs is the average temperature rise of steel (in °F); ΔTb is the average temperature rise of brick (in °F); *p* is the maximum allowable internal pressure (psi); and *D* is the inside diameter of shell (in.). In the equation, when a term in parentheses is negative, it indicates that the steel tank will expand more than the brick lining or that the stress in the steel shell caused by expansion of the brick lining is less than the maximum allowable Ss. The thickness of the shell then should be determined from the internal pressure.

When the thickness of the shell has been established, compression stress in the brick should be determined by the following equation:

$$\frac{bT_b - T_b}{\dfrac{t_b E_b}{t_s E_s} + 1} S_b = E_b \tag{4.2}$$

Should the compressive stress of the brick exceed a safe value, expansion joints may be provided, the thickness of the brick lining may be increased, or stronger brick may be used. A reduction in the thickness of the steel also will reduce the stress in the brick.

The thickness of the flat bottom should be 0.9 mm minimum or sufficient to limit the deflection to 1 m/10 m of the span. Consideration should be given to warpage of flat bottoms, which may result from welding. This can be reduced by increasing the bottom's thickness.

Shells and heads must be made stiff enough to avoid deformations that will seriously crack the brick lining and allow it to disintegrate or fall. Joints in all surfaces that are to be covered by a membrane lining should be made by butt welding. Lap and riveted joints should not be used. Welds at the inside and outside corners should be finished so that they may be ground to a 0.45-mm minimum radius. Vessels should be hydrostatically tested and repaired as necessary before any lining work is started.

4.5.4 **PREVENTING DEFORMATION**

Since the brick lining cannot sustain even very small deformations without cracking, the shell must be very rigid, which can be achieved by the following measures:

- The design should ensure ample wall thickness.
- High-strength steel with a tensile strength exceeding 500 N/mm^2 should not be used.
- The maximum deflection due to wind load should not exceed 100 H.
- Full measure should be taken of the mass of the brick lining, including packing bed, contents, etc. This can be estimated as follows.
  - Equipment with diameters up to 3000 mm, operating at pressure up to 3 bars and temperatures up to 100°C should be designed for an additional stress of 25-30 N/mm^2.
  - Equipment with diameters greater than 3000 mm, operating at pressures up to 10 bars and temperatures up to 200°C should be designed for an additional stress of 30-50 N/mm^2.
- For operating conditions with pressure above 10 bars and temperature above 200°C, it is recommended that each application should be considered separately.

The following additional measures, aimed at avoiding stress concentrations, will reduce the likelihood of (local) deformations occurring:

- Use cylindrical vessels.
- Use vertical vessels supported on skirts with the same diameter as that of the vessel shell. Using legs is not recommended.
- If horizontal vessels are unavoidable, use extra-wide saddles or longitudinal supports of ample rigidity.
- Avoid flat heads and use hemispherical heads or other types of head with very generous knuckle radii.
- Atmospheric storage vessels with flat bottoms can be brick-lined. The temperature decrease over the brick lining should be kept low since a cooler bottom can cause lifting of the brick lining. Welding the bottom plates to the supports can prevent buckling of the steel bottom.
- Avoid sudden changes in wall thickness. Where a change in wall thickness occurs, the internal surface should be flushed (see Figure 4.3).

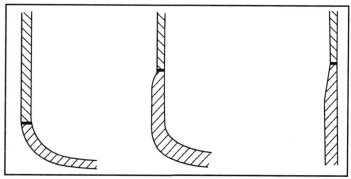

FIGURE 4.3 Recommended shapes for cylindrical seams.

Note: The flange facing should be adapted so it can receive a membrane.

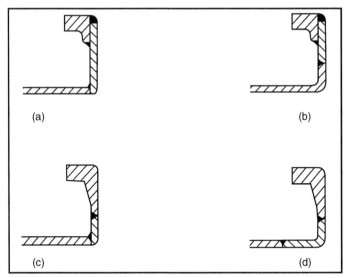

FIGURE 4.4 Nozzle constructions.

- Avoid external stiffening rings or bars attached to the vessel, and avoid internals
- connected directly to the shell.
- Connections for stairs, supports, etc., should be made before the brick lining is applied since welding afterward will damage the chemical-resistant brick lining.
- Minimize the number of nozzles, manholes, and other elements, by combining as many of these as possible into one manhole cover. Nozzles should not protrude into the inside of the vessel; rather, they should be welded either in the corner or onto the bulged-out vessel wall (see Figure 4.4).

Belled-out nozzle openings should be used whenever possible for vessels operating at higher pressures.

4.5.5 SURFACE CONDITIONS

When the vessel to be brick-lined is to have an impervious membrane, which is generally the case, the requirements for the internal surface condition of the shell should be of those ruling for the application of the membrane (see Table 4.3). Air inclusions can be prevented as follows:

- Avoid designs with head curved to the inside (like what is shown in Figure 4.5).
- Provide for correct weld design.
    Preferably in the cylindrical part of the vessel (see Figure 4.6).
    If not possible, then the corner may be made with a profile giving a weld in the cylindrical part and in the bottom; an extra leg may be included to guide rainwater (see Figure 4.7).
- With butt-weld, cone-shaped ends or bottoms, the welds can be made in the corner, provided that they are finished smoothly (see Figure 4.8).
- If corner welds cannot be avoided, air-escape holes are to be created.

Table 4.3 Material of the Anchors

Location	Part	Concrete Surface Temperature (°C)	V-Shaped Studs and Cleats [2]
Radiant section	Floors	N/A	N/A
	Walls	Maximum 600	Carbon steel
		Above 600	Cr-Ni
	Roof, including flue gas branches	Maximum 870	18/8 Cr-Ni
		Above 870	25/20 Cr-Ni
Hot flue ducts and plenum chamber[1]	Unshielded walls and roof	Maximum 870	18/8 Cr-Ni
		Above 870	25/20 Cr-Ni
Convection section	Unshielded walls and roof	Maximum 870	18/8 Cr-Ni
		Above 870	25/20 Cr-Ni
	Shielded walls including bottom part	N/A	Carbon steel
Cold flue ducts	N/A		Carbon steel

- V-welds made from the inside are preferred.
- Sharp edges, weld spatters, and other flaws should be removed.
- Welds should be in-line and be finished smoothly.
- Internal changes should be rounded with a radius of at least 3 mm, but preferably 5 mm.
- Riveted constructions are not acceptable.

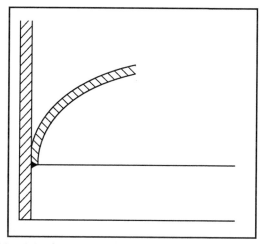

FIGURE 4.5 A form of curved head that is not allowed.

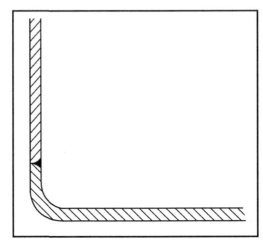

FIGURE 4.6 Recommended joint in a flat bottom.

4.5.6 DESIGN OF METAL EQUIPMENT TO BE LINED WITH CEMENT MORTAR

The steel pipes to be lined should meet the following requirements:

- No holes in the wall (e.g., branches)
- Bending should not exceed 3 mm per 3 m of length
- Maximum out-of-roundness (difference between minimum and maximum dimensions) should not exceed 1.6 mm or 1% of the nominal outside diameter, whichever is the greater

4.5.7 DESIGN OF METAL EQUIPMENT TO BE LINED WITH REFRACTORY

Wall and floor refractories for heaters should be of sufficient thickness to ensure that the exterior face temperature will not exceed 82°C, with a 8 km/h wind velocity and a 40°C ambient temperature. Arch refractory thickness should be in accordance with good design practices, and the resulting casing temperature should be noted on the data sheets.

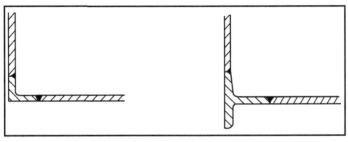

FIGURE 4.7 Recommended construction for flat bottoms in tanks.

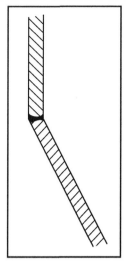

FIGURE 4.8 Recommended joint in a cone-shaped end.

Portions of stacks and ducting that are adjacent to instrumentation and equipment that requires frequent attention should be insulated with a minimum of 138 mm of insulating concrete.

4.5.8 ANCHORING AND REINFORCEMENT LININGS

The anchoring of this type of lining that is 75 mm thick or less should consist of cleats or V-shaped studs with a diameter of 5 mm. These cleats may be made from 5-mm-thick carbon steel or 2-mm stainless steel plate. For the selection of the materials for both types of anchor, see Table 3. For other lining thicknesses, the height and the pitch of the cleats or studs should be adapted.

The anchoring of these linings should be done by 5-mm-diameter V-shaped studs welded to a crimped wall. For the material of the studs, see Table 4.3.

The anchoring of lining to furnace floors is not necessary, except for protruding.

Cleats and studs should be welded to the walls by means of the electric arc welding process using welding electrodes appropriate for the metals concerned.

Welding surfaces must be dry and clean. Cleats and studs should be welded along the whole circumference of their bases, checked by hammer-testing after welding, and rewelded if necessary.

4.5.9 DESIGN OF CONCRETE EQUIPMENT TO BE LINED WITH POLYMERIC MATERIALS

While it is not possible to eliminate shrinkage cracks in concrete, the design should be eliminate structural cracking. Particular attention should be paid to avoiding cracking due to thermal stresses. In the case of loose liners, equipment should be designed to the BS 8110 standard. In the case of bonded liners, equipment should be designed to the BS 8007 standard.

The ability of the lining to accommodate cracking of the concrete will determine the level of design needed. Equipment designed and constructed to class A of BS 8007 is liable to develop cracks up to 0.1 mm wide, and equipment designed and constructed to class B is liable to develop cracks up to 0.2 mm wide.

In the case of class A of BS 8007, the concrete should have a minimum compressive strength of 30 N/mm^2; in that of Class B, the strength should be at least 25 N/mm^2. If necessary, extra reinforcements should be used and construction joints treated so as to promote a bond between adjacent areas of concrete. Expansion joints create special problems in linings and should not be used without consultation between the purchaser and the installer of the lining.

Pipes and fittings should be designed with puddle flanges and cast into the concrete. Where possible, pipes and fittings should either be of the same material as the liner, such that the weld may be made between the liner and the fitting, or designed such that the lining may be carried through the fitting. If this is not possible, the fitting should be designed so that a mechanical joint can be made between the fitting and the lining. In the latter case, the fitting should be of a corrosion-resistant material. The arrangements for the lifting of equipment should be determined at the design stage.

The design of all equipment should allow access during the preparation of the surface and application of the lining, and for the venting of fumes evolved during the operation. In completely enclosed vessels, there should be at least one manhole conforming to BS 470 and one additional branch of not less than 75 mm bore to permit adequate circulation of air.

All corners should be designed to be formed with a 45° fillet with a minimum leg length of 20 mm. All equipment to be placed below ground level or subjected to external water pressure should be provided with a waterproof barrier on the outside of the equipment. All equipment should be designed with a minimum of 20 mm of concrete over reinforcement.

4.5.10 DESIGN OF CONCRETE EQUIPMENT TO BE LINED WITH INORGANIC MATERIALS

The following considerations should be carried out when designing concrete equipment that is to be lined with glass:

- The slopes of the surfaces should be between 3.2 and 6.4 mm per 30 cm and should have attained a minimum compressive strength of 21 MPa (3,000 psi).
- The anchoring system that is used should be placed at the specified center line distance in a diamond pattern with a random orientation of anchor lines. The anchoring system should consist of V-type or longhorn studs.

4.5.11 DESIGN OF CONCRETE EQUIPMENT TO BE LINED WITH BRICK

Special attention should be given to brickwork for concrete equipment since bulging and sagging in the equipment may cause the lining to fail prematurely. Care should be taken to ensure that the inside forms of vessels are adequate to prevent bowing in toward the center.

It is better to have bowing out rather than in. In large rectangular vessels, it is often recommended to bow and batter the inside walls to place the chemical-resistant masonry lining in compression upon thermal expansion. All concrete tanks, regardless of shape, should be reinforced with steel. Where practical, the vessel should be poured monolithically to avoid dry joints, which are potentially weak.

In addition to good engineering design details, it is desirable to have a smooth interior surface when a flexible membrane type lining is to be applied. A wood float finish generally is desirable. Bolt holes, honeycombed surfaces, and other flaws should be filled with a stiff Portland cement mortar. Rough projections or high spots should be removed by chipping or using abrasive stone.

4.5.12 FABRICATION OF METAL EQUIPMENT TO BE LINED WITH ORGANIC MATERIALS

All metal-to-metal joints should be forged by welding. Welds should be homogeneous and free of pores. Welds should be ground smooth and made flush with the parent metal on the side to be covered.

Wherever possible, they should be made from the side to be lined. Where this is not possible, the root should be chipped out and a sealing run should be applied. Internal corner and T-joints should be welded with full penetration. Welds should be ground smooth and concave to the required radius. Welds should be examined according to applicable design codes.

All welds should be continuous on surfaces to be lined. Only butt joints should be permitted on surfaces to be lined. Stitch welding, spot welding, and other noncontinuous welding processes should not be used.

Weld surfaces should be smooth. Some welding procedures provide surfaces of adequate smoothness, but in other cases, surfaces should be ground wholly or partly to remove weld ripples. The grinding process should be done so that the remaining weld does not have sharp edges. Welding procedures should be chosen to avoid porosity on the side of the weld to be lined. It is preferable that capping runs are applied to the lining side to minimize this effect.

All welds should be free of the following surface defects:

- Undercutting
- Cracks
- Porosity
- Any other type of surface cavity
- Lack of fusion

Weld defects that are exposed either on initial inspection or after blast cleaning should be repaired. Repairs should be done by grinding or by welding (with or without subsequent grinding), provided that the requirements for equipment design and fabrication are met. Weld profile details should be generally in accordance with Figures 4.9–4.13.

Filler materials, such as resin, putties, fillers and low melting point solders and brazes, should not be used. Before the equipment is passed for lining, all attachments to be made by welding (such as lagging cleats and lifting lugs) should be complete. All drilling should be completed before the equipment is passed for lining.

When the lining is to be thermoformed at the corners, those corners should be finished to a radius of not less than 5 mm. When the lining is to be welded at the corners, those corners should be finished to a radius of a nominal 1.5 mm.

All slag, anti-spatter compounds or similar materials should be removed. All weld spatter should be removed by chipping and/or grinding. Surface defects such as scores, pitting, and rolling should be removed by grinding or (where necessary) repaired by welding, provided that the requirements of design and construction are met. Rotating parts should be balanced before and after lining.

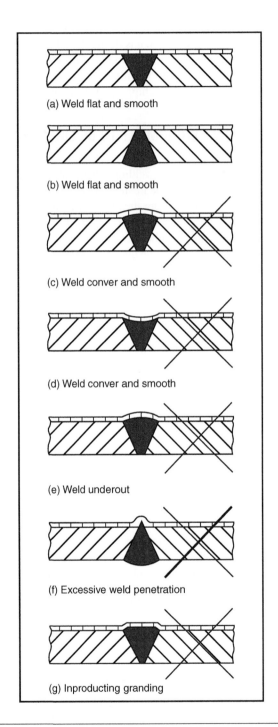

FIGURE 4.9

Weld profile details: Butt welds. The details shown in (a) and (b) are permissible; those shown in (c), (d), (e), (f), and (g) are not permissible.

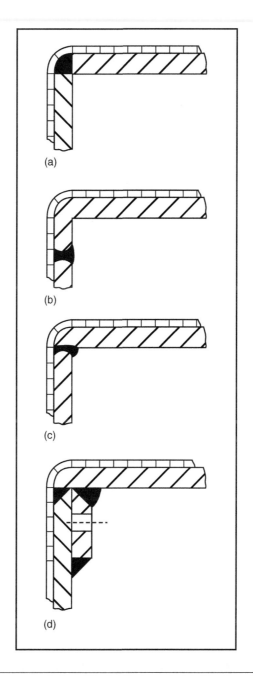

FIGURE 4.10

Weld profile details: External corners and edges. The details shown in (a), (b), (c), and (d) are permissible.

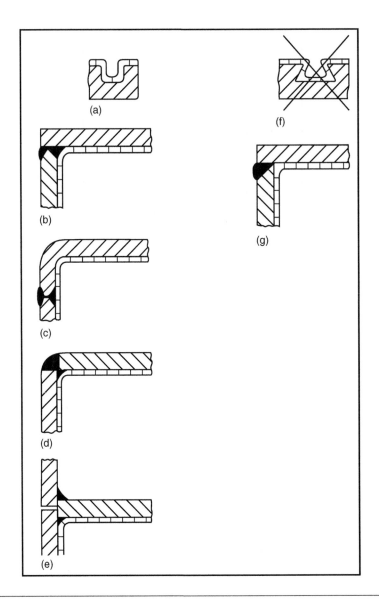

FIGURE 4.11

Weld profile details: Internal corners and edges. The details shown in (a), (b), (c), (d), (e), and (g) are permissible; that shown in (f) is not permissible. The detail shown in (g) is to be used only when the lining is bonded and welded at the internal corners.

4.5.13 FABRICATION OF METAL EQUIPMENT TO BE LINED WITH INORGANIC MATERIALS

In addition to the previous discussions, the following consideration should also be carried out:

Pipe connections to glass-lined equipment should be made only after the vessel has been leveled and securely fastened to a foundation. To avoid stress failures in glassed pipe, the pipe must be adequately

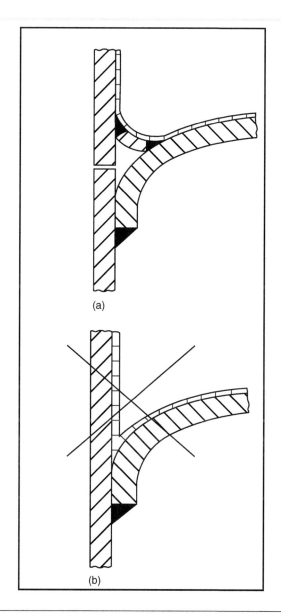

FIGURE 4.12

Weld profile details: Concave heads. The detail shown in (a) is permissible; that shown in (b) is not permissible.

supported by means of pipe hangers, and allowance must be made for expansion of the lines if surface temperature is appreciably above room temperature.

Enough pipe hangers must be provided so that the weight of the pipe and its contents are carried by the hangers rather than on the nozzles of the vessel. The welds should be made as specified in Table 4.4.

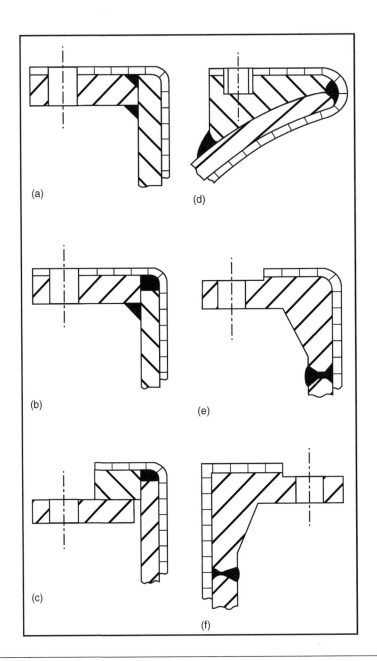

FIGURE 4.13

Weld profile details: Flanges. The detail shown in (f) is to be used only when the lining is bonded and welded at the external corners.

Table 4.4 Weld Types of Glass-Lined Equipment

Type	Comments
	Welding on both sides is also permissible. Angle β on the enamel side may range from 30° to 360° Resulting edges on the enamel side should be rounded
	Angle β on the enamel side may range from 30° to 320° Resulting edges on the enamel side should be rounded
	Roll in under $\alpha = 45° \pm 5°$ Round off edge to $r = S/2$, with r being at least 2 mm or chamfer at less than 45° to $S/2$
	Round off edge to $r = 2$ mm or chamfer accordingly at less than 45°

(Continued)

Table 4.4   Weld Types of Glass-Lined Equipment *(cont.)*

Type	Comments
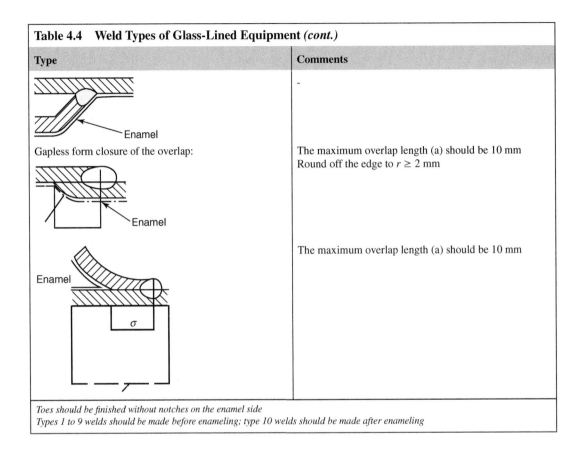	-
Gapless form closure of the overlap:	The maximum overlap length (a) should be 10 mm Round off the edge to $r \geq 2$ mm
	The maximum overlap length (a) should be 10 mm

Toes should be finished without notches on the enamel side
Types 1 to 9 welds should be made before enameling; type 10 welds should be made after enameling

4.5.14 FABRICATION OF CONCRETE EQUIPMENT TO BE LINED WITH ORGANIC MATERIALS

All concrete equipment should be constructed in accordance with the requirements of BS 8007 or BS 8110. Proper curing of the concrete should be ensured by the use of curing membranes wherever necessary. If shuttering is to be removed before 7 days, a curing membrane should be applied. Concrete equipment should be allowed to cure for 28 days before the lining continues.

Equipment that is slip-formed should be bagged as the concrete leaves the formwork, but before the curing membrane is applied. This process will reduce the amount of laitence.

All concrete that has been not cast against shutters or slip-formed should be float-finished. If equipment is to be fitted with loose linings, a steel float or a wooden float may be used. In the case of equipment where the lining is to be bonded, a wooden float finish should be used.

Any steps in the concrete due to misalignment of shutters or surplus material formed because of gaps at joints in shutters should be dressed off and ground smooth.

All holes left after the removal of ties to secure and align formers should be filled. Any surface defects that expose the aggregate should be filled. The material used for filling should be a sand/cement grout with a high cement content or a sand/cement water miscible epoxy resin grout or an epoxy resin mortar.

4.6 SELECTION OF LINING MATERIAL

The choice of the type and the thickness of the lining should be based on the duty for which it is intended (see Table 4.5). For the final selection of lining system and the method of application, which should be made in conjunction with the materials specialist and the lining contractor, the following details should be considered. The designer who selects the lining, with regard to this engineering standard, should verify that such a lining will withstand existing chemical and physical conditions and meet all requirements of the standard.

4.7 PROCESS INFORMATION TO BE ESTABLISHED

Full analysis of the equipment, including constituents present in small and trace quantities, and the details of cleaning operations is needed.

Temperature of material to be handled:

- Normal operating temperature
- Maximum and minimum
- Cycle of temperature variation

Degree of vacuum or pressure:

- Normal operating pressure
- Maximum and minimum pressures
- Cycle of pressure variation

Cycle of operations:

- Whether batch or continuous process

Abrasion and erosion:

- Details of amount, particle size and physical characteristics of the suspended matter together with rates of flow

Mechanical damage:

- Any difficulties expected in the handling and final siting of the equipment or any vibration of equipment and the possibility of mechanical damage

Immersion conditions:

- Constant or intermittent immersion of the lining, and partially or completely filled operation

Special conditions:

- For example, extremes of weather likely to be encountered during the handling, transport and storage of the equipment

4.8 MATERIAL CHARACTERISTICS TO BE CONSIDERED
4.8.1 CHEMICAL RESISTANCE OF THE LINING

Unless documented previous experience demonstrates that a lining will be suitable for a particular duty, appropriate testing should be performed before using it. When testing is required, service conditions should be reproduced accurately (according to ASTM C-581, ASTM D-543, and ASTM D-531).

Where it is not possible to place samples in process streams, service conditions should be simulated. Where it is known that a lining has to withstand an environment where heat transfer is made through the lining, the heat transfer condition should be maintained during the test. Substances including dissolved gasses present in a process stream only in trace quantities should be added to the test liquors.

4.8.2 PHYSICAL AND MECHANICAL CHARACTERISTICS OF LINING

- Available form of material
- Abrasion and erosion aspects
- Shape and weld ability (see Table 4.5)
- Vacuum or pressure resistance
- Thermal expansion of the lining material
- Creep characteristics of the lining material
- Liability of the lining material to stress crack in the particular environment
- Whether or not the lining is to be bonded to the substrate and the temperature limit of the adhesive system when used
- Effect of vacuum on loose linings

4.8.3 OTHER CHARACTERISTICS

- Maintainability of the lining
- Availability of lining material
- Handling of lining material
- Compatibility of lining with the substrate material; efficiently compatible lining material with the substrate; concrete and metal parts (see Table 4.6)
- Thickness of lining; the thickness of finished lining depends upon the material selected and the duties for which it is intended Tables 4.7–4.14 discuss the physical and mechanical characteristics of linings.

4.8.4 COMPLEXITY AND SHAPE OF EQUIPMENT TO BE LINED

The size, configuration, and complexity of the shape of substrate has an important bearing on the suitability of the particular lining material to be used.

Some lining materials are backed with rubber, glass fiber, or other fiber to allow the lining to be bonded to the substrate. There may be limitations on shaping, such as double curvature.

4.8.5 METHOD OF APPLICATION

- The required surface preparation (see Table 4.6)
- The process and equipment should be used for application and installation
- Field or shop application
- Curing condition

No.	Lining Material	Shape	Surface Preparation is Required	Type of Substrate	
				Metal	Concrete
	A- thermoplastic				
1	PE	Sheet	Sa 2½	X	X
2	PE	Nonsheet	Sa 2½	X	—
3	PP	Sheet	Sa 2½	X	X
4	PP	Nonsheet	Sa 2½	X	—
5	PVC	Sheet	Sa 2½	X	X
6	PVC	Nonsheet	Sa 2½	X	—
7	PVDF	Sheet	Sa 2½	X	X
8	PVDF	Nonsheet	Sa 2½	X	—
9	FEP	Sheet	Sa 2½	X	X
10	FEP	Nonsheet	Sa 2½	X	—
11	PTFE	Sheet	Sa 2½	X	X
12	PTFE	Nonsheet	Sa 2½	X	—
13	PFA	Sheet	Sa 2½	X	X
14	PFA	Nonsheet	Sa 2½	X	—
15	E-CTFE	Sheet	Sa 2½	X	X
16	E-CTFE	Nonsheet	Sa 2½	X	—
17	Nylon (polyamid)	Nonsheet	Sa 2½	X	—
18	EVA	Nonsheet	Sa 2½	X	—
19	Fusion-bonded epoxy	Nonsheet	Sa 2½	X	—
	B-thermosetting				
20	Phenol-formaldehyde	All shapes	Sa 2½		
21	Epoxy-phenolic	Nonsheet	Sa 2½	X	—
22	Epoxy	All shapes	Sa 2½	X	—
23	Polyesters	All shapes	Sa 2½	X	X
24	Furanes	All shapes	Sa 2½	X	X
25	Polyurethanes	All shapes	Sa 2½	X	X
26	Polychloroprene	Liquid	Sa 2½	X	X
	C-rubbers			X	X
27	NR	Sheet	Sa 2½		
28	IR	Sheet	Sa 2½	X	X

Table 4.5 Type of the Surface Preparation and Substrate to Be Lined

(Continued)

Table 4.5   Type of the Surface Preparation and Substrate to Be Lined *(cont.)*

No.	Lining Material	Shape	Surface Preparation is Required	Type of Substrate	
				Metal	Concrete
29	MR	Sheet	Sa 2½	X	X
30	NBR	Sheet	Sa 2½	X	X
31	CSM	Sheet	Sa 2½	X	X
32	FKM	Sheet	Sa 2½	X	X
33	CR	Sheet	Sa 2½	X	X
34	Hard and Ebonite rubber	Sheet	Sa 2½	X	X
	D-glass, ceramic				
35	Glass	—	Sa 2½		
36	Porcelain	—	Sa 2½		
37	Ceramic	—	Sa 2½	X	—
	E-bricks			X	—
38	Acid-resistant brick and tile	Fabricated	Sa2	X	X
39	Porcelain tiles (unglazed)	Fabricated	Sa2		
40	Carbon bricks	Fabricated	Sa2	X	X
41	Graphic bricks	Fabricated	Sa2	X	—
42	Silicon carbide bricks	Fabricated	Sa2	X	
	F-cements			X	—
43	Silicate-based cement	—	Sa2	X	—
44	Phenol formaldehyde cement	—	Sa2		
45	Furane resin cement	—	Sa2	X	X
46	Reinforced epoxy cement	—	Sa2	X	X
47	Phenolic furfuraldehyde cement	—	Sa2	X	X
48	Sulfur cement	—	Sa2	X	X
49	Polyester-based cement	—	Sa2	X	X
	G-bitumen and coaltar			X	X
50	Cold-applied bitumen	Liquid	Sa2	X	X
51	Hot-applied bitumen	Solid	Sa2		
52	Cold-applied coal tar	Liquid to paste	Sa2	X	X
53	Hot-applied coal tar	Solid	Sa2	X	X
				X	—

X Means the lining material can be used.
--- Means the lining material shall not be used.

Table 4.6 Working Temperatures for Linings

Lining Material	Maximum Working Temperature (°C)
Polyethylene (PE)	60–65
Polypropylene (PP)	90–100
Polyvinyl chloride (PVC)	60–90
E. CTEF	120–140
PTFE	160–200
PVDF	90–110
PFA	160
FEP	120
Polyamide	80
Fusion-bonded epoxy	80
EVA	—
Phenolformaldehyde	80–150
Epoxy phenolic	90–150
Polychloroprene (liquid)	100
Furan	140
Polyester	70–130
Polyurethane	140
Liquid epoxy lining	90–140
Natural or synthetic polyisoprene	100
Butyl rubber (11R)	120
Nitril rubber (NBR)	110
Chlorosulfonated poly ethylene (CSM)	100
Hard and Ebonite	90
Fluorinated rubber (FKM)	200
Chloroprene rubber (CR)	90
Glass and porcelain	200–260
Ceramic	200
Brick and tiles	65–200
Silicate-based cement	900
Phenolic furfuraldehyde cement	180
Furan resin cement	220
Reinforced epoxy cement	90
Phenol formaldehyde cement	180
Sulfur cement	90
Polyester-based cement	120
Coal tar	32
Bitumen	32

Table 4.7 Specifications of Sheet Thermoplastic Lining

	Sheet Thermoplastics	Shape Ability	Weld Ability	Available Form	Type of Application	Remark
1	PVC	G	G	UPVC sheet PVC sheet	Bond or loose	Unplasticized poly vinyl chloride (UPVC) softens and loses strength as the temperature increases (more than 85°C)
2	PE	B	G	LDPE sheet HOPE sheet	Normally loose	Susceptible to environmental stress cracking in polar organic
3	PP	G	G	Sheets with glass or synthetic fiber or rubber backing	Bond or loose	The backing material can impose limitation on the thermal forming process. Glass fiber backing is not readily formed into complex shapes
4	E. CTFE	B	G	Sheet with a glass fiber backing	Bond or loose	The backing material imposes limitation on the thermoforming process
5	PVDF	G	G	Sheets with glass or synthetic fiber or rubber backing	Bond or loose	The backing material can impose limitation on the thermal process
6	FEP	G	G	Sheet with or without glass fiber backing	Bond or loose	- When FEP is bonded to a substrate the maximum service temperature will be determined by the adhesive used - In the case of glass backed material the complexity of the formed shape may be limited
7	PFA	G	G	Sheet with or without glass and carbon fiber backing	Bond or loose	N/A
8	PTFE	No	G up to 4 mm thickness	Sheet with or without glass fiber backing	Usually loose	The process is very difficult and required special techniques and equipment

G—Good
B—Bad

4.8.6 ECONOMICS OF LINING

The costs that should be considered in selection of lining are mainly the following:

- Cost of lining materials.
- Cost of application.
- Cost of maintenance.

Table 4.8 Specifications of Nonsheet Thermoplastic Lining

	Nonsheet Thermoplastic	Available Form	Type of Application	Remarks
1	PVC	Powder (UPVC) plastisol (PVC)	- Dip - Spray - Fluized bed	Some solvents (e.g. aromatic hydrocarbons), will extract the plasticizer and after evaporation of the solvent, the lining will be hard and liable to develop cracks
2	PE	Powder	- Spray - Fluized bed	All three grades of PE (low. medium, and high density) have susceptibility for environmental stress
3	PP	Powder	- Spray - Fluized bed	Polypropylene-lined pipe and fitting are suitable for conveying corrosive liquid and gases in operating pressures up to 10 bars
4	E.CTFE	Powder	- Spray - Fluized bed	Similar as characteristics noted for PP
5	PVDF	- Dispersion - Powder	- Dip - Spray - Fluized bed	Similar as characteristic noted for PP
6	FEP	- Dispersion	- Dip - Spray	Similar as characteristic noted for PP
7	PFA	Powder	- Spray - Fluized bed	Similar as characteristic noted for PP
8	PTFE	- Dispersion	- Dip - Spray	PTFE lining are resistant to most chemicals but because of the presence of pinholes, are not used for corrosion protection
9	EVA	Powder	- Spray - Fluized bed	EVA has excellent weather resistance. It may be applied to a variety of metals including zinc, without use of primer and adhesion to the substrate is excellent
10	Nylon	Powder	- Spray - Fluized bed	The grades of nylon are available for power coating are Nylon 11 and 12
11	Fusion-bonded epoxy	Powder	- Spray - Fluized bed	Fusion-bonded epoxy coatings are suitable for the interior and exterior of steel pipe lines installed under ground or under water (see API-RP-5L7 and AWWA 213-85)

- Extra cost may be required for installation and transportation due to the possibility of mechanical damage.
- Personnel safety equipment.
- In cases where major shutdowns can be avoided through the use of a certain lining, the economic advantages of that lining will be clear; other cases will require detailed economic evaluations.

Table 4.9 Specification of Thermosets Plastic Lining

Thermosetting Resin	Curing	Available Form	Type of Application	Remarks
Phenol - Formaldehyde	Stoved	Liquid	Spray	This lining has good resistance to erosion. For lining application, several coats are necessary in order to achieve the specified thickness. When defects are discovered, the lining shall be cleaned and abraded to wherever resin is to be applied
Epoxy-phenolic	Stoved	Liquid	Spray	Similar as characteristic is noted for phenol-formaldehyde
Polyester	Cold Curing	Liquid	Spray brushing Trawling	The different types of polyester used for lining include isophthalic, terephthalic, orthophtalic bisphenol. vinyl ester and HET acid. Polyester is used with or without reinforced material
Epoxy	Cold Curing	Liquid	Spray	Typical grades of epoxy are used for lining: Amine-cured solvent containing, polyamide-cured solvent containing, high solid epoxy, solvent-free epoxy, coal tar epoxy
Furane	Cold Curing	Liquid	Brush, roller trowel, spray	This resin shall be used with reinforcing fillers. The final coat is unreinforced resin
Polyurethane	Cold curing backed	Liquid	Spray	The typical grades of polyurethane are used for lining: Backed, air-dried and coal tar polyure-thane. All polyurethane systems are sensitive to moisture
Polychloro prene	Cold curing	Liquid elastomer	Spray	Several coats are necessary in order to achieve the specified thickness. At 15°C, the cure time of lining is approximately 7 days. This lining has good resistance to abrasion and water up to 100°C

4.9 THERMOPLASTIC RESIN LINING

Thermoplastic materials are available in two types: Sheet-applied and nonsheet-applied. Nonsheet applied polymers, which come in the forms of powder and granules, should not be used to line concrete and should be used with only metallic surfaces. Sheet-applied thermoplastic polymers are suitable to line concrete and metallic surfaces, and can be used in bonded or loose forms.

When the lining is to be bonded to the substrate, the type of adhesive, the maximum service temperature, and the conditions required for application and curing should be conferred. The minimum bond strength between the lining and the substrate should be 3.5 N/mm^2 in direct shear and 5 N/mm^2 width in peeling at a test temperature of 20°C when tested in accordance with BS 490: Part 10.4. The adhesive should be capable of maintaining a bond at the design temperature, and after cycling between the ambient and maximum surface temperatures, the applicator should provide the evidence of the suitability of the adhesive.

Table 4.10 Specifications of Rubber Lining

Rubber	Hardness (Shore Durometer) ASTM D2240	Formability	Curing	Remarks
NR or IR	50°—Type A	1—Unvulcanized sheet	1—Use an autoclave for the vulcanization of lined equipment	Lining compounds based on NR, IR, and SBR have less abrasion resistance than other types of rubber linings
SBR	50°—Type A			
CR	60°—Type A			Polychoroprenes give lining compounds greater resistance to heat, ozone, and sunlight than other rubbers
11R	60°—Type A	2—Prevulcanized sheet	2—The adhesive is used for bonding is cold curing	This lining has very good resistance to heat and low permeability to gases
NBR and XNBR	80°—Type A			These linings have excellent resistance to swelling by mineral oils and fuels. Polymers of high acrylonitrile to butadiene ratio have the best resistance, as well as lower gas permeability. Higher butadiene ratios have better low-temperature properties. XNBR is normally used for its outstanding abrasion resistance
CSM	80°—Type A			CSM has excellent resistance to heat and ozone
Hard rubber and Ebonite	80°—Type A			These linings have a higher resistance to chemicals than soft rubbers based on the same polymer types. This resistance again generally increases with decrease in unsaturation of vulcanized polymer
FKM	80°—Type A			These materials have excellent resistance to oil and fuels
BR	N/A			Polybutadiene rubber is normally used in combination with polyisoprene or carboxylated nitrile rubber to produce linings with superior abrasion resistance compared to the individual rubbers; also to improves the lower-temperature properties of nitrile rubber
EPR and EPDM	--			These linings have very good resistance to ozone

When a lining is bonded to the substrate, it may be the choice of adhesive that is the limiting factor. A typical maximum service temperature for bonded lining is 100°C. Except when stated otherwise, the maximum service temperature quoted applies to linings that are not bonded to the substrate.

The selection of the type of lining to be used should be based on the function for which it is intended. Unless previous experience demonstrates that a lining will be suitable for a particular purpose, appropriate testing should be carried out (see ASTM D 543 and ASTM C 581).

Table 4.11 Specifications of Bitumen and Coal-Tar Lining

Lining Material	Form Available	Type of Application	Working Temperature (°C)	Remarks
Cold-applied bitumen	Solvent-based liquid	1—Brushing 2—Spraying	4–32	Solvent-based bitumen linings are suitable for the protection of ferrous metals, galvanized surfaces, concrete, and brick. Emulsion-based bitumen should not be used to protect steel. These linings are neither intended to withstand hot conditions, nor to resist contamination by mineral oils or paint solvents
Hot-applied bitumen	Solid (molten for use)	1—Spraying 2—Mapping 3—Trowelling (For floor) 4—Trough method (for pipe)[1]	4–32	The bitumen in this lining used in molten state. For lining metallic or concrete surfaces, one or more coats of primer should be used prior to hot-applied lining. This lining has high resistance to moisture and good adhesion to steel
Cold-applied coal tar	Thin liquid to heavy paste	1—Spraying 2—Brushing 3—Dipping	4–32	This lining is not recommended for atmospheric exposure and sunlight. Coal tar linings are suitable for water and sewage lines and equipment. This lining has excellent adhesion to ferrous metals, slightly etched galvanized iron and steel, concrete, and brick. It is not recommended where a high degree of resistance to abrasion is required
Hot-applied coal tar	Solid (molten for use)	1—Brushing 2—Spraying 3—Trough method (for pipe)[1] 4—Feed-line method (for pipe)[2] 5—Mapping	4–32	Hot-applied coal tar lining should be used only on metal surfaces and usually is applied over a coal as a primer. This lining provides protection of buried steel water pipelines and structures that are immersed in water

[1]In detailed method, the pipe should be rotated and molten enamel should be introduced into pipe by pouring it through the full length of the pipe.
[2]In the retracting-weir or feed-line method, pipe should be rotated and molten enamel should be supplied to the weir or feed line from a reservoir through supply pipes.

Only virgin polymer should be used for the production of the liner; a maximum of 0.2% wt of additives is permitted. A large amount of additives is allowed if electrically conductive properties are required. Additives or coloring agents should be finely homogenized.

Unless specified otherwise by the purchaser, the specification and quality of lining materials should be approved by the purchaser before application. As the lining should be inspected visually for defects, selection of the bonding liner should be avoided in situations where the shape of equipment does not

Table 4.12 Specifications of Ceramic Lining

Lining Material	Types	Type of Application	Remarks
Silicate-based	1—Alkali-alumina borosilicate 2—Barium, crown 3—Water-soluble silicate	1—Spraying 2—Dipping 3—Brushing (soluble type) 4—Packed cementation 5—Trowelling (soluble type)	Linings with silicate base ceramic, with or without added refractories, have the greatest industrial usage of all ceramics. These linings are used for turbines, exhaust manifolds, heat exchangers, and combustion chambers. Soluble silicates are used for lining aluminum, copper, steel, stainless steel, and magnesium
Oxide-based	Without reinforcement	Flame spraying (up to 6 mm)	Thermal shock resistance of flame-sprayed linings decreases with increasing thickness
		Trowelling (up to 25 mm)	Mineral bonding material and expanded-metal reinforcements are used for ceramic lining with trowelling
Carbide-based	N/A	1—Flame spray 2—Air spray 3—Packed cementation	The principal use of carbide-based ceramic linings is for wear and seal applications, in which the great hardness of carbides is an advantage. This type of lining is used for jet engine seals, compressor and turbine blades, and plug gauges
Silicide-based	N/A	Cementation	This type of lining is used to protect refractory metals against oxidation. Silicide linings generally embrittle the metals (ferrous and nonferrous) to which they are applied, but they should not be used to repair lined metals
Phosphate-bonded		1—Air spraying 2—Trowelling	These materials are formed by the chemical reaction of phosphoric acid and a metal oxide. Phosphate-bonded linings are used to protect metals against heat in high temperatures

Table 4.13 Specifications of Brick, Tile, and Glass Lining

Lining Material	Types	Type of Application	Remarks
Glass	1—High-silica glass 2—Boro-silicate glass	1—Slushing[1] 2—Spraying 3—Hot-dusting[2]	Glass linings are used to protect carbon steel, high-tensile steel, cast iron, and stainless steel equipment against corrosive environments in temperatures up to 260°C
Brick and tile	1—Acid-resistant brick 2—Porcelain, tile 3—Carbon brick 4—Graphite brick	Installation of three layers of membrane, brick, and cement	Chemical-resistant brick linings are multi-layer systems supported by a shell, consist of an impervious membrane to prevent the corrosive medium from reaching the shell and a layer of chemical-resistant brick in chemical-resistant cement. This lining is used to protect steel and concrete vessels and columns
Porcelain lining	1—Alkali resistant 2—Acid resistant 3—Hot-water-resistant 4—Regular-blue-black enamels	1—Slushing[1] 2—Spraying 3—Hot dusting[2]	Porcelain enamels are used for sheet steel, cast iron, and aluminium parts to improve appearance and to protect the metal surface against chemicals and hot water. Porcelain enamels have specific electrical properties and thermal shock capability and are used to line chemical reactors and heat exchangers

[1]*Slushing and spraying are wet processes. Slushing consists of two methods of dipping and pouring.*
[2]*Hot dusting consists of shifting glass dust onto a preheated metal surface.*

Table 4.14 Specifications of Cement Mortar and Refractory Lining

Lining Material	Types	Type of Application	Remarks
Chemical-resistant cement mortar	1—Silicate cements 2—Resinous cements	1—Centrifugally spinning (for pipes) 2—Line troweling 3—Shotcrete 4—Hand troweling	Silicate cement linings are resistant to acids, gases, chlorine solutions, and some salts and solvents Resinous cement, which consists of furane cement, phenolic cement, polyester cements, epoxy cements, and sulfur cements, has more resistance to chemicals than silicate cement
Refractory cement	N/A	1—Shotcrete (gunning) 2—Casting	Refractory cement linings are used to insulate furnaces, boilers, flue ducts, and steel stocks. This type of lining is applied to equipment fabricated in metals
Refractory brick	N/A	Installation of brick lining	Brick refractories are used to insulate reformer furnaces, and where refractory cement linings are unsuitable, this type of lining is normally used in conjunction with refractory cement linings

Table 4.15 Lining Thicknesses

Lining Material	Thickness (mm)	Lining Material	Thickness (mm)
Polyethylene (PE)	0.4–2	For pipe-Min. Rubbers: For vessel-Min	3 5
Polypropylene (PP)	0.2–0.75	Reinforcement Thermosets: Un-reinforcement	0.6–5 0.2–0.4
Plasticized PVC	0.6–10	Glass	0.3–2
Unplasticized PVC	0.25–0.75	Reinforcement Ceramic: Un-reinforcement	2–50 0.01–6
Ethylene-chlorotrifluoroethylene (E-CTFE)	0.2–0.75	Brick	50-80
Polytetrafluoroethylene (PTFE)	0.015–0.05	Chemical-resistant cement	4–13
Polyvinylidene fluoride (PVDF)	0.075–0.8	Refractory cement	40-200
Perfluoroalkoxy (PFA)	0.05–0.3	Cold-applied coal tar	0.1–2.7
Fluorinated ethylene propylencopolymer (FEP)	0.025–0.17	Hot-applied coal tar	1.6–3.2
Polyamid (nylon)	0.25–0.75	Cold-applied bitumen	1.6–3.5
Fusion-banded epoxy	0.25–0.75	Hot-applied bitumen	3–6.5
Ethylene vinyl acetate (EVA)	0.35–1.25	N/A	N/A

allow for visual inspection during installation (i.e., in case of welded joint pipeline with interior diameters smaller than 610 mm).

4.9.1 THICKNESS REQUIREMENTS

The thickness of the finished thermoplastic lining will depend upon the material selected and the functions for which it is intended. The maximum thickness and the minimum thickness should be specified by the designer. If necessary, the material should be capable of being thermoformed and welded to give joints that are free of pinholes.

The thickness of linings based on nonsheet applied material varies considerably with the particular plastic selected (see Table 4.16).When the thickness of the lining is less than 400 μm, it is difficult to obtain linings that are pinhole-free. Furthermore, even if linings are free of imperfections, consideration has to be given to the possibility of damage to them during their lives. If the corrosion rate of the material is low, then the thickness of the lining is not a critical factor. If the corrosion rate is high, then thin linings should not be used because of the risk of severe corrosion through a pinhole. If sheet pipes are to lined with thermoplastic material, the thickness of the thermoplastic liner should be in accordance with Table 4.17.

4.9.2 APPLICATION OF THERMOPLASTIC LININGS

Application of thermoplastic linings, requirements, and inspection of the works should be in accordance with standards.

Table 4.16 Thickness pf Finished Thermoplastic Linings

Nonsheet Thermoplastics	Thickness (mm)
PVC	UPVC: 0.25–0.75 PVC: 0.6–10
PE	0.4–2*
PP	0.2–0.75*
E-CTFE	0.2–0.75*
PVDF	0.075–0.8*
FEP	0.025–0.17*
PFA	0.05–0.3*
PTFE	0.015–0.05
EVA	0.35–1.25
Nylon	0.25–0.75*
Fusion-bonded epoxy	0.25–0.75

Exact thickness to be defined according to lining function.

Table 4.17 Dimensions of Liners and Flares

Normal Pipe Size (mm)	Minimum Liner Thickness (mm)						Minimum Flare Diameter (mm)
	PP	PDVF	FEP	PTFE	PTFE*	PFA	
25	3.2	3.8	1.9	1.5	3	1.9	47.6
40	3.8	4.0	1.9	1.6	3.9	1.9	68.3
50	4.3	4.3	1.9	1.7	3.9	2.0	87.3
80	4.4	4.4	2.3	2.4	4.0	2.2	117.5
100	5.2	5.2	2.4	2.7	4.5	2.4	150.8
150	5.5	5.5	3.3	3.7	5.0	3.3	201.0
200	5.5	5.5	3.3	4.0	5.0	N/A	255.6
250	6.3	N/A	3.5	5.0	6.5	N/A	311.2
300	6.3	N/A	3.5	5.3	8.1	N/A	365.1
350	7.2	N/A	3.5	6.0	N/A	N/A	423.5
400	7.2	N/A	3.5	6.3	N/A	N/A	470.0
450	N/A	N/A	N/A	7.5	N/A	N/A	N/A

Special thicknesses for vacuum or heavy-duty service (chlorine, bromine, etc.) are only applicable when specified in a purchase order.

4.10 THERMOSETTING RESIN LININGS

Thermosetting resins used for the lining are divided into two groups: cold-curing thermosetting resin and stoved-thermosetting resin. Cold-curing resins are suitable to line concrete and metallic surfaces, and stoved resin should only be used to line metallic surfaces. Stoved-thermosetting resin linings are used to protect against corrosive environments, prevent the contamination of products, and provide surfaces that do not get soiled easily or that can be cleaned easily.

The minimum pipe diameter for liquid epoxy lining should be 610 mm to permit inspection and repair of the internal lining by entering the pipe. The actual stoving temperature is specified by the lining supplier, but it normally ranges from 150°C to 200°C depending on stoving time.

All cold-curing resin linings are poor thermal conductors, so they reduce heat transfer. Most of these linings offer good resistance to erosion by suspended particles and to buildup of deposits (see Tables 4.18 and 4.19).

Table 4.18 Resistance of Various Types of Thermoset Epoxy Linings in Chemical Environments

Corrosive		Epoxy Amine Cured Coating	Epoxy Ester Coating (Not Recommended for Immersion)	Coal-tar Epoxy
Acids:				
Sulfuric, 10%		R	LR	R
	50%	LR	NR	R
	78%	NR	NR	N/A
Hydrochloric,	10%	R	LR	LR
	20%	LR	LR	N/A
	35%	NR	NR	N/A
Nitric,	10%	NR	NR	N/A
	20% +	NR	NR	N/A
Phosphoric,	10%	R	R	R
	85%	NR	LR	LR
Acetic. glacial		NR	NR	N/A
Fatty acids		NR	R	R
Boric acid		N/A	N/A	R
Water:				
Tap		R	R	R
Distilled		R	R	LR
Sea		R	R	R
Alkalis:				
Sodium hydroxide,	20%	R	NR	R
	70%	R	NR	LR
Ammonium Hydroxide,	10%	LR	LR	N/A
Aluminum hydroxide		N/A	N/A	R
Potassium hydroxide		N/A	N/A	R
Oxidizing agents		NR	NR	

(Continued)

Table 4.18 Resistance of Various Types of Thermoset Epoxy Linings in Chemical Environments *(cont.)*

Corrosive	Epoxy Amine Cured Coating	Epoxy Ester Coating (Not Recommended for Immersion)	Coal-tar Epoxy
Fats and oils:			
Mineral	R	R	N/A
Animal	R	R	N/A
Vegetable	R	R	N/A
Gases:			
Chlorine (wet)	LR	LR	LR
Ammonia	LR	R	N/A
Carbon dioxide	R	R	R
Aldehydes	LR	LR	N/A
Amines	LR	LR	N/A
Solvents:			
Alcohol ethyl and above	R	LR	N/A
Aliphatic Hydrocarbons	R	R	N/A
Aromatic hydrocarbons	R	R	LR
Ketones	LR	NR	N/A
Ethers	LR	NR	N/A
Esters	LR	NR	N/A
Chlorinated hydrocarbons (general)	LR	NR	N/A
Carbon Tetrachloride	R	LR	N/A
Diethyl ether	R	NR	N/A
Salts:			
Sodium chloride	R	R	R
Sodium phosphate	R	R	R
Copper sulfate	R	R	N/A
Ferric chloride	R	R	R
Wetting agents (ionic and non-ionic)	R	R	N/A
Miscellaneous:			
Hydrogen peroxide, 30%	NR	NR	N/A
Sodium hypochloride	NR	NR	N/A
Chromic acid	NR	NR	N/A
Perchloric acid	NR	NR	N/A

Note: R = Recommended; NR = Not Recommended; LR = Limited Recommendation.

Cold-curing thermoset resins may be used on metallic structures with or without reinforcement regarding the lining thickness required. Some lining materials are unacceptable due to incompatibility with petroleum products or product additives. Acceptable thermosetting materials should be tested in accordance with ASTM-D-543 and ASTM- C-581.

For lining underground storage tanks with thermosetting materials, see API-RP-1631. For lining of water pipelines with liquid epoxy systems, see AWWA-C-210-84. Unless previous experience

Table 4.19 Resistance of Various Types of Polyurethane Linings in Chemical Environments (at Ambient Temperatures)

Chemicals	Air Dry Polyurethane	Baked Polyurethane	Coal-tar Polyurethane
Acids. Mineral:			
Hydrochloric, 10%	LR	LR	LR
Hydrochloric, 37%	LR	LR	N/A
Sulfuric, 10%	LR	LR	LR
Sulfuric. 70%	NR	LR	N/A
Nitric, 10%	LR	LR	N/A
Nitric, 70%	NR	NR	N/A
Phosphoric. 10%	R	R	LR
Phosphoric. 85%	LR	LR	N/A
Chromic, 10%	LR	R	N/A
Chromic, 50%	NR	LR	N/A
Hydrofluoric, 48%	LR	LR	N/A
Hypochlorous	LR	NR	N/A
Acids. organic:			
Acetic, 10%	LR	R	N/A
Glacial acetic	NR	NR	N/A
Anhydride acetic	R	R	N/A
Formic	NR	LR	N/A
Lactic	R	R	R
Cresylic	NR	NR	N/A
Oleic	R	R	N/A
Oxalic	R	R	N/A
Maleic	R	R	R
Stearic	R	R	N/A
Benzene sulfonic	R	R	N/A
Fatty acids	R	R	R
2-ethyl butyric acid	R	R	N/A
Citric acid	N/A	N/A	R
Boric acid	N/A	N/A	R
Acid Salts:			
Aluminum sulfate	R	R	N/A
Ammonium chloride, nitrate, sulfate	R	R	R
Calcium chloride, nitrate, sulfate	RR	R	N/A
Zinc chloride, nitrate, sulfate	R	R	N/A
Ferric chloride	N/A	N/A	R
Magnesium chloride	N/A	N/A	R
Alkalis:			
Aluminum hydroxide,	N/A	N/A	R
Ammonium hydroxide: dilute	LR	R	N/A
Ammonium hydroxide,			
cone.	LR	R	N/A

(Continued)

Table 4.19 Resistance of Various Types of Polyurethane Linings in Chemical Environments (at Ambient Temperatures) *(cont.)*

Chemicals		Air Dry Polyurethane	Baked Polyurethane	Coal-tar Polyurethane
Calcium hydroxide		LR	R	N/A
Potassium hydroxide		LR	R	LR
Sodium hydroxide,	15%	LR	LR	LR
Sodium hydroxide	50%	LR	LR	LR
Alkaline Salts:				
Sodium bicarbonate		R	R	R
Sodium carbonate		R	R	N/A
Sodium sulfide		R	R	N/A
Sodium sulfite		R	R	R
Trisodium phosphate		R	R	R
Sodium nitrate		R	R	N/A
Oxidizing agents:				
Sodium hypochlorite		NR	NR	N/A
Potassium-permanganate		NR	NR	N/A
Sodium chlorate		LR	LR	N/A
Hydrogen peroxide		LR	LR	N/A
Gases:				
Chlorine, dry		LR	R	N/A
Chlorine, wet		LR	R	N/A
Ammonia		LR	R	N/A
Hydrogen sulfide		R	R	N/A
Carbon dioxide		N/A	N/A	R
Solvents:				
Ethyl acetate		NR	R	N/A
Butyl acetate		NR	R	N/A
Acetone		NR	RL	NR
Methyl ethyl ketone		NR	R	N/A
Methyl isobutyl ketone		NR	R	N/A
Cyclohexanone		NR	RL	N/A
Isophorone		NR	R	N/A
Methyl alcohol		NR	R	N/A
Ethyl alcohol		NR	R	N/A
Fatty alcohols		R	R	N/A
Glycols		R	R	R
Trichloroethylene		NR	R	LR
Perchlorethylene		LR	R	LR
Carbon tetrachloride		LR	R	LR
Methylene chloride		NR	LR	N/A
Ethylene dibromide		LR	LR	N/A
Toluene		R	R	LR
Benzene		R	R	N/A
Aromatic hydrocarbons		R	R	N/A
Ethyl ether		R	R	N/A

Table 4.19 Resistance of Various Types of Polyurethane Linings in Chemical Environments (at Ambient Temperatures) *(cont.)*

Chemicals	Air Dry Polyurethane	Baked Polyurethane	Coal-tar Polyurethane
Aliphatic hydrocarbons	R	R	N/A
Gasoline	R	R	N/A
Jet fuel	R	R	N/A
Orthodichlorobenzene	R	R	N/A
Carbon disulfide	R	R	N/A
Dimethyl formamide	NR	NR	N/A
Turpentine	R	R	N/A
Xylene	N/A	N/A	LR
Other:			
Cutting oils	R	R	N/A
Vegetable oils	R	R	N/A
Lubricating oils	R	R	N/A
Diester lubricants	R	R	N/A
Stryrene monomer	R	R	N/A
Glycerin	R	R	N/A
Pyridine	NR	N/A	N/A
Detergents	R	R	N/A
Formaldehyde, 37%	R	R	
Distilled water	R	R	LR
Tap water	R	R	R
Salt water	R	R	LR
Condensate water	R	R	N/A
Fruit juices	R	R	N/A
Milk products	R	R	N/A
Phenol in alcohol	NR	LR	N/A
Sewage waste	R	R	N/A
Hydrazine	LR	N/A	N/A
Glyoxal	R	R	N/A
Anionic wetting agents	R	R	N/A
Acetonitrile	LR	R	N/A
Butyraldehyde	R	R	N/A
Monoethanolamine	LR	LR	N/A

Note: R = Recommended; NR = Not Recommended; LR = Limited Recommendation.

demonstrates that a lining will be suitable for a particular purpose, appropriate testing should be carried out (see ASTM-D-543 and ASTM-C-581).

The selection of the type and thickness of thermosetting lining for a special service should be based on widely accepted standards. The resistances of various types of thermoset epoxy and polyurethane linings in chemical environments are shown in Tables 4.18 and 4.19. The selection of type of lining to be used should be based on the function for which it is intended (see Tables 4.18 and 4.19), as well as the shape of the equipment.

4.10.1 THICKNESS

The dry film thickness of the finished lining will depend upon the type of material selected and the duties for which it is intended. The designer should specify the thickness of the finished lining.

The dry film thickness of the finished thermosetting resin lining without the use of reinforced material (i.e., thin layer) should not thicker than 400 μm, as there is a tendency for the material to crack if it is thicker. It normally should not be thinner than 200 μm to minimize the incidence of pinholes.

The thickness of finished thermosetting resin linings with using of inert fillers and/or reinforcing agent (thick layer) such as mica, glass flake, glass fiber, and carbon black range from 600 μm up to 5 mm. For thick, cold-curing thermosetting resins, typically the liner is 2.5–3.5 mm thick and has an average makeup of 20–25% glass fiber and 75–80% resin. For good long-term lining performance, the temperature should be limited to 60°C.

4.10.2 APPLICATION OF THERMOSETTING RESIN LININGS

For metal equipment, the lining process should start as soon as possible after surface preparation is complete and before any visible rusting occurs. Unless maintained in a dehumidified atmosphere, application of the lining should commence within 4 h. If signs of rusting occur, then the surface should be prepared again to meet the required standard.

For concrete equipment, lining should not proceed until at least 28 days after the concrete was cast and when the free water content is down to the level specified. Where necessary, surfaces should be primed to promote a bond with the lining material. Once a primer has been applied, the equipment should be kept clean and the lining process should start as soon as possible in compliance with manufacturer recommendations.

Note: The primer should be pigmented to facilitate uniform application and to assist in establishing full coverage of the surface to be lined. In the case of steel equipment with large surface areas, holding primers may be used to hold the blast-cleaned surface, provided that the holding primer is compatible with the lining material and does not interfere with the adhesion of lining.

4.10.3 STOVED THERMOSETTING RESINS

Several coats are necessary to achieve the thickness stipulated, and each coat should be allowed to air-dry before application of the next coat. Any intermediate stoving should be at a lower temperature than the final stoving temperature, such that curing does not proceed beyond the stage that impairs intercoat adhesion. All external angles and edges should be strip-coated by applying a thin coat before the rest of the surface is coated.

Before the final stoving takes place, the lining should be tested for continuity. If the continuity of the lined equipment meets the stated standards, final stoving can proceed; if not, further coats should be applied locally or over the whole surface until that occurs. When extra coats are applied, the final thickness should not exceed the specified limit.

Note: This procedure is necessary because once these materials have been cured at the final stoving temperature, it is not possible to obtain the same level of intercoat adhesion.

4.10.4 **THICK LAYER, COLD-CURING THERMOSETTING RESINS**

The lining process should be appropriate to the grade of material selected for the lining.

Cold-curing thermosetting lining are applied by the following methods:

- Hand laid-up mat linings
- Spray-up chopped lining
- Flake glass lining
- Troweled mortar lining

Hand laid-up mat lining consists of two to three layers of glass fiber mat laid up over a blasted primed substrate and finish with a coat of resin. Typically, the liner is 2.5–3.5 mm thick and its makeup averages 20–25% glass fiber and 75–80% resin. For good long-term lining performance, the temperature should be limited to 60°C.

4.10.5 **FLAKE GLASS LINING**

The advent of flake glass lining extended the service temperature range to 71–82°C. It also produced a higher-quality (and premium-priced) lining system. The glass flakes are blended with a suitable resin system selected for the service conditions. Flake glass lining should be applied in three coats (according to manufacturer recommendations). A formulation of 0.75–1 mm thickness of lining used in gas and vapor service and 1.5–2 mm system used in severe corrosive conditions for gas, vapor, and submerged environments. The latter is a system that can be used to line large vessels and tank and flue gas desulfurization systems. Flake glass lining cannot be used for structural repairs. Edges and sharp corners should be radiused and the flake glass lining covered with a mat lining.

4.10.6 **TROWELED MORTAR LINING**

The filled resin lining stabilized with a light roving can be used in lining of sumps, trenches, concrete tanks, and vessels. The troweled mortar lining, although providing much better abrasion resistance than a flake glass lining, does not possess heat resistance to excursions. Silica-filled epoxy liners can be used in the linings of concrete trenches and sumps.

4.11 **RUBBER LINING**

The storage of corrosive or abrasive solutions or suspensions requires that the metal surface of storage tanks, large pipes, and holding vessels be lined with a material that resist such activity. Vulcanized rubber that is securely adhered to the tank or other metal and concrete surface imparts such resistance.

Rubbers are elastomeric polymers with reactive sides along their molecular chain that enable cross-linking with each other. The cross-linking process is called *vulcanization*. This process is induced by heat or vulcanization agents or by a combination of both.

The physical and chemical properties of vulcanized rubbers vary widely according to the type of rubber used and the amount and type of filler and vulcanization agent present in the compound. Some rubbers are available both as hard rubber (ebonites) and as soft rubber. For one rubber, with increasing hardness, chemical resistance increases and abrasion resistance decreases.

Rubber linings are generally designed for vulcanization at elevated temperatures but similar rubber linings containing accelerated systems capable of promoting vulcanization at ambient temperatures are also available. Certain rubber linings can be applied in the form of prevulcanized sheets.

The minimum thickness of rubber lining should be 4.5 mm in vessels and 3 mm in piping. Rubber should be calendared to contain at least four plys for each 3 mm of thickness.

The adhesive system used for bonding the rubber sheet to the substrate should depend upon the type of rubber and the method of vulcanization and may consist of one or more individual layers. It should be the responsibility of the applicator to select an adhesive system that will provide an adequate bond between rubber and substrate.

Note: In cases where there is a likelihood of severe chemical reactions, abrasion, or mechanical damage, the thickness may be increased to 6 mm or greater. When the rubber is used as a membrane under chemical-resistant brick lining, the thickness should not be less than 5 mm. Up to these thicknesses, linings are applied as one single layer except in the case of a composite buildup; i.e., soft rubber base with an ebonite top layer. Thicknesses greater than 6 mm are normally applied in two or more layers. The applicator should provide the company with evidence of the suitability of the adhesive.

The application of rubber linings should be carried out by experienced contractors. Properly applied rubber lining will resist vacuum in the order of 130 m bars. Shop-vulcanized rubber linings generally have a better resistance to vacuum than in situ vulcanized ones. Rubber-lined equipment should not be stored in extreme temperature conditions, such as below 0°C or above 49°C. Avoid sudden changes in temperature.

Rubber-lined equipment may be protected for extended periods of time by storing the tank partially filled with a diluted solution. When recommended by the rubber-lining manufacturer, a 5% sulfuric acid, 5% sodium carbonate solution, or a weak salt solution is an ideal storage medium to help keep the lining flexible, to minimize expansion and contraction, and to keep the air (ozone) from prematurely deteriorating the lining surface. Do not permit the liquid contained within to freeze.

It is recommended that all rubber-lined plants and equipment should be thoroughly inspected at the end of the first year of service. If its condition is satisfactory, then the inspection period after that can be longer than one year, but one-year inspection periods are still preferred.

The temperature range of a particular rubber or compound lining will depend upon the chemical environment to which it is exposed and the composition of the rubber compound. The designer should specify the hardness of the lining material, and it should conform to the value specified on the requisition within a tolerance of ±5°. A minimum of three readings per square meter should be taken.

4.11.1 DESIGN ASPECTS FOR RUBBER LININGS

For the final selection of the type of rubber lining to be used and the method of application, which should be made in conjunction with the materials specialist and the lining contractor, the following details should be considered:

- Products to be handled—A full analysis should be made, including components present in trace quantities. It is imperative to determine the latter, as they may have a deleterious effect on the life span of the lining. For certain services, it may be important to determine whether contamination or discoloration of the material by the liner can occur.

- Temperature—for all products to be handled, determine the normal operating temperature, maximum and minimum temperatures, and the cycle of temperature variation.
- The degree of vacuum or pressure—determine the normal operating pressure, maximum and minimum pressures, and the cycle of pressure variation.
- Determine whether the cycle of operations is a continuous or a batch process.
- Abrasion and erosion—Details of the amount, particle size, and physical characteristics of suspended matter, together with flow rates.
- Immersion conditions—Constant or intermittent immersion of the lining, and partially or completely filled operation. The contractor should be prepared to supply the specifications, including a reference number for the approved rubber compound, and samples of the vulcanized rubber sheet for test and reference purposes.

4.11.2 THICKNESS OF RUBBER

Generally, the thickness of the lining should be a minimum of 3 mm. In cases where there is likelihood of severe chemical reaction, abrasion, or mechanical damage, the thickness may be increased to 6 mm or greater.

Up to 5 mm thicknesses, linings are applied as one single layer except in the case of a composite buildup; i.e., a soft rubber base with an ebonite top layer. Generally, the minimum thickness should be 4.5 mm for vessels and 3 mm for piping.

The thickness of the lining applied to a metallic substrate should be determined with a suitable thickness meter and should conform to the specified thickness within a tolerance of $\pm 10\%$. A minimum of three measurements per square meter should be made. The thickness of the rubber lining applied to concrete or any other nonmagnetic surface should be determined.

4.11.3 HARDNESS OF RUBBER

The hardness of a rubber indicates its chemical resistance and mechanical strength. In general, for one rubber composition, the chemical resistance increases and the mechanical properties (e.g., abrasion resistance) decrease with increasing hardness (see Table 4.20).

Table 4.20 Typical Values for the Different Rubbers Used for Lining Purposes					
Hard NR	Rubber	60°	Type	D	Shore durometer
Soft NR	Rubber	50°	Type	A	Shore durometer
SBR	Rubber	50°	Type	A	Shore durometer
CR	Rubber	60°	Type	A	Shore durometer
MR	Rubber	60°	Type	A	Shore durometer
NBR	Rubber	80°	Type	A	Shore durometer
CSM	Rubber	80°	Type	A	Shore durometer
FKM	Rubber	80°	Type	A	Shore durometer
Ebonites		80°	Type	D	Shore durometer

The hardness is determined by measuring the penetration of a specified indentor under a certain load. Various types of indentors and loads are used. In general, it is common to express hardness in Durometer A or Durometer D readings in accordance with ASTM D 2240.

4.11.4 APPLICATION OF RUBBER LINERS

The application of rubber liners, adherence to requirements, and inspection of the works should be in accordance with widely accepted standards.

4.12 GLASS AND PORCELAIN LINING
4.12.1 GLASS LINING (VITREOUS ENAMELING)

Glass linings are used to protect ferrous metal equipment such as reactors, storage tanks, pipes, and valves in against corrosive environments. Their resistance to strong alkalis depends on temperature (see Table 4.21). At room temperature, practically any alkaline concentration may be safely handled. At the boiling point, concentration having a pH value between 10 and 12 will cause glass etching, and for this reason, that is not recommended. Glass linings should only be used for metals.

Carbon steel (ASTM-A-285-Grade A and B), high-tensile steel, cast iron, stainless steel (Type 430) can be glassed successfully. Items made of cast steel do not glass well.

There are two types of glasses that have the high level of universal corrosion resistance required by the chemical industry:

- High-silica glass, while being the most acid-resistant glass, is very difficult to apply to steel.
- Borosilicate glass is the type used to coat most vessels and equipment for severe chemical service.

Borosilicate glasses are combination of metal oxides, which should be dry-mixed and fused at 1,370°C and then should be chilled suddenly. Glass in this quenched condition is known as *frit*.

The glass frit is prepared for use in the production in two forms: slip and dust. In the spraying and slushing methods of application, the glass lining slip form of glass frit should be used, and in the hot-dusting method, the dust form of glass frit should be used.

4.12.2 GLASS LINING THICKNESS

The thickness of glass linings should be selected based on the function for which they are intended and with reference to the following general types of glassed steel substrate:

- High-voltage tested glass lining or alkali-borosilicate (with a thickness of 1–2 mm) for very severe corrosive service.
- Low-voltage tested glass lining (with a thickness 0.3–1.25 mm) for mild corrosive service.
- Visual-tested glass lining (0.38–0.63 mm) should not be used for corrosive environment, but it is used in handling stored materials.

The upper and lower thickness limits for water heaters should be 0.15 and 0.4 mm. When local increases in thickness are technically unavoidable, they should be limited. The lining thickness should never be greater than 1 mm.

Table 4.21 Corrosion Resistance of Glassed Steel to Alkaline Media

Alkali	Concentration (%)	Temperature (°C)	Liquid Phase Corrosion Rate (μm/year)
NaOH	1	21 65 100	25 125 1,600
	5	21 65 100	25 250 4,025
	10	21 65 100	25 250 4,375
	20	21 65 100	25 550 5,100
Na$_3$PO$_4$	1	21 65 100	25 175 1,000
	5	21 65 100	25 225 2,350
	10	21 65 100	25 150 2,350
	20	21 65 100	25 175 1,975
Na$_2$CO$_3$	1	21 65 100	25 100 450
	5	21 65 100	25 100 725
	10	21 65 100	25 100 100
	20	21 65 100	25 125 1,325

4.12.3 APPLICATION OF GLASS LINING

Application of glass linings should be in accordance with standards.

Porcelain enamels are glassy inorganic coatings that are applied primarily to fabricated sheet steel, cast iron, or aluminum parts to improve their appearance and to protect the metal surface. For standard specifications of steel sheets for porcelain linings, see ASTM-A-424-80.

Table 4.22 Fired Porcelain Enamel Thickness

Type of Metal	Thickness (mm)
Aluminum	0.065–0.125
Sheet steel	0.1–0.225
Cast iron	1–1.78

Alkali-resistant, acid-resistant, hot water–resistant and regular blue-black enamels are four types of porcelain enamel that are used to protect metal surfaces in corrosive environments. In porcelain enamels applied to most metals, silica, alumina and the oxides of boron, sodium, magnesium, calcium and potassium are the basic ingredients.

The raw material is fused at a temperature above 420°C and then chilled to produce frits. The porcelain frit is prepared in two forms: slip and dust. The optimum thickness of porcelain enamel depends on the substrate metal and the service requirements of the part (see Table 4.22).

Normally, porcelain enamels are selected for products or components where there is a need for one or more of the special service requirements that porcelain enamel can provide, such as chemical resistance, corrosion, protection, weather resistance, specific mechanical or electrical properties, and thermal shock capability. Porcelain enamels are used for the lining of chemical reactors, heat exchangers, induction heating coils, jet engine components, and transformer cases.

Preparation prior to porcelain enameling and lining are the same as for glass lining. The porcelain enamels should be fired at 1,010°C.

The hard surfaces of enamel coatings have low coefficients of friction and are highly resistant to abrasion, rating 3½ to 6 on Moh's scale of mineral hardness. (The Moh's scale is a hardness scale for minerals in which 1 represents the hardness of talc; 2, gypsum; 3, calcite; 4, fluorite; 5, apatite; 6, orthoclase; 7, vitreous pure silica; 8, quartz; 9, topaz; 10, garnet; 11, fused zirconium oxide; 12, fused alumina; 13, silicon carbide; 14, boron carbide; and 15, diamond). Being smooth and impervious to absorption, they minimize product adhesion and are readily cleaned. Porcelain enamels are essentially glassy and are electrically nonconductors with dielectric strength in the range of 100–450 volts per 0.025 mm depending upon composition.

4.13 CERAMIC LINING

Ceramic coating and lining, by strict definition, are known as *super porcelains* based on silicates and oxides. However, by extension, high-temperature coating based on carbides, silicides, borides, nitrides, cermets, and some other inorganic materials also have come to be referred to as *ceramic coatings and linings.*

Ceramic coatings and linings are applied to metal, concrete, and brick surfaces to protect them against oxidation, corrosion, and erosion at room temperature and at elevated temperatures. Special coatings and linings have been developed for specific uses, such as chemical resistance, and prevention of hydrogen diffusion. Some of the applications in which ceramic coated metals are

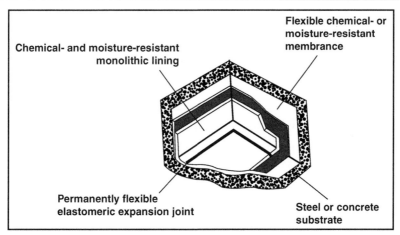

FIGURE 4.14 The surface layer.

employed are furnace components, heat-treating equipment, chemical-processing equipment, and heat exchangers.

Ceramic lining can be applied in two systems: single-layer and dual-layer. Single-layer linings consist of a layer of castable or gunitable ceramic with or without an anchoring system. Single-layer ceramic linings should be used only on metallic surfaces for specific uses such as wear, chemical resistance, and electrical resistance in heat-treating equipment and chemical processing equipment.

Dual-layer linings consist of a layer of membrane protected by a surface layer of erosion-resistant castable ceramic. The surface layer may be supported by either Y- or V-studs or retained in hex mesh supported by 50-mm square washers and studs (see Figure 4.14). The dual-layer system is used to line both steel and concrete equipment. Membrane choices include asphaltics, resins, and synthetic elastomers. Dual-layer systems of ceramic linings should be used to store sodium chlorate ($NaClO_3$), hot HCl acid, titanium dioxide (TiO_2) slurry, ethylbenzene, aluminum chloride, and benzene.

Schematic of a chemical-resistant dual-lining system that provides double protection to the substrate in the form of a flexible membrane and a rigid surface layer. The flexible, corrosion-resistant membrane is applied in direct contact with steel or concrete substrates. It is then covered by a monolithic ceramic lining, which provides protection over a broad pH range and against high temperatures.

4.13.1 CERAMIC LINING THICKNESS

The thickness of ceramic linings depends on the type of material and method of lining, as follows:

- Silicate linings are applied by spraying, brushing, and dipping at the usual thickness of 0.02–0.12 mm, and with reinforcement of up to 25 mm thickness.
- Oxide linings are applied at the usual thickness of 0.04–6 mm, and with reinforcement of up to 25 mm thickness.

- Carbide linings are applied at the usual thickness of 0.04–0.3 mm.
- Silicide linings are applied in thicknesses of 0.01–0.1 mm.
- Phosphate-bonded linings are applied with reinforcement of up to 50 mm thickness.
- Cermet linings are applied in thicknesses of 0.02–0.7 mm.

4.14 CHEMICAL-RESISTANT BRICK AND TILE LINING

A chemical-resistant brick lining is used for the internal protection of steel and concrete types of process vessels, columns, tanks, trenches, etc. It is a multilayer system supported by a shell to provide rigidity and strength. Generally, it consists of an impervious membrane to prevent the corrosive medium reaching the shell; and one or more layers of chemical-resistant brick and tile, laid in a chemical-resistant cement (see Figure 4.15). For more information about acid-proof vessels with membrane and brick linings, see National Association of Corrosion Engineers (NACE) publication.

Brick-lined equipment should be installed such that a complete inspection of the outer surface is always possible. Flat-bottom steel vessels, therefore, should be supported on beams so that an inspection of the bottom can be made.

Chemical-resistant lining should be regularly inspected for defects. The main defects are spalling of the bricks or tiles, erosion effects, cracks in the lining, and degradation of chemical resistant lining materials.

When a defect is detected, repairs should be carried out immediately in order to prevent serious attack to the concrete or steel substrate. Brick linings should be carefully treated and protected

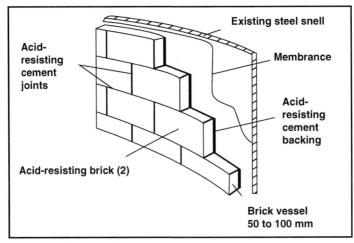

FIGURE 4.15 Chemical-resistant brick and tile lining.

against damage by traffic loads, impact, and local chemical and thermal attacks (steam, leaking flanges, etc.).

4.14.1 TYPE OF BRICK LINING

Chemical-resistant brick lining is a multilayer system consisting of the following layers:

- Impervious membrane
- Chemical-resistant cement
- Layer of brick or tile

These layers are discussed next.

4.14.2 IMPERVIOUS MEMBRANE

It is almost always desirable to use a continuous and impervious membrane lining between the vessel shell and the chemical-resistant brick. This will protect the vessel shell from chemical attack by solutions that permeate the brick lining through pores or cracks. Materials employed for this purpose are sheet elastomers and plastics, as well as coatings that vary in thickness and chemical resistance.

It is desirable to provide brickwork that is sufficiently thick to hold membrane temperatures to about 65°C maximum. When there is no suitable impervious membrane available, even for moderate corrosion and temperature conditions, a brick lining should be used only when there is convincing evidence that a crack-free lining can be applied and maintained e.g., by prestressing the brick lining (i.e., heating and pressing the equipment before the cement is fully cured). In this case, joint cement will cure in the expanded condition and will not contract on cooling down.

The compatibility of some membranes with substrate material is listed in Table 4.23.

4.14.3 CHEMICAL-RESISTANT CEMENT LAYER

Silicate-based cements or synthetic resin-based cements are used to cement the bricks (see Table 4.24). The chemical resistances of these cements are not as universal as that of brick, so careful selection with respect to chemical resistance is required (see Table 4.25).

Table 4.23 Compatibility of Membranes with Substrate Material

SURFACE TO BE LINED	MEMBRANE					
	Hard Soft rubber	Poly-isobutylene	Reinforced Epoxy	Polyurethane	Lead (Sheet and Melted)	Asphaltic Bitumen
Concrete	---	X	X	X		X
Steel	X	X	X	---	X	---

X = Can be used,
--- = Cannot be used.

Table 4.24 Membranes for Use with Brick Linings

Corrodent	Soft Natural Rubber		Neoprene		Plasticized PVC		Asphalt		Lead	
	24°C	65°C	24°C	65°C	24°C	65°C	24°C	65°C	24°C	65°C
Acids										
Acetic, 10%	F	P	F	P	F	P	P	P	P	P
Acetic glacial	P	P	P	P	P	P	P	P	F	P
Benzoic sulfonic, 10%	E	E	G	F	E	E	E	E	G	F
Benzoic	E	E	E	E	E	E	E	E	P	P
Boric	E	E	E	E	E	E	E	E	G	G
Butyric 100%	F	P	F	P	P	P	P	P	P	P
Chloroacetic, 10%	P	P	F	P	F	P	P	P	P	P
Chromic, 5%	P	P	P	P	G	F	G	F	E	E
Chromic, 10%	P	P	P	P	E	F	G	F	E	E
Citric, 10%	E	E	E	E	E	E	E	E	E	E
Fatty acids, 100%	P	P	F	P	G	F	P	P	G	G
Fluosilicic, 40%	E	E	G	G	E	E	P	P	-	-
Formic, 90%	G	F	G	F	F	P	F	P	F	P
Hydrobromic, 48%	G	F	F	P	E	F	E	G	P	P
Hydrochloric, 37%	E	G	F	P	E	F	E	E	P	P
Hydrocyanic, 25%	G	G	E	E	E	E	E	E	G	F
Hydrofluric, 40%	G	F	G	F	E	P	F	F	F	P
Hypochlorous, 10%	E	G	P	P	E	P	P	P	P	P
Lactic, 25%	E	E	E	E	E	E	E	E	P	P
Maleic, 25%	G	G	G	F	G	F	G	F	P	P
Nitric, 5%	P	P	F	P	E	E	G	G	P	P
Nitric, 10%	P	P	F	P	E	E	G	G	P	P
Nitric, 20%	P	P	F	P	E	F	G	F	P	P
Oleic, 100%	F	P	F	P	E	F	F	P	P	P
Oxalic, 20%	G	G	G	G	E	F	E	E	P	P
Perchloric, 40%	G	F	G	F	E	P	-	-	-	-
Phosphoric, 85%	E	E	E	E	E	E	E	E	E	G
Picric, 10%	F	P	F	P	P	P	P	P	G	G
Stearic, 100%	F	P	F	F	E	F	F	P	G	F
Sulfuric, 50%	E	E	E	E	E	E	E	E	E	E

Table 4.24 Membranes for Use with Brick Linings *(cont.)*

Corrodent	Soft Natural Rubber		Neoprene		Plasticized PVC		Asphalt		Lead	
	24°C	65°C	24°C	65°C	24°C	65°C	24°C	65°C	24°C	65°C
Sulfuric, 70%	F	P	F	P	E	P	F	P	E	E
Sulfuric, 87%	P	P	P	P	P	P	P	P	G	P
Oleum, 110% (sulfuric)	P	P	P	P	P	P	P	P	P	P
Alkalis										
Ammonium hydroxide, 28%	G	G	G	G	E	F	G	F	G	F
Calcium hydroxide, SAT.D	E	E	G	G	E	F	G	F	G	F
Potassium hydroxide, 25%	E	E	E	E	E	G	G	F	G	F
Sodium hydroxide, 25%	E	E	E	E	E	G	G	F	G	F
Acid Salts										
Alum 10%	E	E	E	E	E	E	E	E	E	E
Ammonium chloride, nitrate, sulfate	E	E	E	E	E	E	E	E	G	E
Copper chloride, nitrate, sulfate	E	E	E	E	E	E	E	E	E	E
Ferric chloride, nitrate, sulfate, 10%	E	E	E	E	E	E	E	E	E	E
Nickel chloride, sulfate, 10%	E	E	E	E	E	E	E	E	E	P
Stannic chloride, 100%	E	E	G	E	F	E	E	E	E	E
Zinc chloride, nitrate, sulfate, 10%	E	E	E	E	E	E	E	E	G	F
Alkaline Salts										
Barium sulfide	E	E	E	E	E	E	E	P	E	E
Sodium bicarbonate, 10%	E	E	E	E	E	E	E	E	E	E
Sodium carbonate, 10%	E	E	E	E	E	E	G	F	E	E
Sodium sulfide	E	E	E	E	E	E	F	P	G	G
Trisodium phosphate, 10%	E	E	E	E	E	E	P	P	P	P
Neutral Salts										
Calcium chloride, nitrate, 10%	E	E	E	E	E	E	E	E	G	F
Magnesoum chloride, nitrate, sulfate, 10%	E	E	E	E	E	E	E	E	G	F
Potassium chloride, nitrate, sulfate, 10%	E	E	E	E	E	E	E	E	G	F
Sodium chloride, nitrate, sulfate, 10%	E	E	E	E	E	E	E	E	G	F

(Continued)

Table 4.24 Membranes for Use with Brick Linings *(cont.)*

Corrodent	Soft Natural Rubber		Neoprene		Plasticized PVC		Asphalt		Lead	
	24°C	65°C	24°C	65°C	24°C	65°C	24°C	65°C	24°C	65°C
Gases										
Chlorine (dry)	G	F	P	P	E	F	E	E	E	E
Chlorine (wet)	P	P	P	P	E	F	E	F	G	G
Hydrogen sulfide	G	F	E	E	E	E	E	E	E	G
Sulfur dioxide (dry)	G	F	E	E	E	E	E	E	E	G
Sulfur dioxide (wet)	E	E	E	E	E	G	E	E	E	G
Organic Materials										
Acetone, 100%	G	F	G	F	P	P	P	P	E	E
Alcohol, methyl, ethyl, 96%	E	E	E	E	E	F	E	E	E	E
Aniline	F	P	F	P	P	P	P	P	E	E
Benzene	P	P	P	P	P	P	P	P	E	E
Carbon tetracholoride	P	P	P	P	P	P	P	P	G	G
Chloroform	P	P	P	P	P	P	P	P	G	G
Ethyl acetate	P	P	P	P	P	P	P	P	E	E
Ethylene dichloride	P	P	P	P	P	P	P	P	E	E
Formaldehyde, 37%	P	P	G	P	G	E	F	F	P	P
Phenol, 5%	-	-	-	-	P	P	P	P	E	G
Trichloro ethyliene	P	P	P	P	P	P	P	P	G	G
Bleach Solutions										
Chloride dioxide acid solutions	P	P	P	P	E	E	P	P	F	P
Chloride dioxide neutral solutions	P	P	P	P	E	E	P	P	F	P
Chloride dioxide alkaline to pH 11	P	P	P	P	E	E	P	P	P	P
Sodium hypochloric acid, neutral or alkaline to pH 11 (5% sol)	P	P	F	P	E	P	F	F	P	P
Sodium hypochloric above pH 11 (5% sol)	P	P	P	P	E	P	P	P	P	P

Notes:
E: Excellent
G: Good, occasionally used
F: Fair, for limited service only
P: Poor, not recommended

Table 4.25 Resistance of Resin-Cements in Chemical Environments

| Corrosion Media | Resistance of Resins-Cement | | | | | Silicate |
	Furan	Phenolic	Polyester	Epoxy	Sulfur	Cement
Acids:						
Acetic, diluted	R	R	LR	R	R	R
Acetic, glacial	R	LR	NR	NR	LR	R
Chromic	NR	LR	LR	NR	R	R
Citric	R	R	N/A	R	R	R
Fatty acids	R	R	R	R	LR	R
Hydrochloric	R	R	R	R	R	R
Hydrofluoric	R	R	LR	R	R	NR
	(with carbon filler)	(with carbon filler)		(carbon filled)		
Hypochlorous			R		NR	R
Nitric	LR	LR	N/A	NR	N/A	R
Nitric, 2%	NR	N/A	LR	NR	R	R
Nitric, over 2%	N/A	N/A	NR	N/A	N/A	R
Nitric, over 5%	N/A	N/A	N/A	N/A	LR	R
Nitric, under 5%	N/A	NR	N/A	N/A	R	R
Phosphoric	N/A	LR	R	N/A	R	R
Sulfuric, 50%	R	R	R	R	R	R
Sulfuric, 80%	R	R	NR	R	LR	R
Sulfuric, concentrated	LR	LR	NR	NR	NR	
	NR	NR		NR		
Alkalis:						
Ammonium hydroxide				NR	NR	NR
Calcium hydroxide	R	NR	NR	R	NR	NR
Sodium hydroxide, 1%	R	LR	NR	R	NR	NR
	R	NR		R		
				(carbon filled preferred)		
Sodium hydroxide, 10%	LR*	NR	NR	R*	NR	NR
Sodium hydroxide, 25%	LR*	NR	NR	R*	NR	NR
Gases:						
Chlorine	NR	LR	LR	NR	NR	R
Bromine	NR	NR	LR	NR	N/A	N/A
Sulfur dioxide	R	R	R	R	R	R
Hydrogen sulfide	R	R	R	R	R	R
Oils:						
Animal	R	R	R	R	N/A	N/A
Vegetable	R	R	R	R	N/A	N/A
Mineral	R	R	R	R	N/A	N/A

(Continued)

Table 4.25 Resistance of Resin-Cements in Chemical Environments *(cont.)*

| Corrosion Media | Resistance of Resins-Cement | | | | | Silicate |
	Furan	Phenolic	Polyester	Epoxy	Sulfur	Cement
Oxidizing agents:						
Hydrogen peroxide	N/A	N/A	R	LR	N/A	N/A
Hydrogen peroxide, 3%	R	LR	N/A	N/A	N/A	N/A
Bleach	LR	N/A	N/A	NR	N/A	N/A
Acid bleach	N/A	LR	R	N/A	N/A	N/A
Alkaline bleach	N/A	NR	NR	LR	N/A	N/A
Perchloric acid, 10%	R**	R**	R**	R**	R	R
Concentrated oxidizing agents, all types	NR	NR	NR	NR	NR	NR
Chlorine solutions	LR	LR	R	NR	NR	R
Chlorine dioxide solutions	NR	NR	R	NR	NR	R
Salts:						
Alum	R	R	R	R	R	R
Copper sulfate	R	R	R	R	LR	R
Ferric salts	R	R	R	R	R	R
Magnesium sulfate	R	R	R	R	R	R
Potassium permanganate	R	R	R	R	N/A	N/A
Silver nitrate	R	R	R	R	N/A	N/A
Sodium bisulfite	R	R	R	R	N/A	N/A
Sodium carbonate	R	LR	LR	R	LR	NR
Sodium chloride	R	R	R	R	R	R
Sodium hypochlorite	LR	NR	LR	NR	NR	NR
Sodium nitrate	R	R	R	R	R	R
Sodium sulfate	R	R	R	R	R	R
Sodium sulfite	R	R	R	R	LR	NR
Solvents:						
Alcohols	R	R	R	R	R	R
Glycols and glycerin	R	R	R	R	N/A	N/A
Acetone	R	R	NR	LR	NR	R
Aniline	NR	NR	NR	LR	NR	R
Methylene chloride	NR	NR	NR	LR	N/A	N/A
Carbon tetrachloride	R	R	LR	LR	NR	R
Chloroform	R	R	LR	LR	NR	R
Ethyl acetate	R	R	NR	R	NR	R
Gasoline	R	R	R	R	N/A	N/A

*For caustic alkali service, carbon-filled mortars should be used. In the lower range of temperatures, chemical resistance is usually satisfactory, but at high temperatures, different mortars vary in caustic alkali resistance and simulated field tests should be made.
**Handling of perchloric acid in organic materials is hazardous. The danger of explosion is ordinarily the reason that resin mortars are not used.

Note:
R = Recommended
NR = Not Recommended
LR = Limited Recommendation

Table 4.26 Purpose of Chemical-Resistant Cement in Brick Lining

Purpose	Chemical-resistant Cement		
	Silicate Base Cement	**Phenolic Furfur-Aldehyde Cement**	**Furan Cement**
Joint	×	×	×
Membrane	—	×	—
Reinforced Epoxy Cement	**Phenol Form-aldehyde Cement**	**Sulfur Cement**	**Polyester Cement**
×	×	×	×
×	×	—	—

Note:
× Can be used
— Cannot be used

Sometimes it will also be necessary to consider the erosion resistance of the cement. Hydraulic cements should not be used for chemical-resistant brick lining for process equipment. Some chemical-resistant cements can also be used as membranes (see Table 4.26).

4.14.4 SILICATE-BASED CEMENTS

Silicate-based cements are two-component systems. They consist of a sodium or potassium silicate solution mixed with inert fillers (e.g., quartz flour).

Silicate-based cements do not have much resistance to erosion, especially in hot water, steam, and alkali services where washing out of the cement from the joints may occur. The use of resin-based cements for these conditions has to be considered. Silicate-based cements should not be used as a membrane (see Table 4.26) and they do not adhere to rubber membranes.

4.14.5 PHENOL-FURFURALDEHYDE CEMENT

Phenol-furfuraldehyde cements consist of phenol-formaldehyde resin and furane derivates with an inert filler. The cements are supplied as two components: a syrup (the resin solution) and an inert powder (containing also a part of the reactive agent), which should be mixed thoroughly and immediately before use.

In cases where resistance to hydrofluoric acid is required, graphite is used as a filler instead of sand or barites. They are erosion-resistant and free of pores when properly applied. When this substance is used as a membrane, the surface must be coated by a suitable primer before the cement is applied directly to steel or concrete.

4.14.6 FURANE CEMENTS

They are supplied as two components: a powder and a syrup. When these two substances are mixed correctly, they create a cement with excellent adhesive qualities. The syrup cures to a hard, solid resin when suitable catalysts are added.

Furane-cement has a good chemical resistance (see Table 4.25). If a filler such as graphite is added, resistance to hydrofluoric acid is achieved. In addition, such cements are erosion-resistant and free of pores when properly applied.

4.14.7 POLYESTER CEMENTS

Cements based on unsaturated polyester resin are supplied in two or more components which include liquid resin, catalyst, accelerator, and filler that are mixed immediately before use. The addition of inert fillers, such as graphite, to the cement extends its resistance even to hydrofluoric acid, and its resistance to alkalis becomes greater (see Table 4.25). They possess good erosion resistance.

4.14.8 EPOXY CEMENTS

Cements based on epoxy resin are supplied in two or more components (such as liquid resin, cold curing agent, and filler or fiberglass reinforcements). Curing starts immediately after mixing. Epoxy cements have very good adhesion to metallic and concrete substrate. In metallic process equipment, they can be used as a membrane, if necessary, with glass-fiber reinforcement. When applied to a concrete substrate, no prior treatment other than proper surface cleaning is necessary.

If a filler such as graphite is added, resistance of hydrofluoric acid is obtained. Epoxy cements with glass-fiber reinforcement can be used for marine steel structures, especially in splash zones (see NACE standards).

4.14.9 PHENOLIC (PHENOL-FORMALDEHYDE) CEMENTS

Phenolic cements are prepared by combining liquid phenolic formaldehyde resin suitably modified with an acid-containing filler like carbon or silica. The mixing ratio is about two to one powder-to-liquid for the carbon filled and 2½ to one for the silica-filled cements. The chemical resistance of phenolic cement is shown in Table 4.25.

4.14.10 SULFUR CEMENTS

Plasticized sulfur cements essentially are mixtures of sulfur and inert fillers with minor amounts of plasticizers. Sulfur cements are applied as hot melt materials. They can be used with silica or carbon fillers. The major advantage of sulfur cements over the resin cements is that they cost less.

4.14.11 BRICK AND TILE LAYER

There are several types of chemical-resistant brick and tile, and they can be used for lining of steel and concrete equipments. Most brick and tile materials have good resistance to organic chemicals, so resistance to inorganic chemicals often determines the final choice.

Porous bricks offer low resistance to penetration of liquids, gases, or both. They have high thermal conductivity and good thermal shock resistance. Porosity needs to be considered, particularly in the case of crystallizing liquids, where there is a potential danger of volume change destroying the brick.

Dense bricks resist the penetration of liquids. Thermal conductivity is low, so high thermal gradients can occur in the brick and temperature shock will lead to thermal spalling. Erosion resistance of the bricks and tiles also need to be considered. The brick and tile should have a rough surface and should not be glazed, which will ensure good adhesion to the cement.

Arched bricks are required when the width of the side joint at the back of the standard brick is greater than 1.5 times the joint width at the front of the standard brick. Console bricks are sometimes used in the inner course of bricks to support the inside.

The amount of temperature decrease through the brick in a lined steel tank depends on the temperature gradient between the atmosphere and the process temperature and the total thickness of the brick lining. This is subject to calculation as follows. A rule of thumb allowance of 8.5°C per 25 mm of thickness of fire clay or shale type bricks and 5.5°C per 25 mm of thickness of carbon brick has been found adequate for design purposes for process temperatures up to about 200°C. Temperature decrease through the brick in a lined concrete tank will be somewhat less due to the damaging effect of the concrete shell. In almost all instances, the total thickness of brick used should be sufficient to limit the temperature at the membrane lining to 65°C. In some cases, even lower temperatures may be desirable.

4.15 THICKNESS OF THE BRICK LINING

The thickness of the brick lining is determined by the number of layers of bricks, by the dimensions of the brick, and by how the brick sits within the layer. When the brick lining functions as a heat barrier, the minimum thickness is determined by the maximum permissible membrane temperature (see Table 4.27), the steel wall temperature (maximum 100°C), or the maximum temperature gradient of the brick.

Too thin a lining may crack open and tear away because of high stresses due to internal pressure and thermal gradient, even when cements with swelling properties are used. Too thick a lining may result in spalling and crushing of the weaker adjacent layer because of high thermal and tensile stresses.

In general, a lining will comprise two staggered layers of bricks, with the layers overlapping both horizontally and vertically. This arrangement is suitable for severe corrosive conditions, although for 90°C hydrochloric acid, good results have been obtained with one layer of bricks laid radially. At higher temperatures, however, this may cause spalling.

Table 4.27 Thickness and Maximum Temperature Working of Brick-Lining Materials

Layers	Thickness of Lining (mm)	Maximum Temperature (°C)
Carbon steel vessel wall	10	100
Hard natural rubber (80°C, shore D)	5	80
Soft natural rubber (65°C, shore A)	5	80
Butyl rubber	5	110
Polyisobutylene	3	70
Glass-fiber-reinforced epoxy	4	130
Lead	6	70
Silicate-based cement	5–10	900
Synthetic-resin-based cement	5–10	180
Acid-resistant brick (normally applied)	65	65
Carbon brick (unimpregnated)	40	200
Porcelain tile	40	65

4.16 REFRACTORY LINING

4.16.1 REFRACTORY BRICK AND SHAPE LINING

Brick refractories are used for refractory lining of floors and burner areas of boilers and furnaces, and where refractory concrete lining alone is unsuitable. Refractory brick lining is normally used in conjunction with refractory concrete lining.

4.16.2 MATERIALS

Refractory brick and tile classification is as follows:

- High alumina fire brick containing more than 50% wt. alumina (see ASTM C 27)
- Insulating fire brick containing 50% wt. alumina or less (see ASTM C 155)

    Bricks should have flat surfaces within the following limits:

- **Concavity:** Not more than 3 mm in 30 cm and pro rata
- **Convexity:** Not more than 1.5 mm in 30 cm and pro rata

4.16.3 DIMENSIONS

The dimensions of brick should be specified by the purchaser. The bricks should not depart from the stated dimensions by more than 2%.

4.16.4 APPLICATION

Application of refractory brick lining should be in accordance with chemical resistance brick lining.

4.16.5 REFRACTORY CONCRETE LINING

Refractory concrete lining is used to line radiant and convection sections of furnace, waste heat boilers, flue ducts, and steel stacks. These materials should not be applied in freezing or excessively hot weather unless precautions are taken to maintain the ambient temperature of refractory steel and mixing equipment as close to 15°C as possible during mixing and application, and for a minimum of 24 hours thereafter. Neither admixture of any kind nor live steam should be used for this purpose. The vessel is to be placed in an erect position in the field.

Only personnel who are thoroughly familiar and experienced in application of refractory should apply lining. Skilled tradespeople such as brick layers and nozzle operators should have at least one year's previous experience of work on jobs of a similar nature.

4.16.6 TYPES OF REFRACTORY CONCRETE AND USAGE

Refractory concrete lining material consists of a mixture of calcium aluminate cement and lightweight aggregates. It should be supplied by the manufacturer as a dry mix. After the addition of water, it should be suitable for application by gunning or casting in the same way as ordinary concrete. Two type of refractory concrete materials are used for lining of heating equipment: lightweight and medium-weight.

4.16.7 LIGHTWEIGHT (LW) CONCRETE

Lightweight (LW) concrete should be a mixture of calcium aluminate cement with aggregates (e.g., perlite, vermiculite, blast-furnace slag, clay, diatomite, fly ash, shale, or slate). LW concrete should be used as a lining material for radiant sections and convection sections of furnaces and waste heat boilers, except as stated otherwise by the company and for hot flue ducts; e.g., ducts between radiant sections and convection sections of furnaces and waste heat boilers.

4.16.8 MEDIUM-WEIGHT (MW) CONCRETE

Medium-weight (MW) refractory concrete should be a mixture of calcium aluminate cement with aggregates (e.g., blast-furnace slag, clay, diatomite, fly ash, shale, or slate). It should be used as a lining material for furnace header boxes and cold flue ducts (e.g., for ducts from convection sections of waste heat boilers to the stack, for steel stacks, and for convection sections of furnaces and waste heat boilers where shot cleaning will be applied).

The water to be added to the dry mix should be clean, cool, and potable. During cold weather the water temperature should not be less than 5°C. Refractory concrete for lining steel walls should be applied using anchors. Anchoring of lining to furnace floors is not necessary, except for protruding parts.

4.16.9 THICKNESS OF LINING

The thickness of refractory cement lining should preferably be between 40 mm and 200 mm. The minimum thickness of castable refractories in convection sections of fired heaters should be 75 mm. The tolerance of lining thickness should be ±5 mm.

4.16.10 APPLICATION OF LINING

Refractory concrete lining is applied by gunning or casting. The application methods, requirements, and inspection of works should take place in accordance with standards.

4.17 CEMENT-MORTAR LINING

Cement-mortar lining is mainly used to line steel, ductile-iron, and cast-iron pipe and tubing for the purpose of shielding the steel from chemical corrosive attacks encountered in the handling of water and oil field brines and the like. It is intended for the lining of new and old pipes and fittings.

Cement-mortar-lined and -coated steel pipe combines the physical strength of steel with the protective qualities of cement mortar. The lining, applied centrifugally, creates a smooth, dense finish that protects the pipe from tuberculation and provides a measure of corrosion protection. The smooth interior surface provides a high flow coefficient, which is maintained for a long period of time. In addition, the cement-mortar creates a tough, durable, and rugged coating that forms an alkaline environment where oxidation or corrosion of the steel is inhibited.

Note, however, that soft, corrosive water, as well as prolonged contact with heavily chlorinated water, may be damaging to cement-mortar linings. When this environment is anticipated, further studies may be necessary to determine the suitability of this type of lining.

4.17.1 MATERIALS

Cement meeting the requirements of ASTM-C-150 Type I or II (Portland cement) is required for lining pipes and fittings carrying water but cement conferring to ASTM-C-150 Type III [high sulfate-resistant (HSR) similar to API-RP 10E with zero tricalcium aluminate (C3A)] should be used for lining oil field pipe and tubing. This cement is suitable for exposure to corrosive waters, regardless of the soluble sulfate concentration. When API Class C (API specification 10) cement with zero C3A is unavailable, other cements may be used, but only upon specific approval by the company.

Moderate-sulfate-resistant (MSR) cements (having less than 8% C3A) may be used when the corrosive fluid contains less than 1,000 ppm soluble sulfates. These may be either API Class B (MSR) or Class C (MSR) cement.

HSR cements (having less than 3% C3A) should always be used when the corrosive water contains 1,000 ppm or higher soluble sulfates. These may be either API Class B (HSR) or Class C (HSR) cement.

The purchaser of cement-lined pipe may require the applicator to furnish a written statement of the chemical composition of the cement for approval. Similarly, the applicator may require the cement manufacturer to furnish a written statement of the chemical composition of the cement for the approval by the applicator or the purchaser of the cement-lined pipe. The method for determining the C3A content of the cement selected should be by mutual agreement among the cement manufacturer, the applicator and the purchaser of the cement lined pipe.

The cement used to line and fill pipe should be uniform and free of lumps. Sand should be at least 95% quartz (silicon dioxide) and should not lose more than 0.5% on ignition.

The sand should meet ASTM-C-33 standards and conform to screen analysis as specified in Table 4.28.

Fly ash and raw or calcined pozzolans should conform to ASTM-C-618 and API Spec. 10, Section 11. The loss on ignition for fly ash, conforming to ASTM-C-618, should not exceed 4%. The purchaser may require the applicator to furnish satisfactory reports of analyses that certify that the pozzolanic materials meet the ASTM requirements set forth in 16.1.8.

Pozzolanic material should consist of siliceous or a combination of siliceous and aluminous material in a finely divided form that in the presence of moisture will react with calcium hydroxide, at ordinary temperatures, to form compounds possessing cement properties. For mixing the lining material, clear potable water should be used.

Table 4.28 Screen Analysis of Sand Used for Cement-Mortar Lining

Opening (mm) (U.S. Series)	Percent Retained on Each Sieve
1.19 (16)	0
0.590 (30)	0–5
0.420 (40)	15–25
0.297 (50)	49–60
0.177 (80)	20–40
0.149 (100)	0–5
> 0.149 (> 100)	0–2

Chemical admixtures may be used by agreement between the applicator and purchaser. Cement, sand, and pozzolans should be kept clean, dry and free of contaminants. They should be stored in separate bins, tanks, or other containers sealed off from weather and contamination.

Before mixing, cement and pozzolans should be passed through a 6.4 mm or less mesh screen to remove lumps and other extraneous matter. All materials should be handled and transferred in such a manner that all foreign matter is excluded.

Cement-mortar should be composed of cement, sand, and water, well mixed and of proper consistency to obtain a dense, homogeneous lining that will adhere firmly to the pipe surface. To improve the workability, density, and strength of the mortar, admixtures conforming to the latest edition of ASTM-C-494 may be used if the contractor wishes, provided that the ratio of admixture to cement does not exceed that used in the qualification tests of ASTM-C-494. The soluble chloride ion (Cl^-) content of the cement-mortar mix should not exceed 0.15% of the cement weight. No admixture should be used that would have a deleterious effect on potable water flowing in the pipe after the lining has been placed.

4.17.2 CEMENT-MORTAR LINING THICKNESS

Cement-mortar lining should be uniform in thickness except in joints or other discontinuities in the wall. In selecting the lining thickness, it is important to consider the dimensions of any tools or instruments that must pass through the pipe. In considering the clearance for such items, the allowable eccentricity of both lining and pipe should be considered. Tables 4.19 and 4.20 list the variations in lining thicknesses being applied in the shop and on site along the range of tolerances permitted.

4.17.3 APPLICATION OF CEMENT-MORTAR LINING

Application of cement-mortar lining, either in the shop or on site, should be as specified by the company. The application method and requirements of the finished work should be in accordance with standards. The contractor is required to furnish an affidavit that all materials and work furnished under the company's order will comply or have complied with the applicable requirements of the standards.

4.17.4 TYPICAL PROBLEMS EXPERIENCED WITH CEMENT-LINED TABULAR GOODS

Off-specification materials, improper mix properties, or poor application techniques can result in defective cement lining, which can fail by mechanical or chemical mechanisms.

4.17.5 CHEMICAL ATTACKS

Cement linings cannot withstand strong acids and will be deteriorated by acid environments. They should not be used where the pH of the fluid is below 5.0. Cement linings have been successfully used in the following environments:

- When chemical attack can cause holes, grooves, or thinning of the lining
- When chemical attack can cause softening or shrinkage of the cement lining from the pipe wall

However, cement linings are known to deteriorate after reaction with water containing high concentrations of any of several critical individual ions or a mixture of these ions. Recognized as troublesome are high concentrations of sulfate, sulfide, chloride carbon dioxide, or mixtures of sulfide and oxygen.

4.17.6 **VOIDS**

A *void* is a place in the pipe where the cement lining is not continuous. Voids occur during the spinning process when the lining does not distribute to cover all the steel pipe.

4.17.7 **THICKNESS OF LINING**

Cement lining of uniform circumferential thickness that is too thick or too thin along the length of the pipe is a result of any of the following:

- Too little or too much cement mix
- Use of improper sizing cone
- Poor initial longitudinal distribution of cement mixture
- Loss of cement from the ends of the pipe during spinning
- Buildup of cement at the ends of the pipe during spinning

Cement lining that is too thin on one side and too thick on the other side of the pips is a result of any of the following:

- Trying to apply lining to bent or out-of-round pipe
- Pipe-spinning rollers not being aligned
- Pipe-spinning assembly being too short for pipe length
- Excessive lateral (sidewise) movement during spinning
- Cement slurry that is too thick or too thin for the spinning speed

4.17.8 **SAGS**

If the forces of hoop strength and adhesion in the fresh cement lining are not greater than the force of gravity, the lining will sag or pull away from the steel at some points along the top of the pipe.

4.17.9 **DETECTIVE ENDS**

Types of end defects include the following:

- Lining end not being located as specified
- Lining end not being square
- Lining end being chipped or cracked
- Lining thickness being not as specified
- Threads, welding bevel, not cleaned of cement
- Lining separated from steel pipe
- Lining end not being perpendicular to the longitudinal axis of the pipe

If the ends of the cement-lined pipe are not as recommended, a corrosion-resistant joint is difficult to make. The following are major causes of lining failure:

- Foreign material and poor mixing
- Visible bubbles on the surface of the set cement lining
- The appearance of spots or small rough lumps in the fresh lining

4.17.10 CRACKS

Most cracking occurs when the lining is allowed to dry out during curing, transportation, or storage. Each joint of lined pipe should be inspected for cracks after curing, transportation, storage, and handling. The intent in cement lining is always to produce a crack-free lining; however, cracking can occur. Acceptance or rejection of pipe with cracked lining should be left to the discretion of the inspector of the purchaser and the inspector of the applicator.

4.18 SHOP-APPLIED CEMENT-MORTAR LINING

When specified by the company, the cement-mortar lining should be applied in the shop. The lining materials should be as specified by the company based on the duty for which it is intended.

The quality control inspector should ensure that the lining plant is operated and maintained so that no foreign material is introduced into the cement lining or its components during handling, mixing, storage, or at any other time. This person should also ensure that all components are thoroughly mixed in the correct proportions. Foreign material or poor mixing are indicated by the appearance of spots or small rough lumps in the fresh lining.

Wire-fabric reinforcement (ASTM-A-185 or ASTM-A-497) or ribbon-mesh reinforcement should be applied to the interior of fittings larger than 610 mm (24 in). In addition, they should be secured at frequent intervals by tack welding to the pipe, by clips, or by wire.

4.18.1 THICKNESS

Cement-mortar lining should be uniform in thickness, except at joints or other discontinuities in the pipe wall. The lining thickness of pipes should be at least as listed in Table 4.29, along with the range of tolerances permitted or as specified by the purchaser.

Lining thickness should be determined at intervals that are frequent enough to ensure compliance. It should be measured while the mortar is wet.

Table 4.29 Cement-Mortar Lining Thickness of Shop-Applied Pipe

Nominal Pipe Size (mm)	Lining Thickness (mm)	Tolerance (mm)
50 − 75	4	−0.8 + 3.2
100 − 250	6	−1.6 + 3.2
280 − 580	8	−1.6 + 3.2
600 − 900	10	−1.6 + 3.2
> 36	13	−1.6 + 3.2

4.18.2 INSTALLATION

Cement mortar used for the inside joints should be composed of a minimum of 1 part cement to not more than 2 parts sand, by weight, dry-mixed, and moistened with sufficient water to permit packing and troweling without crumbling. Sand should be graded within the limits for plaster sand conforming to ASTM C35. Water should be clean and free of injurious quantities of organic matter, alkali, salts, and other impurities, (potable water). If permitted by the purchaser, workability of the mortar may be improved by replacing not more than 7%, by weight, of the cement with hydrated lime, or by replacing not more than 20%, by weight, of the cement with pozzolan.

Inside joints of mortar-lined pipe should be filled with cement mortar and finished off smooth and flush with the inside surface of the pipe by troweling or equivalent means. Before placing the joint mortar material against the surfaces of the lining, the surfaces should be carefully cleaned, have all soap removed, and then be wetted to ensure a good bond between the lining and the joint mortar. The pipeline should not be put into service until the mortar has cured for a minimum of 24 h. When pipe is 560 mm in diameter and larger, the joints should be finished smooth with the inside surface of the lining by troweling.

When the pipe is smaller than 560 mm, the joint should be finished by placing a sufficient amount of the joint mortar in the bell end of the section against the shoulder of the lining just before installing it in the line. When the section has been laid in place, the joint should be finished by pulling a rubber ball or the equivalent through the joint to finish it off smooth with the inside surface of the lining.

4.19 FIELD-APPLIED CEMENT-MORTAR LINING

When specified by the company, the cement-mortar lining should be applied in place. The lining materials should be as specified by the company based on the duty for which it is intended. Mortar should be mixed long enough to obtain maximum plasticity. The mortar should be used before initial set.

4.19.1 THICKNESS OF LINING

The lining should be uniform in thickness within the allowable tolerance, except at joints or deformations in the pipeline. Cement-mortar lining thickness should be in accordance with at least the requirements set out in Table 4.30, unless otherwise specified by the purchaser. If a cement-mortar lining thickness greater than 13 mm is required by the company, the contractor has the option of applying the mortar lining in multiple applications.

The thickness of cement-mortar lining over the top of rivet heads and lockbar longitudinal seams of steal pipe should not be less than 3.2 mm unless otherwise specified by the purchaser.

4.19.2 APPLICATION OF LINING

Application requirements of the cement-mortar lining for pipeline are as follows. The lining of all straight pipe sections and long radius bends should be accomplished by a machine that progresses uniformly through the pipe, applies cement-mortar against the pipe surfaces, and is provided with an attachment for mechanically trowelling the mortar to obtain a smooth transitions over joints.

Table 4.30 Cement-Mortar Lining Thickness for Pipelines in Place

Type of Pipe

Old Cast Iron and Ductile Iron		Old Steel		New Cast Iron and Ductile Iron		New Steel	
Diameter (mm)	Nominal Thickness of Lining* (mm)	Diameter (mm)	Nominal Thickness of Lining* (mm)	Diameter (mm)	Nominal Thickness of Lining* (mm)	Diameter (mm)	Nominal Thickness of Lining* (mm)
100–250	3.2	100–300	6.4	100–250	3.2	100–300	4.8
300–900	4.8	350–560	7.9	300–900	4.8	350–900	6.4
> 900	6.4	600–1,500	9.5	> 900	6.4	1,070–1,500	9.5
		> 1,500	12.7			1700–2,300	11.1
						> 2,300	12.7

That the lining of bends, specials, and areas adjacent to valves should be machine-sprayed and hand-troweled or, where machine placing is impractical, should all be performed by hand methods.

Application of field-applied cement mortar linings, requirements, inspection of the works, and performance criteria should be in accordance with the standards.

4.20 BITUMEN (ASPHALT) AND COAL-TAR LINING

Bitumen and coal-tar materials are suitable for protecting internal surfaces of steel pipes, fittings, tanks, vessels, and cementitious concrete or brick work equipment (see Tables 4.31 and 4.32).

Bitumen and coal-tar linings are available in two groups: cold-applied (solvent base) and hot-applied materials. Hot-applied linings have more resistance than cold-applied linings. These linings are used for the transport and storage of fluids with temperatures between 4°C and 32°C.

Bitumen and coal-tar lining materials should not be applied to surfaces with a temperature below 4°C to avoid poor wetting of the surface and slow drying of the lining. The lining should never be applied to a surface having a temperature below the dewpoint of the surrounding atmosphere because of the possibility of entrapping moisture beneath the coating. A lining material should never be applied to a surface with a higher temperature than that recommended by the coating manufacturer because blistering and sagging of the lining can result.

For contact-potable water and air conditioning systems, special linings should be specified, which on drying impart no taste or odor to the water or air. For contact with potable water only, bitumen material conforming to BS 3416 should be used.

The excellent resistance to water and the durability of coal-tar lining are attested by the fact that they have provided many years of trouble-free service on ships, floating dry docks, lock gates, tanks, water lines, penstocks, cooling towers, and sewage disposal plants. The coal-tar lining should not be used in contact with foods and potable water. Coal-tar linings are not recommended for atmospheric

Table 4.31 Resistance of Coal Tar and Bitumen Lining in Chemical Environments

Media	Type of Lining		Media	Type of Lining	
	Hot-applied	Cold-applied		Hot-applied	Cold-applied
Acids:			Gases:		
Sulfuric, 10%	LR	NR	Chlorine	NR	NR
Sulfuric, 50%	NR	NR	Ammonia	LR	LR
Hydrochloric, 10%	LR	NR	Carbon dioxide	R	R
Hydrochloric, 20%	NR	NR	Sulfur dioxide	LR	LR
Nitric, 10%	LR	NR	Hydrogen sulfide	R	R
Nitric. Conc	NR	NR			
Phosphoric, 10%	LR	NR	Solvents:		
Phosphoric, 85%	NR	NR	Alcohols	LR	LR
Acetic, 10%	LR	NR	Aliphatic hydrocarbons**	LR	LR
Acetic, glacial	NR	NR	Aromatic hydrocarbons	NR	NR
Fatty acids	NR	NR	Ketones	NR	NR
Water:			Ethers	NR	NR
Tap	R	R*	Esters	NR	NR
Distilled	R	R*	Chlorinated hydrocarbons	NR	NR
Sea	R	R			
			Salts:		
Alkalis:			Sodium chloride	R	R
Sodium hydroxide, 10%	R	LR	Calcium chloride	R	R
Sodium hydroxide, 70%	NR	NR	Sodium sulfate	R	R
Ammonium hydroxide, 10%	R	LR	Sodium bisulfite	R	R
Calcium hydroxide, slaked lime slurry	R	R	Sodium carbonate	R	R
Ammonium salts	R	R	Sodium nitrate	R	R
			Alum	R	R
Oxidizing Agents:			Sodium sulfite	R	R
Concentrated solutions	NR	NR	Sodium acid sulfate	R	R
Dilute solutions	LR	R	Zinc chloride	R	R
			Sodium metassilicate	R	R
			Sodium bichromate	R	R
Fats and oils:			Miscellaneous:		
Mineral	LR	LR	Sodium hypo-chlorite	LR	LR
Animal	LR	LR			
Vegetable	LR	LR			

R = Recommended
NR = Not recommended
LR = Limited recommendation
* Coal tar is not recommended.
** Bitumen is not recommended.

Table 4.32 Maximum Allowable Service Temperature for Coal Tar and Bitumen Lining in Chemical Environments

Corrosive Media	Suggested Maximum Operating Temperature (°C)	Corrosive Media	Suggested Maximum Operating Temperature (°C)
Alum	66	Magnesium sulfate	66
Aluminum chloride	66	Mercuric acetate	66
Aluminum nitrate	66	Methyl alcohol	66
Aluminum potassium sulfate	66	Nickel chloride	66
Aluminum sulfate	66	Nickel nitrate	66
Ammonium chloride	66	Nickel sulfate	66
Ammonium hydroxide	27	Oxalic acid	66
Ammonium nitrate	66	Phosphoric acid	66
Ammonium sulfate	66	Phosphorous acid	66
Barium chloride	66	Phosphorous trichloride	66
Barium hydroxide	66	Phthalic acid	66
Barium nitrate	66	Potassium bicarbonate	66
Benzoic acid	66	Potassium carbonate	66
Boric acid	66	Potassium chloride	66
Cadmium chloride	66	Potassium cyanide	66
Cadmium nitrate	66	Potassium ferricyanide	66
Cadmium sulfate	66	Potassium ferrocyanide	66
Calcium bisulfite	66	Potassium hydroxide, 30%	27
Calcium chloride	66	Potassium nitrate	66
Calcium hydroxide	27	Potassium sulfate	66
Calcium nitrate	66	Salicylic acid	66
Chlorine gas, dry	66	Silver nitrate	66
Chlorine gas, wet	27	Sodium acetate	66
Citric acid	66	Sodium bicarbonate	66
Copper chloride	66	Sodium carbonate	66
Copper nitrate	66	Sodium chloride	66
Copper sulfates	66	Sodium cyanide	66
Ethyl alcohol	66	Sodium hydroxide, 30%	27
Ethylene glycol	66	Sodium nitrate	66
Glycerin	66	Sodium potassium tartrate	66

(Continued)

Table 4.32 Maximum Allowable Service Temperature for Coal Tar and Bitumen Lining in Chemical Environments *(cont.)*

Corrosive Media	Suggested Maximum Operating Temperature (°C)	Corrosive Media	Suggested Maximum Operating Temperature (°C)
Gold cyanide (auric cyanide)	66	Sodium sulfate	66
Hydrochloric acid, 10%	27	Sodium sulfite	66
Hydrochloric acid, conc.	66	Sodium thiosulfate	66
Hydrocyanic acid	66	Sodium thiosulfite	66
Hydrogen sulfide gas, dry	66	Sulfur dioxide gas, dry	66
Hydrogen sulfide gas, wet	66	Sulfur dioxide gas, wet	66
Iron chlorides (ferric and ferrous)	66	Sulfur trioxide gas, dry	66
Iron nitrates (ferric and ferrous)	66	Sulfur trioxide gas, wet	66
Iron sulfates (ferric and ferrous)	66	Sulfurous acid	66
Lactic acid	66	Tannic acid	66
Lead acetate	66	Tartaric acid	66
Lead nitrate	66	Urea	66
Magnesium chloride	66	Zinc chloride	66
Magnesium hydroxide	66	Zinc nitrate	66
Magnesium nitrate	66	Zinc sulfate	66

exposure since they tend to alligator and crack upon prolonged exposure to direct sunlight, although they still afford some level of protection.

Bitumen linings should not be used where contamination of petroleum products is expected. All materials should be applied in compliance with the manufacturer's instructions. Care should be exercised to ensure that there is no mixing of material from different sources or of different types unless examination shows that the final product has satisfactory properties. In particular, it should be recognized that the chemical and physical characteristics of coal-tar-based coatings differ from those of bitumen-based coatings, and that the two kinds of coatings should not be blended in protective linings. It is also essential to clean the plant thoroughly when the use of bitumen lining materials follows that of coal-tar lining materials and vice versa.

4.20.1 THICKNESS

The lining thickness varies depending on the function for which it is intended, but it should comply with the requirements specified in Table 4.33.

Table 4.33 Thickness of Coal Tar and Bitumen Lining		
Lining Material	**Total Thickness (mm)**	**Number of Coats**
Cold-applied coal tar	1–2.6	2–3
Hot-applied coal tar	1.6–3.2	3–4
Cold-applied bitumen	1.6–3.5	3–4
Hot-applied bitumen	2–4.5	2–3

4.21 CLADDING OF PRESSURE VESSELS

This section covers the requirements for design, fabrication, process, repair, and inspection of pressure vessels constructed of steel with corrosion-resistant integral cladding and weld metal overlay cladding.

Cladding involves a surface layer of sheet metal usually put on either by rolling two sheets of metal together, centrifugal casting (extrusion and explosion), or a thin liner is spot-welded to the walls of a steel equipment. A higher alloy rod is necessary for welding clad parts to avoid dilution of the weld deposit and loss of corrosion resistance (see Table 4.34).

Cladding presents a great economic advantage in that the corrosion barrier or expensive material is relatively thin and is backed up by inexpensive steel. Costs might be astronomical if the entire wall were made of highly corrosion-resistant material. Cladding material is chosen for its corrosion resistance in particular conditions, such as seawater, sour gas, and high temperature. Specific uses include heat exchangers, reaction and pressure vessels, furnace tubes, tubes, and tube elements for boilers, scrubbers, and other systems involved in the production of chemicals.

4.22 SELECTION OF CLADDING MATERIALS

4.22.1 CLAD PLATE

Integral clad plate bonded by the rolling or expansion method should be homogeneously made so that the material has a quality and thickness as specified. The clad plate to be used for pressure vessels should meet one of the following standard specifications or the equivalent:

- ASME SA 263, "Corrosion-Resisting Chromium Steel-Clad Plate, Sheet,
- and Strip"
- ASME SA 264, "Stainless Chromium-Nickel Steel-Clad Plate, Sheet, and
- Strip"
- ASME SA 265, "Nickel-Base Alloy Clad Steel Plate"
- ASME B 432, "Copper and Copper Alloy Clad Steel Plate:

The integral clad plate bonded by the explosion method should not be used for hydrogen service vessels operating at elevated temperatures and high pressures at a hydrogen partial pressure 7 kg/cm^2

A and over (hereinafter referred to as *hydrogen service*). However, its use for removable parts such as manhole covers and tube sheets may be permitted.

4.22.2 WELD METAL OVERLAY CLADDING

Weld metal overlay cladding should be mainly used for hydrogen service vessels as stated previously. When the hydrogen service vessel is to be operated in a special condition such as the hydrocracking process and hydro-desulfurization process (hereinafter referred to as *special hydrogen service*), the weld metal overlay cladding should be strictly adopted for all inside surfaces of the vessel.

4.23 DESIGN

4.23.1 NOZZLES AND MANHOLES

Nozzles with sleeve linings should be flanged and sized to a minimum of 38.1 mm nominal pipe size (NPS). All nozzles and manholes should be fabricated with integral cladding or weld metal overlay cladding where possible. However, nozzles sized 100 mm NPS and less may be fabricated with sleeve lining which behind space should be vented to atmospheric with a 3.2-mm nominal pipe thickness (NPT) tell-tale hole and the hole is filled with grease.

The flange facings and neck connection weld of shell to nozzles and manholes should be lined with weld metal overlay cladding, as shown in Figure 4.16. All nozzles and manholes for special hydrogen service vessels should be employed with the weld metal overlay cladding. In the case, the raised-face type with a spiral wound gasket should be recommended for the flange connection. Where the ring-joint type of facing is unavoidably employed, the corner radius of the ring-joint groove of the flange facing should be made as large as possible, and the final weld metal overlayer of the groove should be made after the final post-weld heat treatment (PWHT) is complete.

4.23.2 INTERNAL ATTACHMENT

Welding of internal support lugs and rings to be directly attached to the shell or head of a special hydrogen service vessel should be performed with full penetration prior to PWHT, and the final weld overlayer around the support lugs and rings should be applied after PWHT, as shown in Figure 4.17.

4.23.3 FABRICATION

Welding electrodes for cladding materials should be used as described in Table 4.34.

4.23.4 CLAD DISBONDING PREVENTION

Preventives of weld metal overlay clad disbonding, which may occur during operation of the special hydrogen service vessel, should be provided in the vessel fabrication.

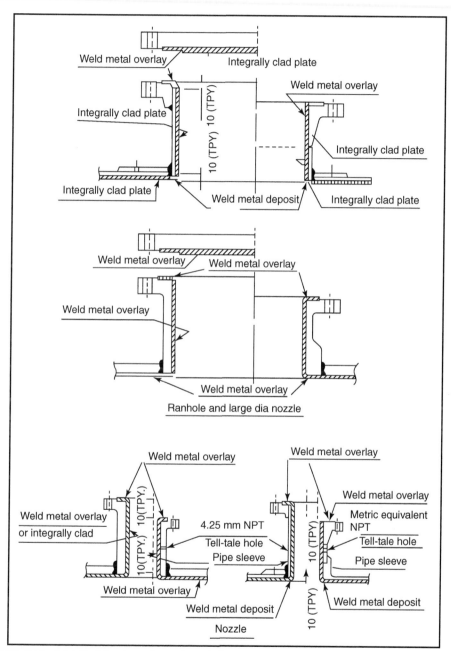

FIGURE 4.16 Typical detail of lined nozzles and manholes.

Table 4.34 Welding Electrodes for Cladding Materials

Asme/Astm Spec.	Cladding Material	AWS-Welding Electrode Specifications*	
		Weld to Base Metal	Weld with Cladding Metal
SA 263	Type 405 or 410S	E 310 - XX or E 309 - XX	E 310- XX or E 309 - XX
SA 264	Type 304		E 308 - XX
	Type 304 L		E 308 L - XX
	Type 316	E 310-Mo-XX or E 309 - Mo - XX	E 316-XX
	Type 316 L		E 316L-XX
	Type 321 or 347	E 310-XX or E 309-xx	E 347 - XX
SA 265	UNS 04400 (Ni - Cu Alloy)	ENiCu-7	ENi Cu-7
B 432	UNS C 70600 UNS C 72200 UNS C 71500 (Cu-Ni Alloy)	ENiCu-7 or ENi-1	Ni Cu - 7 or ECuNi

The AWS electrode designation xx is 15–16.

4.23.5 THICKNESS

No corrosion allowance should be added to the base plate. The thickness of the cladding material, which is made from integral cladding or weld metal overlay cladding, should not be included in the calculations of wall thickness. The thickness of the integral clad and weld metal overlay clad materials should be considered as corrosion allowance.

4.23.6 CLADDING PROCESS

The principal cladding techniques include cold roll bonding, hot roll bonding, hot pressing, explosion bonding, and extrusion bonding All of these involve some form of deformation to break up surface oxides, create metal-to-metal contact, and heat the material to accelerate diffusion.

The techniques differ in the amount of deformation and heat used to form the bond and in the method of bringing the metals into intimate contact. Cold and hot roll bonding apply primarily to sheet (less than 5 mm thick), but explosion bonding is usually restricted to thicker gages (up to several centimeters). The material should be free of injurious defects, have a workmanlike appearance, and conform to the designated finish.

Stainless chromium-nickel steel and corrosion-resisting chromium steel plates should be blast-cleaned and pickled (see ASTM A 240) prior to cladding. Nickel-based alloy plate should be furnished in the as-rolled condition prior to cladding unless otherwise specified.

The welding procedure and the welders or welding operators should be qualified in accordance with Section IX of the ASME Code.

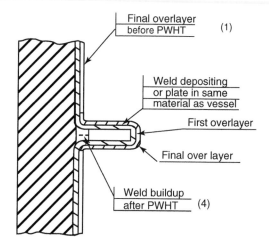

(a) Heavy-duty support

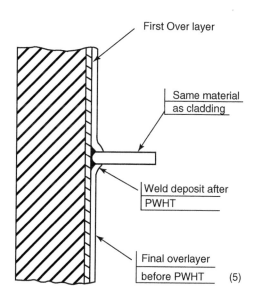

(b) Light-duty support

FIGURE 4.17 Lined details of internal support.

4.23.7 **HEAT TREATMENT**

Unless otherwise specified or if the purchaser and the manufacturer agree to do otherwise, all clad plates should be furnished in the normalized or tempered condition, or both. For example, all austenitic stainless-steel clad plates should be given a heat treatment that involves heating them to the proper temperature for the solution of the chromium carbides in the cladding, and then individually air cooling each plate. For base methods of air-hardening low-alloy steels, this heat treatment should be followed by a tempering treatment. When plates over 25 mm in thickness are to be cold-formed, the purchaser may specify that they be heat-treated for grain refinement of the base metal.

4.23.8 **REPAIR OF CLADDING**

Unless otherwise specified, the material manufacturer may repair defects in cladding by welding provided that the following requirements are met:

- Prior approval is obtained from the purchaser if the repaired area exceeds 3% of the cladding surface.
- The defective area is removed and the area prepared for repair is examined by a magnetic particle method or a liquid penetrate method to ensure that all defective areas have been removed. The method of testing and the acceptance standard should be agreed upon by the purchaser and the manufacturer.
- The weld will be deposited in a suitable manner so that its surface condition is equivalent in corrosion resistance to that of the alloy cladding.
- The repaired area is examined by the liquid penetrate method.
- The location and extent of the weld repairs, together with the repair procedure and
- examination results, are transmitted as a part of the certification.

All repairs in Corrosion Resistance Alloy (CRA) nickel type 410, or repair penetrating into the base steel, should be stress-relieved to eliminate residual stresses.

Note: For stainless steel clad, at the request of the purchaser or the inspector, plates should be heat treated again following repairing by welding.

4.23.9 **INSPECTION AND TEST METHODS**

The inspection and testing should be done in accordance with Table 4.35.

4.23.10 **MATERIAL INSPECTION**

The materials of integral clad plate to be used for vessels should be inspected by ultrasonic examination to check the quality of the bonding in accordance with ASME SA 578, "Straight-Beam Ultrasonic Examination of Plain and Clad Steel Plates for Special Applications," and the acceptance level should be S6, as specified in the standard. However the band having a width not less than 75 mm along with cut edge should have no defects.

Table 4.35 Scope of Inspection and Testing

Inspection Item	Division of Work	
	PR	**MFR**
1. Material inspection	R	Tr & S
2. Welding inspection 2.1 Magnetic particles Examination for weld buildup and weld deposit of heavy-duty support for special hydrogen service vessels	W/R	Tr & S
2.2 Liquid penetrate examination 2.2.1 Weld overlay and groove weld 2.2.2 Internal attachment weld on the weld overlay or clad metal 2.2.3 Entire weld surface of final layer and gasket seating surface	W/R W/R W/R	Tr & S Tr & S Tr & S
2.3 Ultrasonic examination 2.3.1 For cladding of general service vessels 2.3.2 For cladding of special hydrogen service vessels 2.3.3 For weld buildup and weld deposit for heavy-duty support	R W/R W/R	Tr & S Tr & S Tr & S
2.4 Ferrite check for austenitic stainless steel weld overlay	R	Tr & S
2.5 Chemical analysis for weld overlay	R	Tr & S
3. Inspection for completed vessel 3.1 Leak test for weld of pipe sleeve	R	Tr & S

PR: The purchaser
MFR: The manufacturer
R: Verify by reviewing the manufacturer's inspection/testing record
W: Witness inspection/testing
Tr: Manufacturer's own inspection/testing with the record to be prepared
T: Manufacturer's own inspection/testing
S: Submission of manufacturer's inspection/testing record

4.23.11 MAGNETIC PARTICLE EXAMINATION (MT)

For vessels in special hydrogen service, the entire surface of the buildup weldment and weld deposit for heavy-duty support should be checked by magnetic particle examination before the placement of the first layer of weld metal overlay cladding. This is done to confirm that no cracks or other defects exist. The method of MT is in accordance with Appendix 6 in ASME Code Sect. VIII Div. 1.

4.23.12 LIQUID PENETRATE EXAMINATION (PT)

Liquid penetrate examination should be performed at the following portions of weldments to check that no cracks or other defects exist:

- All final weld metal overlay surfaces and final layer of groove welds for clad plate after PWHT.
- All final welds of internal attachments on the weld metal overlay or clad metal.

  The method of PT in accordance with Appendix 8 in ASME Code Sect. VIII Div. 1.

4.23.13 ULTRASONIC EXAMINATION

Ultrasonic examination for clad vessels in general service should be performed as follows:

1. All surfaces of integrally bonded clad plate after cold forming when the resulting extreme fiber elongation is more than 5% from as-rolled condition, as well as all hot forming, should be examined to detect gross lack of bonding, if applicable.
2. All overlay weld deposits, including everything but the joint of the clad plate, should be examined to detect gross lack of fusion.
3. The acceptance level should be S6, as shown in ASME SA 578.

For clad vessels in special hydrogen service, ultrasonic examination should be performed in the following manner. All surfaces of weld metal overlay cladding and all weld metal overlay deposits of butt joints should be examined from the counter side of the clad surface in accordance with the requirements of paragraph T-543 in ASME Code Section V, Article 5. after PWHT. The acceptance standard should be as follows:

- There should be no imperfection interpreted to be cracks, or lacking bond.
- There should be no indication exceeding 40% of the reference level by cathode ray tube (CRT).
- Indications of amplitude of 20–40% by CRT should be recorded when the indication has the maximum length of 25 mm and over before PWHT, the defect portion should be repaired in accordance with the approved welding procedure specifications.

The entire surface of weld buildup and weld deposit for heavy duty support should be examined to detect any lack of fusion. The acceptance level should be level I in ASME SA 578.

4.23.14 FERRITE CHECK

Austenitic stainless steel weld metal overlay cladding should be subject to a ferrite check. The ferrite content of the weldment at 1.5 mm from the surface of the overlay, as shown in Figure 4.18, should be limited to 3–10% value of the Schaeffler diagram. Where the inert gas shielded welding process is employed, the De Long diagram and the nitrogen analysis evaluation should be applied.

A ferrite check should be performed for each overlay welding procedure. The ferrite check sample may be made from test pieces.

4.23.15 CHEMICAL ANALYSIS

Chemical analysis of weld metal overlay should be performed by taking samples from 1.5 mm in depth from the surface, as shown in Figure 4.18, to ensure that the chemical composition conforms to the requirements of the applicable material specification in ASME Section II, Part C.

4.23.16 AIR TEST

Air testing for the welding of pipe sleeves should be performed before PHWT if it is to be conducted. No leaks should be permitted.

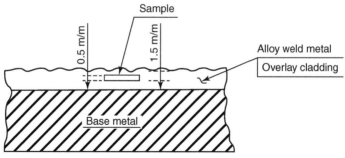

FIGURE 4.18 Sample for chemical analysis

4.24 **CLADDING OF PIPELINE WITH CRA (NICKEL-BASED)**

Clad pipe consists of a composite material with CRA on the inside and carbon steel on the outside (see Table 4.36). CRA selection is based on a battery of tests using in-service coupon testing, ASTM G 28, ASTM A 262 E, and ASTM G 48 testing for intergranular corrosion pitting, crevice corrosion testing, and other kinds of tests.

4.24.1 **PIPE**

Forming is the commonly used process for pipe, but seamless clad pipe is generally made by metallurgical bonding using co-extrusion, explosion bonding of the liner, or centrifugal casting. Clad thickness should be considered as corrosion allowance and thickness of the outer pipe is calculated from pressure and mechanical considerations.

4.24.2 **FITTINGS**

Bimetallic fittings can be produced by forming from clad pipe or clad plate using closed-die forming or hot extrusion processes. Generally, however, the quantities are not enough to interest the manufacturer.

In such a situation, manual/automatic weld overlay on a standard carbon steel fitting is the most viable route. For fittings made from clad pipe or plate, the same requirements as are applied to the pipe can be used. However, for manual/automatic weld, compatibility of the weld material with pipe outer material must be considered. In addition, the wall thickness and tolerances on the outer wall have to be matched. Further, to get good uniform coverage with weld overlays, a minimum of two overlay passes should be specified. The deposits with two overlays generally make the total thickness of the fitting more than the total thickness of the pipe. This problem can be prevented by specifying the inside taper on the end of the fittings to ensure proper matching to the pipe's beveled end. The intent is to taper the inside of the outer wall and not build a taper on the overlay thickness, because that reduces the corrosion resistance and extends the life of the fitting (see Figure 4.19).

Table 4.36 Commonly Used Outer and Inner Layers

Outer Pipe	Inner Cra Liner Pipe
	Super-austentic stainless steel 904
ASTM A 106 Gr. B	Duplex stainless steel 2205
API 5L Gr. B	Alloy 625
API 5LX Gr. 42	Alloy 825
API 5LX Gr. 60	Alloy B-2
ASTM A516 Gr. 60/65	Alloy C-276
	Monel 400

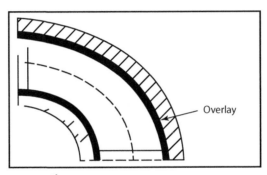

FIGURE 4.19 Overlaid elbow cross section.

4.24.3 FLANGES

To obtain a proper overlay on a flange face, the raised face should be machined off. To achieve a dilution-free overlay on the raised face, undercut the flange face by at least 1.58 mm. After machining the flange face, overlay of the inside and face of the flange should be conducted in the same way as for the fittings. Finally, the proper gasket seating flange finish should be prepared by remachining the flange face.

4.24.4 BALL VALVES

Applying overlay on the wet parts of ball valves is a viable alternative at higher pressure classes and usually sizes larger than 100 mm. Overlays of 316 stainless steel, duplex stainless steel, inconel 825, and other metals have been successfully applied. The valve bodies to be overlaid are first machined over size, and then overlay material is applied. After overlaying, heat treatment as required is carried out, and then surfaces are machined to their final dimensions. The overlay should be thick enough to provide an effective erosion barrier, while simultaneously thin enough to be economical when placed over solid alloy valves. Additionally, while deciding on the overlay thickness, the possibility of the delusion of carbon steel into the surface of overlay material should also be considered.

4.24.5 OTHER VALVES

Most commonly used valves, such as gate, globe and checks, have not yet been introduced into clad construction. However, special oil field-type choke valves and through-conduit gate valves have been manufactured in clad construction.

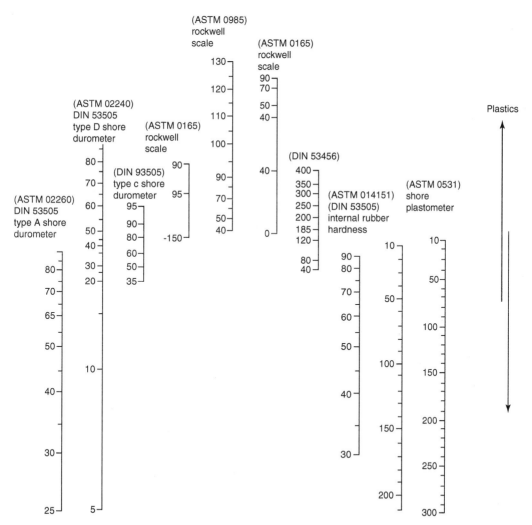

FIGURE 4.20 Hardness comparison chart for plastics and rubbers.

4.24.6 WELDING

Dilusion of carbon steel and CRA into each other is a major problem, resulting in crack formation and reduced corrosion resistance. Welding clad piping systems must be taken into account with the base material and the CRA material requirements in the selection of electrodes. It is essential to provide a weld deposit that will match the strength of the base material (in this case, CRA) and be compatible with welding onto CRA material. It is good practice to have a minimum of two passes for the overlay. The filler material for the carbon steel outer pipe can be steel weld. The weld procedure specification, procedure qualifications, and welder qualifications should clearly specify the requirements for chemistry of material, shield gas composition, preheat, interpass parameters, hardness testing requirements in overlay, heat affected zone and base metal liquid penetrant examination, ultrasonic examination, mechanical properties, impact testing, microstructure and macrostructure examination, etc. The procedure should also address the question of using coupons for testing.

4.25 EFFECT OF LINING MATERIALS ON WATER QUALITY

When used under the conditions for which they are designed, materials in contact with or likely to come into contact with potable water should not constitute a toxic hazard, should not support microbial growth, and should not give rise to an unpleasant taste or odor, cloudiness, or discoloration of the water.

Concentrations of substances, chemicals, and biological agents leached from materials in contact with potable water, as well as measurements of the relevant organoleptic/physical parameters, should not exceed the maximum values recommended by the World Health Organization (WHO).

CONSTRUCTION GUIDELINES FOR LINING

5

This chapter discusses the minimum requirements for application and installation of organic and inorganic lining for various types of equipment. It includes requirements for preparation of material for use, application, drying or curing, transportation, installation, repair of lining, and inspection of lined parts. It applies to equipment fabricated in metal or concrete and to both bonded and loose linings.

5.1 SELECTION OF PROCESS AND MATERIAL

The final selection of the type and thickness of the lining, as well as the method of application, should be made in consultation with the materials specialists and lining experts.

The following details should be included on the requisition sheet of the equipment concerned:

- Materials to be handled
- Temperature minimum, maximum, normal
- Degree of vacuum or pressure
- Cycle of operations
- Abrasion and errosion aspects
- Immersion conditions
- Storage

5.2 APPLICATOR QUALIFICATIONS

Lining installation should be done only by an applicator who is qualified by sufficient commercial experience in installing the type of lining proposed. To satisfy this requirement, the applicator should submit a list of applications and installations, and where such data is available, the service conditions and reported performance records, that he or she has handled.

5.3 INSPECTOR QUALIFICATIONS

All inspections, whether for the company or the applicator, should be performed only by a qualified inspector.

An inspector should have the following minimum qualifications:

- Complete knowledge of the job specifications and their requirements
- A practical knowledge of all phases of lining application work, including:
  - Pre-application surface finish requirements, such as grinding of welds, sharp edges, etc.
  - Surface preparation
  - Lining application techniques and workmanship
  - Lining materials
  - Continuity, thickness, cure tests, and tolerances of standards
  - Equipment and tools used in all phases of lining application work
- Adequate experience and training in the inspection of lining applications and in the use of the instruments utilized for the inspection and evaluation of lining applications

The specifications should stipulate if final acceptance of the work is to be made by a duly appointed representative of the user. If so, it is wise for that representative to be a qualified inspector.

5.4 QUALIFICATION TESTING

The full program of qualification testing is required before a manufacturer will be allowed to deliver for the first time. It may be required that the qualification testing of a certain make be repeated completely or in part, because of time elapsed or new developments, or other reasons. Regardless, any changes in the design and method of manufacture of parts will require new or additional qualification tests.

The qualification testing should be performed on products with representative diameters. It should be done by the manufacturer and witnessed and certified by an independent authority. Alternatively, testing and certification may be performed by an independent testing organization. This should be confirmed by submitting a certificate stating the test results.

Qualification testing program to which a manufacturer is subjected before his delivery to ensure that he can fulfill the requirements.

5.5 RESPONSIBILITY FOR INSPECTION AND TESTING

The manufacturer of the equipment and contractor of the lining are responsible for all inspections and tests specified, and they should provide a certificate of inspection and testing. The contractor should provide the opportunity for the purchaser to inspect all materials and application procedures, both during and after the installation work, and he or she should cooperate fully. In cases where the installation contractor has been responsible for determining the design required to meet a particular performance, performance tests may be arranged after the installation work for been completed or has become fully operative (see BS 6466). The stage of inspection should be according to, but not limited to, the following:

- The equipment as fabricated
- Surface cleanliness
- Application techniques
- Visual appearance of lining
- Adhesion of lining

- Continuity of lining
- Thickness of lining
- Cure evaluation of lining

Note that no test instrument that is destructive or detrimental to the integrity of an applied lining should be used for any test.

5.6 SHEET-APPLIED THERMOPLASTIC RESIN LINING

This section specifies requirements for the lining of equipment using sheet thermoplastics.

The thermoplastics that can be used for sheet lining include:

- Polyethylene (PE)
- Polypropylene (PP)
- Polyvinyl chloride (PVC)
- Polyvinylidene fluoride (PVDF)
- Fluorinated ethylene propylenecopolymer (FEP)
- Perfluoroalkoxy (PFA)
- Polytetrafluoroethylene (PTFE)
- Ethylene-chlorotrifluoroethylene (E-CTFE)

Sheet thermoplastics are applied to equipment fabricated in metal or concrete and in both bonded and loose lining. Requirements for lining design, including fabrication of the equipment, selection of lining, state of preparation necessary for the surface to be lined, and thickness of lining, should be according to "Lining standards."

The thermoplastic sheets are installed in vessels in one of two ways: they are either combined with a backing material (usually fiberglass, cloth, or carbon fiber) and bonded to the wall with an epoxy- or rubber-based adhesive; or they are left loose, attached only at flanges, as a gasket would be. In both cases, the sheets are welded together at their edges, with a rod of like material employed to seal the joints.

The inspector may stop lining operations when conditions such as weather, improper surface preparation, improper application procedure, or poor material performance indicate that an inferior lining will result. In case of bonding liner, the operation should be suspended by either the contractor or the company inspector when local relative humidity exceeds 90%, if the atmospheric temperature is less than 7°C, or if the temperature in the vicinity of the lining operations is less than 4°C. If the atmospheric temperature in the vicinity of the lining operation goes below 4°C, supplementary artificial heat may be used to keep the proper temperature.

5.7 PREPARATION

Preparation of material for use with sheet thermoplastic are generally manufactured by extrusion. The width, length, and thickness of the sheeting should be as specified by standards included as part of the contract if the manufacturer applies the lining.

5.7.1 PREPARATION OF SURFACE TO BE LINED

For metals, all grease oil, temporary protective and chalk should be removed from the surface to be lined. All welds should be ground smooth to prevent mechanical damage to the liner, and equipment to which liners are to be fitted should be free of loose rust and dirt. The flange edge covered by the loose liner should be rounded off.

All surfaces to which a lining is to be bonded should be maintained at a temperature at least 3°C above the dew point throughout the preparation and lining processes. If there is a risk that this condition will not be maintained owing to ambient conditions or a change in ambient conditions, dehumidifying and heating equipment should be used as needed. In addition, all surfaces to which liners are to be bonded should be blast-cleaned.

In the case of carbon steel and cast iron, the standard should be Sa 2½, as defined in ISO 8501. The average surface profile of blast-cleaned surfaces (measured peak to valley) should be 150 μm for sheet-applied thermoplastics.

When thermosetting, resin-based adhesives are to be used to bond the lining to the equipment, a minimum temperature of 10°C should be maintained from the start of the lining process until the adhesive is fully cured. All dust, residue, and debris left on the surface after blast cleaning should be removed by brushing or vacuum cleaning.

For concrete, any external corners not formed with a chamfer should be rubbed down to a radius not less than 5 mm. Equipment to which loose liners are to be fitted does not require further surface preparation, provided that all surfaces are clean and smooth.

All surfaces to which a lining is to be bonded should be treated to remove laitance and shutter release agents. The specified method for this operation is blast cleaning. The blast-cleaning process should be controlled so that all laitance is removed without exposing the profile of the aggregate. After blast cleaning, all dust and debris should be removed.

Following the removal of laitance from concrete surfaces to which linings are to be bonded, any small holes exposed should be filled.One material recommended for this purpose is a smooth paste made from a water-miscible epoxy resin, cement, and a fine filler.

All surfaces to which a lining is to be bonded should be maintained at a temperature at least 3°C above the dew point during the lining processes. If there is a risk that this condition will not be maintained owing to ambient conditions or a change in ambient conditions, dehumidifying and heating equipment should be used.

When thermosetting resin-based adhesives are to be used to bond the lining to the equipment, a minimum temperature of 10°C should be maintained from the start of the lining process until the adhesive is fully cured.

5.8 APPLICATION OF LINING

Following the blast cleaning of metal surfaces to which a liner is to be bonded, the clean surface should be primed as soon as possible—in any case, not more than 4 h later and before any visible rusting occurs.

In the case of concrete equipment to which linings are to be bonded, lining should not proceed until the free water content is down to a level compatible with the adhesive system used. Whether the lining is to be bonded or not, sheets should be cut and where necessary, hot-formed so that there is a tight fit between the liner and substrate. Wherever possible, corners in the plastic sheets should be hot-formed so that welding in corners is avoided.

In the case of pipework, special expansion techniques may be employed in the fitting of the liner. When linings are being bonded to the substrate, the material should be fitted in such a way that air is not trapped behind the sheet. When adhesives are used for bonding, the batch should be used within its pot life or else discarded.

The welding process used for joining the lining should be appropriate to the material and circumstances. The fabricator of the lining should provide evidence that the operators engaged on lining the vessels are capable of making good welds. When required, the applicator of the lining should provide type welds for testing. These welds should have at least 85% of the strength of the parent material when tested in accordance with BS 2782: Methods 320A to 320F. Welds made in a lining forming a mating surface should be ground flat.

5.8.1 BONDING

Sheet thermoplastics are installed in equipment in one of two ways: bonded and loose, as specified by the designer. In both cases, edges with a rod of like material are employed to seal the joints.

5.8.2 BONDED SHEET-APPLIED PROCESS

The sheets are combined with a backing material (usually fiber glass, cloth, or carbon fiber) and bonded to the wall with an epoxy- or rubber-based adhesive. Rubber-based compounds should not be used when organic solvents are present, or when the process temperature exceeds 100°C. In these cases, epoxy-based adhesive should be used.

Unless stated otherwise, bonded liners should be specified when the equipment has a complex shape (dished heads or a conical section, for example), it has internal components, or when there are wide swings in the equipment's operating temperature (37–50°C in magnitude).

5.8.3 LOOSE SHEET-APPLIED PROCESS

In the loose sheet-applied process, the sheet is attached only at flanges, as a gasket would be. Where the shape of the equipment is relatively simple, process temperature are not cyclic, and pressures are low, nonbonded liners offer users the same chemical resistance as bonded liners, and at a lower cost. The savings are realized by eliminating the costs of the backing material, adhesive, and labor required to bond the liner. In a vacuum, loose liners will quickly collapse, so metal rings, lined with the same polymer as the vessel walls, should be used to hold them in place.

Unless specified otherwise by the designer, when deciding whether to install a bonded or disbonded liner in a pressurized vessel, use 3.5 bar(g) as the dividing line. If the vessel's working pressure is below 3.5 bar(g), a loose liner can safely be used, provided that the vessel's interior has a relatively simple shape, and processing temperature does not vary widely. Above 3.5 bar(g), a bonded liner should be used.

5.8.4 SELECTION OF PROCESS

The lining process should be appropriate to the grade of material selected for the lining.

There are two types of PE sheets available: low-density polyethylene (LDPE) and high-density polyethylene (HDPE). PE sheets are not normally used as a bonded liner. They are easy to weld, but complicated shapes are not easily formed.

PP is used with a glass or synthetic fiber or rubber backing, so it can be bonded to substrates. The backing material can impose limitations on the thermal-forming process. Glass fiber-backed PP is not readily formed into complex shapes, but it can be welded easily.

There are two types of PVC sheets available: unplasticized PVC (UPVC) and plasticized PVC. UPVC softens and loses strength as the temperature increases. When fully bonded to a substrate, it may be used as a lining material up to 85°C, UPVC can be shaped easily and can be welded. Plasticized PVC is usually bonded to the substrate and can be shaped and welded.

PVDF is used with a glass or synthetic fiber or rubber backing, so it can be bonded to substrates. The backing material can impose limitations on the thermal forming process. It is readily welded.

FEP is used with a glass fiber backing to provide a bondable surface. When it is bonded to a substrate, the maximum service temperature will be determined by the adhesive used. FEP can be thermoformed, but in the case of glass-backed material, the complexity of the formed shape may be limited. In addition, FEP can be welded.

PFA can be thermoformed and welded. It can be used as a loose or bonded liner.

E-CTFE is used with a glass-fiber backing, so it can be bonded to a substrate; the backing material may impose limitations on the thermoforming process. E-CTFE can be welded.

PTFE sheets cannot be thermoformed, but when heated, it can be swaged over flanges. PTFE sheets up to 4 mm thick may be welded. It is usually used as a loose liner, although it is possible to etch the surface to make bonding to the substrate possible. It is also available with a glass-fiber backing. The process is very difficult and requires special techniques and equipment. Note that PTFE sheets are sliced from a large block that has been produced either by ram extrusion or a combination of compression molding and sintering. For thicknesses that exceed 6 mm (0.25 in), PTFE sheets are molded.

5.8.5 DRYING

After application of sheet lining (if necessary), the lined surface should be dry in accordance with the manufacturer's recommendations.

5.9 TRANSPORTATION AND STORAGE

Lined equipment should be stored in a covered area or a protected compound. When necessary, linings should be shielded from direct sunlight. All branches, manholes, and other openings should be protected from mechanical damage by using wooden blanks or other suitable material.

In the case of some linings that are flared over flanges, wooden planks should be fitted as soon as flaring is complete. These planks should be left in place until just before the mating flange is bolted in position.

If for any reason flange joints in such equipment are broken, the wooden planks should be fitted until such time as the joint is remade. It is recommended that such joints are broken only at ambient temperatures.

Lifting should be arranged so that chains and other lifting aids do not come into contact with lined surfaces. High local loads on lined surfaces should be avoided. Loose fittings should not be placed inside lined equipment while it is being transported.

The responsibility for arranging the transport of lined equipment will vary. Whoever is responsible (whether it be the company, fabricator, or applicator) should instruct the carrier about any handling precautions.

5.10 REPAIR OF LINING

When it is necessary to replace part of the lining, the sheet used for rectification should be of the same grade as that used for the original lining. In the case of linings bonded to the substrate, the adhesive used for replacing parts of the lining should be that specified for the original lining.

If more than one fault is found in a weld, and those faults are less than 50 mm apart, then for the purpose of rectification, the faults should be considered as one large defect. After all rectification is finished, the lining should be subject to inspection as appropriate, particularly with regard to continuity testing.

5.11 INSPECTION AND TEST METHODS
5.11.1 VISUAL APPEARANCE OF LINING

The surfaces of the plastic liner before and after application should be free of defects, such as blisters, cracks, scratches, dents, nicks, or sharp tool marks, which would be expected to affect the performance of the liner. The absence of these defects needs to be detected visually, with a dye penetrant, or with continuity testing (wet sponge or spark testing). The plastic liner should fit snugly to the steel housing, and no entrapments should be present between the plastic liner and the internal surface of the substrate.

The adhesive should be capable of maintaining a bond at the design temperature and after cycling between the ambient and maximum surface temperatures. If the completed lining shows blistering, cracking, mechanical damage, or unallowable variation in thickness, the lining should be rejected and entirely removed and redone at the contractor's expense.

5.11.2 ADHESION OF LINING

As much as possible, adhesion testing of the applied lining should be avoided because the testing is destructive and the lining has to be repaired afterward. All linings should be inspected visually for evidence of lack of adhesion to the substrate. If required, test plates should be used to demonstrate that the process employed provides a lining with the required level of adhesion. These test plates should be made of the same material as the substrate, and the lining process should be the same as that employed for the lining of the equipment and done under the same conditions and at the same time. The minimum bond strength between the lining and the substrate should be 3.5 N/mm^2 in direct shear and 5 N/mm^2 in the width of the peel at a test temperature of 20°C when tested in accordance with BS EN ISO 252.

5.11.3 CONTINUITY OF LINING

The lining should be tested for pinholes. For noncorrosive conditions, local repairs are permitted if the number of pinholes exceeds that specified. If local repairs are not possible, the lining should be removed and replaced. The appropriate method for the continuity testing of sheet thermoplastic lining is spark testing.

5.11.4 THICKNESS OF LINING

The thickness of the finished lining will depend upon the material selected and the duties for which it is intended. If necessary, the material should be capable of being thermoformed and welded to give joints that are pinhole free.

5.12 **INSTALLATION**

At all times during the installation of lined equipment, the contractor should use every precaution to prevent damage to the lining. No metal tools or heavy objects should be permitted to come into contact unnecessarily with the lining. Any damage to the equipment from any cause during the installation and before final acceptance by the company should be repaired by and at the expense of the contractor.

When making up lined pipe, a certain amount of judgment must be used because of the coating that is applied to the pin end and in the joint area of the coupling. For stabbing, a plastic stabbing guide must be used to guide the pin end directly into the middle of the coupling, to eliminate any contact of the pin end with the coupling's top edge. Because of the coating in the threads, initial makeup normally produces higher than normal torque values, but subsequent make and breaks of the same connection is more in line with published torque values. For this reason, the initial makeup of lined pipe should first be made by position, plus-or-minus one thread from hiding the last scratch to ensure coating-to-coating overlap of the pin end with the coupling while monitoring the torque. Keep in mind that maximum torque values may be observed until the coating is removed from the threads.

The installation of plastic-lined piping is performed in a way that is similar to that for flanged steel piping with respect to supporting, thermal expansion, etc. Lined piping is mainly used in aboveground installations.

5.12.1 **FLANGE CONNECTIONS**

The piping should be installed in such a way that no damage is caused to the flared or molded liner facing. The use of 1-mm-thick smooth metal guide plates is recommended when making connections or when installing individual sections in an existing line. Flange facings should be cleaned prior to installation.

A lap-joint type flange may be used on one end of each straight pipe to enable bolt alignment during installation, especially in combination with glass-lined piping or equipment; the other flange may be rigid to the pipe. In the case of a totally plastic-lined system, all flanges can be rigid to the pipes and fittings.

5.12.2 **FLARING IN THE FIELD**

Whenever possible, prefabricated lined pipe lengths provided with flanges with shop-flared or molded ends should be used. To adapt an existing plastic-lined pipe to the designed length, spacers or distance pieces lined with the same polymer can be used. Solid spacers should be used only to a maximum thickness of about 6 mm.

Flaring should be avoided, but when absolutely necessary, it may be performed only in a workshop with equipment supplied by the pipe manufacturer. The instructions should be provided by the manufacturer and must be closely followed.

Special attention has to be given to the following:

- Cutting of steel tube (for swaged pipes only)
- Flaring temperature
- Cooling rate after flaring

The flaring tool should not be removed before the pipe has cooled to room temperature. For swaged pipes, only threaded flanges with straight or tapered threads are used for field fabrication. Flanges should be fully tightened and secured by tack welding to prevent inadvertent turning, which will damage the liner.

For that reason, flanges must have tapered threads or be of the threaded-socket type. The pipe end should be provided with sufficient threading to accommodate the flange.

A perforated metal ring must be used as a backup gasket for pipes made in accordance with the swaging method. Obviously, no backup gasket is required when a socket flange with a fully radiused edge is used.

Immediately after the flaring operation, a wooden or metal flange protector is installed to prevent mechanical damage prior to installation and to keep the flared end in position.

5.12.3 BOLTING

Flange bolts should be tightened with a torque wrench, using greased bolts and nuts, to the values specified on the piping by the pipe manufacturer in a criss-cross manner. Excessive bolt loading may damage the plastic flange face. The use of appropriate spring washers between the nut and the flange is recommended. Bolts should be retorqued after a service period of 24 h.

5.12.4 WELDING

Under no circumstances should welding be performed on lined piping, nor should it be used as a welding ground, as this will cause irreparable damage to the liner.

5.12.5 VENTING SYSTEM

Care should be taken that the venting system does not become blocked by paint or other deposits. Regular inspection of the vent holes is recommended. No sharp tools should be used to clean the vent holes. When lined piping is insulated, the use of vent-hole extensions is strongly recommended.

5.12.6 DISASSEMBLING

Plastic-lined steel pipes and fittings should be dismantled from an existing pipeline only at temperatures below 40°C; this prevents retraction of the plastic flange face. Immediately after disconnection, a flange protector should be installed on each flange face.

5.13 NONSHEET-APPLIED THERMOPLASTIC RESIN LINING

This section specifies the requirements for the lining of equipment using nonsheet-applied thermoplastic resins. The materials considered include thermoplastic powders, plastisols, etc.

The thermoplastics that can be used for nonsheet-applied lining include some polymers and the following:

- Polyamides (nylon)
- Ethylene vinyl acetate (EVA)
- Fusion-bonded epoxy resin

Although it is not a thermoplastic, fusion-bonded epoxy resin powder is included in this category since the same application techniques are used.

Requirements for design and fabrication of the equipment state necessary preparation for the surface to be lined. The applicator should supply test pieces, such as panels, to which the lining has been applied and which will serve as reference samples.

Nonsheet-applied thermoplastic resin should not be used to line concrete and be used only for metal surfaces. These resins should be applied with spraying, dipping, or fluidized bed processes.

The lined equipment should be identified with signs indicating the presence of interior lining, stating its specific type, and prohibiting welding or burning on the exterior surface. Upon requests, the applicator of the lining should provide a certificate of inspection and testing. The stages of inspection for the equipment should be as follows:

1. After preparation
2. After primer application
3. After lining application

5.13.1 PREPARATION

Substances that have been mixed for application must be used within time limits as specified by the manufacturer. Such materials not used in the designated period will be discarded. Contractors should handle, store, mix, and apply the lining in strict accordance with the manufacturer's specifications or as directed by an authorized representative of the lining material manufacturer.

5.13.2 PREPARATION OF SURFACES TO BE LINED

Surface preparation of metal prior to lining should be in accordance with standards. The average surface profile of blast-cleaned surfaces for nonsheet-applied thermoplastic lining should be 75 μm.

5.13.3 APPLICATION METHODS

The lining process should start as soon as possible after blast cleaning is complete and before any visible rusting occurs. Unless maintained in a dehumidified atmosphere, application of the lining should commence within 4 h. If signs of rusting appear, then the surface should be prepared again to the required standard.

Where necessary, surfaces should be primed in order to promote a bond with the lining material. The primer should be pigmented to facilitate uniform application and to assist in establishing full coverage of the surface to be lined. Once the primer has been applied, the equipment should be kept clean and the lining process should continue as soon as possible.

The lining process should be appropriate to the grade of material selected for the lining and the article to be lined. Nonsheet-applied thermoplastics should be used only for the lining of metallic surfaces.

5.13.4 **LINING PROCESSES**

Three of the most widely used lining processes are described in this section. Included are Rotomold techniques for special cases.

This method of dipping into a liquid or plastisol consists of dipping a heated and primed metal article into a tank of liquid or plastisol. The heat allows a layer of material to deposit and fuse with the primer. At a predetermined time, the article is removed from the tank and cured in an oven at a carefully controlled temperature.

This fluidized bed process involves dipping a heated and (if so specified) primed article into a tank of fluidized plastic powder for a predetermined time. The powder sinters and fuses into a homogeneous and adherent coating. For some materials, a post-cure heating cycle may be needed to achieve optimum properties of the lining. Fluidizing is achieved by passing air at low pressure through the tank. The air causes the powder to behave as if it were a liquid, reducing its resistance to items entering it and ensuring full coverage of the components.

The spraying technique involves spraying powder or liquid plastic onto a metal object that is either preheated or subsequently heat-dried. Principal materials applied by spraying include the fluoroplastics, thermoplastics, and fusion-bonded powder coatings. Application takes place by electrostatic or pressure spraying.

In addition to the above techniques, the Rotomold process can also be used. The piece of equipment to be lined is used as the mold for a Rotomolded lining by charging a preweighted amount of powder resin into a hollow mold. The part is then placed into a heated oven and rotated, depositing a seamless, pinhole-free liner to the interior of the part. This produces a perfect conformation of the liner to the vessel wall, eliminates the need for expensive internal tooling such as that required for transfer molding, and allows lining of parts that cannot be lined in any other way.

5.13.5 **SELECTION OF PROCESS**

The lining process should be appropriate to the grade of material selected for the lining.

PE powder for lining is available in three grades: LDPE, medium-density polyethylene, and HDPE. The lining is applied by spraying or the fluidized bed process.

Suitably formulated PP powders for lining are available in several grades, including homopolymers, copolymers, and mixtures with other polyolefins. Linings based on unmodified polypropylene will not normally bond to the substrate, whereas those based on powders formulated with special additives will. Improved adhesion can be achieved by using a liquid primer.

The grades of nylon available for powder lining are Nylon 11 and 12. Linings based on powder will bond to substrates.

Two forms of PVC are used for lining: UPVC and plasticized PVC. The kind of UPVC used for lining is normally a copolymer of PVC and PVA. UPVC is normally applied as a powder and bond to substrates. Plasticized PVC when used for lining is applied as a plastisol.

Some solvents (e.g., aromatic hydrocarbons) will extract the plasticizer, and after evaporation of the solvent, the lining will be hard and liable to crack.

An EVA copolymer can be applied to a variety of metals (including zinc) without the use of primer. The adhesion of EVA to the substrate is excellent.

A PVDF lining may be applied as a dispersion lining or by powder lining. Reinforced lining systems are available. Priming of substrates is necessary to achieve adhesion. A processing temperature (curing) of 250°C is required.

PFA is a copolymer of tetrafluoroethylene and perfluoropropyl vinyl ether. Nonsheet PFA lining should be applied as a powder. A processing temperature (curing) of 350°C is required.

PTFE is not melt processable, and it can be applied only as a dispersion lining. The lining produced are thin and are not pinhole-free. It is necessary to use etching primers in order to promote adhesion. The lining should consist of at least two coats, uniformly applied via spraying. The priming coat should be applied immediately after the cleaning of the surface. Application by brushing is allowed only for areas inaccessible to a spray gun. Each coat should be sintered separately. This should be performed by heating the entire piece of equipment between 380°C and 400°C, followed by gradual cooling to ambient temperatures.

An FEP lining may be applied as a dispersion lining. A processing temperature of 360°C is required.

E-CTFE linings may be applied by fluidized bed or electrostatic spraying. Adhesion to carbon steel surfaces is good, but in the case of other metals, a primer may be required to obtain good adhesion.

The processing temperature of 360°C to 400°C is required. These linings are available in thicknesses that correspond to standards. A lining thickness of up to 5 mm can be achieved by rotomolding.

Fusion-bonded epoxy resins are available in three grades of powder suitable for application by fluidized bed, electrostatic, and normal powder spray. The equipment that has been cleaned should be preheated so that the surface temperature at the entrance of the lining station is between 220°C and 245°C, or as specified by the manufacturer.

Graduated meltable temperature indicators should be used to measure the temperature. The epoxy powder should be applied to the preheated pipe, fitting, or vessel by methods approved by the manufacturer at a uniform cured-film thickness. After the lining has cured according to the manufacturer's recommendations, it may be cooled with air or water spray to a temperature below 93°C for inspection and repair. The practices for unprimed internal fusion bonded epoxy coating of line pipe are recommended in API-RP-5L7. For more information about using of fusion bonded epoxy for the lining of steel water pipelines, see AWWA C 213.

5.14 CURING

For thermally cured materials (namely plastisols and fusion-bonded epoxy resins), the state of cure should be tested as described in the next sections.

5.14.1 PLASTISOLS

The hardness of the linings should be tested in accordance with widely accepted standards. It should be within 5 international rubber hardness degrees (IRHDs) of the hardness of the material.

5.14.2 FUSION-BONDED EPOXY RESINS

The lining should be tested for solvent resistance. This test should be carried out by laying a cloth soaked in methyl isobutyl ketone (MIBK) over selected areas of the lining for 3 min. After this time, the coating should show no signs of softening when scratched with a fingernail.

Another method of estimating the cure is to carry out an impact test to meet the ASTM G 14-77 standards. If this test is to be used, the impact resistance should be specified by the material manufacturer.

In the case of disputes about the curing state of fusion-bonded epoxy resin linings, samples should be taken and the state measured by an independent authority using differential scanning calorimetry.

5.15 REPAIR OF LINING

Defective coatings should be repaired by the applicator. This include coatings with thicknesses below the specified minimum defects disclosed by the holiday detector, as well as obvious defects resulting from mechanical damage to the coated surface.

Rectification of faults should be done using the same material as originally used for the lining. Any other materials should be used only with the company's written consent.

Damage to linings attributable to the contractor's operations should be repaired by the contractor at no cost to the company. When the company inspector determines that the damaged area is large enough to require relining, it should be done at no cost to the company. If the company inspector approves, small areas may be touched up with a brush or by spray.

Damage to linings attributable to the mill or from equipment defect repairs should be repaired by the contractor at the sole expense of the equipment manufacturer. Should the company inspector thinks relining is required, it should be done. Small areas, if approved by inspector, should be touched up with a brush or spray at no additional cost if the inspector approves. The inspector should approve all relining and repairs to linings, and the contractor should maintain records of such work in order to receive reimbursement for it. If the number of defects is large, and the damage covers a large surface area, the lining should be removed completely and the work redone.

5.16 INSPECTION AND TEST METHODS
5.16.1 APPEARANCE OF LINING

The lining should be inspected visually for blisters, flaws, sagging, and inclusions of foreign material. Any defects in the lining should be removed and the lining replaced. The repaired area should be subjected to a full inspection.

The plastic liner should fit snugly to the steel housing, and no entrapments should be present between the plastic lines and the internal surface of the substrate.

5.16.2 ADHESION

The lining should be inspected visually for evidence of lack of adhesion to the substrate and, where applicable, lack of intercoat adhesion. As far as possible, adhesion testing of the applied lining is to be avoided because it can be destructive, requiring the lining to be repaired.

In the rare cases where testing is required, test plates should be used to demonstrate that the process employed provides a lining with the required level of adhesion. These test plates should be of the same material as the substrate, and the lining process should be the same as that employed for lining the equipment and done under the same conditions and at the same time. For thick linings such as plastisols, the adhesion test should be that given in Section 10.4 of BS 490.

If evidence of lack of adhesion to a substrate or lack of intercoat adhesion is found, the lining should be removed totally or partially, depending on the problem area. The area should be relined and subjected to full inspection.

5.16.3 THICKNESS OF LINING

A survey of the thickness of the lining should be made. Of the instruments available, those that operate on a single probe electromagnetic or eddy current principle should be used in most situations. The choice of instrument in other cases is determined by the nature of the substrate. The instruments should be calibrated against reference plates at least twice a day.

Attention is drawn to the fact that thickness measurements of thin films in corners and on curved surfaces of small radii may not be accurate. The thickness of the lining of all pipes and fittings should be measured at both ends at the facing of each flange and meet the requirements of the standards. A reduction of 10% at the flange facing due to flaring is acceptable, but variations in wall thickness of more than 20% are not allowed.

5.16.4 CONTINUITY OF LINING

The lining should be tested for pinholes. If the number of pinholes exceeds the number specified, local repairs should be performed. if local repairs are not possible, the lining should be removed and replaced.

5.16.5 HYDROSTATIC PRESSURE TEST

A total of 10% of the pipes and fitting furnished under construction standard should be subjected to a hydrostatic pressure test. The company should be contacted for those cases where testing would result in damage of the pipe ends caused by the end caps. Alternatively, an underwater test may be carried out as agreed with the company.

Installed piping systems should be pressure-tested with water at ambient temperatures at a pressure of 1.5 times the maximum working pressure for a period of at least 4 h. No weeping at flanges or through vent holes (if any) is allowed during this test. Owing to variations in ambient temperature, the pressure may fluctuate. Care should be taken that the test pressure does not exceed the lowest-rated element in the system.

5.16.6 THERMAL CYCLE TESTING

Pressure/temperature testing and vacuum testing will be required only for fluoropolymer-lined piping intended for vacuum, pressure, or heavy-duty service and will be specified as such in the company order.

5.17 STOVED THERMOSETTING RESIN LINING

This section specifies requirements for the lining of equipment using stoved thermosetting resins.
The materials considered include:

- Phenol-formaldehyde resin
- Epoxy-phenolic resin

Stoved thermosetting resins should not be applied to equipment manufactured in cast iron or concrete. The requirements for the design and fabrication of the equipment, the state of preparation necessary for the surface to be lined, and the thickness of the lining should be according to standards.

The thermosetting stoved resin-based linings are used for a number of duties, including protection against corrosive environments, prevention of contamination of products, and provision of surfaces that do not foul easily or that can be cleaned easily. The linings are normally applied as solvent solutions that may contain pigments, and then are stoved at elevated temperatures to remove solvents and cure the resin. The actual stoving temperatures are specified by the lining supplier, but they are normally in the 150–200°C range; and therefore, it is usual for the stoving to occur in an oven in the applicator's works. Heat curing in situ is possible, but it presupposes adequate insulation of the vessel and careful temperature control.

Linings based on these resins are generally less than 300 μm in thickness and though they can be applied completely free of pinholes on relatively small vessels. It is very difficult to obtain linings that are pinhole free on large surface areas. Furthermore, even if the applied linings are free of imperfections, consideration should be given to the possibility of damage during operation.

Before selecting a lining of this type, therefore, some knowledge is required of the likely rate of corrosion of the substrate and the causes of such corrosion. If the corrosion rate is low, there may not be a problem; but if it is high and due to simple solution of the metal, then caution is advised because of the danger of severe corrosion through a pinhole. On the other hand, if a high corrosion rate is due to an effect such as erosion or corrosion, then thin linings can be an effective barrier.

5.17.1 PREPARATION OF MATERIALS FOR USE

All materials furnished by the contractor should be of the specified quality. To prepare the lining material, follow the instructions provided by the manufacturer.

5.17.2 PREPARATION OF SURFACE TO BE LINED

Stoved thermosetting lining should be used only for metallic equipment (with the exception of cast iron, where it should not be used).

5.18 APPLICATION METHODS

The lining process should start as soon as possible after blast cleaning is complete and before any visible rusting occurs. Unless maintained in a dehumidified atmosphere, application of the lining should commence within 4 h. If signs of rusting occur, then the surface should be prepared again to the required standards.

Several coats are necessary in order to achieve the thickness stipulated and each coat should be allowed to air-dry before the next coat is applied. Any intermediate stoving should be at a lower temperature than the final stoving temperature, such that curing does not proceed beyond the stage that impairs intercoat adhesion. All external angles and edges should be strip coated by applying a thin coat before the rest of the surface is coated.

Before the final stoving takes place, the lining should be tested for continuity. If the continuity of the lined equipment meets the quality agreed at the tender stage, final stoving can proceed; if not, further coats should be applied either locally or over the whole surface until the specified standard is reached. When extra coats are applied, the final thickness should not exceed the specified limit. This procedure is necessary because once these materials have been cured at the final stoving temperature, it is impossible to achieve the same level of intercoat adhesion.

5.19 CURING

The basic reasons for heating backing-type resin are to remove volatile solvent and products of polymerization from lining and tank, and to accomplish polymerization or curing to obtain chemical and permeation resistance. If the lining has been properly heated for the correct period of time, it should be polymerized completely until little or no solvent odor is present.

The control required to ensure that the necessary minimum metal temperature is obtained and that the maximum temperature, as specified by the resin manufacturer, is not exceeded at any given spot in the equipment.

After the linings have been fully cured, it is possible to achieve only a mechanical bond when further coats are applied.

5.20 TRANSPORTATION AND STORAGE

The lined equipment should be stored and transported with the following additional requirements:

- When lined pipe is to be transported by truck, a flatbed trailer must be used with at least three bolsters between each layer of pipe; the bolsters are aligned vertically above the previous layer to provide even support. The load should be tied down to prevent any bending, shifting, or other movement of pipe. All lined pipe should be transported with closed thread protectors on both ends, preferably made of plastic or a plastic-lined steel (composite).
- To stack the lined pipe, wooden bolsters should be placed between each layer, directly above the pipe racks, with each layer being blocked to prevent shifting.
- If the lined pipe must be drifted, the yard personnel must use a wooden, PTFE, or plastic drift to prevent damage to the lining.

5.21 REPAIR OF LINING

When defects and damage are discovered, the coating should be cleaned and abraded wherever resin is to be applied. The area to be recoated should extend at least 50 mm in all directions beyond the defect. The extent of repair will determine the method of cure.

If the defects are small, then local heating via infrared or hot air should be used to cure the repair. For extensive repairs, all the equipment should be heated, in which case care should be taken to ensure that the original coating is not overcured. Use of cold curing, thermosetting resins for repairs should not be permitted without the purchaser's written consent.

5.21.1 INSPECTION

After application of the final coat, but before the cure cycle is complete, the lining should be visually examined for blisters, flaws, and other imperfections. If such defects are found, then the lining should be removed in the affected area and replaced before final stoving.

5.21.2 ADHESION

The lining should be inspected visually for evidence of lack of adhesion of the coating to the substrate and for lack of intercoat adhesion. If evidence of lack of adhesion to a substrate or of intercoat adhesion is found, the lining should be removed totally or partially, depending on the problem area. The area should be relined and subjected to full inspection before final stoving.

5.21.3 THICKNESS OF LINING

A survey of the thickness of the lining should be made before final stoving. Of the instruments available, those that operate on a single-probe electromagnetic or eddy current principle should normally be used. The choice should be determined in other cases by the nature of the substrate. The instruments should be calibrated against reference plates at least twice a day. In fact, the thickness measurements of thin film in corners and on curved surfaces of small radiuses may not be accurate.

5.21.4 LINING CONTINUITY

The lining should be tested for pinholes before final stoving. If the number of pinholes does not exceed the number in the design specifications, final stoving can proceed. If the number of pinholes exceeds this number, further coats should be applied to the lining, either locally or over the whole surface, depending on the location and number of pinholes. After final stoving, the lining should be tested again for pinholes by spark testing or wet sponge testing according to the required standards.

5.21.5 TEST OF CURING

In the solvent wipe test, the lining should be examined for the curing state by placing a rag soaked in a solvent [either methyl ethyl ketone (MEK) or acetone] on the lining for 3 min. After this, the lining should show no sign of softening when scratched with a fingernail. Any apparent softening indicates that the lining is not fully cured and further curing should be performed.

The lining should be visually inspected for overcuring, and any overcured linings should be rejected. The appearance of blisters after the final stoving and considerable darkening of the color of the lining compared with the normal color are evidence of overcuring.

Most heat-curing phenolic and epoxy-phenolic resin linings will show a marked degree of color change between an inadequately and an adequately cured applied film. A set of cure color standards should be used to evaluate the degree of completeness of bake cure. The material should be well cured, and overcured and undercured linings should be rejected.

Variations of formulation, including amounts and types of pigments, will affect the characteristic color. Thus, a set of bake cure color standards are required for each specific material as supplied by each manufacturer. Code symbols should be used to identify each specific material and the degrees of color for each set of bake cure color standard.

5.22 COLD-CURING THERMOSETTING RESIN LINING

The construction standard concerning cold-curing thermosetting resin lining specifies minimum requirements for the lining of equipment using cold curing thermosetting resin. They apply to equipment fabricated in metal or concrete.

The lining materials considered include:

- Liquid epoxies
- Polyesters
- Furanes
- Polyurethanes
- Polychloroprene

Although not thermosetting resins, liquid elastomeric linings (such as polyurethanes) and solvated elastomeric linings (such as polychloroprene) are included since the same application techniques are used. The resins may contain fillers, reinforcing agents, or both.

Requirements for design and fabrication of the equipment, the state of preparation necessary for the surface to be lined, and the thickness of the lining should be according to standards. All materials furnished by the contractor should be of the specified quality. The entire operation should be performed by and under the supervision of experienced persons who are skilled in the application of liquid epoxy linings.

5.22.1 PREPARATION OF MATERIAL FOR USE

All materials furnished should be of the specified quality. To prepare the primer and finish coat of lining materials for application, follow the instructions provided by the manufacturer. Mix only the amount of material that will be used within its pot life.

5.22.2 PREPARATION OF SURFACES TO BE LINED

For metals, all grease, oil, temporary protective, and chalk should be removed from the surface to be lined. Degreasing should be performed using vapor degreasing equipment or as recommended in the standards. All surfaces to be lined should be maintained at a temperature at least 3°C above the dew point throughout the preparation and lining processes. If there is a risk that this condition will not be maintained owing to ambient conditions or a change in ambient conditions, dehumidifying and heating equipment should be used.

All surfaces to be lined should be blast-cleaned. In the case of carbon steel and cast iron, the standard of blasting should be as defined in ISO standard 8501 and listed in Table 5.1. Only nonmetallic grit should be used for blast-cleaning aluminum and its alloys. The average surface profile of the prepared substrate, measured peak-to-trough, should be 40 μm.

Special treatment may be required for other metals as follows:

- All dust, residue, and debris left on the surface after blast-cleaning should be removed by brushing and vacuum cleaning.
- Before abrasive blasting of the steel tank, it should be rechecked with the combustible gas indicator to ensure that no flammable vapors have entered it.

Table 5.1 Standard of Blasting

Types of Lining	Grade of Lining	Degree of Cleanliness	Average Surface Profile, Measured Peak-to-Trough (µm)
Polyester	A, B, C	Sa 2½	75
Polyester	D	Sa 2½	150
Epoxy	A and B	Sa 2½	40
Epoxy	C and D	Sa 2½	75
Epoxy	C and D, with hot spray applied	Sa 2½	150
Epoxy	E	Sa 2	75
Furane	All	Sa 2½	75
Polyurethane	Spray	Sa 2½	75
Polyurethane	Trowel	Sa 2½	75
Polychloroprene	All	Sa 2½	75

For concrete, remove forming oil with detergent before blasting. Any external corners not formed with a chamfer should be rubbed down to a radius of not less than 3 mm. All surfaces to be lined should be treated to remove laitance and shutter release agents. The specified method for this operation is blast cleaning. The blast-cleaning process should be controlled so that all laitance is removed without exposing the profile of the aggregate. After blast cleaning, all dust and debris should be removed.

Notes:

1. An alternative method of removing laitance that is sometimes used is acid etching. This process is only really applicable to horizontal surfaces. Furthermore, the presence of shutter oils will reduce the effectiveness of acid etching. The thickness of the laitance of a concrete surface varies considerably, and it is very important that acid is allowed to dwell on the surface for long enough to remove all the laitance.

   When acid etching is used, the next operation should to wash the concrete with water, followed by a drying process. Acid etching is not suitable when the equipment is to be lined with polyurethane.

2. Removal of laitance on concrete invariably leaves a surface that contains a large number of small holes of varying diameter and depth.

   Unless the lining material will fill or effectively bridge the holes remaining after the removal of laitance, then these holes should be filled before the lining process begins. One material recommended for this purpose is a smooth paste made from a water-miscible epoxy resin, cement, and a fine filler.

   After removal of laitance, all surfaces to be lined should be maintained at temperatures at least 3°C above the dew point throughout the preparation and lining processes. If there is a risk that this condition will not be maintained owing to ambient conditions or a change in ambient conditions, dehumidifying and heating equipment should be used.

   Fill and patch all holes with polymer mortar or putty. It is not possible to apply lining over any gaps.

5.23 APPLICATION METHODS

The lining process should be appropriate to the grade of material selected for the lining

5.23.1 POLYESTER LININGS

All polyester resins may be applied in a variety of ways, and it is usual to grade according to the thickness of the applied lining typical grades. The criteria are as follows:

- *Reinforced linings up to 1 mm thick.* These systems consist of a resin reinforced with mica, carbon, or small-diameter glass flakes. A resin primer is normally used. The systems are applied in two coats and are used for light corrosive duties and for situations where protection of the product from contamination is necessary.
- *Reinforced linings from 1–2 mm thick.* These systems are made from resins reinforced with glass flakes of up to 3 mm in diameter. A resin primer is used. The systems are applied either in two coats or in two coats with a special topcoat to provide a smoother surface. These systems provide a tough, corrosion-resistant lining.
- *Laminate reinforced linings.* These systems are up to 5 mm thick and normally consist of the following:
  - A resin-based primer
  - Glass fiber reinforcement (chopped strand mat or chopped fiber) thoroughly wetted with resin
  - A resin-based topcoat
- *Laminate-reinforced linings with screed.* These systems are up to 5 mm thick and normally consist of the following:
  - A resin-based primer
  - A screed of resin and inert filler up to 2 mm thick
  - Glass fiber reinforcement laid on top of the screed and thoroughly wetted out with the resin
  - A resin-based topcoat applied by trowel and roller

5.23.2 LAID BY HAND MAT LININGS

These mat linings consist of two layers of glass-fiber mat laid over a blasted and primed substrate and finished with a coat of resin. Typically, the liner is 2.5–3.5 mm thick and averages 20–25% glass and 75–80% resin. This type of lining is used extensively in immersion services on steel and concrete substrates.

With all systems applied to metal substrates, a resin-based primer should be sprayed or brushed onto the substrate immediately after preparation. The resins should be catalyzed immediately before application, and thorough mixing with a mechanical mixer is essential.

Systems that incorporate a layer or layers of glass fiber reinforcement should be rolled to ensure proper consolidation. Care should be taken during the rolling operation to fully wet the glass and expel all air. Adjacent pieces of reinforcement should be overlapped by not less than 50 mm. The edges should be worked out by brushing with a stippling action. If more than one layer of glass reinforcement is used, then all joints should be staggered through the thickness of the laminate.

When the outer layer is reinforced with chopped strand material, then an additional layer of resin and surface tissue should be applied. The final coat should consist of a flow coat of resin. This resin should normally contain 0.4–0.6% paraffin wax (with a melting point of 55–60°C) to prevent loss of styrene and minimize air inhibition of the cure. The work should be scheduled so that good adhesion is obtained between successive layers in adjacent areas. To attain this, any coat should be applied before the previous coat has reached a curing state that would prevent good intercoat adhesion.

If work is interrupted so that one layer is fully cured before work is complete, then the surface of the resin of the previous coat should be removed by grinding. If the previous coat contains glass fibers, then these should be exposed in the grinding process.

For good long-term lining performance, limit the temperature to 60°C. Avoid temperature shocks. For instance, do not drain a 60°C solution from the tank and turn a 15°C water hose on the lining to wash out the vessel.

Good results have been obtained with a high elongation resin (10%) prime coat followed by a 4% elongation lay-up. For polyester lining of under ground storage tanks, see API-STD-1631.

5.23.3 GLASS FLAKE LININGS

To apply a flake glass-polyester lining over a steel surface, the prepared surface should be primed with the lining manufacturer's primer by spraying or brushing. The prime coat should be cured for the amount of time that meets the manufacturer's recommendations.

Prior to applying the intermediate coat, the primer coat should be tack free but should not have set any longer than 8 weeks (nor less than 8 h minimum). In addition, the primer should be checked for styrene sensitivity. If it does not exhibit styrene sensitivity, it must be abraded by sanding, grinding, or sandblasting. At least 75% of the original surface must be uniformly removed.

Any exposed steel should be reprimed. The primed surface should be wiped down with styrene prior to lining to ensure removal of all dust and other forms of contamination. The application of the liner intermediate coat should not proceed if either of the following conditions exists:

- The relative humidity in the work area is greater than 90%.
- The surface temperature is less than 3°C above the dew point of the air in the work area.

The colors of the intermediate coat and the topcoat should be different (for example, an ultramarine base and a pink topcoat) so that you can ensure by looking that coverage has been obtained. The liner intermediate coat should be applied by trowel to a minimal thickness of 0.75–1 mm. The surface of the liner intermediate coat should be applied with a short-nap paint roller dampened with styrene. Avoid excessive application of styrene to prevent softening of the lining.

Prior to the topcoat application, the intermediate coat should be tack free and have set less than 14 days (the manufacturer's instructions might differ from this).

If the intermediate coat has set more than 14 days, it should be abraded by sanding, grinding, or sandblasting. At least 75% of the original surface must be uniformly removed. More important, the intermediate coat should be checked for styrene sensitivity. If the styrene is sensitive, good intercoat adhesion will exist. If blasting had been done in the vessel after application of the intermediate coat, or if the surface is otherwise visually dirty, the intermediate coat should be wiped with styrene to ensure removal of dust and other contaminants.

The topcoat should be applied by trowel to a thickness of 0.75–1 mm or as specified by the designer. Sharp edges should be protected with multiple layers of glass mat–reinforced polyester extending a minimum of 100 mm on the adjacent flat surfaces. Layers should be built up with a 450-g/m² mat, which should be sized, silane-finished, and dried. A minimum of two layers that are 1.1 mm thick will be applied. Then a layer of 0.25 mm is placed above the lining. A bisphenol resin will be used.

All reinforced linings are finished with a wax coat containing approximately 0.2% reinforcing strips. Some manufacturers cover the mat with flaked glass, ending with a 4.5–5-mm thickness on sharp edges and adjacent surfaces.

5.23.4 TROWELLED MORTAR LININGS

Trowelled mortar linings, filled polyester linings stabilized with light roving, are very important to use in pumps, trenches, concrete tanks, and vessels. These can be further modified to provide additional abrasion resistance, which in many cases is extremely important.

Next, a silica-filled base coat is applied to the primed substrate. A light roving is embedded in it and rolled with resin until soaked through. A topcoat follows and, if abrasion is a problem, is enhanced with a corundum filler. The surface may be rolled with styrene to provide a glazed, slick surface.

5.23.5 EPOXY LININGS

Typical grades of epoxy materials used for lining are as follows:

- *Amine-cured solvent containing systems.* These systems can be based on liquid epoxy resins, but more often, a solid resin is used, which tends to give a better performance. The hardener is usually in the form of an amine adduct. Sensitivity to temperature and humidity depends upon the type of hardener that is used. The lining is applied as a multicoat system, which may well include a primer.
- *Polyamide-cured solvent-containing systems.* In these systems, a polyamide resin is used as the hardener in place of an amine. They are not often used as lining materials, as their general properties are substantially reduced compared to amine-cured systems. They are generally sensitive to conditions of high humidity.
- *High-solid epoxy systems.* These systems are similar to the first two on this list, except that the solvent content is very low. Nearly all of them are based on liquid epoxy resins. Hardeners will be either an amine, aromatic, or aliphatic polyamine, or a polyamide. Sensitivity to temperature and humidity depends on the type of hardener that is used. The lining is usually applied in two coats.
- *Solvent-free epoxy systems.* These systems are similar to high-solid epoxy systems, except that they are based on liquid epoxy resin with no solvents present. When the hardener is an aromatic polyamine, a three-pack system consisting of a base, a hardener, and an accelerator is used. These systems are suitable for use with food and drinks, provided that a suitable nontoxic and taint-free resin is used. These resins also may be reinforced with glass flake or glass fiber to provide additional strength.
- *Coal-tar epoxy systems.* These are two-pack systems containing either coal tar or coal-tar pitch. The epoxy may be liquid or solid resin. The amounts of solvent are small or nonexistent. Corrosion-resistant properties vary markedly depending on the coal tar/epoxy resin ratio, and it is important that this is specified. For instance, where good resistance to sulfuric acid is required, a proportion of at least 40% epoxy resin is required.

Hardeners such as amines, amine adducts, and polyamides are commonly used, depending on temperature and humidity. The systems usually include two or three coats.

Some of the resin systems are moisture sensitive, and in the case of site work, it may be necessary to install and operate dehumidifying equipment during the lining process. The temperature of the mixed coatings and of the substrate at the time of application should not be lower than 13°C. It is permissible to heat the substrate by in-line heaters or other means to 49°C to facilitate application and curing.

The lining system should consist of a liquid, two-pack, chemically cured, and rust-inhibitive epoxy primer and one or more coats of a liquid, two-pack epoxy finish coat. The primer and finish coats should be from the same manufacturer. Alternatively, the lining system may consist of two or more coats of the same epoxy material without the use of a separate primer.

The total system should be applied in two or more coats and a total dry film thickness of between 365 μm and 635 μm. Both the primer and finish coats should be spray-applied in accordance with the manufacturer's recommendations. Airless spraying or a centrifugal wheel is the preferred method of application.

After mixing, the primer should be applied without thinning to a dry film thickness of 25–40 μm. A minimum of 4 h drying time and a maximum drying time as recommended by the manufacturer is required before the finish coat is applied. If more than the maximum drying time elapses, the surface must be reprimed either by removing the primer or abrading its surface to roughen it.

When the resin system contains solvents, the interval between coats should be sufficient to allow the solvent to evaporate. The epoxy finish coat(s) should be applied over the dry primer or first coat as recommended by the manufacturer. If more than one coat is applied, the second coat should be applied within time limits recommended by the manufacturer to prevent delamination between coats. If the recommended period between coats is exceeded, a recommended repair procedure should be obtained from the coating manufacturer.

For more information about using liquid epoxy lining systems with steel water pipelines, see AWWA C 210.

5.23.6 FURANES

All surfaces to which furane resin linings are to be applied should be primed before application. The primer should be fully cured before application as well. The components of the resin system should be mixed thoroughly and applied within the time limit specified by the manufacturer of the lining.

Application by brush, roller, trowel, or special spray equipment is permissible. When glass fiber reinforcement is used, the coats should be rolled well to ensure a void-free laminate.

More than one coat of resin is usually applied, and the work should be scheduled so that good adhesion is obtained between successive layers in adjacent areas. To achieve this, any coat should be applied before the previous coat has reached a state of cure which would prevent good intercoat adhesion. When a final topcoat of unreinforced resin is applied, the thickness should not exceed 0.5 mm.

The curing of furane resin linings should be in accordance with the resin manufacturer's recommendations. When the curing conditions involve heating, hot air should be used. No part of the lining should be heated above 50°C during the early stages.

5.23.7 POLYURETHANE

All the polyurethane systems used for lining process equipment should have multiple components. The components should be mixed thoroughly before application. When polyurethane systems used for lining contain solvents, the interval between coats should be sufficient to allow the solvents to evaporate.

The work of applying successive coats should be scheduled so that the next coat goes on before the preceding coat is fully cured. Otherwise, poor intercoat adhesion will result.

All the polyurethane systems used for lining are sensitive to moisture. A small amount of moisture will accelerate the curing of the lining. Excessive moisture has an adverse effect, and it may be necessary to control the moisture content of the atmosphere during the application of the lining.

Where concrete equipment is to be lined, the free water content of the concrete is particularly important.

5.23.8 POLYCHLOROPRENE

Special primers should be used to promote a bond between substrates and liquid polychloroprene. The liquid rubber contains solvents, and the lining process requires the application of a number of coats. The process should be scheduled with enough time to allow the solvents from one coat to evaporate completely before the next coat is applied.

The work of applying successive coats should be scheduled so that the next coat is applied before the preceding coat has fully cured. Otherwise, poor intercoat adhesion will result.

The curing of the lining depends upon the type of curing agent and the temperature. At 15°C, the cure time is approximately 7 days. The process may be accelerated by the application of heat.

5.23.9 CURING

After application, the lining should be cured in accordance with the manufacturer's recommendations. There is no suitable procedure for evaluating the adequateness of chemically cured linings at the time of application. However, the hardness of the applied coating has been used as a guide, in that it may be proportional to the degree of curing achieved.

5.23.10 REPAIR OF LINING

When rectification of faults is to be made in linings that are fully cured, special attention should be paid to the problems of achieving adhesion between new resin and the cured lining. In the case of linings based on polyester resins, the first step should be to remove the surface wax over a patch that extends 50 mm beyond the area to be repaired. With all linings, a patch that extends 25 mm beyond the area to be repaired should be ground to remove the gloss.

When priming of the substrate is an essential part of the system, the first step after preparation should be to establish whether the primer is intact. If the primer is damaged, then it should be repaired before the rest of the work proceeds.

Rectification of faults should be done with the same material as was originally used for the lining. If other materials are used, the purchaser must consent in writing.

After all rectification work is completed, the lining should be subject to inspection as appropriate, particularly with regard to continuity testing. Repairs should be electrically inspected using continuity testing.

5.23.11 INSPECTION

The lining should be inspected visually for blisters, flaws, sagging, and inclusions of foreign material. Defects should be removed and the lining replaced. If the number of defects is large and covers a large surface area, the lining should be removed completely and the work of lining redone.

Visual inspection should be as per ASTM D 2563 standards, as follows:

- Acceptance Level I for aggressive environments
- Acceptance Level II for all other environments

Next, a survey of the thickness of the lining should be made. Of the available instruments, those that operate on a single-probe electromagnetic or eddy current principle should normally be used. The choice should be determined in other cases by the nature of the substrate. The instruments should be calibrated against reference plates at least twice a day.

Notes:

- For determining the thickness of reinforced thermosetting resin linings, see ASTM D 3567.
- The thickness measurements of thin film in corners and on curved surfaces of small radii may not be accurate.
- If concrete is the substrate, it may be appropriate to monitor the thickness of the linings by using wet film thickness gauges.

5.23.12 ADHESION OF LINING

The lining should be inspected visually for evidence of lack of adhesion to the substrate and, where applicable, lack of intercoat adhesion. As far as possible, adhesion testing of the applied lining is to be avoided because the testing is destructive, requiring the lining to be repaired.

If required, test pieces should be used to demonstrate that the process employed provides a lining with the required level of adhesion. These test pieces should be of the same material as the substrate and the lining process should be the same as the one employed to line the equipment and done under the same conditions and at the same time.

5.23.13 CONTINUITY TESTING

The lining should be tested for pinholes. Two main types of instruments are used for continuity testing, and are described next.

For wet sponge testing, it is usual for wet sponge probes operating on low voltage to be used for linings up to 350 μm thick. The sponge probe of the instrument should be wetted with a 3% solution of sodium chloride, to which a small amount of wetting agent (detergent) has been added. If the substrate is austenitic steel, then a 3% solution of ammonium sulfate should be used instead of the salt solution.

The sponge should be moved across the surface in a systematic way so the whole of the surface is examined. The speed of travel should be controlled so that time is allowed for imperfections in the lining to wet through. When a pinhole is discovered, the position of the hole should be clearly marked.

Before proceeding with further testing, the surface of the lining adjacent to the pinhole should be dried thoroughly to avoid tracking back to the pinholes that already have been discovered.

The finished lining should be tested using a high-voltage, high-frequency spark tester set for the voltages listed in Table 5.2.

A probe should be moved continuously over the surface of the lining at a speed not exceeding 100 mm/s. Applying the spark to one spot for any appreciable length of time should be avoided, as prolonged exposure to the spark can damage the lining. When a defect in the lining is discovered, it should be clearly marked. When concrete is the substrate, the only method of testing for pinholes

Table 5.2 Voltage Levels	
Glass flake–reinforced linings	**2,500 to 5,000 V, depending on composition**
Lay-up and spray linings	5,000 volts to 1.5 mm thickness 10,000 volts to 3 mm thickness

is by spark testing using an A/C-type instrument. For continuity testing, the surface of all but small pieces of equipment should be divided by chalk lines or other suitable marks into smaller areas of about 1 m^2.

Local repairs are permitted if the number of pinholes does not exceed 5 per square meter. If local repairs are not possible, the lining should be removed and replaced.

When curing lining, the following points are important:

- The lining should be examined for the state of cure.
- The hardness of the lining is a good indication of the cure, so it should be the minimum specified.
- The lining should be tested for solvent resistance using the specified solvent.
- This test should be carried out by laying a cloth soaked in the specified solvent over selected areas of the lining for 3 min. After this time, the lining should show no sign of tackiness.
- For linings based on polyester resins, any wax should be removed from the surface before applying this test.
- If the lining is found to be undercured and the appropriate curing schedule has been followed, then remedial action should be undertaken. If this treatment is not successful, then the lining should be removed and replaced.

The average of several hardness readings made with the Barcol impressor on a reference sample, should at least be equal to the minimum specified by the lining manufacturer based on the type and material of lining selected. A minimum of one reading should be made for each 10 m^2 of lining installed and for each opening (i.e., nozzle or manway). If the required hardness is not obtained, remedial procedures should be undertaken as mutually agreed by both company and contractor.

Reference samples prepared by the applicator and meeting all quality requirements should be supplied to the company prior to execution of the work, to the following specifications:

- Samples should be 150 mm × 150 mm.
- One sample should be prepared of the complete system, to be used as a standard by the inspector for quality and finish of the completed work.
- For all systems that require a seal coat, an additional sample should be prepared without this finish coat. This sample should be used by the inspector as a standard and calibration for hardness tests, thickness measurements, and spark test.

5.23.14 TRANSPORTATION AND STORAGE

Lined equipment should be stored and transported with the following additional requirements. When lined pipe is transported by truck, a flatbed trailer should be used with at least three bolsters between each layer of pipe, with the bolsters aligned vertically above the previous layer to provide even support. The load should then be tied down to prevent any shifting, bending, or movement of the pipe. All pipe should be transported with closed-end thread protectors on both ends, preferably made of plastic or plastic-lined steel (composite).

For storage, the pipe should be placed on at least three racks or wooden sills evenly spaced to keep it around 48 cm off the ground. To stack the pipe, wooden bolsters should be placed between each layer, directly above the pipe racks, with each layer being blocked to prevent shifting. When movement of the pipe is required on the racks, bars or similar objects should never be placed in the pipe's inside diameter (ID). If thread protectors come loose, the threads and lining should be inspected for damage, and the thread protector reinstalled prior to any movement.

When running the pipe, it is important to select the best tools available, especially slips, power tongs, backup tongs, and slip-grip elevators. Equipment with as much surface contact as possible to the pipe (e.g., full wraparound tongs) should be used. When the tubing is lifted onto the rig floor, the pin-end thread protector must be in place to protect the threads and lining that covers the chamfer and (typically) the first two threads. It is removed just prior to stabbing for makeup.

5.24 **RUBBER LININGS**

This section specifies minimum requirements for the lining of equipment using cold rubber. The material considered include natural and synthetic rubbers applied as unvulcanized or prevulcanized sheet. Brushed or sprayed rubber linings are not discussed here.

The rubbers can be used for lining include the following:

- Natural polyisoprene
- Synthetic polyisoprene
- Butyl rubber
- Nitril rubber
- Chloro-sulfonated polyethylene
- Hard or Ebonite rubber
- Fluorinated rubber
- Chloroprene rubber (CR)

Rubber linings are applied to equipment and construction fabricated in metal or concrete. They are mainly used for two purposes: corrosion protection and abrasion protection. In brick-lined equipment, they form an impermeable membrane between the brick lining and load-bearing construction. Rubber linings are used economically for chemical services in temperature ranges from −20 to 200°C. In many cases of high-temperature service, protective brick linings are used to reduce the actual temperature of rubber linings and extend the expected service life by enough time to justify the added expense. Alternating from one chemical service to another is not recommended.

Linings should not be applied when the temperature of the surface is less than 3°C above the dew point of the air in the work area, the relative humidity is higher than 80%, or both.

The applicator of the lining should provide a certificate of inspection and testing when requested.

5.24.1 **PREPARATION OF MATERIALS FOR LINING**

Unvulcanized lining materials should be produced using a calender, extruder or roller die. When linings are produced by calendaring, multi-ply construction should be employed. The minimum number

Table 5.3 A Guide to the Number of Plies Related to a Given Thickness

Thickness of Lining (mm)	Minimum Number of Plies
3 to 5	3
6	4

of plies employed in the manufacture of a finished sheet depends upon the rubber compound. A guide to the number of plies related to a given thickness is given in Table 5.3.

When roller die or extrusion methods are used for the production of lining materials, then single-ply or multi-ply sheets are permissible.

With prevulcanized lining materials, the unvulcanized lining sheet should be prepared and vulcanized by heating (normally under pressure in an autoclave) or by a rotary vulcanization process (see also ASTM D-3182).

5.24.2 PREPARATION OF UNVULCANIZED RUBBER SHEETS

Unless otherwise specified, condition the sheeted compound for 1–24 h at 23°C (±3°) at a relative humidity not greater than 55%. For maximum precision, condition for 1 to 24 h in a closed container or in an area controlled at 35 ±5% relative humidity to prevent the absorption of moisture from the air.

Place the sheeted compound on a flat, dry, and clean metal surface and cut pieces that are 4.5 ±1.5 mm shorter in width and length than the corresponding dimensions of the mold cavity. Mark the direction of the milling on each piece. The mass of a 150-×-150-mm sheet or a 150-×-75-mm sheet to be vulcanized in the molds should be as shown in Table 5.4.

A film of suitable material, such as a nonlubricated, 0.1-mm-thick aluminum foil, may be placed both above and below the sheet in the mold to prevent contamination with materials remaining in the mold from previous cures. The mass of the unvulcanized sheet should be reduced to compensate for the thickness of the foil.

5.24.3 PREPARATION OF SURFACE TO BE LINED

Metal surfaces to be lined should be smooth and free of pitting, cavities, porosity, and other surface irregularities. The surface also should be free of oil, grease, and other foreign matter. All surfaces of carbon steel and cast iron to be lined should be blast-cleaned. The standard of blasting should be Sa 2½ as defined in Swedish standard SIS 05 59 00.

After this operation, the surface roughness should have a peak-to-valley height of 50–75 μm. The thoroughly cleaned surface should be pretreated before the lining is applied. In the case of metals, except for carbon steel and cast iron, methods of preparation of the substrate should promote an acceptable bond between the substrate and the lining. Immediately after the surface treatment of the metallic substrate, the grit, dust, and other materials are removed and a layer of adhesive primer with a dry-film thickness of approximately 30 μm should be applied.

Unless maintained in a dehumidified atmosphere, application of the primer should commence within 4 h. Should signs of rusting occur, then the surface should be prepared again to the required standards.

Table 5.4 Mass of Unvulcanized Rubber Sheet (in Grams)

Density of Compound	150 mm × 150 mm	150 mm × 75 mm
0.94	52 ± 3	26 ± 1.5
0.96	53	27
0.98	54	27
1.00	55	28
1.02	56	28
1.04	57	29
1.06	58	29
1.08	59	30
1.10	60	30
1.12	61	31
1.14	62	31
1.16	63	32
1.18	64	32
1.20	65	33
1.22	66	33
1.24	67	34
1.26	68	34
1.28	69	35
1.30	70	35

All surfaces to be lined should be maintained at a temperature of at least 3°C above the dew point throughout the preparation and lining processes. If there is a risk that this condition will not be maintained owing to ambient conditions, dehumidifying and heating equipment should be used.

Unless otherwise stated, all parts which are not rubber-lined should be derusted and painted with one coat of a suitable epoxy resin-based primer. This should be performed after vulcanization of the rubber.

With concrete, any external corners not formed with a chamfer should be rubbed down to a radius of not less than the thickness of the rubber to be used for lining. Lining of concrete equipment should not proceed until at least 28 days after the concrete was cast and the free water content is down to a level of less than 5% wt.

Notes:
- Satisfactory adhesion of rubber lining to concrete depends upon the water content of the concrete. It is desirable that the free water content be less than 5%. If the concrete mix used is carefully controlled, it is possible that after a curing period of 28 days, the water content will be down to 5%, but in other conditions, the required curing period may be longer.

- There are no reliable methods for measuring the absolute water content of concrete, although moisture meters may be useful in determining problem areas. It is advisable to line a test area and evaluate the adhesion before proceeding with a complete lining.

Surfaces to be lined should be treated to remove laitance and shutter release agents. Of the two possible methods for this operation, blast cleaning is preferred. This process should be controlled so that all laitance is removed and exposure of the profile of the aggregate kept to a minimum. After blast cleaning, all dust and debris should be removed.

The alternative method of removing laitance is acid etching. This process is more difficult to control on vertical and overhead surfaces, and the presence of shutter oils will reduce its effectiveness. The thickness of the laitance of a concrete surface varies considerably, and it is very important that acid be allowed to dwell on the surface for a sufficient length of time to remove all laitance. When acid etching is used, the next operation is washing of the concrete with water, followed by drying.

Unless the lining material will fill or effectively bridge the large number of small holes (of varying diameter and depth) that invariably remain in the concrete surface following the removal of laitance, then these holes should be filled with any appropriate mortar before the work of lining begins. One material recommended for this purpose is a smooth paste made from a water-miscible epoxy resin and a fine filler and cement.

After removal of laitance, all surfaces to be lined should be maintained at a temperature at least 3°C above the dew point throughout the preparation and lining processes. If there is a risk that this condition will not be maintained owing to ambient conditions or a change in ambient conditions, dehumidifying and heating equipment should be used.

5.24.4 APPLICATION METHODS

Prepare all metal surfaces to be lined in accordance with standards. Apply one coat of adhesive primer immediately after blasting to prevent rusting. Apply additional coats of adhesive primer, if necessary, as specified by the lining manufacturer.

Apply the required number of coats of intermediate or tie adhesives, or both, as specified by the lining manufacturer. Allow sufficient drying time between the adhesive coats so the coat being applied does not lift up the preceding one.

Apply the type and thickness of the lining specified using a minimum number of sheets and splices consistent with good lining practice. Overlap the edges of the sheets approximately 50 mm unless restricted by dimensional tolerances. The rubber lining sheets may be washed with a recommended solvent and allowed to dry before application.

During the application, roll the sheets and carefully stitch all the seams and corners to eliminate all trapped air between the lining and adhesive-coated surfaces, so that there is full contact with all coated areas (see also ASTM D 3486). Skive the edges of all the sheets at a 45° angle (minimum) from the top surface to the bottom of the sheet. Use a closed-skive construction, commonly known as a *down skive,* wherever possible. This is required where the lining is a combination of hard-face stock and soft cushion. Open-skived splices may be used when specified by the lining manufacturer.

Prior to vulcanization, inspect all the lined surfaces for blisters (trapped air), pulls, or lifted edges at seams and surface defects. After lining, also check any special dimensional tolerances required.

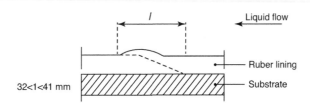

FIGURE 5.1

Overlap bevel joint (one layer).

The scope of this procedure covers the calibration of the equipment and the use of the equipment to determine if there are leaks and their location on sheet linings. It is essential that personnel be instructed in the application procedures to be adopted when entering rubber-lined vessels. A clearance certificate should always be obtained from the appropriate authority before doing so.

The following points should be observed:

- Personnel should not wear studded boots or other footwear likely to cause damage. Rubber-soled shoes preferably should be worn.
- In cases where solid deposits have to be removed, the use of metal spades or other tools must be avoided. Wooden or lined implements should be used.
- The ends of ladders or scaffolding likely to come in contact with rubber linings should be covered in such a way that damage is avoided. Swinging air lines or hoses can also puncture rubber linings. Metallic ends, therefore, should be covered to prevent this from occurring.
- In large ebonite-lined vessels, precautions to avoid successive flexing should be taken and walkways laid if necessary.

5.24.5 JOINTS

Overlap bevel joints (see Figure 5.1) should be used when joining separate sheets of unvulcanized rubber. The total contacting surface between the sheets should be at least four times the sheet thickness, but should not exceed 32 mm at any point. Where applicable, overlaps should follow the direction of the liquid flow.

When the total lining thickness is built up from more than one layer, only the joints in the top layer should be of the overlap bevel type, the under layers being flush jointed as shown in Figure 5.2.

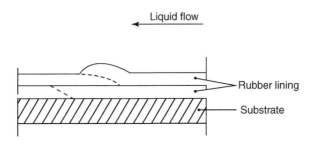

FIGURE 5.2

Overlap bevel joint (two layers).

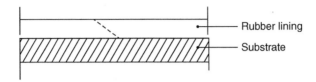

FIGURE 5.3

Flush joint.

The relatively weak flush joint (Figure 5.3) is applied when the lining is used as a base for chemical-resistant brick lining. Joints in the different layers should be staggered.

Joints between rubber pipe linings and the rubber on the flange facing should not protrude so as to restrict the bore of the pipe or to prevent efficient sealing between the flange faces of adjacent lengths. All scarf joints should be closely inspected. Any separation of the joint should be investigated (see Figure 5.6). If the extent of the separation is small, then the rubber should be ground back. If the separation of the rubber is extensive, then the joint and rubber adjacent to the joint should be removed and replaced.

5.24.6 GASKETS

To prevent the gasket and lining bonding together, the rubber flange facing should be lightly rubbed with colloidal graphite.

5.24.7 FLANGES

Rubber lining on pipe flange connections should be as shown in Figure 5.4. The flanges for vessels are shown in Figure 5.5

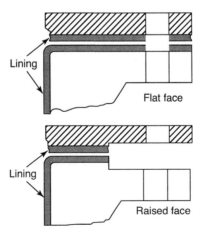

FIGURE 5.4

Rubber-lined pipe flange joints.

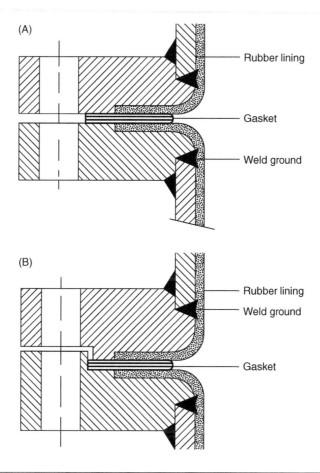

FIGURE 5.5

Rubber-lined flanged connections for vessels. (a) Normal connection. (b) Self-centering connection.

5.24.8 **VULCANIZING**

Vulcanization of rubber linings should be performed in shop or in situ as described next.

A typical lining procedure for shop vulcanization involves the following steps. Immediately after preparation of the substrate, an adhesive primer layer is applied. After evaporation of the solvent, and no longer than 96 h thereafter, the precut unvulcanized rubber sheets are applied. Care is taken to position the sheets accurately without inclusion of air and with joints. The next step is to vulcanize the rubber. This is carried out in an autoclave usually at a pressure of 4-6 bars and a temperature of 140–160°C.

When the equipment to be lined is of such dimensions that it cannot be placed in an autoclave, or there is too high a risk of damaging the shop-applied rubber lining, or the equipment is made from concrete, in situ lining can be performed. In this process, after surface preparation of the substrate, the unvulcanized sheets are applied similarly.

For the subsequent vulcanization, different heat treatment methods are employed—either steam, hot water, or hot air at atmospheric pressure. Alternatively, precatalyzed rubber sheets may be used,

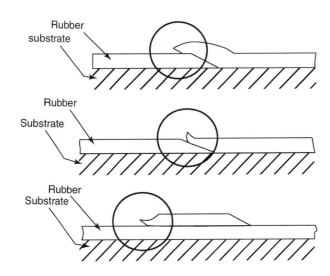

FIGURE 5.6

Unacceptable faults in joints in rubber linings.

these cure naturally at ambient temperatures. Usually, such sheets consist of a completely vulcanized top layer to which a layer of catalyzed and partially vulcanized rubber is laminated. The lining should not be applied when the temperature of surface is less than 3°C above the dew point of the air in the work area, the relative humidity is higher than 80%, or both.

5.24.9 METHOD OF VULCANIZATION

Vulcanization of the lining should be carried out by one of the following methods:

- Autoclave vulcanization
- Using equipment as its own autoclave
- Steam or hot air vulcanization at ambient pressure
- Hot water vulcanization
- Self vulcanization at ambient temperature

The method employed depends upon the design and size of the equipment. It may be necessary to shield the equipment to reduce heat losses that otherwise would lengthen the duration of vulcanization when using methods (b), (c), or (d).The duration of vulcanization will depend upon the method used and the composition of the lining material.

With the agreement of the applicator, interruption of vulcanization should be permitted to detect and repair any faults present. The equipment should then undergo further heat treatment to complete the vulcanization.

5.24.10 AUTOCLAVE VULCANIZATION

The equipment should be placed in an autoclave which is then heated to the required temperature and pressure, which usually at a pressure of 4–6 bar and temperature of 140–160°C..

5.24.11 **USING EQUIPMENT AS ITS OWN AUTOCLAVE**

With all outlets sealed and a steam trap condensate drain attached to a convenient outlet to ensure continuous removal of condensate, saturated steam should be injected until the equipment is pressurized to a predetermined pressure. The pressure should be within the design pressure limits of the equipment. Precautions should be taken against failure of the steam supply since, in such cases, condensation can cause a vacuum and collapse the vessel.

5.24.12 **STEAM OR HOT AIR VULCANIZATION AT AMBIENT PRESSURE**

With outlets covered to reduce steam losses and provisions made to drain condensate from the equipment, the steam should be injected until the vulcanization temperature is attained. This temperature should be maintained for the required time period. Attention is drawn to regarding failure of steam supply.

Note: Hot air may be used as an alternative to steam in some cases, provided that the temperature and heat input can be achieved.

5.24.13 **HOT WATER VULCANIZATION**

With all outlets below the top flange sealed off and the equipment partially filled with water, steam should then be injected into the water, raising the water level and temperature. If the water reaches boiling point before the equipment is full, further water should be added to attain the required level. The temperature should then be maintained for the required time period while maintaining the water level.

5.24.14 **SELF-VULCANIZATION AT AMBIENT TEMPERATURES**

Rubber linings should be specially designed so that they are capable of vulcanizing under ambient conditions. The time to vulcanize is temperature-dependent; at temperatures below 15°C, it may be necessary to use supplementary heating in order to reduce vulcanization times to an acceptable period. After vulcanization, inspect all lined surface for blisters, open seams pinholes, lifted edges and surface defects.

5.24.15 **TRANSPORTATION**

Rubber-lined equipment and piping should not be transported or assembled if the ambient temperature is below or is likely to drop below 0°C. The objects should be handled with care; hoisting should be performed using nonmetallic slings. In particular, branches, openings, and flange facings should be protected adequately since these are vulnerable.

Lifting should be arranged so that chains and other lifting aids do not come into contact with lined surfaces. Loose fittings should not be placed inside lined equipment while it is being transported. Responsibility for arranging transport of lined equipment will vary, and whoever is responsible (purchaser, fabricator, or applicator) should instruct the carrier about the precautions in handling.

5.24.16 **STORAGE**

Lined equipment should be stored under cover or in a protected compound.

All branches, manholes, and other openings should be protected from mechanical damage by using wooden blanks or other suitable material. Between delivery and use, store lined vessels away from direct sunlight, heat, or outdoor seasonal weathering.

Flexible linings may be stored outdoors, provided that the vessels are covered with protective tarpaulins and are not subjected to extreme temperature conditions, such as below 0°C or above 49°C. Avoid sudden changes in temperature. Tanks stored or used in the outdoors may be painted a light color on the outside to reflect heat.

To ensure protection, store semi-hard and especially bone-hard lined equipment in a way that does not allow it to be subjected to extremely cold climatic conditions. If this occurs, thermal stress and expansion may introduce cracking.

Rubber-lined equipment may also be protected for extended periods of time by storing the tank partially filled with a diluted solution. When recommended by the rubber-lining manufacturer, a 5% sulfuric acid, 5% sodium carbonate solution, or weak salt solution make ideal storage media to help keep the lining flexible, to minimize expansion and contraction, and to keep the air (ozone) from prematurely deteriorating the lining surface. Do not permit the liquid contained within to freeze.

Shelter large rubber-lined equipment that cannot be filled with a solution under a suitable structure to protect it from the direct rays and heat of the sun. Provide sufficient air space between the tank and covering to allow for circulation. For small tanks that can be stored inside, cover any open tops and outlets with plywood, or other suitable material and store them away from steam coils or other high-temperature sources. Inspect any stored vessel prior to being put into service. Do not carry on any welding nor any other activity requiring intense heat in the vicinity of a lined tank.

When tanks might be stored outside, take care to ensure good weather ability of the paint. Primer paints are not designed to withstand prolonged atmospheric weather conditions. It is essential that the contractor, the purchaser, or both issue instructions to those responsible for installation on the handling procedures, and special reference is made to the need to wear soft, clean footwear when entering lined equipment and the need to protect lined surfaces from ladders and scaffold poles.

5.25 INSTALLATION OF RUBBER-LINED PARTS

During the installation of rubber-lined equipment, care should be taken to avoid damage of the rubber lining. Due attention should be paid to the susceptibility of linings to collapse when subjected to vacuum and to the risk of damage by local overheating.

Properly applied rubber linings will resist vacuum in the order of 130 mbars absolute. Shop-vulcanized rubber linings have in general a better resistance to vacuum than in situ vulcanized ones.

Rubber-lined equipment and piping is installed in a way that is similar to that of carbon steel equipment. No welding is allowed on lined equipment and piping, nor should they be used as a welding ground.

For connecting rubber-lined equipment and piping, only flanged connections may be used. The rubber-lined flange facings should not be damaged. Rubber-lined piping is used only in aboveground installation.

If the equipment to be lined is located outdoors, it will be necessary, in certain latitudes and seasons, to protect the equipment from the weather and to provide the heat necessary to maintain a temperature above 15°C. In any event, the equipment must be properly protected against rain and in warm weather from direct exposure to the sun. A satisfactory installation cannot be made if conditions are such as to induce condensation on surfaces prior to cementing and lining. Nor can lining be properly cured unless equipment is protected against excessive heat loss during cure.

5.26 REPAIR OF LINING

Repair of damaged rubber linings can be performed in situ or at the contractor's work using an autoclave for vulcanization. The repaired area should be checked for adhesion, thickness, hardness, and continuity in accordance with widely accepted standards. The quality of the repairs should be certified.

5.26.1 SPOT REPAIRS

The total amount of shop repairs allowable in equipment should not be more than 100 cm^2/m^2 of lined surface. No lining repairs are allowed in piping, on flange facings, or on nozzle necks of equipment.

The damaged rubber is cut away, the exposed edges of the lining are beveled and the surface is roughened. Depending on the size of the damage and the service conditions, repairs of equipment can be made.

5.26.2 DAMAGE IN NONCRITICAL SERVICES

For vessel linings not in direct contact with the process liquid, a vulcanized rubber sheet, cut to fit the spot to be repaired, is glued with an appropriate adhesive. A hot-air device can be used to make the sheet sufficiently flexible for handling. Alternatively, if the chemical resistance is not impaired, repairs can be performed with cements based on synthetic resins, such as epoxy or phenol formaldehyde.

5.26.3 MINOR DAMAGE IN CRITICAL SERVICES

For vessel linings in direct contact with the process liquid, an unvulcanized rubber sheet, cut to fit the spot to be repaired, is glued with an appropriate adhesive. Subsequently, heat equivalent to the normal vulcanizing conditions is applied by pressing a heating element against the rubber. Areas to be repaired should not be greater than the heating surface of this element. Alternatively, if chemical resistance is not impaired, repairs can be carried out with a glass-fiber reinforced epoxy resin system cured with hot air in order to yield optimum properties, or with a phenol formaldehydebased cement.

5.26.4 MAJOR DAMAGE IN CRITICAL SERVICES

Most rubber-lining manufacturers have a method utilizing a hot-water vulcanizing or self-vulcanizing type of rubber for repairing large areas in contact with the process fluid. Such rubber compounds have been modified with respect to their accelerator system so that vulcanizing at 90°C or at ambient temperatures is possible. These procedures can also be followed for repairs of equipment, which has been in service. However, the defective and adjacent lining should be thoroughly neutralized, cleaned, and dried.

5.26.5 COMPLETE REPAIR

Provided that previously used plant and equipment is structurally sound and that it has not been subjected to extensive corrosion, relining is normally possible. It is, however, particularly important that the lining contractor should inspect such items and agree that they are in a suitable condition to undertake the work.

When it is necessary to replace a rubber lining in existing equipment, the old lining will have to be removed by one of the following methods:

- Burning—Most rubbers may be removed by this procedure; but although effective, it is subject to certain disadvantages:
    - Metal structures may warp as a result of the heat generated.
    - Noxious or toxic fumes will be generated, and in the case of ebonites and the essentially noncombustible polychloroprene, they will be particularly objectionable and may constitute a health hazard.

It is essential, therefore, that care be taken to ensure that anti-pollution laws are not infringed by this procedure.

- Heating—This is a more acceptable alternative to burning and involves raising the temperature by externally heating the equipment so that the bond to the metal is weakened. The heating process is followed by mechanical removal of the rubber.
- Mechanical removal—The most effective method in instances where adhesion to the metal is still good is to cut through the rubber so that it is divided into narrow strips. These can then be most readily removed by application of a mechanical chisel to the rubber or metal interface.

After removal of the old lining, the metal surface should be prepared in the normal way. Inspection to check the vessel's mechanical suitability for the duty also should be performed at this stage if necessary.

When the old lining has been removed and inspection has confirmed that the item is mechanically sound and in a suitable condition for relining, the standard procedures should be followed. However, if there is any reason to suppose that there has been any impregnation of the metal by chemicals, it is sound practice to sweat the metal in live steam before shot-blasting. Where noticeable corrosion has occurred, relining should be avoided.

5.26.6 INSPECTION AND TESTS

Before proceeding, the applicator should ensure that all materials to be used in the lining process are visually examined and where appropriate, physically and practically tested to confirm that they are in an acceptable condition. For the unvulcanized rubber sheeting, this should include the following:

- Visual examination to ensure that it has no imperfection that could significantly affect the performance of the finished vulcanized lining
- Visual and physical examination to ensure that the actual thickness is within ±10% of the specified thickness
- A test to ensure that the sample taken from each roll achieves the correct hardness when vulcanized in accordance with the customary practice of the applicator

5.26.7 VISUAL APPEARANCE OF THE LINING

Inspection for visible defects should be carried out over the entire surface of the lining in a good light, with attention paid to any areas of mechanical damage, cuts, blisters, lack of adhesion, and poor jointing. If the applicator finds a defect prior to vulcanization, it should be removed and the area overlaid with unvulcanized rubber of the same type as the original unvulcanized rubber on which it is bonded. After subsequent vulcanization, the area concerned should be considered identical to a lap joint or seam and thus fully acceptable.

All scarf joints should be closely inspected. To repair defects found after vulcanization, the surface of the lining local to the defect should be prepared by abrading. Adhesive and a patch of unvulcanized rubber, both of the same type as used for the lining, should be applied. The size of this patch should not be greater than 25 mm in any direction beyond the defect. Next, the patch should be vulcanized. After rectification, the repaired area should be tested for pinholes. Minor wrinkles and surface markings which have no significant effect on the performance of the lining should be acceptable under the responsibility of the purchaser, the contractor, or both.

After vulcanization defects such as blisters, blow holes, small cracks, or scuffed areas in the lining may be found. When such faults are discovered after vulcanization, the defect may be patched.

5.26.8 TESTS

The extent of vulcanization should be checked by a hardness test. If actual hardness levels measured indicate that further vulcanization is required, the lining should be retested after such vulcanization.

The hardness of a rubber is a property indicative of its chemical resistance and mechanical strength. The hardness is determined by measuring the penetration of a specified indentor under a certain load. Various types of indentors and loads are used. In general, it is common to express hardness in Durometer A or Durometer D readings in accordance with ASTM D 2240 or BS ISO 7619-2.

The hardness should conform to the value specified on the requisition within a tolerance of ±5°. A minimum of three reading per square meter should be taken.

5.26.9 CONTINUITY OF LINING (HIGH-FREQUENCY SPARK TEST)

Testing for pinholes and other discontinuities should be performed at the following times:

- When the rubber has been applied and before vulcanization
- After interrupted vulcanization
- After complete vulcanization
- After any remedial work

Before testing begins, the surface should be made dry and free of dirt.

No sparks should be produced when the liner, applied on a metallic substrate, is tested with a direct-current apparatus approved by the company using a voltage determined by the following formula:

$6 (1 + \text{thickness in mm}) = \text{kV}$

which should not exceed 30 kV.

This voltage can be adjusted for high carbon black–filled (soft) rubbers to approximately 3 kV per mm thickness (exact voltage to be determined on a test sample). Antistatic linings on metallic substrates should be checked with wet sponge testing, which uses a low-voltage holiday detector.

The presence of continuous pores in rubber linings applied on concrete substrates can only be determined visually unless a conductive layer has been applied to the concrete. The probe should be moved continuously over the surface of the lining at a speed not exceeding 100 mm/s. Applying the spark to one spot for any appreciable length of time should be avoided.

5.26.10 WET SPONGE TESTING

In the case of rubber linings that are normally at least 3 mm thick and may contain scarf joints up to 32 mm in length, wet sponge testing is not recommended. This is because the time that may be required for the electrolyte to penetrate the hole in the rubber or a leak in joint may be rather long. By the time the electrolyte reaches the substrate, the probe may no longer be in a position to complete the circuit. Furthermore, in some cases when the rubber contains a cut, it is possible that the electrolyte will not penetrate that cut under the conditions of the test.

5.26.11 BOND STRENGTH TESTS

In nondestructive testing, the adhesion between the rubber lining and the substrate should be homogeneous and without any defects. This may be investigated by lightly tapping the rubber lining with an appropriate wooden hammer. At areas where the adhesion is broken, a hollow sound will occur. This is no quantitative test method.

In destructive testing, the bond strength between soft rubber and a substrate using test samples should be measured by BS 903, Part A 21- Section 21.1, or ASTM D 429 Method B. Test samples should consist of pieces of substrate, similar to that used in the manufacture of the equipment, to which has been applied rubber of the same mix and thickness used. For the main lining, for preparation of pieces for test purposes from products, see ASTM D 3182.

There is no quantitative test for measuring the bond strength of hard rubber or Ebonite to substrates. A quantitative assessment of the bond can be made by chipping the lining on a test plate with a chisel 25 mm wide at a time when it should be possible to loosen only small pieces of hard rubber or ebonite at each blow.

Minimum load figures for bond strengths obtainable with various types of rubber, vulcanized by different methods, should be as given in Table 5.5.

5.26.12 THICKNESS TESTS

The thickness of the lining applied on a metallic substrate should be determined with a suitable thickness meter and should conform to the specified thickness within a tolerance of ±10%. A minimum of three measurements per square meter should be made. The thickness of the rubber lining applied on concrete or any other nonmagnetic surface should be determined destructively.

5.26.13 PRESSURE/VACUUM TESTING

If appropriate, equipment and piping should be tested with water at a pressure equal to the test pressure mentioned in the related design code and at the maximum allowable service temperature for the particular lining. Alternatively, it should be vacuum-tested at 130 m bars absolute at ambient temperatures. These conditions should be maintained for a period of 1 h. At the end of the test, the lining should be visually inspected. No blisters, cracks, or other surface irregularities should be permitted. Thereafter, the lining should pass the high-voltage spark test.

Table 5.5 Test Loads for Adhesion of Vulcanized Soft Rubber to Carbon Steel and Concrete Substrates

Type of Rubber	Carbon Steel			Concrete (kN/m)
	Pressure Vulcanization (kN/m)	Vulcanization by Hot Water or Steam at Atmospheric Pressure (kN/m)	Self-vulcanization at Ambient Temperature (kN/m)	
Natural or synthetic polyisoprene	3.5	2.7	2.7	1
Styrene-butadiene-rubber (SBR)	3.5	2.7	2.7	1
Chloroprene rubber (CR)	3.5	2.7	2.7	1
Butyl rubber (IIR)	3.5	2.7	2.7	1
Ethylene-propylene rubber (EPR)	3.5	2.7	2.7	1
Nitrile rubber (NBR)	3.5	2.7	2.7	1
Butadiene and/or blends (BR)	3.5	2.7	-	1
Chlorinated rubber (CSM)	2.7	2.7	2.7	1

5.26.14 OTHER TESTS

In addition to previous methods, the following test methods applicable in general to vulcanized rubber, should be utilized as required by the purchaser:

- Tension test; ASTM D 412 ASTM D
- Aging test; ASTM D 573 and D 865 ASTM D
- Immersion test; ASTM D 471 ASTM D
- Abrasion test; ASTM D 2228 ASTM D
- Chemical resistance test; ASTM D 3491 ASTM-D 3491

5.27 BITUMEN, ASPHALT, AND COAL-TAR LINING

This section specifies the following requirements for the lining of equipment using bitumen and coal-tar-derived materials:

- Bitumen and coal-tar materials are suitable for protecting internal surfaces of steel pipes, fittings, vessels, and cementitious or brickwork equipment (see also ISO 5256).
- Bitumen and coal-tar base linings are available for hot or cold application.
- Cold-applied linings consist of liquid solutions of bitumen or coal tar in volatile solvents.

- Hot-applied linings consist of bitumen or coal-tar pitch and filler. It is melted before use and applied in molten form.
- For certain applications as specified by the designer, bitumen and coal tar can be reinforced with mineral fillers or to produce heavy-consistency products suitable for application by trowel or heavy-duty spray.

The applicator of the lining should provide a certificate of inspection and testing when requested.

5.28 PREPARATION

5.28.1 PREPARATION OF MATERIALS FOR USE

Where thinning of cold-applied lining is necessary (e.g., as a first coat on very porous surfaces), only the manufacturer's recommended thinners should be used. For hot-applied lining, the enamel should be heated in suitable agitated heating kettles equipped with accurate and easily read recording thermometers. These thermometers should be checked and adjusted by the inspector whenever necessary. Both solidified and molten enamel should be maintained moisture and dirt free at all times prior to, and at the time of, heating and application.

The solidified enamel charge should be melted and brought up to application temperature, which should be in accordance with the recommendations of the manufacturer. Excess enamel remaining in a kettle at the end of any heating should not be included in a fresh batch in an amount greater than 10% of the batch. Kettles should be emptied and cleaned frequently as required. The residual material removed in cleaning the kettles should not be blended with any enamel.

5.28.2 PREPARATION OF SURFACES TO BE LINED

Complete removal of mill scale, heat treatment scale, previous coatings, and paint, loose dirt, grease, oil, salt, and other substances that could be harmful to the adhesion of lining to steel should be performed by surface preparation. The preparation should be carried out to achieve a quality of at least Sa 2½ as in ISO 8501.

Immediately before the application of the lining, the surface should be free of all traces of abrasive and dust. If metallic surfaces are not to be coated immediately after cleaning, a suitable protective film should be applied to prevent corrosion.

Concrete or other nonmetallic surfaces should be clean and free of oil and any dust or powdery material prior to application of lining.

5.28.3 APPLICATION METHODS

Bitumen and coal-tar lining can be applied in the factory or other workplace on each pipe, fitting, or vessel. Cold-applied bitumen and coal-tar linings are readily applied by brushing, spraying, or dipping as appropriate by the job. For lining porous surfaces such as concrete, one coat of primer should be applied by brush prior to the application of the main lining. Hot-applied bitumen and coal-tar lining can be applied by trowel, spray, rotating or other methods. Considerable skill is required in all methods of application. The materials are heated as needed in a kettle near the application site.

For vertical surfaces, the material is daubed on with a stiff brush, covering small rectangular areas with short strokes and overlapping to form a continuous lining. In weld areas, the brush strokes should be in the direction of the weld; a second coat then should be applied in the opposite direction.

For horizontal surfaces, the material can be poured on and then trowelled out. If unevenness occurs where a smooth surface is required, it may be permissible to direct a blowtorch onto the surface and finish by trowelling.

For lining of pipe, the centrifugal casting method should be used. When hot-applied coal tar is used as a protective interior lining of steel water pipelines, the temperature of the water must not exceed 32°C, and its flow rate must be sufficient to prevent stagnation.

The lining process should be appropriate to the type of lining material.

5.28.4 COLD-APPLIED BITUMEN LINING

Bitumen or solutions are readily applied by brushing or spraying and are often used as priming coats for the heavy-duty materials which can be applied hot or cold at the works or on site. When lining the interior of tanks, use a forced air supply to disperse the solvent and prevent residual solvent condensing on the lining and then washing off.

For lining metallic surfaces, one or more coats of synthetic primer should be followed by at least two coats of bitumen-based lining material to obtain the specified thickness. For lining cementitious or brickwork surfaces, two or more coats of bitumen-based solution should be applied to obtain the specified thickness.

For lining of drinking water tanks or cisterns, two or more coats of solvent-based bitumen lining should be applied to interior surfaces to obtain the specified thickness. The lining material should meet the requirements of BS 3416 regarding the health hazards.

5.28.5 HOT-APPLIED BITUMEN LINING

For the lining of metallic surfaces, one or more coats of cold-applied bitumen primer should be followed by hot-applied bitumen coatings. The application should be performed by rotating the pipe and introducing the lining material in molten state, or by mat, trowel, or spray to vertical surfaces (e.g., vessels).

If the involved parties agree, protection at certain types of joints may be effected by the application of a thick, cold-applied bitumen coating. Hot melt bitumen, when used as a tank bottom covering or as an impervious membrane for acid-proof brick floor lining, can be poured and leveled by squeegeeing in successive coats.

5.28.6 COLD-APPLIED COAL-TAR LINING

All cold-applied coal-tar linings can be applied by brush, spray, or dip. Usually, only two or three coats are applied to obtain the desired dry film thickness. Drying time between coats will range between 16 to 96 h, depending upon the temperature, humidity, and air velocity over the coated surface.

Cold-applied linings generally require no primer. They require from one day to several weeks for the solvent to evaporate and the film to harden.

5.28.7 HOT-APPLIED COAL-TAR LINING

With hot-applied coal-tar linings, all blasted surfaces should be cleaned from dust and grit and should be primed immediately following blasting and cleaning. The use of coal-tar primer that has become fouled with foreign substances or has thickened through evaporation of the solvent oils is not permitted.

In some cases, the application of the primer should be by hand brushing, spraying, or other suitable means and should be in accordance with instruction for application supplied by the manufacturer of the primer.

Spray-gun apparatus to be used should include a mechanically agitated pressure pot and an air separator that will remove all oil and moisture from the air supply. Suitable measures should be taken to protect wet primer from contact with rain, fog, mist, spray, dust, or other foreign matter until completely hardened and enamel-applied.

In cold weather, when the temperature of the steel is below 4°C, or at any time when moisture collects on the steel, the steel should be warmed to a temperature of approximately 30–40°C, which should be maintained long enough to dry the pipe surface prior to priming. To facilitate spraying and spreading, the primer may be heated and maintained during the application at a temperature of not more than 50°C.

The minimum and maximum drying times of the primer, as well as the period between application of primer and application of coal-tar enamel, should be in accordance with instructions issued by the manufacturer of the primer. If the enamel is not applied within the allowed maximum time after priming, the part should be reprimed with an additional light coat of primer, or the entire prime coat should be removed by reblasting and the part reprimed.

During cold weather, when metal surface temperature is below 4°C, or during rainy or foggy weather, when moisture tends to collect on cold surfaces, enameling should be preceded by warming of the primed part. Warming should be done by any method that will heat the part uniformly to recommended temperature without injury to primer. The steel temperature of the part should not exceed 70°C.

The application of the enamel to the internal surface of all pipes other than specials should be by centrifugal casting by either the trough method (see note 1) or the retracting-weir or feed-line method (see note 2). On odd shapes of flat surfaces, the hot-applied coal-tar lining is applied over the dry primer by hand-daubing in shingle fashion using a dauber or on horizontal surfaces with a glass mop.

Notes:

1. In the trough method, the pipe should be rotated and molten enamel should be introduced into the pipe by a pouring trough that extends the full length of the pipe.
2. In the retracting-weir and feed-line methods, the pipe should be rotated and molten enamel should be supplied to the weir or feed line from a reservoir through supply pipes and maintained at application temperature by means of insulation and the use of suitable methods of heating both reservoir and supply line.

5.28.8 APPLICATION OF COAL-TAR ENAMEL TO ENDS OF PIPE SECTIONS

When pipe sections are to be joined together by field welding, a band that is free of protective materials should be left on the inside and outside if any surfaces at the ends of the sections. This band should be 15 cm wide, or as per the specifications, to permit the making of field joints without injury to the lining and coating.

When pipe sections are to be joined with mechanical couplings, bands free of protective materials should be left on the exterior surface if any at the ends of the sections. This band should be 15 cm wide, or as per the specifications, to permit joint makeup. The interior enamel lining should extend to the pipe end.

For bell-and-spigot ends with rubber gaskets, the interior enamel lining should extend from the end of the pipe at the spigot end to the holdback in the bell end. The exterior coating should extend from the lip of the bell to the holdback on the spigot end. The exposed steel surfaces on the inside of the bell and the outside of the spigot end should be given a coating of synthetic primer to a dry film thickness of 0.06 mm ±0.01 mm. For joints other than those specified in the standards, the length of pipe to be left bare at the ends should agree with the specifications.

5.29 DRYING
5.29.1 COLD-APPLIED LINING

Typical drying times for cold-applied linings containing fast-curing solvent can vary considerably. The manufacturer's recommendations for curing times should be followed.

5.29.2 HOT-APPLIED LINING

Water used for chilling the enamel lining following centrifugal casting should not be applied until the enamel has hardened sufficiently to prevent water marks.

5.29.3 TRANSPORTATION

The lined equipment should be handled so as not to cause damage either to the bevels or the lining and coating. During transport of parts to the lined works storage site, all appropriate precautions should be taken to avoid damage to the parts and any lining and coating.

5.29.4 REPAIR OF LINING

Lining material used should be compatible with the previously applied lining. Damaged and nonadherent lining should be removed before effecting a repair.

After preparing and priming exposed surfaces, the lining should be built up, pore free, to the full thickness by troweling or swabbing the molten lining, followed by smoothing to the original contour of the bore. Careful warming of the metal and edges of the existing lining may be necessary to achieve satisfactory adhesion in hot-applied enamels.

5.30 INSPECTION AND TEST METHODS
5.30.1 LINING THICKNESS

The lining thickness should be measured by employing a nondestructive method permitting determination of the lining thickness with an uncertainty of measurement that is not greater than 10%.

5.30.2 ADHESION

Assessment of the adhesion should only be made at a pipe or vessel wall temperature of at least +10°C. Parallel incisions should be made in the lining using a knife. Following this, attempts should be made to lift the strips from the pipe or vessel in order to determine the adhesive strength of the lining. The assessment can only ever be qualitative.

5.30.3 LINING CONTINUITY

The contractor should electrically inspect all interior linings applied or repaired by hand daubing and subjected to traffic or personnel entering the pipe or vessel, or that otherwise exhibit any evidence of physical damage. Any defect in the coating and lining should be satisfactorily repaired at the contractor's expense.

The inspection, which is to be performed on each part, is intended to reveal imperfections in the lining, but not to test the resistance to electrical breakdown of a lining free of such imperfections. The primary input wattage should be no higher than 20 W, and the minimum pulses at crest voltage should be 20/s. The operating voltage of the detector should never exceed 15,000 V.

During measurement, the electrodes (e.g., metal brushes) should be in close contact with the lining surface, since any air gap would falsify the results. The existence of imperfections is indicated by the sound of a sparkover or by the signals emitted by the instrument.

5.31 GLASS AND PORCELAIN LININGS

This section specifies the requirements for the lining of equipment using glass and porcelain lining materials. Glass linings are applied to equipment fabricated in steel, cast iron, or stainless steel. Porcelain enamels are applied to fabricated sheet steel and cast iron in two coat of ground coat and covercoat. For aluminum, neither ground coats nor adherence-promoting oxides are required. Single-coat systems are used for most applications.

The method of application for porcelain and glass lining is either wet or dry process (see also NACE-6H-160). The requirements for design and fabrication of the equipment include the state of preparation necessary for the surface to be lined.

5.31.1 PREPARATION

The glass frit is prepared for use in the glass lining process in two forms: slip and dust. To prepare slip, the frit is ground in porcelain-lined ball mills with specific amounts of water and a suspending agent such as clay. After a period of grinding, a slurry results that is composed of finely ground glass held in suspension by the clay. By the proper use of water and electrolytes, the specific gravity and viscosity are adjusted to obtain the necessary flow and spraying properties. Dust is prepared by dry-grinding the frit in a ball mill to the required fineness.

5.31.2 PREPARATION OF SURFACE TO BE LINED

After the metal has been fabricated and is ready for glassing, it should be placed in the furnace, brought up to a temperature of about 730–900°C (normalizing). After normalizing, the metal should be cleaned and blasted with suitable abrasives.

Table 5.6 Application Method of Glass Lining on Metal

Method of Lining	Type of Lining Material	Base Metal	Notes
Wet process Slushing (dipping or poured) Spraying	Glass slip Glass slip	Sheet steel and cast iron Cast iron	For glass lining of sheet steel, the ground coat and cover coat should be applied with the wet process
Dry process	Glass dust	Cast iron	For glass lining of cast-iron parts, the ground coat should be applied with the wet process and cover coat can be applied by either the wet or dry process

5.31.3 APPLICATION METHODS

Glass material after preparation should be applied in two different coats (a ground coat and a cover coat) over fabricated sheet steel and cast iron and stainless steel. The ground coat is the first coat of glass that is applied to the metal surface. It is formulated specifically for the purpose of promoting adherence to the base metal and is usually not a high corrosion-resistant glass. All surfaces exposed to view or surfaces that come in contact with corrosive media should be covered by a continuous ground coat or appropriate enamel layer which should be smoothly finished.

Cover coat enamels are applied over the ground coat to improve the appearance and proper ties of the coating. They can also applied directly to properly prepared decarburized steel and aluminum substrates.

Two basic methods are used to apply glass enamels to base metal. These include dry-process and wet-process enameling. The ground coat of glass lining usually is applied with the wet process, and the cover coat can be applied with either the wet or dry process (see Table 5.6).

5.31.4 WET PROCESS

The wet process consists of two methods of lining: slushing and spraying (see ISO 28764).

5.31.5 SLUSHING

Slushing consists of either dipping the item to be coated into a container of slip or pouring the slip over the metal surface. The dried coating is then fired in a furnace. This method is most suited for intricate shapes and pipe.

5.31.6 SPRAYING

Spraying is the method used to coat large vessels and accessories. The slip is sprayed onto a clean metal surface, allowed to dry, and the item is then placed in a furnace and the coating fused down. Subsequent coats of glass are sprayed over the fired glass surface and fused.

5.31.7 DRY PROCESS (HOT-DUSTING)

Hot-dusting consists of shifting (powdering) glass dust on one side of a preheated metal surface that has been ground-coated with the wet process. The item is immediately replaced in a furnace and the glass fused down. The process is repeated on the hot item until the desired thickness of glass is reached. This method is mainly used on cast iron items, such as valves and fittings (see ISO 28764).

5.32 CURING

After the spray-dusted or slushed coating is thoroughly dried, the item is placed in a furnace, brought to the firing temperature recommended by the material manufacturer, soaked, and removed from the furance to cool. The ground coat is fired at a higher temperature than the cover coat. Between coats of glass, all radii and rough areas on the lining are ground before applying the next coat.

The heating (firing) procedure used in glassing steel must be carefully controlled. For this reason, the design of the equipment must be such that there are no sudden changes in metal thickness, such as in very heavy flanges.

5.33 TRANSPORTATION

5.33.1 GLASS-LINED EQUIPMENT

During shipment, glass-lined equipment is fastened to skids by means of metal straps or by bolting to shipping legs. It is good practice to keep the equipment on these skids until the vessel has been moved to its final location. Items externally glassed should be shipped completely boxed and properly cushioned. Size permitting, vertical tanks are best handled in the upright position.

When hoisting, it is recommended that a four-leg bridle sling be attached to the skid under the body of the tank. If choker slings are desired, do not hoist an unjacketed vessel using only one sling, as this results in too much load concentration on the vessel. In using two choker slings, it is good practice to distribute the load over a large area. An unjacketed tank should be wrapped with 2.5×15.2-cm wood lagging on the lower half of the tank before applying the choker slings. Larger vessels are shipped with lifting loops welded on.

Under no condition should a vessel be lifted by attaching slings to drive supports or nozzles.

5.33.2 GLASS-LINED PIPE

Lined piping should be handled with due regard to the fragility of the lining, which should include prevention of shock and excessive loads. The ends of lined piping should be protected with covers to prevent the entry of foreign matter and the lined surfaces from being damaged. Welding or arc striking directly on lined pipe should not be allowed.

5.34 INSTALLATION

Do not weld on any metal that has been glassed. When welding in the vicinity of a glassed surface, be sure to protect the glass from flying sparks and weld spatter. During welding on jackets and other accessories, precautions must be taken to prevent high local heat import.

Pipe connections to glass-lined equipment should be made only after the vessel has been leveled and securely fastened to a foundation. To avoid stress failures in glassed pipe, the pipe must be adequately supported by means of pipe hangers and allowance must be made for expansion of the lines if surface temperature is appreciably above room temperature. Enough pipe hangers must be provided so that the weight of the pipe and its contents are carried by the hangers rather than on the nozzles of the vessel.

The high firing temperatures, as well as the stresses introduced by glass coating, tend to distort nozzles on vessels. Likewise, long lengths of pipe tend to bow. The first step in installing pipe is to rotate the pipe or turn it end to end to find the best fit-up. If the alignment is extremely poor, glass-lined wedges or porcelain spacers should be employed. Small misalignments can be compensated for by shimming gaskets.

The tightening of split flanges and bolts should be done carefully and evenly using a torque wrench. Uneven and excessive tightening can cause the glass to spall off of radii. A glassed flange with four bolts or more should be tightened evenly with alternately, diametrically opposite "1, 3, 2, 4" tightening of the bolts. A glass-surfaced, flat-faced flange should not be bolted to a raised-face flange because of the hazard of snapping the glassed surface about the fulcrum edge of the raised-face flange.

U-bolts should not be used on the pipe support. Where U-bands are used, the tightening torque should be approximately 1 kg/m. All glassed flange joints require gaskets. The standard gasket has a combination of hardboard sheet and resilient, semihard material enclosed in a Teflon envelope.

Where warpage is such that the joint cannot be easily sealed, the gasket must be built up or shimmed. To properly shim a gasket, the operator should have special instructions, as the gasket can easily be damaged. Where pressures are excessive, shroud rings or metal reinforced gaskets may be installed to keep the gasket from blowing out.

For installation of glass-lined water heaters, see DIN 4753-Part 3.

5.35 REPAIR OF LINING

Because glassed equipment is unique in its method of fabrication, repair techniques common to other solid materials of construction cannot be used. Repair materials other than glass must be used in such cases, even though they may not have the nearly complete inertness to chemical attack, as does the original glass lining. A change in chemical conditions (severe or mild chemical service), therefore, may require a change in the repair materials within any given chemical process.

5.35.1 SEVERE CHEMICAL SERVICE

Temporary repairs consist of cements applied directly to the prepared surface in the form of air-drying liquids or puttylike mixtures. Only one group of silicate (or ceramic) cements has been found to have sufficient adhesion for this type of repair. Other cements, such as the furan resin and polyesters, have sufficient chemical and temperature limitations, but because of their lack of adherence to glass, they should be used only with suitable metals. Thus, the latter cements involve permanent rather than temporary repairs.

For maximum adherence and serviceability, the silicate cements require a 24-h application time, including setting and acid treatment. They are resistant only to strong acids and thus should not be used with dilute acids, water, or alkaline solutions. The maximum temperature limitation is in the 175–185°C range.

Permanent repairs consist of metal patches in the form of discs, plates, sleeves, caps, boots, and other elements held on by means of studs and nuts and separated from the glass by a suitable gasket (usually polytetrafluorethylene). Some suitable cement is necessary to prevent seepage. The metal selected must be satisfactory for the chemical conditions involved. Materials in current use include:

- Tantalum
- Silver
- Hastealloy steel
- Zirconium
- Nickel
- Titanium
- Molybdenum
- Stainless steels

The chemical limitations of this second type of repair must be determined by the metal selected. Under the proper mechanical and chemical considerations, such repairs are suitable up to 230°C. **Caution:** Two different metals may set up galvanic cells when immersed in the same continuous electrolyte. Abnormal deterioration of one or both of the metals may result.

5.35.2 MILD CHEMICAL SERVICE

Temporary repairs with silicate cements may be used under mild service conditions if the acidic concentration is suitable. The rather long installation period limits the use of this type of repair for mild services except in emergency situations. Since mild service is normally understood to be less than 52°C, there is no maximum temperature limit for these cements under mild service conditions. When special techniques are employed, epoxy compounds may be used to repair glass-lined equipment under mild service conditions.

Permanent repairs for this service are the same as for severe chemical service, except that the gasket may not be used. Generally, the less expensive metals are selected. The temperature of this type of repair is limited by the maximum equipment operating temperature which is usually 50°C.

5.35.3 INSPECTION AND TEST METHODS

The visual inspection of all vitreous enameled parts should be performed under diffused artificial illumination from daylight fluorescent tubes of between 30 and 50 lumens per square foot.

The quality of the finish should comply with the following requirements:

- Cracks—The lining surface should contain no cracks.
- Flaking—The lining should not have flaked off any lined surfaces. The quality of the finish may show the following imperfections subject to the conditions stated.
- Hairlines and strain lines—Hairlines or strain lines of the ground coat showing through the topcoat with no break or crack in the lining are permitted, provided that they do not detract from the general appearance of the appliance.
- Tears (beads)—these are permitted, provided that they do not detract from the appearance or function of the appliance in service.

- Runs (drain lines)—these are permitted, provided that they do not detract from the appearance or function of the appliance in service.
- Specks and inclusions—Specks showing on or through the enamel surface are permitted, provided that they are not concentrated in one area and are not greater than 0.75 mm in diameter, and provided that they do not detract from the general appearance of the appliance.
- Pinholes—There should be no holes in the lining surface that can be shown to extend to the base metal. For pinholes that do not penetrate the ground coat, the requirements given for specks and inclusions apply.
- Blisters—These are permitted, provided that they are unavoidable due to the particular design or fabrication of the article and that they do not detract from the appearance or function of the appliance in service.
- Depressions and raised areas—Smooth, well-covered depressions or raised areas in the enamel surface are permitted, provided that they are small and widely spaced.
- Orange peel (ripple)—this is permitted, provided that it does not detract from the general appearance of the appliance.

5.35.4 THICKNESS TEST

The thickness of glass lining should be measured as specified in DIN EN ISO 2178. Measurements should be made at five different points for each square meter of the inside surface of the vessel. The result should be stated as the maximum, minimum, and average values.

5.35.5 CONTINUITY TEST (ELECTROSTATIC TEST)

Proper initial inspection prior to assembling and field inspection after the assembling of glass-lined equipment is very important. Visual inspection normally suffices, especially if a satisfactory preventive maintenance schedule is being followed. When unusual circumstances prevail or questionable areas are apparent, the electrostatic inspection should be used.

The value of the test voltage on initial inspection of glass-lined chemical equipment is 20 KV, DC. This voltage definitely ensures a minimum sound glass thickness in addition to guaranteeing a continuous glass lining. A value of 5 KV, DC (60 cycles) is used to inspect storage vessels and field inspection of chemical equipment where only a minimum thickness of glass is required.

5.35.6 RESISTANCE TO IMPACT STRENGTH

Resistance to impact strength should be tested in accordance with ISO 4532 with a spring-load of 10 N.

5.35.7 RESISTANCE TO ABRASION

Resistance to abrasion should be tested in accordance with BS 1344: Part 4.

5.35.8 RESISTANCE TO THERMAL SHOCK AND HEAT

Resistance of glass-lined equipment to thermal shock should be tested in accordance with BS 1344 Part 1, and resistance to heat should be tested in accordance with BS 1344: Part 7.

Table 5.7 Chemical Resistance of Glass-Lined Equipment	
Resistance to Citric Acid	**ISO 28706-1**
Resistance to products of combustion containing sulfur compounds	BS 1344: Part 3
Resistance to alkali	BS 1344: Part 6
Resistance to detergent	BS 1344: Part 5

5.35.9 CHEMICAL RESISTANCE

Chemical resistance of glass-lined equipment should be tested as described in Table 5.7.

5.35.10 LEAK TESTING

Leak testing for the completed piping should be conducted by using air or N^2 gas in the following conditions (the hydrostatic pressure test should not be required):

- Test pressures should be 1.1 times the design pressure.
- The test pressure should be maintained for 10 min.

5.36 CERAMIC LINING

This section specifies the requirements for the lining of equipment using ceramics. The ceramic types that can be used for lining include silicate-based, oxide-based, carbide-based, silicide-based, phosphate-bonded material, and cermets.

Ceramic lining applies to equipment and construction fabricated in metal, concrete, and brick (see Table 5.8).

5.36.1 DIPPING

Dipping can be used for all parts, including rivets or spot-welded assemblies, except for those assemblies in which the faying surface would be inadequately covered by the slurry.

5.36.2 FLAME SPRAYING

Flame spraying can be applied to work pieces in a wide range of sizes and shapes. The three methods of flame spraying to the substrate surface are combustion flame spraying, plasma-arc flame spraying, and detonation-gun spraying.

The first two methods utilize coating materials in powder or rod form; detonation-gun spraying uses only powder materials. After preparation and prior to ceramic lining, the sprayed metal coating, masking tape, rubber or sheet metal, depending on the severity of the surface roughening operation should be applied. Sprayed molybdenum or nickel-chromium-alloy undercoating can be used in thicknesses ranging from 0.05 to 0.3 mm to provide an optimum physical bond for the ceramic lining.

Table 5.8 Application Methods of Ceramic Lining on Metals

Application Method	Type of Lining Material	Surface Preparation of Metallic Surfaces
Air-spraying	Silicate-based, phosphate-bonded	Degreasing and blasting or pickling
Flame spraying: Combustion flame spraying	Silicate-based Silicide-based Oxide-based Carbide-based	Blasting and metallic Spray undercoat
Plasma arc-flame spraying	Oxide-based	Blasting and metallic
Detonation-gun spraying	Carbide-based	Spray undercoat
Dipping	Silicate-based	Degreasing and pickling or blasting
Troweling	Oxide-based (colloidal) Phosphate-bonded Silicate-based (soluble)	Degreasing and blasting
Cementation: Packed cementation	Carbide-based Silicate-based Silicide-based	Power tool cleaning Degreasing and water blasting
Fluidized bed cementation	Silicide-based	Etching and blasting

5.36.3 CEMENTATION

Pack cementation and the fluidized-bed process are two types of cementation processes employed in ceramic lining. These processes are used to produce impervious, oxidation-resistant coating for nickel-based, cobalt-based, and vanadium-based alloys, and refractory metals.

5.36.4 TROWEL LINING

Troweled linings are used for furnaces, hot-gas ducts, and certain repair patches on other ceramic linings for relatively short service exposure. Surface roughening is accomplished by blasting or degreasing, or by attaching reinforcements such as wire mesh, corrugated metal, angular clips, or honeycomb structures. Reinforcement usually is required for surfaces having a finish of less than 6 μm.

5.36.5 CERAMIC LINING ON CONCRETE SUBSTRATE

For single-layer linings, the ceramic material should be applied over an anchoring system. With dual-layer linings, the membrane should be applied to the recommended thickness, ensuring that all studs are completely coated and that the linings are free of pinholes.

5.36.6 CERAMIC LINING ON BRICK SUBSTRATE

The preferred method of application in this case is to gunite the material. Guniting allows the material to be applied under pressure, this enabling the material to be packed into the 13-mm open joints and allowing the system to be supported by the studs or keying into the joints.

5.37 CURING

Methods of curing should be agreed upon by the purchaser and applicator.

5.38 TRANSPORTATION

Ceramic-lined equipment should always be handled carefully. Minor impact and bending normally have no effect on linings.

5.39 REPAIR OF LINING

Repairs of damaged ceramic lining should be performed by the contractor.

5.40 INSPECTION AND TESTS

5.40.1 VISUAL INSPECTION

Although visual inspection or comparison is of only limited usefulness, many plants prepare samples of lining with surface defects that are known to be deleterious to the protective value and service life of the lining and use these samples as visual comparators.

5.40.2 CONTINUITY OF LINING

High-temperature testing and fluorescent-penetrant testing are the test procedures for determining lining continuity and oxidation resistance.

5.40.3 HARDNESS AND STRUCTURE TESTING

The microscope is a useful tool for observing the structure of linings, and hardness testing gives a direct measure of the inter particle bond strength. Accepted Vickers hardness values of aluminium oxide deposited by various methods are 600–800 for flame-sprayed linings, 700–1,000 for plasma-sprayed linings, and 1,000–1,200 for detonation-gun-sprayed linings.

5.41 BRICK AND TILE LINING

This section specifies the requirements for chemical-resistant brick and tile lining for process equipment. However, it does not cover refractory brick lining.

The chemical-resistant brick lining is a multilayered system supported by a shell.It consists of an impervious membrane to prevent the corrosive medium reaching the shell, and one or more layers of chemical-resistant brick laid in a chemical-resistant cement.

Brick lining applies to equipment fabricated in metal or concrete. Extra care is required for the operation of brick-lined equipment, particularly during startup and shutdown, when its operational limitations should be considered. Depending on the selection criteria, it may be sensitive to changes in pressure, temperature, and acid concentrations, which could cause damage or even collapse of the lining.

The applicator of the lining should provide a certificate of inspection and testing.

5.42 **PREPARATION**

5.42.1 **PREPARATION OF MATERIALS FOR USE**

Bricks and tiles should be stored near the job site under the same temperature conditions as the equipment to be lined (between 20°C and 25°C) for approximately 48 h before using, to avoid temperature and humidity changes during the lining process. The bricks and tiles should be clean.

5.42.2 **PREPARATION OF EQUIPMENT AND SURFACE TO BE LINED**

All steel equipment should be hydraulically pressure-tested in accordance with the applicable code to confirm adequate strength and liquid tightness before the application of membrane and brick linings (at least 48 h). All loose and foreign materials, particularly oil and grease, should be removed. When necessary to remove mill scale, rust, or other contaminants, blast cleaning to Sa 2 in accordance with SIS 05 59 00 should be performed.

All projections and welds should be ground smooth, weld spatter removed, and all corner welds ground to a 4.7-mm minimum radius. The prime coat of the adhesive system (for bonding membrane) should be applied immediately after blasting to prevent rusting.

Concrete equipment should be water-tested to ensure liquid tightness before commencement of the brick lining. Defects that could enable water to enter the vessel should be repaired with synthetic resin injection or other methods.

To avoid air inclusions and to ensure sound attachment of the lining, the concrete substrate should be free of loose sand, dust, laitance, oil, grease, or other contaminants. This can be achieved by means of blast cleaning or mechanical steel brushing; pinholes should be opened by blasting.

The moisture content of the concrete should not exceed 5% by volume. Generally, this may be reached after 28 days of hardening and drying. The moisture content of the substrate should be checked regularly during the installation of the lining. Measuring equipment should be calibrated, and the method of establishing moisture content should be approved.

Concrete surfaces that have been attacked by chemicals should first be prepared by neutralizing or, if necessary, by local replacement of concrete. Repairs should be carried out after consultation with the specialist.

Small defects (up to a depth of approximately 50 mm) should be sealed with a quartz-filled epoxy mortar (with a composition of 75% volume quartz and 25% volume resin). Larger repairs should be performed with nonshrink, cement-based mortars.

The surfaces of repaired defects should be smooth and flush with the surrounding surfaces. The final surfaces should be smooth and even, without any sharp edges. The walls and floor should not bulge inward because that could cause the brick lining to break away as a result of uneven expansion during the operation.

5.43 METHOD OF LINING

The design of brick linings should ensure that the thermal, chemical, and mechanical effects of operation do not cause cracks to develop, thereby invalidating the lining as a corrosion barrier. Bricklaying is to be performed with care. The only method for inspection and testing of the applied brick lining is by visual examination.

Since bricks have to be laid by hand, the dimensions of the equipment to be brick-lined should allow sufficient room for a person to work inside it with reasonable freedom. The minimum free diameter recommended is 600 mm.

Linings for large flat surfaces will need to be thicker than curved surfaces because in the latter case, the curvature contributes to the overall strength. It will be necessary to provide supports in cases where the lining cannot be supported by the contour of substrate, except where the mass of the brick lining is low enough to rely on the adhesion of the cement.

Inaccurate shaping of the substrate will cause unexpected compressive stresses and buckling of the brick lining. Stirred vessels or vessels with flat bottoms should have shaped bottom linings made of straight bricks. The vessel wall lining should not rest on the bottom lining unless more than two layers are applied on the bottom. Direct passage to the vessel wall should be avoided as much as possible.

The internals should not be supported by the vessel wall through the brick lining. Instead, they should rest on the brick lining by incorporating special console bricks (see Figure 5.7).

Supporting grids for an internal packing bed can be from corrosion-resistant metal or alternatively fabricated in situ from hexagonal ceramic elements jointed together with cement (see Figure 5.8).

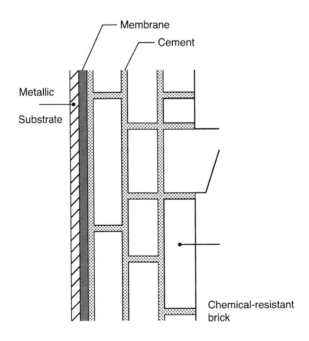

FIGURE 5.7

Support ring of preshaped brick (internal structure).

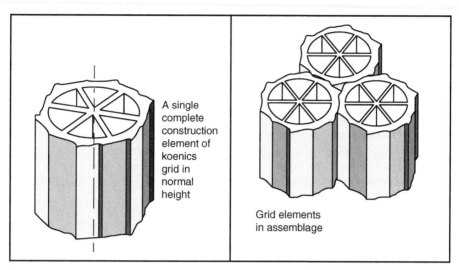

A single
complete
construction
element of
koenics
grid in
normal
height

Grid elements
in assemblage

FIGURE 5.8

Support grids assembled from constructional elements.

Metal grids are easily removable. However, there are some disadvantages to their use, including vulnerability to mechanical damage by shock or concentrated loads, the small free passage area, and (especially for larger diameters) the low load-bearing capacity.

Ceramic element grids can be used for large diameters providing high load-bearing capacities with good free passage areas. The disadvantages to using them are that the vessel needs to be divided into different parts, sometimes requiring extra manholes; and because of the large openings, intermediate layers may be needed (e.g., stacked rings followed by dumped rings). The necessity to support the grid during construction and hardening of the cement joints also needs to be considered.

Internal elements independent of the lining, such as inlet pipes, spray nozzles, distribution trays, and hold-down trays, should be installed so that they do not impede thermal movement of the brick layers, either by resting on them or by being locked into the brick lining. They should be fabricated from materials with the appropriate chemical, thermal, and mechanical properties.

The brick lining should be constructed in such a way that the lining does not tear away from the vessel wall. Since the coefficient of thermal expansion of steel is about twice that of the brick lining, the steel shell will tend to expand more than the lining, thus introducing tensile stresses in the lining. This effect is increased by internal pressure. Since the tensile strength of brick and cement is low, as is the bond strength of cement, cracks will develop unless special precautions are taken.

With low operating temperatures, benefits can be derived from making the lining thicker so that the expansion of the steel shell is equal to that of the outer layer of bricks, or to the average expansion across the brick lining. At higher operating temperatures, tensile stresses may become excessive and a correspondingly thick brick lining would add considerably to the cost. In practice, good results are being obtained with relatively thin brick linings because the membrane that is generally used has a low heat conductivity that keeps the steel shell temperature low, and it has also a high coefficient of thermal expansion so that it is able to compensate for the difference in thermal expansion between the steel shell and brick lining. However, the maximum allowable temperature for the membrane should not be exceeded. A lead membrane is not recommended for this application.

The minimum diameter that can effectively be brick-lined is 600 mm. Smaller diameters for lining should be prefabricated in short sections and assembled with flanges.

The contractor responsible for the brick lining should ensure before the process begins that the specified lining thickness, together with the required final internal dimensions (after lining), can be realized and that the dimensions of the object to be brick-lined are correct. The temperature of equipment to be brick-lined should be kept between 18°C and 22°C. Higher and lower temperatures will influence the correct curing of the cement. When the equipment to be lined is at a temperature above 22°C, the cement should be mixed in small quantities in some other location and kept between 15°C and 20°C before use. Temperatures that are too high influence the pot life of cement unfavorably.

For conditions below 15°C, the equipment's temperature should be raised, preferably by electric heating, to avoid uncontrolled moisture development. Painting of the outside surfaces of equipment with a high-reflecting white coating will reduce the uneven heating effect of the sun. A lightweight shield will reduce the effects of sun, rain, and wind on the surface.

Condensation is not allowed on the steel or concrete substrate, the membrane, or the installed layers of the lining. The substrate temperature, therefore, should always be at least 3°C above dew point; relative humidity should not exceed 85%.

The surface temperature and relative humidity of the air should be controlled by electric heaters and air-drying equipment. A daily record of the working conditions should be kept. Cement should remain free of contact with water and vapor.

5.44 APPLICATION OF LINING FOR METALLIC EQUIPMENT
5.44.1 MEMBRANE LAYER APPLICATION

Membranes should be continuous, watertight, and sufficiently flexible, for which a proven installation procedure and careful application are necessary. The liquid tightness should be tested after installation; the procedure and test equipment should be approved by the company. When applied on a metal substrate, the membrane can be spark-tested by using a direct electric current that will produce a spark wherever the membrane is not liquid-proof. The voltage should not exceed 30 kV and is based on the following formula:

6 (1 + thickness in mm) = kV

Antistatic linings on metal substrates should be inspected with the wet sponge test (a low-voltage holiday detector). The membranes should also be visually inspected for air inclusions (blisters), cracks, and other imperfections.

5.44.2 LEAD MEMBRANE

For the adhesion of lead to a steel substrate, the steel surface should be lightly tinned with a 0.02–0.05-mm-thick layer that should be free of pores. To obtain good adhesion, within 24 h before application of the tin, dirt, grease, and rust should be removed thoroughly by blast cleaning or etching with hydrochloric acid.

The requirements for the installation and testing of lead membrane linings should be in accordance with DIN 28058. The tightness of the lining should be verified with the sulfuric acid indicator test. For this purpose, the lead surface is primed with a solution of 20% sulfuric acid that is washed away with clean water after 3 h. If there are pores in the lead lining, they will be filled with the acid solution. If a mixture of water and Congo red having a pH indicator in the 3–5 range is then applied and allowed to dry, the pores will show up blue against the red surface.

The minimum thickness of a homogeneous lead lining should be 6 mm, with a tolerance on thickness of 0 to 25%. This can be checked with a magnetic layer thickness meter (nondestructive), or by taking a measurement after locally melting the lead (destructive). Poor adhesion of the lining will reduce the heat conductivity, which could cause further detachment of the lead lining due to temperature changes and pressure/vacuum variations during operation. The adhesion should be ultrasonically tested from the outside (steel) side of the equipment, using examples of both correct bonding and poor adhesion for interpretation.

5.44.3 POLYISOBUTYLENE (THERMOPLASTIC) MEMBRANE

The application of a polyisobuylene membrane should be in accordance with the manufacturer's instructions, which generally would be as follows:

Separate polyisobutylene sheets of 3 mm minimum required thickness are joined together by welding to form the membrane. This is attached to the substrate with an adhesive (glue). The separate sheets and the substrate are both coated with the adhesive but the overlap of the sheets to be joined, 30 mm wide approximately, should be kept free of adhesive.

Between about 1 h and 12 h, depending on the temperature and type of adhesive, the sheet of membrane material will be ready for sticking to the substrate. To avoid air inclusion, the sheets should be positioned from the center to the sides using a suitable wooden tool to avoid damage; preheating of the sheets will facilitate their installation.

The separate sheets should then be jointed at the overlap by welding in accordance with the manufacturer's instructions, which is generally done by roughening with sandpaper, cleaning the weld areas with a suitable solvent and welding with hot-air welding equipment, the air being directed by a tapered mouthpiece at a temperature between 300°C and 350°C. When the surfaces become soft, they are pressed together with a roller, and the seam should almost disappear. Vertical seams should be welded and rolled downward to release any remaining solvent.

For severe chemical conditions, the weld seams should be reinforced by welding an additional strip over the completed seam. For equipment with angular corners, the membrane should be reinforced with a welded, corner-shaped patch covering a small triangle on all three sides, forming a corner that is stuck or welded over the membrane.

Small damaged parts of the membrane can be repaired by welding patches of the same material over the spot. Larger damaged parts should be removed and the substrate prepared, after which a new piece can be inserted with adhesive and welded to the surrounding material.

Thermoplastic membranes should be inspected by spark testing, or with a wet sponge, low-voltage tester. Checking adhesion by careful knocking on the surface can be difficult. Therefore, testing after installation may be carried out by filling the equipment with water and raising its temperature to 70°C during a period of 24 h. After draining, imperfections in the adhesion will be seen as blisters or bulges on the surface of the lining.

5.44.4 REINFORCED EPOXY MEMBRANE

Glass fiber–reinforced epoxy resin (thermosetting material) should be used for membrane with the minimum required thickness of 4 mm. Dosing and mixing of the epoxy resin components and installation of the alternative layers of resin and glass fiber should be done in accordance with the manufacturer's instructions. The clean, prepared surface of the substrate should first be primed with the selected epoxy resin, and any remaining holes and cracks in concrete surfaces should be filled with resin.

The laminate is applied "wet-in-wet," and any resin layer that is allowed to cure completely should be lightly blast-cleaned or roughened with sandpaper before the next layer of the laminate is applied. Resin and glass fiber are applied to the surface, and the resin is evenly distributed by rolling and pressing. The glass-fiber reinforcement is soaked completely, and air is removed.

The glass fabric reinforcement should be installed with an overlap of between 25 and 50 mm. Mixing equipment should be calibrated for the quantity of the components to be mixed; dosing and mixing should be carefully performed in accordance with the manufacturer's instructions. To obtain good adhesion of the membrane to a brick lining, the final and sealing layer of thermosetting resin is treated with silver sand.

5.44.5 BRICK AND TILE LAYER APPLICATION

To prevent damage to the bottom membrane, the brick lining of the bottom should be finished first. Brick linings in vertical equipment are built up ring upon ring, with the bricks placed tightly against the membrane. If practical, the brick lining for horizontal equipment should be installed with the cylindrical part of the equipment placed in a vertical position.

For horizontal vessels, both the axial and circumferential joints should be in line. For vertical equipment, the vertical joints need not necessarily be in line. For linings consisting of more than one layer of brick or tile, the joints of the layers should be staggered. Normally, the same cement is used for bedding against the membrane and for the axial and circumferential joints. Wedges should be used for joints that are to be filled later with a different cement. The installation rules are equally applicable for alumina-based, acid-resistant bricks and tiles, porcelain tiles, carbon and graphite bricks, and special ceramic lining materials.

Joints between the bricks of chemical-resistant brick linings should be as small as possible to obtain good strength and resistance. The joint space given hereinafter should be strictly observed, particularly when the swelling properties of the cement involved have been determined in the design (see Table 5.9).

Table 5.9 Joints of Chemical-Resistant Brick Lining with Cement

Cement	Bed Joints	Axial and Circumferential Joints	Joints to Be Filled Afterward
Silicate-based cement	5 mm maximum × 8 mm	3 mm	7 mm
Synthetic resin-based cement	5 mm maximum. × 8 mm	5 mm* (for prestressed constructions, the design instruction for axial joints should be adhered to)	7 mm; joint depth 15 mm

Air inclusions in the cement behind or between the bricks should be prevented. The cement should be placed on the brick being laid, as well as on the surface to be lined, and when the same cement is used for bed, axial, and circumferential joints, against the side of the installed bricks. The joints are then filled with the positioning of the brick to achieve a homogeneous filling of joints, the surplus cement being removed immediately. When curing with hot, dry air is to be applied, the curing process for steel vessels can be started during application of the bricks as follows.

The metal wall temperature should be maintained as high as possible; i.e., at approximately 35–40°C during the application of the brick lining. To avoid obstruction of the lining work, and so as not to influence the pot life of the cement, heat should be applied on the outside of the metal walls. Drying of the bed cement layer should be controlled so that it does not harden too quickly.

5.44.6 CHEMICAL-RESISTANT CEMENT LAYER

The dosing and mixing of components should be strictly in accordance with the manufacturer's instructions. Using other procedures could disturb the chemical reactions and curing, giving different chemical properties to the cement. Improvement of the processability of the cement by the application of a modified mixing ratio is allowed only within the manufacturer's limits or with the approval of the company.

Different types or qualities of cements should never be mixed. Mixers and tools should be kept clean and dry to prevent contamination of the cement.

5.45 APPLICATION OF LINING FOR CONCRETE CONSTRUCTION
5.45.1 EXPANSION JOINTS

Expansion joints are the weakest parts in chemical-resistant brickwork and tiling. Therefore, they should be installed outside zones of chemical attack. If this is impossible, it is recommended that they be located to minimize the chance of aggressive products permeating them.

At the expansion joint, the reinforced concrete should have a 10-mm-wide gap that should be filled with semirigid polyurethane foam, insulation cord, or other appropriate material. The concrete fill applied on top of the concrete to provide the required stop for drainage should have a gap at the same location and of the same width. This joint should be sealed with a plastic (e.g., polyisobutylene foil or other suitable material). The foil should be adhesive-bonded to the substrate; the adhesive should be a bituminous or rubber type.

The membrane and a layer of brick or tile should then be applied, keeping the joint open. Finally, a rubber or plastic seal should be fitted with a suitable adhesive in the joint to prevent penetration of liquid.

5.45.2 MEMBRANE LAYER APPLICATION

Prior to the application of a membrane or a coat of primer, the concrete should be at least 28 days old and should be free of dust, oil, grease, and other contaminants. Asphaltic bitumen membranes should be applied to a primed surface by squeegee or brush until it is smooth, even, and free of irregularities. The surface of the membrane should be sanded for good adhesion of the subsequent cement layer [e.g., by brushing with a solution of bitumen and spreading quartz sand (0.7–1.2 mm grain size) onto the bitumen coating while it is still tacky]. The minimum dry film thickness (DFT) should be 20 mm. The main thermoplastic membrane material is polyisobutylene. It should be adhesive-bonded to the cleaned concrete surface. The sheet should be joined either by adhesive bonding or welding; vulcanizing is not required. The minimum DFT should be 3 mm, and other thermoplastic membranes should not be used.

For specific chemical conditions, cold-cured epoxy-resin-based membrane should be used, if necessary with glass-fiber reinforcement. The clean and rough concrete surface should be given an epoxy-resin-based primer and within 24 h after application of the primer. The epoxy resin, and the glass-fiber reinforcement, if any, should be applied the minimum DFT of this membrane should be 2.5 mm. Applying an epoxy membrane during rain or at temperatures below 10°C is forbidden.

5.45.3 BRICK AND TILE LAYER APPLICATION

Bricks and tiles should be clean and dry and should be at a temperature of at least 15°C when being applied. If a brick lining has to be applied in winter, provisions should be taken to protect the area where the bricklaying takes place from cold, rain, snow, and other conditions.

Tiles used for floor, trenches, and neutralization pits should be at least 30 mm thick. For walls in pump houses, the minimum thickness of tiles should be 20 mm. For narrow joints, the bricks or tiles should fit correctly, which requires that they should be selected at sites with regard to their squareness and dimensions. Vertical parts should be lined first, and then the horizontal parts. Acid-resistant bricks or tiles should be applied to pump foundations before the bricks or tiles are laid on the adjoining floors.

5.45.4 CEMENT LAYER APPLICATION

Cements should be mixed in accordance with the supplier's instructions. The tools and mixer should be clean and dry. Different types of cements should never be mixed. The cements should not be applied under freezing conditions.

5.45.5 HYDRAULIC CEMENT

Hydraulic cement can be used as a bedding mortar (with a cement/sand ratio that is usually 1:3 by volume) for tiles in mildly aggressive conditions. If layers of a hydraulic mortar are applied to make slopes, they should be kept wet during curing (for about one week) to obtain the optimum strength and to avoid hair cracks. A hydraulic cement that is delivered in paper bags (50 kg) should be used within 8 h of opening the bag.

5.45.6 SILICATE CEMENT

Silicate cement may be used as a bedding material, applied on an asphaltic bitumen membrane. The joints between bricks and tiles should then be sealed with a resin-based cement.

Four days after application, the brickwork should be washed with dilute acid; e.g., a 10%-wt. solution of hydrochloric acid. This treatment is important, since the alkali hydroxide formed during curing is detrimental and would eventually destroy the joint. Silicate cements do not adhere to rubber membranes.

5.45.7 CEMENT BASED ON PHENOLFURFURALDEHYDE RESIN

Cements with a phenofurfuraldehyde resin base are supplied as two components: a liquid and a powder, which should be mixed thoroughly and used immediately. The cements are used for both embedding and sealing of the joints between bricks and tiles. The rate of setting and curing of the cement is influenced by temperature. At 15–20°C, the mortar starts to set in about 4 h and cures in 1–2 days. At lower temperatures, the mortar starts to set and cure at a slower rate. If the temperature falls below 15°C, consideration may be given to accelerating the curing by heating. However, care should be taken to ensure that the temperature does not exceed 80°C; otherwise, the difference in expansion between the substrate and the top surface may adversely affect adhesion.

In order to give the cement its full chemical resistance, especially with caustic alkalis, the cement requires a heat treatment at 80°C for 24 h after it has fully cured. Contact with water or water vapor during curing should also be avoided. The heating, therefore, should be performed by using electric heaters. It is essential that during curing, the cement does not come into contact with free alkali, since this alkali would tend to neutralize the acid catalyst. Consequently, the concrete floor should be primed with two coats of a suitable primer when these cements are used as a membrane. The primer should be in accordance with the resin cement manufacturer's recommendations.

5.45.8 CEMENTS BASED ON FURANE RESIN

In general, the properties of furane cement resemble those of the cements based on phenolic resin, but curing at high temperatures to obtain full chemical resistance is not necessary, and they are somewhat easier to apply.

Cements based on furane resin cannot be applied directly to concrete. When a membrane of this cement is to be applied, the concrete should be pretreated with a primer in accordance with the resin cement manufacturer's instructions. For application of these cements, the same rules apply as for the application of cements based on phenol-furfuraldehyde resin.

5.45.9 CEMENTS BASED ON POLYESTER RESIN

Cements based on unsaturated polyester resin are supplied in the form of a powder and a liquid resin, which should be mixed immediately before use. These cements should not be mixed or applied under freezing conditions. They are self curing at 15–20°C, and a complete cure at this temperature can be obtained in 24 h. The curing time and the pot life are affected by temperature. Contact with water or water vapor during curing should be avoided.

5.45.10 CEMENTS BASED ON EPOXY RESIN

These cements are generally supplied as a paste of puttylike consistency with a liquid curing agent. After the two components have been mixed, the cement cures within 1 h at temperatures of 10–30°C. The curing time is affected by temperature. The cement can be used for continuous floors and for embedding and sealing purposes. Contact with water or water vapor during curing should be avoided.

5.45.11 JOINTS

The cement layer between the bricks or tiles and the membrane should have a thickness of about 5 mm. The joints between bricks or tiles should be as small as practicable, preferably not more than 3 mm wide. However, wider joints of 5–7 mm will be required; e.g., for certain hot-pour jointing materials and when joints between the bricks and tiles are to be sealed. When rejoining may be required after a period of service, the joint should be made 5 mm wide. The width of the joints should be consistent over the full depth of the joint and free of cavities.

5.46 CURING AND PRESTRESSING

Chemical-resistant brick linings cannot resist high tensile and bending stresses. Proper curing and prestressing, for steel vessels only, should result in compressive stress in the brick lining while the carbon steel remains under modest tension. To obtain good chemical resistance, the cement should be completely cured. However, curing or prestressing treatment should not begin within 8 days after installation of the brick lining is complete, and it should be finished within 6 to 8 weeks after the installation.

5.46.1 CURING WITH ACIDIC LIQUID

Curing with acidic liquid is applied for brick linings operating at ambient temperatures. Considering the time restraints, the vessel is filled with an acidic liquid. After 3 weeks, it is emptied, washed with water, and inspected.

5.46.2 CURING WITH DRY, HOT AIR

Curing with dry, hot air is applied for brick linings operating at conditions up to 80°C and 1 bar(g). The curing should preferably begin with application of heat at the installation stage.

Considering the time restraints, dry, hot air is introduced in the bottom of the dry and closed equipment. To control proper curing of the cement, especially during the initial period, direct flame heaters should not be used. The air temperature should be raised at 5–7°C/h and the pressure simultaneously increased for equipment that will operate under pressure.

The final curing conditions should be maintained for 16 h, followed by a reduction of 3–5°C/h, reducing the pressure at the same time. Care should be taken that metal wall shrinkage is not less than that of the lining material. When returned to ambient temperature and pressure, the brick lining should be inspected.

5.46.3 PRESTRESSING WITH ACIDIC LIQUID

Prestressing with acidic liquid is a wet-curing process suitable for brick linings operating above 80°C and 1 bar(g) under severe chemical conditions. Considering the time restraints, the vessel is either filled with acidic liquid or the liquid is circulated. The liquid is heated to gradually raise the wall temperature to the operating temperature, while the pressure is raised to the test pressure. Circulating liquids should thoroughly wet all parts of the brick lining, and after maintaining the operating conditions for several hours to cure the cements completely, the temperature and pressure are slowly reduced to the ambient level.

5.46.4 CURING LIQUIDS

In general, diluted acids are used for curing; a solution with a pH value between 2 and 5 is recommended for synthetic resin–based cements. For silicate-based cement, a 5–10% solution of sulfuric acid is suitable; hydrochloric, phosphoric, or acetic acid solutions are also acceptable.

Solutions of calcium chloride, sodium bisulfite, or calcium bisulfite should not be used for silicate-based cements since they require washing to remove insoluble salts that could react with the lining cement. The curing liquid is brought to the required concentration at an ambient temperature in a separate vessel of rubber-lined, thermoplastic, or glass fiber–reinforced plastic material, the liquid is then pumped or circulated to the equipment.

For equipment that will operate under pressure, a curing liquid pressure of 0.5 bar (g). is required. The liquid temperature should be raised with steam injection (either directly or via a sparger) or by circulation through a heat exchanger at a rate of 5–7°C/h until it reaches the operating temperature. To limit expansion of the lining during heating, the temperature differential in the equipment should remain within 15°C. The pressure should be raised simultaneously with the temperature to reach the required operating pressure at about 100°C. Heating with open steam will raise the pH value with the dilution of the liquid; this should be corrected by the draining and addition of fresh liquid to keep the pH below 5.

The final curing conditions should be maintained for 72 h, followed by cooling at a rate between 3 and 5°C/h while gradually reducing the pressure. Care should be taken that the rate of metal wall shrinkage is not less than that of the lining material. Boiling of the liquid should be prevented by controlling the ratio between pressure, temperature, and liquid concentration. After cooling to ambient conditions, the vessel should be left for 3 h. Then, after draining and washing, the lining should be inspected.

The following data should be recorded during prestressing and curing:

- Temperature of the liquid in equipment or liquid inlet and outlet temperature
- Metal wall temperature at three representative points
- Ambient temperature
- Pressure inside the equipment.

Notes:
- Steam injection (either directly or via a sparger) is generally more economical than a heat exchanger circulation system.
- The lining contractor should provide a calculation showing the stresses expected during prestressing and curing.

5.46.5 TRANSPORTATION

Chemical-resistant bricks and tiles should be stored adequately protected for dry and cool conditions at a relative maximum humidity of 70% and a temperature between 5 and 20°C in a frost-free warehouse. When stored during installation in the open air, the material should be stacked on wooden pallets and covered with tarpaulins.

To prevent possible cracking, deformation, and disbanding caused by shock or vibration to the brittle lining materials, the equipment should not be transported or handled after the brick lining is applied. Equipment to be lined with brick should be properly installed before the lining is applied.

If transport of brick-lined equipment or parts thereof (e.g., small vessels, pipe sections, ducting, etc.) cannot be avoided, the design and execution should make allowance for more rigid construction, adequate lifting points, additional internal and external supports, and temporary studs for rigid fixing during transport. The equipment should be completely cured and prestressed before handling.

In the case of unforeseen moving of equipment, stiffening rings or structures should be designed and applied. However, welding on lined equipment should be avoided wherever possible.

5.47 REPAIR OF LINING

5.47.1 REPAIR OF LINING FOR METALLIC SUBSTRATE

To avoid mechanical damage, special protective provisions should be made for cleanout, scaffolding, and brick lining repair activities. Nozzles and manholes should be opened only when required for access, and the required working climate should be realized.

For the execution of local repairs, the remaining lining should be properly supported when bricks have to be removed. Shocks and vibrations of the surrounding brick lining should be avoided.

Even minor defects of brick linings should be consistently repaired to prevent the defects from spreading to deeper layers of the brick, the membrane, and the substrate. This includes the replacement of dissolved or washed-out cement from joints.

Scraping to sound material and subsequent filling with fresh cement is sufficient when the damage is not too deep. When the erosion nearly equals the thickness of the final brick layer of the lining, the affected area should be completely removed and replaced by a new layer to secure the bedding adhesion to other layers.

To repair a leak, disbanding, wide cracks, fallout of bricks, or severe spalling, all the affected material should be removed as far as required to do the following:

- Repair or replace part of the locally affected metal substrate.
- Replace the leaking and affected part of the membrane with an appropriate weld to
- sound membrane material.
- Allow replacement of brick lining rejected by the inspector, and ensure complete and proper bonding to the remaining brick lining.

All surfaces of substrate, membrane, and brick lining should be thoroughly cleaned and dried before any replacement work commences. The adhesion between cement and wet or dirty bricks will be significantly lower than with clean and dry bricks.

The original brick configuration should be maintained upon replacement. Welding on brick-lined equipment should be avoided, since most membranes will be permanently damaged and the brick lining will be affected. The thermal expansion of the substrate during welding will loosen the brick lining, with the possibility of future leakage.

If welding cannot be avoided, the brick lining and membrane should be locally removed up to a minimum distance of 500 mm from the weld. After welding, surface preparation, and drying, proper replacement of membrane and brick lining should take place as previously described.

5.47.2 REPAIR OF LINING FOR CONCRETE SUBSTRATES

Chemical-resistant linings should be regularly inspected for defects. They should be carefully treated and protected against damage by traffic loads, impact, and local chemical and thermal attacks (steam, leaking flanges, etc.). When a defect is detected, repairs should be carried out immediately in order to prevent serious damage to the concrete substrate.

The main defects are spalling of the bricks or tiles, erosion effects, cracks in the lining, and degradation of the chemical-resistant lining materials.

Spalling of the brick lining might be due to any of the following:

- Inadequate brick quality (e.g., composition, porosity)
- Exposure to exceptional operating conditions (e.g., thermal, chemical, or other loads more severe than those foreseen)

Local spalling might be due to impact by a falling object. Impact by mechanical load should always be avoided.

Damaged areas or spots should be repaired by replacing them with new material, either of the same quality or of a quality that is wholly compatible, with respect to physical and chemical properties, with the adjacent original material. If the effects of erosion or attack by chemicals are slight, the joints can be repaired by scraping out to sound material and filling with fresh cement. If the depth on the scraped-out joint is 75% or more of the thickness of the brick layer, all the cement in the joint should be removed and replaced, if necessary by relaying the bricks.

When cracks in the lining are present, they should be opened completely to establish the condition of the membrane and/or the substrate. Degradation of the lining materials may indicate an excessive chemical attack. The chemical conditions that caused the degradation should be ascertained. If defects other than those described above are found, the cause of these other defects should be ascertained to avoid further attack of the concrete construction.

For repair, a sufficient number of bricks should be removed to restore the brick lining configuration.

5.47.3 INSPECTION

The only method for inspection and testing of applied brick lining is visual examination. The equipment should be inspected by experienced personnel in accordance with this manual. Equipment to be assembled from parts should be checked for correct assembly before installation of the lining.

Test pressures for brick-lined equipment should be limited to 10% above the operating pressure, to prevent unacceptable deformations. This equipment should be inspected at regular intervals, observing any local obligations, and also whenever any leak or product contamination occurs.

The inspection should be restricted to visual observation, with consideration of the following:

- The general condition of the brick lining.
- The color of the bricks.
- The level of cement in the joints; excessive chemical attack (e.g., by fluorides) could reduce the thickness of bricks or tiles, which may be indicated by protuberance of the joints.
- Regular shape of the brick lining; disbanding of bed joints could cause irregularities.
- The position of all bricks and tiles, with no loose or displaced parts.
- Cracks; deformation of the equipment due to lack of or improper pretreatment can cause irregularities.
- Spalling; generally, distributed spalling could result from incorrect composition and porosity characteristics of the bricks, or too severe operational conditions caused by frequent temperature or pressure changes. Local spalling could result from direct impact of liquid or vapor jets, causing rapid temperature changes, impact, and the effect of boiling on the interface level.
- Cement condition in the joints; erosion, dissolving or washing out; e.g., for silicate-based cements caused by steam, hot water, or chemical attack.
- Lining in and around nozzles and manholes; when design, location, material selection, installation, special treatment, and operations are correctly done, only minor repairs should be expected.
- When disbonding and spalling of bricks are noticed, this should be further investigated by careful hammer testing.

If defects (e.g.,leakages, disbonding, wide cracks, missing bricks, severe spalling, material reduction, or open joints) are demonstrated by visual inspection or hammer testing, a further thorough examination is necessary. This may require locally opening up the brick lining, depending on the severity of the damage.

Advice from specialists based on visual inspection, hammer testing, and further examination (including laboratory analysis) should be considered when deciding whether the damage could be the result of material selection, materials supplied, the design of equipment and lining, specifications, protection of materials during storage and installation, curing, or operations. There also may be other considerations, such as local conditions, factors in inspection, operating reports, and other factors that may help the specialist to decide on the type and extent of any repairs that may be required.

5.47.4 STARTUP AND OPERATION

Brick-lined equipment should only be operated after the following has occurred:

- Complete curing of the cement
- Required prestressing is completed
- Release and approval by inspector

The equipment should be brought into service very gradually, by slowly increasing the temperature and pressure to the operation requirements. The same careful handling is required when the equipment is taken out of operation. Temperature stresses caused by inexpert handling could destroy the acid-resistant brick lining completely.

Accurately written operating instructions should be established based on information to be supplied by the brick lining contractor for equipment with critical limitations of temperature and pressure during startup, operations, or both. Brick-lined equipment standing idle should be protected against frost, especially for those cases were moisture can reach the lining.

5.48 CEMENT-MORTAR LINING

This section specifies requirements for chemical-resistant cement-mortar lining for process equipment. It does not discuss refractory cement lining.

For cement lining applied to equipment fabricated in metal or concrete, the applicator should exercise diligence to provide uniform linings, without thick or thin areas. Adequate and properly spaced hold-downs on the application machinery should be used. New and used parts should be free of mill varnish, oil, paraffin, corrosion products, mill scale, thread lubricant, or any other foreign material when the wet cement mix is introduced for lining.

For more information about cement-mortar linings for water pipelines, whether applied on site or at the shop, see AWWA C104, AWWA C205, and AWWA C602.

5.48.1 PREPARATION

Mortar for the lining should be composed of cement, sand, and water that have been well mixed and are of a consistency such that a dense, homogeneous lining is produced. Unless otherwise specified by the purchaser, the mortar may also include admixtures and pozzolanic materials.

The approximate proportions of cement and sand in the mortar for the lining should be 1 part of Portland cement to 1½ parts of sand by volume; the proportion of sand to cement should be not more than 3 parts sand to 1 part cement by weight. The exact proportions should be determined by the characteristics of the sand used. Pozzolanic material can be substituted for a part of the Portland cement if desired; the proportion should be approximately 1 part pozzolanic material to 5 parts Portland cement by volume. Admixtures (i.e., resin and additives), if added, should be used in strict compliance with the manufacturer's recommendations.

The water content should be the minimum quantity that produces a workable mixture, with full allowance made for moisture on the interior of the pipe surfaces. Slump tests should be made periodically on freshly mixed mortar immediately prior to the mortar being conveyed to the lining machine. The test should be made in accordance with ASTM C143/C143M. Nominal slumps of cement-mortar mixes for application of lining are indicated in Figures 5.9 and 5.10.

Water for mixing mortar should be clean and free of mud, oil, and injurious amounts of organic material or other deleterious substances. Potable water should be used when available.

Mortar should be mixed long enough to obtain maximum plasticity. The mortar should be used before initial setting has taken place. The soluble chloride-ion (Cl-) content of the cement-mortar mix should not exceed 0.15%, expressed as a percentage of cement weight.

The minimum temperature of the wet mix should be maintained at not less than 10°C, and the maximum temperature should not exceed 32°C. Do not place cement mix in pipe when the ambient temperature is less than 4°C.

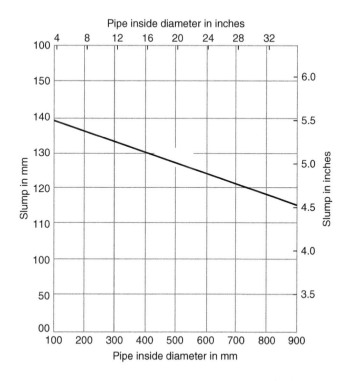

FIGURE 5.9

Nominal slumps of cement-mortar mixes for application of pipe linings using pump feed.

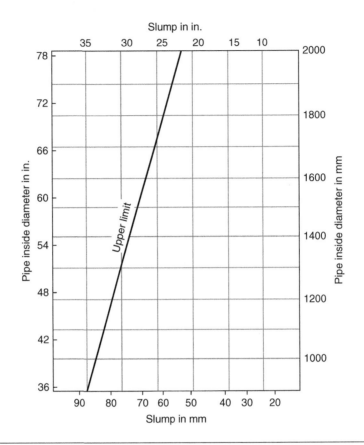

FIGURE 5.10

Slump limits of cement-mortar mixes for application of pipe linings using mechanical feed.

5.48.2 PREPARATION OF SURFACES TO BE LINED

The steel surfaces to be cement-lined should be cleaned to remove rust, oil, scale, and previously applied paint to the Sa 2 degree of Swedish standard ISO 8501 that could interfere with the adherence of the cement mortar. After the interior of the pipe or tubing has been cleaned, a visual inspection of each joint is made by the applicator to ensure that no scale, sand, or other foreign matter is left in the pipe and that the pipe or tubing can be properly cement-lined.

The concrete or masonry surfaces to be cement-lined should be cleaned to remove all unsound and loose material by chipping, scarifying, abrasive blasting, or water blasting. Abrasive-blasted surfaces do not require chipping to remove paint, oil, grease, and other contaminants, and provide a roughened surface for the proper bonding of shotcrete.

5.48.3 METHOD OF LINING

Cement mortar should be composed of cement, sand, and water, be well-mixed, and have the proper consistency to obtain a dense, homogeneous lining that will adhere firmly to the substances surface.

Cement-mortar lining should be applied by spinning, mechanical placement (line traveling), the pneumatic process (shotcrete or gunite), and hand troweling.

For shop application of cement lining of pipes, centrifugal spinning should be used. For field application of cement lining of pipes, mechanical placement (line traveling) should be used.

Vessels, miters, angles, bends, reducers, and other special elements should be lined with mortar by hand troweling, mechanical placement, and pneumatic placement. Hand troweling should be used for lining pipelines where other methods of application are impractical, such as sharp bends, or areas closely adjacent to valves.

All defects, including but not restricted to sand pockets, voids, oversanded areas, blisters, and cracking as a result of impact should be cut out and replaced by hand or pneumatic placement. The lining should be cured in such a manner as to produce a properly hydrated mortar lining that is hard and durable. The cure may be affected by the application of seal material to the still moist lining.

5.48.4 PNEUMATIC PLACEMENT (SHOTCRETE OR GUNITE PROCESSES)

The use of gunite without reinforcement should be limited to small or confined areas for gunite thicknesses not exceeding 20 mm. The details of such installation should be approved by the purchaser. During guniting operations, the equipment should be grounded in locations where there is a possibility of flammable materials being present. In addition, protection should be provided to prevent damage to adjacent structures.

The reinforcement should be placed not less than 13 mm from the surface to be gunited, and there should be a minimum of 20 mm between the reinforcement and final surface of the gunite. Where the gunite thickness does not permit these requirements, the reinforcement should be placed midway between the steel surface and the final gunite surface.

The reinforcement should be fastened by welding or by tie wires to anchoring devices welded to the surfaces to be gunited. Such devices may be any combination of rods, clip angles, studs, or nuts on edge. Spacing of these points should be a maximum of 450 mm in both directions.

Dampen absorptive substrate surfaces prior to the placement of shotcrete to facilitate bonding and to reduce the possibility of shrinkage cracking from premature loss of the mixing water. There should be brooming or scarifying of the surface of freshly placed shotcrete to which additional layers of shotcrete are to be bonded after hardening. The surface should be dampened just prior to the application of succeeding layers.

First, all corners and any area where rebound cannot escape or be blown free should be filled with sound material. The corners between the web and the flanges of structural steel must be completed before application to the flat area. Do not place shotcrete if drying or stiffing of the mix takes place at any time prior to delivery to the nozzle. Do not use rebound or previously expended material in the shotcrete mix.

5.48.5 FINISHING

Provide one of the following final surface finishes when specified:

- Broomed
- Floated

- Troweled
- Sponge floated
- Flash finish

Avoid troweling and starting moisture curing within a relatively short period after placement of shotcrete.

5.48.6 MECHANICAL PLACEMENT FOR LINING PIPELINES (LINE TRAVELING)

The lining of pipelines should be applied in one pass or more by a machine traveling through the pipe and distributing the mortar uniformly across the full section and long radius bends of the pipe. The discharge should be from the rear of the machine so that the newly applied mortar will not be marked.

The rate of travel of the machine and the rate of mortar discharge should be mechanically regulated to produce a smooth surface and uniform thickness throughout. The mortar should be densely packed and adhere wherever applied; there should be no injurious rebound.

5.48.7 PROCEDURE FOR RIVETS AND OPEN JOINTS

In steel pipe that is 600 mm and larger in diameter, mortar may be applied by hand ahead of the lining machine for uniform thickness over the line of rivet heads. Open joints should be packed with mortar lining where necessary to provide a smooth surface across the joint. Such mortared areas should be moist and free of surface checking before the machine lining proceeds.

5.48.8 FINISHING

The lining machine should be provided with attachments for mechanically troweling the mortar. The trowel attachment should be such that the pressure applied to the lining will be uniform and produce a lining of uniform thickness with a smooth finished surface, free of spiral shoulders.

5.48.9 SPIN OR CENTRIFUGAL PROCESS (FOR LINING OF SHOP-APPLIED PIPES)

Straight sections of pipe should be lined by use of a spinning machine specifically designed and built for the purpose of rotating the pipe section and centrifugally applying cement-mortar linings to the interior of steel pipe.

The mortar should be mixed in batches. The amount of cement and sand entering into each batch should be measured by weight. The quantity of water entering the mixer should be measured automatically by an adjustable device, or it should be measured in another way that ensures that the correct quantity of water is being added.

When required to prevent distortion or vibration during the spinning, each section of pipe should be suitably braced with external or internal supports appropriate to the equipment. For the application of linings by a spinning machine, the entire quantity of mortar required for completion of the lining of the section of pipe should be placed without interruption.

After the mortar has been distributed to a uniform thickness, the rotation speed should be increased to produce a dense mortar with a smooth surface. Provisions should be made for the removal of surplus water by air blowing, tilting of the pipe, or other methods.

The application of cement-mortar lining to miters, angles, bends, reducers, and other special sections, the shape of which precludes application by the spinning process, should be accomplished by mechanical placement, pneumatic placement, or hand application. Then it should be finished to produce a smooth, dense surface.

5.48.10 REINFORCEMENT

Wire-fabric reinforcement or ribbon-mesh reinforcement should be applied to the interior of fittings larger than 610 mm and should be secured at frequent intervals by tack-welding to the pipe by clips or wire. The wires placed with 50-mm spacing on the 50 mm × 100 mm fabric should extend circumferentially around the fitting. Repaired areas of machine-applied linings at miters, pipe ends, outlets, and other cuts made in the lining for fabrication of the fitting need not be reinforced if the width of the repair area does not exceed 300 mm. Repairs for widths exceeding 150 mm should be bonded to the steel and adjacent faces of the lining with a bonding agent.

5.48.11 HAND-TROWELED APPLICATION FOR PIPELINES

In pipe that is 600 mm and larger in diameter, lining should be performed by hand in places where machine placing of cement lining is impractical, such as sharp bends, specials, or areas closely adjacent to valves. The engineer may permit the correction of any defect by hand application.

If necessary, areas to be lined should be moistened with water immediately prior to placing the hand-applied mortar. Steel-finishing trowels should be used for the application of cement by hand, except at bends. The outer edges of hand-traweled areas may be brushed to reduce the abutting offset. All hand-finishing work in a section of pipeline should be completed within 24 h after completion of the machine application of mortar lining to that section.

5.48.12 CURING OF CEMENT LININGS

Methods of curing should be defined by the company in terms of job requirements. The following methods of steam and atmospheric curing have been found generally successful.

5.48.13 STEAM CURING

Begin steam curing between 2 and 4 h after final spin occurs. High atmospheric temperatures and low humidities may shorten the time requirements, whereas low atmospheric temperatures and high humidities may lengthen the time requirements.

Leave the lining in the steam-curing chamber at 57–74°C for not less than 18 h. The purchaser may require a longer curing time; if so, it should request that.

During heating and cooling, do not increase or decrease the temperature of the steam curing chamber at a rate of more than 0.6°C/min. Place end caps on the cement-lined pipe before placing in the steam chamber. Do not expose the cement-lined pipe for more than 2 h after spinning without end caps in place.

Leave the end caps in place until the pipe is installed in the field system. Keep the cement lining moist and protect it from freezing until delivered to the purchaser, and keep the cement lining moist following steam curing. The storage time following cure and before installation should be at the discretion of the company.

5.48.14 MOIST (ATMOSPHERIC) CURING

Keep the cement lining moist at all times during the curing process. Use airtight end caps to seal the pipe ends to retain moisture in the lining. Do not expose the cement lining for more than 2 h after spinning without end caps in place. Keep the caps in place until the pipe is installed in the field service.

Maintain the curing temperature at not less than 10°C for at least 8 days. Atmospheric- temperature-cured, cement-lined pipe should not be transported or installed for at least 8 days after application of the lining.

5.48.15 MEMBRANE CURING

Membrane curing, if any, should consist of the complete encapsulation of the coating by the application of curing compound that will retain the moisture of the applied cement lining.

5.49 TRANSPORTATION

5.49.1 TRANSPORTATION OF EQUIPMENT

Do not drop the equipment on to or off of the transporting vehicle. Tie-downs should be used to ensure that the pipe will not shift during shipment. Any tie-down can result in a lining damage, so the pipe should be properly protected. All impact should be avoided.

5.49.2 TRANSPORTATION OF LINED PIPE

Cement-lined pipe should always be handled carefully. Minor impacts and bending normally have no effect on the lining. The following practices are recommended to prevent lining damage during handling:

- Always keep the airtight end caps in place. Loss of end caps permits the lining to dry out, which can result in severe conditions
- Always load or unload by hand when practical, whenever loading or unloading.
- Do not drop the pipe onto or off of the transporting vehicle.
- Do not roll the pipe onto or off of the transporting vehicle in such a manner that it bangs into other pipes.

Pipe should always be supported to avoid undue flexure when being transported. Supports should be placed every 1.2 m for pipe 152 mm and smaller and every 1.8 m for larger pipe.

Dunnage should always be placed within 0.6 m of the ends to avoid banging the ends together. When trucking, a flat-bed trailer provides the best support. Pipe that cannot be loaded by hand may be

Table 5.10 Number of Tiers of Lined Pipes

Nominal Pipe Size		Maximum Number of Tiers
mm	**in.**	
50	(2)	10
76	(3)	10
100	(4)	8
152	(6)	6
200	(8)	4
250–400	(10–16)	3

picked up at the midpoint if the ends sag no more than 0.6 m. If sagging exceeds 0.6 m, a 3-point (bar-sling) arrangement should be used.

Tie-downs should be used to ensure that the pipe will not shift during shipment. Any tie-down can damage the lining, so the pipe should be properly protected. All impacts should be avoided. Hooks or other devices that insert into the ends of the pipe should never be used. Lined pipe should never be dropped after completing a welded joint.

The pipe should never be bent to such an extent that the metal is deformed or lining damage will result. The recommended maximum bending angle is 5°, or 76 mm in 30 m when lowering pipe into a trench or similar operation.

The number of tiers of lined pipes should not exceed the values given in Table 5.10.

5.50 INSTALLATION AND JOINING

5.50.1 INSTALLATION AND JOINING OF EQUIPMENT

The appropriate seal material should be placed at the joints of cement-lined pipes.

5.50.2 INSTALLATION AND JOINING OF LINED PIPE

For preventing corrosion damage at the joints of cement-lined pipes, the following techniques should be used:

1. The lengths of pipe should be butted together and checked for alignment and good contact of the cement lining and pipe ends.
2. The appropriate seal material should be applied in accordance with widely accepted standards.
3. Lines smaller than 100 mm should be tack-welded at the 12 o'clock and 3 or 9 o'clock positions. The first welding pass should start opposite the chosen second tack weld and continue in the direction of the 6 o'clock position. This is necessary to prevent a gap between the pipe ends opposite the first welding pass. Pipe with diameters of 100 mm and larger should be tack-welded at the 4, 8, and 12 o'clock positions.

4. The pipe should be joined by the shielded metal-arc welding process. The arc should not come in direct contact with the cement lining or seal material. Starts and stops should be staggered to prevent starting or stopping more than once in the same place. Welding slag should be cleaned from all weld passes. Welding materials used should be controlled and approved by the company in accordance with the current list of welding consumables published by internationally acknowledged bodies.

5.51 SEAL MATERIALS

5.51.1 MAGNESIUM OXIDE/GRAPHITE/HYDRAULIC CEMENT

Only magnesium oxide/graphite/hydraulic cement is suitable for use in welded joints. The dry compound should be mixed thoroughly with clean water in a clean container by stirring with a clean welding rod. Pot life of the mixture is approximately 1 h. Only enough compound should be prepared in each batch to last for 45 min. to 1 h. Additional water should not be used to thin the mixture after it has started to set. The mixing and application container should be thoroughly washed with clean water before mixing another batch of compound.

The compound can best be applied to the cement at the ends of the pipe by squeezing it from a plastic squeeze-type bottle or applicator. A continuous layer or bead of compound should be applied to the cement liner on the ends of both lengths of pipe to be welded. Enough should be used to form a small continuous bead of excess compound in the welding groove when the pipe ends are butted together. Using too much compound will cause a larger bead of excess compound inside the pipe, which is unnecessary and wasteful. The compound mixture should be thin enough to be squeezed through the nozzle of the applicator, but not thin enough to run off the pipe after it is applied.

5.51.2 PLASTIC MATERIALS

For plastic materials, follow the same procedure as described above, except that a mixing procedure applicable to the material selected should be used. Many compounds are available, but several are not resistant to heat. The user should make sure that the compound selected is suitable for the intended service. The pot life of most catalytically cured materials is limited and differs for various compounds.

5.51.3 ASBESTOS GASKETS

The pipe ends to be joined should be tack-welded together and the asbestos gasket placed in a V formation. The V should be closed by moving the pipe lengths and tack-welding the pipe at the top. Lined-up clamps should be installed and the welding completed.

The thickness of the gasket should be 0.8 mm for thin wall pipe and 1.6 mm for greater wall thicknesses. The outside diameter (OD) of the gasket should be equal to the ID of the steel pipe minus 1.5–2.5 mm. The ID of the gasket should be equal to the ID of the lining, within a tolerance of zero to 1 mm.

5.51.4 CAULKING

Pipe with a diameter large enough for a person to enter it can be caulked at the joint. Numerous compounds are available, but a polysulfide rubber is most widely used. The user should make sure that the compound selected is suitable for the intended purpose.

5.51.5 THREADED AND COUPLED PIPE

For coating exposed threads, many compounds are available. The user should make sure that the compound selected is suitable for the intended service. The compounds include the following:

- Magnesium oxide/hydraulic cement compounds will lock the joint if they are included in the makeup portion of the joint.
- Epoxy materials will lock the joint if included in the makeup portion of the joint.
- Polysulfide rubber has been reported as having low resistance to strong H2S and strong acid environments.
- Coal-tar epoxy.
- Coal-tar modified with vegetable pitch.
- Polyurethane elastomer.
- Polychloroprene molded and cured inside the coupling or rubber rings has been used to seal the exposed collar area.

5.52 REPAIR OF LINING

All defects, including but not restricted to sand pockets, voids, oversanded areas, blisters, and cracking as a result of impact should be cut out and replaced by hand or pneumatic placement at the same thickness as required for the mortar lining. In the case of water pipelines, temperature and shrinkage cracks in the mortar lining that are less than 1.6 mm in width need not be repaired. Cracks wider than 1.6 mm need not be repaired if it can be demonstrated to the satisfaction of the purchaser that the cracks will heal autogenously under continuous soaking in water. The autogenous healing process may be demonstrated by any procedure that keeps the lining of the pipe continually wet or moist. Pipe used in the demonstration should be representative of the pipe to be supplied, and water for the moistening of the pipe should be chemically similar to the water to be carried in the pipeline.

Defective or damaged areas of linings may be patched by cutting out the defective or damaged lining to the metal so that the edges of the lining not removed are perpendicular or slightly undercut. The cutout area and the adjoining lining should be thoroughly wetted, and the mortar applied and troweled smooth. After any surface water has evaporated, but while the patch is still moist, it should be cured.

For more information about the repair procedure of welded steel, cement-lined pipe leakage, see API RP 10E.

5.53 INSPECTION AND REJECTION (FOR LINED PIPE)

To ensure good practices of application of cement lining, the types of inspection that should be performed are as follows:

- Quality control should be done by the applicator during manufacture.
- Shop inspection by an inspector nominated by the company should be performed at the manufacturer's works.
- Inspection of lined equipment are generally the same as lined pipe, except all lined equipment should be inspected.

5.54 INSPECTION DURING APPLICATION

The quality control inspector should check, inspect, and test for the following causes of lining failures described next.

5.54.1 VOIDS

A *void* is a place in the pipe or equipment where the cement lining is not continuous. Each pipe should be inspected immediately after the application by looking through the pipe from each end (for pipe smaller than 600 mm ID), or using a strong light on the other end (for pipe larger than 600 mm ID),. Voids appear as dark places in the shiny surface of the wet lining. They should be repaired immediately to achieve the required thickness, and then the cured pipe should be reinspected for voids before the company will accept it.

5.54.2 LINING THICKNESS

Cement thickness at and near each end of the pipe should be determined by an internal caliper and measuring scale. Cement thickness near the center of the pipe should be measured by cutting the pipe and examining the new ends.

Poor lateral distribution of the cement mix can sometimes be detected by looking through the pipe at the fresh-spun lining using a strong light at the other end. The appearance of concentric rings indicates poor lateral cement distribution.

5.54.3 SAGS

If the forces of hoop stress and adhesion in the fresh cement lining are not greater than the force of gravity, the lining sags or pulls away from the pipe surface at some point along the top of the pipe. Each joint should be inspected before and after curing by looking through the pipe from each end, using a strong light at the other end. Sags appear as large smooth lumps in the lining at the top of the pipe.

5.54.4 DEFECTIVE ENDS

The types of end defects include:

- Lining end not being located as specified
- Lining end not being perpendicular to the longitudinal axis of pipe
- Lining end not being square
- Lining end being chipped or cracked
- Lining thickness not being as specified
- Threads or welding bevel and land not being cleaned of cement
- Lining being separated from the steel pipe surface

Both ends of each joint of cement-lined pipe must be inspected before and after curing for the defects mentioned above. If the ends of the cement-lined pipe are not as specified, it is very difficult to make a corrosion-resistant joint in the pipe. Therefore, end defects should be a major cause for rejection. Cement-lined pipe is damaged most often at the ends during transportation and handling.

Therefore, the ends should also be inspected before the pipe is installed. Any repaired ends should meet specifications.

5.54.5 FOREIGN MATERIAL AND POOR MIXING

The quality control inspector should ensure that the lining plant is operated and maintained so that no foreign material is introduced into the cement lining or its components during handling, mixing, storage, or at any other time. This person should also ensure that all components are thoroughly mixed in the correct proportions. Foreign material or poor mixing are indicated by the appearance of spots or small rough lumps in the fresh lining.

5.54.6 CRACKS

The detection of cracks and the correct evaluation of the detrimental effect of cracks is the most difficult part of inspection. Unless the pipe is large enough for a person to enter, lining cracks are difficult to see unless viewed from a point perpendicular to the surface of the lining. The best inspection tool consists of a light source and a mirror on the end of a rod which can be inserted into the lined pipe. The inspector can look into the end of the pipe and see the illuminated surface of the lining in the mirror. Under some conditions, the cracks in cement lining either "heal" or do not lead to corrosion of the pipe because of the inherent alkalinity of the cement. However, if the fluid is corrosive (e.g., salt water) and if pressure surges are present, even hairline cracks can cause corrosion failure of the pipe. Most cracking occurs when the lining is allowed to dry out during curing, transportation, and storage. The quality control inspector should ensure that the lining is still moist after inspection and that airtight end caps are placed and maintained on the pipe. The lining of each pipe should be inspected for cracks after curing, transportation, storage, and handling.

5.55 REFRACTORY LINING

This section specifies the requirements for refractory brick or refractory concrete lining for process equipment.

5.55.1 PREPARATION

When not properly stored, factory-prepared dry mix may absorb moisture, which will cause setting. Such material will contain lumps and should be discarded. If the stored mix becomes hot, the rate of moisture absorption will accelerate upon cooling. Therefore, the mix should always be stored in a cool and dry atmosphere.

If bags containing dry mix are stored in the open air (allowed for a maximum of 2 days only). They should always be placed on a ventilated platform off the ground and protected by a covering such as a tarpaulin. In addition, they should arranged such that water cannot come into contact with any of the bags. Care should be taken to avoid the possibility of ground moisture causing a high humidity underneath the covering.

Materials should not be piled directly on the ground or floor. Pallets, planking, or a combination of the two must be used.

If the dry mix has been stored for more than 6 months, its quality should be checked before use. In general, checking of the bulk density and compressive strength will be sufficient.

5.55.2 MIXING

Water for mixing the refractory should be cool, free of impurities, and from a chemical stand point good enough to drink. Mixing with water should preferably be performed in a mixing machine such as a paddle-type mixer which empties itself completely on discharge. The duration should be kept to the minimum necessary for thorough mixing, usually 3–4 min. Prolonged mixing should be avoided.

Mixers should be clean and free of old mortar, Portland cement, lime, and dirt to avoid contamination of the castable materials. Before commencing, mixers should be checked for mechanical reliability to prevent breakdown while the operation is in progress.

The size of the batch for hand mixing should be limited to a volume that will permit thorough mixing and pouring within 15 min. from the moment mixing has started, particularly at high ambient temperatures. The freshly mixed concrete should be of a soft puddling consistency so that no tamping is required. Tamping consistencies may be used on some work, such as on sleep slopes, to eliminate the cost of form work.

Remember that the use of excessive amounts of mixing water will reduce cold (unfired) strength and will cause excessive shrinkage. In cold weather, the water temperature should not be less than 5°C.

The surface to be lined should be thoroughly sand-blasted with sharp, abrasive sand or wire-brushed to remove all unbounded scale, slag, rust, and other foreign materials. All sand-blasting dust should be removed before lining material is applied.

5.56 APPLICATION METHODS

During application, special care should be taken that all irregular areas and corners are completely filled and that voids or air pockets are precluded. If any surface finish is required, the surface should simply be leveled off with a screed or a wood float. The surface should not be troweled to a slick finish.

In cold weather, the concrete and the surface to which it will be applied should be kept at a temperature above 5°C, during both application and curing. This should be accomplished by providing an enclosure in which a temperature above 5°C is maintained during mixing, application, and curing. Linings containing water should be protected against freezing; however, live steam should not be used for this purpose.

Expansion joints in the lining should be provided every 1.5–2.0 mm in the horizontal and vertical directions. Straight-through joints filled with ceramic fiber board or ceramic fiber blanket (see BS 6466) should be acceptable where local experience has shown them to be satisfactory. Ceramic fiber blankets should be compressed to around 60% of its nominal thickness and contained in a compressed state (such as in an expanded metal cage). Straight-through joints are not recommended for unshielded walls or roofs of the radiant sections. In linings of steel stacks, the carving of horizontal shrink slits 1 mm wide and one-fifth of the lining thickness deep should be carried out at distances of approximately 2.0 m.

The application method for refractory brick lining is the same as for chemical resistance brick lining. For refractory brick lining, refractory bricks are used in conjunction with refractory concrete.

5.57 REFRACTORY CEMENT LINING

Application of refractory cement lining may be either by casting or gunning (shotcreting), either in place or be prelining. For gunning with a dry gun, 30% more than the design quality calculated is required; for casting, it is 10%.

5.57.1 **REFRACTORY CASTING METHODS**

Casting in a horizontal position permits easy inspection and requires neither skilled labor nor any special equipment. This method can be used only for new work; its interference with steel structural work, as well as the special care required while lifting and assembling the lined sections or panels, are serious drawbacks. Sufficient space will be required for handling the sections or panels (see Figures 5.11 and 5.12).

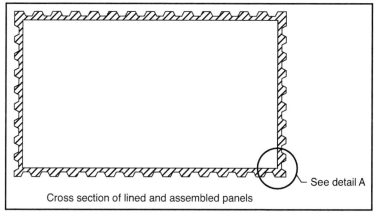

Cross section of lined and assembled panels

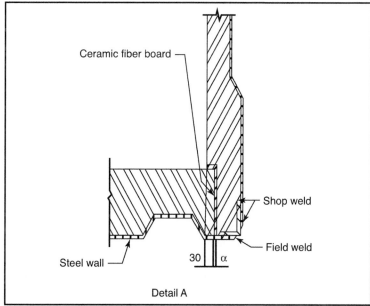

Detail A

FIGURE 5.11

Horizontal casting on separate panels at ground level and subsequent assembly (anchoring not shown).

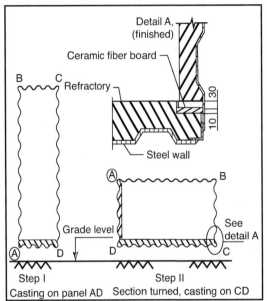

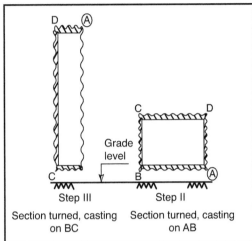

FIGURE 5.12

Horizontal casting inside preassembled panels (anchoring not shown).

5.57.2 CASTING IN A VERTICAL POSITION

Casting in a vertical position should be restricted to minor repairs and to those cases where gunning or horizontal casting is not feasible. This method requires shuttering, which should be a maximum of 500 mm high (see Figure 5.13). The shuttering should be made reasonably watertight, and the inside surfaces should be sufficiently oiled, but not excessively.

Note: Dry wooden shuttering may need treatment to prevent the absorption of mixing water from the fresh concrete. The concrete should be cast evenly along the length of the shuttering and consolidated in order to obtain a homogeneous composition. Skilled labor and close supervision is essential.

5.57.3 FORMS (SHUTTERING) FOR CASTING

Both wood and metal forms used for casting hydraulic-setting castable refractories should be rigid and strong to prevent movement due to the pressures and loads that develop during placement. The forms should be coated with acid-free oil to prevent wood forms from absorbing water from the castable and to allow the form (whether metal or wood) to part easily when the forms are removed.

Forms for casting tubular lines or stacks may be tubular collapsible steel or tubular concrete forms, such as "so no voids," which are constructed of heavy cardboard and water proofed on the OD. The refractory contractor should be responsible for furnishing accurately made forms that are suitable for the work being handled. All form designs should be submitted in complete detail for approval. Wooden or cardboard forms are to be carefully removed after the castable has set. Under no circumstances are these forms to be burned off because in so doing, high temperatures are flashed onto the face of the still green castable, usually resulting in damage to the face of the lining, or even the entire lining.

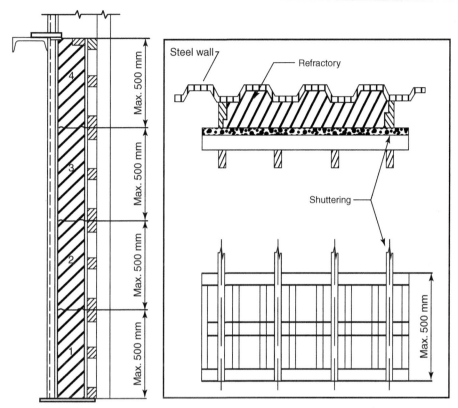

FIGURE 5.13

Vertical casting of refractory (anchoring not shown).

5.57.4 **CASTING PROCESS**

Pour refractory behind forms using funnels and sheet metal tubes as required. Cast refractory should be applied in a manner that will minimize material segregation. All materials should be thoroughly worked into place and air bubbles expelled within 15 min. from the moment that mixing starts. Each batch from the mixer should be so placed that the full thickness of the lining will be reached, using temporary weirs as necessary.

On no account should another layer be added later to complete the lining. Once application has started, it should proceed without interruption until the entire lining of the part has been put in place. If an unavoidable interruption does occur, the wet edge of the lining should be cut back at a right angle to the surface to create an edge of full thickness. All materials ahead of the cut should be removed and discarded.

If any surface finish is required, the surface should be leveled off with a screed or a wood float. The surface should not be troweled to a slick finish.

To make certain that refractory is placed free of voids, use magnetic vibrators or a ramrod. Do not excessively vibrate or ram; agitate only sufficiently to ensure good flow. Care must be taken not to vibrate or ram after initial set occurs, which is approximately 15–20 min. after mixing.

5.57.5 REFRACTORY GUNNING METHOD

Gunning (shotcreting) should normally be done with a dry gun. The addition of the correct quantity of water in the dry mix at the spray gun and the even mixing of the concrete are completely dependent on the skill and attentiveness of the gun crew. Skilled labor and close supervision should be imperative.

This method should be employed after erection of the steel structure is finished. It is suitable for applying linings in horizontal, vertical, and overhead positions. Once gunning has started, the application of the lining should be continuous until the whole lining or section is completed.

Work should always proceed along the wet edge of the band just finished and reach full lining thickness as quickly as possible before proceeding to another panel. Vertical walls should be gunned in horizontal bands, working upward from the bottom of the area to be lined.

In those cases where the size of the unit precludes successive band placement before initial setting has taken place, or where the gunning will be interrupted for more than 10 min., the working edge should be cut back to full thickness at a right angle to the steel wall. All materials ahead of the cut should be removed and discarded.

On jobs where the gunning of the lining will not be a continuous operation, the day's work should not be terminated until a unit or section is finished. This means that the work should cease only at natural stopping points of complete subsections. The tolerance of the lining thickness with the gunning method should be ±5 mm.

Sections of the lining below the minimum thickness should be cut out entirely and replaced. At no stage should additional material be placed over previously applied material to build up the required thickness.

Where trimming will be necessary, it should be done by edge-cutting with a trowel; surface smoothing by trowel should be avoided. Only qualified personnel thoroughly familiar and experienced with the pneumatic application of refractory concrete should be employed for this work. Make certain that all gunning equipment is in first-class condition and has been thoroughly cleaned before gunning operations begin.

Gun nozzles should be suitable for handling the refractory material being applied, and spare nozzles or repair parts should be on hand for immediate use. The hose lengths between the gunning machine and the equipment being lined should be as short as possible. Excessive lengths of hose are more vulnerable to plugging, and as a result, more material is lost in cleaning.

All tools, mixing equipment, and water storage vessels should be kept clean and free of contaminating materials such as Portland cement, lime, and sodium silicate before starting the gunning operation. Contamination will affect setting and strength.

Refractory materials should not be applied in freezing or excessively hot weather unless precautions are taken to maintain refractory, steel, and mixing equipment at as close to an ambient temperature of 15°C as possible during mixing and application, and for at least 24 h thereafter. No admixtures of any kind, nor live steam, should be used for this purpose.

The specified thickness of the lining should not be obtained by building up the material, nor should a thickness less than specified be applied and allowed to set. If possible, gunning should be continued until the full thickness has been applied. When it is necessary to discontinue gunning at an intermediate point for more than 20 min., the insulating layer of a dual lining or a monolithic lining should be cut back immediately to the steel wall with a steel trowel. This cut should be made at a right angle to the steel wall. At locations where the full thickness has been applied, all materials beyond the cut should be discarded. Before gunning is resumed, each of these terminated sections must be thoroughly cleaned and wetted to damp out any suction. Failure to carry out this operation may cause lack of adhesion, cracking, or a powdery surface.

Material gunned with either too much or too little water should be immediately removed and the area regunned. Rebound material or material that has been removed from plugged gunning machines or hoses should be discarded and should not be reused under any circumstances. Rebound material should not be permitted to accumulate at any point where lining has already been applied. If it has accumulated in areas to be lined, it should be completely removed prior to applying the lining.

5.58 CURING AND DRYING

During the curing period, the lining should be kept cool to avoid cracking or damage. This can best be accomplished by using a water spray, which should be applied as soon as the surface of the lining is hard enough to permit spraying without washing out the cement.

An indication that spraying can start should normally be obtained when the cement will no longer stick to a wet hand placed lightly on the concrete surface. In hot and dry weather, it may be found necessary to start spraying two or three h after the concrete has been placed. The water spray should merely consist of a fine mist; a water stream should be avoided.

The horizontal parts of the lining should normally be covered with wet cloths or jute sacks. Once the water spraying has been started, it should be continued, preferably without interruption, for at least 24 h. If curing is done in an ambient temperature below 15°C, curing time in excess of 24 h is required. At no time during the curing period should the intervals between water spraying be more than 30 min.

Note: A liquid membrane-forming compound may be applied as an alternative to curing of the lining by water spray, where local experience has shown this to give good results. The compound should be applied within 30 min. after the installation of the lining to the whole of its exposed surface in sufficient thickness to prevent evaporation of water from the lining. The color of the compound should contrast with the lining to facilitate full coverage.

After curing, the concrete should be permitted to dry under atmospheric conditions as long as possible, but not less than 24 h. After this drying period, the insulating refractory concrete should be inspected and any cracks 1.5 mm wide or more should be repaired.

When equipment with newly installed linings is put into service (started up), the temperature in the equipment should be increased gradually to expel excess moisture slowly. The following heating schedule may serve as a guide:

1. Raise the temperature slowly to 300°C (measured at the flue duct) at a rate of not more than 15°C/h, and maintain this temperature for about 24 h.
2. Raise the temperature to 500°C in not less than 12 h.
3. Continue to heat to the furnace's operating temperature at a rate of approx. 50°C/h.

This heating schedule should be adapted as necessary to provide a compromise if two or more different refractory materials have to be heated simultaneously. The schedule should be registered by temperature recorders in conjunction with the installed thermocouples.

5.59 TRANSPORTATION

Refractory-lined equipment that includes piping should not transported until at least 24 h after the curing operation has been completed. Rigging should be such that flexure and distortion are prevented to avoid damage to the lining.

5.59.1 INSTALLATION

For pipes 620 mm and greater with weld seams or joints that require welding after lining is applied, leave a 76–84-mm unlined section on each side of seam. After welding and inspection of the weld, thoroughly clean the steel shell and apply the refractory lining.

5.59.2 REPAIR OF LINING

If linings have to be repaired, the concrete at the damaged area should be entirely removed down to the steel wall in such a way that a dovetail is obtained. Where needed, additional cleats or V-studs should be applied.

Before installing the new lining to a repair area, the area should be cleaned by blowing with an air hose to remove all loose particles. The newly exposed concrete surface should then be thoroughly wetted and lining material of the composition as originally used should be applied.

Normal curing and drying are again required. Any voids or dry-filled spaces will produce a dull sound. Such spots should be chiseled out completely to the vessel steel wall and laterally to sound refractory materials. The area removed should at least enough to expose three studs. The sides of the cut should be tapered toward the equipment wall, forming a keyhole to form a stronger patch.

5.59.3 INSPECTION AND TEST METHODS

Prior to refractory lining application, prepare and test the samples. Mixing and application of the field application should not proceed until samples are satisfactory. After application of lining, inspect the part.

5.59.4 PREAPPLICATION TESTS (TEST SAMPLE)

Sample mixtures should be made of the refractory to be tested in strict accordance with the applicable refractory specifications before proceeding to mix and apply for field installation. Three samples should be prepared.

The sample mix should be water-cured or membrane-cured with applicable specifications for a 114 mm × 114 mm × 65 mm sample. Mixing and application for field installation should not proceed until samples are satisfactory.

Any changes in materials or materials supplied from different manufacturers and changes in the water source should require new sample mixture testing. Samples should be prepared in accordance with ASTM C860 and compression tested in accordance with ASTM C93.

5.59.5 INSPECTION OF LINED SUBSTANCE

Following initial curing, but before drying, test the refractory lining by striking it with a ball-peen hammer at about 30-cm intervals over the entire surface.

5.60 **SAFETY**

5.60.1 **HANDLING AND APPLYING LINING MATERIALS**

Mixing can present many problems. Organic lining materials should be stored in safe, well-ventilated areas where sparks, flames, and the direct rays of the sun can be avoided. Containers should be kept tightly sealed until they are ready for use. Warning tags should be placed on toxic materials. The recommended safety rules for mixing operations include the following:

- Use protective gloves..
- Keep the face and head away from the mixing container.
- Use protective face cream.
- Avoid splash and spillage, as well as inhalation of vapors.
- Use eye-protection goggles.
- Mix all materials in well-ventilated areas away from sparks and flames.
- Use low-speed mechanical mixers whenever possible.
- Clean up spillage immediately.
- Check the temperature limitations of materials being mixed.

Many materials are dangerous at high temperatures. Protective devices and equipment required for application of lining materials are determined by the type of lining, as well as by the environment. The lining manufacturers should provide complete mixing and application instructions, including safety requirements. Unless definite information regarding explosion and toxicity hazards inherent in the material are provided by the manufacturer, a written request for such data should be made before starting the lining process. Records of previous applications using similar materials also should be examined.

Protective face cream is recommended for all spraying operations. Goggles should be worn wherever possible.

5.60.2 **HEALTH HAZARDS OF MATERIALS**

A lining material may be considered a health hazard when its properties are such that it can either directly or indirectly cause injury or incapacitation (either temporary or permanent) from exposure by contact, inhalation, or ingestion.

Degrees of health hazard are ranked according to the probable severity of injury or incapacitation. Materials that on brief exposure could cause death or major residual injury even if prompt medical treatment is given include the following:

- Materials that can penetrate ordinary rubber protective clothing
- Materials that under normal conditions or under fire conditions give off gases that are extremely toxic or corrosive through inhalation or through contact with the skin

Materials that after brief exposure could cause serious temporary or residual injury even if prompt medical treatment is given include the following:

- Materials giving off highly toxic combustion products
- Materials corrosive to living tissue or toxic by skin absorption

Materials that on intense or continued exposure could cause temporary incapacitation or possible residual injury unless prompt medical attention is given include the following:

*   Materials giving off toxic combustion products
*   Materials giving off highly irritating combustion products
*   Materials that under either normal conditions or fire give off toxic vapors

Materials that on exposure can cause irritation, but only minor residual injury, even if no treatment is given include the following:

*   Material that under fire conditions give off irritating combustion products
*   Materials that cause irritation to the skin without destruction of tissue
*   Materials that on exposure of fire conditions offer no hazard beyond that of ordinary combustible material

5.60.3 FLAMMABILITY HAZARDS OF MATERIALS

A lining material may be considered a flammability hazard when it will burn under normal conditions. Degrees of hazard are ranked according to the susceptibility of materials to burning.

Materials that will rapidly or completely vaporize at atmospheric pressure and normal ambient temperature, and that are readily dispersed in air and burn include the following:

*   Any liquid that takes liquid form under pressure and has a vapor pressure not greater than 1 atm at 38°C
*   Materials that may form explosive mixtures with air and that are readily dispersed in air, such as mists of flammable or combustible liquid droplets
*   Materials that can be ignited under almost all ambient temperature conditions

The materials that produce hazardous atmospheres with air under all ambient temperatures and are readily ignited include the following:

*   Materials having a flash point of 38°C (100°F) or below and having a vapor pressure not greater than 1 atm (14.7 psia) at 38°C (100°F).
*   Materials that ignite spontaneously when exposed to air.
*   Materials that must be moderately heated or exposed to relatively high ambient temperature before ignition can occur. Materials of this type are those having a flash point above 38°C (100°F) but not greater than 93°C (200°F).
*   Materials that must be preheated before ignition can occur. These materials are those that will support combustion for 5 min. or less at 815°C (1,500°F).

5.61 TOXICITY OF MATERIALS

Virtually all solvent solution linings are highly flammable in liquid form, and vapors released in the process of application are explosive in nature if they are concentrated in sufficient volume in closed or restricted areas. Even vapors from ordinary enamels and oil paint may be accumulated in such density as to result in explosion if a source of ignition is present. Generally speaking, however, solvents used in solvent solution coatings are more volatile and dangerous than those used in conventional paints or linings (see Tables 5.11 and 5.12).

Table 5.11 Threshold Limit Values for Commonly Used Solvents

Solvent	Maximum Allowance Concentration[1]	Flash Point Temperature (°C)[2]
Acetone	1,000	−17.8
Benzene	25	−11.1
n-butanol	100	28.9
Butyl cellosolve	50	60
Carbon tetrachloride[3]	10	N/A
Cyclohexane	400	−17.2
Ethylene dichloride	100	13.3
Ethanol	1,000	12.8
Gasoline	500	7.2
Methanol	200	11.1
Naphta, coal-tar	200	37.8–43
Naphta, petroleum	500	<-17.3
Perchloroethylene	200	N/A
Isopropanol	400	11.7
Stoddard solvent	500	37.8–43
Toluene	200	4.4
Trichloroethylene	100	N/A
Turpentine	100	35
Xylene	200	17.2

[1]Maximum allowance concentration of vapor in air (or hygienic standard), in parts per million by volume, for breathing during a continuous 8 h workday. These figures should be used as guides for evaluating exposures; they are subject to change when better information is available.
[2]TAG closed-cup tester: From "Flammable Liquids and Gases," National Fire Codes, Vol. 1.
[3]Because of cumulative health hazards, use of this cleaning solvent is not recommended.

A wide variety of solvents are used in the formulation of many linings now available. These include, but are not limited to, the ketone group, which includes such solvents as acetone, MEK, MIBK, ethyl amyl ketone (EAK), and cyclohexane; and the hydrocarbons group, which includes toluol, toluene, xylol, and xylene. In many instances, the ketones act as the solvents that hold the resins in solution; the hydrocarbons are used primarily as diluents to control the viscosity or flow qualities of the lining. Most linings have a combination of solvents from each group. The epoxy and synthetic rubber linings use mostly solvents of the hydrocarbons group, although ketones may be used in limited quantities.

All of these solvents are highly flammable and should be handled with the greatest care. Solvents, as well as other components of many modern solution linings, present other hazards that must be guarded against at all times. Solvents of all groups are toxic to varying degrees and may cause serious effects to those working with them unless appropriate precautions are taken. Excessive breathing of concentrated

Table 5.12 Threshold Limit Values for Fumes, Mists, and Dusts

Hazard	Maximum Allowance Concentration[1]
Lead	0.2
Dust (no free silica)	50.0
Silica:	
High (above 50% free SiO)	50
Medium (5-50% free SiO)	20
Low (below 5% free SiO)	50
Total dusts below 5% free SiO	50
Chromic acids and chromates as CrO3	0.1

[1]*Maximum allowable concentration (or hygienic standards), in milligrams per cubic meter of air, for breathing during a continuous 8 h workday.*

solvent vapors may cause dizziness or nausea, excessive drying or irritation of the mucous membranes, and, in rare cases, allergic reactions with the skin.

The epoxides used in epoxy linings and compounds are particularly irritating to the skin, and some people are seriously affected by allergic reactions if proper hygiene is not practiced while these materials are in use. Common reactions include swelling around the eyes or lips and rashes of the skin. Some epoxy linings have polyamides as curing agents, which react similar to a mild acid on mucous membranes.

Misting and spraying of lead-based paints may be dangerous if the vapors are inhaled. Minute particles of lead may cause lead poisoning if exposure is not avoided. Good personal hygiene is necessary to control ingestion of lead from contamination by cigarettes or food.

The following basic safety precautions should govern the use of all linings:

- Always provide adequate ventilation.
- Guard against fire, flames, and sparks, and do not smoke while working.
- Avoid breathing of vapors or spray mist.
- Use protective skin cream and other protective equipment.
- Practice good personal hygiene.

For details on preparation for opening the tank and the safety aspects of tank cleaning, entry, and lining, see API STD 1631.

5.62 **LINING CONTINUITY AND TESTS**
5.62.1 **DISCONTINUITY DEFINITIONS**

A *lining discontinuity* is any point or area at which a lining film does not cover or hide the substrate of the lining film and which usually will be penetrated by the exposure medium. All discontinuities or latent discontinuities (those which may develop later during service) are to be avoided. Holidays, skips, and misses are gross discontinuities caused by faulty workmanship. In common usage in the industrial lining field, the term *holidays* has become synonymous with *discontinuities.*

  Pores, voids, fish-eyes, pits, and *pinholes* are the names given to various types of small cavities or holes in a lining film, all of which are discontinuities that usually will be penetrated by the exposure medium. These are commonly acknowledged to be caused by (1) application shortcomings, (2) imperfections of the substrate surface, (3) a contaminant in the lining film, (4) a contaminant on the substrate, (5) too-rapid release of solvents or products of reaction such as water, or (6) a shortcoming of the film-forming properties of the lining material.

  Blisters are faults of the lining film caused by too-rapid release of solvents or products of reaction or entrapped air. Blisters often will occur after the applied lining has been exposed to high temperatures. They may or may not become discontinuities, as defined above.

5.62.2 DESCRIPTIVE TERMINOLOGY

The degrees of continuity are termed as Conditions A, B, and C, which are as follows:

* Condition A—Pinhole free. The applied lining film should be absolutely continuous.
* Condition B—Relatively pinhole free. The applied film should contain only a negligible number of points of minor discontinuity. No more than two points of discontinuity should occur within an area having a radius of 15 cm, as measured from a point of discontinuity (i.e., a pinhole). No gross discontinuity (i.e., larger than a pinpoint) should be allowed.
* Condition C—Commercially continuous. The applied film should contain only a small number of points of discontinuity. No more than two points of discontinuity should occur within an area having a radius of 15 cm, as measured from a point of discontinuity (i.e., a pinhole). No more than 40% of the total number of allowable points of discontinuity should occur within any area equal to 25% of the total area being lined.

  The total allowable number of discontinuity points for all specified thicknesses of linings in all three conditions defined above are given in Table 5.13.

  Unless specified otherwise by the company, the degree of continuity of the linings should be condition A.

Table 5.13 Total Number of Allowable Points of Discontinuity			
Surface Area Being Lined (m²)	**Condition A**	**Condition B**	**Condition C**
0.9	0	1	5
0.9–4.5	0	2	10
4.5–9	0	5	20
9–45	0	10	30
45–90	0	15	50
90–450	0	25	75

5.62.3 **USE OF CONTINUITY TESTS**

The ideal time to test a lining film for continuity is at that time during application when most of the solvents and reaction products have been released from the film, but before the film has acquired a degree of cure that will hinder proper correction of any faults that may be located during inspection. With high-bake cured lining, the most practical time for performing tests would appear to be just before the final bake, or just before application of the last coat. However, testing just before the final bake adds to the cost because another heating and drying operation is necessary to harden the lining film to allow for traffic contact (e.g., walking on the surface by the inspectors). Whether made before the final bake or before application of the last coat, testing will not always ensure that the lining film quality will remain the same after the final bake has been completed.

Most chemically converted linings are more readily touched up than are high-bake cured lining and thus can be tested after completion of the cure. Some, however, develop such a high amount of chemicals that proper adhesion of the touch-up lining to the cured film cannot be achieved.

Wet or damp sponge continuity tests should not be used without having determined that the lining will not be detrimentally affected by water at that stage of the application. The tests and instruments used for continuity testing have definite limitations for quantitative or qualitative determinations as to the number of pinholes that exist. In addition, time exposure in service and abuse of the lining may later expose pinholes or other lining deficiencies.

5.62.4 **CONTINUITY-TESTING EQUIPMENT**

The approach to continuity testing is to consider the lining as an electrical insulator and search for holes by trying to make electrical contact with the substrate through the lining. There are two main methods of searching:

- Wet sponge testing, which consists of a wet sponge soaked in an electrolyte as one probe of a low-voltage circuit, the other being the earth return.
- Spark testing, in which a discharge from a high-frequency source or a direct current spark is used to find the fault

5.62.5 **WET SPONGE TESTING**

The instruments used for this type of testing are simple. Normally, there are two circuits powered by low-voltage batteries. The primary circuit carries two probes, one of which is connected to the metal substrate. The second is connected to a sponge that is soaked in a 3% solution of electrolyte to which a small amount of a wetting agent has been added. The sponge is passed over the lining. If the lining contains a pinhole or some other discontinuity, the electrolyte penetrates the defect to the metal substrate, current flows in the primary circuit, which triggers the secondary circuit and the alarm operates.

The sensitivity of this type of instrument depends upon the resistance below which current can be detected in the primary circuit. The voltage at which the instrument operates is not an important factor. The resistivity of 3% sodium chloride solution (a commonly used electrolyte) is less than 100 ohm/cm, whereas the resistivity of tap water is about 1,000 ohm/cm. Therefore, it is important that sponges are soaked in an electrolyte.

It is simple to calculate the minimum size hole that can be detected in a given thickness of lining for a specific sensitivity of the instrument in question. If an instrument has a sensitivity of 106 ohm, current will be detected in the primary circuit if the resistance is less than 106 ohm. Taking the resistivity of the electrolyte as 100 ohm/cm., then it will be possible to find a hole approximately 15 μm in diameter in a 200-μm-thick lining.

If the instrument is of the right sensitivity, a hole 1 μm in diameter will be found, providing that the hole fills with electrolyte. With this type of instrument, all holes are significant because if the electrolyte can penetrate it, then so can a corrodant. In fact, with conditions of varying temperature and pressure in service, the working conditions can be more searching than the test. However, it is possible to produce an instrument that is too sensitive. For example, if the primary circuit of an instrument will operate up to a resistance of 500 M, then in humid conditions, if the operator sing a probe touches the metal substrate, it will trigger the alarm.

Some instruments are available that give a varying signal depending on resistance. This means that the signal has to be interpreted. It is preferable that the instrument responds on a go or no-go basis.

The sensitivity of the test instrument will be specified by the company. There has been evidence in the past that similar models of instrument vary considerably in sensitivity. Therefore, a certificate of performance should be consulted for individual instruments.

Instruments of the wet sponge type can be used for testing a wide range of linings, but they are not considered suitable for use with some partially cured resin linings. Remember that before attempting to repair defects found by wet sponge testing, it is necessary to remove traces of electrolyte and dry the lining.

5.62.6 SPARK TESTING

There are two types of spark testing equipment in general use:

- High frequently with an AC source
- Direct current (high voltage)

The modes of operation of the two types are very different, as described in the next sections.

5.62.7 HIGH-FREQUENCY TEST EQUIPMENT

In high-frequency testing instruments, a Tesla coil is used to generate a high-frequency discharge. Models are available that operate on supply voltages of 240 V, 110 V, or 50 V. The voltages at which they discharge can vary between 5,000 and 50,000 V. Normally, output is controlled, but the actual output cannot be recorded on a meter. However, it is possible to measure the voltage for any set position of the control by the method laid down in BS 358—namely, measuring the gap that the spark will jump between 20-mm-diameter spheres. The voltage of the discharge varies, but the peak voltage for any setting of the instrument will not be exceeded.

The instruments have a single electrode. When the electrode is held on or close to the lined surface, there is a corona discharge. When a fault is present, the discharge concentrates at that point and the fault is readily identified. It is possible to survey a large area very quickly with a small probe. When linings are being examined, for ease of operation, it is possible to use an extended probe so that a band (e.g., 150 mm wide) can be examined in one sweep. Tests have shown that the risk of damage by high-frequency spark testing is remote if the time that the probe is allowed to dwell on any one spot is short.

Breakdown of a lining by spark testing is by erosion until a critical thickness is reached. During the time that the surface of a lining is exposed to a spark in test conditions, the material lost by spark erosion is negligible. The only time the critical thickness of a lining is likely to be reached is when there is a bubble with a thin skin in the lining. Such imperfections are undesirable.

5.62.8 DIRECT CURRENT SPARK TESTERS

This type of instrument may be powered by mains or a rechargeable battery. The output voltage varies with the make, but a common range is 1,000 V to 20,000 V. Most instruments have a voltage control and a built-in voltmeter and this is the model recommended for use. In the case of models operating on battery power, the operator should check frequently that full-output voltage is available. Batteries need recharging at frequent intervals.

With DC instruments, it is necessary to have an earth return. Unlike the high-frequency instrument, there is a discharge only when the electrode is close enough to a defect for the DC spark to bridge the gap to earth. When contact is made, a spark can be seen, but most instruments of this type incorporate an audio signal. The energy of the spark from a DC instrument is greater than that from a high-frequency instrument, and it is possible for defects to be enlarged by the action of the spark.

5.62.9 TESTING

Whichever type of instrument is selected for continuity testing, it is absolutely essential that the whole of the lining is surveyed. The spark from a high-frequency tester ionizes the air, and the spark from such a source will jump a much larger gap than is the case with DC sparks. It is particularly important that electrodes used with DC instruments conform to the surface contour; otherwise, defects may be missed. It may be difficult to meet this requirement when using very wide probes. It is essential that inspectors have an understanding of how the instruments work and the need to survey the linings in a controlled fashion. Because of the need to ensure that continuity testing is carried out to the required standard, it is preferable that the testing is done by a person other than the operators who applied the lining.

5.63 APPLICATION OF RUBBER SHEET LINING IN PIPE
5.63.1 CEMENTING

Apply one coat of primer as soon as possible after blasting to prevent oxidation. Apply additional coats of primer, intermediate, or tie adhesives as specified by the manufacturer. Allow sufficient drying time between coats so the coat being applied does not lift up the preceding coat.

5.63.2 LINING APPLICATION

Whenever calendared or extruded flat sheet lining is supplied, the tube should be formed by using longitudinal skived butt splices. The spliced tube's outside circumference should be slightly smaller than the inside circumference of the pipe to be lined.

Whenever unvulcanized extruded seamless elastomer tubes are supplied, the tubes should have a slightly smaller outside circumference than the inside circumference of the pipe to be lined.

For bleeder strings, apply twisted multifilament string or yarn lengthwise in two to four locations to allow proper gas venting between the lining and the pipe. These strings may be used at the applicator's option.

Enclose the tube in a fabric liner and attach a tow rope. Pull the tube into the pipe with a slow, constant pull. The fabric liner facilitates the positioning of the tube in the pipe and prevents premature bonding.

5.64 TUBE INFLATION
5.64.1 SMALL- AND MEDIUM-DIAMETER PIPE

Remove the fabric liner and expand the elastomer tube against the pipe wall by using pressurized air. A mechanical extension and flange arrangement may be used for the pipe ends so that a minimum of 70 kPa(g) internal pressure can be maintained in the expanded tube for at least 5 min.

5.64.2 LARGE-DIAMETER PIPE

Pipe too large to feasibly line by inflating a tube and large enough to allow personnel to enter should be rubber-lined in the same manner as tanks or duct work. Bleeder strings may be used, at the applicator's option, to facilitate the escape of gases during cure.

Remove extension and flare excess stock over the flange face and trim flush. Apply a covering to the full face of flange. Skive the ID of the flange stock to slightly less than the ID of the lining and stitch firmly to tube stock. On larger pipes, the flange stock may be lapped into the pipe lining instead of the skive used on smaller pipes. Rubber should be removed from bolt holes after curing by means of a knife, reamer, or other suitable tool.

5.64.3 CURING

Cure rubber-lined pipe as specified by the rubber lining manufacturer.

5.64.4 LINING INSPECTION AND SPARK TESTING

Special equipment is usually required so that the spark tester can reach all areas inside long lengths of pipe.

5.64.5 IDENTIFICATION AND PROTECTION

Identify each piece by stamping on a ground area so that numbers will remain visible as described next.

"Rubber Lined—Do Not Cut or Weld." Protect the linings on flange faces during shipment or storage by covering with plywood or another suitable material.

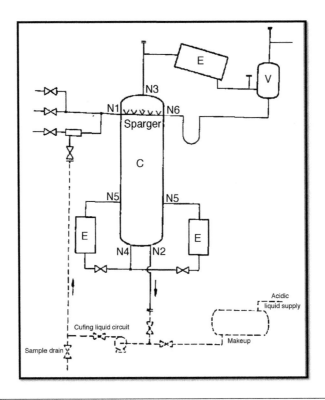

FIGURE 5.14

Typical example for curing the brick lining installed in a column

5.65 CURING OF BRICK LINING WITH ACIDIC LIQUID

The circulation of the acidic liquid to the steam sparger is maintained by a pump. Heat is supplied as steam to the sparger. The acidic liquid is prepared at ambient temperature in a separate vessel (not shown in Figure 5.14). To spray the acidic liquid against the equipment wall, a spray pipe is installed in nozzle N1 (see Figure 5.14 for details).

5.66 CURING OF TANK LINING

This section gives the heating equipment requirements, methods, and control equipment to obtain drying and curing of tank lining materials. This section is limited to the interior tank lining of the dispersion or solvent solution type (i.e., baking, chemically converted, and air-dried) where resistance to immersion is the primary factor.

Only three basic types of lining materials are considered. The first is air-dried materials (e.g., vinyls, vinyl copolymers, and vinylidene chloride), the second is chemically converted materials (i.e., catalyzed epoxies, catalyzed phenolics, catalyzed furanes, and combinations of these), and the third is baking-type materials (i.e., high-baked phenolics, phenolic epoxies, and urethanes). All three of these types must be heated before they are placed in service to ensure optimum chemical and permeation resistance. Only the extent and degree of heating varies with each type.

5.67 **REASONS FOR HEATING LINING MATERIALS**

The basic reasons for heating each of the three types of lining materials are given here for each type.

5.67.1 **AIR-DRIED TYPE**

The reasons for using air-dried lining materials are as follows:

- To remove volatile solvent (or vehicle in case of emulsions) from lining to ensure optimum film density and increase chemical and permeation resistance
- To remove solvent or vehicle from the vessel, to prevent solvent wash and to decrease or eliminate residual odor, so that the curing process can proceed

5.67.2 **CHEMICALLY CONVERTED TYPE**

The reasons for using chemically converted lining materials are as follows:

- To remove volatile solvent and products of polymerization from the lining and the tank
- To accelerate and ensure complete polymerization or cure, to obtain maximum chemical and permeation resistance

5.67.3 **BAKING TYPE**

The reasons for using the baking type of lining materials are as follows:

- To remove volatile solvent and products of polymerization from the lining and the tank
- To accomplish polymerization or curing to obtain chemical and permeation resistance

In each case, one reason for heating is to remove volatile solvent or water not only from the lining, but also from the vessel. Unless it is removed from the tank during the heating process, a portion of the solvent or water will be reabsorbed into the lining material and the heating process will have been all but wasted. Severe examples of this are called *solvent wash,* a redilution of the lining material especially prevalent with thermoplastic materials.

Therefore, when *heating* is mentioned in this section, a type of heating is meant that utilizes a high volume of air circulation over the surface of the film and maximum air change in the tank itself.

5.67.4 **HEATING EQUIPMENT**

Many types of heating units are capable of giving a high-volume (1.4–3.76 m^3/s) stream of hot air. They fall into one of four categories: gas-fired, oil-fired, steam, or electric.

Each of these categories can be subdivided into direct- or indirect-fired. To prevent intercoat contamination, as well as surface dirt and possible degradation of lining film, indirect-fired heating(eliminating products of combustion) is preferred. Indirect-fired heating is appreciably less efficient. The sizable heat loss usually prohibits indirect fire for higher bakes (above 120°C, or even above 66°C in large vessels). When direct-fired units must be used, considerable effort should be made to provide as clean heating air as possible. Direct-fired oil heaters are dirty compared to most direct-fired gas heaters. Sources of gas vary as to combustion products, and the gas-air combustion mixture must be adjusted carefully to achieve the cleanest flame.

For heating to a metal temperature of 82–93°C (as in the force-drying or force-curing of air-dried or chemically converted lining materials, and some between coat-drying of high-baked materials), a high-volume, low-temperature, indirect-fired unit normally is preferred. Steam-operated units generally are adequate; however, gas- or oil-fired units can be used effectively. Steam-, gas-, or oil-fired units are preferred in the field or on large tanks; the electric as well as the gas and oil fired for oven type cures (when heating or baking of the applied lining is accomplished by placing the entire tank in a high volume, hot air oven).

5.67.5 METHOD OF HEATING

There are no definitive rules as to a precise method of heating. Each heating project must be engineered beforehand to determine three points:

- What the metal temperature must be and how long it should be maintained to remove solvent or water from the film
- What the metal temperature must be and how long it should be maintained to complete polymerization in the case of chemically converted or baking-type materials
- The amount of air circulation or frequency of air change required to remove the solvent or moisture from the tank

After these three points are established, it is necessary to determine what equipment will be required, where to place it, how to obtain a uniform distribution of heat and air circulation over the surface of the lining, how to obtain a uniform metal temperature with no hot or cold spots, and whether insulation is necessary, how much, and of what type. Influencing each of these factors is the capacity of each lining material to take heat. This capacity must be determined from the volatility and boiling point of each solvent or blend of solvents in the material. This allows the time-temperature scale to be determined for the coating system. Excessive heat or rate of heat application can result in solvent entrapment or entrapment of polymerization products with resultant blistering or boiling of the finished film. Often, this blistering or boiling is not readily observable; but it causes a marked increase in permeability of the film and is thus detrimental to proper performance.

The following factors are significant in determining the method of heating, equipment to be used, and time-temperature rates to achieve proper curing:

- Tank location (inside a building, as opposed to exposed to the weather)
- Tank supports (steel saddles or on a sand base) and the amount of heat that thus will be lost
- Atmospheric conditions (hot or cold still air or moving air)
- Tank dimensions
- Tank construction (internal baffles that might interfere with circulation and various metal masses in the same tank)
- Hazardous conditions (explosive fumes in the area)

- Whether the tank is insulated already and the type of insulation used
- Accessibility for blowers and heaters
- Filter losses in a dirty atmosphere

The problem of designing a permanent, large-capacity oven for minimum temperature variation throughout is great. Complications arise when it is necessary to have a large volume of air movement or air change, coupled with the necessity of having to accomplish this without knowing specifically the size of the piece or pieces to be heated and without knowing the mass of metal involved.

The design problem is great for an oven when the type and degree of insulation can be specified and fixed and when it is possible to use necessary ducting, baffles, outlets, and fans to circulate the heated air, as well as for removing the cooler air. Design problems are even greater when it is necessary to heat a mass of metal (such as a tank) by introducing the heated air through a restricted and fixed number of openings without the help of internal baffles, ducting, and fans.

There is little or no correlation between air temperature and metal temperature during the heating of a tank because the primary concern for proper curing is the minimum lining temperature. For all practical purposes, the exterior metal temperature and the minimum lining temperature are identical. Therefore, when a temperature is mentioned in connection with the curing of a lining material, it should be in reference to the metal temperature alone.

In addition to the minimum lining or metal temperature, attention must be paid to the maximum metal temperature. Overheating can be as detrimental as underheating or undercuring. Overheating of air-dried and some chemically converted coatings is difficult to detect by physical inspection. On the other hand, baking-type coatings are less sensitive and certainly have many more failures from underbaking than from overbaking, though they can be charred. Unfortunately, the temperature range between the maximum and minimum limits often is small.

An effort has been made to develop an empirical formula for computing heater capacities required to obtain varying metal temperatures in various-sized tanks, but the task has been abandoned because of the many uncontrollable variables that prevent correlation between test results and field experience. The factors necessary to complete proper curing of linings are based primarily on knowledge and skills derived from experience. They include proper calorie capacity, proper heater design, method, type and degree of insulation required, and ways and means of setting up ducting and curing time. These processes still must be considered an art rather than a science, and they are found principally with those few organizations specializing in the application of the three types of tank-lining materials discussed here.

5.68 TEMPERATURE CONTROLS

The controls required to ensure that the necessary minimum metal temperature is obtained and that the maximum temperature is not exceeded at any given spot in the tank consist mainly of surface temperature–measuring devices such as controller pyrometers on the heater, surface contact pyrometers, surface thermometers, temperature indicating crayons, and thermocouples connected to direct reading potentiometers.

Recorders in connection with these instruments would be advantageous. Physical proof would be available after completion of the heating cycle to show that the maximum temperature were not exceeded and that the minimum temperature was obtained and held for the specified period. However, these records would be of little value because they would record the temperatures of a few isolated points rather than the entire surface. The use of recorders, therefore, increase the coat of the applied lining. Such recorders would be another expensive piece of equipment that would have to be amortized against each tank lining project without giving much benefit.

The only value of surface temperature–measuring devices is to assist the applicator performing the heating or curing operation in determining when minimum temperatures are obtained and when an area is approaching the maximum temperature permissible. As these readings are obtained, the applicator must take whatever action is necessary. This person must know, by experience, where the critical areas are in any given tank (where the greatest heat loss will be, thus the area with the lowest temperature, or where the hot spots will be), and must check these areas constantly during the entire heating operation. Insufficient attention to this phase of the operation or a mistake in judgment could cause complete failure, requiring the applicator to remove the lining and start over again.

5.69 TESTING AND INSPECTION

Determining whether the cure or force-dry has been accomplished on an applied lining material by test methods is almost impossible. Therefore, considerable emphasis must be placed on proper engineering of the heating operations and on the controls used during the heating process. Spasmodic work is being done to develop hardness test standards for the three types of materials, but to date, no success has been reported.

Three simple tests are outlined next that can be made during final inspection of a lining to determine whether the material is free of solvent, cured, and ready for service.

5.70 AIR-DRIED LINING MATERIALS

If air-dried linings have been properly heated with high-volume hot air for sufficient time, there should be little or no solvent odor in the case of solvent-based materials. There should be no excessive humidity in the case of water-based materials.

5.71 CHEMICALLY CONVERTED MATERIALS

If the chemically converted lining has been heated properly at a high enough temperature and with sufficient air volume for the proper length of time, the lining should be polymerized completely, with little or no solvent odor present. The lining material should be hard and solid.

If the lining has been sufficiently cured, a solvent-soaked rag rubbed on the surface should not soften the lining or remove any coloring. More conclusive indication can be obtained by rubbing the lining with fine sandpaper. If the lining is properly cured, the sandpaper should not gum up. This test, however, is far from being foolproof.

5.72 **BAKING-TYPE MATERIALS**

If the baking-type lining has been properly heated for the correct period of time, it should be polymerized completely, with little or no solvent odor present. In most cases, a change in appearance or color occurs, although this may be only minor and difficult to observe.

Some applicators have successfully used prepared color comparison strips for each phase of the color change that occurs during the curing of this type of lining. A good indication can be obtained by testing the lining being cured with the appropriate strip. With experience, this technique can become a valuable guide.

BAKING-TYPE MATERIALS

ENGINEERING GUIDELINES FOR PROTECTIVE COATINGS IN BURIED AND SUBMERGED STEEL STRUCTURES

The main task of protective coatings is to prevent or minimize external corrosion of buried or submerged steel structures. The coating isolates metal from contact with the surrounding environment. Since a perfect coating cannot be guaranteed, cathodic protection is used in conjunction with coatings to provide the first line of defense against corrosion. And since a properly selected and applied coating should provide 99% of the necessary protection, it is of utmost importance to know the advantages and disadvantages of available coatings. The right coating material, properly used, will make all other aspects of corrosion control relatively easy. The number of coating systems available requires careful analysis of the many desired properties for an effective pipe coating.

Therefore, optimum selection and proper application of protective coatings is of paramount importance. Over an extended period of time, a protective coating deteriorates as a result of contact with moisture, oxygen, chemicals, fluctuating temperatures, abrasion, pressure, and many other possible factors. Proper and timely maintenance is required to get the optimum performance from a protective coating.

The selection and application of maintenance coating is more complicated than it is in initial construction. Climatic conditions, chemical exposure, available time, budget, health and safety, and the grade of surface preparation have a significant amount of influence on the planning of optimum design coating. Finding out about the operating and installation conditions is the beginning of the process of selecting the best coating system. The available steel sources and job location may limit the coatings available to each project. Selecting a high-quality applicator is the most important consideration—and frequently, it is also the most neglected. Second only to the coating and applicator, inspection at the coating mill, and especially on the job site during construction, is vital to ensuring that a high-quality pipe coating system has been installed.

This chapter covers the minimum requirements for the design and selection of coating systems for the protection of pipes, storage tanks, and piling systems that are to be buried or submerged in water. Corrosion protection of steel structures in the oil, gas, and petrochemical industries, including refineries, chemical and petrochemical plants, gas plants, oil exploration and production units, is explored in detail. Included here are the essential elements involved in surface preparation, selection of coating systems, and repairing of coating defects. Note that the coating of stainless steel, galvanized steel, and nonferrous alloys under external corrosive conditions is subject to the approval of the design engineer.

Essentials of Coating, Painting, and Lining. http://dx.doi.org/10.1016/B978-0-12-801407-3.00006-7

6.1 FIELD OF APPLICATION

This section deals generally with the following structures to be coated, and mainly involving buried and submerged steel pipes:

1. The types of pipes to be coated to which standards apply, including both welded and seamless pipes of nonalloy steel used for the conveyance of gas and fluids.
2. The types of fittings to be coated to which standards apply; mainly bends, tees, reducers, and collars.
3. Valves and insulating joints; the standard involved here applies to all buried valves and insulating joints.
4. Storage tanks that should be externally coated and buried underground or submerged in water; in all cases, the storage tank should considered as a pipe that is closed at both ends, requiring external protective coating at overall surfaces.

Many materials have the basic corrosion control properties, but only a few of them meet the overall needs for protecting buried, submerged, or aboveground steel structures against corrosion.

6.2 PURPOSE OF COATING

Coatings prevent the corrosion of buried and submerged structures in the following ways:

- They inhibit corrosion by providing an adhesive film with a high resistance to ionic transport.
- They reduce the current requirements for cathodic protection by providing an electrically insulating film.
- They assist in the uniform distribution of cathodic protection current.

Although the initial coating procedure can be expensive, coatings will lead to a considerable reduction in cathodic protection power consumption, saving money in the long run. Coatings are considered an integral part of any cathodic protection system, and in most situations, they provide the main part of any corrosion protection system. Cathodic protection provides backup corrosion protection of the structure at points where failure of or damage to the coating has occurred. In compact structures, many combinations of coating systems are used. In situ repairability should be a significant consideration when choosing the coating system.

6.3 COATINGS AND CATHODIC PROTECTION
6.3.1 INFLUENCE OF COATINGS ON CATHODIC PROTECTION REQUIREMENTS

Although it is technically possible to protect buried or immersed steel structures and pipelines without coatings by applying cathodic protection only, it is seldom desirable to do so because the large current required to do this is expensive, and it is often difficult to arrange the anodes to give a uniform current distribution. A good, high-insulation coating greatly reduces the current required to maintain the required steel-to-soil potential and provides a more uniform spread of current from the anodes. A protective coating, therefore, should always be applied to any buried and immersed structure or pipeline that needs to be cathodically protected.

Table 6.1 Design Current Densities for Different Pipeline Coatings (Operating Temperatures up to 30°C)

Coating Type	Pipeline Life (Years)		
	0–5	**5–15**	**15–30**
	Current Density (mA/m²)		
Asphalt bitumen and coal-tar, 6 mm asphalt mastic	0.040	0.100	0.200
Fusion-bonded epoxy Liquid epoxy Coal-tar epoxy	0.010	0.020	0.05
Polyethylene Polypropylene	0.002	0.005	0.010
Plastic tape (laminated)	0.040	0.100	0.200

Note: The current densities given in Table 6.1 already include the current requirements due to the expected coating breakdown during the life of the pipeline.

6.3.2 INFLUENCE OF CATHODIC PROTECTION ON COATINGS

The current required to protect a structure or pipeline is approximately proportional to the area of bare steel (see Table 6.1), so theoretically, cathodic protection should be unnecessary when the steelwork is perfectly coated. In practice, however, coatings are often damaged in transport or during laying, or they may contain imperfections such as pinholes. Even in low-corrosivity soils, the slightest discontinuity in the protective coating may result in severe local corrosion. As a result, in corrosive conditions, even coated structures or pipelines should be given cathodic protection.

Pipeline coatings of bitumen, coal-tar, or epoxy coal-tar are not much affected by properly applied cathodic protection. However, a potential more negative than -2.0 V with reference to a copper/copper sulfate electrode may damage the coatings by causing hydrogen evolution on the steel surface.

Cathodic protection of painted or metal-sprayed and painted structures should be planned carefully because oil-based paints may be saponified by the alkalinity that develops at the cathodically protected surface; sprayed aluminum or zinc may be attacked in a similar way. The surface potential, therefore, should be maintained as closely as possible to the value needed for protection, and overprotection should be avoided.

The recommended off-potential limits for underground coatings (to Cu/CuSO4 half-cell) are as follows:

- Epoxy powder fusion-bonded—1.5 V
- Epoxy coal-tar (not included in the standards)—1.5 V
- Hot-applied enamel (coal-tar and bitumen) —2.0 V
- Polyethylene (two layers)—1.0 V
- Plastic tape—1.5 V
- Polyethylene (three layers)

6.4 COATING DESIGN AND DESIRABLE CHARACTERISTICS OF COATINGS

6.4.1 EFFECTIVE ELECTRICAL INSULATOR

Since soil and salt corrosion is an electrochemical process, a pipe coating has to stop the current by isolating the structure from the environment.

6.4.2 EASE OF APPLICATION

The coating material must be suitable and properly applied. Many excellent pipe coatings require exacting application procedures that are difficult to achieve consistently. Consistent quality may be obtained with a coating system that is least affected by any variables that might exist. Coating application specifications and good construction practice, combined with proper inspection, contribute to the quality of the finished coating system.

6.4.3 APPLICABLE TO PIPING WITH A MINIMUM OF DEFECTS

The ability to be applied to piping with only minimal problems correlates with ease of application. No coating is perfect, which is why cathodic protection is required.

6.4.4 ADHESION TO METAL SURFACE

Coating adhesion is important to eliminate water migration between the metal substrate and the pipe coating. The coating adhesion ensures permanence and the ability to withstand handling during installation without losing effectiveness.

6.4.5 RESIST DEVELOPMENT OF HOLIDAYS

Once the coating is buried, two elements that may destroy or degrade it are soil stress and environmental contaminants. Soil stress, which occurs in certain soils that are alternately wet and dry, creates tremendous forces that may split or cause thin areas. Adhesion, cohesion, and tensile strength are important properties to evaluate in order to minimize this problem. The coating's resistance to chemicals, hydrocarbons, and acidic or alkaline conditions has to be known in order to evaluate performance in known contaminated soils.

6.4.6 HANDLING, STORAGE, AND INSTALLATION

The ability of a coating to withstand damage is a function of its impact, abrasion, and ductile properties. Pipe coatings are subjected to a great deal of handling between application and backfill. While precautionary measures of proper handling, shipping, and stockpiling are recommended, coatings vary in their ability to resist damage. Outside storage requires resistance to ultraviolet rays and temperature changes. These properties must be evaluated to ensure proper performance.

6.4.7 CONSTANT ELECTRICAL RESISTIVITY

Since corrosion is an electrochemical reaction, a coating with a high electrical resistance over the life of the system is important. The percentage of initial resistance drop is less indicative of the quality of the pipe coating than is the overall level of electrical resistivity.

6.4.8 **RESISTANT TO DISBONDING**

Since most pipelines are eventually cathodically protected, the coating must withstand cathodic disbondment. The amount of cathodic protection is directly proportional to the quality and integrity of the coating. Considering interference and stray current problems, this becomes a most important requirement. Cathodic protection does two things: (1) it drives water through a coating that would ordinarily resist penetration; and (2) it may produce hydrogen at the metal surface where current reaches it, and the hydrogen breaks the bond between the coating and melt surface. Cathodic protection cannot render a coating completely resistant to damage, so it is very important to choose a coating that minimizes these effects. The ASTM G8 test for cathodic disbonding of pipeline coatings, commonly known as the *salt crack test,* measures a coating's resistance to damage by cathodic protection.

6.4.9 **EASE OF REPAIR**

Recognizing that some damage may occur and that the weld area must be field-coated, compatible field materials are required to make repairs and complete the coating after welding. Manufacturers' recommendations should be followed. Variables in conditions influence what materials are chosen.

All nine of the abovementioned characteristics are important when deciding what pipe coating to use. The factors described in the next sections should also be considered.

6.4.10 **TYPE OF SOIL OR BACKFILL**

Soil conditions and backfill influence the coating system and thickness that are specified. Soils are rated by their shrink-swell factor (i.e., soil stress). High shrink-swell soils can damage conventional coatings. Ideally, trenches should be free of projections and rocks, permitting the coating to bear on a smooth surface. When backfilling, rocks and debris should not strike the pipe coating. The following ASTM tests are recommended to measure the resistance to penetration of the pipe coating if it is set on stones in the trench:

- ASTM D 785, "Method of Test for Rockwell Hardness of Plastics and Electrical Insulating Materials"
- ASTM D 5, "Method of Test for Penetration of Bituminous Materials"
- ASTM D 2240, "Method of Test for Indention Hardness of Rubber and Plastics by Means of a Durometer"

The following ASTM tests are recommended to measure the resistance against damage by rock in back fill:

- ASTM G 13, "Limestone Drop Test"
- ASTM G 19, "Falling Weight Test"

Soil stresses on pipe coatings may be evaluated by ASTM D 427, "Method of Test for Shrinkage Factory of Soils."

6.4.11 **ACCESSIBILITY OF PIPELINE**

When a pipeline is inaccessible or in a marine environment, the best system should be selected, regardless of initial cost. Performance under similar conditions for at least five years or well-designed laboratory tests on new products are the best criteria to determine the right type of coating.

6.4.12 OPERATING TEMPERATURE OF PIPING

Surface temperature and environmental conditions must be considered because, once buried, a coating experiences wet heat, which is more detrimental than dry heat and harms its effectiveness. ASTM G 8, "Cathodic Disbonding of Pipeline Coatings," is a modified disbondment test that determines resistance to elevated temperatures.

6.4.13 AMBIENT TEMPERATURES DURING CONSTRUCTION AND INSTALLATION

Temperatures during construction and installation are often more critical than during operations. For instance, some thermoplastic systems, such as mastics, tapes, or enamels, may become brittle in freezing temperatures (polyethylene coating systems, however, have been field bent at −40°C). Extra care in handling, transport, and storage is needed under extreme conditions.

6.4.14 GEOGRAPHICAL AND PHYSICAL LOCATION

Pipe source and coating plant location often determine the coating, or at least are factors that affect the cost. Severe environments, such as river crossings, pipe inside casings, exceptionally corrosive soils, high-soil stress areas, and rocky conditions require special attention. For large projects in remote areas, the economics may favor a railhead or field coating site.

6.4.15 HANDLING AND STORAGE OF COATED PIPE

Handling, shipping, and stockpiling are important in the selection process. Some coatings require special handling and padding, and all require careful handling. Most underground coatings are not designed for aboveground use and are affected by excessive aboveground storage. Coal-tar and asphalt enamel and mastic coatings are protected from ultraviolet deterioration by whitewash or craft paper. In polyethylene, the addition of 2.5% carbon black is the most satisfactory deterrent.

Stock should be rotated (on a first-in, first-out basis) to minimize potential problems. Long-term storage requirements determine what coating is selected.

6.4.16 COSTS

Evaluation of pipe coating properties with these considerations assists in selection. The most misunderstood factor is costs. In the economics of pipe coating, the end has to justify the means. The added cost of coatings and cathodic protection has to pay for itself through reduced operating costs and longer life. True protection costs include not only the initial costs of coating and cathodic protection, but also installation, joint coatings, and repairs. Field engineering and facilities to correct possible damage to other underground facilities may add costs, possibly outweighing the initial costs of the pipe coating.

6.4.17 CURRENT DENSITY REQUIREMENTS

The current density required to protect a buried structure depends on the type and performance of the coating in question. Table 6.1 gives minimum design values for new construction projects. The current density values in the table are related to the total pipeline surface area and take into account coating deterioration during the refereed life of the pipeline. It is assumed that pipeline construction is carried out in a manner to avoid coating damage during construction and operation.

For protection of pipelines with elevated operating temperatures, the minimum design current densities given in Table 6.1 should be increased by 25% for each 10°C increase in temperature above 30°C.

6.5 **GENERAL REQUIREMENTS**

Table 6.2 gives the typical properties for coating.

Table 6.2 Typical Properties of Representative Coating Systems for Compact Structures

Coating System*	Coating Site	Ease of On-site Application	Structure Pretreat-ment (for steel)	Coating Thickness mm	Susceptibility to Damage From		
					Soil♣ Stresses	Cathodic Disbond-ment	Impact
Coal-tar Enamel	Field (over-the-ditch) yard	Difficult	Wire brush or blast γ	3–6	Medium	Medium	Medium
Extruded Polyethylene	Shop	N/A	Blast	2.5–3.5	Low	Low	Medium
Fusion-bonded polyethylene	Shop	N/A	Blast	2.5–3.5	Low	Low	Low
Fusion bond Epoxy	Site and shop	Difficult	Blast	0.35–0.45	Low	Low	Low
Asphalt Enamel	Field and yard	Difficult	Wire brush or blast γ	3–6	Medium	Medium	Medium
PVC, polyeth-ylene backed laminated tape	Field	Easy	Wire brush or blast γ	1.5–3.0	High	Medium	High
Petrolatum Tapes	Field	Easy and highly con-formable	Wire brush γ	3–6	High	Not applicable	Medium
Heat-shrink Sleeves	Field	Medium	Blast	1.0–3.0	Low	Medium	Medium
Coal-tar epoxy	Field and yard	Medium	Blast	0.3–0.6	Low	Medium	Medium

Metalliferous primers should not be used in coating systems with structures requiring cathodic protection.
♣*Properties resulting from soils that produce stresses (e.g., clay).*
γ*It is good practice to blast-clean surfaces prior to coating application to ensure maximum adhesion. Wire brush pretreatment, which may leave mill scale on a steel surface, may leave the structure in a condition susceptible to stress-corrosion cracking; thus it is inferior to blast-cleaned surfaces.*
Used on site welded joints; difficult to repair.
Note:
The properties tabulated in this table relate only to the basic standard coating for each system. Coating performance can vary substantially from these values and depends on the characteristics of the actual system used.

6.6 COATING SCHEDULE

6.6.1 TYPES OF COATING

With reference to the previous considerations, only the following types of coatings have been selected for the purpose of this engineering standard. Among these, the desired coating systems should be selected in accordance with the information discussed in the following sections for a particular underground or submerged structure, including offshore risers and piling systems.

6.6.2 BITUMINOUS COATINGS

Bituminous coatings include coal-tar and bitumen (asphalt) enamels that are applied in the molten state. These coatings are applied on the site.

6.6.3 EXTRUDED POLYETHYLENE COATINGS

Two systems are available for extruded polyethylene coatings. One is a polyethylene sleeve shrunk over a hot-applied asphalt mastic adhesive. The first is a dual extrusion where a butyl adhesive is extruded onto the blast-cleaned pipe, followed by multiple fused layers of polyethylene. The second is a three-layer, multistage process where an epoxy powder primer is sprayed onto the hot blast-cleaned pipe, followed by a fused copolymer adhesive and then by multiple fused layers of polyethylene. The latter utilizes multiple extruders in a proprietary method, which obtains the maximum bond with minimum stress. These coatings are applied in the factory.

6.6.4 FUSION-BONDED EPOXY POWDER COATINGS

Fusion-bonded epoxy coatings are applied to preheated pipe surfaces of 204–260°C with or without primer. These coatings should be applied in the factory.

6.6.5 PLASTIC TAPE

Prefabricated, cold-applied plastic tape is normally applied as a three-layer system consisting of primer, corrosion-preventive tape (on the inner layer), and a mechanical protective tape (on the outer layer). These coatings are generally applied in the field.

6.6.6 CONCRETE COATINGS

Cement mortar coatings are usually used for cast-iron pipes and for shielded areas where cathodic protection cannot be used effectively. It is relatively expensive, but it results in a strong, durable coating for speciality applications. Such coatings are used as negative buoyancy and armor protection over ordinary coatings in marine environments.

6.7 CHARACTERISTICS OF SPECIFIED COATINGS

This section describes the main characteristics of the desired coatings. The special characteristics of these coatings are listed in each individual subsection. The conventional coating to be used on a lined pipe should be determined regarding the coating properties given next.

6.7.1 BITUMINOUS ENAMEL (ASPHALT AND COAL-TAR)

The characteristics of bituminous enamel are as follows:

- Chemically inert.
- Highly moisture-resistant for coal tar (less for bitumen).
- Very good electrical resistivity.
- Brittle at low temperatures; sags at high temperatures.
- Both yard and over-ditch coating is possible.
- Dangerous fumes result from the high application temperatures required when using coal tar.
- Service temperatures range from −10 to 70°C, depending on the type.

6.7.2 EXTRUDED COATINGS

The characteristics of extruded coatings are as follows:

- Good chemical resistance.
- Low water absorption.
- Extremely good electrical resistivity.
- Flexible.
- Suited to plant application only.
- Service temperatures range from -40 to 80°C.

6.7.3 FUSION-BONDED EPOXY

The characteristics of fusion-bonded epoxy are as follows:

- Extremely good chemical resistance.
- Polyamide-catalyzed epoxies have better water resistance than amide or amine-adduct-cured epoxy.
- Extremely good electrical resistivity.
- Flexible.
- Plant application only.
- High-impact and abrasion strength.
- Excellent adhesion to steel.
- The proper method of application is electrical deposition; it is the only sure way to prevent pinholes and voids in the coating.
- Service temperatures range from −70 to 120°C.

6.7.4 PLASTIC TAPE

The characteristics of plastic tape are as follows:

- Good chemical resistance.
- Good electrical resistiance.
- Both yard and over-ditch application is possible.
- Low impact and abrasion strength.

- Subject to pressure deformation from rocky backfill and damage during transportation from plant to field.
- Poor resistance to aromatic hydrocarbons.
- Service temperatures range from -20 to $55°C$, depending on the type.

6.7.5 FACTORS AFFECTING THE DESIGN OF PIPELINE COATING

The major influences on the design of pipeline coatings are:

- Diameter and length of pipe
- Service temperature of the pipe's internal media
- Pressure and frequency of expansions and contractions
- Soil resistivity, soil analysis, and soil stress potential
- Statistical condition of the pipeline route and right of way
- Availability of materials and cost
- Limitation of access to work
- Repair access and frequency of prospected repair
- Condition of manpower requirement at the site
- Transport and handling of the pipe sections
- Availability of electricity beside the pipeline for permanent impressed current-cathodic protection installation
- Provision of access equipment
- Special safety and security regulations that may limit the coating design
- Protection conditions of the adjacent pipeline

6.8 SELECTION OF COATING TO BE USED

The coating selected for a specific application ideally should have both the lowest cost per meter of pipe and the desired characteristics of good electrical and mechanical strength and long-term stability under the environmental conditions and cathodic protection. In order to select the optimum coating system, the factors described in the next subsections should be considered.

6.8.1 FUNCTION

The questions to consider concerning the function of the coating system include the following:

- What is the main function of the structure?
- What are the second functions of the structure?

6.8.2 LIFE

The questions to consider concerning the life of the coating system include the following:

- How long is it required to fulfill this function?
- What is the desired amount of time until the system first has to be maintained?

6.8.3 **ENVIRONMENT**

The questions to consider concerning the environment that the coating system must function within include the following:

- What is the general environment at the site of the structure?
- What localized effects exist or are expected to occur later?
- Is the structure buried or immersed? If so, is it immersed in seawater or buried in a seabed?
- Is sulfate-reducing bacteria present?
- Are other types of bacteria present?
- Is the existing amount of soil stress likely to remain consistent, or will it change?
- What other factors may affect the structure (e.g., surface temperature and abrasion, service temperature and fluctuation, and service pressure and fluctuation)?

6.8.4 **SPECIAL PROPERTIES**

What special properties are required of the coating (e.g., coefficient of friction)?

6.8.5 **HEALTH AND SAFETY**

The questions to consider concerning health and safety include the following:

- What problems need to be taken into account during initial treatment?
- What problems need to be taken into account during maintenance treatment?

6.8.6 **TOLERANCE**

The questions to consider concerning the tolerance of the coating system include the following:

- Does the coating need to be able to tolerate indifferent surface preparation?
- Does the coating need to be able to tolerate departures from specifications?
- Does the coating need to be able to tolerate indifferent application techniques?

6.8.7 **COATING SYSTEMS**

The questions to consider concerning the coating system itself include the following:

- What coating systems are suitable?
- Are these systems readily available?
- Are the system elements mutually compatible?
- Which facilities should be required?

6.8.8 **COATING FACILITIES**

The questions to consider concerning the coating facilities include the following:

- Are the coating facilities readily available:
  - For factory application?
  - For site application?
  - For field application?

- Do they cover all sizes and shapes of fabrication?
- Do they permit speedy application?
- Do the facilities permit working in a way that meets adequate standards?

6.8.9 COMPATIBILITY WITH ENGINEERING AND METALLURGICAL FEATURES

The questions to consider concerning compatibility with engineering and metallurgical features include the following:

- Is the design and jointing of the structure compatible with the preferred coating technique?
- Does surface preparation (blasting) or application of coating affect the mechanical properties of the steel in any way that matters?
- Is the system compatible with cathodic protection?

6.8.10 DELAYS

The questions to consider concerning delays include the following:

- What delays should be allowed between fabrication and the first protective coating?
- What delays should be allowed between application of primer and coating?
- What delays should be allowed between application of coating and installation?
- What delays should be allowed between final coating and repair?

6.8.11 TRANSPORT, STORAGE, AND HANDLING

The questions to consider concerning transport, storage, and handling include the following:

- How well does the coating withstand excessive or careless handling?
- How well does the coating withstand abrasion and impact?
- How well does the coating withstand early stacking?
- How well does the coating withstand exposure to seawater during transit?
- How well does the coating withstand exposure to sunlight?

6.8.12 EXPERIENCE

It is important to evaluate the consistent performance of the coating.

6.8.13 EXPORTING/IMPORTING

What special precautions should be taken when the steelwork is exported or imported?

6.8.14 MAINTENANCE

The questions to consider concerning maintenance include the following:

- Is the deterioration of the coating rapid and serious if maintenance is delayed?

- What access is there going to be for effective maintenance?
- What is the possibility of effective maintenance?

6.8.15 COSTS

The considerations concerning costs include the following:

- The approximate costs of
  - the basic system
  - any additional items
  - transport
  - access
- What are the approximate costs of maintenance?

6.8.16 CATHODIC PROTECTION

The questions to consider concerning cathodic protection include the following:

Is there a specific need for restricting cathodic protection current to the absolute minimum? For example, locations where cathodic protection current sources can be installed may be limited and widely spaced, necessitating the best possible current distribution.

Is electricity available to the structure for impressed current cathodic protection systems to be installed?

Are there any restrictions for impressed current systems (e.g., lack of electricity, location, etc.)?

Are there any restrictions regarding soil resistivity, availability of galvanic anodes, etc.?

6.8.17 ACCESS

Will all or part of the structure be installed where it is not readily accessible (such as river crossings, swampland installations, submarine locations and other similar situations)?

Each coating system considered should be evaluated carefully in terms of the preceding items. All application and performance characteristics of each coating must be determined, particularly with respect to limitations beyond which good performance cannot be expected.

A relatively simple coating system may be fully adequate if, for instance, a pipeline is to be installed in a rock-free soil that is not subject so soil stress; if application and installation conditions are reasonably dry and not subject to temperature extremes; if the pipeline operating temperature is not appreciably above the soil temperature; and if pipeline accessibility is reasonable, with no unusual limitations on cathodic protection installations. Typically, a single-layer standard pipeline enamel with felt wrapper or pipeline plastic tapes could do an excellent job.

On the other hand, a coating as described above might not be satisfactory in adverse conditions. Under rocky conditions, a coating system that will resist impact damage and penetration by steady pressure should be specified. If soil stress is a problem, materials that will resist distortion under such conditions should be used. If ambient temperatures are extreme, materials that will not become brittle and crack at low temperatures should be used. If high temperatures are the problem, a material should be selected that will not soften and be easily damaged during handling. If the pipeline, once installed, will be essentially inaccessible for maintenance work, it may be essential to have the best coating available.

The choice between the use of yard-coated pipe and over-the-ditch coating procedures is largely economic. Factors involved include location of the coating plant with respect to the pipeline right of way (which will influence shipping costs) and whether the pipeline project is large enough to justify the cost of using over-the-ditch coating equipment. The cost of over-the-ditch coating can vary considerably with the type of coating being used, as some materials require more equipment and larger crews than others. With some coating materials, establishment of centrally located railhead field coating plants may be justified on large projects. In any event, the choice is best based on a cost analysis for the type of project being planned.

6.8.18 COATING SELECTION CRITERIA

In summing up this subject, the following minimum criteria should be used to select the coating system:

- Resistance to deterioration when exposed to corrosive media
- High dielectric resistance
- Resistance to moisture transfer and penetration
- Applicable with a minimum of defects
- Resistant to bacteria and microbial growth
- Good adhesion to metallic surfaces
- Resistance to mechanical damage during handling, storage, and installation
- Resistance to cathodic disbonding
- Ease of repair
- Retention of physical properties with time
- Conditions during shipping, storage construction, and installation
- The level of inspection and quality control during coating application
- Cost and availability
- Service-proven experience
- Low water absorption
- Compatibility with the type of cathodic protection to be applied to the system in the case of submarine pipes
- Compatibility with the system operating temperature
- Sufficient ductility to minimize detrimental cracking
- Resistance to future deterioration in submerged environment

6.8.19 FIELD JOINT COATINGS

Coated pipe sections connected by welding, mechanical coupling, or both by means of valves or other underground appurtenances are considered field joints. Coating of field joints must be equal to or better than the coating on the pipeline and should be compatible with the main coating.

If materials requiring primer are used, the primer may be hand-applied in a uniform coat. Curing or drying time must be in accordance with the manufacturer's specifications.

Coating materials must be applied substantially free of voids, wrinkles, and air or gas entrapment. This may require the use of materials that will conform to the shape or irregular appurtenances, such as valves. Petrolatum tape coating should be used for irregular shops, such as bare valves and fittings, when applicable.

A new coating must overlap and adhere to existing material. The overlap must be sufficient to allow for shrinkage of both new and existing coating (e.g., 10 cm on each side). When hand-applied tape is selected for field joints, it should be used with 50% overlap on its own.

Field joint coating systems should be suitable for field application. They should be fast and easily applicable, and they should not require special attention.

6.8.20 TYPES OF FIELD JOINT COATINGS

With yard-applied coatings, the coating of many field joints has to be carefully selected and applied with sufficient overlap to ensure that the whole length of the pipeline is protected. For the protection of the joint, many suitable coverings are available. The recommended systems are given in Table 6.3.

Belowground unburied valves should be coated with asphalt mastic to a minimum thickness of 3.5 mm.

Table 6.3 List of Preferred Materials for Coating Welded Joints

Type	Possible Combinations of Different Types of Coating Each Side of the Weld		Choice of Coating First Choice	Second Choice
A	Fusion-bonded epoxy	Fusion-bonded epoxy	Fusion-bonded epoxy	Two-component liquid epoxy
B	Two-component liquid epoxy (if any)	Two-component liquid epoxy (if any)	Two-component liquid epoxy	Hand-applied laminated tape with 50% overlap
C	Fusion-bonded epoxy	Two-component liquid epoxy (if any)		
D	Coal-tar or bituminous enamel	Two-component liquid epoxy	Two-component liquid epoxy Overlap sealed with hand-applied laminated tape	
E	Coal-tar or bituminous enamel	Two-component liquid epoxy (if any)		
F	Coal-tar or bituminous enamel	Coal-tar or bituminous enamel		
G	Polyethylene plastic tape	Fusion-bonded epoxy	Heat-shrinkable tape	Cold-applied laminated tape with self-adhesive tape overwrap
H	Coal-tar or bituminous enamel	Polyethylene		
I	Polyethylene	Multicomponent liquid (if any)		
J	Polyethylene	Polyethylene		

Notes:
Bare or painted pipe or fittings should be coated with hand-applied tape before the relevant butt joint coating is applied.
When the butt weld to be coated is on a pipeline that will operate at less than 30% specified minimum yield strength (SMYS) and less than 20°C, the use of joint coatings other than those detailed in this table may be considered.
Polyethylene refers to both two- and three-layer polyethylene coatings.
In cases where the operation temperature of the pipeline is about 80°C, then the special heat-shrinkable tape with multicomponent liquid epoxy primer should be used. The material specifications should be approved by the company, and the field application of materials should be in accordance with manufacturer instructions.

6.9 COATING APPLICATION

The external coating should be applied according to standards.

The procedure normally includes the following:

- Handling and treatment of coating materials
- Surface preparation
- Temperature, air humidity, and time lags between steps in the coating process
- Testing methods
- Acceptance criteria
- Repair procedure following the attachment of cathodic protection cables, padeyes, etc.
- Handling and transport of coated pipes
- Quality control and inspection
- Coating repair
- Reporting procedure

6.9.1 STATUS OF COATING

The quality control reports should include the following:

- Acceptance criteria according to the coating specifications
- Surface preparation data
- Temperature and humidity measurements
- Number of coats and total dry film thickness
- Adhesion data
- Holiday detection
- Information on the location of reinforcement in the coating

A preproduction test is to be carried out at the coating yard to demonstrate that the coating can be adequately applied under the prevailing conditions.

6.9.2 FIELD JOINT COATING

Field joint coating should be applied according to an approved procedure of a similar nature. The field joint coating should be compatible with the pipe coating.

6.9.3 REPAIR AND REJECTION

Criteria for acceptance, repair, and rejection of coating before burial or submersion of pipe are to be stated.

6.9.4 SURFACE PREPARATION

At the time of application of the coating, the steel surface to be coated should be dry and free of all contaminants (such as previous coatings, paint, loose dirt, grease, oil, salt, and other materials) that could be harmful to the surface preparation or to the adhesion of the coating to the steel. The surface of the steel should be prepared in accordance with standards. The prepared surface should be Sa 2½ to SIS 055900 for all the coating systems specified in this engineering standard.

6.10 DESCRIPTION OF COATING SYSTEMS
6.10.1 BITUMINOUS COATINGS

Enamels are formulated from coal-tar pitches or petroleum asphalts hot-applied (blown bitumen) and have been widely used as a protective coating for many years. Coal-tar and asphalt enamels are available in various grades, and they form a corrosion coating combined with glass wool to achieve mechanical strength for handling. These materials should meet the requirements of relevant standards, and 295 enamel coatings have been the workhorse coatings of the industry and provide efficient, durable corrosion protection.

Bituminous coating systems may be used within a service temperature range of −10 to 70°C. When temperatures fall below 4°C, precautions should be taken to prevent cracking and disbonding during field installation. Enamels are affected by ultraviolet rays and should be protected by craft paper or whitewashing. Enamels also are affected by hydrocarbons. A barrier coat is recommended when contamination exists. This coating can be used on all sizes of pipe.

Enamel coatings are low-cost coatings whose protective properties depend on film thickness. They have good resistance to dilute acids and alkalis, salt solutions, and water, but they are not resistant to vegetable oils, hydrocarbons, and other solvents. They may become brittle in cold weather and soften in hot weather. Enamel-coated articles should not be stacked (also see Table 6.4).

Enamel coatings are not suitable for aboveground structure and piping; it should only be used for underground and subsea structures and pipelines. The coatings may be applied in a coating yard or over-the-ditch, as appropriate for the job. The designer should specify the method of application. Coating at a yard is likely to produce the best results, assuming that proper control is exercised and that subsequent transport, handling, and joint coating are carried out with care.

The cost of materials for hot-applied coatings is usually relatively low, whereas the cost of application is relatively high. These coatings generally should not be used for buried pipelines and structures if the operating temperature is above 60°C, or above 70°C in case of subsea pipelines unless special enamel coating is specified.

Enamel coatings are widely used for submarine pipelines alone or under the concrete weight coating. Enamel coating thickness should be at least 6 mm.

Bituminous enamel, glass-fiber-reinforced coatings should be used for coating line pipes and networks buried in normal soil, except when the soil is contaminated with hydrocarbons or other solvents (see Table 6.4), or when the temperature of the pipeline contents exceeds 50°C when the pipeline is buried in consolidated fill.

Recently, coal-tar enamel use for buried pipes and structures has decreased for the following reasons:

- Fewer suppliers
- Environmental and health hazard regulations
- Increased acceptance of other coatings, such as extruded polyethylene and fusion-bonded epoxy coatings

Table 6.4 Comparison of Bituminous Enamel (Asphalt) Characteristics with Coal-Tar Enamel

Use of Resistant Coatings	Coal-tar Enamel (Hot-applied)	Bituminous Enamel (Hot-applied)
Temperature resistance	Poor	Poor
Abrasion resistance	Fair	Fair
Bacteria and fungus resistance	Good	Poor
Chemical resistance	Good	Good
Hardness	Fair	Poor
Acid-oxidizing	NR	NR
Nonoxidizing	Good	Good
Organic	NR	NR
Alkali	Good	Good
Salts: Oxidizing	NR	NR
Nonoxidizing	Seawater OK	Seawater OK
Solvent: Aliphatic	NR	NR
Aromatic	NR	NR
Oxygenated	NR	NR
water:	Excellent	Good
Moisture permeability	Low	Fair
Petroleum products	Good	NR
Flexibility	Good	Good
Root penetration	Good	Poor
Soil resistance	Excellent	Excellent
Weather and UV light resistant	NR	Good
Principal hazard-application	Coal-tar fumes	

NR, not recommended

6.10.2 DESCRIPTION OF BITUMINOUS COATING SYSTEMS

The bituminous coating system consists of a cold-applied primer coat, which should be selected in conjunction with the bitumen or coal-tar derived coating material with which it should be compatible. Coal-tar primer or bitumen primer should be used with coal-tar enamel or bituminous enamel, respectively, and is suitable only for site or yard application. Fast-drying synthetic primer can be used both with coal tar and bituminous enamel and is suitable for site application and field or ditch application. The primer should apply at the thickness specified by the manufacturer with reference to the relevant standards for each primer.

One or more coats of bituminous coal-tar enamel or bituminous enamel are built up to form the thickness required for the type of protection.

One or more reinforcements of glass fiber mat as inner wrap, embedded in each protective layer.

One protective layer of coal tar or bitumen-saturated fiber glass mat as an outer wrap.

One solar protective layer, in the case of coal-tar enamel, with the following formula as a whitewash to prevent excessive heating of the coating by solar radiation.

All whitewash should be mixed as described next:

Whitewash ingredients should include water, 3.8 l boiled linseed oil, 68 kg processed quicklime, and 4.5 kg salt. Add the salt to water, then add quicklime and linseed oil slowly and simultaneously, and then mix thoroughly. Next, allow the mixture to stand at least 3 days before it is used.

In certain special cases [for example, the nature of backfill, rocky areas, environmental temperatures, or working temperatures (i.e., about 50°C)], an additional mechanical protections as rockshield under the concrete coating and concrete slabs or rockshield may be specified by the designer with reference to the job requirements.

6.10.3 TYPES OF COATING SYSTEM

Two types of coating system are generally specified for bituminous coating as follows.

Single-coat systems consist of the following:

- One coat of primer
- One coat of bituminous enamel
- One layer of glass-fiber inner wrap
- One layer of glass-fiber outer wrap

Double-coat systems consist of the following:

- One coat of primer
- One coat of bituminous enamel
- One layer of glass-fiber inner wrap
- One coat of bituminous enamel
- One wrap of glass-fiber outer wrap

Single-coat systems are usually used for field (over-the-ditch) coating applications and double-coat systems are used for yard applications.

6.10.4 CHARACTERISTICS OF COAL-TAR AND BITUMINOUS ENAMELS

Table 6.4 gives a comparison of bituminous enamel (asphalt) characteristics with coal-tar enamel.

Coal tar is harder than asphalt enamel, but asphalt's weatherability is better. However, proper asphalt enamel can be used for underground waterlines and gas pipelines, but coal tar for oil-processing areas and oil and gas pipelines.

Despite of bituminous enamel, coal-tar enamel, due to the presence of polynuclear carcinogen hydrocarbon compounds, is toxic to vegetable and sea animals. Therefore, it makes a good coating for pipelines in seabeds and forest environments.

6.11 TEMPERATURE LIMITATIONS (THE APPLICATION AND SERVICE TEMPERATURES)

6.11.1 APPLICATION TEMPERATURE

For application of enamel coatings by flooding or other means, the temperature of the coating material should be such that the viscosity is controlled to give the thickness of coating required, but not so high as to cause excessive fuming. No grade of material should be heated above the maximum application temperature given in Table 6.5.

6.11.2 SERVICE TEMPERATURE

In general, materials with higher softening points or lower penetrations are intended for use in higher temperature conditions. Coal-tar service (operating) temperature requirements are met by modifying the combined materials with various plasticizers. Grade 105/15 or bitumen grade A at normal and lower-than-normal ambient temperatures.

Coal-tar grade 105/8 and bitumen grade B are suitable at ambient temperatures in both low, temperate, and hot climates. Coal-tar grade 120/5 should be designated for use at elevated service temperatures up to 70°C, or up to 115°C in the case of offshore pipelines when an additional concrete anti-buoyancy coating material is used. Under these conditions, a degree of hardening will occur early in use. The manufacturer of the product should be consulted about its suitability under particular conditions.

6.11.3 APPLICATION AND INSPECTION PROCEDURE

Surface preparation should be by blast-cleaning to SIS 05 5900 Grade Sa 2½, preceded by removal of surface contamination. The prepared surface should be primed with appropriate primer (for field coating application, only synthetic primers should be used). The primer should be applied at the rate

Table 6.5 Application Temperatures

Grade of Coating Material		Maximum Application Temperature	Service Temperature
Coal-Tar	Bitumen		
		(°C)	(°C)
105/15	A	250	Notes: For service temperatures higher and lower than normal (0–35°C) only synthetic primer should be used Bituminous enamel grade C can be used for service temperature 0–60°C and coal-tar enamel grade 120/5 can be used for service temperatures 0–80°C for buried structures, or up to 115°C in the case of offshore pipelines when an additional concrete weigh coating is used
105/8	B	250	
120/5	C	260	

recommended by the manufacturer, with reference to the relevant standards for the specified primer, and the recommended maximum and minimum rates. It should be allowed to dry to a uniform film free of bubbles and discontinuities.

The primed surface should be enamel-coated only within the time limits recommended by the manufacturer and should be free of dust, moisture, and other contaminants before flood coating. The flood coating of enamel should be applied with an approved machine that also has been equipped for spiral wrapping of the inner and outer wrap.

The first flood coat of enamel should have the inner wrap pulled in so that it does not touch the surface of the steel pipe and is embedded in the middle 50% of the enamel thickness. The second flood coat should have the outer wrap pulled on and securely bonded, without wrinkles, to the enamel. The two flood coats may be combined if the company approves. Each wrap should overlap by at least 25 mm.

The enamel should be applied at a temperature not exceeding that specified in the standards. It should be melted in a boiler fitted with mechanical agitators and should be continuously stirred. All other aspects of enamel handling, melting, and application should be as specified in the standards.

The coating should terminate 250 mm or be cut back 100 mm for sizes up to size DN (500 mm) and 150 mm for sizes over DN (500 mm) from each end of each length of pipe. It should be neatly trimmed to a 45° bevel. The finished thickness of the coating should average 5 mm, with a minimum of 4 mm and a maximum of 6 mm. The minimum thickness over seam or spiral welds may be relaxed to 3 mm, provided that the coating satisfies holiday detection requirements.

After inspection and repair of defects, the coating should be covered with weather-resistant white-wash or similar approved solar protection coating if the coating is applied on site. The coated pipe should be suitably marked to identify the grade of enamel employed.

For either field or yard application of coating, use the procedures described next.

Inspection should include the following points:

- Monitoring the particle size, cleanliness, and mix of the blast-cleaning media
- Visual checks in good light, after blast-cleaning, of the pipe surface for steel defects and occluded grit
- Control of temperature and freedom from moisture of the pipe surface before priming and before flood coating
- The enamel melting and application temperatures
- The location of the inner wrap in the thickness of the enamel
- The adhesion of the coating to the pipe and to the outer wrap
- Overall holiday detection, including testing of repairs
- The adhesion or bond test should be as required in standard

6.11.4 HANDLING AND STACKING

All coated pipe should be handled and transported according to standards. The contractor should ensure that pipe is not handled under unsuitable temperature conditions.

Stacking of coated pipe should be limited to such a height that neither flattening nor indentation of the coating occurs.

6.11.5 FIELD REPAIR, JOINTS, AND FITTINGS

Field coating repair and coating of joints, fittings, and special sections should be performed by using hand-applied laminated tape and its primer. The tape should be wrapped with 50% overlap.

For irregular shapes, such as valves and fittings which are buried, petrolatum tape should be used. Belowground unburied valves should be coated with asphalt mastic to a minimum thickness of 3.5 mm.

6.11.6 STANDARD COATING MATERIALS

The standard coating materials should be as follows:

- Synthetic primer
- Bituminous primer
- Coal-tar primer
- Hot-applied bituminous enamel
- Hot-applied coal-tar enamel
- Inner wrap
- Outer wrap
- Rock shield
- Hand-applied plastic tape coating

6.11.7 CATHODIC PROTECTION CHARACTERISTICS

Bituminous coatings have good electrical resistance and need a low cathodic protection current (see Table 6.1). The recommended design current density for 15–30 years of service life is 200 $\mu A/m^2$ of the external pipe surface.

These coatings are more resistant to cathodic disbonding than other coatings. The recommended potential limits for underground coating (to Cu/CuSO4 half-cell) is −2 volts.

6.12 EXTRUDED POLYETHYLENE COATING

Extruded polyethylene coating has been available since 1956. Its growth and acceptance has been remarkable since that time. Its initial problems of stress cracking and shrinkage have been minimized by improved quality and grade of high-molecular-weight polyethylene resins.

There are two systems available for the coating of line pipes. One is an extruded polyethylene sleeve, shrunk over a primed pipe by the crosshead extruded method. The other is a dual extrusion (via the side extrusion method), where a butyl adhesive (soft primer) or polyethylene copolymer (hard primer) is extruded onto the blast-cleaned pipe followed by multiple fused layers of polyethylene. The latter utilizes multiple extruders in a proprietary method, which obtains maximum bond with minimum stress.

In both methods, the pipe is normally preheated to between 120°C and 180°C, depending on the type of adhesive primer. The sleeve type is available on 130–610-mm pipe, while the dual extrusion is presently available on 63.5–260-mm pipe. The accepted standard to which pipe is coated with these types of polyethylene coatings is DIN 30670.

Improved adhesion and cathodic disbonding resistance can be achieved by priming the pipe surface with an epoxy-based layer, on top of which the adhesive layer and polyethylene coating are being applied. This new-generation, three-layer corrosion protection system with three-layer polyethylene coating should be installed according to standards.

6.12.1 CHARACTERISTICS OF POLYETHYLENE COATINGS

Polyethylene coatings are durable and their penetration and impact resistance are better than the resistance of hot-applied (asphalt or coal-tar enamel) coatings are, therefore, less prone to mechanical damage during transport, handling, storage, and laying. They also exhibit great electrical resistance, which allows for low cathodic protection current requirements throughout a long period of service (see Table 6.1). Pigmenting the material with carbon black has eliminated ultraviolet degradation problems resulting from lengthy exposure to sunlight.

Polyethylene coatings are not recommended for pipelines operating above 65°C or 80°C (depending on the grade of polyethylene).

Polyethylene coatings have good bendability (1.9° per pipe diameter length at −40°C). Swelling may occur in hydrocarbon environments.

These coatings are applied only in the yard at a thickness that depends on the pipe diameter (see Table 6.6).

6.12.2 APPLICATION PROCEDURE

Both crosshead and side extrusion procedures preheat a bare pipe prior to blast cleaning to SIS 050590 Grade Sa 2½ with the sleeve type coating. The adhesive undercoating is applied by flood-coating the hot material over the pipe before it passes through an adjustable wiper ring that controls the thickness. After adhesive primer is applied, the pipe passes through the center of the crosshead die where polyethylene is extruded in a cone shape around the undercoating and pipe. Right away, the polyethylene is quenched with water to shrink it around the undercoating and pipe. Following electrical inspection, the pipe ends are trimmed for cutting back, and the coated pipe is stockpiled.

In the dual-extrusion (side-extrusion) system, the blast-cleaned pipe is rotated at a calibrated rate. The first of two extruders applies a film of adhesive primer (soft or hard) of predetermined width and thickness, fusing the film to the rotating pipe in two layers. While the primer is still molten, high-molecular-weight polyethylene is applied from the second extruder in multiple layers of a predetermined

Table 6.6 Minimum Thickness for Polyethylene Coatings		
	Minimum Thickness (mm)	
Diameter of Pipe (mm)	**Standard**	**Reinforced**
Up to 250	2.0	2.5
250 to 500	2.2	3.0
500 to 800	2.5	3.5
800 and over	3.0	3.5

thickness, producing a bonded coating 2–3.5 mm thick. Water quenching, electrical inspection, thickness measurement, visual inspection, and cutback is completed prior to stocking.

For polyethylene copolymer adhesive, the system requires a high temperature (i.e., 200°C) for the application of adhesive primer. For three-layer polyethylene coating, the side-extrusion system equipped with an electrostatic spray gun to allow a powder epoxy primer to be applied to the cleaned pipe prior to application of adhesive primer. Application of three-layer polyethylene coating should be according to standards.

6.12.3 HANDLING AND STACKING

All coated pipes should be handled and stored at the coating factory and handled, transported, and stored for installation according to standards.

6.12.4 FIELD JOINTS AND SPECIALS

Field joints and specials should be coated either by polyethylene shrink tape or sleeve or by cold-applied tape to be put in place with 50% overlap over its primer.

Polyethylene shrink tapes and sleeves have both advantages and disadvantages over conventional cold-applied tape. Their advantages are that they are self-tensioning and resistant to direct sunlight. Their disadvantages are that they require a source of heat (a flame torch) for application, which is a major issue; field construction crews must be skilled to apply the heat shrink tape and sleeve properly; the application process is slow and time-consuming; and they are more expensive.

For irregular shapes, such as valves and fittings which are buried, petrolatum tape should be used.

6.12.5 CATHODIC PROTECTION CHARACTERISTICS

Due to high electrical resistance, coatings need low cathodic protection current throughout a long period of service. The recommended design current density for 15–30 years of service life is 10 $\mu A/m^2$ of the external pipe surface. The recommended potential limits for underground coatings (to Cu/CuSO4 half-cell) is -1.0 V.

6.12.6 FUSION-BONDED EPOXY COATING

Fusion-bonded powder epoxy coatings were introduced in 1959 and have been commercially available since 1961. This is a thin-film coating and can be applied to small and larger diameter pipes (19–1,600 mm). This coating is widely used for land-based pipelines operating at elevated temperatures.

Fusion-bonded powder epoxy coatings have good mechanical and physical properties and may be used aboveground or belowground. On aboveground installations, to eliminate chalking and to maximize service life, topcoats should be used with a urethane paint system. Of all the pipe coating systems, the fusion-bonded epoxy resin system is most resistant to hydrocarbons, acids, and alkalies.

Perhaps the main advantage of fusion-bonded powder epoxy coatings is that because they cannot cover up apparent steel defects due to their lack of thickness, they permit excellent inspection of the steel surface before and after coating. The number of holidays that occur is a function of the surface condition and thickness of the coating, and increasing the thickness of the applied coating will minimize this occurrence.

6.12.7 CHARACTERISTIC OF FUSION-BONDED EPOXY COATING

Despite its low film thickness (350–450 μm), fusion-bonded epoxy coating displays many desirable characteristics that are not found in the traditional pipe coating systems. For example, this system is tough, has great flexibility, and provides good adhesion to the steel pipe, along with extremely good chemical resistance. In terms of its high dielectrical strength, very small quantities of current for complete cathodic protection are required.

An extra benefit from epoxy pipe coating that is not easily achieved with other coating systems is the ability to withstand a relatively high temperature (approximately 100°C) for an extended length of time without damage, provided that the environment is dry. Some epoxy thin-film systems can even withstand wet environments at this elevated temperature. It is currently the best practice to specify this type of coating for land-based pipelines operating at temperatures 65°C.

A shortcoming of the coating system is its increased sensitivity to sharp impact damage, which requires careful attention during transportation, field handling, and pipe laying. Fortunately, impact damage does not normally cause disbonding outside the damage area and can be readily repaired by hot-melting or with liquid epoxy resins.

Experience has shown that proper surface preparation prior to the application of the epoxy resin powder is extremely important with this coating. To obtain a satisfactory coating, furthermore, it is absolutely necessary that good quality control during the application process is strictly performed.

Fusion-bonded epoxy coating should be ordered only against detailed specifications covering both the epoxy resin materials and their application. It should be applied in accordance with ANSI/AWWA C213.

The coating should be applied to a minimum thickness of 350 μm, and a maximum of 450 μm of coating applied outside these limits should be rejected and reprocessed.

6.12.8 APPLICATION AND INSPECTION PROCEDURE

The coating is plant-applied by applying epoxy resin powder by means of multiple electrostatic guns on to a blast-cleaned (to Sa 2½ to SIS 505900) and preheated pipe (approximately 230–240°C). The pipe surface should be free of protective oil, lacquer, and mill primer. The pipe surface should also be as free as possible from scale, silvers, laminations, and similar defects.

The pipe surface should be blast-cleaned. The cleaning media should be selected to achieve a surface profile of 40–80 μm. The appropriate blend of shot and grit to achieve this profile is necessary.

The powder epoxy should be used in accordance with ANSI/AWWA C215.

6.12.9 HANDLING AND STACKING

All coated pipes should be handled and stored at the coating factory in accordance with ANSI/AWWA C215. One shortcoming of the coating system is its increased sensitivity to sharp impact damage, which requires careful attention during transportation, field handling, and pipe laying.

Transportation, field handling, and storing for installation should be done in accordance with standards.

6.12.10 **FIELD JOINTS AND FITTINGS**

Pipe joints and fittings should be coated by hot-melting or with liquid epoxy resins. The materials should be in accordance with ANSI/AWWA C215.

6.12.11 **CATHODIC PROTECTION CHARACTERISTICS**

Due to high electrical resistance, the coating needs low cathodic protection current over a long period of service. The recommended design current density for 15–30 years of service life is 50 $\mu A/m^2$ of the external pipe surface. The recommended "off"-potential limits for underground coating (to Cu/CuSO4 half-cell) is -1.5 Volts.

6.12.12 **PLASTIC TAPE COATING SYSTEM**

Cold-applied plastic tape coating systems are applied as three-layer systems consisting of primer, corrosion-preventive tape (on the inner layer) and a mechanical protective tape (on the outer layer). This system is recommended for temperatures up to 57°C. This temperature is a limitation for normal tape, but there are special tape systems available for temperatures up to 93°C.

The primer's function is to provide a bonding medium between the pipe surface and the adhesive or sealant on the inner layer. The inner-layer tape consists of a plastic backing and adhesive. This layer protects against corrosion, so it has to provide high electrical resistivity and low moisture absorption and permeability, along with an effective bond to the primed steel.

The outer-layer tape consists of a plastic film and an adhesive of the same type of materials used in the inner-layer tape. The purpose of the outer-layer tape is to provide mechanical protection to the inner-layer tape, and also to resist the elements during outdoor storage.

6.12.13 **CHARACTERISTICS OF COATING SYSTEMS**

Plastic tape coating systems have the following characteristics:

- A three-layer coating system is applied in normal construction conditions. This coating system is applied cold to a prepared pipe surface.
- The coating can be applied by hand to small-diameter pipes and small pipe sections, but it normally would be applied by machine.
- The coating can be easily applied in the field.
- The coating is suitable for operation temperatures ranging from −34°C to 57°C.

6.12.14 **DESCRIPTION OF COATING SYSTEM**

The plastic tape coating system consists of the following:

- One primer coat, properly applied. The material specifications should be in accordance with standards.
- One protective layer (inner wrap), which should be applied to the primed steel pipe. The material specifications should be in accordance with standards.
- One mechanical protective layer (outer wrap) to be applied over protective layer. The material specifications should be in accordance with standards. The spiral overlap of each layer should be 2.54 cm.

6.12.15 **APPLICATION AND INSPECTION PROCEDURE**

Prior to the application of primer, the pipe should be prepared to Sa 2½ according to SIS 505900 by blast cleaning.

In the field application of coating, the pipe is welded together beside the canal. Then the surface preparation, priming, and wrapping is performed continuously over the ditch. The coating is inspected simultaneously, and the approved coated pipeline will be buried.

In the yard and field application of coating, the pipes are surface-cleaned and primed at the yard. The primed pipes are transported to the field, jointed, cleaned of contamination, reprimed, wrapped, inspected, and buried.

Field coating repairs and coating of fittings and special sections should be performed using hand-applied laminated tape and its primer. The tape should be wrapped with 50% overlap.

6.12.16 **CATHODIC PROTECTION CHARACTERISTICS**

Due to good electrical resistance, the coating needs low cathodic protection current (see Table 6.1). The recommended design current density for $15 - 30$ years of service life is 200 $\mu A/m^2$ of the external pipe surface.

Although plastic tape coatings have certain advantages and are relatively easy to apply, many problems have arisen with this system in practice. One major drawback of tapes is their sensitivity to disbonding, particularly at the overlaps. As a result, cathodic protection currents are easily shielded, rendering the cathodic protection system ineffective so that corrosion can proceed unabated. For this reason, the use of tape coatings should be limited to special cases where other coatings cannot be selected.

Recommended potential limits for underground coating (to Cu/CuSO4 half-cell) is -1.5 V.

6.13 **CONCRETE**

Mortar lining and coating have the longest history of protecting steel or wrought-iron coating and cast iron from corrosion. When steel is encased in concrete, a protective iron oxide film forms. So long as the alkalinity is maintained and the concrete is impermeable to chlorides and oxygen, corrosion protection is obtained. See AWWA C205 for a detailed reference on concrete coatings.

Today, concrete used as corrosion coating is limited to internal lining. The external application is applied over a corrosion coating for armor protection and negative buoyancy in marine environments. A continuously reinforced concrete coating has proved to be the most effectively controlled method.

Materials including water, sand, and heavy aggregate and cement are mixed in the application plant. Then they are conveyed by belt to the throwing heads, where controlled-speed belts and brushes deposit the mixture onto the coated pipe surface. The rotating pipe is moved past the throwing heads to receive the specified thickness of concrete. Simultaneously, the galvanized wire reinforcement is applied with an overlap. To increase tensile strength and to improve impact resistance, additional layers of wire of steel fibers may be specified. Welded wire cages are an alternative method of reinforcement. Other application methods include forming or molding of concrete in place and applying it to the pipe by means of a plastic film.

6.13.1 CONCRETE WEIGHT COATINGS

Concrete weight coatings are normally applied to offshore pipelines, river crossings, and marsh lines to maintain lateral and vertical stability. The amount of concrete is determined by the calculated required submerged weight of the pipeline, which is also known as *negative buoyancy*.

Most frequently, the concrete is applied by impingement over an anticorrosion coating of asphalt or coal-tar enamel. This design has demonstrated good short- and long-term characteristics. Combined with properly selected tensioners on a lay barge, this design has also been successfully installed offshore in many areas.

There must not be any electrical contact between the pipe and the reinforcement, as this may make subsequent cathodic protection of the pipe difficult or even impossible.

Application methods for concrete coatings other than by impingement are being developed to resolve problems resulting from applying the coating over anticorrosion, fusion-bonded epoxy coating. Current experience with these applications is limited.

6.14 COATING OF SUBSEA PIPELINES

Subsea pipelines are defined as lines that are laid in or on the seabed. This section covers the requirements for coating against corrosion the external surfaces of pipelines that are welded and joint-coated on a lay barge, followed by laying pipe over the stringer. It also discusses pipelines laid by reel barge or by pulling into the sea or across creeks, estuaries, rivers, and canals.

All coatings on such subsea pipelines should be compatible with concrete or bituminous weight coating and with normal levels of cathodic protection. The protective potential is no less than $-1.30\,$V, measured against a silver/silver chloride half-cell.

Hot-applied coal-tar enamel glass fiber reinforced coatings should be used to coat subsea pipelines, which may or may not be weight-coated. It should not be used in the following circumstances:

- The temperature of the pipeline contents exceeds 70°C.
- The pipeline is to be laid from a reel.
- The pipeline is to be laid by pulling or placing and is not to be concrete weight-coated.

Epoxy powder coating should be used when one or both of the following conditions apply:

- The temperature of the pipeline contents is too high for coal-tar enamel but does not exceed 95°C (200°F). This includes pipeline risers.
- The pipeline is to be laid from a reel.

Pipelines that are laid by pulling or placing and are not concrete weight-coated (this includes prefabricated spool pieces), should be coated with epoxy powder. Plastic tape coating should not be used for subsea pipelines.

When designing coating for subsea pipelines, the factors described in the next sections should be considered.

6.14.1 CONCRETE SLIPPAGE

For submarine applications, some coatings (e.g., fusion-bonded epoxy, polyethylene, and polypropylene) will normally need an intermediate coating to provide increased friction to avoid slippage between concrete and coating during pipe-laying. For lay barges with a single tensioner, precautions may be needed to avoid breakage and slippage of the concrete at the ends of the pipe. This might be achieved with temporary infill blocks, or it could incorporate stronger longitudinal reinforcing wire. (Any dimensional irregularities at the end of the coating, such as "bell ends," will exacerbate this problem.) The exposed end portion of anticorrosion coating may become too short or even disappear if slippage does occur.

6.14.2 ANTICORROSION COATING DAMAGE

An intermediate "barrier" layer may be needed to prevent damage from the concrete impingement process.

6.14.3 CHOICE OF ANTICORROSION COATING

There are normally many factors involved in the choice of coating for a particular pipeline. The above two potential problem areas may need to be taken into account when making this decision.

6.14.4 PIPE DIMENSIONS AND STIFFNESS

Pipes with large diameter/wall thickness (D/t) ratios have a tendency to become oval when loaded externally. They also may buckle at the field joint area when concrete-coated pipes are installed. Large concrete thickness can make this occur.

FUNDAMENTALS OF PROTECTIVE COATING CONSTRUCTION

This chapter gives the minimum requirements for the initial construction and maintenance of coating of steel pipelines. It focuses on corrosion protection of pipelines in the oil, gas, and petrochemical industries, underground facilities of gas transmission and distribution system, marine and subsea facilities and where applicable. Inspection and test methods for quality control are also specified, and construction requirements of the following coating systems for pipelines are discussed:

- Cold-applied tape coating
- Fusion-bonded epoxy coating
- Hot-applied enamel coating
- Concrete coating
- Polyethylene coating

7.1 GENERAL CONDITIONS FOR APPLICATION

Weather is an externally important element that must be taken into consideration during any coating application. Attention should be paid to the temperature and humidity limits of each coating.

The coating should not be applied in the following environments:

- Under windy conditions
- When the temperature is less than 3°C above the dew point
- On wet or damp surfaces, as adhesion will be affected
- When the ambient temperature is less than 4°C

7.2 COATING MATERIALS

The coating materials should be certified to establish that they comply with all the provisions contained in the relevant standards for coating, and prior to application. The contractor may make any investigations necessary, by testing or other means, to ensure compliance.

When the contractor is responsible for supplying coating materials, they should be certified by the manufacturer to be in accordance with the requirements cited in the relevant standards for the coating. The contractor should obtain and retain all certificates and manufacturers' data sheets and the latter should be made available for examination on request. The company may make any investigation

Essentials of Coating, Painting, and Lining. http://dx.doi.org/10.1016/B978-0-12-801407-3.00007-9

necessary, by way of testing, batch sampling, manufacturing and factory inspection, to satisfy itself that the contractor has complied with all requirements.

7.3 MATERIAL HANDLING AND STORAGE

The materials should be stored in the manufacturer's original packaging under ventilated conditions and away from direct sunlight. Where required and applicable, they should be kept in air-conditioned storage. In addition, they should be handled in such a way that they do not suffer any damage.

All coating materials consigned to the coating site should be properly stored in accordance with the manufacturer's instructions at all times to prevent damage and deterioration prior to use. Materials should be used in the order in which they are delivered.

7.4 IDENTIFICATION OF COATING MATERIALS

All materials supplied for coating operations should include the following information:

- The manufacturer's name and address
- The material and order number
- The batch number
- Date of manufacture and shelf life
- Directions for mixing and thinning with solvents, if any
- Directions for handling and storing the coating materials
- Material safety data sheet

7.5 PIPE IDENTIFICATION

All identification markings, whether internal or external to the pipes, should be carefully recorded before surface preparations begin. The date that the coating was finished and the coating factory markings, including pipe identification, should be legibly marked on the coating surface of each pipe.

7.6 PROTECTION OF WELD END PREPARATION

Weld end preparations should be protected from mechanical damage during handling, storage, surface preparation, and the coating process. The methods used should also ensure that no damage occurs to the internal surface of the pipe. Protection during handling and storage should be in accordance with relevant standards. Weld end preparations should be protected from coating during the application process by a method approved by the company.

For technical reasons, the ends of the pipes should be free of any coating layer (cut back) over a length of 100 mm for sizes up to size DN (500 mm) inclusive and over a length of 150 mm for sizes over DN (500 mm), unless specified otherwise by the company.

The uncoated ends of pipes should not exceed 150 mm unless specified otherwise by the company. They should be protected against atmospheric corrosion by a temporary paint or primer that is easily removable by brushing.

7.7 **SURFACE PREPARATION**

The method of surface cleaning and surface preparation should be specified by the contractor as part of the coating procedure qualifications. The surface preparation for the coating application should be performed in accordance with the standards.

When oil, grease, or other contaminants are present, they should be removed with a suitable solvent without spreading them over the surface. For pipes that have been subjected to contamination, the contaminant should be removed by washing either with potable water or an approved chemical cleaner. If a chemical cleaner is used, subsequent washing with potable water will be necessary. The pipe should be dried before blast cleaning.

Pipes should be blast-cleaned to a minimum of an Sa 2½ finish to ISO 8501-1. The blast profile should be between 40 and 100 μm in thickness, unless specified otherwise by an individual coating system measured by the standard method.

The metal surface should be inspected immediately after blast cleaning, and all slivers, scabs, and other flaws revealed by blast cleaning that are detrimental to the coating process should be removed using a method approved by the company. After the removal of defects, the remaining wall thickness should comply with the relevant pipe specifications. Any pipe found to have defects that exceed the levels permitted in the relevant pipe specifications should be set aside for examination by an authorized company representative, and no subsequent action should be taken without the agreement of the company.

Directly before coating, any dust, grit, or other contaminants should be removed from the pipe surface by a method established as acceptable by the relevant coating procedure test and recorded in the relevant coating procedure. When rust blooming or further surface contamination has occurred, the pipe should be cleaned again in accordance with standards and blast-cleaned again, also in accordance with standards. Coating should take place before any further contamination or rust blooming appears.

7.8 **COATING PROCEDURE TESTS**

The coating process should comply with the procedure established in the coating procedure qualifications. Any changes in coating materials, pipe dimensions, the pipe manufacturing process, or the coating process may necessitate a new coating procedure approval tests, at the discretion of the company. In addition, approved procedure tests should be confirmed as proving tests, at intervals of not more than 1 year, for each type of coating material used by the contractor and for each size of pipe and pipe manufacturing process as requested by the company.

7.9 **INSPECTION AND TESTING (QUALITY CONTROL)**

The quality control system should include at least the requirements listed in Table 7.1. All inspections and testings listed in Table 7.2 should be made by the contractor and witnessed and certified by the inspector.

After examination or testing, should the inspector find that any pipe has not been cleaned or coated in accordance with this construction standard, the contractor must remove the coating that is considered defective or inadequate, and to reclean and recoat the pipe until the inspector approves.

Table 7.1 Minimum Quality Control Requirements

Requirements

Check the cleanliness of components immediately prior to cleaning

Monitor the size, shape, and cleanliness of the blast-cleaning material and process

Check visually, in good light, the surface of the components for metal defects, dust, and entrapped grit

Check the component surface blast profile

Check for residual contamination of component surfaces

Check temperature control of the component surface by an agreed method

Check the weather conditions

Check the coating thickness

Check the coating continuity

Check the coating adhesion

Check the cure of coating

Supervision to ensure the adequate and proper repair of all defects

Check on the coating appearance

Check for damage to weld and preparations

Table 7.2 Coating Requirements and Test Methods for Coating Procedure Approval Tests

Approval Tests	Hot-Applied Enamel	Cold-Applied Tape	Fusion-Bonded Epoxy Coating	Concrete
Surface preparation	1) Visual inspection			
	2) Acceptable limit: As specified in 5.7			
Coating thickness	AWWA C 203	AWWA C 214	AWWA C 213	AWWA C 205
	Minimum requirements: As specified by the design date			
Porosity (holiday)	AWWA C 203	AWWA C 214	AWWA C 213	Not applicable
	Acceptable limit: As specified in 5.15.4			
Adhesion	AWWA C 203	AWWA C 214	AWWA C 213	Not applicable
Impact resistance	AWWA C 203	AWWA C 214	AWWA C 213	Not applicable
Elongation/shear adhesion	Not applicable	AWWA C 214	AWWA C 213	Not applicable
Temperature resistance	AWWA C 203	AWWA C 214	AWWA C 213 (Hot water resistance)	Not applicable
Penetration resistance	AWWA C 203	AWWA C 214	AWWA C 213	Not applicable
Insulation resistance	AWWA C 203	AWWA C 214	Not applicable	Not applicable
Cathodic disbonding	AWWA C 203	AWWA C 214	AWWA C 213	Not applicable

The inspector should have access at any time to the construction site and to those parts of all plants that are concerned with the performance of work under this construction standard. The contractor should provide the necessary inspection tools and instruments for the inspector, as well as the normal facilities necessary for inspection.

7.10 QUALITY SYSTEM

The contractor should set up and maintain such quality assurance and inspection systems as are necessary to ensure that the good and service supplied comply in all respects with the requirements of construction standard. The company will assess such systems against the recommendations of the applicable parts of ISO 9002 and should have the right to undertake such surveys as are necessary to ensure that the quality assurance and inspection system are satisfactory.

The company should undertake the inspection and testing of goods and services during any stage at which the quality of the finished good may be affected and also to undertake inspection or testing or raw materials, purchased pipes and components, or both.

7.11 TEST CERTIFICATES

The contractor should issue the required certificates in accordance with DIN 50049 (Para.: 2.3-1986) for all coating production tests. For all tests witnessed by the inspector, a certificate should be prepared and issued by the contractor and certified by the inspector in accordance with DIN 50049 (Para.: 3.1.A-1986).

7.12 PROCEDURE QUALIFICATION

Before bulk coating of pipes commences, the requirements of should be met and a detailed sequence of operations to be followed on the coating of components should be submitted to the expert for checking the compliance with this construction standard and formal approval. The company should also specify which coated pipes are to be subjected to testing in order for the coating procedure to be approved formally. No coated pipes should be dispatched to the company, nor should the coating process be completed, until the procedure has been approved in writing by the company.

7.12.1 COATING PROCEDURE SPECIFICATIONS

The coating procedure specifications should incorporate all of the following (but not limited to them):

- The coating systems to be used, along with the appropriate data sheets
- Cleaning of pipes and components, as well as the method of cleaning
- Cleaning medium and technique
- Blast-cleaning finish, surface profile, type of abrasive, and surface cleaning in the case of blast cleaning
- Dust removal
- Coating application methods

- Preheat time and temperature, if any
- Powder spray, if any, including use of recycled material
- Curing and quenching time and temperature
- Post cure time and temperature
- Repair technique
- Technique used to strip coating

7.12.2 COATING PROCEDURE APPROVAL TESTS

A batch of 10–20 pipes of any specific pipe mill should be selected by the inspector and coated by the contractor in accordance with the approved coating procedure specifications, with the coating operations being witnessed by the inspector. Three pipes from the coated pipes should be selected by the inspector and subjected to the complete set of tests. Testing should be witnessed by the inspector, and a full set of records should be presented to the company for consideration.

Bulk coating of pipes should not commence until all short- and long-term test results have been approved officially by the company, unless the contractor takes responsibility should any long-term testing fail. All test methods should be in accordance with Table 7.2.

7.12.3 SHORT-TERM APPROVAL TESTS

To test the thickness, at least 12 measurements should be made in accordance with Table 7.2 at locations uniformly distributed over the length and periphery of each pipe selected for the test. 50% of these measurements should be made along and over the longitudinal weld seam, if any.

When testing porosity, each pipe being examined should be holiday-detected over 100% of its coated surface. No defect should be observed when tested in accordance with Table 7.2.

The adhesion test should be carried out on each pipe at five locations uniformly distributed over the length and periphery of the pipe. In this respect, the requirements and the method of the testing should comply with Table 7.2.

Long-term approval tests should be performed, when specified by the company, on sections taken from all three coated pipes selected for the testing. The adhesion test should be carried out at five different locations on five test sections, but after 30 days keeping in the hot air. No change in the mean force necessary to pull the coating off must occur.

Cathodic disbanding, temperature resistance, penetration resistance, impact resistance, insulation resistance, and elongation should be tested for compliance with Table 7.2.

7.13 PRODUCTION COATING REQUIREMENTS
7.13.1 COATING PROCESS

The production coating process should be approved in accordance with standards. The thickness of the coating should comply with the values specified when tested in accordance with the method specified in Table 7.2.

Where pipe is to be concrete-coated, the person to specify the correct thickness of anticorrosion coating to be applied should be identified.

7.14 INSPECTION AND REJECTION OF FINISHED COATING

The inspection of finished coating should be in accordance with the standards. The quality and values to be achieved should be the same as those identified in the standards.

7.14.1 COATING COLOR AND APPEARANCE

Coating color and appearance should be uniform and free of runs, sags, blistering, roughness, foaming and general film defects.

7.15 COATING REQUIREMENTS AND TEST METHODS

After formal approval of all short- and long-term tests, the contractor will be authorized to begin bulk production and should perform the routine inspection and tests in accordance with standards during coating production. All the inspection and tests witnessed by the inspector should be certified.

7.15.1 SURFACE PREPARATION

This test should be carried out on each individual pipe as detailed in Table 7.2. Every pipe that does not comply with the minimum requirements must be rejected for repreparation.

7.15.2 THICKNESS

Unless specified otherwise, thickness testing should be carried out on each individual coated pipe. Every pipe that does not comply with the minimum requirements specified should be rejected for subsequent stripping and recoating. Should two consecutive pipes fail to satisfy the requirement, the cause should immediately be determined. If the cause is not discovered after four consecutive attempts, the coating process should be stopped to perform a full investigation, including checking all the pipes back to the preceding acceptable pipe.

7.15.3 POROSITY

Each individual line pipe should be checked for holidays over 100% of its coated surface in accordance with Table 7.2. Up to two holidays per pipe length will be allowed on a maximum of 5% of coated pipe lengths during each 8-h production shift.

Any individual pipe with more than two holidays should be rejected for subsequent stripping and recoating. If more than two holidays per pipe length are detected on two consecutive pipes, the cause should immediately be determined. If the cause is not resolved after four consecutive attempts, the coating process should be stopped so that a full investigation can be conducted. All holidays detected on nonrejected pipes should be repaired in accordance with standards and satisfactorily retested.

7.15.4 ADHESION

This test should be carried out three times during each 8-h production shift on each individual lined pipe. It should be performed in accordance with Table 7.2 and at the ends of the pipe coating surface. If the coating adhesion at any location is below the desired requirements, the pipe should be rejected for subsequent stripping and recoating; in this case, the second consecutive pipe should be checked. Should two consecutive pipes fail to satisfy the requirements, the cause should immediately be investigated. If the cause is not found after four consecutive attempts, the coating process should be stopped so that a full investigation can take place, which should involve checking all pipes back to the preceding acceptable pipe.

7.16 DEFECT RATE

Should any test in any production shift show a rejection rate of more than 10% for 50–457 mm and 5% for 508–1,420 mm of coated pipes, then every pipe in that shift should be individually subjected that test. In these cases, the contractor should conduct an investigation to establish the cause of the defect at the same time. The cost of retrieval and any expenses incurred as a result of additional examinations should be borne by the contractor.

7.17 REPAIRS

Repairs of holidays and damaged areas due to destructive testing should be made in accordance with the procedures specified for each individual coating system. All repairs should be retested for holidays in accordance with Table 7.2.

After field repair of coating, and before lowering the pipe into the ditch, the pipe should be tested for holidays once more. All defective places should be plainly marked and repaired using the identified methods.

7.18 STRIPPING OF COATING

Rejected coating should be removed only by the procedures specified. The process should cause no mechanical damage to the pipe, and the steel temperature should not exceed 250°C.

7.19 TRANSPORTATION, HANDLING, AND STORAGE
7.19.1 HANDLING AND STORAGE REQUIREMENTS

All coated pipes should be handled and stored in such a manner as to prevent damage to the pipe walls, the weld end preparations, and the coating.

Nylon slings or protected hooks that do not damage pipe ends should be used for loading, unloading, and stacking the pipes. Coated pipes should be stored off the ground at all times. Storage may be affected by the use of battens suitably covered with soft material such as rubber sheeting. These pipes may only be stacked to a height such that no flattening of the coating occurs. The pipes should be separated from each other with sufficient and proper dunnage.

During long storage, the coating should be protected from contact with petrol, oil, or grease, as some of these substances can cause swelling in the coating layer or damage the coating otherwise.

The contractor should clean the end cutback of concrete coated pipes, to remove all concrete spatter and any deleterious material, prior to loading out. The method of cleaning should be submitted for approval by the company.

The contractor should rebevel concrete coated pipe ends only when specifically instructed to do so by the company. The procedure for rebeveling should be agreed with the company prior to it being carried out. The contractor should not carry out repairs to the pipe other than rebeveling, minor filing, or grinding.

Concrete-coated pipes should not be stocked more than 4 m high. Pipes fitted with anodes or arresters should form a top tier and should be laid out individually on top of the pipe stacks at the frequency required for loadout and laying.

7.19.2 TRANSPORTATION LOADING

The loading operations should be witnessed and certified by the inspector. The coated pipes should be loaded on trucks. The coating manufacturer should provide all necessary means (such as saddles, battens, etc.) for safe transporting of the coated pipes.

7.20 COLD-APPLIED TAPE COATINGS

The application of cold-applied tape coatings should be a continuous, three-step operation starting with a properly prepared pipe surface. The three steps following immediately one after the other should consist of the following:

*   Priming
*   Inner-layer tape application
*   Outer-layer tape application

The coating materials should be stored and applied at the temperature recommended by the manufacturer. It should be applied by one of the following methods:

*   Over the ditch application—In this procedure, the pipe is welded together beside the ditch. Then the surface preparation, priming and wrapping is performed continuously over the ditch. The coating is inspected simultaneously and the approved coated pipeline will be buried.
*   Field application—In this procedure, the pipes are prepared and primed in the yard. The primed pipes are transported to the field, jointed, cleaned of contamination, reprimed, wrapped, inspected, and buried.

7.21 SURFACE PREPARATION PRIOR TO COATING

Preheating to remove oil, grease, and mill scale may be used provided that all pipes are preheated in a uniform manner to avoid distortion. The blast profile should have the profile of 40–75 mm in thickness or as specified by the tape manufacturer.

Blast-cleaned pipe surfaces should be protected from conditions of high humidity, rain, or surface moisture. No pipe should be allowed to flash-rust before coating. For pipe sizes bigger than 750 mm, each longitudinal weld seam should be primed and covered with 100 mm strip of inner-layer tape prior to placing the coating materials.

7.22 COATING APPLICATION
7.22.1 PRIMER APPLICATION

The primer should be applied in a uniform, thin film at the coverage rate recommended by the manufacturer. The primer should be thoroughly mixed and agitated as needed during application to prevent settling. The primer may be applied by spraying, or other suitable means to cover the entire exterior surface of the pipe. After application, the primer coat should be uniform and free of floods, runs, sags, drips, and bare spots. The primed pipe surface should be free of any foreign substances such as sand, grease, oil, grit, rust particles, or dirt. The state of dryness of the primer prior to application of the inner-layer tape should be in accordance with the manufacturer's recommendations.

7.22.2 APPLICATION OF INNER-LAYER TAPE

The inner-layer tape should be applied directly onto the primed pipe surface with mechanical coating and wrapping equipment with suitable tape dispensing ability. The inner-layer tape should be spirally applied with an overlap width of 25 mm and application tensions as recommended by the manufacturer. When applied to spirally welded pipe, the direction of the tape spiral should essentially conform to the weld spiral. The applied tape should be tight, wrinkle-free, and smooth. When a new roll of tape is started, the end lap should be overlapped on the previous roll by at least 150 mm measured circumferentially, and it should be smooth and placed so that it maintains the continuity of the inner-layer coating.

7.22.3 APPLICATION OF OUTER-LAYER TAPE

The outer-layer tape should be applied over the inner-layer tape using the same type of mechanical equipment used to apply the inner-layer tape. The overlap of the outer-layer tape should not coincide with the overlap of the inner-layer tape. The outer layer may be applied at the same time as the inner layer.

7.22.4 CUTBACKS

Cutbacks should be a minimum of 150 mm from the ends of the pipe or as specified by the company. The cutbacks may be a straight edge for the total thickness of the coating, or they may be tapered as specified by the company.

7.23 COATING OF WELDED FIELD JOINTS

The joints should be coated with cold-applied tapes. Welded joints should be cleaned of mud, oil, grease, and other foreign contaminants; and the exposed metal in the weld zone should be wire-brushed by hand or power-brushed so as to remove all corrosion. The adjacent cold-applied tape coating should be cleaned of all foreign matter and should be dry at a distance of 50 mm from the cutback zone. For this purpose, the weld zone is the uncoated area that results when two joints with coating cutback are assembled in the field in accordance with standards.

7.24 MECHANICAL COUPLING AND PIPE ENDS

Where rubber-gasketed joints or mechanical couplings are used, the coating may extend to the ends of the pipe, but the coating thickness on the pipe surfaces that receive the rubber sealing gaskets should not be in excess of that recommended by the manufacturer of the sealing device. If coating the pipe to the ends will interfere with the proper seating of the seal, the coating should be removed at the amount required by the type of joint used.

7.25 COATING REPAIR

All holidays detected, such as damaged or flawed areas or misaligned lapping, should be repaired by peeling back and removing the outer and inner layers from the damaged area. The holiday area should then be brushed with primer, and either a length of inner-layer tape should be wrapped around the pipe to cover the defective area, or a patch of inner-layer tape should be applied directly to the defective area as specified by the company. The minimum lap at the damaged area should be 100 mm all around. The repaired area should be tested with a holiday detector after the repair is completed. If no holidays are found, the repaired area should be covered with outer-layer tape with a minimum lap of 100 mm beyond the inner-tape patch.

7.26 FUSION-BONDED EPOXY COATING

Epoxy powder coating should be applied in the shop to the pipe, and the supplier must been informed of this. Throughout the cleaning and coating, 75 mm of pipe ends should be masked, and it should be suitably protected from mechanical damage at all times.

The average coating layer thickness should be 400 μm, with a minimum of 350 μm and a maximum of 450 μm, unless otherwise specified by the company.

7.26.1 COATING APPLICATION

The pipe should be preheated to the temperature (between 220°C and 245°C) recommended by the epoxy powder manufacturer. The heating equipment should be suitably controlled to maintain uniform temperatures and to achieve the required uniform temperature along the entire surface of the pipe joint. On no account should the pipe be heated in excess of 260°C. Prior to coating, the pipe temperature should be checked with approved and calibrated instruments to ensure that the temperature complies with the established coating procedure.

Oxidation of the steel prior to coating or other apparent oxide formation is unacceptable. If such oxidation should occur, the pipe should be cooled to the ambient temperature and recleaned.

The epoxy powder should be applied by electrostatic spraying to create a coating that is uniform in color, gloss, and thickness and free of any runs, sags, and holidays. The production coating process should be performed strictly in accordance with the qualified coating procedure and specifications. Failure to comply with the above requirements should result in the coating being rejected and completely removed and the pipe being recoated at no expense to the company.

Recycling of the powder is allowed up to a proportion of 30%, well mixed with fresh powder, provided that the powder has passed through a magnetic filter to remove all metallic dust and through a sieve to remove all aggregates. This is to ensure that the particle size of the recycled powder is within the acceptable limits of fresh, unused powder as specified by the manufacturer. Powders of different batches should never be mixed. Traceability of the coating material on each individual pipe should be ensured.

After application of the epoxy powder, the coating should be allowed to cure in accordance with the qualified coating procedure before being quenched in water to permit handling, inspection, and, if required, repair.

The pipe should be allowed to cool to at least 60°C and the coating thickness determined with an approved gauge.

7.27 FIELD COATING OF WELDED JOINTS
7.27.1 EPOXY APPLICATION

The weld area should be heated to a temperature of 260°C using a circumferential induction heating coil of sufficient size, width, and power to provide the required heat in the weld zone and 50 mm back under the fusion-bonded pipe coating. Immediately afterward, the weld should be coated with a powder coating with a minimum thickness of 380 μm using air or electrostatic spraying. The powder coating should attain full performance properties in accordance with standards when cured. Application should be done as rapidly as possible to prevent premature cooldown of the heated zone. The welded-joint coating should overlap the original pipe coating by no less than 25 mm.

The joint coating should be cured from the residual heat remaining in the heat zone. No quenching or forced cooling should be allowed, and the heat zone should be protected from adverse weather conditions, such as rain or high winds, that would cause cooling at high rates.

On completion of the coating operation, the joint coating should be inspected for continuity. Holidays should be repaired according to relevant standards. Inspection and repair may begin after the heat zone has cooled sufficiently.

The exterior of field-welded joints may be coated with hot-applied tape, cold-applied tape, liquid epoxy (AWWA C 210), or polyethylene shrink sleeve, or as otherwise specified or approved by the company.

7.27.2 COATING OF CONNECTIONS AND APPURTENANCES

Fusion-bonded coatings can be applied to articles preheated to between 150°C and 245°C. The temperatures should meet the manufacturer's recommendations. The articles may be heated by any controllable means that does not leave a residue or contaminant on the surface to be coated. Care should be exercised to ensure that the articles or parts thereof can sustain the preheating without damage.

Fusion-bonded epoxy coatings should be uniformly applied to the specified surfaces at the specified thicknesses by fluid bed, electrostatic spraying, or air spraying according to the manufacturer's recommendations. The choice of the method of application depends on the size, shape, and configuration of the article to be coated. Mechanical thread system parts, such as nuts and bolts, can have this coating applied, provided that the thickness of the coating will not gall or impede the normal operation of the thread system. At the option of the company, thread systems may be left bare and protected only with rust-preventing oil. In this case, the oil should be removed after final field assembly by means of solvent cleaning, and the threads should be coated using liquid epoxy. The company should also specify the coating requirements for flange faces or other appurtenances.

If it is necessary to post-cure fusion-bonded coating, the coated articles should be heated immediately after the coating goes on according to the coating manufacturer's recommendations until total curing is achieved. Once the coating is applied, it should be visually inspected for blisters, bubbles, voids, and other discontinuities. The coatings should also be electrically inspected for holidays. Inspection and repair may begin after the article has cooled sufficiently. Holidays and imperfections detected by electrical inspection or visual means should be repaired.

7.28 COATING REPAIR
7.28.1 MINOR DEFECTS

Pipes that require repair due to scars, coating imperfections, and other small defects (less than 25 mm in diameter) should be repaired using materials from the same manufacturer as the fusion-bonded epoxy as follows:

- Areas of pipe that need spot repairs should be cleaned to remove dirt, scale, and damaged coating using surface grinders, files, or sanders. The adjacent coating should be roughened, and all dust should be wiped off.
- A two-part, solid, liquid epoxy patching compound or a hot-melt adhesive patching compound should be applied on the prepared areas to a minimum thickness of 305 μm. The freshly coated area should be allowed to cure in accordance with the manufacturer's recommendations prior to handling and storage. Liquid epoxy should not be applied if the pipe temperature is 13°C or less, except when the manufacturer's recommended heat curing procedures are followed.

7.28.2 MAJOR DEFECTS

Pipe sections with coating defects, such as partial coating, unbonded coating, inadequate film thickness, or holidays in excess than 1 per 0.9 m^2 pipe should be reprocessed. The applicator should demonstrate the integrity of the techniques used to repair defects by carrying out adhesion, hot water resistance, and holiday detection testing in accordance with Table 7.2.

7.29 HOT-APPLIED ENAMEL COATING

Two types of coating system are generally specified for hot-applied coating. Single-coat systems consist of the following:

- One coat of primer
- One coat of hot-applied enamel
- One layer of glass fiber inner wrap
- One layer of glass fiber outer wrap

Double-coat systems consist of the following:
- One coat of primer
- One coat of hot-applied enamel
- One layer of glass fiber inner wrap
- One coat of hot-applied enamel
- One layer of glass fiber outer wrap

The coating may be applied in a coating yard, field, or over-the-ditch procedure. To reduce surface oxidation in sunlight and sagging due to excessive heat of yard-applied coatings during storage, the application of a coat of whitewash can be specified (see AWWA C203). When concrete application is to follow enamel coating, this step is normally omitted.

7.29.1 COATING APPLICATION

Hot-applied coating may be applied by one of the following methods:

- Yard application—In this procedure, the pipes are surface-prepared and primed, enameled, and wrapped at the yard or factory. The coated pipes are transported to the field and jointed. The welded joints, connections, fittings, and special sections should be coated.
- Field application—In this procedure, the pipes are surface-prepared and primed at the yard. The primed pipes are transported to the field, jointed, cleaned of contamination, reprimed, coated, inspected, and buried.
- Over-the-ditch—In this procedure, the uncoated pipeline should be surface-prepared, primed, and joined together and then coated over the ditch.

During over-the-ditch coating work, care must be taken to handle the pipes so that they will remain clean and dry. Foreign matter and dirt, as well as moisture, will reduce their coating effectiveness.

7.29.2 PRIMER APPLICATION

The prepared surfaces should be dry at the time the primer is applied, and no primer should be applied during rain or fog unless the pipe to be primed is protected from the weather by suitable housing. At the option of the contractor, the application of the primer should be by hand brushing, spraying, or other suitable means and should be in accordance with application instructions supplied by the manufacturer of the primer. The apparatus to be used to apply primer should be acceptable to the inspector. Spray-gun apparatuses to be used should include a mechanically agitated pressure pot and an air separator that will remove all oil and free moisture from the air supply.

After application, the priming coat should be uniform and free of floods, runs, sags, drips, holidays, or bare spots. Any bare spots or holidays should be recoated with primer. All runs, sags, floods, or drips should be removed by scraping and cleaning, and the cleaned area should be retouched or remedied by reblasting and repriming. Suitable measures should be taken to protect wet primer from contact with rain, fog, mist, spray, dust, or other foreign matter until it has completely hardened and enamel applied over it.

In cold weather (when the temperature of the steel is below 7°C), or at any time when moisture collects on the steel, the steel should be warmed to a temperature of approximately 30–38°C, which should be maintained long enough to dry the pipe surface prior to priming. To facilitate spraying and spreading, the primer may be heated and maintained during the application at a temperature of not more than 49°C.

The minimum and maximum drying times of the primer and the period between application of primer and application of enamel should be in accordance with instructions issued by the manufacturer of the primer. If the enamel is not applied within the allowed maximum time after priming, the pipe should be reprimed with an additional light coat of primer or the entire prime coat should be removed by reblasting and the pipe reprimed as determined by the company.

7.29.3 PREHEATING OF PRIMED PIPES

During cold weather, when the pipe surface temperature is below 7°C, or during rainy or foggy weather, when moisture tends to collect on cold pipes, enameling should be preceded by warming of the primed pipes. This should be done by any method that will heat pipe uniform to the recommended temperature without injury to primer. The steel temperature of pipe should not exceed 70°C.

7.30 PREPARATION OF ENAMEL FOR USE

7.30.1 PREPARATION OF BITUMINOUS ENAMEL

Bituminous enamel should be heated in kettles approved by the engineer and equipped with accurate and easily read thermometers. The enamel should not be overheated, nor should it be kept in the kettle for an excessive period of time. Operating kettles should be completely emptied at least once each day and cleaned, when necessary, before another charge of unmelted enamel is added. The exception to this is that, in the practice of field patching, the engineer may permit continuous use of a heating kettle not exceeding 200 l capacity.

The enamel should be maintained moisture- and dirt-free at all times before and during the time of heating and application. In loading the kettles, the enamel should be broken into pieces not exceeding 10 kg each.

In heating the enamel, the charge should be melted and brought up to the application temperature without damage to the enamel and in a manner approved by the engineer. The hot enamel should be thoroughly and continuously stirred. A patching kettle that has a capacity of 200 l or less should be thoroughly stirred with an iron paddle at intervals not exceeding 15 min.

7.30.2 PREPARATION OF COAL-TAR ENAMEL

Coal-tar enamel should be heated in suitable agitated heating kettles equipped with accurate and easily read recording thermometers. The thermometers will be checked and adjusted by the inspector whenever necessary. The charts, therefore, should constitute a basis for accepting or rejecting any enamel due to improper heating, handling, or both.

The maximum temperature of the supply or operating kettles to which the enamel may be heated and the maximum time that the enamel may be held in the kettles at application temperature should be in accordance with the instructions supplied by the manufacturer. Enamel that has been heated in excess of the maximum allowable temperature or that has been held at application temperature for a period longer than that specified should be rejected. Fluxing the enamel will not be permitted.

Excess enamel remaining in a kettle at the end of any heating should not be included in a fresh batch if it would be greater than 10% of the batch. Kettles should be emptied and cleaned as often as required. The residual material removed in cleaning the kettles should not be blended with any enamel.

A minimum of 50% of the original enamel penetration at 25°C should be retained in the applied enamel. This minimum penetration should be evidence of satisfactory melting and handling practice. The inspector should periodically take samples of the enamel as it is being applied to the pipe. If the penetration is less than 50% of the original enamel penetration at 25°C in any kettle, the inspector may reject the enamel. The cost of such testing should be borne by the company if the penetration exceeds 50% of the original value, or by the contractor if it is less. While the enamel is being applied, all kettles should be equipped with suitable screens to exclude particles of foreign matter or other deleterious materials that could cause flaws in the finished coating.

7.30.3 APPLICATION OF ENAMEL

The enamel should be applied by pouring on the revolving pipe and spreading to the specified thickness. Enamel should be applied so that each spiral resulting from the spreading operations should overlap the preceding spiral, producing a continuous coat that is free of defects, skips, or holidays. The thickness of coating should be 2.4 mm, and the allowable variation in thickness should not exceed ±0.8 mm. Where the protrusion of the weld seam interferes with this thickness, the thickness of the enamel above the weld seam should meet the minimum thickness specified.

7.30.4 APPLICATION OF INNER AND OUTER WRAP

Inner and outer wrap should be mechanically applied in a continuous end-feed machine or in a lathe-type machine or by other suitable wrap application equipment. In single-wrap system, only one layer of glass-fiber outer wrap should be applied. In double-wrap system, one layer of glass-fiber inner wrap should be applied, followed by one coat of hot-applied enamel. Finally, one layer of glass-fiber outer wrap should be applied.

If low-porosity outer wrap, which has been stored, is applied under ambient, high-humidity conditions, "gassing",(that is, the formation of craters or voids in the enamel beneath the outer wrap) may occur. To prevent gassing, apply a film of outer-wrap saturant or hot enamel to the underside of the outer wrap before it is drawn into the enamel on the pipe.

Because of their composition, some outer wraps are required to be stored in a dry, covered area and in the original packaging until immediately prior to use. Check with the manufacturer for proper storage and stacking requirements.

The outside surfaces of the outer wrap are to be given a coat of whitewash, water-emulsion latex paint or a kraft-paper finish coat following final inspection. The spirally wrapped kraft paper should be tack-bonded with enamel at frequent intervals.

The coated pipe should not be rolled or supported on its enameled and wrapped surface until it has cooled and hardened enough to avoid deformation of the coating system.

7.31 COATING OF JOINT, FITTING, AND CONNECTIONS
7.31.1 PIPE SECTIONS TO BE FIELD-WELDED

When pipe sections are to be joined together by field welding, a band that is free of protective materials should be left at the ends of the sections. This band should be at least 250 mm from each end to permit field joints to be formed without injury to the coating. The welded joints should be cleaned and dried, primed, and then coated.

For irregular shapes such as valves and fittings, the buried petrolatum tape should be used. The special primer should be applied by brush and then petrolatum tape should be wrapped over the parts.

The belowground unburied valves should be coated with asphalt mastic. The cold-applied asphalt mastic should be applied by brush over the valves.

7.31.2 PIPE SECTIONS TO BE JOINED WITH MECHANICAL COUPLING AND BELL-AND SPIGOT ENDS

In sections to be joined by mechanical coupling, a band free of protective materials should be left on the ends of the sections. This band should be of sufficient width to permit joints to be made up.

For bell-and-spigot ends with rubber gaskets, the coating should extend from the lip of the bell to the hold-back on the spigot end. The exposed steel surfaces on the inside of the bell and the outside of the spigot end should be given a coating of fast-drying synthetic primer to a dry-film thickness of 0.06 mm ±0.01 mm, or bitumen primer or coal-tar primer as appropriate.

7.32 COATING REPAIR

7.32.1 DAMAGES, FLAWED AREAS, AND HOLIDAYS

These areas should be repaired using materials made by the manufacturer of the material originally used to coat the pipe. At no time should damaged enamel be repaired by putting enamel over loose or damaged enamel where the damage goes down to the metal or where the bond of the enamel has been destroyed. Damaged areas are generally categorized into three types: pinpoint or bubble, exposed metal, or extensive damage. Repair of these types of damage is discussed in the following sections. Repairs are to be made using materials meeting the original specification. Repaired areas should be inspected in accordance with AWWA C 203.

7.32.2 REPAIR OF PINPOINT OR BUBBLE-TYPE DAMAGE

Remove dirt, foreign matter, kraft paper, and outer wrap with a sharp knife, taking care not to damage the surrounding enamel. Pour correctly heated enamel over the prepared area to the specified thickness and then cover it with a patch of outer wrap.

7.32.3 REPAIR OF EXPOSED-METAL TYPE DAMAGE

For damaged areas that are approximately 100 mm by 100 mm in size, repair them by removing dirt, foreign matter, and all disbonded enamel, and then bevel the surrounding edges. Clean the metal surface properly, using wire brushes if required, and reprime bare areas. After primer has dried, apply correctly heated enamel over the prepared area to the specified thickness and then cover it with a layer of outer wrap.

7.32.4 EXTENSIVE DAMAGE

Extensive damage means defects such as partially uncoated, unbonded enamel; severe cracking; excessive holidays; or inadequate film thickness. Enameled pipe with these flaws should be reprocessed.

7.32.5 ELECTRICAL INSPECTION

Repaired areas should be electrically inspected in accordance with Table 7.2.

7.33 CONCRETE COATING OF LINE PIPE

This section discusses the minimum requirements for the application of reinforced concrete coating to the exterior surface of steel-lined pipe (which has been coated previously with anticorrosion coating) for use in marine environments, estuaries, rivers, wadis, and swamps.

Concrete coating thicknesses less than 25 mm are not covered, however. In such cases, the coating should be applied in accordance with AWWA C 205. The materials used for concrete coating, which can include cement, aggregates, water, and additives (if required), should be specified.

Concrete coating should not be applied when the pipe, coating, or air temperature exceeds 35°C or falls below 4°C. Pipes should be handled at all times in a suitable manner to avoid damage to pipe attachments, coating, or concrete.

7.33.1 REINFORCEMENT STEEL

Steel reinforcement should be provided to limit spalling and control cracking of the concrete. The minimum circumferential reinforcement should be 0.5% of the longitudinal cross-sectional area of the concrete coating. The minimum longitudinal reinforcement should be 0.05% of the transverse cross-sectional area of the concrete coating (see Figure 7.1).

The reinforcement provided should be either the steel-welded, wire-fabric type or the cage type. The selection of the type of reinforcement should be made by the company as stated in the specifications.

The use of galvanized steel poultry netting (or chicken wire; e.g., to ASTM A-390) is generally not recommended. However, it may be used in certain situations, such as small river crossings, where external forces on the concrete would be minimal.

The minimum distance between the reinforcement and the anticorrosion coating should be 10 mm. When more than one layer of reinforcing is used, there should be spaces of a minimum of 10 mm between layers. Contacts should not be made with anodes or buckle/crack arresters, and if the reinforcing passes over arresters, the minimum gap should be 10 mm.

Reinforcing should terminate 20 mm (±5 mm) from the end of the concrete coating and adjacent to the anodes. There should be no electrical contact between reinforcing and pipe or between reinforcing and anodes. The reinforcements should be free of oil, grease, and dirt.

The steel-welded wire fabric should be galvanized. The chemical and physical properties should be in accordance with standards. The minimum diameter of reinforcing wire should be 1.6 mm. The longitudinal overlap should be at least 30 mm. For concrete thicknesses of up to 50 mm, one wrap of reinforcing should be used. For thicknesses of 50–120 mm, two layers should be considered. The reinforcement should be positioned at least 10 mm below the outer concrete surface.

The cage type reinforcement (Figure 7.2) should be positioned within the middle third of the concrete coating with a concrete cover of at least 9 mm. The material used should be hard-drawn wire that suits either BS 4483 or ASTM A 615 M. The reinforcement should be positioned within the middle third of the concrete coating with a concrete cover of at least 9 mm. The diameter of the circumferential and longitudinal bars should be calculated from the required percentage of reinforcing, and it should be at least 3 mm.

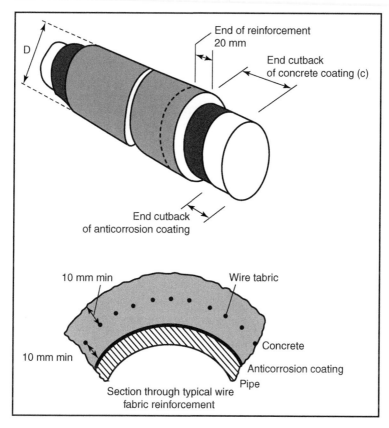

FIGURE 7.1

The longitudinal reinforcement.

The contractor should propose a welding procedure that meets a national or international code or standard to be approved by the company.

The contractor should carry out shear testing of four welds each month in accordance with BS 4483 or ASTM A 185. The welding at intersections of the cage should result in a shear strength specified by BS 4483 or ASTM A 185.

The spacing of the longitudinal bars should be between 50 and 250 mm; not less than four longitudinal bars (with approximately equal spacing) should be provided. The circumferential bar spacing should be not more than 150 mm.

Cages should have two loops spaced 50 mm apart at each pipe end and adjacent to anodes and arresters. Close-fitting arresters may be designed to have the concrete around them; in such cases, this requirement does not apply.

Where a lap is required, one cage should have its longitudinal wires extended to ensure a minimum overlap of 200 mm with the longitudinal wires of the other cage, and with the circumferential wires of each cage having at least 25 mm between them. Lap wires should be bent down as necessary to maintain alignment.

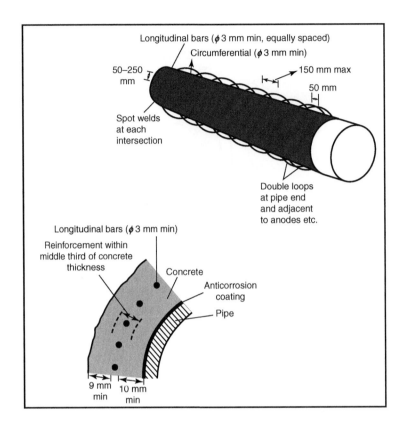

FIGURE 7.2

The cage type of reinforcement.

Cages should be rigidly held concentric to the pipe at the correct location by electrically insulating plastic or concrete spacers. The type of spacer should be proposed by the contractor for approval by the company. Spacers should have flush bases to prevent indentation into the anticorrosion coating.

7.34 COATING PROCESS

7.34.1 PROPORTIONING OF MATERIALS

The contractor will be permitted to select any proportion of materials that will produce the requirements of concrete strength and density/submerged weight. The mix design needs to be demonstrated by prequalification testing and be witnessed and approved by the company prior to the start of coating. Any approved mix design should not be changed without retesting and approval by the company. Mix designs submitted to the company for approval should contain at least the following information:

- The proportions and weights of the component materials used for the mix
- The water/cement ratio (not to exceed 0.45 by weight)

- The grading of the aggregates accompanied by appropriate curves
- The test results of the mix for strength, density, and water absorption
- The measures taken to minimize drying shrinkage

7.34.2 **APPLICATION OF CONCRETE COATING**

Prior to application of concrete coating, anticorrosion coating should be applied to the pipe, and then it must be inspected and tested with a holiday detector specified by standards for each individual coating. Any pipes containing coating defects should be put aside or repaired in accordance with standards. The testing should take place just prior to the fitting of the reinforcement steel and the attachment of anodes or buckle/crack arresters.

Special care should be taken with pipes to which sacrificial anodes, buckle/crack arresters, or both have been fitted. Anode attachment and cable-to-pipe connectors should be inspected to ensure that they are secure and properly coated with an anticorrosion coating determined by the user of the holiday detector.

The concrete should be applied to the pipe by either the impingement or the compression method. The concrete should be placed on the pipe within 30 min of the water being added to the mix. Each pipe should be coated in a continuous operation in such a manner that the thickness of coating is applied uniformly, smoothly, without corrugations, and concentric with the steel pipe.

If more than one application is required to produce a coating of the specified thickness, then the time allowed between the first and the second coat should not exceed 30 min. If a period exceeding 30 min does occur, the previous coating layer should be removed and the entire pipe recoated.

Both ends of each pipe (i.e., the bare pipe and the anticorrosion coating) should be completely free of concrete or any other foreign matter. The concrete should be terminated either square or tapered at an angle to the pipe surface, as specified in the scope of work. The outer surface of each anode (except the ends) should be kept free of concrete.

Each concrete-coated pipe should be weighed immediately after the coating operation and its submerged weight established and recorded.

7.34.3 **LIMITATIONS**

Concrete coating should not be applied when the pipe, coating, or air temperature exceeds 35°C, or when there is evidence of flash setting of concrete. If the ambient temperature falls below 4°C, the contractor may request to use a "Winter Concrete Application" procedure. Prior to use, this procedure should be qualified by the contractor in the manner specified by the company.

The use of recycled rebound material is allowed. Secondary mixing of the freshly batched concrete with the recycled material should follow immediately and should produce a homogeneous cohesive mixture. The amount of recycled material used should not exceed 10% of the total mix by weight. When breaks in the operation exceed 30 min for any reason, then the recycled material not previously added to the mix should be discarded and removed from the coating area.

7.34.4 **CURING**

Immediately after coating is completed, the pipe should be removed from the coating machine and transferred, after weighing and measuring the diameter, to the storage yard for curing. The coated pipe should be gently lifted and transported to the curing yard and placed, so as to prevent cracking or damage to the concrete, in a single layer.

Curing should be performed either by using water curing, steam curing, curing by sealing compounds, or polyethylene wrapping. The exposed surfaces of the concrete should be protected during curing from any adverse effects of sunshine, drying winds, rain, or running water.

The curing process should continue until a minimum compressive strength of 14 N/mm^2 has been achieved (as demonstrated by the qualification tests and any subsequent core tests), after which the pipe can be lifted, transported, or stacked.

If the curing process involves steam or other warm, humid air, then it should be demonstrated that the process will have no deleterious effects on the concrete. Under no circumstances should the pipe wall be allowed to reach a temperature that would cause any damage to the anticorrosion coating (i.e., 60°C), or as advised otherwise by the company.

Prior to use, the contractor should present the curing facilities and procedure to the company for approval.

7.34.5 WATER CURING

Water curing should consist of wetting and moistening the concrete coating, starting not later than 6 h after the coating process has completed. The concrete coating should be kept continuously moist by intermittent spraying for a period of at least 7 days. The interval between spraying should not be more than 24 h. At temperatures below 4°C, suitable precautions should be taken to prevent damage due to freezing.

7.34.6 STEAM CURING

Curing by steam should not start sooner than 3 h after concrete application has completed. Concrete-coated pipes should be enclosed in plastic or a similar covering that will maintain steam circulation.

Steam circulation should start at ambient temperatures and should be controlled to give a temperature gradient of approximately 10°C/h, up to a maximum steel/coating temperature of 60°C. The pipes should be held under steam curing for at least 6 h and then allowed to cool for a similar period. The contractor should demonstrate by prequalification that the curing time used is sufficient to meet the required strength levels.

7.34.7 CURING BY SEALING COMPOUNDS (MEMBRANE)

Sealing compounds should meet the requirements set out in ASTM C 309-89. The material should be stored, prepared, and applied in strict conformity with the printed instructions supplied by the compound manufacturer. The compound should be nontoxic and nonflammable and should not react with any constituent of the concrete, the reinforcement, the anodes, the protective coating, or the pipe steel.

Unless otherwise specified by the compound manufacturer, membrane sealing compounds should be sprayed over the concrete surface within 6 h after the concrete has been applied, and it should remain in place for at least 7 days. The materials should not be applied at a temperature of less than 4°C. Sealing compounds should not be sprayed on any part of the anode surface.

7.34.8 CURING UNDER POLYETHYLENE WRAPPING

Wrapping in plastic film (e.g., polyethylene film), should take place by a mechanical apparatus immediately after the coating is applied to the pipe and preferably before the pipe is removed from the concrete coating apparatus. A light spray of water should be applied before applying the plastic film. It should have a minimum thickness of 0.1 mm, and the overlap of the sheet should not be less than 25% of the sheet width. Polyethylene film should be in accordance with ASTM C 171. The polyethylene wrapping should be removed prior to loading out of pipe.

7.35 INSPECTION AND TESTING

The company may have a full- or part-time representative at the contractor's plant to monitor sampling, testing, calibrating, and curing during production. The contractor should supply details of all sampling and testing procedures and should receive the company's approval prior to use.

Strength, density, and water absorption tests should be carried out by a third party proposed by the contractor and approved by the company unless the contractor's own facilities and procedures have been deemed acceptable.

7.35.1 MISCELLANEOUS INSPECTIONS/TESTS

When steel wire fabric reinforcement is used, at least two pipes per shift should be selected at random and checked that the distribution of the reinforcement is as specified in standards. The method of checking should be to remove a small area of concrete coating (200 mm long × 100 mm wide) by using a high-pressure water jet. Any pipe found satisfactory after inspection should be reinstated.

An electrical resistance check should be made between the pipe and the inner reinforcement on at least two pipes per shift. This should be done in the period between concrete application and curing using an ohmmeter. The resistance should be at least 1,000 ohms.

Pipes with unsatisfactory distribution of reinforcement, or which fail the electrical resistance test, should be stripped and recoated after the proper reinforcement has been installed. In this event, three additional pipes from the same batch should also be checked to gain acceptance of the batch.

The diameter measurements immediately after concrete coating has been applied should satisfy the following two requirements:

- The radial distance between any high and low point of the coating surface should not exceed 8 mm in any 500-mm length of pipe, nor 5 mm in the last 1-m of each end.
- The diameter of each coated pipe should be measured using a girth tape in six positions spaced at approximately equal intervals, with the two end measurements being approximately 600 mm from the ends of the concrete coating. All diameter measurements should be recorded, and the mean value per pipe should be used to calculate the mean concrete thickness. The concrete thickness and acceptance tolerance should be as stated in the scope of work.

Coated pipes outside the stated tolerances should be repaired or stripped and recoated at the discretion of the company.

All concrete-coated pipes should be visually inspected and examined for damage, cracks, or other defects. All cores removed for strength testing should be visually examined for voids and other defects.

Note that circumferential surface cracking of the concrete, with a crack width less than 5 mm, should not be considered a defect. Longitudinal surface cracks of any width and less than 250 mm in length should not be considered defects, but holes with 10-mm nominal diameter should be drilled at the crack tips to prevent crack propagation. Longitudinal cracks in excess of 1,000 mm in length or extending the full depth of the coating are not acceptable, and coating with such cracks should be removed from the entire length of the pipe. Longitudinal cracks with a length between 250 mm and 1,000 mm should be repaired.

Circumferential annular cracking that is visible at pipe ends is not acceptable. Any such cracking noted should be reason to consider the concrete suspect and liable to rejection. An investigation should be made, including a back check of pipe produced prior to and following the suspect pipe. Any concrete coating noted with annular cracking of the concrete should be destructively tested by the contractor. This testing should take the form of saw cutting and removal of sections of the concrete from the pipe to determine whether the concrete is a homogeneous mass throughout the length of the pipe.

Surface damage will not be considered a defect if all of the following is true:

- The total surface area of damage per pipe is less than 0.1 m^2
- The maximum depth does not exceed 20% of the coating thickness
- The remaining concrete is sound

Damage at the ends of the concrete need not be repaired if the damaged area is less than one-third of the circumference, for a length less than 200 mm.

7.35.2 STRENGTH AND DENSITY TESTING

The compressive strength and density of the applied concrete after 7 and 28 days should be determined by two different methods of testing:

- Taking test cubes from the transfer belt between the concrete batching plant and the application machine (or from the mixer outlet for continuous mixing equipment)
- Taking of core samples from the actual concrete coating on the pipes

Acceptance should be based only on core sample strength results after 28 days.

7.36 REPAIRS AND REJECTION

Concrete-coated pipes may be rejected by the company when any of the previously stated acceptance criteria are not met. In that case, the coating should be completely removed rather than partially repaired. Damage, cracks, and core holes should be repaired.

Concrete used for repairs should contain the same proportions of dry constituents as the original coating, but the water/cement ratio may be higher to improve workability. Prior to repairing, the defective area should be thoroughly cleaned by removing concrete coating sufficiently to clean under the reinforcing bars and the concrete edges should be cut so as to form a holding key. All repairs should be moist-cured for a minimum period of 48 h.

The contractor should demonstrate that the procedures for repair application and curing will result in an acceptable coating, with the compressive strength of repair being at least equal to the strength of the coating. All repair methods should be subject to approval prior to use.

Repairs should be affected by removing the concrete to expose uninjured reinforcements throughout the damaged areas. Any damaged reinforcement should be replaced.

The edges of the removed area should be undercut to provide a key lock for the repair material. A stiff mixture of cement, aggregate, and water should be troweled into and through the reinforcement, and built up until the surface is even and smooth with the original coating around the repair. The coating resulting from the repair should be equal in strength to the coating originally applied.

Any repair made by the concrete applicator should be appropriately positioned and allowed to cure properly for a minimum of 36 h before further handling. However, if a shield is used over the repair and allowed to remain in place, the pipe may be handled immediately.

7.36.1 DAMAGED AREAS

Damaged areas should be repaired when they are considered defects. Areas of damage not in excess of 0.8 m^2 in any 3-m length of pipe (restricted to two such areas in any 12-m length of pipe) may be repaired by hand-patching if repairs are carried out while the concrete is in a "green" state (i.e., within 4 h of coating application and prior to curing).

If the contractor wishes to repair areas on cured concrete, a guniting process should be used and the procedure should be approved by the company prior to use. Areas where coating has been damaged during coating application (in excess of 0.8 m^2 but not more than 25% of the total coating area) should be repaired immediately before the pipe is removed from the coating applicator by reapplication of the spalled areas using the application machine used to coat the pipe. If the damage to concrete is in excess of 25% of the total coating area, repairs should not be allowed. All coating should be removed and the pipe recoated.

The contractor should propose a procedure for removing rejected coating and should receive the company's approval prior to use.

7.36.2 CRACKS

Longitudinal cracks longer than 250 mm should be repaired as specified here, and in addition, the ends of each crack should be drilled with a hole with a 10-mm nominal diameter to prevent crack propagation. The bottom of this hole should be 7–10 mm from the anticorrosion coating.

Repairs to cracks should be made by opening the crack to 25 mm wide by chiseling. The prepared area should be undercut to form an adequate key for the repair material. The repair material should be troweled or impinged into the prepared area and smoothed down to the level of the original coating. Prior to repairing, the prepared area should be inspected by the company for damage to the coating or metal surface.

7.36.3 CORE HOLES/INSPECTION AREAS

Core holes should not be refilled without prior inspection and approval by the company. They then should be hand-filled with a stiff mix similar to that used in the coating process. Prior to use, the procedure should be proposed in detail by the contractor and approved by the company.

7.37 POLYETHYLENE COATING APPLICATION

Application requirements for the following three types of shop-applied polyethylene coatings for the exterior of steel pipe are considered next; the type of coating should be specified by the company:

- **Type 1:** The extruded polyethylene sleeve, shrunk over a primed pipe by the cross-heat-extruded method. This type of coating should be applied in accordance with DIN 30670.

- **Type 2:** Dual extrusion (the side extrusion method), where the soft primer is extruded onto the blast-cleaned or epoxy-primed pipe followed by multiple fused layers of polyethylene.
- **Type 3:** Powder sinter coating (fusion-bonded polyethylene coating) consists of polyethylene powder distribution from an applicator running the length of the pipe onto the continuously rotating heated pipe below it.

7.37.1 FUSION-BONDED, LOW-DENSITY POLYETHYLENE COATING (POWDER SINTERING)

The standard specifies the requirements for applying fusion-bonded low-density polyethylene coating of pipes and fittings that are installed in locations that are not subject to weather exposure. Polyethylene materials complying with standards are intended to be applied to pipes and fittings where the pipeline's operating temperatures are not normally more than 60°C nor less than −40°C.

7.37.2 COATING THICKNESS

Unless specified otherwise by the company, the minimum coating thickness of the fusion-bonded polyethylene should be as specified in Table 7.3.

7.37.3 APPLICATION OF COATING

Immediately prior to the application of polyethylene compound, the surface to be coated should be free of deleterious contaminants and surface defects. With this process, the surface-prepared steel pipes are heated by a direct gas flame or hot-air heated oven to 320–360°C to obtain good adhesion of the polyethylene to metal.

The hot pipe is placed beneath a chute, running the length of the pipe, from which a polyethylene powder with particle size of no more than 400 μm falls continuously (or intermittently as required) onto the rotating pipe. The coated pipes are cooled by blowing cold air through them after having been put onto end supports to keep them clear of the ground to prevent damage to the still soft polyethylene coating.

Table 7.3 Coating Thickness

Nominal Size of Pipe, mm	Minimum Coating Thickness (mm)	
	Normal	Heavy (see Note[1])
≤100	1.8	2.5
>100 ≤ 250	2.0	2.5
>250 ≤ 499	2.2	3.0
≥500 ≤ 749	2.5	3.5
≥750	3.0	3.5

[1]*Where additional protection is required (e.g., because of rocky backfill expansive or clays), consideration may be given to specifying the heavy thickness.*
Note: *When the specified minimum coating thickness exceeds 45% of the pipe wall thickness, the company should consult the applicator.*

During coating, the beveled ends of the pipes should be protected against mechanical damage and against contamination with coating material. The coating should be applied in such a way as to obtain a smooth outer surface compliant with standards.

If, after being heated, a pipe or fitting is allowed to cool below the coating temperature without being coated, the surface should be reprepared by abrasive blast cleaning before being reheated for coating.

7.37.4 PIPE ENDS

At the pipe ends, the coating should be cut back over a length of 150 mm ±20 mm unless otherwise specified. At the cutback, the coating edge should be shaped to form a bevel angle of 30–45°. For stainless steel pipes, nonferrous or stainless steel mechanical tools should be used.

7.37.5 END TREATMENT OF COATING

The junction at the ends of the pipe coating and the pipe surface should be tapered to a feathered edge over a distance not less than five times the coating thickness, and then sealed with lacquer.

7.37.6 PROTECTION OF WELDED JOINTS ON POLYETHYLENE-COATED STEEL PIPES (COATING OF FIELD JOINTS)

The protection of the uncoated area of steel pipe adjacent to the welded joints can be achieved by means of thermo-shrinkable polyethylene tubing, polyethylene powder, and wrapping with plastic adhesive tape (cold-applied tape).

7.37.7 FINISHED COATING SYSTEM REQUIREMENTS

The finished coating should comply with standards, except that the cathodic disbonding can be up to 15 mm longer.

7.37.8 REPAIRS

If a flaw is located, the polyethylene coating should be repaired to the satisfaction of the company. The application of a hot spatula is a satisfactory method for small repairs (up to 650 mm²), provided that there is a residual layer of polyethylene still adhering strongly to the steel surface.

Apart from repairs resulting from destructive testing, the number of repairs should not exceed three per pipe or fitting. Except for repairs resulting from destructive testing, the area of any single repair should not exceed 0.1 m², and the length of such repairs should not exceed either the nominal pipe diameter in the circumferential direction or twice the nominal pipe diameter in the longitudinal direction.

7.38 COATING APPLICATION OF FITTINGS AND CONNECTIONS

The application method of cold-applied primer and prefabricated tape, polyethylene shrink sleeves, and asphalt mastic coating for fittings, connections, and special parts of pipelines should be as described in the next sections. The selection of type of coating for connections and fitting should comply with requirements.

7.39 COLD-APPLIED PRIMER AND PREFABRICATED TAPE

7.39.1 SURFACE PREPARATION

Prior to the application of primer and prefabricated tape, the surfaces of connections and fittings should be prepared at least to St3 in accordance with ISO 8501-1 and then coatings can be applied. Welded joints should be allowed to cool prior to surface preparation.

7.39.2 COATING APPLICATION

A uniform and continuous coat of primer should be applied to prepared surfaces in accordance with the manufacturer's recommendations. The primer coverage and curing or drying time should be sufficient to ensure an effective bond between the substrate and the coatings. Priming application should be limited to that amount that can be wrapped during the same workday; otherwise, the steel must be reprimed.

Prefabricated tape should be applied by hand spirally to the primed surface with the following considerations in mind:

- The total applied thickness of the tape should not be less than 500 μm using the overlap of 50%.
- Tape laps should not be used unless the pipe circumference is less than the length of the tape.
- Maximum and minimum temperatures for application and handling specified by the manufacturer should be followed.

7.39.3 COATING REPAIR

All damages, flawed areas, holidays, or misaligned lapping should be repaired by peeling back and removing the tape layers from the affected areas. The repair areas should be brushed with primer, and then a patch of tape covering a minimum of 100 mm around the affected area should be wrapped around the pipe, or by applying a patch of tape as specified by the company. The repaired areas should be tested with a holiday detector after the repair is completed.

7.39.4 INSPECTION OF COATED SURFACES

After wrapping has been completed, the contractor should conduct an electrical inspection of all wrapped surfaces with an electrical holiday detector. Any defect in the wrapping should be satisfactorily repaired at the expense of the contractor.

7.40 POLYETHYLENE HEAT-SHRINKABLE COATINGS

Heat-shrinkable coatings are made of radiation cross-linked polyethylene and are available in the following types:

- Tubular type—These coatings are installed before joining the pipe ends by sliding the coating from a free end of the pipe onto the area to be coated.
- Cigarette wrap—These coatings are wrapped circumferentially around the pipe area to be coated. Each wraparound coating is provided with either a separate or a built-in closure to secure the overlap during shrinking.

- Wraparound tape—These coatings are helically wrapped around the pipe area to be coated with an overlap of 50%.

All three of these coating types are better than conventional cold-applied tapes because they are self-tensioning and resistant to direct sunlight, and some brands can be used at elevated temperatures up to 90°C. Their major disadvantage is that they require a blowtorch for application. Skilled field construction crews are needed to apply the heat-shrink sleeves properly. Shrink sleeves are extensively being used for the field-joint protection of polyethylene and fusion-bonded epoxy-coated pipes.

7.40.1 SURFACE PREPARATION

Surfaces that have been blast-cleaned and primed in a mill or shop before being shipped to the field location may be cleaned using solvent wash and wire brushing or other approved means at the time the heat-shrinkable coating is applied.

7.40.2 COATING APPLICATION

The application procedure used for each type of coating system should comply with the manufacturer's specifications. Maximum and minimum temperatures for application and handling specified by the manufacturer should be followed.

Where the heat-shrinkable coating joins with a mill-applied coating, it should bond to and overlap the mill coating by at least 50 mm. The supplied width should be a minimum of 127 mm wider than the exposed steel area to be covered. When this coating is applied over pipe coated, coal-tar enamel coating, kraft paper or whitewash should be removed from the area.

7.40.3 COATING REPAIR

All damaged or flawed areas, holidays, and misaligned lapping should be repaired by using heat-shrinkable coating or cold-applied primer and prefabricated tape. The damaged area should be covered with at least 50 mm of overlapping around the damaged area by using either a precut patch or wraparound coating. After the repair is completed, the area should be tested with a holiday defector as described previously in this chapter.

7.40.4 INSPECTION OF COATED SURFACES

After heat-shrinkable coating has been properly applied, the contractor should conduct an electrical inspection of all wrapped surfaces with an electrical holiday detector. If no spark is seen, the coating has passed the electrical inspection test. The primary input power should be no higher than 20 W, and the minimum pulses at crest voltage should be 20 Hz.

The operating voltage of the detector should be determined by the following formula:

$$V = 1250 \sqrt{t}$$

where V is the inspection voltage and t is the total coating system thickness, in mils.

7.41 ASPHALT MASTIC COATING

Mastic coating should not be used with a primer. The material should be such that it will bond firmly to a clean steel surface. It should be cold-applied by brush.

7.41.1 APPLICATION OF MASTIC FOR BELOWGROUND VALVES AND FITTING

All belowground valves and fittings that are not already coated should be thoroughly cleaned, and brush-coated. The mastic should be applied in two or three coats to a total dry film thickness of at least 3 mm. Each coat should be allowed adequate time to dry, set, or harden before handling or before a further coat is applied. (12 h minimum for each coat). The mastic should be applied evenly and be free of voids, runs, and missed spots.

7.42 INSTALLATION OF COATED PIPE AND BACKFILLING

At all times during construction of the pipeline, the contractor should use every precaution to prevent damage to the protective coating on the pipe. No metal tools or heavy objects should be permitted to come into contact unnecessarily with the finished coating. Workers must not walk on the coating except when absolutely necessary, and in such cases, they should wear shoes with rubber or other footwear that will not damage the coating. Any damage to the pipe or the protective coating during the installation of the pipeline and before final acceptance by the company should be repaired in accordance with this construction standard to the satisfaction of the company, and at the expense of the contractor.

7.42.1 BENDING

Cold field bending should be done using padded machines designed for bending coated pipe. The field bends should be uniform and should not exceed 1½ degrees for each section of pipe length equal to the pipe diameter.

7.42.2 PROTECTION DURING WELDING

A 460-mm-wide strip of heat-resistant material should be draped over the top half of the pipe on each side of the coating holdback during welding to avoid damaging the coating by hot weld splatter. No welding ground should be attached to the coated part of the pipe.

7.42.3 APPLICATION OF ROCK SHIELD

Before hoisting the pipe, the contractor should apply rock shield to the coated pipe when specified by the company.

7.43 HOISTING

Pipe should be hoisted from the trench side to the trench by using wide-belt slings. Metal chains, cables, tongs, or other equipment likely to cause damage to the coating should not be permitted, nor should dragging or skidding of the pipe. The contractor should allow inspection of the coating on the underside of the pipe while it is suspended from the slings. Coated pipe should be handled with nylon sling choker belts or the equivalent. Caliper clamps should not be used. Any damage should be repaired in accordance with standards.

7.44 PIPE BEDDING AND TRENCH BACKFILL

Backfilling should be conducted at all times in such a manner as to avoid abrasion or other damage to the coating on the pipe. Unless otherwise specified by the company, the requirements described next should be met.

Where the trench traverses rocky ground containing hard objects that could penetrate the protective coating, a layer of screened earth or sand (not less than 150 mm thick) or other suitable bedding should be placed at the bottom of the trench prior to installation of the pipe. Placement of backfill around the exterior of the coated pipe should be done only after the inspector has accepted the exterior coating.

7.45 BACKFILLING

Backfilling should be conducted at all times in such a manner as to avoid abrasion or other damage to the protective coating on the pipe. Unless otherwise specified by the company, the requirements presented in the next sections should be provided.

7.45.1 PLACING OF BACKFILL

Placing of backfill around exterior-protected pipe should be done only in the manner approved by the inspector.

7.45.2 INSTALLATION AND TYPE OF BACKFILL

Immediately after pipe is placed and aligned in the trench, and before the joint is completed, loose backfill should be placed about the pipe, except at field joints, to a depth of 150 mm above the pipe. This backfill should be free of large stones, frozen lumps, trash, or material that may decay.

7.45.3 ROCKS AND HARD OBJECTS

If rocks or other hard objects are present in the backfill along any section of the pipeline, the backfill should be screened before being placed around the pipe. Alternatively, if the contractor wishes, suitable waste backfill from other parts of the line may be transported to and placed about the pipe in such sections.

7.45.4 METHODS OF COMPACTION

Compaction of backfill in the trench should be by flooding, puddling, tamping, jetting, or another method agreeable to both the company and the contractor. Rodding with metal rods or other metal tools that will come in contact with the pipe coating should not be permitted.

MATERIALS AND CONSTRUCTION FOR THREE-LAYER POLYETHYLENE COATING SYSTEMS

8

This chapter deals with factory- and field-applied polyethylene coatings for the external surfaces of buried or submerged steel pipes. The coating may be two or three layers as specified by the user.

Two-layer polyethylene coating consists of adhesive primer and polyethylene and is suitable for operating temperature up to 50°C. Three-layer polyethylene coating consists of epoxy primer, adhesive primer, and polyethylene and is intended for operating temperatures up to 80°C.

8.1 REQUIREMENTS FOR FACTORY-APPLIED COATING MATERIALS AND TESTING METHODS

This section specifies materials for use in three-layer polyethylene coating systems. It deals specifically with the properties, minimum requirements, and related testing methods to establish their suitability for use.

Material specifications and the testing method for two-layer polyethylene coating systems should be in accordance with DIN 30670.

8.2 COATING MATERIALS
8.2.1 SELECTION OF COATING MATERIALS

The coating materials supplied should be certified by the manufacturer to ensure that all the materials and equipment comply with the provisions described in this part of the chapter prior to application. The purchaser may make any investigations necessary, via testing, batch sampling, manufacturing, and factory inspection to verify the contractor's compliance.

The contractor is responsible for ensuring that the range of values for any material under consideration will be capable of providing a finished product that complies with relevant standards when related to the specific mode of operation to be used.

It is intended that these specifications should be used to encourage and stimulate the development of progressively better external pipeline coatings. Thus, where certain minimum performance values are stated, should future coating material testing yield better results than the specification requirements, then these new values may be adopted as the minimum requirements (if they are economically viable) and the specification upgraded accordingly.

Essentials of Coating, Painting, and Lining. http://dx.doi.org/10.1016/B978-0-12-801407-3.00008-0

8.2.2 IDENTIFICATION OF MATERIALS

All materials supplied for coating operations should be suitably marked with the following information:

- Manufacturer's name and address
- Material and order number
- Batch number
- Date of manufacture and stable working shelf life (including storage condition limits)
- Directions for mixing and/or thinning with solvents as required
- Directions for handling and storing of the coating materials
- Any necessary information and warnings

The contractor should require that the material manufacturer supply certificates confirming that the tests detailed here have been carried out on the batches supplied and that the materials meet this standard. These certificates should be made available for examination on request.

8.3 EPOXY POWDER

8.3.1 BASIC PROPERTIES AND TESTS FOR THE RAW POWDER

Epoxy powder is a thermosetting material used as a primer in a three-layer polyethylene coating system for steel pipes. It should be specifically formulated and designed so it is suitable for electrostatic application and can improve adhesion of the coating system and provide maximum cathodic disbanding resistance. To ensure an acceptable coating, the contractor should obtain from the manufacturer specified and qualified ranges of values for all the properties listed in Table 8.1.

8.3.2 INFRARED SCAN

An infrared spectrogram, preferably made using a standard potassium bromide (KBr) disk, should be obtained from a first batch of the powder epoxy using an approved method. This should subsequently be used for comparison with another type of spectrogram.

Table 8.1 Typical Raw Epoxy Powder Properties			
Property	**Unit**	**Test Method**	**Typical Value**
Gloss at 60° angle	%	DIN 67530	65 ± 5
Gel time	s	DIN 55990-T8	43 ± 10
Density	g/cm³	DIN 55990-T3	1.5
Particle size	%	DIN 55990-T2	90 between 10 to 80 μm
Moisture content	% weight	Acceptable method to company	0.5 Max
Shelf life at 30°C and % 60 humidity	month	N/A	12 Min
Theoretical coverage	g/m²	Acceptable method to company	90 g for 60 μm DFT
Note: *The test for raw epoxy powder properties is under the responsibility of the manufacturer.*			

8.3.3 THERMAL ANALYSIS

Thermal analysis data for each batch should be made available by showing, with the use of a differential scanning calorimeter (DSC), the glass transition (GT) of the raw powder and the enthalpy of the curing powder. The glass transition temperature (GTT) of the fully cured powder also should be quoted. The reference curve should be provided as part of the production data sheet. The limiting values of ΔH, TG1, and TG2 should be identified by the manufacturer.

8.4 BASIC PROPERTIES AND TESTS FOR CURED COATINGS

Testing of cured coatings should be carried out on a laboratory-coated steel plate, and the thickness of the coating must be 60 μm. Prior to coating, the steel surface should be blast-cleaned to a level of Sa 2½ to Swedish standard SIS 05 5900 and surface profile with a height of approximately 50 μm from peak to valley.

8.4.1 APPEARANCE

The coating should have a uniform appearance.

8.4.2 THERMAL ANALYSIS

The GTT should be measured on a prepared laboratory sample for recording and checking purposes. The typical value of it should be greater than 100°C.

8.4.3 FLEXIBILITY/BENDING

The flexibility/bending should be measured in accordance with DIN 53152, with a typical value of less than 5 mm.

8.4.4 HARDNESS

The hardness of the cured epoxy film should be more than 85 bucholtz when tested in accordance with DIN 53155.

8.4.5 IMPACT RESISTANCE

The impact resistance of the cured epoxy film should be at least 120 kg/cm at 20°C, in accordance with the ASTM G14 test method. The test panel thickness should be 3 mm.

8.5 BASIC PROPERTIES AND TESTS FOR BONDING AGENTS (ADHESIVES)

This adhesive is used as the second layer in a three-layer polyethylene coating system for steel pipes. It should be an ethylene base copolymer, specifically formulated and designed to be suitable for extrusion applications. The contractor should obtain from the manufacturer specified values for all the

Table 8.2 Physical Properties of Adhesive

Property	Unit	Test Method	Value
Density	g/cm³	DIN 53479	0.900–0.950
Melting index (2.16 kg/190°C)	g/10 min	DIN 53735	0.5–8 or as suitable for application As PE (topcoat)
Elongation	%	DIN 53455	95 (min)
Melting point	°C	DSC	9 (typical)
Co-monomer content	%		

Note: *The testing for raw epoxy powder properties is under the responsibility of the manufacturer.*

properties listed in Table 8.2, and ensure that the adhesive manufacturer should carry out tests 1 and 2 in the table for each batch, and the remaining tests should be carried out as type tests at least twice for each order.

The adhesive should be uncolored and made of raw, top-quality material to provide the following properties:

- Excellent peeling resistance
- Excellent mechanical strength
- Excellent thermal stability
- Strong adhesion to fusion-bonded epoxy film, as well as to steel surfaces
- Homopolar bond with the polyethylene top coat (third layer)

8.6 BASIC PROPERTIES AND TESTS FOR POLYETHYLENE

Polyethylene is a thermoplastic resin for use as a topcoat in a three-layer polyethylene coating system for steel pipes. It should be specifically formulated and designed for extrusion application.

The user should obtain from the manufacturer specified values for all properties listed in Table 8.2.

The polyethylene should be made of raw, top-quality material to provide the following properties:

- Excellent peeling resistance
- Excellent mechanical strength
- Excellent thermal stability
- Excellent impact resistance
- Excellent penetration resistance
- Strong adhesion to the adhesive layer
- Excellent stability against ultraviolet rays

The polyethylene should be adequately weather-resistant and stable for fabrication and use. For these purposes, the nature and quantity of the antioxidant should be in accordance with BS 3412-76.

8.7 COLOR

The polyethylene should be uniform in color and free of obvious foreign matter. The color should be blackened with carbon black; the carbon black characterization, dispersion, and content should be in accordance with BS 3412-76.

Table 8.3 Physical Properties of Black Polyethylene

Property	Unit	Test Method	Value
1. Density (in black)	g/cm³	DIN 53479	0.946 min
2. Melting index (2.16/kg 190°C)	g/10 min	DIN 53735 ISO 1133D	0.3 (typical)
3. Elongation	%	DIN 53455	600–700
4. Tensile strength at yield	N/mm²	DIN 53455 or ISO R 527	15 min
5. Tensile strength at break	N/mm²	DIN 53455 or ISO R 527	25 min
6. Hardness	Shore. D	DIN S3S05 or ISO 868	55 min
7. Vicat softening point	°C	DIN 53460	115 min
8. Melting point	°C	DSC	125
9. Low-temperature brittleness	°C	ASTM D 746	−70 no fracture
10. Stress-cracking resistance (methyl -ethylceton)	h	ASTMD 1693 or ISO 4599	>1,000
11. Carbon black content	%	ASTM D 1603	2–2.5 min
12. Dielectric strength	kV/mm	IEC 243	30 min
13. Fungus bacteria	N/A	ASTMD 3173 and ASTMD 3180	Pass no growth
14. 01 Tin 200°C	°C/min	EN728	30 min

8.8 QUALITY SYSTEMS

The contractor should set up and maintain such quality assurance and inspection systems as are necessary to ensure that the goods or services supplied comply in all respects with the requirements of this standard. The user should have the right to undertake inspection or testing of raw materials or purchased components before application.

8.9 MATERIALS
8.9.1 ACCEPTABLE MATERIALS

Only polymer systems conforming to this standard should be considered for use as coatings applied in accordance with this section. The contractor should be responsible for the conformity with the requirements of this construction standard, and should obtain and retain all certificates and manufacturer's data sheets. Certificates should be made available on request.

8.9.2 IDENTIFICATION OF COATING MATERIALS

The contractor should ensure that all materials supplied for coating operations are clearly marked with the following information:

- Manufacturer's name, trademark, and address
- Name of material, order number
- Batch number
- Date of manufacture and expiration date for use
- Safety data sheet
- Technical data sheet

8.9.3 STORAGE OF COATING MATERIALS

To ensure that the properties of all coating materials are maintained in compliance with the relevant elements described previously, all the materials consigned to the coating plant should be properly stored in accordance with the manufacturer's recommendations at all times to prevent damage and deterioration prior to use. Materials should be used in the order in which they are delivered.

8.9.4 PIPE IDENTIFICATION

All identification markings, whether internal or external to the pipes, should be carefully recorded before surface preparations begin. The date of the coating finish and the coating factory markings, including pipe identification, should be legibly marked on the coating surface of each pipe.

8.10 PROTECTION OF PIPE END PREPARATION

Pipe end preparations should be protected from mechanical damage during handling, storage, surface preparation, and the coating process. The methods used should also ensure that no damage occurs to the internal surface of the pipe. The pipe end preparations should be protected from coating during the process by a suitable method approved by the user.

For technical welding reasons, the ends of the pipes should be free of any coating layer (cut back) over a length of 100 mm up to pipe size DN 500 mm inclusive and over a length of 150 mm for sizes over DN 500 mm, unless specified otherwise by the user. The uncoated ends of pipes should be temporarily protected against atmospheric corrosion by a temporary paint easily removable by brushing.

8.11 SURFACE PREPARATION

The method of surface cleaning and surface preparation should be specified by the contractor as part of the coating procedure qualifications and should take into account the requirements in accordance with standards. Where oil, grease, or other contaminants are present, they should be removed, without spreading them over the surface, with a suitable solvent. For pipes, which have been subjected to contamination, the contaminant should be removed by washing either with potable water or an approved chemical cleaner. If a chemical cleaner is used, subsequent washing with potable water will be necessary. The pipe should be dried before blast cleaning. All processes should be in accordance with standards.

Pipes should be blast-cleaned to a minimum of Sa 2½ finish to SIS 05 5900. The blast profile should have a height of between 40 μm and 75 μm, measured by an agreed method. The blast-cleaning medium used should be according to standards.

The metal surface should be inspected immediately after blast cleaning and all slivers, scabs, and other flaws made visible by blast cleaning and detrimental to the coating process should be removed using a method approved by users. After the removal of defects, the remaining wall thickness should comply with the relevant pipe specification. Any rectified areas should be blast cleaned to meet the requirements of the standards.

Any pipe found to have defects that exceed the levels permitted in the relevant specifications should be set aside for examination by an authorized user representative and no subsequent action taken without the agreement of the user. Directly before coating, any dust, grit, or other contaminants should be removed from the pipe surface by a method established as acceptable by the relevant coating procedure testing and recorded in the relevant coating procedure.

Where rust blooming or further surface contamination has occurred, the pipe should be cleaned again in accordance with standards and blast cleaned again in accordance with coating should take place before any further contamination or rust blooming appears.

8.12 COATING PROCEDURE TESTS

The coating process should comply with the coating procedure qualifications. Any changes in coating materials, pipe dimensions, the pipe manufacturing process, or the coating process may, at the discretion of the user, necessitate a new coating procedure approval test. Additionally, approved procedure tests should be confirmed as proving tests, at intervals of not more than 1 year for each type of powder, adhesive, and polyethylene used by the contractor and for each size of pipe and pipe manufacturing process as requested by the user.

8.13 INSPECTION AND TESTING (QUALITY CONTROL)

The quality control system should include as a minimum the requirements listed in Table 8.4. All inspections and testings should be made by the contractor and witnessed and certified by the inspector. Should the inspector find out upon examination or testing that any pipe has not been cleaned or coated in accordance with this standard, the contractor should be required to remove any coating that is considered defective or inadequate, and to reblast and recoat the pipe to the requirements for approval of the inspector.

The inspector should have access at any time to the construction site and to those parts of all plants that are concerned with the performance of work under this standard. The contractor should provide

---

Table 8.4 Minimum Quality Control Requirements

Requirements

1. Check cleanliness of pipes immediately prior to blast cleaning
2. Monitor size, shape, and cleanliness of the blast-cleaning material and process
3. Check visually, in good light, the surface of the pipes for metal defects, dust, and entrapped grit
4. Check the pipe surface blast profile
5. Check for the residual contamination of pipe surface
6. Control the temperature of the pipe surface (pipe temperature shall not exceed 250°C)
7. Check recycled coating material for contamination and moisture
8. Check the coating thickness (first and second layers and total)
9. Check the temperature control of the quenching system
10. Check the coating adhesion
11. Perform holiday detection of 100% of the surface area of all coated pipes
12. Perform supervision to ensure the adequate and proper repair of all defects
13. Check on coating color and appearance, e.g., uniformity and flow
14. Check for damage to pipe end preparations

---

the necessary inspection tools and instruments for the inspector, as well as normal facilities necessary for inspection.

8.14 QUALITY SYSTEMS

The contractor should set up and maintain such quality and inspection systems as are necessary to ensure that the goods and services supplied comply in all respects with the requirements of this standard. The user should assess such systems against the recommendations of the applicable parts of ISO 9002 and should have the right to undertake all necessary surveys to ensure that the quality assurance and inspection systems are satisfactory. In addition, the user should have the right to undertake inspection and testing of the goods and services during any stage of manufacturing at which the quality of the finished goods may be affected, and to undertake the inspection or testing of raw materials, purchased pipes, or both.

8.15 COATING PROCEDURE QUALIFICATIONS

Before bulk coating of pipes commences, the requirements of standards should be met and a detailed sequence of operations for the coating of pipe should be presented to the user so that compliance with this standard and formal approval can be achieved. The user should also specify which coated pipes are to be subjected to testing for formal approval of the coating procedure. No coated pipes should be dispatched to the user, and no coating process should be performed until the coating procedure has been approved in writing by the user.

8.15.1 COATING PROCEDURE SPECIFICATIONS

The coating procedure specifications should incorporate full details of the following (but is not limited to these items):

- The coating system to be used, with appropriate data sheets
- Pipe cleaning
- Blast-cleaning medium and technique
- Blast-cleaning finish, surface profile, and surface cleaning
- Dust removal
- Preheating time and temperature
- Powder epoxy, adhesive, and polyethylene, including the use of recycled material
- Curing and quenching time and temperature
- Repair technique
- Coating stripping technique

8.16 COATING PROCEDURE APPROVAL TESTS

A batch of 10–20 pipes of any specific pipe mill should be selected by the inspector and coated by the contractor in accordance with the approved coating procedure specifications, with the coating operation witnessed by the inspector. Three of the coated pipes should be selected by the inspector and subjected to a complete set of tests. The testing should be witnessed by the inspector, and a full set of records should be presented to the user for consideration.

Bulk coating of pipes should not commence until all short- and long-term test results have been approved officially by the user unless the contractor takes responsibility for the failure of any test. All test methods should be in accordance with Table 8.6.

8.17 SHORT-TERM APPROVAL TESTS
8.17.1 THICKNESS

For the purpose of checking thickness, at least two measurements should be made in accordance with Table 8.6 at locations uniformly distributed over the length and periphery of each pipe selected for the

Table 8.5 Minimum Coating Thickness			
Pipe Diameter (mm)	Powder Epoxy Resin (First Layer) (mm)	Adhesive (Second Layer) (mm)	Polyethylene (Third Layer)
Up to DN 250	0.060	0 30	2.5
DN 250 up to DN 500	0.060	0 30	3.0
>DN 500	0.080	0.35	3 (normal) or 3.5 (reinforced) when specified

Table 8.6 Coating Requirements and Test Methods for Coating Procedure Approval Tests

Tests/Inspection	Test Methods and Requirements
1. Surface preparation	Visual inspection
2. Coating thickness	Electromagnetic thickness gauge is used; it should be calibrated daily with the standard calibrated plates
3. Porosity	DIN 30670 No defect at 25 kV
4. Adhesion	DIN 30670, Method I Acceptable limit: min 23°C, 8 kg/cm min 80°C, 2 kg/cm
5. Impact resistance	DIN 30670 Acceptable limit: 5 Jul/mm
6. Elongation	DIN 30670 Acceptable limit: Minimum of 200% for extruded coating
7. Indentation (hardness)	DIN 30670 Acceptable limit: 0.3 mm
8. Thermal cycle resistance	30°C, 1 h 1 cycle: 60°C 1 h Number of cycles: 100 Acceptable limit: No cracking
9. Environmental stress-cracking resistance	ASTM D 1693 Acceptable limit: No cracking after 300 h
10. Thermal aging	DIN 30670 Acceptable limit: ±35% change in melting index value
11. Specific electrical test	DIN 30670 Acceptable limit: 108 Ωm, 2 Min
12. Cathodic disbonding	ASTM G8 Acceptable limit: 5 mm

test; they should be checked for compliance with Table 8.5. In addition, 50% of these measurements should be made along and over the longitudinal weld seam, if any.

8.17.2 POROSITY

Each pipe selected for the test should be checked for holidays over 100% of its coated surface and checked for compliance with Table 8.6.

8.17.3 ADHESION

Adhesion testing should be carried out on each pipe at five locations uniformly distributed over the length and periphery of the pipe. In this respect, the mean force necessary to pull off the coating should be as listed in Table 8.6. None of these tests can fail.

8.17.4 LONG-TERM APPROVAL TESTS

The tests should be performed on sections taken from all three of the coated pipes selected for approval. An adhesion test should be carried out at five different locations on five test sections in accordance with relevant standards, but after 30 days in the air at 80°C. No change in the mean force required to pull off the coating can occur.

For cathodic disbanding, the test sections should be checked for compliance with the specifications in Table 8.6.

For environmental stress-cracking resistance, the test sections should be checked for compliance with the specifications in Table 8.6.

For thermal cycle resistance, the test sections should be notched with a length of 30 mm and a depth of 0.3 mm and then checked for compliance with the specifications in Table 8.6.

For impact resistance, the test sections should be tested and checked for compliance with the specifications in Table 8.6.

For thermal aging, the test sections should be tested and checked for compliance with the specifications in Table 8.6.

For elongation, the samples taken from the three pipes should be tested and checked for compliance with the specifications in Table 8.6.

For specific electrical resistance, the samples taken from the three pipes should be tested and checked for compliance with the specifications in Table 8.6.

For indentation resistance, the samples taken from the three pipes should be tested and checked for compliance with the specifications in Table 8.6.

8.18 PRODUCTION COATING REQUIREMENTS

8.18.1 COATING PROCESS

The production coating process should be carried out using a procedure that is in accordance with the standards. The thickness of each layer and the total thickness should comply with the values in Table 8.5 when tested in accordance with DIN 30670.

8.19 INSPECTION OF FINISHED COATING

The inspection of the finished coating should be in accordance with the standards. The quality and values to be achieved should be the same as those identified in the standards.

8.19.1 CHECK ON COATING COLOR AND APPEARANCE

The color and appearance of the coating should be uniform and free of runs, sags, blistering, roughness, foaming, and general film defects.

8.20 COATING REQUIREMENTS AND TEST METHODS

After formal approval of all short- and long-term tests by the user, the contractor will be authorized to commence the bulk production. The contractor should perform routine inspection and testing in accordance with the standards during coating production. All the inspection and tests witnessed by the inspector should be certified.

8.20.1 THICKNESS

This test should be carried out three times during each 8-h production shift and each time on four consecutive pipe lengths in accordance with DIN 30670. Every pipe that does not comply with the minimum requirements of Table 8.6 should be rejected for subsequent stripping and recoating. Should two consecutive pipes fail to satisfy the requirement, the cause should be investigated immediately. If the cause is not resolved after four consecutive pipes, the coating process should be stopped so that a full investigation can be conducted that involves checking all pipes since the last acceptable pipe.

8.20.2 POROSITY

Each individual line pipe should be holiday-detected over 100% of its coated surface in accordance with DIN 30670. Up to two holidays per pipe length is allowed for repair on a maximum of 5% of the coated pipe lengths during each 8-h production shift.

Any individual pipe with more than two holidays should be rejected for subsequent stripping and recoating. If more than two holidays per pipe length are detected on two consecutive pipes, the cause of the high holiday rate should immediately be investigated. If the cause is not resolved after four consecutive pipes, the coating process should be stopped so that a full investigation can take place. All holidays detected on nonrejected pipes should be repaired in accordance with satisfactorily retested.

8.20.3 ADHESION

This test should be carried out three times during each 8-h production shift, each time on one individual line pipe. The test should be carried out at room temperature and at two ends of the pipe coating surface and checked for compliance with Table 8.6. If the coating adhesion at any location is below the requirements laid out in Table 8.6, the pipe should be rejected for subsequent stripping and recoating. In this case, the second consecutive pipe should be checked. Should two consecutive pipes fail to satisfy the requirement, the cause should be investigated immediately; if the cause is not resolved after four consecutive pipes, the coating process should be stopped so that a full investigation can take place involving checking all pipes back to the last acceptable pipe.

8.20.4 DEFECT RATE

Should tests during any production shift show a rejection rate of more than 10% for 50–457 mm and 5% for 508–1,420 mm of coated pipes for any single test, then every pipe in that shift should be individually subjected to that test. In such cases, the contractor should simultaneously conduct an investigation to establish the cause of the defect. The cost of retrieval and any additional expenses incurred as a result of this additional examination should be borne by the contractor.

8.21 **HANDLING AND STORAGE REQUIREMENTS**

All coated pipes should be handled and stored in such a manner as to prevent damage to the pipe walls, the weld end preparations, and the coating. Nylon slings or protected hooks, which do not damage pipe ends, should be used for loading, unloading, and stacking.

The coated pipes should be stored at all times off the ground. Storage may be affected by the use of battens that are suitably covered with soft material such as rubber sheeting. The coated pipes may be stacked only to a height such that no flattening of the coating occurs. In this respect, the formula given in API RP 5L5 should be used for the calculation of static load stress.

The pipes should be separated from each other with sufficient and proper dunnage. During long storage, the polyethylene coating should be protected from contact with petrol, oil, or grease, as some of these substances can cause swelling.

8.22 **TRANSPORTATION LOADING**

The loading operations should be witnessed and certified by the inspector. The coated pipes should be loaded on trucks. The contractor should provide all necessary means, such as saddles, battens, or other elements, that ensures safe transport of the coated pipes.

8.23 **STRIPPING OF COATING**

Rejected coating should be removed only by the procedures specified by the user. The process should cause no mechanical damage to the pipe, and the steel temperature should not exceed 250°C.

8.24 **HOLIDAY DETECTION OF COATINGS**

Holiday detection should be carried out using equipment approved by experts to be used on the surfaces in question, which are at ambient temperatures and free of moisture. For all coatings (except adhesive tape), the operating voltage should be 125 V per 25 μm of coating thickness (e.g., 2 kV for 400 μm).

The rate of travel of the probe over the surface should be a maximum of 300 mm/s. All holidays should be repaired.

For fusion-bonded epoxy coating systems, carbon-impregnated neoprene or rolling spring types of electrodes should be used. For other coatings, the splayed brush electrode may also be used, if authorized by the user.

The brush and carbon-impregnated neoprene types of electrode should be of the curved type, conforming to the contour of the pipe. All holidays, imperfections, and damaged areas should be identified with a waterproof marker.

All markings should be sufficiently distant from the holiday, imperfection, or damaged area to allow surface preparation and patching to take place without detriment to the adhesion of the coating. All holiday detectors should be calibrated at the start of every workday and at other times when requested by the user.

Table 8.7 Coating of Butt Joints—Material Application Charts

Existing Coating	Butt Joint	Existing Coating
FBE	FBE	FBE
MCL	MCL	MCL
FBE	MCL	MCL
CTE	MCL + TAPE or H-TAPE	FBE
CTE	MCL + TAPE or H-TAPE	MCL
CTE	TAPE or H-TAPE	CTE
PE	TAPE or H-TAPE	FBE
CTE	TAPE or H-TAPE	PE
PE	TAPE or H-TAPE	MCL
PE	TAPE or H-TAPE	PE

8.25 COATING OF BUTT JOINTS—MATERIAL APPLICATION CHARTS

Table 8.7 shows Coating of Butt Joints for various existing coatings. Key to abbreviations and markings are given below:

- FBE—Fusion-bonded epoxy coating
- MCL—Multicomponent liquid coating
- TAPE—Hand-applied laminate tape epoxy
- H-TAPE—Heat-shrinkable tape
- CTE—Coal-tar enamel
- PE—Polyethylene

Note: Bare or painted pipe or fittings should be coated with tape before the relevant butt joint coating is applied.

8.26 APPROVAL OF FIELD JOINT AND FITTING COATING EQUIPMENT AND APPLICATORS

All processes and equipment for applying field joint coating should be approved by the user prior to applicator qualification trials.

8.26.1 FIELD JOINT COATING APPLICATORS

Written procedures and drawings must be submitted prior to undertaking qualification trials, providing all details of the method of working and parameters to be used. These procedures must have the acceptance of the user or the user's representative before trials commence.

In this procedure, 10 welded joints should be coated under the supervision of the user or the user's representative under simulated or actual field conditions.

All operators who carry out successfully the coating including heating when required applications should be deemed to be qualified to carry out production work on any contract, using the same process and equipment within 12 months of completing the trial.

Additional and subsequent operators should demonstrate their ability to coat the joint within the specification requirements on production work, which should be supervised or carried out by previously qualified operators in the ratio of four skilled workers to two unskilled ones.

8.26.2 REQUIREMENTS FOR COATING MATERIALS

The coating materials within the three-layer coating system should be approved by the user before application. The polymeric adhesive and polyolefin compound should be supplied by the same manufacturer as the coating materials.

The manufacturer should ensure that the quality of each type of coating material is in compliance with the requirements of this part of the standards. The qualification should be repeated if there are changes in the material composition, in the production process that influence the material processing behavior, and in the production facility.

The supplier should obtain from the coating material manufacturer technical data sheets showing at least the properties described in Tables 8.1, 8.2, and 8.3. A certificate of analysis (COA) should be issued by the manufacturer of each component. The manufacturer should supply an inspection certificate for each batch.

Each batch of coating material should be accompanied by a COA according to EN 10204, 3.1.B, which states that all the tests have been carried out and the results are in accordance with the manufacturer's product specifications and requirements of Tables 8.8, 8.9, and 8.10

8.27 COATING MATERIAL

8.27.1 EPOXY MATERIAL

The applicator should use epoxy material that is in compliance with Table 8.8.

8.27.2 ADHESIVE MATERIAL

The applicator should use adhesive material that is in compliance with Table 8.9.

8.27.3 POLYOLEFIN (HDPE/PP) TOPCOAT MATERIAL

The applicator should use high-density polyethylene/polypropylene (HDPE/PP) material that is in compliance with Table 8.10.

Table 8.8 Requirements for the FBE Powder Primer

No.	Properties	Unit	Requirements	Test Method
1.	Density	g/cm³	As per manufacturer's specification ±0.05	ISO 8130-2
2.	Gel time	s	Within 20% of manufacturer's specification	ISO 8130-6
3.	Particle size: Maximum powder retained on 150-μm mesh Maximum powder retained on 250-μm mesh	%	3.0 0.2	CSA Z 245.20-02
4.	Specific coating resistance after 100 days of exposure in 3% NaCI solution @ 23°C	ohm.m	$>10^8$	NF A 49-710
5.	3.0° flexibility test @ 23°C, 0°C & −20°C		No cracking	CSA Z 245.20-02
6.	Cathodic disbondment after 28 days@ 65°C in 3% NaCI solution at −1.5 volts (calomel electrode) potential, initial defect diameter Do = 6 mm	mm	7 (maximum)	NF A 49-710
7.	Moisture content (max)	% by mass	0.5	ISO 21809-1 (Annex K)
8.	Degree of cure (differential thermal analysis)	°C	$-2°C \leq \Delta Tg \leq +3°C$	ISO 21809-1 (Annex D)
9.	Glass transition temperature (Tg_2) (DSC Analysis)	°C	≥ 95 (for 3LPE) ≥ 120 (for 3LPP)	ISO 21809-1 (Annex D)
10.	Water resistance (1,000 h @ 80 °C)	N/A	No blistering, swelling $< 5\%$, loss of hardness $< 10\%$	ASTM D 870
11.	Adhesion to pipe surface	N/A	Max. rating 2	CSA Z 245.20-02

Notes:
The epoxy powder should not contain any calcium carbonate.
Determination of the particle size by a MALVERN instrument instead of the sieving method is acceptable.

8.28 REQUIREMENTS FOR FACTORY-APPLIED COATING

8.28.1 COATING SYSTEM QUALIFICATIONS

The applicator should apply coating materials that are qualified as per the requirements. The qualification process should be repeated if the coating line, coating materials, or application procedures are modified.

Table 8.9 Requirements for the Adhesive Material (Co-Polymeric or Grafted Adhesive in Pellet or Powder Form)

No.	Properties	Unit	Requirement		Test Method
			PE	PP	
1.	Density	Kg/m³	Within 1% of manufacturer's specified nominal		ISO 1183
2.	Melt flow rate (190°C, 2.16 Kg)	g/10 min	Within 20% of manufacturer's specified nominal		ISO 1133
3.	Oxidation induction time @ 210°C	min	≥ 20	≥ 20	EN 728 ISO 11357
4.	Tensile strength at break @ 23°C	MPa	≥ 15	≥ 20	ISO 527
5.	Ultimate elongation 2 23°C	%	≥ 600	≥ 400	ASTM D 638 ISO 527
6.	Hardness	Shore D	≥ 47	≥ 55	ISO 868
7.	Melting point (D.S.C)	°C	≥ 120	≥ 160	ISO 3146
8.	Vicat softeninig temperature A/50 (10 N)	°C	≥ 95	135	ISO 306
9.	Water content (maximum)	%	0.1	0.1	ISO 15512

Table 8.10 Minimum Requirements for Polyethylene and Polypropylene Topcoat

No.	Properties	Unit	Requirement		Test Method
			HDPE	PP	
1.	Density, compound	g/cm³	≥ 0.946	≥ 0.09	ISO 1183
2.	Melt flow rate	g/10 min	0.15–0.8 (190°C, 2.16 Kg)	0.5–4 (230°C, 2.16 Kg)	ISO 1133
3.	Oxidation induction time	Minutes	≥ 30 @210°C	≥ 30 @220°C	EN 728
4.	Tensile strength at break @ 23°C	MPa	≥ 18	≥ 24	ISO 527
5.	Carbon black content	%	2.0–2.5	N/A	ASTM D 1603 ISO 6964
6.	Elongation at break @ 23°C	%	≥ 600	≥ 400	ISO 527
7.	Hardness	Shore D	≥ 55	≥ 60	ISO 868
8.	Melting point	°C	≥ 125	≥ 160	ISO 3146
9.	Vicat softening temperature A50/(10N)	°C	≥ 120	≥ 140	ISO 306
10.	ESCR	h	≥ 1000	N/A	ASTM D 1693 (Condition B)
11.	Volume resistivity	Ohm.cm	$\geq 10^{16}$	$\geq 10^{16}$	ASTM D 257
12.	Water content	%	≤ 0.05	≤ 0.05	ISO 15512
13.	UV resistance and thermal aging	%	$\Delta MFR \leq 35$	$\Delta MFR \leq 35$	ISO 21809-1 (Annex G)

8.28.2 APPLICATION PROCEDURE SPECIFICATIONS

Prior to the start of coating production, the applicator should prepare an application procedure specification (APS), including all of the following:

- Incoming inspection of pipes and pipe tracking
- Data sheets for coating materials, including any materials to be used for coating repairs
- Data sheets for abrasive blasting materials
- Certification, receipt, handling, and storage of materials for coating and abrasive blasting
- Preparation of the steel surface, including control of environmental parameters, methods and tools for inspection, and testing of surface preparation
- Coating application, including tools and equipment needed for the control of process parameters essential for the quality of the coating
- Layout sketch or flow diagram for the coating plant
- Methods and tools and equipment for inspection and testing of the applied coating
- Repairs of coating defects and any associated inspection and testing
- Stripping of defective coating
- Preparation of coating cutback areas
- Marking and traceability
- Handling and storage of pipes
- Any special conditions for dispatch of coated pipes including protection of pipe ends
- Documentation

The APS should cover all items associated with quality control as defined in this part of the standard. It should be approved by the purchaser prior to the start of production. It should be available to the purchaser's inspector on request at any time during production. The coating work and associated inspection and testing should be carried out in accordance with the APS.

8.29 COATING SYSTEM APPLICATION

8.29.1 SURFACE PREPARATION

Prior to the start of grit blast cleaning, the steel should be checked that it is free of grease and other such contamination. If such contamination is found, it should be cleaned as per SSPC SP1. Chemical cleaning solutions and any remaining detergents should be removed.

Grit blast cleaning should be carried out by machine as per ISO 8504 part 2, to a cleanliness standard of Sa 2½ in accordance with ISO 8501/1. The abrasive used should be steel or chilled shot and iron grit conforming to ISO 11124 parts 1–4. Abrasive material that becomes worn or dirty should be replaced via frequent small additions of fresh grit. The compressed air for blasting should be free of water and oil. Adequate separators, filters, and traps should be provided as needed. Suppliers should submit for approval a procedure to periodically verify recycled grit quality in accordance with ISO 11125 parts 1–7.

The surface profile should be within the range of 60–100 μm, and should be measured according to ISO 8503/4 (measured with a 2.5-mm cutoff) or using a replica tape system. The surface profile should be checked once every hour, and the value recorded in the inspection report.

Table 8.11 Maximum Elapsed Time	
Relative Humidity (RH)	**Maximum Elapsed Time**
RH > 80%	2 h
70% < RH ≤ 80%	3 h
RH ≤ 70%	4 h

The steel surface should be visually inspected immediately after blast cleaning, and all defects should be removed by filling and grinding. The blast-cleaned surface should not be contaminated with dust, metal particles, hydrocarbons, water, or foreign matter that could be detrimental to the applied coating. Testing for dust contamination should be carried out in accordance with ISO 8502/3. Requirements for dust contamination should be at a maximum of class 2. The coating process should commence after completion of blast cleaning of the steel surface. The total elapsed time between the start of blasting of any line pipe and the heating to the specified temperature should not exceed the specifications given in Table 8.11.

Any pipe surface not processed within this table should be completely reblasted before coating. To avoid risk of condensation, the temperature of the environment and the steel surface must be 3°C higher than the dew point.

8.29.2 CHEMICAL PRETREATMENT

After blast cleaning and before application of the epoxy primer, soluble salts and chloride contamination on the steel surface should be checked for using an approved salt detector instrument measuring conductivity, SCM400, or the equivalent. The soluble salt content should not excess 2 $\mu g/cm^2$. Line pipes with chloride contamination in excess of 2 $\mu g/cm^2$ should be subject to chemical pretreatment using an approved phosphoric acid solution.

The surface to be coated should be heated to a temperature of 45–65°C and treated with a low-pressure (0.5–2.0-bar) spray application of a 10% v/v solution of an approved acid washing material and process. A uniform pH of 1 or less should be maintained over the entire surface of the treated area. The acid-washed pipe surface should remain wet for approximately 20 s and then rinsed with clean potable water before its starts to dry out.

High-pressure water rinses at 50–70 bars should be used to remove any treatment residue. The wetted surface of the rinsed pipe should have a pH of 6 or greater.

Water must be clean, with less than 200 ppm of dissolved solids and < 50 ppm chlorides, and a conductivity of < 100 μm Siemens. The water should not be reused. The chloride level should be retested.

8.29.3 CHROMATE PRETREATMENT

The contractor should ensure that the temperature of the substrate is maintained between 45–60°C and the chromate solution temperature does not exceed 45°C. The diluted solution of chromate (typically 10% v/v in clean potable water) should be applied to the blast-cleaned steel surface by a suitable method that results in a completely wetted surface with a uniform film of chromate solution remaining on the surface. Any drainage concentration, drips, and other issues should be alleviated by wiping or other suitable means. The contractor should be fully responsible for adherence to local regulations and material safety sheets for using a chromate solution.

8.29.4 COATING APPLICATION

Before the application, the pipe should be uniformly heated in accordance with the manufacturer's specified temperature limits, but the temperature should not exceed 270°C. The surface temperature of the pipe should be monitored and controlled within the limits recommended by the powder manufacturer.

Oxidation of the steel prior to coating or other apparent oxide formation is not acceptable. If such oxidation occurs, the pipe should be set aside and recleaned.

Fusion-bonded epoxy (FBE) coating should be applied to the pipe surface by electrostatic spray with the pipe at earth potential and the epoxy powder charged to high potential.

The pipe temperature should be checked and recorded periodically by temperature-indicating sticks (three to six times per pipe joint) or a recording pyrometer. If a pyrometer is used, it should be checked for error not less that every 4 h against a calibrated temperature-measuring instrument. Prior to starting the FBE powder application, the recovery system should be thoroughly cleaned of any unused powder.

Recycled powder is permitted only when collected and processed through a closed system, which controls the ratio of recycled to virgin powder. The ratio of recycled powder to virgin powder should not exceed 20%.

When the coating application on a pipe length is discontinuous, the pipe should be rejected, stripped of its coating, and prepared for recoating. The minimum thickness of the FBE layer at any point should be 200 μm.

For adhesive layer-coating, the contractor should ensure that the rollers push the adhesive film to eliminate any air entrapment or voids. The adhesive layer should be applied before the gel time of the FBE has expired by using either the cross-head or side extrusion technique. Application of the adhesive should not be permitted after the FBE has fully cured. The contractor should establish to the satisfaction of the purchaser or the purchaser's representative that the adhesive is applied within the gel time window of the FBE and at the temperature specified by the FBE manufacturer. The contractor should state the proposed minimum and maximum time intervals between FBE and the adhesive applications at the pipe temperature range and overlap. The minimum thickness of the adhesive should be 150 μm or in accordance with the coating material manufacturer's recommendations.

High-density polyethylene/polypropylene (HDPE/PP) should be applied over the adhesive immediately after the adhesive layer. The extrusion process should be controlled such that no air is trapped between the adhesive and polyethylene/polypropylene layer or internally within the polyethylene/polypropylene. The tension applied to the polyethylene/polypropylene layer should be kept at a minimum level to prevent additional stresses being built into the topcoat. The polyethylene/polypropylene topcoat should be applied to have a minimum three-layer coating total thickness, as specified in Table 8.12.

Coating should be removed by cutting back from the pipe ends. The amount of cutback required for different pipe diameters should be as follows:

- For pipe sizes up to 20 inches: 100 mm.
- For pipe sizes 20 inches and larger: 150 mm.

Cutback includes full removal of both polyethylene/polypropylene and of the adhesive; the epoxy primer should be left at least 10 mm from the edge of the polyethylene cutback. The cutback end should

Table 8.12 Minimum Total Coating Thickness

Pipe Diameter mm	Powder Epoxy Resin (First layer) (mm)	Adhesive (Second layer) (mm)	Total Thickness, mm	
			3LPE	3LPP
Up to DN 500 (20)	0.2 (200)	0.15 (150)	2.5	2.5
DN 500 (20) up to DN 900 (36)	0.2 (200)	0.15 (150)	3.0	2.5
> DN 900 (36)	0.2 (200)	0.15 (150)	3.5	3

be tapered to an angle of no more than 30° from the metal surface. This taper of the coating should preferably be achieved by using rotating metal brushes; it should not in any way cause cuts on the metal or lifting or disbanding of the tapered part. A butyl primer should be applied by brush, sealing the tapered part and the nearby metal zone for a length of about 20 mm.

The pipe bar ends should be properly cleaned and free of fabrication residue. A temporary protective agent (Varnish) should be applied to the pipe bar ends and should be easily removable by brushing. The increased cutback length (up to 30 cm) on a maximum of 2% of the coated line pipes may be acceptable, subject to the purchaser's approval. In this case, the applicator should supply the suitable size and quantities of heat-shrinkable sleeves and relevant accessories.

8.30 QUALITY CONTROL REQUIREMENTS

The minimum tests to be carried out and the frequencies in the qualification (preproduction) and production stages are specified in Tables 8.14 and 8.15. All the tests listed in these tables should be addressed in the APS.

Table 8.13   Properties of the Applied Coating

Properties	Unit	Test Method	3LPE	3LPP
Appearance and continuity	N/A	ISO 21809-1 (Annex B)	Uniform color, free of defects and discontinuities, delaminations, separations, and holidays	
Impact strength at $23 \pm °C$ e	J/mm	ISO 21309-1 [Annex E)	> 7	> 10
Indentation resistance at $23 \pm 2°C$ e at maximum class temperature	mm	ISO 21309-1 (Annex F)	≤ 0,2 ≤ 0,4	≤ 0,1 ≤ 0,4
Elongation at break at $23 \pm 2°C$ e	%	ISO 527	≥ 400%	≥ 400%
Peel strength	N/mm	ISO 21809-1 (Annex C)	≥ 15 at 23°C ≥ 3 at 80°C	≥ 25 at 23°C ≥ 4 at 90°C or at maximum operating temperature if above 90°C

(Continued)

Table 8.13 Properties of the Applied Coating (*cont.*)

Properties	Unit	Test Method	3LPE	3LPP
Degree of cure of the epoxy	°C	ISO 21809-1 (Annex D)	$-2°C \leq \Delta T_g : \leq 3°C$	
Product stability during extrusion of the PE/PP top layer process	%	ISO 1133	$\leq 20\% \Delta$ MFR for 3LPE $\leq 35\% \Delta$ MFR for 3LPP (virgin compounded granulate before extrusion/ extruded foil after extrusion of the same batch)	
Flexibility at 0°C	N/A	ISO 21309-1 (Annex I)	No cracking at 2.5°/ pipe diameter length	
UV Resistance and Thermal aging	%	ISO 21809-1 (Annex G)	Max. ±35% change of MFI	
Porosity (Air entrapment)	N/A	See 4.5	No voids or air entrapment	
Average radius of cathodic disbond-ment at: 23°C/28 days;−1,5,V 65°C/48h;−1.5.V Maximum opera-tion Temperature/28days/1.5.V	mm	ISO 21809-1 (Annex H)	≤ 7 ≤ 7 ≤ 15	
Specific Electrical Insulation Resistance	$\Omega.m^2$	DIN 30670 DIN 30678	$> 10^{10}$	

8.31 SOME TEST PROCEDURES

8.31.1 RESIDUAL MAGNETISM

Measurements of residual magnetism should be made using a Hall-effect gaussmeter or other type of calibrated instrument approved by the coater. The gaussmeter should be operated according to the manufacturer's instructions. The accuracy should be verified at least once per day.

Four readings at spots 90° apart should be taken around the circumference of each end of the line pipe. The average of the four readings should not exceed 25 gauss, and no single reading should exceed 30 gauss with a Hall-effect gaussmeter, or equivalent values with other types of coater-approved instrument.

The coater should submit for approval a procedure to verify the residual magnetism on the line pipe. Any line pipe that does not meet the above requirements should be considered defective. All defective line pipes should be degaussed and remeasured, or segregated.

8.31.2 ADHESION OF FBE PRIMER

The adhesion test for FBE primers should be carried out on dummy pipes in accordance with the procedure described next. The acceptance criteria should be at a rating of 1–2.

1. Inscribe a V-cut with two 20-mm lines intersecting at approximately 5 mm from their ends at a 30–45°.
2. Insert the blade of the a strong-pointed knife at the point of the V-cut at a 45° angle to the surface. Then, with an upward flicking action, attempt to dislodge the coating within the V. If little or no coating is removed, repeat this action within the V at least four times to confirm the integrity of the coating.

Table 8.14 Requirements for Inspection of Surface Preparations

Properties	Test Method	Requirement	Frequency Qualification	Frequency Production
Surface condition before blasting	Visual inspection	Free of contamination	Each pipe	Each pipe
Surface condition after blasting	Conductive measurement, ISO 8502-9	Salt content maximum of 20 mg/m^2	Each pipe	5 pipes at start of production and 1 pipe/shift
Environmental conditions	Calculation	As determined at time of measurement	Once	Every 4 h
Pipe temperature before blasting	Thermocouple	Minimum of 3°C above the dew point	Once	Every 4 h
Shape and properties of abrasive	Visual + Certification ISO 11124 resp. ISO 11126	Conformity to certificate, compliance to manufacturing/working procedures	Once	Once per shift
Water-soluble contamination of abrasives	ASTM 4940	Conductivity maximum of 60 mS/cm	Once	Once per shift
Surface roughness of blasted surface (R$_z$)	ISO 8503-4	60 mm to 100 mm	5 pipes	Every 1 h
Visual inspection of blasted surface	ISO 8501-1	Grade Sa 2½	Each pipe	Each pipe
Presence of dust after dust removal	ISO 8502-3	Maximum of class 2	5 pipes	Every 1 h
Chromate solution temperature	Thermometer	<45°C	Once	Once per shift
Chromate concentration (titration test) and uniformity/color appearance	N/A	As per manufacturer's instructions	Once	Once per shift
Pipe condition prior to coating	Monitoring	No rust; pipe temperature at least 3°C above the dew point	Continuously	Continuously
Temperature of extruded adhesive and polyolefin	Thermometer	Compliance to APS	Once	Every 1/h
Preheating temperature before coating	Thermometer	Compliance to APS	Each pipe	Every 5th pipe

8.31.3 HOT-WATER SOAK TEST (FOR FBE PRIMERS)

An adhesion test as detailed next should be carried out on two pipes coated only with an FBE layer. A coated sample of 200 mm x 100 mm machined from a pipe ring of the coated pipe should be immersed in tap water of 800°C for 24 h. The bare edges of the sample should be coated to prevent ingress of moisture beneath the coating.

Table 8.15 Requirements for Inspection and Testing of Applied Coating

Properties	Test Method	Requirement	Frequency Qualification	Frequency Production
Minimum epoxy thickness	ISO 2808	200 mm	At startup	Once per shift
Minimum adhesive thickness	ISO 2808	150 mm on pipe body	At startup	Once per shift
Degree of cure of the Epoxy	ISO 21809-1 (Annex D) ISO 11357-2	$-2°C \leq \Delta T_g \leq +3°C$	At startup	Once per shift
Appearance	Visual	Uniform color, free of defects and discontinuities, declamations and separations	Continuously	Continuously
Holiday detection	ISO 21809-1 (Annex B)	No holidays at 25 kV.	Each pipes	Each pipe
Total thickness of coating	ISO 21809-1 (Annex A)		5 pipes	Every 10 pipes
Impact resistance	ISO 21809-1 (Annex E)		3 pipes	Once per PE/PP batch
Peel strength @ 23°C @ T_{max}[a]	ISO 21809-1 (Annex C)		5 pipes 3 pipes	Every 4 h once per 100 pipes
Indentation resistance @ 23°C @T_{max}	ISO 21809-1 (Annex F)		Once	Once per PE/PP batch
Elongation at break	ISO 527		Once	Once per PE/PP batch
Cathodic disbondment test @23°C/28 days/−1.5V @ 65°C/2 days/−1,5V @80°C/28 days/−1.5V.	ISO 21809-1 (Annex H)		Once Once N/A	Once per 4,000 pipes once per week once per order
Flexibility	ISO 21809-1 (Annex I)		Once	N/A
Residual magnetism		Average: 25 gauss Max: 30 gauss	Each pipe	Each pipe
In-process degradation of PE/PP	ISO 1133	Δ MFR max. 35% for PP; 20% for PE between raw and extruded material	Once	Once per PE/PP batch
Thermal cycle resistance 1 cycle: −30°C/1 h 1 cycle: +70°C/1 h No. of cycles: 100	N/A	See Note "b"	No crack	Thermal cycle resistance 1 cycle: −30°C/1 h 1 cycle: +70°C/1 h No of cycles: 100

Table 8.15 Requirements for Inspection and Testing of Applied Coating *(cont.)*

Properties	Test Method	Requirement	Frequency Qualification	Frequency Production
Porosity (Air entrapment)		No voids or air entrapment	5 pipes	Every 4 h
Bonding of coating on longitudinal weld		No voids or air entrapment	Once	Once per week
Hot-water soak test (for FBE) hours @ 80°C		No disbondment or blistering	2 pipes	Once per FBE batch
Adhesion of FBE		Refusal to peel or a cohesive failure	2 pipes	Once per FBE batch
Specific Electrical Insulation Resistance	DIN 30670 DIN 30678	$\geq 10^{10}$ $\Omega.m^2$	Once	Once per PE/PP batch
UV Resistance and Thermal aging	ISO 21809-1 (Annex G)	Max. ± 35% change of MFI	Once	Once per PE/PP batch
Cutback	Measuring	100 ± 5 mm up to 20" 150 ± 10 mm > 20"	Each pipe	Each pipe
Coating repairs	Visual/Holiday detection	No holidays	Once for demonstration	Each defect

Notes:
a) For peel testing, a sample of coated pipe should be conditioned (or heating the internal surface of the pipe at the test location) for at least 1 h at the test temperature. The surface temperature should be monitored during the test and the results reported.
b) The test sections should be notched with a length of 30 mm and a depth of 0.3 mm before testing.

Immediately after 24 h of exposure, the coated sample should be removed and allowed to cool to the ambient temperature. Then the coating adhesion should be tested by the following method:

1. Using a sharp knife, make two incisions of approximately 15 mm in length through the steel surface to form an X, with an angle of intersection of approximately 30°.
2. Commencing at the intersection, an attempt should be made to lift the coating from the steel substrate using the bland of the knife.

Refusal of the coating to peel or a cohesive failure that is entirely within the coating in the absence of excessive voids caused by foaming constitutes passing the test. Partial or complete adhesion failure between the coating and the metal substrate constitutes failure. Cohesive failure caused by voids in the coating leaving a honeycomb structure on the steel surface is also a failure. In addition, the coating should not show any tendency toward disbanding or blistering. A slight discoloration of the coating is acceptable.

8.31.4 BONDING OF THE COATING ON THE LONGITUDINAL WELD

The coating application process should ensure a fully satisfactory interlayer bonding along the entire circumference of the lined pipe. This should particularly apply to the longitudinal weld and the region adjacent to the edges of the final weld cap. Voids and interlayer bond checks should be verified by means of peeling tests or partial stripping of coating along the longitudinal weld. Voiding or air entrapment at any location, including alongside the weld, should not be allowed.

8.31.5 POROSITY (AIR-ENTRAPMENT) TEST

After adhesion (peeling) testing is complete, all the peeled strips of coating should be examined for porosity (air entrapment). No air entrapment should be allowed.

8.31.6 RETESTING

If any required test should fail, the coater should test two additional pipes, one pipe before the failed one and one after it. If the follow-up tests are successful, all pipe that has been coated since the last acceptable test should be considered acceptable except for the failed pipe, which must be rejected. However, if the follow-up tests also fail to meet the requirements of this specification, all pipes coated since the last acceptable test must be rejected.

8.32 REPAIR OF COATING DAMAGE

All defects detected by the holiday detector and damaged coating areas less than 50 cm^2 resulting from destructive testing or mechanical accidents should be repaired at the applicator's expense according to the following stipulations:

- If one individual damaged area is greater than 50 cm^2, the pipe should be rejected, stripped, cleaned, and recoated.
- A maximum of three repairs per line pipe is allowed, but in no case should the total repaired areas exceed 100cm^2.
- A maximum of 2% of damaged line pipes is allowed. If the amount of damaged line pipes is greater than this percentage, the applicator should perform immediate corrective action.
- Damage to the polyethylene topcoat layer that does not decrease the total coating thickness should not be repaired.

All the repaired areas should be subjected to the following checks:

- Visual inspection on 100% of repairs
- Holiday detection on 100% of repairs
- Thickness testing on 100% of repairs

All the repairs carried out on defects with exposed steel should be recorded by the applicator.

Recording should include the line pipe number, the dimensions of any damage, and the results of all the checks. The applicator should prepare a detailed coating repair procedure, including all types of expected defects and the coating stripping procedure for lined pipes. The coating repair and stripping procedures should be approved and qualified by the purchaser before production starts.

TECHNICAL GUIDELINES FOR MATERIAL SELECTION—PART 1

This chapter covers the minimum requirements for the composition, analysis, properties, storage life and packaging, inspection, and labeling of a wide variety of paint and coating materials.

9.1 ASPHALT MASTIC (COLD-APPLIED)

9.1.1 COMPOSITION

The ingredients and proportions of asphalt mastic should be as specified in Table 9.1 and the following criteria. The mastic based on the specified ingredients should be uniform, stable in storage, and free of grit and coarse particles. No waste, coal tar, or coal-tar solvents may be added.

The asphalt mastic should contain a maximum of 40% by weight of inorganic fillers and minimum and 60% by weight of the vehicle. The vehicle should consist of 55 ±10% by weight of asphalt and 45 ±10% by weight of petroleum solvents.

The fillers used should be suitable inorganic minerals (such as mica, silicate, or slate flour) and short-fiber asbestos. Sand and other harsh abrasive materials should not be used. The fillers (exclusive of asbestos) should be fine enough to pass through a 0.043-mm sieve opening. Not more than 15% by weight of the total mastic should be retained on a 0.043-mm sieve opening, as determined by washing the mastic through the screen with a suitable solvent.

Table 9.1 Composition

Ingredients	Required		Ingredient Standards	
	Minimum Wt%	Maximum Wt%	ASTM Method	U.S. Federal Standard
Filler (40 wt.% maximum) inorganic filler contains short-fiber asbestos and free inert material	-------	100	-------	-------
Vehicle (60 wt.% minimum) Asphalts	45	65	-------	-------
Mineral spirit thinner	35	55	D235	TT-T-291 Grade 1 or equivalent

9.1.2 VEHICLES

The bitumens should consist of asphalts, either naturally occurring or petroleum-derived, and they may include up to 50% by weight of gilsonite (natural asphalt); any petroleum asphalt used should be a straight-run, steam, or vacuum-refined residue and should be air-blown in accordance with good commercial practice.

The volatile solvents should consist of any suitable aliphatic petroleum solvents with a minimum flash point of 32°C, such as white spirit thinners.

9.1.3 ANALYSIS

The mastic should conform to the composition (analysis) requirements given in Table 9.2.

9.1.4 PROPERTIES

The mastic should meet the requirements of Table 9.3 and the following criteria:

- Odor—The odor should be normal for the materials permitted (as per ASTM Standard D1296).
- Color—The color should be black.

Table 9.2 Analysis

| Characteristics | Requirements | | ASTM Method | U.S. Federal Standard No. 141 |
	Minimum Wt%	Maximum Wt%		
Filler (wt.%)	N/A	40	D1856	4021*
Volatiles (wt.%)	N/A	38	D2369	N/A
Nonvolatile bitumens calculated by difference (wt.%)	27	N/A	N/A	4053
Pigment% by volume of nonvolatile	N/A	25	N/A	4311

Use benzene and extract at least five times.

Table 9.3 Properties

| Characteristics | Requirements | | ASTM Method |
	Minimum	Maximum	
Density Kg/l	0.96	1.14	D 1010
Flash point of solvents, (°C)	32	N/A	D 3278

- Compatibility—There should be no incompatibility of any of the ingredients of the mastic when two volumes of it are mixed slowly with one volume of mineral spirits (U.S. Federal Standard No. 141, Method 4203).

9.1.5 SKINNING

There should be no skinning in a three-quarters-filled closed container after 48 h when tested in the standard manner specified in U.S. Federal Standard No. 141, Method 3021.

9.1.6 WORKING PROPERTIES

The mastic should be capable of being applied by brush and trowel in addition to spraying with standard high-pressure spray equipment of the types recommended for the application of mastic. The consistency should be such that it may be sprayed at a temperature of 10°C with a tank pressure of not more than 550 kPa gauge and no more than 550 kPa gauge of air pressure on the gun nozzle. A coating thus applied should be uniform, continuous, and free of any pinholing.

9.1.7 SAG TEST

A 3-mm, wet-film coating of the mastic should not sag when suspended vertically for 1 h immediately after spraying, and there should be no flowing and piling up of the coating on the lower half of the panel.

9.1.8 DRYING TIME

A wet film of coating that is sufficient to deposit a 1,600-μm-thick dry film should set to touch in 4 h and should reach practical hardness by 36 h at a test temperature of 21–24°C and 45%–55% relative humidity.

9.1.9 HARDNESS

The dry, cured mastic should have a penetration of not more than 1.0 mm at 25°C, 2.0 mm at 35°C, 7.5 mm at 66°C, and 15 mm at 82°C when tested according to ASTM Standard D5 using a 50 g weight.

The coating is considered to have reached practical hardness when the film pressure between the thumb and the fingers gives a slightly touchy condition, but the film is not ruptured and none of the coating appears on the fingers.

9.1.10 ACID AND ALKALI RESISTANCE

Spray a coating of the mastic on both sides of degreased 10×30-cm cold-rolled steel panels to give a dry film thickness of approximately 1,600 μm. After the coating has dried for 24 h, bake for 10 days at 52°C and allow to cool at room temperature.

Seal each panel around the edges with a high-melting-point wax by dipping each edge approximately 66 mm into a molten solution of the wax and then immerse it in test solutions of acids and alkalis to a depth of 15 cm for 30 days at 24°C or 10 days at 52°C. On removal, the panel should be rinsed in tap water, wiped dry with a soft rag, and observed for any marked change in appearance. The coating should then be removed by solvent cleaning and the surface examined for evidence of pitting and rusting. The panel should show no sign of attack after immersion for 30 days in a 10% solution of hydrochloric acid, sulfuric acid, and sodium hydroxide. On removal of the coating by solvent cleaning, the surface should be free of any pitting or rusting.

9.1.11 CORROSIVITY

Precleaned, polished strips of copper, aluminum, and steel of any convenient size, but not less than 12 mm × 76 mm, should be immersed completely in a bath of the mastic at a temperature of 77–82°C for 24 h. The strips should then be removed, rinsed with a petroleum naphtha solvent, and dried. They should be examined for evidence of staining or corrosion. Steel strips should show no trace of corrosion at all, and copper and aluminum strips should show no more staining and corrosion than similarly prepared strips that have been heated for 24 h in an oven at 77–82°C but not immersed in the compound.

9.1.12 FIRE RESISTANCE

Spray and condition a coating of the compound on three steel panels that are 7.6 cm × 15 cm. Suspend one of the test panels vertically in a shielded hood. A laboratory burner with the air supply shut off and the flame regulated to 5 cm in length should be placed under the panel so that the lower end of the panel is 2.5 cm into the flame. The flame should be allowed to remain under the test panel for 20 s. After the flame is withdrawn, the time that the flaming continues should be observed. This test should be run three times and the results averaged. The dry coating may char, but it should not support combustion for more than 10 s when the flame source is removed.

9.1.13 RESISTANCE TO FLOW AT ELEVATED TEMPERATURES

Spray and condition a coating of the compound; parallel lines spaced 1 cm apart should be drawn across the width of the surface. The panel should then be suspended in a vertical position in an oven at 160–166°C for 24 h. Upon removal, observe the surface of the coating for shifting of the lines. There should be no evidence of flow or creep.

9.1.14 SALT-SPRAY RESISTANCE

Prepare three steel panels, and then expose them for 450 h to a salt spray or fog as prescribed in ASTM Standard B 117, "Salt Spray (Fog) Testing." On removal from the salt (fog) cabinet, observe the surface for evidence of rusting and pinholing. The coating should be tested for lifting by inserting a spatula under the coating along one of the waxed edges and noting whether it lifts easily from the panels or adheres firmly. Lifting of the coating as a solid film or in an area of 26 cm² or more should be grounds for rejection. The coating should then be removed by solvent cleaning and the panel surface observed for pitting and rusting. The coating should show no signs of blistering or excessive lifting. Upon removal of the coating, the steel surface should be free of pitting or rusting.

9.1.15 COLD-TEMPERATURE ADHESION

A 1,600-μm dry film coating of mastic should be prepared and tested as specified in U.S. Federal Specification TT-C-520, "Coating Compound, Bituminous, Solvent Type, Underbody (for Motor

Vehicles)"; it should be subjected to successive slamming at angles of 70°, 80°, and 90° in the slamming machine specified. Any loosening, flaking, or chipping should not exceed 5% of the total area at −23°C.

9.1.16 ABRASION RESISTANCE

Spray and condition a coating of the mastic and lay the panel flat in a sandblasting cabinet. The coating should then be subjected to 20 passes of the blast from a suction-type sandblast gun conforming to the following dimensions:

- Diameter of air orifice: 6 mm
- Diameter of nozzle: 12 mm
- Inside diameter (ID) of sand hose: 3 cm
- Length of sand hose: 3 m
- Air pressure at a nozzle of 6–7 Kg/cm2 gauge

The sandblast gun should traverse the 10-cm width of the panel in not less than 2 s at a distance of 2.5 cm from the surface of the coating. After 20 passes with the sandblast gun, the area should be examined for evidence of film failure. The coating should not be cut through to bare metal at any point.

9.1.17 WATER VAPOR TRANSMISSION

The water vapor permeability of a 1,600-μm-thick dry coating of mastic that has been prepared on a porous base and cured as in Section 6.10 should not be over 0.4 grams per square meter per 24 h under a vapor pressure differential of 11.8 mm of mercury at 25°C. The test should be made with standard permeability glass cups, each of which contains 40 cc of water, and then covered with a sample of the dry coating. The units should be sealed with wax and placed in a room or container held at 50% relative humidity and at a temperature of 25°C. At different time intervals, the cells are weighed separately. The weight loss is taken as the amount of water vapor (weight per 24 h per area) lost by diffusion through the coating after equilibrium in water vapor transmission has been established.

9.1.18 STORAGE LIFE AND PACKAGING

The mastic should show no thickening, curdling, gelling, or hard caking when tested as specified in U.S. Federal Standard No. 141, Method 3011, after it has been stored for 24 months from the date of delivery (unless otherwise specified by the company).

The packaging should meet the relevant requirements of ASTM D3951 (88) unless otherwise specified by the purchaser. All materials supplied under this specification should be subject to timely inspection by the purchaser or the purchaser's authorized representative. The purchaser should have the right to reject any materials found to be defective under this specification. In the case of disputes, the arbitration or settlement procedure established in the procurement documents should be followed. Samples of any or all ingredients used in the manufacture of this paint should be supplied upon the purchaser's request, along with the supplier's name and identification of the material.

Unless otherwise specified, the methods of sampling and testing should be in accordance with U.S. Federal Test Method Standard No. 141 or applicable methods of the American Society for Testing and Materials (ASTM).

9.1.19 DIRECTIONS FOR USE

Asphalt cold-applied mastic is used to protect steel from corrosion in severe surroundings. It will perform best if applied over blast-cleaned or pickled steel; however, it may be used over well-cleaned steel if all rust scale, loose rust, loose mill scale, and loose or nonadherent paint are removed. Oil and grease should be removed as well. For severe exposure, apply over a thoroughly dry, rust-inhibitive primer. The following directions for use should be supplied with each container of mastic:

1. Mix the mastic thoroughly before use.
2. Apply the mastic using high-pressure spray equipment. It can be applied by brush as well; in these cases, apply it with a daubing action. Apply it to the specified film thickness, or if none is specified, to at least 1,600 μm dry or approximately 3,200 μm wet.

The surface to be painted should be dry; the surface temperature should be at least 3°C above the dew point and the air temperature should be over 4°C. Do not paint outdoors in rainy weather or if freezing is expected before the paint dries. Under normal conditions, this mastic will dry for recoating in 24 h, but it will remain soft for long periods.

9.1.20 SAFETY PRECAUTIONS

The following safety instructions should be supplied with each container of mastic:

- Mastic is hazardous because of its flammability and potential toxicity, so proper safety precautions should be observed. Safe handling practices are required, including, but not limited to, the provisions of SSPC-PA Guide 3, "A Guide to Safety in Paint Application," and to the following specifications:
  - Keep mastic away from heat, sparks, and open flame during storage, mixing, and application. Provide sufficient ventilation to maintain vapor concentration at less than 25% of the lower explosive limit.
  - Avoid prolonged or repeated breathing of vapors or spray mist, and prevent contact of the paint with the eyes or skin.
  - Clean hands thoroughly after handling mastic and before eating or smoking.
  - Provide sufficient ventilation to ensure that vapor concentrations do not exceed the published permissible exposure limits. When necessary, supply appropriate personal protective equipment and enforce its use.
- This mastic may not comply with some air pollution regulations because of its hydrocarbon solvent content.
- Ingredients in this paint that may pose a hazard include asbestos, hydrocarbon solvent, and asphalt. Applicable regulations governing safe handling practices should apply to the use of this mastic.

9.2 MATERIAL AND EQUIPMENT STANDARDS FOR VINYL PAINT (ALUMINUM) AS INTERMEDIATE AND TOPCOAT (FINISH)

This section covers the minimum requirements for composition, analysis, properties, storage life and packaging, inspection, and labeling of a ready-to-mix vinyl paint (aluminum) to be used as an intermediate coat and topcoat (finish).

9.2.1 COMPOSITION

Details of the composition, ingredients, and proportions should be as specified in Table 9.4. The paint based on the specified ingredients should be uniform, stable in storage, and free of grit and coarse particles. Beneficial additives such as antiskinning agents, suspending agents, or wetting aids may be added. The aluminum paste should be packaged separately unless otherwise specified in the procurement documents.

This paint should contain approximately 14% by volume of nonvolatile, film-forming solids (pigment and binder).

Table 9.4 Composition

Ingredients	Typical Composition		Ingredient Standard ASTM
	Wt.%	**Vol.%**	
Pigment (10 ±1%)			
Aluminum paste	10.0	6.3	D 962
			Type 2
			Class B
Vehicle (90 ±1%)			
Vinyl resin A[1]	7.5	5.2	N/A
Vinly resin B[2]	7.5	5.1	N/A
Dioctyl phthalate[3]	1.5	1.2	N/A
MIBK[4]	36.7	42.6	D 1153
Toluene	36.7	39.6	D 362
Total	100.0	100.0	

[1]*Vinyl resin A should be a hydroxyl-containing vinyl chloride acetate copolymer. It should contain 89.5–91.5% vinyl chloride, 5.3–7.0% vinyl alcohol, and 2.0–5.5% vinyl acetate. The inherent viscosity of the resin (ASTM Standard D 1243, Method A) at 20°C should not be less than 0.5.*

[2]*Vinyl resin B should be a carboxyl-containing vinyl chloride acetate copolymer. It should contain 85–87% vinyl chloride, 12–14% vinyl acetate, and 0.5–1.0% maleic acid. The inherent viscosity of the resin (ASTM Standard D 1243, Method A) at 20°C should not be less than 0.48.*

[3]*Dioctyl phthalate di-2-ethylhexyl phthalate should be commercial material that conforms to the following requirements:*
- *Specific gravity at 25°C: 0.980–0.9861*
- *Refractive index at 25°C: 1.4830–1.6859*

[4]*When specified in the procurement documents, suitable high-boiling vinyl solvent may be substituted for a portion of the methyl isobutyl ketone (MIBK) to make the paint more amenable to application in hot weather or by brushing.*

Table 9.5 Analysis of Paint

| Characteristics | Requirements | | ASTM Method | U.S. Federal Standard No. 141 |
	Minimum Wt.%	Maximum Wt.%		
Pigment	6.0	7.5	D 1208	4021*
Volatiles	75.0	78.5	D 2369	N/A
Nonvolatile vehicle	15.5	17.5	N/A	4053
Uncombined water coarse particles and skins, as retained on sieve opening	N/A	0.5	D 1208	4081
Standard 0.44 mm (325 mesh screen)	N/A	0.25	D 185	4092

Using extraction mixture "C" (1:1 toluene and acetone)

9.2.2 ANALYSIS

The paint should conform to the composition (analysis) requirements laid out in Table 9.5.

9.2.3 PROPERTIES

The paint should meet the requirements specified in Table 9.6. The odor should be normal for the materials permitted (ASTM Standard D 1296). Before mixing with the aluminum paste, the vehicle should be clear. The color after mixing should be typical of aluminum paint (dull aluminum luster).

Table 9.6 Paint Properties

| Characteristics | Requirements | | ASTM Method | U.S. Federal Standard No. 141 |
	Minimum	Maximum		
Paint consistency viscosity* shear rate 200 (rpm)				
Kreb units	54	60	D 562	N/A
Grams	75	97	D 562	N/A
Density (Kg/l)	0.9	0.97	D 1475	N/A
Drying time (min):				
Tack-free	N/A	15	D 164D	4061
Dry hard	N/A	60	D 1640	4061

Viscosity 48 h or more after manufacturing.

There should be no evidence of incompatibility of any of the ingredients of the paint when two volumes of the paint are slowly mixed with one volume of thinner consisting of 85% toluene and 15% MIBK by volume (U.S. Federal Standard No. 141, Method 4203).

The paint should show good adhesion when tested as follows:

1. Apply one 25-μm coat of dry film of the mixed paint to a clean steel panel free of rust or scale, and to a similar panel pretreated with Wash Primer U.S. MIL-P-15328, "Primer (Wash) Pretreatment," or SSPC paint 27, "Basic Zinc Chromate-Vinyl Butyral Wash Primer" at a thickness of 12.5 μm. Also apply it to a panel similar to the preceding one, but over which has also been applied one coat of MIL-P-15929, "Primer Coating, Shipboard, Vinyl-Red Lead (for Hot Spray)," at a thickness of 25 μm)
2. After drying for 24 h, the film on each panel should be subjected to a knife test to determine whether the paint exhibits good adhesion to the undercoats and to the steel.

9.2.4 WORKING PROPERTIES

The paint should be easily applied when tested in accordance with U.S. Federal Standard No. 141, Methods 4331 and 4541. It should show no streaking, running, or sagging after drying.

9.2.5 STORAGE LIFE AND PACKAGING

While in a full, tightly covered container, the paint should show no thickening, curdling, gelling, or hard caking when tested as specified in U.S. Federal Standard No. 141, method 3011, after storage for 24 months from date of delivery (unless otherwise specified by the company).

The packaging should meet the relevant requirements of ASTM D3951 (88).

All materials supplied under this specification should be subject to timely inspection by the purchaser or the purchaser's authorized representative. The purchaser should have the right to reject any materials supplied that are found to be defective under this specification. In case of disputes, the arbitration or settlement procedure established in the procurement documents should be followed.

Samples of any or all ingredients used in the manufacture of this paint should be supplied upon request by the purchaser, along with the supplier's name and identification of the materials. Unless otherwise specified, the methods of sampling and testing should be in accordance with U.S. Federal Test Method Standard No. 141 or applicable methods of the ASTM.

9.2.6 DIRECTIONS FOR USE

Vinyl (aluminum) paint is intended for use over vinyl butyral wash primer or as a finish coat over vinyl chloride-acetate copolymer paint. It is supplied in two components, to be mixed just before using.

Gradually add small portions of the liquid to the aluminum paste. Mix the paste thoroughly until it is smooth and free of lumps. Then gradually add the remainder of the liquid, stirring constantly. It is recommended that the paint be mixed by a mechanical mixer.

The paint should be thinned as necessary with solvent containing not more than 85% toluene and 15% MIBK or methyl ethyl ketone (MEK). The amount of thinning will depend upon application methods and conditions and may be as high as 25–33% by volume. When required, this paint may be tinted to a contrasting color by the addition of a stable tinting pigment dispersed in a vinyl chloride-acetate copolymer resin solution.

Apply by conventional air spray. Brushing may be used in small areas. The surface to be painted should be dry, above 2°C, and not less than 3°C above the dew point. Do not paint outdoors in rainy weather. Apply so as to obtain a minimum dry film thickness of 25 μm.

A wet film of paint should be deposited on the surface when spraying; the spray gun should be adjusted so that proper atomization is obtained, but no dry powder is deposited onto the surface. The nozzle should be held about 150 mm from the surface during application.

If application is to be made by brush, the brush should be heavily loaded with paint; apply it quickly and smoothly. Avoid excessive brushing, and do not go back over the surface until the paint is thoroughly dry.

At temperatures between 16°C and 27°C, dry at least 1 h between coats and 72 h before immersion. Varying atmospheric conditions and degrees of ventilation in confined spaces may allow shorter or require longer drying times. This paint is not to be used as a priming coat next to bare steel.

Paints are hazardous because of their flammability and potential toxicity; therefore, proper safety precautions should be observed. Safe handling practices are required and should include, but not be limited to, the provisions of SSPC-PA Guide 3, "A Guide to Safety in Paint Application," and to the following instructions:

- Keep paint away from heat, sparks, and open flame during storage, mixing, and application. Provide sufficient ventilation to maintain vapor concentration at less than 25% of the lower explosive limit.
- Avoid prolonged or repeated breathing of vapors or spray mist, and prevent contact of the paint with the eyes or skin.
- Clean hands thoroughly after handling paint and before eating or smoking.
- Provide sufficient ventilation to ensure that vapor concentrations do not exceed the published permissible exposure limits. When necessary, supply appropriate personal protective equipment and enforce its use.

This paint may not comply with some air pollution regulations because of its hydrocarbon solvent content. Ingredients in this paint that may pose a hazard include hydrocarbon solvent. Applicable regulations governing safe handling practices should apply to the use of this paint.

9.3 RED LEAD, IRON OXIDE, RAW LINSEED OIL, AND ALKYD PRIMER

This section covers the minimum requirements for the composition, analysis, properties, packaging, inspection, and labeling of red lead, iron oxide, raw linseed oil, and alkyd primers.

9.3.1 COMPOSITION

The ingredients and proportions of this primer should be as specified in Table 9.7. The primer based on the specified ingredients should be uniform, stable in storage, and free of grit and coarse particles.

Table 9.7 Composition of Red Lead, Iron Oxide, Raw Linseed Oil, and Alkyd Primer

Ingredients	Typical Composition		Ingredient Standard	
	Wt.%	Vol.%	ASTM	U.S. Federal
Pigment (77.5 Wt.% Minimum)				
Red lead (97% Pb_3O_4)	56.3	16.8	D 83	
Red or brown iron oxide (85% Fe_2O_3)	18.4	11.0	D 3722*	N/A
Aluminum stearate	0.3	0.8	N/A	MIL-A-15206
Vehicle (25 Wt.% Maximum)				
Raw linseed oil	14.0	39.8	D 234	TT-L-215
Alkyd resin solids	5.2	12.9	N/A	TT-R-266, Grade II
Mineral spirit thinner	5.8	18.7	D 235	TT-T-291 Grade I
Driers	N/A	N/A	D 600 CLASS B	N/A
Total	100.0	100.0		

Either red or brown oxide (natural) should also comply with the following specifications:
- *Water solubles: 0.3% Max.*
- *Coarse particles on 0.044 mm sieve opening (325 mesh screen): 0.1% Max.*
- *Moisture and other volatile matter: 0.2% Max.*
- *Organic matter: None*

No rosin or rosin derivatives may be used. Beneficial additives such as antiskinning agents, suspending agents, or wetting aids may be added. This primer should contain approximately 82% by volume of nonvolatile, film-forming solids.

9.3.2 ANALYSIS

The paint should conform to the composition (analysis) requirements laid out in Table 9.8.

9.3.3 PROPERTIES

The paint should meet the requirements of Table 9.9 and the following specifications:

- Odor—The odor should be normal for the materials permitted (ASTM Standard D 1296).
- Color—The color should be typical of the specified mixture of red lead and red iron oxide.
- Compatibility—There should be no evidence of incompatibility of any of the ingredients of the primer when two volumes of the primers are slowly mixed with one volume of mineral spirits in accordance with U.S. Federal Standard No. 141, Method 4203.
- Skinning—There should be no skinning in a three-quarters-filled, closed container after 48 h when tested in the standard manner specified in U.S. Federal Standard No. 141, Method 3021.

Table 9.8 Analysis of Red Lead, Iron Oxide, Raw Linseed Oil, and Alkyd Prime

Characteristics	Requirements		ASTM Method	U.S. Federal Standard No. 141
	Minimum Wt.%	Maximum Wt.%		
Pigment	75	N/A	D 2371	4021
Volatiles	N/A	6	D 2369	N/A
Nonvolatile vehicle calculated by difference	19	N/A	N/A	4053
Uncombined water	N/A	0.5	D 1208	4081
Coarse particles and skins, as retained on standard 0.044 mm sieve opening (325 mesh, screen)	N/A	N/A	N/A	N/A
Rosin or rosin derivatives	N/A	0	D 1542	N/A

Table 9.9 Properties of Red Lead, Iron Oxide, Raw Linseed Oil, and Alkyd Primer

Characteristics	Requirements		ASTM Method	U.S. Federal Standard No. 141
	Minimum	Maximum		
Viscosity* shear rate 200 (rpm)				
Grams	190	230	D 562	N/A
Krebs	80	87	D 562	N/A
Density (Kg/l)	2.64	N/A	D 1475	N/A
Fineness of grind, hegman units (Mic)	5.0 (40)	N/A	D 1210	N/A
Drying time, (h)	N/A	24		4061
Flash point (°C)	38	N/A	D 3278	N/A
Sag resistance, (μm)	150	N/A	D 2801	4494

Viscosity 48 h or more after manufacture.

The paint should be easily applied by all three methods (brush, spray, and roller) when tested in accordance with U.S. Federal Standard No. 141, methods 4321, 4331, and 4541. The paint should show no streaking, running, or sagging after drying.

9.3.4 STORAGE LIFE AND PACKAGING

While in a full, tightly covered container, the primer should show no thickening, curdling, gelling, or hard caking when tested as specified in U.S. Federal Standard No. 141, Method 3011, after storage for 12 months from the date of delivery.

9.3.5 INSPECTION

All materials supplied under this specification should be subject to timely inspection by the purchaser or the purchaser's authorized representative. The purchaser should have the right to reject any materials supplied that are found to be defective under the standard. In the case of disputes, the arbitration or settlement procedure established in the procurement documents should be followed.

Samples of any or all ingredients used in the manufacture of this primer should be supplied upon the purchaser's request, along with the supplier's name and identification of the material. Unless otherwise specified, the methods of sampling and testing should be in accordance with U.S. Federal Test Method Standard No. 141 or applicable methods of the ASTM.

9.3.6 DIRECTIONS FOR USE

Red lead, iron oxide, raw linseed oil, and alkyd paint is intended for use as a primer over hand-cleaned steel in atmospheric exposure; it will perform better if the steel is power tool–cleaned, blast-cleaned, or pickled. All rust scale, loose rust, loose mill scale, and loose or nonadherent paint should be removed. Oil and grease should be removed to the fullest extent practical, as residues of oil and grease remaining on the surface will result in decreased paint performance.

Mix paint thoroughly before use. If the pigment has settled, pour off most of the liquid into a clean container and then thoroughly mix the pigment with the remaining liquid, taking care to scrape all the pigment off the bottom of the can. Gradually add the poured-off liquid and mix thoroughly. This may be made easier by transferring the contents to a larger container or by pouring the primer to and from another container. Examine the bottom of the container for unmixed pigment and screen the paint before applying.

Thin the paint only if necessary, using only mineral spirits. For brush application under normal conditions, no thinning should be required. For spray application, add up to 1 l of thinner per 8 l of primer when necessary.

Apply by brush or spray to the specified film thickness, or if none is specified, to at least 50 μm (dry) or approximately 75 μm (wet). The surface to be painted should be dry; the surface temperature should be at least 3°C above the dew point; and the temperature of the air should be over 4°C. Do not paint outdoors in rainy weather or if freezing temperatures are expected before the paint dries. Allow the primer at least 24 h of drying time in good weather before recoating.

9.3.7 SAFETY PRECAUTIONS

Paints are hazardous because of their flammability and potential toxicity; so proper safety precautions should be observed. Safe handling practices are required and should include, but not be limited to, the provisions of SSPC-PA Guide 3, "A Guide to Safety in Paint Application," and to the following criteria:

- Keep paint away from heat, sparks, and open flame during storage, mixing, and application. Provide sufficient ventilation to maintain vapor concentration at less than 25% of the lower explosive limit.
- Avoid prolonged or repeated breathing of vapors or spray mist, and prevent contact of the paint with the eyes or skin.
- Clean hands thoroughly after handling paint and before eating or smoking.

- Provide sufficient ventilation to ensure that vapor concentrations do not exceed the published permissible exposure limits. When necessary, supply appropriate personal protective equipment and enforce its use.
- This paint may not comply with some air pollution regulations because of its hydrocarbon solvent content.
- Ingredients in this paint that may pose a hazard include red lead, hydrocarbon solvent, and lead drier. Applicable regulations governing safe handling practices should apply to the use of this paint.

9.4 RED LEAD, IRON OXIDE, AND ALKYD INTERMEDIATE PAINT

This section covers the minimum requirements for the composition, analysis, properties, storage life, packaging, inspection, and labeling of red lead, iron oxide, and alkyd intermediate paints.

9.4.1 COMPOSITION

The ingredients and proportions of the paint should be as specified in Table 9.10. The paint based on these ingredients should be uniform, stable in storage, and free of grit and coarse particles.

Table 9.10 Composition of Red Lead, Iron Oxide, and Alkyd Intermediate Paint

Ingredients	Typical Composition		Ingredient Standard	
	Wt.%	Vol.%	ASTM	U.S. Federal Standard
Pigment: (58.4 Wt.% ±2%)				
Red lead (95% Pb_3O_4)	25.1	4.9	D 83	N/A
Red or brown oxide (85% Fe_2O_3)	13.4	5.6	D 3722*	N/A
Magnesium silicate	19.9	12.4	D 605	
Vehicle: (41.6 Wt.% ±2%)				
Alkyd resin solids	23.1	37.1	N/A	TT-R266, Grade II
Mineral spirit thinner	18.5	40.0	D 235	TT-T-291, Grade I
Driers	N/A	N/A	D 600	N/A
			CLASS B	
Totals	100.0	100.0		
Water solubles		0.3% Max		
Coarse particles on 0.044 sieve opening (325 Mesh sieve)		0.1% Max		
Moisture and other volatile matter		0.2% Max		
Organic matter		None		

Either red or brown iron oxide (natural) shall also comply with the following:

Table 9.11   Analysis of Red Lead, Iron Oxide, and Alkyd Intermediate Paint

| Characteristics | Requirements | | ASTM Method | U.S. Federal Standard No. 141 |
	Minimum Wt.%	Maximum Wt.%		
Pigment	56	N/A	D 2371	4021
Volatiles	N/A	22	D 2369	N/A
Nonvolatile vehicle	19	N/A	N/A	4053
Calculated by difference uncombined water	N/A	1.0	D 1208	4081
Coarse particles and skins, as retained on standard 0.044-mm sieve opening (325 mesh screen)	N/A	1.0	D 185	N/A
Rosin or rosin derivatives	N/A	0	D 1542	

No rosin or rosin derivatives may be used. Beneficial additives such as antiskinning agents, suspending agents, or wetting aids may be added. This intermediate paint should contain approximately 60% by volume of nonvolatile film-forming solids (pigment and binder).

9.4.2 ANALYSIS

The paint should conform to the composition (analysis) requirements listed in Table 9.11.

9.4.3 PROPERTIES

The paint should meet the requirements of Table 9.12 and the following specifications:

- Odor—The odor should be normal for the materials permitted (ASTM Standard D1296).
- Color—The color should be typical of the specified mixture of red lead and iron oxide.

Either red or brown iron oxide (natural) should also comply with the following:

- Compatibility—There should be no evidence of incompatibility of any of the ingredients of the paint when two volumes of the paint are slowly mixed with one volume of mineral spirits (U.S. Federal Standard No. 141, Method 4203).
- Skinning—There should be no skinning in a three-quarters-filled, closed container after 48 h when tested in the standard manner specified in U.S. Federal Standard No. 141 Method 3021.

The paint should be easily applied by all three methods (brush, spray, and roller) when tested in accordance with U.S. Federal Standard No. 141, Methods 4321, 4331, and 4541. The paint should show no streaking, running, or sagging after drying.

The film prepared as described in U.S. Federal Standard No. 141, Method 6221, after baking for 24 h at 93°C should show no cracking when suddenly chilled to 0°C and quickly bent sharply on itself through 180° over a 3-mm mandrel. The film on the bent part of the panel should show satisfactory adhesion. The gloss should be dull (ASTM Standard D523).

Table 9.12 Properties of Red Lead, Iron Oxide, and Alkyd Intermediate Paint

| Characteristics | Requirements | | ASTM Method | U.S. Federal Standard No. 141 |
	Minimum	Maximum		
Paint consistency:				
Viscosity* shear rate 200 (rpm)				
Grams	120	220	D 562	N/A
Kreb units	65	85	D 562	N/A
Density (Kg/l)	1.6	N/A	D 1475	N/A
Fineness of grind, microns+	65	N/A	D 1210	N/A
Fineness of grind, hegman units	3			
Drying time (h)	N/A	18	N/A	4061
Flash point (°C)	38	N/A	D 3278	N/A
Sag resistance (μm)	152	N/A	D 2801	4494

rpm = round per minute
*Viscosity 48 h or more after manufacture.
+gage depth rounded to nearest 5 mm.

9.4.4 STORAGE LIFE AND PACKAGING

While in a full, tightly covered container, the paint should show no thickening, curdling, gelling, or hard caking when tested as specified in U.S. Federal Standard No. 141, Method 3011, after storage for 24 months from the date of delivery (unless otherwise specified by the company). The packaging should meet the relevant requirements of ASTM D3951-88.

9.4.5 INSPECTION

All materials supplied under this specification should be subject to timely inspection by the purchaser or the purchaser's authorized representative. The purchaser should have the right to reject any materials supplied that are found to be defective under the standard. In the case of disputes, the arbitration or settlement procedure established in the procurement documents should be followed.

Samples of any or all ingredients used in the manufacture of this paint should be supplied upon the purchaser's request, along with the supplier's name and identification of the material. Unless otherwise specified, the methods of sampling and testing should be in accordance with U.S. Federal Standard No. 141 or applicable methods of the ASTM.

9.4.6 DIRECTIONS FOR USE

Red lead, iron oxide, and alkyd intermediate paint is intended for use as an intermediate coat over rust-inhibitive primers on structural steel, over itself, or over other oleoresinous paints. It also may be used as a finish coat as desired, and it may be used for priming steel that has been pickled or blast-cleaned. All oil, grease, dust, and loose or nonadherent paint should be removed, as residues of oil and grease remaining on the surface will result in decreased paint performance.

Mix paint thoroughly before use. If the pigment has settled, pour off most of the liquid into a clean container. Thoroughly mix the pigment with the remaining liquid, taking care to scrape all the pigment off the bottom of the can. Gradually add the poured-off liquid and mix thoroughly. This may be made easier by transferring the contents to a larger container or by pouring the paint to and from another container. Examine the bottom of the container for unmixed pigment, and screen paint before applying.

Thin paint only if necessary, using only mineral spirits. For brush application under normal conditions, no thinning should be necessary for spray application, add up to 1 l of thinner per 8 l of paint.

Apply by brush or spray to the specified film thickness, or if none is specified, to at least 38 μm (dry) or approximately 75 μm (wet). The surface to be painted should be dry; the surface temperature should be at least 3°C above the dew point, and the temperature of the air should be over 4°C. Do not paint outdoors in rainy weather or if freezing temperatures are expected before the paint dries. Allow paint at least 18 h of drying time in good weather before recoating.

9.4.7 SAFETY PRECAUTIONS

Paints are hazardous because of their flammability and potential toxicity, so proper safety precautions should be observed. Safe handling practices are required and should include, but not be limited to, the provisions of SSPC-PA Guide 3, "A Guide to Safety in Paint Application," and to the following specifications:

- Keep paint away from heat, sparks, and open flame during storage, mixing, and application. Provide sufficient ventilation to maintain vapor concentration at less than 25% of the lower explosive limit.
- Avoid prolonged or repeated breathing of vapors or spray mist, and prevent contact of the paint with the eyes or skin.
- Clean hands thoroughly after handling paint and before eating or smoking.
- Provide sufficient ventilation to ensure that vapor concentrations do not exceed the published permissible exposure limits. When necessary, supply appropriate personal protective equipment and enforce its use.
- This paint may not comply with some air pollution regulations because of its hydrocarbon solvent content.
- Ingredients in this paint that may pose a hazard include red lead, hydrocarbon solvent, and lead drier. Applicable regulations governing safe handling practices should apply to the use of this paint.

9.5 WHITE ALKYD PAINT FOR TOPCOAT (FINISH)

This section covers the minimum requirements for the composition, analysis, properties, storage life, packaging, inspection, and labeling of white alkyd paint for topcoat (finish).

Table 9.13 Composition of White Alkyd Paint

Ingredients	Typical Composition	Ingredient Standards	
	Wt%	ASTM Method	U.S. Federal
Pigment: (47.5 ±2.5 Wt. %)	N/A	N/A	N/A
Titanium dioxide (Minimum)	67.0	D 476	N/A
Reinforcing and tinting pigments (Maximum)	33.0	D 602	N/A
	N/A	D 605	N/A
Totals	100.0	N/A	N/A
Vehicle: (52.5 ±2.5 Wt.%)	N/A	N/A	N/A
Alkyd resin solids	46.8	N/A	TT-R-266
Mineral spirit thinner	53.2	D 235	TT-T-291
	N/A	N/A	Grade I
Total	100.0	N/A	N/A

9.5.1 COMPOSITION

The ingredients and proportions of white alkyd paint should be as specified in Table 9.13. The paint based on the specified ingredients should be uniform, stable in storage, and free of grit and coarse particles; no rosin or rosin derivatives may be used. Beneficial additives such as antiskinning agents, suspending agents, or wetting aids may be added.

This paint should contain approximately 48% by volume of nonvolatile, film-forming solids (pigment and binder).

Table 9.14 Analysis of White Alkyd Paint

Characteristics	Requirements	ASTM Method	Federal Standard No. 141
	Wt %		
Pigment (Minimum)	45.0	D 2371	4021
Volatiles (Maximum)	30.0	D 2369	N/A
Nonvolatile vehicle (minimum) calculated by difference	23.5	N/A	4053
Uncombined water (Maximum)	1.0	D 1208	4081
Coarse particles and skins as retained on standard 0.04-mm sieve (325 mesh screen) (Maximum)	1.0	D185	N/A
Rosin or rosin derivatives	0	D 1542	N/A

9.5.2 ANALYSIS

The paint should conform to the composition (analysis) requirements of Table 9.14.

9.5.3 PROPERTIES

The paint should meet the requirements of Table 9.15 and the following specifications:

- Odor—The odor should be normal for the materials permitted (ASTM Standard D1296).
- Color—The color should be white to U.S. Federal Standard 595 No. 17886.
- Compatibility—There should be no evidence of incompatibility of any of the ingredients of the paint when two volumes of the paint are slowly mixed with one volume of mineral spirits (U.S. Federal Standard No. 141, Method 4203).
- Skinning—There should be no skinning in a three-quarters-filled, closed container after 48 h when tested in the standard manner specified in U.S. Federal Standard No. 141, Method 3021.

The paint should be easily applied by all three methods (brush, spray, and roller) when tested in accordance with U.S. Federal Standard No. 141, Methods 4321, 4331, and 4541. The paint should show no streaking, running, or sagging after drying.

The film prepared as described in U.S. Federal Standard No. 141, Method 6221, after baking for 24 h at 93°C should show no cracking when suddenly chilled to 0°C and quickly bent sharply on itself

Table 9.15 Properties of White Alkyd Paint			
Requirements			
Characteristics		**ASTM Method**	**U.S. Federal Standard No. 141**
Paint consistency			
Viscosity* shear rate 200 rpm			
Grams	120		
Kreb units Minimum	65	D 562	N/A
Grams	220		
Kreb units Maximum	85	D 562	N/A
Density Kg/l	1.34	D 1475	N/A
Fineness of grind, μm$^+$	25		
Fineness of grind, hegman units	6	D 1210	N/A
Drying time, h			
Dry hard (Maximum)	18	N/A	4061
Flash point, °C (Minimum)	30.0	D 3278	N/A
Sag resistance μm (Minimum)	152	D 2801	4494
*Viscosity 48 h or more after manufacture +Gauge depth rounded to nearest 5 mm.			

through 180° over a 3-mm mandrel. The film on the bent part of the panel should show satisfactory adhesion. The gloss should be no less than eggshell (ASTM Standard D523).

9.5.4 STORAGE LIFE AND PACKAGING

When kept in a full, tightly covered container, the paint should show no thickening, curdling, gelling, or hard caking when tested as specified in Federal Standard No. 141, Method 3011, after storage for 24 months from date of delivery (unless otherwise specified by the company). The packaging should meet the relevant requirements of ASTM D 3951(88).

All materials supplied under the standard should be subject to timely inspection by the purchaser or the purchaser's authorized representative. The purchaser should have the right to reject any materials supplied that are found to be defective under the standard. In the case of disputes, the arbitration or settlement procedure established in the procurement documents should be followed.

Samples of any or all ingredients used in the manufacture of this paint should be supplied upon the purchaser's request, along with the supplier's name and identification of the material. Unless otherwise specified, the methods of sampling and testing should be in accordance with U.S. Federal Test Method Standard No. 141 or applicable methods of the ASTM.

9.5.5 DIRECTIONS FOR USE

White alkyd paint is intended for use as a topcoat (finish) over rust-inhibitive primers on structural steel, over itself, or over other oleoresinous paints. All oil, grease, dust, and loose or nonadherent paint should be removed, as residues of oil and grease remaining on the surface will result in decreased paint performance. If the undercoat is damaged, the steel should be spot-cleaned and -primed with rust-inhibitive primer.

Mix paint thoroughly before use. If the pigment has settled, pour off most of the liquid into a clean container. Thoroughly mix the pigment with the remaining liquid into a clean container, taking care to scrape all the pigment off the bottom of the can. Gradually add the poured-off liquid and mix thoroughly. This may be made easier by transferring the contents to larger container or by pouring the paint to and from another container. Examine the bottom of the container for unmixed pigment, and screen paint before applying.

Thin paint only if necessary, using only mineral spirits. For brush application under normal conditions, no thinning should be required. For spray application, add up to 1 l of thinner per 8 l of paint.

Apply by brush or spray to the specified film thickness, or if none is specified, to at least 38 μm (dry) or approximately 75 μm (wet). The surface to be painted should be dry; the surface temperature should be at least 3°C above the dew point and the temperature of the air should be over 4°C. Do not paint outdoors in rainy weather or if freezing temperatures are expected before the paint dries. Allow paint at least 18 h of drying time in good weather before recoating.

9.5.6 SAFETY PRECAUTIONS

Paints are hazardous because of their flammability and potential toxicity, so proper safety precautions should be observed. Safe handling practices are required and should include, but not be limited to, the provisions of SSPC-PA Guide 3, "A Guide to Safety in Paint Application," and to the following specifications:

- Keep paint away from heat, sparks, and open flame during storage, mixing, and application. Provide sufficient ventilation to maintain vapor concentration at less than 25% of the lower explosive limit.
- Avoid prolonged or repeated breathing of vapors or spray mist, and prevent contact of the paint with the eyes or skin.
- Clean hands thoroughly after handling paint and before eating or smoking.
- Provide sufficient ventilation to ensure that vapor concentrations do not exceed the published permissible exposure limits. When necessary, supply appropriate personal protective equipment and enforce its use.

This paint may not comply with some air pollution regulations because of its hydrocarbon solvent content. Ingredients in this paint that may pose a hazard include hydrocarbon solvent and lead drier. Applicable regulations governing safe handling practices should apply to the use of this paint.

9.6 COLORED ALKYD PAINT FOR TOPCOAT (FINISH)

This section covers the minimum requirements for the composition, analysis, properties, storage life, packaging, inspection, and labeling of alkyd paint for topcoat (finish).

9.6.1 COMPOSITION

The ingredients and proportions of colored alkyd paint should be as specified in Table 9.16. The paint based on the specified ingredients should be uniform, stable in storage, and free of grit and coarse particles; no rosin or rosin derivatives may be used. Beneficial additives such as antiskinning agents, suspending agents, or wetting aids may be added.

This paint should contain approximately 45% by volume of nonvolatile film-forming solids (pigment and binder).

Table 9.16 Composition of Colored Alkyd Paint

Ingredients	Typical Composition	
	Nominal% by Volume	Ingredient Standards
Pigment	Fade-resistant chemical-resistant colored pigment	BS 5493
Solids (nominal %) minimum	45	BS 5493
Nonvolatile vehicle and pigment		
Volatiles (Maximum)	55	ASTM D 2369
Total	100%	

Table 9.17 Analysis of Colored Alkyd Paint

Characteristics	Requirement		
	Wt%	ASTM	U.S. Federal Standard No. 141
Volatiles (Maximum)	30	D 2369	N/A
Nonvolatile vehicle (Minimum)			
Calculated by difference	70	N/A	N/A
Uncombined water (Maximum)	1.0	D 1208	4081
Coarse particle and skins, as retained on standard 0.45-mm sieve opening (325 mesh screen) (Maximum)	1.0	D 185	N/A

9.6.2 ANALYSIS

The paint should conform to the composition (analysis) requirements listed in Table 9.17.

9.6.3 PROPERTIES

The paint should meet the requirements laid out in Table 9.18 and the following specifications:

- Odor—The odor should be normal for the materials permitted (ASTM Standard D 1296).

Table 9.18 Properties of Colored Alkyd Paint

Characteristic	Requirements		
		ASTM Method	U.S. Federal Standard No. 141
Paint consistency viscosity* shear rate 200 (rpm)	N/A	N/A	N/A
Grams kreb units (Minimum)	120 65	D 562	N/A
Grams kreb units (Maximum)	220 85	N/A	N/A
Fineness of grind, μm	25	D 1210	N/A
" "Hegman units	6	N/A	N/A
Drying time, (h)	N/A	N/A	N/A
Dry hard (Maximum)	18	N/A	4061
Flash point (°C) (Minimum)	30.0	D 3278	N/A
Sag resistance, μm (Minimum) U.S. federal 4494	152	D 2801	N/A

*Viscosity 48 h or more after manufacture.

Table 9.19 Color of Colored Alkyd Paint

Paint Color	Color No. to BS 381 C
Arctic blue	112
Sea green	217
Brilliant green	221
Canary yellow	309
Light straw	384
Middle brown	411
Signal red	537
Light orange	567
Light gray	631

- Color—The color should be as specified in purchase order with reference to Table 9.19.
- Compatibility—There should be no evidence of incompatibility of any of the ingredients of the paint when two volumes of the paint are slowly mixed with one volume of mineral spirits (U.S. Federal Standard No. 141, Method 4203).
- Skinning—There should be no skinning in a three-quarters-filled, closed container after 48 h when tested in the standard manner specified in U.S. Federal Standard No. 141, Method 3021.

The paint should be easily applied by all three methods (brush, spray, and roller) when tested in accordance with U.S. Federal Standard No. 141, Methods 4321, 4331, and 4541. The paint should show no streaking, running, or sagging after drying.

The film prepared as described in U.S. Federal Standard No. 141, Method 6221, after baking for 24 h at 93°C, should show no cracking when suddenly chilled to 0°C and quickly bent sharply on itself through 180° over a 3-mm mandrel.

The film on the bent part of the panel should show satisfactory adhesion. The gloss should be no less than eggshell (ASTM Standard D 523).

9.6.4 STORAGE LIFE AND PACKAGING

When in a full, tightly covered container, colored alkyd paint should show no thickening, curding, gelling, or hard caking when tested as specified in U.S. Federal Standard No. 141, Method 3011, after storage for 24 months from the date of delivery unless otherwise specified by the company. The packaging should meet the relevant requirement of ASTM D3951 (88).

9.6.5 INSPECTION

All materials supplied under the standard should be subject to timely inspection by the purchaser or the purchaser's authorized representative. The purchaser should have the right to reject any materials supplied that are found to be defective under the standard. In the case of disputes, the arbitration or settlement procedure established in the procurement documents should be followed.

Samples of any or all ingredients used in the manufacture of this paint should be supplied upon the purchaser's request, along with the supplier's name and identification of the material. Unless otherwise specified, the methods of sampling and testing should be in accordance with U.S. Federal Test Method Standard No. 141 or applicable methods of the ASTM.

9.6.6 DIRECTIONS FOR USE

Colored alkyd paint is intended for use as topcoat (finish) over rust-inhibitive primers on structural steel, over intermediate paint, over itself, or over other oleoresinous paint. All oil, grease, dust, and loose or nonadherent paint should be removed, as residues of oil and grease remaining on the surface will result in decreased paint performance. If the undercoat is damaged, the steel should be spot-cleaned and -primed with rust-inhibitive primer.

Mix paint thoroughly before use. If the pigment has settled, pour off most of the liquid. Thoroughly mix the pigment with the remaining liquid, taking care to scrape all the pigment off the bottom of the can. Gradually add the poured-off liquid into a clean container and mix thoroughly. This may be made easier by transferring the contents to a larger container or by pouring the paint to and from another container. Examine the bottom of the container for unmixed pigment, and screen paint before applying.

Thin paint if necessary, using only mineral spirits. For brush application under normal conditions, no thinning should be necessary. For spray application, add up to 1 l of thinner per 8 l of paint.

Apply by brush or spray to the specified film thickness or if none is specified, to at least 38 μm (dry) or approximately 75 μm (wet). The surface to be painted should be dry; the surface temperature should be at least 3°C above the dew point; and the temperature of the air should be over 4°C. Do not paint outdoors in rainy weather or if freezing temperatures are expected before the paint dries. Allow paint at least 18 h of drying time in good weather before recoating.

9.6.7 SAFETY PRECAUTIONS

Paints are hazardous because of their flammability and potential toxicity, so proper safety precautions should be observed. Safe handling practices are required and should include, but not be limited to, the provisions of SSPC-PA Guide 3, "A Guide to Safety in Paint Application," and to the following specifications:

- Keep paint away from heat, sparks, and open flame during storage, mixing, and application. Provide sufficient ventilation to maintain vapor concentration at less than 25% of the lower explosive limit.
- Avoid prolonged or repeated breathing of vapors or spray mist, and prevent contact of the paint with the eyes or skin.
- Clean hands thoroughly after handling paint and before eating or smoking.
- Provide sufficient ventilation to ensure that vapor concentrations do not exceed the published permissible exposure limits. When necessary, supply appropriate personal protective equipment and enforce its use.

This paint may not comply with some air pollution regulations because of its hydrocarbon solvent content. Ingredients in this paint that may pose a hazard include hydrocarbon solvent, and lead drier. Applicable regulations governing safe handling practices should apply to the use of this paint.

9.7 CHLORINATED RUBBER PAINT FOR TOPCOAT (FINISH)

This section covers the minimum requirements for the composition, analysis, properties, storage life packaging, inspection, and labeling of chlorinated rubber topcoat (finish) paint.

9.7.1 COMPOSITION

The ingredients and proportions of chlorinated rubber paint should be as specified in Table 9.20. The paint based on the specified ingredients should be uniform, stable in storage, and free of grit and coarse

Table 9.20 Composition of Chlorinated Rubber Paint for Topcoat

Ingredients	Typical Composition		Ingredient Standards U.S. Federal
	Wt.%	Vol.%	
Pigment (23±5 Wt.%)	N/A	N/A	N/A
Colored or tinting	N/A	N/A	N/A
Pigments[1]	26.8	9.1	N/A
Extender pigments[1]	0.3	0.2	N/A
Vehicle (77±5 Wt.%)	N/A	N/A	N/A
Chlorinated rubber[2]	23.2	19.2	N/A
Chlorinated resin	5.6	4.8	MIL-C-429
			Type II
Chlorinated plasticizer	9.2	10.9	MIL-C-429
			Type I
Plasticizer[3]	2.3	3.5	N/A
Solvent[4]	31.6	51.1	N/A
Vehicle stabilizers[5]	0.7	0.8	N/A
Vapor phase stabilizers[6]	0.2	0.3	N/A
Suspending aids[7]	0.1	0.1	N/A
Totals	100	100	N/A

[1]*Lightfast and compatible chemically resistant colored pigments should be used to provide a tint or color when desired. Extender pigments must be compatible and chemically resistant.*
[2]*Chlorinated rubber should contain approximately 66% by weight chlorine. The viscosity (based on a solution of 20% by weight concentration in toluene at 20°C) should fall in the range of 9 to 14 cP, when measured according to ASTM Standard D445. Up to 50% by weight of the amount of chlorinated rubber can be in the viscosity range of 17–25 cP, but spray application is more difficult.*
[3]*Alternative plasticizers can be used, provided that they are compatible, high-quality, and chemically resistant.*
[4]*The solvent should consist of aromatic or a blend of aromatic and aliphatic hydrocarbons, with the aliphatic portion limited to 25% by weight, and should have a minimum kauri butanol value of 75 up to 10% by weight of turpentine (ASTM Standard D13) or other high-boiling, aromatic-type solvents may be added to improve application properties.*
[5]*The vehicle stabilizer should be a mixture of four parts of zinc oxide and one part pentaerythritol. Other suitable chloride acceptors, such as an epoxidized vegetable oil, may be used.*
[6]*A vapor phase stabilizer, such as propylene oxide, can also be used. It must be a high-grade commercial material suitable for the intended purpose.*
[7]*The suspending aid should be hydrogenated castor oil or montmorillonite mineral.*

particles. No rosin or rosin derivatives may be used. Beneficial additives such as antiskinning, suspending agents, or wetting aids may be added.

This paint should contain approximately 35% by volume of nonvolatile film-forming solids.

9.7.2 ANALYSIS

Chlorinated rubber paint should conform to the composition(analysis) requirements of Table 9.21.

9.7.3 PROPERTIES

Chlorinated rubber paint should meet the requirements of Table 9.22 and the following specifications:

- Odor—The odor should be normal for the materials permitted (ASTM Standard D 1296).
- Color—The color should be as specified in purchase order with reference to Table 9.23.
- Compatibility—There should be no evidence of incompatibility of any of the ingredients of the paint when one volume of paint is slowly mixed with one volume of xylene (U.S. Federal Standard No. 141, Method 4203).

Solvent blends should be checked in an unpigmented film deposited on glass. The dried film must be clear and bright. The flexibility and permeability of the trial film should also be compared with that obtained from the typical formulation of the paint.

The paint should be easily applied by all three methods (brush, spray, and roller) when tested in accordance with U.S. Federal Standard No. 141, Methods 4321, 4331, and 4541. The paint should show no streaking, running, or sagging after drying.

9.7.4 STORAGE LIFE AND PACKAGING

When in a full, tightly covered container, the paint should show no thickening, curdling, gelling, or hard caking when tested as specified in U.S. Federal Standard No. 141, Method 3011, after storage for

Table 9.21 Analysis of Chlorinated Rubber Paint for Topcoat

| Characteristics | Requirements | | ASTM Method | U.S. Federal Standard No. 141 |
	Minimum Wt.%	Maximum Wt.%		
Pigment	18.0	28.0	D 1208	4021
Volatiles	N/A	53.0	D 1208	N/A
Nonvolatile vehicle	20.0	N/A	N/A	4053
Uncombined water	N/A	0.25	D 1208	4081
Coarse particles and skins, as retained on 0.045-mm sieve opening (standard 325 mesh screen)	N/A	0.05	D 185	N/A
Rosin or rosin derivatives	N/A	0	D 1542	N/A

Table 9.22 Properties of Chlorinated Rubber Paint for Topcoat

Characteristics	Requirements		ASTM Method	U.S. Federal Standard No. 141
	Minimum	**Maximum**		
Paint consistency viscosity* shear rate 200 (rpm)				
Grams	120	190		
Kreb units	65	80	D 562	N/A
Density (kg/l)	1.0	1.4	D 1475	N/A
Fineness of grind, μm	25	N/A	D 1210	N/A
" " Hegman units	6.0			
Drying time:				
Set to touch, min	15	N/A	N/A	4061
Dry hard (h)	1	N/A	N/A	4061
Flash point (°C)	26.7		D 3278	N/A

*Viscosity 48 h or more after manufacture.

Table 9.23 Color of Chlorinated Rubber Paint for Topcoat

Paint Color	Color No. to BS 381 C
Arctic blue	112
Sea green	217
Brilliant green	221
Canary yellow	309
Light straw	384
Middle brown	411
Signal red	537
Light orange	567
Light grey	631
White	595

24 months from the date of delivery unless otherwise specified by the company. The packaging should meet the relevant requirement of ASTM D3951.

9.7.5 INSPECTION

All materials supplied under this specification should be subject to timely inspection by the purchaser or the purchaser's authorized representative. The purchaser should have the right to reject any materials supplied that are found to be defective under this specification. In the case of disputes, the arbitration or settlement procedure established in the procurement documents should be followed.

Samples of any or all ingredients used in the manufacture of this paint should be supplied upon the purchaser's request, along with the supplier's name and identification of the material. Unless otherwise specified, the methods of sampling and testing should be in accordance with U.S. Federal Test Method Standard No. 141 or applicable methods of the ASTM. Quality control procedures, documents, and certificates should be presented to the company for review and approval.

9.7.6 DIRECTIONS FOR USE

Chlorinated rubber topcoat paint is intended for use as an intermediate or finish coat over rust-inhibitive chlorinated rubber primer or other suitable primer on structural steel. Before applying, remove all moisture, oil, grease, dirt, and loose or nonadhering paint. Sound old coatings that are compatible with this chlorinated rubber paint may remain, but damaged areas or areas of poor adhesion must be spot-cleaned and -primed.

Preferred primers are chlorinated rubber, modified chlorinated rubber, chemically cured epoxy, zinc-rich, and others specifically recommended by the manufacturer. Where a zinc-rich primer is used and the system is considered for use with a water immersion service, a seal coat between the primer and this intermediate or topcoat paint may be required to eliminate blistering. Contaminated prime coats should be cleaned by appropriate methods before application of succeeding coats.

The surface to be painted should be dry and the surface temperature should be at least 3°C above the dew point. Do not paint outdoors if precipitation, dew, or condensation is expected before the paint dries.

At temperatures of 15–27°C and relative humidities of 40–60% allow at least 3 h of drying time between coats. Allow at least 48 h of drying time after the last coat is applied before placing in a water immersion service. If the temperature is below 16°C, or if the relative humidity is above 60%, allow 4–7 days before placing in a water immersion service.

Mix paint thoroughly before use. If simple stirring is inadequate, pour off most of the liquid into a clean container. Thoroughly mix the pigment with the remaining liquid, taking care to scrape all the pigment off the bottom of the can. Gradually add the poured-off liquid and mix thoroughly. This may be made easier by transferring the contents to a larger container or by pouring the paint to and from another container. Examine the bottom of the container for unmixed pigment, and screen paint before applying.

Generally, thinners are not used for brush application. For spray application, the coating may be thinned with xylene for up to 1-1/2 pints per gallon of unthinned paint. A by-volume blend of 80% minimum of xylene and 20% maximum of an aliphatic solvent having an evaporation rate faster than that of xylene may be used instead of straight xylene.

Apply by brush or spray to the specified film thickness of, if none is specified, to at least 38 μm (dry) or approximately 100 μm (wet). When application is by spraying, the equipment and operator technique should be properly adjusted to prevent dry spraying and to deposit a wet film of paint on the substrate. Clean the equipment with xylene or the reducing thinner both before and after use.

9.7.7 SAFETY PRECAUTIONS

Paints are hazardous because of their flammability and potential toxicity, so proper safety precautions should be observed. Safe handling practices are required and should include, but not be limited to, the provisions of SSPC-PA Guide 3, "A Guide to Safety in Paint Application," and to the following specifications:

- Keep paint away from heat, sparks, and open flame during storage, mixing, and application. Provide sufficient ventilation to maintain vapor concentration of less than 25% of the lower explosive limit.
- Avoid prolonged or repeated breathing of vapors or spray mist, and prevent contact of the paint with the eyes or skin.
- Clean hands thoroughly after handling paint and before eating or smoking.
- Provide sufficient ventilation to ensure that vapor concentrations do not exceed the published permissible exposure limits. When necessary, supply appropriate personal protective equipment and enforce its use.

This paint may not comply with some air pollution regulations because of its hydrocarbon solvent content. Ingredients in this paint, if so formulated, that may pose a hazard include lead and chromate pigments, hydrocarbon solvent, and plasticizers. Applicable regulations governing safe handling practices should apply to the use of this paint.

9.8 CHLORINATED RUBBER PAINT FOR INTERMEDIATE COAT

This section covers the minimum requirements for the composition, analysis, properties, storage life and packaging, inspection, and labeling of chlorinated rubber intermediate coat paint.

9.8.1 COMPOSITION

The ingredients and proportions of chlorinated rubber paint for intermediate coat should be as specified in Table 9.24. The paint based on the specified ingredients should be uniform, stable in storage, and free of grit and coarse particles. No rosin or rosin derivatives may be used. Beneficial additives such as antiskinning, suspending agents, or wetting aids may be added.

This paint should contain approximately 35% by volume of nonvolatile film-forming solids.

9.8.2 ANALYSIS

The paint should conform to the composition (analysis) requirements given in Table 9.25.

9.8.3 PROPERTIES

The paint should meet the requirements of Table 9.26 and the following specifications:

- Odor—The odor should be normal for the materials permitted (ASTM Standard D1296).
- Color—The color, or a contrasting shade, should be obtained by using compatible, chemically resistant tinting pigments.
- Compatibility—There should be no evidence of incompatibility of any of the ingredients of the paint when one volume of paint is slowly mixed with one volume of xylene (U.S. Federal Standard No. 141, Method 4203). Solvent blends should be checked in an unpigmented film deposited on glass.

Table 9.24 Composition of Chlorinated Rubber Paint for Intermediate Coat

| Ingredients | Typical Composition | | Ingredient ASTM | Standards U.S. Federal |
	Wt. %	Vol. %		
Pigment:	N/A	N/A	N/A	N/A
Titanium dioxide	4.0	1.3	D 476 Type IV	N/A
Tinting pigments[1]	0.4	0.3	N/A	N/A
Extender pigments Magnesium silicate	15.6	7.3	D 606	N/A
Mica	6.4	2.9	D 607	N/A
Vehicle:	N/A	N/A	N/A	N/A
Chlorinated rubber[2]	20.1	15.5	N/A	N/A
Chlorinated hard resins	5.0	4.0	N/A	MIL-C-429, Type II
Chlorinated plasticizers	10.0	11.0	N/A	MIL-C-429, Type I
Plasticizer[3]	N/A	N/A	N/A	N/A
Solvents[4]	37.0	56.2	N/A	N/A
Vehicle stabilizer[5]	0.6	0.7	N/A	N/A
Vapor phase stabilizers[6]	0.2	0.3	N/A	N/A
Suspending aids[7]	0.7	1.5	N/A	N/A
Totals	100.0	100.0	N/A	N/A

[1]Lampblack or lightfast and compatible, chemically resistant colored pigments should be used to provide a tint or color when desire. When this paint is to be used in a paint system to the interior of potable water vessels, the ingredients used should not be deemed objectionable for such use by the U.S. Environmental Protection Agency (EPA) or any agency that would have jurisdiction over such matters.
[2]The chlorinated rubber should contain approximately 66% by weight chlorine. The viscosity (based on solution of 20% by weight concentration in toluene at 20°C) should fall in the range of 9 to 14 cP, when measured according to ASTM Standard D445. Up to 50% by weight of the amount of chlorinated rubber could be in the viscosity range of 17–25 cP, but spray application is more difficult.
[3]Alternative plasticizers can be used, provided that they are compatible, high-quality, and chemically resistant.
[4]The solvent should consist of aromatic or a blend of aromatic and aliphatic hydrocarbons, with the aliphatic portion limited to 25% by weight, and should have a minimum kauri butanol value of 75. Up to 10% by weight of turpentine (ASTM Standard D13) or other high-boiling aromatic-type solvents may be added to improve application properties.
[5]The vehicle stabilizer should be a mixture of four parts zinc oxide and one part pentaerythritol. Other suitable chloride acceptors, such as an epoxidized vegetable oil, may be used.
[6]A vapor-phase stabilizer, such as propylene oxide, can also be used. It must be a high-grade commercial material suitable for the intended purpose.
[7]The suspending aid should be hydrogenated castor oil or montmorillonite mineral.

The dried film must be clear and bright. The flexibility and permeability of the trial film should also be compared with that obtained from the typical formulation of paint.

The paint should be easily applied by all three methods (brush, spray, and roller) when tested in accordance with U.S. Federal Standard No. 141, Methods 4321, 4331, and 4541. The paint should show no streaking, running, or sagging after drying.

Table 9.25 Analysis of Chlorinated Rubber Paint for Intermediate Coat

Characteristics	Requirements		ASTM Method	U.S. Federal Standard No. 141
	Minimum Wt. %	Maximum Wt. %		
Pigment and extender	16.0	30.0	D 1208	4021
Volatiles	N/A	54.6	D 1208	N/A
Nonvolatile vehicle	22.0	N/A	N/A	4053
Uncombined water	N/A	0.25	D 1208	4081
Coarse particles and skins, as retained on 0.045-mm sieve opening (standard 325 mesh screen)	N/A	0.05	D 185	N/A
Rosin or rosin derivatives	N/A	0	D 1542	N/A

Table 9.26 Properties of Chlorinated Rubber Paint for Intermediate Coat

Characteristics	Requirements		ASTM Method	U.S. Federal Standard. No. 141
	Minimum	Maximum		
Paint consistency viscosity shear rate 200 (rpm)				
Grams	165	350		
Kreb units	75	100	D 562	
Density (kg/l)	1.2	1.4	D 1475	
Fineness of grind, μm	40	N/A	D 1210	N/A
Drying time:				
Set to touch, min	15	N/A	N/A	4061
Dry hard h	1	N/A	N/A	4061
Flash point (°C)	26.7	N/A	D 3278	N/A

9.8.4 STORAGE LIFE AND PACKAGING

When in a full, tightly covered container, the paint should show no thickening, curdling, gelling, or hard caking when tested as specified in Federal Standard No. 141, Method 3011, after storage for 12 months from the date of delivery. The packaging should meet the relevant requirement of ASTM D3951-88.

9.8.5 INSPECTION

All materials supplied under this specification should be subject to timely inspection by the purchaser or the purchaser's authorized representative. The purchaser should have the right to reject any materials supplied that are found to be defective under this specification. In the case of disputes, the arbitration or settlement procedure established in the procurement documents should be followed.

Samples of any or all ingredients used in the manufacture of this paint should be supplied upon the purchaser's request, along with the supplier's name and identification of the material. Unless otherwise specified, the methods of sampling and testing should be in accordance with U.S. Federal Test Method Standard No. 141 or applicable methods of the ASTM.

9.8.6 DIRECTIONS FOR USE

Chlorinated rubber paint is intended for use as an intermediate coat or topcoat over rust-inhibitive chlorinated rubber primer or other suitable primer on structural steel. Before applying, remove all moisture, oil grease, dirt, and loose or nonadhering paint. Sound old coatings that are compatible with this chlorinated rubber paint may remain, but damaged areas or areas of poor adhesion must be spot-cleaned and -primed.

Preferred primers are chlorinated rubber, modified chlorinated rubber, chemically cured epoxy, zinc-rich, and others specifically recommended by the manufacturer. Where a zinc-rich primer is used and the system is considered for a water immersion service, a seal coat between the primer and this intermediate or topcoat paint may be required to eliminate blistering. Contaminated prime coats should be cleaned by appropriate methods before application of succeeding coats.

Mix paint thoroughly before use. If simple stirring is inadequate pour off most of the liquid into a clean container. Thoroughly mix the pigment with the remaining liquid, taking care to scrape all the pigment off the bottom of the can. Gradually add the poured-off liquid and mix thoroughly. This may be made easier by transferring the contents to a larger container or by pouring the paint to and from another container. Examine the bottom of the container for unmixed pigment, and screen paint before applying.

Generally, thinners are not used for brush application. For spray application, the coating may be thinned with xylene up to 1.5 l per 8 l of unthinned paint. A by-volume blend of 80% minimum of xylene and 20% maximum of an aliphatic solvent having an evaporation rate faster than that of xylene may be used instead of straight xylene.

Apply by brush or spray to the specified film thickness, or if none is specified, to at least 75 μm (dry) or approximately 225 μm (wet). When application is by spraying, the equipment and operator technique should be properly adjusted to prevent dry spraying and to deposit a wet film of paint on the substrate. Clean the equipment with xylene or the reducing thinner both before and after use.

The surface to be painted should be dry and the surface temperature should be at least 3C above the dew point. Do not paint outdoors if precipitation, dew, or condensation is expected before the paint dries.

At temperatures of 15–27°C and relative humidities of 40–60%, allow at least 3 h of drying time between coats. Allow at least 48 h of drying time after the last coat is applied before placing in a water immersion service. If the temperature is below 16°C, or if the relative humidity is above 60%, allow 4–7 days before placing in a water immersion service.

9.8.7 SAFETY PRECAUTIONS

Paints are hazardous because of their flammability and potential toxicity, so proper safety precautions should be observed. Safe handling practices are required and should include, but not be limited to, the provisions of SSPC-PA Guide 3, "A Guide to Safety in Paint Application," and to the following specifications:

- Keep paint away from heat, sparks, and open flame during storage, mixing, and application. Provide sufficient ventilation to maintain vapor concentration at less than 25% of the lower explosive limit.
- Avoid prolonged or repeated breathing of vapors or spray mist, and prevent contact of the paint with the eyes or skin.
- Clean hands thoroughly after handling paint and before eating or smoking.
- Provide sufficient ventilation to ensure that vapor concentrations do not exceed the published permissible exposure limits. When necessary, supply appropriate personal protective equipment and enforce its use.

This paint may not comply with some air pollution regulations because of its hydrocarbon solvent content. Ingredients in this paint, if so formulated, that may pose a hazard include lead and chromate pigments, hydrocarbon solvent, and plasticizers. Applicable regulations governing safe handling practices should apply to the use of this paint.

9.9 CHLORINATED RUBBER-INHIBITIVE PRIMER

This section covers the minimum requirements for the composition, analysis, properties, storage life and packaging, inspection, and labeling of chlorinated rubber-inhibitive primer.

9.9.1 COMPOSITION

The ingredients and proportions of chlorinated rubber-inhibitive primer should be as specified in Table 9.27. The primer based on the specified ingredients should be uniform, stable in storage, and free of grit and coarse particles. No rosin or rosin derivatives may be used. Beneficial additives such as antiskinning agents, suspending agents, or wetting aids may be added.

This primer should contain a minimum of 32% by volume of nonvolatile film-forming solids (pigment and binder).

9.9.2 ANALYSIS

The primer should conform to the composition (analysis) requirements of Table 9.28.

9.9.3 PROPERTIES

The primer should meet the requirements of Table 9.29, and the following specifications:

- Odor—The odor should be normal for the materials permitted (ASTM D1296).
- Color—The color should be consistent with the choice of inhibitive pigment and should be obtained by using compatible, chemically resistant tinting pigments.
- Compatibility—There should be no evidence of incompatibility of any of the ingredients of the primer when one volume of the primer is slowly mixed with one volume of xylene (U.S. Federal Standard No. 141, Method 4203).

Table 9.27 Composition of Chlorinated Rubber-Inhibitive Primer

Ingredients	Typical composition		Ingredient ASTM	U.S. Federal Standards
	Wt. %	Vol. %		
PIGMENT: (39 ± 1 Wt. %)				
Rust-inhibitive pigments[1]	26.3	5.3	N/A	N/A
Tinting pigments[2]	0.3	0.2	N/A	N/A
Extender pigments	12.1	6.4	N/A	N/A
Barytes			D602	N/A
Magnesium			D605	N/A
Mica			D607	N/A
Vehicle: (61 ±10Wt. %)				
Chlorinated rubber[3]	12.3	10.8	N/A	N/A
Chlorinated resins	3.3	3.0	N/A	MIL-C-429, Type II
Chlorinated plasticizers	6.7	8.4	N/A	MIL-C-429, Type I
Plasticizer[4]	N/A	N/A		N/A
Solvents[5]	37.0	64.0	N/A	N/A
Vehicle stabilizers[6]	0.4	0.5	N/A	N/A
Vapor phase stabilizers[7]	0.1	0.2	N/A	N/A
Suspending aids[8]	1.5	1.2	N/A	N/A
Totals	100.0	100.0		

[1]The rust-inhibitive pigments should be of good quality and such that when it is incorporated in the vehicle portion of this specification paint, the proposed paint will perform as well as or better than the paint made from the control formulation at each inspection period in the tests.

[2]Lightfast and compatible, chemically resistant colored pigments should be used to provide a tint or color when desired.

[3]The chlorinated rubber should contain approximately 66% chlorine by weight. The viscosity (based on a solution of 20% by weight concentration in toluene at 20°C) should fall in the range of 9–14 cP when measured according to ASTM Standard 445. Up to 50% by weight of the amount of chlorinated rubber could be in the viscosity range of 17–25 cP, but spray application is more difficult.

[4]Alternative plasticizers can be used, provided they are compatible, high-quality, and chemically resistant.

[5]The solvent should consist of aromatic or a blend of aromatic and aliphatic hydrocarbons, with the aliphatic portion limited to 25% by weight and should have a minimum kauri butanol value of 75. Up to 10% by weight of turpentine (ASTM Standard D 13) or other high-boiling, aromatic-type solvents may be added to improve application properties.

[6]The vehicle stabilizer should be a suitable chloride accepter, such as expoxidized vegetable oil.

[7]A vapor phase stabilizer, such as propylene oxide, can also be used. It must be a high-grade commercial material suitable for the intended purpose.

[8]The suspending aid should be hydrogenated castor oil or montmorillonite mineral.

Table 9.28 Analysis of Chlorinated Rubber-Inhibitive Primer

Characteristics	Minimum Wt. %	Maximum Wt. %	ASTM Method	U.S. Federal Standard No. 141
Pigment	29.0	49.0	D1208	4021
Volatiles	N/A	46.0	D1208	N/A
Nonvolatile vehicle	15.0	N/A	N/A	4053
Uncombined water	N/A	0.25	D1208	4081
Coarse particles and skins as retained on standard 0.045-mm sieve (325 mesh screen)	Opening	0.05	D185	N/A
Rosin or rosin derivatives	N/A	0	D1542	N/A

Table 9.29 Properties of Chlorinated Rubber-Inhibitive Primer

Characteristics	Minimum	Maximum	ASTM Method	U.S. Federal Standard No. 141
Primer consistency:				
Viscosity shear rate 200 (rpm)				
Grams	140	225	D562	N/A
Kreb units	70	90	D562	N/A
Density (kg/l)	1.4	1.7	D1475	N/A
Fineness of grind, (μm)	65	N/A	D1210	N/A
” ” Hegman units	3			
Drying time:				
Set of touch, (min)	15	N/A	N/A	4061
Dry hard, (h)	1	N/A	N/A	4061
Flash point, (°C)	26.7	N/A	D3278	N/A

Note: Viscosity 48 h or more after manufacture.

Solvent blends should be checked in an unpigmented film deposited on glass. The dried film must be clear and bright. The flexibility and permeability of the trial film should also be compared with that obtained from the control formulation primer. See Table 9.29.

The primer should be easily applied by all three methods (brush, spray, and roller) when tested in accordance with U.S. Federal Standard No. 141, Methods 4321, 4331, and 4541. The primer should show no streaking, running, or sagging after drying.

When subjected to the tests, the performance of the primer should be equal to or better than the control formulation primer given in Table 9.29.

The primer should be capable of being applied to produce a dry film thickness of at least 38 μm without running or sagging. Depending upon the volume of the solid content, this will require a wet film thickness of approximately 125 μm.

Panels prepared for the working properties test under Section 6.5 may be used for these measurements. A minimum of five measurements should be taken to calculate the average dry film thickness.

A salt fog test should be performed in accordance with ASTM Standard B 117. Panels should be 8 cm × 16 cm × 1/3 cm cold-rolled steel and sand-blasted in accordance with SSPC-SP 10, "Near-White Blast Cleaning," using 20–40 mesh silica sand.

The primer should be applied by spraying to a 75 μm (dry) film thickness, approximately 225 μm (wet) film. After 1 week of drying at 21–27°C, the panels should be scored as indicated in ASTM Standard B 117.

Panels should be exposed in the cabinet at a 30° angle from the vertical. Panels should be tested in duplicate and examined at 96-h intervals for a period of 500 h.

A flexibility test should be run in accordance with U.S. Federal Standard No. 141, Method 6221. Panels should be 8 cm × 16 cm, 20-gauge cold-rolled steel, cleaned in accordance with U.S. Federal Standard No. 141, Method 2011, Procedure D, followed by Procedure A.

Table 9.30 Composition of Control Paint

| Ingredients | Typical Composition | | Ingredient ASTM | Standards U.S. Federal |
	Wt. %	Vol. %		
Basic lead silico-chromate	30.0	11.5	D1648	N/A
Barytes	13.0	4.7	D602	N/A
Chlorinated rubber[1]	12.5	11.6	N/A	N/A
Chlorinated resins	7.4	7.2	N/A	MIL-C-429, Type II
Chlorinated plasticizer	4.1	5.5	N/A	MIL-C-429, Type I
Modified hydrogenated castor oil	1.5	2.7	N/A	N/A
Epoxidized soya oil	0.9	1.2	N/A	N/A
Propylene oxide	0.1	0.2	N/A	N/A
Xylene	23.7	44.1	D364	N/A
Ethylene glycol monoethyl ether acetate	6.3	11.3	D343	N/A
Total	100.0	100.0		

[1]The chlorinated rubber should contain approximately 66% by weight chlorine. The viscosity (based on a solution of 20% by weight concentration in toluene at 20°C) should fall in the range of 9–14 cP, when measured according to ASTM Standard D 445. Up to 50% by weight of the amount of chlorinated rubber could be in the viscosity of 17–25 cP, but spray application is more difficult.

The primer should be applied by spraying to a **75** μm (dry) film thickness, and approximately 225 μm (wet) film. After 1 week of drying at 21–27°C, the panel should be bent over a 4-cm mandrel and should show no cracking or loss of adhesion.

Since exterior exposure environments vary, exposure tests must be based on comparisons with the control formulation paint. The proposed paint should provide at least as much protection as the control formulation primer. Panels should be 16-cm channel steel in a 30-cm-long section and exposed at a 45° angle at south exposure with horizontal flanges.

Each panel should have a weld bead approximately 8 cm long placed about 8 cm from the left end of the panel. The primer should be applied by spray to a 75-μm (dry) film thickness and approximately 225-μm (wet) film. After 1 week of drying at 21–27°C, the coated steel should be scored in an "X" pattern 8 cm from the right end. Each stroke of the "X" must be 10 cm long. Then the panels should be placed on exposure. After 60 days of exposure, both the proposed and control primers should show no blistering or rusting on the face and only minor blistering or rusting at the weld and score.

In the fresh-water immersion test, panels should be 8 cm × 16 cm × 1/3 cm cold-rolled steel prepared according to SSPC-SP 10. "Near-white Blast Cleaning," using 20–40 mesh silica sand. The primer should be applied by spray to 75-μm (dry) film thickness, and approximately 225-μm (wet) film. Half of the immersed area and half of the unimmersed area should be coated with the selected topcoat system. After 1 week of drying at 21–27°C, the panels should be immersed halfway in distilled water at 38°C. Panels should be examined every 24 h for blistering, leaching, rusting, and loss of adhesion. Both the proposed and control primers, should survive 500 h of immersion with no other defect than a slight discoloration (examine after a 1-h drying period).

9.9.4 STORAGE LIFE AND PACKAGING

When in a full, tightly covered container, the primer should show no thickening, curdling, gelling, or hard caking when tested as specified in U.S. Federal Standard No. 141, Method 3011, after storage for 12 months from the date of delivery at a temperature of 10–43°C. The packaging should meet the relevant requirement of ASTM D3951 (88).

9.9.5 INSPECTION

All materials supplied under this specification should be subject to timely inspection by the purchaser or the purchaser's authorized representative. The purchaser should have the right to reject any materials supplied that are found to be defective under this specification. In the case of disputes, the arbitration or settlement procedure established in the procurement documents should be followed.

Samples of any or all ingredients used in the manufacture of this paint should be supplied upon the purchaser's request, along with the supplier's name and identification of the material. Unless otherwise specified, the methods of sampling and testing should be in accordance with U.S. Federal Test Method Standard No. 141 or applicable methods of the ASTM.

9.9.6 DIRECTIONS OF USE FOR CHLORINATED RUBBER-INHIBITIVE PRIMER

Chlorinated rubber-inhibitive paint is intended for use as a primer on structural steel. Before applying, remove all moisture, oil, grease, dirt, and loose or nonadhering paint. Rust, rust scale, mill scale, and other flaws should be removed by abrasive blast cleaning in accordance with SSPC-SP 6,"Commercial

Blast Cleaning," or SSPC-SP 8, "Pickling." Sound old coatings that are compatible with this chlorinated rubber primer may remain, but damaged areas or areas with poor adhesion must be spot-cleaned before priming. When the coating system is intended for immersion, the surface should be cleaned in accordance with SSPC-SP10. "Near-white Blast Cleaning, "or SSPC-SP 8.

This primer is intended to be followed by an intermediate paint such as chlorinated rubber intermediate coat paint and/or a finish coat conforming to chlorinated rubber topcoat paint. The entire system is designed to provide maximum chemical resistance. In more severe environments, a second prime coat is recommended.

Mix primer thoroughly before use. If simple stirring is inadequate, pour off most of the liquid into a clean container. Thoroughly mix the pigment with the remaining liquid, taking care to scrape all the pigment off the bottom of the can. Gradually add the poured-off liquid and mix thoroughly. This may be made easier by transferring the contents to a larger container or by pouring the paint to and from another container. Examine the bottom of the container for unmixed pigment, and screen paint before applying. Generally, thinners are not used for brush application. For spray application, the coating may be thinned with xylene up to 1.5 l per 8 l of unthinned paint. A by-volume blend of 80% minimum of xylene and 20% maximum of an aliphatic solvent having an evaporation rate faster than that of xylene may be used instead of straight xylene.

Apply by brush or spray to the specified film thickness, or if none is specified, to at least 38 μm (dry) or approximately 125 μm (wet). When application is by spraying, the equipment and operator technique should be properly adjusted to prevent dry spraying and to deposit a wet film of paint on the substrate. Clean the equipment with xylene or the reducing thinner both before and after use.

The surface to be painted should be dry, and the surface temperature should be at least 3°C above the dew point. Do not paint outdoors if precipitation, dew, or condensation is expected before the paint dries.

At temperatures of 15–27°C and relative humidities of 40% to 60%, allow at least three h of drying time between coats. Allow at least 48 h of drying time after the last coat is applied before placing in a water immersion service. If the temperature is below 16°C, or if the relative humidity is above 60%, allow 4–7 days before placing in a water immersion service.

9.9.7 SAFETY PRECAUTIONS

Paints are hazardous because of their flammability and potential toxicity, so proper safety precautions should be observed. Safe handling practices are required and should include, but not be limited to, the provisions of SSPC-PA Guide 3, "A Guide to Safety in Paint Application," and to the following specifications:

- Keep paint away from heat, sparks, and open flame during storage, mixing, and application. Provide sufficient ventilation to maintain vapor concentration at less than 25 % of the lower explosive limit.
- Avoid prolonged or repeated breathing of vapors or spray mist, and prevent contact of the primer with the eyes or skin.
- Clean hands thoroughly after handling primer and before eating or smoking.
- Provide sufficient ventilation to ensure that vapor concentrations do not exceed the published permissible exposure limits. When necessary, supply appropriate personal protective equipment and enforce its use.

This primer may not comply with some air pollution regulations because of its hydrocarbon solvent content. Ingredients in this primer, if so formulated, and which may pose a hazard, include lead and

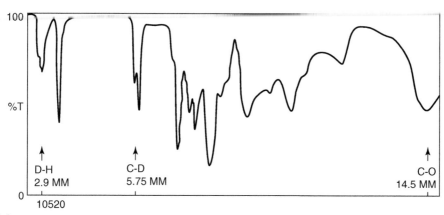

FIGURE 9.1

Infrared specturm of resin.

chromate pigments, hydrocarbon solvent, and plasticizers. Applicable regulations governing safe handling practices should apply to the use of this primer.

9.10 ALKYD PAINT (ALUMINUM) LEAFING AS TOPCOAT (FINISH)

This section covers the minimum requirements for the composition, analysis, properties, storage life, packaging, inspection, and labeling of alkyd paint leafing (aluminum) as topcoat (finish). This paint consists of a two-component container with leafing-type aluminum paste separated from a long oil alkyd varnish vehicle. The aluminum paste is mixed with the alkyds varnish prior to use.

9.10.1 COMPOSITION

The ingredients and proportions of the mixed paint should be as specified in Table 9.31. The paint based on the specified ingredients should be uniform, stable in storage, and free of grit and coarse particles.

Table 9.31 Composition of Mixed Paint			
	Required		
Characteristics	**Minimum Wt%.**	**Maximum Wt%**	**Ingredient ASTM**
Pigment (20.3 ± 0.5 Wt. %)			
Aluminum paste	100	N/A	N/A
Vehicle (79.7 ± 0.5 Wt. %)			
Alkyd varnish solids	50	N/A	N/A
Mineral spirit thinner	N/A	50	D235
Driers	N/A	N/A	D600 Class B

Table 9.32 Analysis of Alkyd Varnish

| Characteristics | Requirements | | ASTM Method | U.S. Federal Standard No. 141 |
	Minimum Wt. %	Maximum Wt. %		
Volatiles	N/A	50	D2369	N/A
Nonvolatile vehicle				
Calcalated by difference	50	N/A	N/A	4053
Rosin or rosin derivatives	N/A	0	D1542	N/A

No rosin or rosin derivatives may be used. Beneficial additives such as antiskinning agents, suspending agents, or wetting aids may be added. This paint contains approximately 40% by volume of nonvolatile, film-forming solids (pigment and binder).

9.10.2 ANALYSIS

The alkyd varnish should conform to the composition (analysis) requirements listed in Table 9.32.

9.10.3 PROPERTIES

The alkyd varnish should meet the requirements of Table 9.33 and the following criteria:

- Odor—The odor should be normal for the materials permitted (ASTM D1296).
- Color—The color should be less than 11 on the Gardner 1933 scale (ASTM D1544).

Table 9.33 Properties of Alkyd Varnish

| Characteristics | Requirements | | ASTM Method | Federal Standard No. 141 |
	Minimum	Maximum		
Viscosity* gardner airbubble				
Viscometer	C	E	D1545	N/A
Density (kg/l)	0.91	0.97	D1475	N/A
Drying time (h)				
Set to touch	N/A	4	N/A	4061
Dry hard	N/A	10	D154	4061
Flash point (°C)	30	N/A	D3278	N/A
*Viscosity 48 h or more after manufacture.				

- Compatibility—There should be no evidence of incompatibility of any of the ingredients of the paint when two volumes of the mixed paint are slowly mixed with one volume of mineral spirits (U.S. Federal Standard No. 141, Method 4203).
- Skinning—There should be no skinning in a three-quarters-filled, closed container after 48 h when tested in the standard manner specified in Federal Standard No. 141, Method 3021.

The paint should be easily applied by all three methods (brush, spray, and roller) when tested in accordance with U.S. Federal Standard No. 141, Method 4321, 4331, and 4541. The paint should show no streaking, running, or sagging after drying.

A dried film of the varnish should be clear, smooth, and glossy. A dried film (with a thickness of 25 µm ± 5 µm) of the varnish should show no cracking when bent over a 3.2-mm mandrel after 17 h of air-drying, plus 24 h of baking at 102–107°C.

Dried films, prepared as in Section 6.8 should resist boiling water for 10 min, and should withstand immersion in distilled water for 24 h. Upon removal after 2 h of drying, the film should show no whitening, blistering, or loss of adhesion, but slight dulling is permissible. After air-drying for 17 h, plus 24 h of baking at 104°C, the mixed paint should show no detrimental film effects after a painted panel is immersed in gasoline for 4 h.

The aluminum paste should comply with the requirements of either U.S. Federal Specification TTP-320 Type II Class B, or ASTM Standard D962 Type 2 Class B, with the exception that the total retained on 0.045 opening (325 mesh) sieve should be within the range of 4–6%. However, the hiding and covering capacity should be equivalent.

9.10.4 STORAGE LIFE AND PACKAGING

When in a full, tightly covered container, the ready-to-mix and already mixed paint should show no gas evolution, thickening, curdling, gelling, or hard caking when tested as specified in U.S. Federal Standard No. 141, Method 3011, after storage for 12 months from date of delivery. The packaging should meet the relevant requirements of ASTM D3951 (88).

9.10.5 INSPECTION

All materials supplied under this specification should be subject to timely inspection by the purchaser or the purchaser's authorized representative. The purchaser should have the right to reject any materials supplied that are found to be defective under this specification. In the case of disputes, the arbitration or settlement procedure established in the procurement documents should be followed.

Samples of any or all ingredients used in the manufacture of this paint should be supplied upon the purchaser's request, along with the supplier's name and identification of the material.

Unless otherwise specified, the methods of sampling and testing should be in accordance with U.S. Federal Test Method Standard No. 141 or applicable methods of the ASTM.

Refer to ANSI Standard Z129.1, "Precautionary Labeling of Hazardous Industrial Chemicals," for information on labeling.

9.10.6 DIRECTIONS FOR USE

Alkyd paint (aluminum) leafing is intended for use as a finish coat over rust-inhibitive primer on structural steel or over other oleoresinous paints. All oil, grease, dust, and loose or nonadherent paint should be removed to the fullest extent practical, as residues of oil and grease remaining on the surface will result in decreased paint performance. If the undercoat is damaged, the steel should be spot-cleaned and -primed with rust-inhibitive primer.

Mix paint thoroughly before use. Add the aluminum paste to the mixing varnish in the ratio of 1 kg of aluminum paste per 4.2 l of the varnish vehicle. To mix the paste with the varnish, add a small amount of the varnish to sufficient aluminum paste in a large container. Thoroughly mix the aluminum paste with the small portion of varnish until a smooth, thin paste is achieved. Gradually add more of the varnish while stirring. Continue adding paste and mixing until all of the varnish is incorporated with the vehicle. Examine the bottom of the container for unmixed paste. Screen paint before applying. Mix only enough for one day's use.

If this paint is furnished in a single component and the pigment has settled, pour off most of the liquid. Thoroughly mix the pigment with the remaining liquid, taking care to scrape all the pigment off the bottom of the container. Gradually add the poured-off liquid and mix thoroughly. This may be made easier by transferring the contents to a larger container or by pouring the paint to and from another container.

Thin paint only if necessary, using only mineral spirits or turpentine. For brush application under normal conditions, no thinning should be necessary. For spray application, add up 1.5 l of thinner per 8 l of paint when necessary.

Apply by brush or spray to the specified film thickness, or if none is specified, to at least 38 μm (dry) or approximately 102 μm (wet). The surface to be painted should be dry; the surface temperature should be at least 3°C above the dew point and the temperature of the air should be over 4°C. Do not paint outdoors in rainy weather or if freezing temperatures are expected before the paint dries.

Allow paint a drying time of at least 24 h in good weather before recoating.

9.10.7 SAFETY PRECAUTIONS

Paints are hazardous because of their flammability and potential toxicity, so proper safety precautions should be observed. Safe handling practices are required and should include, but not be limited to, the provisions of SSPC-PA Guide 3, "A Guide to Safety in Paint Application," and to the following specifications:

- Keep paint away from heat, sparks, and open flame during storage, mixing, and application. Provide sufficient ventilation to maintain vapor concentration at less than 25% of the lower explosive limit.
- Avoid prolonged or repeated breathing of vapors or spray mist, and prevent contact of the paint with the eyes or skin.
- Clean hands thoroughly after handling paint and before eating or smoking.
- Provide sufficient ventilation to ensure that vapor concentrations do not exceed the published permissible exposure limits. When necessary, supply appropriate personal protective equipment and enforce its use.

This paint may not comply with some air pollution regulations because of its hydrocarbon solvent content. Ingredients in this paint that may pose a hazard include hydrocarbon solvents. Applicable regulations governing safe handling practices should apply to the use of this paint.

9.11 ALKYD PAINT (ALUMINUM) NONLEAFING AS INTERMEDIATE

This section covers the minimum requirements for the composition analysis, properties, storage life and packaging, inspection, and labeling of alkyd paint (aluminum) nonleafing as intermediate. This paint employs nonleafing aluminum and is furnished in a single-compartment container.

9.11.1 COMPOSITION

Ingredients and proportions of the paint should be as specified in Table 9.34. The paint based on the specified ingredients should be uniform, stable in storage, and free of grit and coarse particles. No rosin or rosin derivatives may be used. Beneficial additives such as antiskinning agents, suspending agents, or wetting aids may be added.

This intermediate paint contains approximately 40% by volume of nonvolatile film-forming solids (pigment and binder). The paint based on the specified ingredients should be uniform, stable in storage, and free of grit and coarse particles. No rosin or rosin derivatives may be used. Beneficial additives such as antiskinning agents, suspending agents, or wetting aids may be added.

9.11.2 ANALYSIS

The alkyd varnish should conform to the composition (analysis) requirements listed in Table 9.35. The pigment should conform to the composition (analysis) requirements in Table 9.36.

9.11.3 PROPERTIES

The paint should meet the requirements of Table 4 and the following criteria:

- Odor—The odor should be normal for the materials permitted (ASTM Standard D 1296).
- Color—The color should be less than 11 on the Gardner 1933 scale (ASTM D 1544).
- Compatibility—There should be no evidence of incompatibility of any of the ingredients of the paint when two volumes of the ready-mixed paint are slowly mixed with one volume of mineral spirits (U.S. Federal Standard No. 141, Method 4203).

Table 9.34 Composition of Alkyd Paint (Aluminum) Nonleafing

| Ingredients Wt. % | Required | | Ingredient U.S. Federal | Standards |
	Minimum Wt. %	Maximum ASTM		
Pigment: (20.3 ± 0.5 wt.%) aluminum paste:	100	N/A	N/A	N/A
Vehicle: (< 79.7 ± 0.5wt.%) alkyd varnish solids[2]	50	N/A	N/A	TT-R-266 Type I Class A
Mineral spirit thinner	N/A	50	D235	TT-T-291 Type I
Driers	N/A	N/A	D600 Class B	N/A

Table 9.35 Analysis of Alkyd Varnish

| Characteristics | Requirements | | ASTM Method | U.S. Federal Standard No. 141 |
	Minimum Wt. %	Maximum Wt. %		
Volatiles	N/A	50	D2369	N/A
Nonvolatile vehicle				
Calculated by difference	50	N/A	N/A	4053
Rosin or rosin derivatives	N/A	0	D1542	N/A

Table 9.36 Analysis of Pigment (Aluminum Paste)

| Characteristics | Requirements | | ASTM Method |
	Minimum Wt. %	Maximum Wt. %	
Nonvolatile matter at 105–110°C	65	N/A	N/A
Easily extracted fatty oil matter (lubricants)	N/A	3.0	N/A
Total impurities other than fatty and oily matter	N/A	0.7	N/A
Coarse particles and skins as retained on standard 0.045-mm sieve opening (325 mesh screen)	N/A	1.0	D185
Leafing	N/A	None	N/A

- Skinning—There should be no skinning in a three-quarters-filled, closed container after 48 h when tested in the standard manner specified in U.S. Federal Standard No. 141, Method 3021.

The paint should be easily applied by all three methods (brush, spray, and roller) when tested in accordance with U.S. Federal Standard No. 141, Methods 4321, 4331, and 4541. The paint should show no streaking, running, or sagging after drying.

A dried film of the varnish should be clear, smooth, and glossy. A dried film (25 μm +5 μm thickness) of the varnish should show no cracking when bent over a 3.2-mm mandrel after 17 h of air-drying, plus 24 h of baking at 102–107°C.

Dried films should resist boiling water for 10 min and should withstand immersion in distilled water for 24 h. Upon removal after 2 h of drying, the film should show no whitening, blistering, or loss of adhesion, but slight dulling is permissible.

After air-drying for 17 h, plus 24 h of baking at 104°C, the paint should show no detrimental film effects after a painted panel is immersed in gasoline for 4 h.

Aluminum paste for this intermediate paint should be equivalent in fineness to the standard lining grade as defined by ASTM Standard D962, Type 4, Class B.

Table 9.37 Properties of Alkyd Varnish

| Characteristics | Requirements | | ASTM Method | U.S. Federal Standard No. 141 |
	Minimum	Maximum		
*Viscosity gardner				
Airbubble viscometer	C	E	D1545	N/A
Density kg/l	0.91	0.97	D1475	N/A
Drying time, (h) set to touch	N/A	4		4061
Dry hard	N/A	10	D154	4061
Flash point, (°C)	30		D3278	N/A

*Viscosity 48 h or more after manufacture

The aluminum pigment paste should consist of commercially pure aluminum, in the form of fine, polished flakes, and a suitable fatty lubricant or metallic soap lubricant combined with a volatile thinner. It should contain no fillers or adulterants. There should be no appreciable settling of the metallic portion of the paste in the container; i.e., no free liquid should be present. The test methods are those given in ASTM Standard D480 or U.S. Federal Specification TT-P-320.

The aluminum paste should be nonleafing. A total of 2 g of paste mixed with 25 ml of a leaf-testing vehicle (ASTM Standard D480) in a 250-cc beaker should show no more than a trace of leafing on the surface of the vehicle. In doubtful cases, absence of leafing should be confirmed.

The sample of aluminum paste to be tested may be compared with a sample mutually agreed upon by the purchaser and the vendor.

9.11.4 STORAGE LIFE AND PACKAGING

The ready-mixed paint should show no gas evolution, thickening, curdling, gelling, or hard caking when tested as specified in U.S. Federal Standard No. 141, Method 3011, after storage for 12 months from date of delivery or otherwise specified by company. Packaging should meet the relevant requirements of ASTM D3951 (88).

9.11.5 INSPECTION

All materials supplied under this specification should be subject to timely inspection by the purchaser or the purchaser's authorized representative. The purchaser should have the right to reject any materials supplied that are found to be defective under this specification. In the case of disputes, the arbitration or settlement procedure established in the procurement documents should be followed.

Samples of any or all ingredients used in the manufacture of this paint should be supplied upon the purchaser's request, along with the supplier's name and identification of the material. Unless otherwise specified, the methods of sampling and testing should be in accordance with U.S. Federal Test Method Standard No. 141 or applicable methods of the ASTM.

9.11.6 DIRECTIONS FOR USE

Alkyd paint (aluminum) is intended for use as an intermediate on structural steel or over other oleoresinous paints. All oil, grease, dust, and loose or nonadherent paint should be removed to the fullest extent practical, as residues of oil and grease remaining on the surface will result in decreased paint performance. If the undercoat is damaged, the steel should be spot-cleaned and -primed with rust-inhibitive primer.

If the pigment has settled, pour off most of the liquid. Thoroughly mix the pigment with the remaining liquid, taking care to scrape all the pigment off the bottom of the container. Gradually add the poured-off liquid and mix thoroughly. This may be made easier by transferring the contents to a larger container or by pouring the paint to and from another container. Examine the bottom of the container for unmixed pigment, and screen paint before applying.

Thin paint only if necessary, using only mineral spirits or turpentine. For brush application under normal conditions, no thinning should be necessary. For spray application, add up to 1 l of thinner per 8 l of paint.

Apply by brush or spray to the specified film thickness, or if none is specified, to at least 38 μm (dry) or approximately 100 μm (wet). The surface to be painted should be dry; the surface temperature should be at least 3°C above the dew point, and the temperature of the air should be over 4°C. Do not paint outdoors in rainy weather or if freezing temperatures are expected before the paint dries.

Allow paint a drying time of at least 24 h in good weather before recoating. Note: This paint is not intended to be used as a priming coat next to bare steel.

9.11.7 SAFETY PRECAUTIONS

Paints are hazardous because of their flammability and potential toxicity, so proper safety precautions should be observed. Safe handling practices are required and should include, but not be limited to, the provisions of SSPC-PA Guide 3, "A Guide to Safety in Paint Application," and to the following specifications:

- Keep paint away from heat, sparks, and open flame during storage, mixing, and application. Provide sufficient ventilation to maintain vapor concentration at less than 25% of the lower explosive limit.
- Avoid prolonged or repeated breathing of vapors or spray mist, and prevent contact of the paint with the eyes or skin.
- Clean hands thoroughly after handling paint and before eating or smoking.
- Provide sufficient ventilation to ensure that vapor concentrations do not exceed the published permissible exposure limits. When necessary, supply appropriate personal protective and enforce its use.

This paint may not comply with some air pollution regulations because of its hydrocarbon solvent content. Ingredients in this paint that may pose a hazard include hydrocarbon solvent. Applicable regulations governing safe handling practices should apply to the use of this paint.

9.12 ALKYD PAINT (HIGH-BUILD THIXOTROPIC LEAFING ALUMINUM) AS TOPCOAT (FINISH)

This section covers the minimum requirements for the composition, analysis, properties, storage life and packaging, inspection, and labeling of alkyd paint (high-build thixotropic leafing aluminum) as topcoat (finish). This paint contains leafing aluminum pigment and a thixotropic long-oil alkyd in a

Table 9.38 Composition of Alkyd Paint (High-Build Thixotropic Leafing Aluminum)

Ingredient	Required		Ingredient Standards	
	Minimum Wt.%	Maximum Wt.%	ASTM	U.S. Federal
Pigment: (26.2% min)				
Aluminum paste, 65% nonvolatile	100	N/A	D962, Type 2 Class B	TT-P-320 Type II Class 2
Vehicle: (73.8% Max)				
Thixotropic alkyd solids	55.6	N/A	N/A	TT-T-291
Mineral spirit thinner	N/A	44.5	D235	Type 1
Driers	N/A	N/A	D600 Class B	
Additives	N/A	N/A		

single package. Thixotropic paint appears to possess a high viscosity, which is altered because of a reversible gel structure when the paint is agitated. When the paint is stirred, sprayed, brushed, or rolled, the level of viscosity falls rapidly and the paint becomes liquefied. During application, the appearance is similar to other coatings. However, the liquid state will not remain for long. If undisturbed, the paint will regel within a short time.

9.12.1 COMPOSITION

The ingredients and proportions of the paint should be as specified in Table 9.38. The paint based on the specified ingredients should be uniform, stable in storage, and free of grit and coarse particles. No rosin or rosin derivatives may be used. Beneficial additives such as antiskinning agents, suspending agents, or wetting aids may be added.

9.12.2 PERCENTAGE

This paint contains approximately 49% by volume of nonvolatile film-forming solids (pigment and binder).

9.12.3 ANALYSIS

The thixotropic alkyd resin should conform to the composition (analysis) requirements of Table 9.39. The paint should conform to the composition (analysis) requirements of Table 9.40.

Table 9.39 Analysis of Thixotropic Alkyd Resin

| Characteristics | Requirements | | ASTM |
	Minimum	Maximum	
Nonvolatile resin, % by weight of solution	59	61	D1208
Drying oil acids, % by weight of solution	65	N/A	D1398
Phthalic anhydride % by weight of nonvolatile resin	9	N/A	D1306
Acid number of nonvolatile resin	N/A	8	D1639
Color, gardner color standards of 1953	N/A	13	D1544
Polyamide resin. % by weight of nonvolatile resin[1]	2.5	N/A	N/A
Rosin or rosin derivatives	N/A	0	D1542

[1]*Polyamide Resin:*
Softening Point   105–115°C
Amine Value   −2–7
Acid number   −10 maximum
Specific gravity   −0.95–0.98

Table 9.40 Analysis of Paint

| Characteristics | Requirements | | Method | ASTM Standard No. 141 |
	Minimum Wt. %	Maximum Wt. %		
Pigment	17	N/A	D1208	4021
Volatiles	N/A	42	D2369	N/A
Nonvolatile vehicle Calculated by difference	41	N/A	N/A	4053
Uncombined water	N/A	1.0	D1208	4081
Coarse particles and skins, as retained on standard 0.045-mm. sieve opening (325 mesh screen)	N/A	0.5	D185	N/A
Rosin or rosin derivatives	N/A	0	D1542	

9.12.4 PROPERTIES

The paint should meet the requirements of Table 9.41 and the following criteria:

- Odor—The odor should be normal for the materials permitted (ASTM Standard D1296).
- Color—A dried film of the paint should show good leafing, as indicated by a bright aluminum surface.

Table 9.41 Properties of Paint

| Characteristics | Requirements | | ASTM Method | U.S. Federal Standard |
	Minimum	Maximum		
Viscosity* shear rate 200 (rpm)				
Grams	255	300		
Kreb units	90	95	D562	N/A
Density (kg/l)	1.0	1.05	D1475	N/A
Drying time (h):				
Set to touch	N/A	4	D1640	4061
Tack free	10	16	D1640	4061
Flash point (°C)	41	N/A	D3278	N/A
Sag resistance, (mm)	0254	N/A	D2801	4494

Viscosity 48 h or more after manufacture.

- Compatibility—There should be no evidence of incompatibility of any of the ingredients of the paint when two volumes of the paint are slowly mixed with one volume of mineral spirits (U.S. Federal Standard No. 141, Method 4203).
- Skinning—There should be no skinning in a three-quarters-filled, closed container after 48 h when tested in the standard manner specified in U.S. Federal Standard No. 141, Method 3021.

The paint should be easily applied by all three methods (brush, spray, and roller) when tested in accordance with Federal Standard No. 141, Methods 4321, 4331, and 4541. The paint should show no streaking, running, or sagging after drying.

Apply paint quickly with a 4-cm brush to a 10 cm × 30 cm × 0.8 cm, cold-rolled steel panel in a horizontal position. Obtain a wet film thickness of 175–200 μm on as much of the panel as possible. The final brush stroke should be in the 30-cm direction, air-dried for 72 h at 21–24°C and baked for 5 h at 99°C. This panel should show no cracking on the radius of the bend when bent over a 3.2-cm mandrel. The film on the bent part should show satisfactory adhesion.

9.12.5 STORAGE LIFE AND PACKAGING

When in a full, tightly covered container, the paint should show no thickening, gas evolution, curdling, gelling, or hard caking when tested as specified in U.S. Federal Standard No. 141, Method 3011, after storage for 12 months from date of delivery. The packaging should meet the relevant requirements of ASTM D3951 (88).

9.12.6 INSPECTION

All materials supplied under this specification should be subject to timely inspection by the purchaser or the purchaser's authorized representative. The purchaser should have the right to reject any materials supplied that are found to be defective under this specification. In the case of disputes, the arbitration or settlement procedure established in the procurement documents should be followed.

Samples of any or all ingredients used in the manufacture of this paint should be supplied upon the purchaser's request, along with the supplier's name and identification of the material. Unless otherwise specified, the methods of sampling and testing should be in accordance with U.S. Federal Test Method Standard No. 141 or applicable methods of the ASTM.

9.12.7 DIRECTIONS FOR USE

Alkyd paint (high-build thixotropic leafing aluminum) is intended for use as a finish coat over rust-inhibitive primers on structural steel, over itself, or over other oleoresinous paints. It is suitable for outdoor exposure in rural, industrial, and marine environments and for interior use. All oil, grease, dust, and loose or nonadherent paint should be removed to the fullest extent practical, as residues of oil and grease remaining on the surface will result in decreased paint performance. If the undercoat is damaged, the steel should be spot-cleaned and -primed with rust-inhibitive primer.

Mix paint thoroughly before use. This thixotropic paint settles only slightly and moderate hand stirring should be sufficient. Thin paint only if necessary, using only mineral spirits with a minimum flash point at 60°C and a boiling range of 182–215°C. Under normal conditions, no thinning should be necessary. Add up to 1 l of thinner per 15 l of paint when necessary.

Apply by brush or spray to the specified film thickness, or if none is specified, to at least 75 μm (dry) or approximately 175 μm (wet). The surface to be painted should be dry, the surface temperature should be 3°C above the dew point, and the temperature of the air should be over 4°C.

Do not paint outdoors in rainy weather or if freezing temperatures are expected before the paint dries. Allow paint at least 24 h of drying time in good weather before recoating.

9.12.8 SAFETY PRECAUTIONS

Paints are hazardous because of their flammability and potential toxicity, so proper safety precautions should be observed. Safe handling practices are required and should include, but not be limited to, the provisions of SSPC-PA Guide 3, "A Guide to Safety in Paint Application," and to the following specifications:

- Keep paint away from heat, sparks, and open flame during storage, mixing, and application. Provide sufficient ventilation to maintain vapor concentration at less than 25% of the lower explosive limit.
- Avoid prolonged or repeated breathing of vapors or spray mist, and prevent contact of the paint with the eyes or skin.
- Clean hands thoroughly after handling paint and before eating or smoking.
- Provide sufficient ventilation to ensure that vapor concentrations do not exceed the published permissible exposure limits. When necessary, supply appropriate personal protective equipment and enforce its use.

This paint may not comply with some air pollution regulations because of its hydrocarbon solvent content. Ingredients in this paint that may pose a hazard include hydrocarbon solvent. Applicable regulations governing safe handling practices should apply to the use of this paint.

9.13 ALKYD PAINT (BLACK) AS INTERMEDIATE AND TOPCOAT (FINISH)

This section covers the minimum requirements for the composition, analysis, properties, storage life and packaging, inspection, and labeling of alkyd paint (black) to be used as intermediate and topcoat (finish).

9.13.1 COMPOSITION

The ingredients and proportions of alkyd paint (black) as intermediate and topcoat should be as specified in Table 9.42. The paint based on the specified ingredients should be uniform, stable in storage, and free of grit and coarse particles.

No rosin or rosin derivatives may be used. Beneficial additives such as antiskinning agents, suspending agents, or wetting aids may be added.

This paint contains approximately 37% by volume of nonvolatile, film-forming solids (pigment and binder).

9.13.2 ANALYSIS

The paint should conform to the composition (analysis) requirements of Table 9.43.

9.13.3 PROPERTIES

The paint should meet the requirements of Table 9.44 and the following criteria:

- Odor—The odor should be normal for the materials permitted (ASTM D 1296).
- Color—The color should be typical of carbon black.
- Compatibility—There should be no evidence of incompatibility of any of the ingredients of the paint when two volumes of the paint are slowly mixed with one volume of mineral spirits (U.S. Federal Standard No. 141, Method 4203).

Table 9.42 Composition of Alkyd Paint (Black) as Intermediate and Topcoat

	Typical Composition		Ingredient Standards	
	Wt. %	Vol. %	ASTM	U.S. Federal
Pigment (5.5 ±0.5 Wt. %)				
Carbon black	5.5	3.0	D 561	N/A
Vehicle (94.5 ±0.5 wt. %)				
Alkyd resin solids	41.0	34.0	N/A	TT-R-266
Mineral spirit thinner	53.5	63.0	D 235	TT-T-291, Grade I
Driers	N/A	N/A	D 600 Class B	N/A
Totals	100.0	100.0		

Table 9.43 Analysis of Alkyd Paint (Black) as Intermediate and Topcoat

| Characteristics | Requirements | | ASTM Method | U.S. Federal Standard No. 141 |
	Minimum Wt. %	Maximum Wt. %		
Pigment	5.0	6.0	D 2371	4021
Volatiles	N/A	55.0	D 2369	N/A
Nonvolatile vehicle				
Calculated by difference	40.0	N/A	N/A	4053
Uncombined water	N/A	1.0	D 1208	4081
Coarse particles and skins, as retained on standard 0.045-mm sieve opening (325 mesh screen)	N/A	0.5	D 185	N/A
Rosin or rosin derivatives	N/A	0	D1542	N/A

Table 9.44 Properties of Alkyd Paint (Black) as Intermediate and Topcoat

| Characteristics | Requirements | | ASTM Method | U.S. Federal Standard No. 141 |
	Minimum	Maximum		
Viscosity* shear rate 200 (rpm)				
Grams	120	180	D 562	N/A
Kreb units	65	78	D 562	N/A
Density (kg/l)	0.91	N/A	D 1475	N/A
Drying time (h)				
Tack free	N/A	10	N/A	4061
Dry hard	N/A	18	N/A	4061
Flash point, degrees (°C)	38	N/A	D 3278	N/A
Sag resistance, mic	152	N/A	D 2801	4494

Viscosity 48 h or more after manufacture.

- Skinning—There should be no skinning in a three-quarters-filled, closed container after 48 h when tested in the standard manner specified in U.S. Federal Standard No. 141, Method 3021.

The paint should be easily applied by all three methods (brush, spray, and roller) when tested in accordance with U.S. Federal Standard No. 141, Methods 4321, 4331, and 4541. The paint should show no streaking, running, or sagging after drying.

The film prepared as described in U.S. Federal Standard No. 141, Method 6221, after baking for 24 h at 93°C, should show no cracking when suddenly chilled to 0°C and quickly bent sharply on itself through 180° over a 3-mm mandrel. The film on the bent part of the panel should show satisfactory adhesion.

The gloss should be high (ASTM Standard D 523).

9.13.4 STORAGE LIFE AND PACKAGING

When in a full, tightly covered container, the paint should show no thickening, curdling, gelling, or hard caking when tested as specified in U.S. Federal Standard No. 141, Method 3011, after storage for 12 months from the date of delivery. The packaging, should meet the relevant requirements of ASTM D 3951 (88).

9.13.5 INSPECTION

All materials supplied under this specification should be subject to timely inspection by the purchaser or the purchaser's authorized representative. The purchaser should have the right to reject any materials supplied that are found to be defective under this specification. In the case of disputes, the arbitration or settlement procedure established in the procurement documents should be followed.

Samples of any or all ingredients used in the manufacture of this paint should be supplied upon the purchaser's request, along with the supplier's name and identification of the material.

Unless otherwise specified, the methods of sampling and testing should be in accordance with U.S. Federal Test Method Standard No. 141 or applicable methods of the ASTM.

9.13.6 DIRECTIONS FOR USE

Alkyd paint (black) is intended for use as an intermediate or finish coat over rust-inhibitive primers on structural steel, over itself, or over other oleoresinous paints. All oil, grease, dust, and loose or nonadherent paint should be removed, as residues of oil and grease remaining on the surface will result in decreased paint performance. If the undercoat is damaged, the steel should be spot-cleaned and -primed with rust-inhibitive primer.

Mix paint thoroughly before use. If the pigment has settled, pour off most of the liquid into a clean container. Thoroughly mix the pigment with the remaining liquid, taking care to scrape all the pigment off the bottom of the can. Gradually add the poured-off liquid and mix thoroughly. This may be made easier by transferring the contents to a larger container or by pouring the paint to and from another container. Examine the bottom of the container for unmixed pigment, and screen paint before applying.

Thin paint only if necessary, using only mineral spirits. For brush application under normal conditions, no thinning should be necessary. For spray application, add up to 1 l of thinner per 8 l of paint.

Apply by brush or spray to the specified film thickness, or if none is specified, to at least 38 μm (dry) or approximately 100 μm (wet). The surface to be painted should be dry; the surface temperature should be at least 3°C above the dew point, and the temperature of the air should be over 4°C. Do not paint outdoors in rainy weather or if freezing temperatures are expected before the paint dries.

Allow paint at least 18 h of drying time in good weather before recoating.

9.13.7 SAFETY PRECAUTIONS

Paints are hazardous because of their flammability and potential toxicity, so proper safety precautions should be observed. Safe handling practices are required and should include, but not be limited to the provisions of SSPC-PA Guide 3, "A Guide to Safety in Paint Application," and to the following specifications:

- Keep paint away from heat, sparks, and open flame during storage, mixing, and application. Provide sufficient ventilation to maintain vapor concentration at less than 25% of the lower explosive limit.
- Avoid prolonged or repeated breathing of vapors or spray mist, and prevent contact of the paint with the eyes or skin.
- Clean hands thoroughly after handling paint and before eating or smoking.
- Provide sufficient ventilation to ensure that vapor concentrations do not exceed the published permissible exposure limits. When necessary, supply appropriate personal protective equipment and enforce its use.

This paint may not comply with some air pollution regulations because of its hydrocarbon solvent content. Ingredients in this paint that may pose a hazard include hydrocarbon solvents and lead drier. Applicable regulations governing safe handling practices should apply to the use of this paint.

9.14 ACRYLIC SILICON FINISH PAINT FOR TEMPERATURE APPLICATIONS UP TO 200°C

Standard specifications cover the minimum requirements for the composition, properties, storage life and packaging, inspection, and labeling of acrylic silicon paint. The paint is intended for protection of equipment (stacks, cat crackers, boilers, heat exchangers, etc.), that operate up to 200°C. In this case, the steel surface should be treated with organic or inorganic zinc-rich primer.

9.14.1 COMPOSITION

The paint should consist of acrylic silicone resin, pigment, and additives, together with solvents needed to give consistency suitable for application by spray, brush, or other approved methods. The solid content of the paint should not be less than 40% by volume when tested by ASTM D2697.

9.14.2 PROPERTIES

The paint should be suitable for temperature services up to 200°C where color is required. It should withstand thermal shock from 200–230°C and temperature surges to 260°C. The paint should exhibit excellent weathering properties, thereby making it suitable for use as a topcoat in coastal and severe industrial environments (ASTM D-2485, Method B).

- Odor—The odor of the paint should not be offensive, irritating, or putrid during and after application. It should be normal for the materials permitted as ASTM D-1296.
- Color—The color should be as specified by the purchaser, with reference to Table 9.45 or other color.

Table 9.45 Paint color for Acrylic Silicon Finish Paint

Paint Color	Color No. to BS 381 C	Color No. to Rall (Approx)
Arctic blue	112	5024
Sea green	217	6017
Brilliant green	221	6002
Canary yellow	309	1018
Light straw	384	1000
Middle brown	411	8007
Signal red	537	3020
Light orange	557	2000
Light gray	631	7033
Aluminum	N/A	N/A

- Compatability—The paint should be compatible with the thinner. Wet dry films over intermediate should not show any defects.

The paint should be easily applied by brush and spray when tested in accordance with U.S. Federal Standard Methods 4321, 4331. The paint should show no streaking, running, sagging, cracking, chipping, or flaking after drying.

The paint as received should show no evidence of livering, skinning, or hard settling of pigment; the container should not be affected. The material should be easily dispersed in liquid form by stirring by hand to form a smooth, homogeneous paint free of persistent foam when tested in accordance with U.S. Federal Standard test method 3011.1.

The fineness of dispersed pigments in the vehicle system should not be greater than 40 μm according to ASTM D-1210.

When determined in accordance with ASTM D-1475, the density of paint should be within 5% of the value of the manufacturer's specification. When determined in accordance with ASTM D-4287, the density of paint should be within 5% of the value of the manufacturer's specification.

Coating should be applied and cured on a surface prepared carbon on carbon steel plate to Sa 2½ with 35-50 μm (dry) film thiclcness. The adhesion classification should be at least 4B according to ASTM D-3359 method B. The adhesion classification of the complete system applied according to the manufacturer's specification.

Because of diversity of potential service environments, this specification may require the paint to be further exposed and qualified by at least one additional test relating to the intended exposure, as described next.

The paint should have a satisfactory resistance to weathering. This should be clearly and adequately demonstrated by at least two document case histories that confirm satisfactory performance for a minimum of 5 years, at or near its recommended maximum service temperature, in a moderate or more severe environment.

For using paint in system 2C, applicable to chemical and marine atmospheres, the complete system should withstand spray testing according to ASTM B-117 for 300 h, without any blistering of coating or rusting of the coated portion. The system should also withstand humidity testing according to ASTM D-2247 for 300 h, without any blistering of coating and rusting of the coated portion.

9.14.3 STORAGE LIFE AND PACKAGING

The product should meet the standard requirements after storage of at least 12 months from the date of delivery in a full, tightly covered container in normal conditions. The packaging should meet the relevant requirements of ASTM D 3951 unless otherwise specified by the purchaser.

Packing should be accomplished in a manner that will ensure acceptance by a common carrier, at the lowest rate, and will afford protection against physical or mechanical damage during shipment.

Shipment marking information, in addition to the labeling required, should be provided on interior package and exterior shipping containers.

9.14.4 INSPECTION

All materials supplied under the standard should be subject to timely inspection by the purchaser or the purchaser's authorized representative. The purchaser should have the right to reject any materials supplied that are found to be defective under the standard. In the case of disputes, the arbitration or settlement procedure established in the procurement documents should be followed.

The supplier should be responsible for the performance and costs for all laboratory testing as stated in the standard specifications. The supplier should place, at no cost, at the disposal of the purchaser's inspectors all means necessary for carrying out their inspection, specification, or test results to check for conformity of materials with standard specification, checking of marking, and packing and temporary acceptance of materials. Samples submitted to the purchaser will be tested in the purchaser's laboratory or in a responsible commercial laboratory designated by the purchaser.

The supplier should furnish the purchaser with a certified copy of the results of tests made by the manufacturer covering the physical and performance characteristics of each batch of product to be supplied under the standard. The supplier should furnish, or allow the purchaser to collect, samples of the material representative of each batch of product. Certified test reports and samples furnished by the supplier or collected by the purchaser should be properly identified with each lot of product.

Prior to the acceptance of the supplier's material, samples of material submitted by the supplier or collected by the purchaser will be tested by the purchaser. If any sample is found not to conform to standard specifications, materials represented by such sample will be rejected.

The number of samples for testing should consist of 10% of the lot or batch, but in no case should it be fewer than 1 or more than 10 containers. The results of the tests on two specimens (top and bottom) should be averaged for each test specified at the standard specifications to determine conformance with the specified requirements.

A lot or batch should consist of an indefinite number of containers offered for acceptance and filled with a homogeneous mixture of material from one isolated container or filled with a homogeneous mixture of material manufactured by a single plant run (not exceeding 24 h) through the same processing equipment, with no change in ingredient material.

9.14.5 **DIRECTIONS FOR USE**

In addition to the manufacturer's instructions for use, the following directions should be supplied with each container of paint: "This paint is intended for use on primed structural steel. The surface of steel should be prepared in accordance with the applicable standards before applying the primer."

For use on steel surfaces subjected to high temperatures (up to 200°C), the primer used should be organic or inorganic zinc primer. Apply by brush or spray to the specified film thickness, or if none is specified, to at least 38 μm (dry). When application is by spraying, the equipment and operator technique should be properly adjusted to prevent dry spraying and to deposit a wet film of paint on the substrate. Clean the equipment with suitable thinner both before and after use. The surface to be painted should be dry, and the surface temperature should be at least 3°C above the dew point.

9.14.6 **SAFETY PRECAUTIONS**

This paint is hazardous because of its flammability and potential toxicity, so proper safety precautions should be observed. Safe handling practices are required and should include, but not be limited to, the provisions of SSPC-PA Guide 3, "A Guide to Safety in Paint Application," and to the following instructions:

- Keep paint away from heat, sparks, and open flame during storage, mixing, and application. Provide sufficient ventilation to maintain vapor concentration at less than 25% of the lower explosive limit.
- Avoid prolonged or repeated breathing of vapors or spray mist, and prevent contact of the paint with the eyes or skin.
- Clean hands thoroughly after handling paint and before eating and smoking.
- Provide sufficient ventilation to ensure that vapor concentration do not exceed the published permissible exposure limits. When necessary, supply appropriate personal protective equipment and enforce its use.
- This paint may not comply with some air pollution regulations because of its hydrocarbon solvent content.

9.15 **VINYL PAINT (BLACK) AS INTERMEDIATE AND TOPCOAT (FINISH)**

This section covers the minimum requirements for the composition, analysis, properties, storage life and packaging, inspection and labeling of vinyl paint (black) to be used as intermediate and topcoat (finish).

9.15.1 **COMPOSITION**

The ingredients and proportions of vinyl paint (black) should be as specified in Table 9.46. The paint based on the specified ingredients should be uniform, stable in storage, and free of grit and coarse particles. Beneficial additives such as antiskinning agents, suspending agents, or wetting aids may be added.

This paint contains approximately 13% by volume of nonvolatile, film-forming solids (pigment and binder).

Table 9.46 Composition of Vinyl Paint (Black)

Ingredients	Typical Composition		Ingredient Standards
	Wt. %	**Vol. %**	**ASTM**
Pigment: (1.0–1.5%) carbon black	1.0	0.5	D 561
Vehicle: (98.5–99.0%) vinyl resin A[1]	7.5	4.8	N/A
Vinyl resin C[2]	7.5	4.8	N/A
Dioctyl phthalate[3]	3.0	2.7	N/A
Methyl isobutyl ketone[4]	40.5	45.4	D 1153
Toluene	40.5	41.8	D 362
Totals	100.0	100.0	

[1]*Vinyl rosin A should be a hydroxyl-containing, vinyl chloride, acetate copolymer. It should contain 89.5–91.5% vinyl chloride, 5.3–7.0% vinyl alcohol, and 2.0–5.5% vinyl acetate. The inherent viscosity of the rosin (ASTM Standard D1243, Method A) at 20°C should not be less than 0.5.*
[2]*Vinyl rosin C should be a vinyl, chloride-acetate copolymer. It should contain 85–88% vinyl chloride, 12–15% vinyl acetate, by weight. The inherent viscosity of the resin (ASTM Standard D 1243, Method A) at 20°C should not be less than 0.48.*
[3]*Dioctyl phthalate (di-2-ethylhexyl phthalate) should be commercial material that conforms to the following requirements:*
- *Specific gravity @ 25°C: 0.980–0.9861*
- *Refractive index @ 25°C: 1.4830–1.6859*
[4]*When specified in the procurement documents, suitable high boiling vinyl solvent may be substituted for a portion of the MIBK to make the paint more amenable to application in hot weather or by brush.*

9.15.2 ANALYSIS

The paint should conform to the composition (analysis) requirements listed in Table 9.47.

9.15.3 PROPERTIES

The paint should meet the requirements of following specifications:

- Odor—The odor should be normal for the materials permitted (ASTM D 1296).
- Color—The color should be black.
- Compatibility—There should be no evidence of incompatibility of any of the ingredients of the paint when two volumes of the paint are slowly mixed with one volume of thinner consisting of 85% toluene and 15% MIBK by volume (U.S. Federal Standard No. 141, Method 4203).

The paint should be easily applied when tested in accordance with U.S. Federal Standard No. 141, Methods 4331 and 4541. The paint should show no streaking, running, or sagging after drying.

The paint should show good adhesion when tested as follows:

1. Apply one coat with a 25-μm (dry) film thickness of the mixed paint to a panel pretreated with wash primer MIL-P-15328, "Primer (Wash) Pretreatment," or SSPC-Paint 27, "Basic Zinc Chromate-Vinyl Butyral Wash Primer" (12.5-μm, dry film thickness), and to a panel similar to the preceding one, but over which has also been applied one coat of MIL-P-15929, "Primer Coating, Shipboard, Vinyl-Red Lead (for Hot Spray)" (25-μm, dry film thickness).

Table 9.47 Analysis of Vinyl Paint (Black)

| Characteristics | Requirements | | ASTM Method | U.S. Federal Standard No. 141 |
	Minimum Wt. %	Maximum Wt. %		
Pigment	1.0	1.5	D1208	4021
Volatiles	78.0	81.0	D2369	N/A
Nonvolatile vehicle				
Calculated by difference	18.0	20.5	N/A	4053
Uncombined water	N/A	0.5	D 1208	4081
Coarse particles and skins, as retained on standard 0 45-mm sieve opening (325 mesh screen)	N/A	0.25	D185	4092

2. After drying for 24 h, the film on each panel should be subjected to a knife test to determine whether the paint exhibits good adhesion to the undercoats.

9.15.4 STORAGE LIFE AND PACKAGING

When in a full, tightly covered container, the paint should show no thickening, curdling, gelling, or hard caking when tested as specified in U.S. Federal Standard No. 141, Method 3011, after storage for 12 months from the date of delivery. The packaging should meet the relevant requirements of ASTM D3951 (88).

Table 9.48 Properties of Vinyl Paint (Black)

| Characteristics | Requirements | | ASTM Method | U.S. Federal Standard No 141 |
	Minimum	Maximum		
Viscosity* shear rate 200 (rpm)				
Grams	100	150	D 562	N/A
Kreb units	60	72	D 562	
Density (kg/l)	0.88	0.95	D 1475	
Fineness of grind, (μm)	15	N/A	D1210	
Drying time, min				
Tack free	N/A	15	D 1640	4061
Dry hard	N/A	30	D 1640	4061

Viscosity 48 h or more after manufacture.

9.15.5 INSPECTION

All materials supplied under this specification should be subject to timely inspection by the purchaser or the purchaser's authorized representative. The purchaser should have the right to reject any materials supplied that are found to be defective under this specification. In the case of disputes, the arbitration or settlement procedure established in the procurement documents should be followed.

Samples of any or all ingredients used in the manufacture of this paint should be supplied upon the purchaser's request, along with the supplier's name and identification of the materials.

Unless otherwise specified, the methods of sampling and testing should be in accordance with U.S. Federal Test Method Standard No. 141 or applicable methods of the ASTM.

Now we will discuss the directions of use of Vinyl (black) paint. It is not to be used as a priming coat next to bare steel. This paint is intended for use as an intermediate or finish coat over vinyl butyral wash primer or vinyl resin paint. Mix thoroughly before use.

The paint should be thinned as necessary with a solvent containing not more than 85% toluene and 15% MIBK or MEK. The amount of thinning will depend upon application methods and conditions and may be as high as 25–33% by volume of the paint.

Apply by conventional air spraying. Brushing may be used in small areas. The surface to be painted should be dry, above 2°C, and not less than 3°C above the dew point. Do not paint outdoors in rainy weather.

Apply so as to obtain a minimum dry film thickness of 25 μm. A wet film of paint should be deposited on the surface when spraying; the spray gun should be adjusted so that proper atomization is obtained, but no dry powder is deposited on the surface; the nozzle should be held about 152 mm from the surface during application.

If application is to be made by brush, a brush should be heavily loaded with paint. Apply the paint quickly and smoothly, avoid excessive brushing, and do not go back over the surface until thoroughly dry.

At temperatures between 16°C and 27°C, dry for at least 1 h between coats and 72 h before immersion. Varying atmospheric conditions and degrees of ventilation in confined spaces may allow shorter or require longer drying times.

9.15.6 SAFETY PRECAUTIONS

Paints are hazardous because of their flammability and potential toxicity, so proper safety precautions should be observed. Safe handling practices are required and should include, but not be limited to, the provisions of SSPC-PA Guide 3, "A Guide to Safety in Paint Application," and to the following criteria:

- Keep paint away from heat, sparks, and open flame during storage, mixing, and application. Provide sufficient ventilation to maintain vapor concentration at less than 25% of the lower explosive limit.
- Avoid prolonged or repeated breathing of vapors or spray mist, and prevent contact of the paint with the eyes or skin.
- Clean hands thoroughly after handling paint and before eating or smoking.

- Provide sufficient ventilation to ensure that vapor concentrations do not exceed the published permissible exposure limits. When necessary, supply appropriate personal protective equipment and enforce its use.
- This paint may not comply with some air pollution regulations because of its hydrocarbon solvent content.
- Ingredients in this paint that may pose a hazard include hydrocarbon solvents. Applicable regulations governing safe handling practices should apply to the use of this paint.

9.16 SILICONE ALKYD PAINT (WHITE OR COLORED) AS TOPCOAT (FINISH)

This section covers the minimum requirements for the composition, analysis, properties, storage life and packaging, inspection, and labeling of silicone alkyd paint as topcoat (finish).

9.16.1 COMPOSITION

The ingredients and proportions of the silicone alkyd paint should be as specified in Table 9.49.

This paint should contain at least 60–67% by volume of nonvolatile, film-forming solids (pigments and binder). The main pigment for white should consist of titanium dioxide conforming to ASTM-D-476, Type IV, or any combination of colored pigments to obtain the color specified, provided that the paint complies with the requirements of this specification.

The vehicle should consist of silicon-modified medium oil soya alkyd copolymer of the air-drying type, together with suitable thinners, driers, antiskinning agents, wetting agents, dispersing agents, and stabilizers combined, producing a material conforming to all requirements specified herein. The silicone intermediate used in the preparation of the copolymerized resin should be hydroxyl-functional. The vehicle should conform to the composition (analysis) requirements of Table 9.50.

The solvent should be mineral spirits conforming to ASTM-D 235 or U.S. Federal Specification TT-T-291. The driers should conform to ASTM-D 600.

Table 9.49 Composition of Silicone Alkyd Paint

Ingredients	Wt. %	Ingredient Standards
Total solids	50–70	ASTM D2369
Vehicle solids	30–40	N/A
Solvents	30–40	N/A

Note: Place a portion (approximately 10 g) of the vehicle (separated as described in ASTM Standard D2698) in a dropping bottle and weigh to the nearest 0.1 mg. Weigh one of the 60-mm aluminum dishes to fourth-decimal-place accuracy. Transfer a small sample that does not exceed 0.3 g to the dish, and determine its exact weight by loss of weight of the bottle. Dissolve the sample in 2 ml of reagent-grade toluene and dry it in a gravity convection oven at 105 ±2°C for 3 h. After cooling for 30 min, weigh the dish to the nearest 0.1 mg. From the weight of residue in the dish and the weight of the sample taken, calculate the percentage of vehicle solids.

Table 9.50 Vehicle Characteristics

| Characteristics | Requirements | | ASTM Method | U.S. Federal Standard No. 141 |
	Minimum	Maximum		
Copolymer resin solids, percent by weight of extracted vehicle solids[1]	50.0	N/A	N/A	N/A
Silica (SiO$_2$), percent by weight of copolymer resin solids[2]	14.0	N/A	N/A	N/A
Phthalic anhydride, percent by weight of copolymer resin solids	14.0	17.0	D1306	7014[3]
Soya oil acid content (based on solids)	41.0	55.0	D1393	7014[3]
Soya oil	Positive		D2800	N/A

[1]Copolymer resin content of nonvolatile vehicle isopropanol extraction: Weigh 5 g (to the nearest 0.1 mg) of vehicle (separated as in ASTM D2698) into a centrifuge bottle or tube fitted with a cap. Add 50 ml of isopropanol (technical-grade), cap the bottle or tube, and shake vigorously for 2 min. Centrifuge for 15 min at a minimum of 2,000 rounds per minute (rpm). Decant the isopropanol extract and repeat the extraction and condition the bottle or tube in an oven at 135°C for 3 h. Remove the bottle or tube, cool for 30 min at room temperature, and weigh. Calculate the copolymer resin solids using the following formula. The percentage of copolymer resin solids is figured by the formula $(R \times 100)/(SXD)$ where R is the weight of residue (in the bottle or tube), S is the weight of sample (vehicle), and D is the percentage of vehicle solids (see the note in Table 9.49).
[2]Silica content of vehicle: From a stoppered bottle or weighing pipet, weigh accurately by difference about 3 g of the vehicle into a properly ignited and weighed 75-mm porcelain evaporating dish. Dry at 105°C in an oven for 3 h. Place the dried sample in a cold muffle furnace, gradually increase the temperature over a period of 3 h to 800°C, and then maintain this temperature for an additional hour. After cooling in a dessicator, weigh the dish and contents and calculate the percent of silica as follows: The percentage of silica is figured by the formula $A \times 100 /(SxD)$ where A is the weight of ash, S is the weight of sample (vehicle), D is the percentage of vehicle solids (see the note in Table 9.49.
[3]Altered by the substitution of petroleum ether for chloroform.

Table 9.51 Analysis for White Paint

| Characteristics | Requirements | |
	Minimum Wt. %	Maximum Wt. %
Total solids	64	N/A
Pigment solids	31	35
Vehicle solids	37	N/A

9.16.2 ANALYSIS

High-gloss white paint should conform to the composition (analysis) requirement of Table 9.51.

9.16.3 PROPERTIES

The paint should meet the requirements of Table 9.51 and the following specifications:

- Odor—The odor should be normal for the materials permitted (ASTM Standard D1296). The odor of the wet enamel and of the film at any stage of drying should not be obnoxious or objectionable.
- Color—Draw down a coat of paint on a white opaque glass panel using a doctor blade with a 150-μm gap clearance designed to deposit a wet film thickness of approximately 75 μm. After 48 h of drying at 21–24°C and 50% relative humidity, compare the dried film with the standard chip (U.S. Federal Standard No. 595) for white and with BS 381C (see Table 9.52) for other colors in accordance with ASTM Standard D1729 for compliance. If doubt exists as to the color match, an instrumental referee method may be used (ASTM Standard D2244).
- Dilution Incompatibility—There should be no evidence of incompatibility of any of the ingredients of the paint when one volume of the paint is slowly mixed with one volume of mineral spirits (U.S. Federal Standard No. 141, Method 4203). However, slight pigment settling should be permitted.

The paint, as packaged, should be applied easily when tested in accordance with U.S. Federal Standard No. 141, Method 4321. The paint should dry to a smooth, uniform film, free of seeds, runs, sags, or streaks. The dried film should show no discernible brush marks.

Prepare a steel panel in accordance with U.S. Federal Standard No. 141, Method 2011, using the petroleum naphtha ethylene glycol monoethyl ether mixture. Apply the paint to this panel by spraying to a dry film thickness of 23–28 μm. The paint should be easily applied when tested in accordance with U.S. Federal Standard No. 141, Method 4331. The paint should show no running, sagging, or streaking. The air-dried film should show no seeding, dusting, floating, fogging, mottling, hazing, excessive orange peel, or other film defect.

For adhesion, use the panel by air-drying for 18 h, then baking for 2 h at 105 ±2°C. Condition the panel for 1 h under referee testing conditions (see Section 9 of U.S. Federal Standard No. 141). Then score a line through to the metal across the width of the film using a sharp, pointed knife. The film should then be taped perpendicular to and across the score line with waterproof, pressure-sensitive tape, 2 cm wide, conforming to U.S. Federal Standard PPP-T-60 Type IV. Press the tape down with firm pressure. Allow approximately 10 s for the test area to return to room temperature. Grasp the free end

Table 9.52 Colors

Paint Color	Color No. to BS 381 C
Arctic blue	112
Sea green	217
Brilliant green	221
Canary yellow	309
Light Straw	384
Middle Brown	411
Signal Red	537
Light orange	567
Light gray	631
White	595

of the tape and rapidly strip it from the film by pulling back from the panel at approximately 180°. The paint should show no removal of the film or loosening beyond 2 mm on either side of the score line.

Determine the paint's flexibility in accordance with U.S. Federal Standard No. 141, Method 6221. Apply a 5-cm-wide film of paint on a smooth-finish steel panel, prepared in accordance with U.S. Federal Standard No. 141, Method 2011, using the petroleum naphtha ethylene glycol monoethyl ether mixture with a suitable film applicator that will give a dry film thickness of 23–28 μm. The panel should be prepared from new, cold-rolled, rust-free carbon steel which is 250 ±25 μm thick with a Rockwell 15-T maximum hardness of 82 and a finish with surface roughness of 0.2–0.3 μm. Air-dry in a horizontal position for 18 h, and then bake for 168 h at 105 ±2°C. Condition the panel for a half-hour under standard testing conditions (see Section 9 of U.S. Federal Standard No. 141). Bend over a 6-mm mandrel. Examine the coating for cracks over the area of the bend in strong light at seven times diameters magnification. The paint should withstand bending without cracking or flaking.

Determine the hiding power in accordance with ASTM Standard D 2805. Draw down a film using an applicator that will deposit a dry film with 15 μm maximum thickness. Air-dry for 72 h, measure the thickness of the dried film, and then measure the reflectance. Calculate the contrast ratio. A dry film thickness of 25 μm maximum for white paint (with a minimum reflectance of 84%) should give a dry film contrast ratio of 0.95.

9.16.4 STORAGE LIFE AND PACKAGING

Determine the condition of the paint in the container in accordance with U.S. Federal Standard No.141, Method 3011. The paint should be free of grit, seeds, skins, lumps, thickening, and livering

Table 9.53 Properties of Silicone Alkyd Paint

| Characteristics | Requirements | | ASTM Method | U.S. Federal Standard No. 141 |
	Minimum	Maximum		
Flash point. pensky-martens, closed cup, (°C)	30	N/A	D93	N/A
Water, percent by weight of paint	N/A	0.5	D95	4081
Coarse particles and skins, 0.045 standard sieve opening retained on (325 mesh sieve), percent by weight of pigment	N/A	0.1	D185	4092
Viscosity* shear rate 200 (rpm): grams	125	175	D562	N/A
Kreb units	67	77.0	D562	N/A
Fineness of grind, (μm)	25	N/A	D1210	N/A
Drying time:				
Set to touch, (h)	N/A	2	D1640	4061
Dry hard, (h)	N/A	8	D1640	4061

Viscosity 48 h or more after manufacture.

and should show no more pigment settling or caking that can be readily reincorporated to a smooth, homogeneous state.

Reseal the can and then agitate it for 3 min in a paint shaker. On reexamination of the contents, the disclosure of gel bodies, undispersed pigment, or unsatisfactory settling properties is cause for rejection.

Determine skinning after the paint has been in a partially full container for 48 h in accordance with U.S. Federal Standard No. 141, Method 3011 except use a three-quarters-filled, 250-ml, multiple-friction-top can. The paint should show no skinning. Reseal the can and store for 7 days at 60°C, and then look at the paint. The paint should show no livering, curdling, hard caking, or gummy sediment, and it should mix readily to a smooth, homogeneous state.

Determine the storage stability of the package paint in accordance with ASTM Standard D1849 using a standard quart can, allowing it to stand undisturbed for 24 months. The paint should show no skinning, livering, curdling, hard-dry caking, or tough gummy sediment. Evaluate pigment settling or caking, but agitate the can for 5 min on the paint shaker prior to examination. The paint should remix readily to a smooth homogeneous state and must be useable. The consistency of the paint after storage should be 62–82 Krebs Units (ASTM Standard D562). The packaging should meet the relevant requirements of ASTM D 3951 (88).

9.16.5 INSPECTION

All materials supplied under this specification should be subject to timely inspection by the purchaser or the purchaser's authorized representative. The purchaser should have the right to reject any materials supplied that are found to be defective under this specification. In the case of disputes, the arbitration or settlement procedure established in the procurement documents should be followed.

Samples of any or all ingredients used in the manufacture of this paint should be supplied upon the purchaser's request, along with the supplier's name and identification of the material. Unless otherwise specified, the methods of sampling and testing should be in accordance with U.S. Federal Test Method Standard No. 141 or Applicable Methods of the ASTM.

9.16.6 DIRECTIONS FOR USE

Silicone alkyd paint is intended for use as a finish coat over a rust-inhibitive primer (or other suitable primer) and an intermediate coat on structural steel. Before applying, remove all moisture, oil, grease, dirt, and loose or nonadhering paint. Old coatings in good condition that are compatible with this silicon alkyd paint may remain, but damaged areas or areas of poor adhesion must be spot-cleaned and -primed.

Mix paint thoroughly before use. If simple stirring is inadequate, pour off most of the liquid into a clean container. Thoroughly mix the pigment with the remaining liquid, taking care to scrape all the pigment off the bottom of the can. Gradually add the poured-off liquid and mix thoroughly. This may be made easier by transferring the contents to a larger container or by pouring the paint to-and-from another container. Examine the bottom of the container for unmixed pigment, and screen paint before applying.

Thin paint only if necessary, using only mineral spirits. For brush application under normal conditions, no thinning should be necessary. For spray application, add up to 1 l of thinner per 8 l of unthinned paint when necessary.

Apply by brush or spray to the specified film thickness, or if none is specified, to at least 38 μm (dry) or approximately 63 μm (wet). The surface to be painted should be dry; the surface temperature should be at least 3°C above the dew point; and the temperature of the air should be over 4°C. Do not paint outdoors in rainy weather or if freezing temperatures are expected before the paint dries.

The thinner should be made of mineral spirits conforming to ASTM-D235 or U.S. Federal Specification TT-T-291, up to 1 l of thinner may be added per 8 l of unthinned paint. Allow paint at least 24 h of drying time in good weather before recoating.

9.16.7 SAFETY PRECAUTIONS

Paints are hazardous because of their flammability and potential toxicity, so proper safety precautions should be observed. Safe handling practices are required and should include, but not be limited to, the provisions of SSPC-PA Guide 3, "A Guide to Safety in Paint Application," and to the following specifications:

- Keep paint away from heat, sparks, and open flame during storage, mixing, and application. Provide sufficient ventilation to maintain vapor concentration at less than 25% of the lower explosive limit.
- Avoid prolonged or repeated breathing of vapors or spray mist, and prevent contact of the paint with eyes or skin.
- Clean hands thoroughly after handling paint and before eating or smoking.
- Provide sufficient ventilation to ensure that vapor concentrations do not exceed the published permissible exposure limits. When necessary, supply appropriate personal protective equipment and enforce its use.

This paint may not comply with some air pollution regulations because of its hydrocarbon solvent content. Ingredients in this paint that may pose a hazard include lead and chromate-containing pigments and hydrocarbon solvents. Applicable regulations governing safe handling practices should apply to the use of this paint.

9.17 WASH PRIMER (BASIC ZINC CHROMATE-VINYL BUTYRAL)

This section covers the minimum requirements for the composition, analysis, properties, storage life and packaging, inspection, and labeling of basic zinc chromate-vinyl butyral wash primer.

9.17.1 COMPOSITION

The ingredients and proportions of basic zinc chromate-vinyl butyral should be as specified in Table 9.54.

This primer contains approximately 10% by volume of nonvolatile, film-forming solids (pigment and binder).

9.17.2 ANALYSIS

The analysis of the primer should conform to the composition (analysis) requirements of Table 9.55.

Table 9.54 Composition of Basic Zinc Chromate-Vinyl Butyral

Ingredients Per 378.5 lit. Paint	Typical Composition		Ingredient Standards		
	(Kg)	**(l)**	**ASTM**	**U.S. Federal**	**Military Spec.**
Ingredients of resin component (303 l)[1]					
Polwinyl-butyral resin[2]	25.455	23.089			
Zinc chromate (insoluble type)[3]	24.545	6.435			
Magnesium silicate (type A or B)	3.636	1.237			MIL-P-15173
Lampblack	0.273	0.151	D209		
Butyl alcohol, normal	56.818	69.947		TT-B-846	
Isopropyl alcohol	160.455	203.633		TT-l-735	
Water ingredients of acid component (77.5 l)	6.818	6.818			
Phosphoric acid (Class 1)	12.727	7 57		0.0-670	
Water	11.364	11.364			
Isopropyl alcohol:	45.0	56.775		TT-l-735	

[1]The formula of the base is given slightly in excess of 303 l to allow for normal manufacturing loss.
[2]The resin should be polyvinyl, partial butyral resin containing only polyvinyl butyral, polyvinyl alcohol, and polyvinylacetate in the molecule. The resin should contain 18–20% vinyl alcohol and not more than 1% of vinyl acetate. A 6% solution of the resin in methanol should have a viscosity of 12–18 cP at 20°C. The specific gravity (25°/25°C) of the resin should range from 1.05 to 1.15.
[3]The zinc chromate should be of an insoluble type, showing an analysis of 16–19% C_2O_3, 67–72% ZnO, and not more than 1% water soluble salts.

Table 9.55 Analysis of Basic Zinc Chromate-Vinyl Butyral

Characteristics	Requirements		ASTM Method	U.S. Federal Standard No. 141	Military SPEC.
	Minimum	**Maximum**			
Characteristics of resin component: pigment, percent by weight of resin	N/A	N/A	N/A	N/A	N/A
Component:	9.5	11.9	N/A	N/A	MIL-P 15328 D
Volatiles, percent by weight of resin	N/A	N/A	N/A	N/A	N/A
Component:	79.0	81.5	D2369	N/A	N/A
Nonvolatile vehicle, percent by weight of resin	N/A	N/A	N/A	N/A	N/A
Component (calculated by difference)	8.5	10.0		4053	N/A
Ratio of pigment to nonvolatile vehicle by weight	1.07	1.15	N/A	N/A	MIL-P 15828D

(Continued)

Table 9.55 Analysis of Basic Zinc Chromate-Vinyl Butyral *(cont.)*

| Characteristics | Requirements | | ASTM Method | U.S. Federal Standard No. 141 | Military SPEC. |
	Minimum	Maximum			
Coarse particles and skins, as residue retained on 0.045-mm sieve opening (standard no. 325 mesh screen), percent by weight of resin component:	N/A	0.2	D185	4092	N/A
Chromium trioxide (C_2O_3), percent by weight of pigment	13.5	N/A	N/A	N/A	MIL-P 15328D
Zinc oxide (ZnO) percent by weight of pigment	57.9	N/A	N/A	N/A	MIL-P 15328 D
Distillation of 100 g of thinner obtained from resin	N/A	N/A	N/A	N/A	MIL-P 15328 D
Component:					
Initial boiling point. (°C)	79	82	N/A	N/A	MIL-P 15328 D
Temperature at 80 cm^3, point, (°C)	N/A	85	N/A	N/A	N/A
Temperature at 105 cm^3, point, (°C)	116	N/A	N/A	N/A	N/A
End point, temperature (°C)	N/A	120	N/A	N/A	N/A
Volume at end point, cm^3	116	N/A	N/A	N/A	N/A
Characteristics of acid component:	N/A	N/A	N/A	N/A	MIL-P 15328 D
Phosphoric acid, percent by weight of acid component:	15.0	16.5	N/A	N/A	MIL-P 15328 D
Distillation of 150 g of acid component:	N/A	N/A	N/A	N/A	N/A
Initial boiling point, (°C)	75	82	N/A	N/A	N/A
Temperature at 105 cm^3. point, (°C)	N/A	84	N/A	N/A	N/A
Volume at end point, cm^3	105	N/A	N/A	N/A	N/A
Maximum temp. during distillation (°C)	N/A	192	N/A	N/A	N/A

Notes:
The solvent portion of the formulation should conform to the following requirements:
- *Aromatic compounds with eight or more carbon atoms to the molecule, except ethylbenzene (total aromatics less ethylbenzene) should not exceed 1% by volume.*
- *The ethylbenzene content of the solvent should not exceed 1% by volume. Compounds with olefinic or cycloolefinic unsaturation should result in a negative test.*
- *Ketones should not exceed 1% by volume.*

9.18 PROPERTIES

The primer should meet the requirements of the following specifications:

- Odor—The odor of the resin component and of the acid component should be normal for the volatile substances permitted when tested in accordance with ASTM D1296.
- Color—The color of the primer after drying should be characteristic of the pigments specified.

Water should be added to the resin component during manufacture in the exact amount specified. The finished resin component should give a negative test for the presence of excess water when tested.

The presence of excess water in the resin component should be determined by the following laboratory test on the thinner removed from the resin component by distillation. Upon completing the distillation, mix well and remove a 10-cm^3 portion to a 100-cm^3 glass-stoppered graduated cylinder. Add 90 cm^3 of chemically pure (c.p) benzene and shake well. Formation of a cloudy solution indicates the presence of excess water. The thinner removed from properly prepared resin component should result in a clear solution when tested as specified.

The presence of butanol should be determined on the fraction of the distillate from the resin component which distills at 117–119°C. This material should have a refractive index of 1.395–1.398 at 25°C. When 5 ml of this material is placed in a 100-ml glass-stoppered graduated cylinder with 60 ml of distilled water and shaken, a clear homogeneous solution should be formed.

A film of mixed paint should be hard and tough and should adhere tightly to the metal panel. It should be difficult to furrow off with the knife and should not flake, chip, or powder. The knife cut should show beveled edges.

Mix the coating, but omit the standing period. Using a 0.0076-cm film applicator, draw down a 5.08-cm-wide film of the mixed coating on aluminum, steel, and galvanized steel panels, solvent-cleaned as specified in method 2011 of U.S. Federal Standard 141, using the petroleum naptha-ethylene glycol monoethyl ether mixture. Air-dry for 1 h under referee conditions, and then perform a knife test as specified in method 6304 of U.S. Federal Standard 141.

There should be no evidence of incompatibility of any of the ingredients of the mixed coating when tested. Compatibility with the thinner should be determined in accordance with method 4203 of Federal Standard 141. A mixture of 50 cm^3 of mixed primer and 50 cm^3 of isopropyl alcohol conforming to TT-I-735 should be used. The isopropyl alcohol should be added slowly after mixing.

When tested, the acid and resin components should form a smooth, uniform mixture and should show no signs of thickening of gelatin when examined 24 h after mixing. The components should mix readily at any temperature between 4°C and 32°C and should be suitable for spray application within that temperature range.

Add slowly one part by volume of the acid component, with rapid stirring, to four parts by volume of the resin component. Store in a closed glass container for 6 h. Then spray a portion of the mixed material on a solvent-cleaned steel panel to a dry film thickness of 0.00076–0.00127 cm and examine for leveling and evenness of application. Retain the remainder of the mixed material in the closed glass container for 18 additional hours and examine for absence of nonuniformity by the appropriate sections of method 3011 of Federal Standard 141.

A flowout film of the mixed primer, after drying on glass for 24 h, should exhibit a surface that is smooth in appearance and free of defects such as pinholes, coarse particles, skins, or agglomerates of any kind.

Table 9.56 Properties of Basic Zinc Chromate-Vinyl Butyral

| Characteristics | Requirements | | ASTM Method | U.S. Federal Standard No. 141 |
	Minimum	Maximum		
Resin: viscosity shear rate 200 (rpm)				
Gram	110	165		
Kreb units	63	75	D562	
Density (kg/l)	0.38	0.93	D1475	
Fineness of grind (μm)	25	N/A	D1210	
Acid:				
Density	0.90	0.93	D1475	
Primer:				
Dry hard (min)	N/A	30		4061*

Drying time should be determined by method 4061 of Federal Standard 141, except that the primer should be drawn down on a steel panel using a firm applicator. The specified conditions of temperature and humidity should apply only for referee tests in case of disputes. All other tests should be conducted under prevailing laboratory conditions.

Prepare a flowout film of the primer by pouring approximately 15 cm^3 of the mixed primer across a glass panel near the upper edge while the panel is lying flat. Then tilt the panel so that the coating spreads over everything but the upper edge. Next, place the panel in an almost vertical position and allow to drain. After 24 h, examine the film for compliance with relevant standards. Coarse particles, skins, and agglomerates are characterized by being larger than the dispersed pigment in particle size extending beyond the plane of the film.

9.19 STORAGE LIFE AND PACKAGING

9.19.1 CONDITION IN CONTAINER

This primer should be supplied in two components. The resin component should be capable of being remixed to a smooth, uniform consistency. It should not liver, should not exceed 85 Krebs units in viscosity, and should not exceed 1 h of dry hard time (for pretreatment primer). It should not curdle, gel, or show any other objectionable properties for at least 12 months after the date of delivery. The packaging should meet the relevant requirement of ASTM D3951 (88).

9.19.2 INSPECTION

All materials supplied under this specification should be subject to timely inspection by the purchaser or the purchaser's authorized representative. The purchaser should have the right to reject any materials supplied that are found to be defective under this specification. In the case of disputes, the arbitration or settlement procedure established in the procurement documents should be followed.

Samples of any or all ingredients used in the manufacture of this paint should be supplied upon the purchaser's request, along with the supplier's name and identification of the material. Unless otherwise specified, the methods of sampling and testing should be in accordance with U.S. Federal Test Method Standard No. 141 or applicable methods of the ASTM.

9.19.3 DIRECTIONS FOR USE

Basic zinc chromate-vinyl butyral primer is intended to be used primarily on clean steel free of rust and scale or on clean galvanized metal. Four volumes of resin component should be mixed with one volume of acid component just prior to use as described next.

First, break up the pigment settled in the resin component with a wooden paddle, mechanical stirrer, or mixer, and mix to distribute the pigment evenly throughout the resin. After the resin component is thoroughly mixed, slowly pour one volume of the diluent into four volumes of the resin component with constant agitation. Do not pour off the liquid that has separated from the pigment and then add the acid component to the settled pigment to aid mixing. Material that is not mixed properly may gel and be unfit for use.

The resin component should be mixed with the acid component in quantities which will be applied within 6–8 h after mixing. Primer that cannot be used within a maximum of 8 h after mixing with the acid component should be discarded. Screen paint before applying.

Apply the wash primer by spraying or brushing. Spraying is generally the preferred method, but brushing may be desirable when dealing with rough or poorly prepared steel. Roller coating may be used only if specified by experts.

Paint brushes should be clean and dry or, if not, wetted with alcoholic solvents. When sprayed, the primer must be deposited on the surface wet. If dusting is encountered, move the gun closer to the surface; if already within 15 cm of the surface, decrease atomizing air pressure or increase the liquid pressure, or add thinner.

Where thinning is desired, isopropanol (99% grade), normal butyl alcohol, or denatured ethanol should be used. At least 25% thinning is usually necessary to get a uniform application. Use denatured ethyl or isoprophy alcohol to clean the equipment.

Apply to a film thickness of 8–13 μm (dry) or approximately 75–125 μm (wet). Note that at this thickness, which should not be exceeded, the base metal will show through the coating, as evidenced by uneven coloring. This is the normal appearance; do not attempt to hide the base metal completely. When spot-treating, cover only spots free of old paint. A slight overlap of existing paints is generally not harmful, provided that adherence of the wash primer to the old paint is satisfactory and the old paint is not lifted.

The next coat of paint may be applied as soon as the wash primer is dry, usually from a half-hour to 4 h later except when otherwise authorized by the inspector. This wash primer should be applied over clean, dry steel; however, a slightly damp surface may be painted over, provided adequate normal butyl alcohol is used in the thinner. If the surface is excessively wet, the vinyl butyral resin will be thrown out of solution and form a gel, or the dried film will turn white, become brittle, and lack adhesion to the steel.

This wash primer is not intended for use as a shop coat for steel, and it should be recoated with the prime coat of paint before exposure, preferably within 24 h. It is intended for use over clean, dry, descaled steel. It does not work over phosphate-treated steel and should not be used over paint or wetting oils. It must be used on bare metal for best results. If used over mill scale, it may contribute to the mill scale lifting.

This wash primer may be used to bond conventional paints to galvanized surfaces or stainless steel. Solvent cleaning of such surfaces, even though apparently clean, is advisable before applying the wash primer. Adhesion to some types of white rust preventatives on galvanized steel is poor.

Almost all paints will adhere well to this wash primer; exceptions are certain types of vinyl paints and some lacquers. These coatings may require an intermediate or bonding coat. Paints containing alcohol or ketone solvents generally show best bonding.

9.19.4 SAFETY PRECAUTIONS

Paints are hazardous because of their flammability and potential toxicity, so proper safety precautions should be observed. Safe handling practices are required and should include, but not be limited to, the provisions of SSPC-PA Guide 3, "A Guide to Safety in Paint Application," and to the following specifications:

- Keep paint away from heat, sparks, and open flame during storage, mixing, and applications. Provide sufficient ventilation to maintain vapor concentration at less than 25% of the lower explosive limit.
- Avoid prolonged or repeated breathing of vapors or spray mist, and prevent contact of the paint with the eyes or skin.
- Clean hands thoroughly after handling paint and before eating or smoking.
- Provide sufficient ventilation to ensure that vapor concentrations do not exceed the published permissible exposure limits. When necessary, supply appropriate personal protective equipment and enforce its use.

This paint may not comply with some air pollution regulations because of its solvent content. Ingredients in this paint that may pose a hazard include zinc chromate, hydrocarbon solvent, and phosphoric acid. Applicable regulations governing safe handling practices should apply to the use of this paint.

9.20 VINYL RED LEAD PRIMER

Standard specifications cover the minimum requirements for the composition, analysis, properties, storage life, packaging, inspection and labeling of vinyl red lead primer.

9.20.1 COMPOSITION

The ingredients and proportions of red lead primer should be as specified next.

The primer should be composed of 43% by weight of total solids, pigment, resin, plasticizer and solvents as specified. Small amounts of wetting agents, suspension aids, and stabilizers may be used at the discretion of the manufacturer, provided all of the requirements of the specification are met. The pigment should be red lead conforming to type I, grade 97 of U.S. Federal Standard TT-R-191.

The nonvolatile vehicle should conform to the requirements laid out in Table 9.57.

The volatile portion of the primer should be composed only of the materials listed in Table 9.58 and should conform to the requirements shown in the table when tested by the methods specified.

Table 9.57 Vehicle Characteristics

| Material | Percent by Weight | | U.S. Fed Standard Specification |
	Minimum	Maximum	
Vinyl resin*	89	92	TT-T-656
Tricresyl phosphate	8	11	

The resin should be a hydroxyl-containing vinyl chloride-acetate copolymer composed of 89.5–91.5% vinyl chloride, 5.3–7.0% vinyl alcohol, and 2.0–4.0% vinyl acetate. It should have a specific minimum gravity of 1.35 and be furnished as a powdered solid, not less than 98% of which should pass through a standard 0.84-mm sieve opening (No. 20 U.S. Standard Sieve Series). An 18% solution of the resin in MIBK should be on darker than Gardener Color Standard No. 5. Vacuum-dry a film of an acetone solution of the resin isolated as specified by Method 4032 of U.S. Federal Standard No. 141 on a rock salt plate at 70°C, and scan the infrared spectrum from 2.5–15 µm. The spectrum should match the one shown on Figure 9.1, showing the 2.9-, 5.75-, and 14.5-µm bands in relatively the same absorbance ratio as illustrated.

Table 9.58 Volatile (% by Weight)

| Material | Requirements | | | | Ingredient Standard |
| | Composition (g) | | Composition(l)[1] | | |
	Minimum	Maximum	Minimum	Maximum	
MIBK	50	N/A	N/A	N/A	TT-M-268
Methyl n-butyl ketone[2]	N/A	N/A	50	N/A	N/A
MEK	7	10	17	20	TT-M-261
Toluene	N/A	40	N/A	15	TT-T-548
Aliphatic naphtha[3]	N/A	N/A	N/A	15	TT-M-95, Type 1

[1]The methyl n-butyl ketone should contain no more than 5% by volume of branched chain ketones.
[2]The aliphatic naphtha should contain no more than 11% by volume of aromatic hydrocarbons.
[3]The volatile content of composition L should also conform to the following requirements by volume when tested as specified in 4.3.4 of MIL-P-15929C.
a) Solvents with an olefinic or cyclo-olefinic type of unsaturation: 5% maximum.
b) A combination of aromatic compounds with eight or more carbon atoms to the molecule except ethyl benzene: 3% maximum.
c) A combination of ethyl benzene ketones having branched structures or toluene: 20% maximum.
d) Total of a + b + c: 20% maximum.

9.20.2 ANALYSIS

The primer should conform to the composition (analysis) requirements shown in Table 9.59.

9.20.3 PROPERTIES

Vinyl red lead should meet the requirements given in Table 9.60. The color of the primer should be characteristic of the pigment used and should approximate color number 22510 predominantly orange of U.S. Federal Standard No. 595.

Table 9.59 Analysis of Red Vinyl Lead Primer

Characteristic	Requirements	
	Minimum	**Maximum**
Vinyl resin, percent by weight of nonvolatile	39	92
Vehicle tricresyl phosphate percent by weight of nonvolatile vehicle	8	11
Chlorine content, percent by weight of nonvolatile vehicle	45	48
Vinyl acetate, percent by weight of nonvolatile vehicle	2	3
Total solids, percent by weight of primer	43	0.75
Pigment, percent by weight of primer	24	27
Red lead (Pb_3O_4), percent by weight of pigment	96	0.75
Vehicle solids, percent by weight of primer	18	21
Water, percent by weight of primer	0.75	0.5

Table 9.60 Properties of Red Vinyl Lead Primer

Characteristics	Minimum	Maximum	ASTM Standard	U.S. Federal Standard No. 141
Viscosity* shear rate 200 (rpm)				
Kerbs units	72	82	D562	4281
Grams	150	200		
Density (kg/l)	1.15	1.21	D1475	4184
Fineness of grind (μm)	40	N/A	D1210	4411
Hegman units	5	N/A	D1210	4411
Drying time, air dry, min. Set to touch	N/A	15	D1640	4061
Dry hard	N/A	30	D1640	4061
85° specular gloss	40	N/A	N/A	6103

Viscosity 48 h or more after manufacture.

The primer tested as specified next should spray satisfactorily in all respects and should not exhibit running, sagging, or streaking. The dried film should show no dusting or mottling and should present a smooth uniform finish free of seeds. Reduce two parts by volume of the primer with one part by volume of thinner conforming to Table 9.58. Spray a 13-μm (dry) film thickness of pretreatment coating conforming to MIL-P-15328, to a steel panel cleaned with the aliphatic naphtha, ethylene glycol monoethyl ether mixture as specified in method 2011 of U.S. Federal Standard 141.

Allow the pretreatment coating to air-dry for 1 h and then recoat with a 2.3–2.8 μm (dry) film of the primer. Air-dry for 1 h and observe for compliance with the previously specified requirements; then allow to air-dry for 48 h and apply a second coat of the primer to a 23 to 28 μm (dry) film thickness. Condition the panel for knife tests. For referee testing, use automatic application as specified in method 2131 of U.S. Federal Standard 141.

A film of primer, tested as specified next should be hard and tough and should adhere tightly to and not flake or crack from the metal. The film should ribbon or curl from the metal upon cutting, and there should be no separation between the pretreatment coating and the primer and between primer coats.

Allow the panel that has been prepared to air-dry for 48 h. Then perform the knife test as specified in method 6304 of U.S. Federal Standard 141 for compliance with the abovementioned specifications.

9.20.4 STORAGE LIFE AND PACKAGING

When stored in a full, tightly covered container at normal temperatures, the primer should show no thickening, curdling, felling, or hard caking when tested as specified in U.S. Federal Standard No. 141, Method 3011, and should meet the requirements of standard after storage of at least 12 months from the date of delivery. The packaging should meet the relevant requirements of ASTM D 3951 (88) unless otherwise specified by the purchaser.

9.20.5 INSPECTION

All materials supplied under this specification should be subject to timely inspection by the purchaser or the purchaser's authorized representative. The purchaser should have the right to reject any materials supplied that are found to be defective under this specification. In the case of disputes, the arbitration or settlement procedure established in the procurement documents should be followed.

Samples of any or all ingredients used in the manufacture of this paint should be supplied upon the purchaser's request, along with the supplier's name and identification of the material.

9.20.6 DIRECTIONS FOR USE

In addition to the manufacturer's instructions, the following directions for use should be supplied with each container of paint.

This primer is intended for use over a vinyl butyral wash primer. The primer should be thinned as necessary with solvent containing not more than 40% toluene, 50% MIBK, and 10% MEK. The amount of thinning will depend upon application methods and conditions; it may be as high as 25–33% by volume of primer.

Apply by conventional air spray. Brushing may be used in small areas. The surface to be painted should be dry, above 2°C, not less than 3°C above the dew point. Do not paint outdoors in rainy weather. Apply the paint to obtain the specified film thickness. The minimum dry film thickness should be 25 μm.

A wet film of primer should be deposited on the surface when spraying; the spray gun should be adjusted so that proper atomization is obtained such that dry paint (similar to overspraying) is not deposited on the surface. The nozzle should be held about 15 cm from the surface during application.

If application is to be made by brush, the brush should be heavily loaded with primer; apply the paint quickly and smoothly. Avoid excessive brushing, and do not go back over the surface until thoroughly dry.

At temperature between 16°C and 27°C, dry at least 1 h between coats and 72 h before immersion. Varying atmospheric conditions and degrees of ventilation in confined spaces may allow shorter or require longer drying times.

Unless otherwise specified, the methods of sampling and testing should be in accordance with U.S. Federal Test Method Standard No. 141 or applicable methods of the ASTM.

9.20.7 SAFETY PRECAUTIONS

In addition to the manufacturer's instructions for use, the following safety directions should be supplied with each container of paint.

Paints are hazardous because of their flammability and potential toxicity, so proper safety precautions should be observed. Safe handling practices are required and should include, but not be limited to, the provisions of SSPC-PA Guide 3, "A Guide to Safety in Paint Application," and to the following specifications:

- Keep paint away from heat, sparks, and open flame during storage, mixing, and application. Provide sufficient ventilation to maintain vapor concentration at less than 25% of the lower explosive limit.
- Avoid prolonged or repeated breathing of vapors or spray mist, and prevent contact of the paint with the eyes or skin.
- Clean hands thoroughly after handling paint and before eating or smoking.
- Provide sufficient ventilation to ensure that vapor concentrations do not exceed the published permissible exposure limits. When necessary, supply appropriate personal protective equipment and enforce its use.

This paint may not comply with some air pollution regulations because of its hydrocarbon solvent content. Ingredients in this paint that may pose a hazard include hydrocarbon solvents. Applicable regulations governing safe handling practices should apply to the use of this primer.

9.21 MATERIAL AND EQUIPMENT STANDARD FOR COAL-TAR EPOXY POLYAMIDE PAINT AS PRIMER, INTERMEDIATE, AND TOPCOAT (FINISH)

This section covers the minimum requirements for the composition, analysis, properties, storage life and packaging, inspection, and labeling of coal-tar epoxy-polyamide paint as primer, intermediate, and topcoat (finish).

9.21.1 COMPOSITION

The ingredients and proportions of coal-tar epoxy polyamide paint should be as specified in Table 9.61.

This paint contains approximately 71% by volume of nonvolatile, film-forming solids (pigment and binder).

Table 9.61 Composition of Coal-Tar Epoxy Polyamide Paint

| Ingredient | Required | | Typical[3] Composition | |
| | Minimum | Maximum | Components A and B | |
	Wt.%	Wt.%	Wt.%	Vol.%
Component A (82 ± 0.5 wt.%) (80 vol.%)				
Coal tar pitch	33.0	36.0	28.2	29.0
Polyam1de resin	11.0	12.0	9.5	12.4
Magnesium silicate[1]	30.0	33.0	25.8	11.9
Xylene[2]	18.0	21.0	15.4	22.9
Gelling agent and activator	2.5	2.6	2.0	2.4
Accelerator	1.2	1.3	1.1	1.4
Component a totals		100	82.0	80.0
Component B (18 ± 0.5 wt.%) (20 vol.%)				
Liquid epoxy resin	N/A	N/A	18.0	20.0
Totals (components A and B)	100	100	100	100

[1]When specified in the procurement documents, a dark red coating should be furnished in which 50% or more (by volume) of the magnesium silicate is replaced by synthetic red iron conforming to ASTM Standard D3721. The red coating should meet all of the test requirements prescribed for the black coating, except that the nonvolatile content of component A should be an amount reflecting the greater specific gravity of the iron oxide pigment.
[2]In those cases where the specified volatiles are not permitted, the volatile portion of the coal-tar epoxy coating may be replaced with exempt materials to the extent necessary to assure compliance with the applicable regulations. The modified coating should meet all of the test requirements specified herein, except that determination of compliance with the nonvolatile weight content should reflect any difference in specific gravity between xylene and the substituted solvents.
[3]The most favorable composition.

Coal-tar pitch used in this paint is defined as a product obtained from the distillation of high-temperature, crude coke oven tar, which in itself is a product obtained during the destructive distillation of coal in slot ovens operated at a temperature above 700°C. The coal-tar pitch should be composed primarily of a complex combination of three or more membered condensed ring aromatic hydrocarbons.

The epoxy resin should be a diepoxide condensation product of bisphenol-A and epichlorohydrin with terminal epoxide groups. The polyamide resin should be a condensation product of a dimerized fatty acid in polyamines.

Acceptable gelling agents are organic derivatives of magnesium montmorillonite and hydrogenated castor oil. Acceptable activators, if used, are methanol, ethanol, or propylene carbonate. The accelerator should be 2, 4, 6-tri (dimethylaminomethyl) phenol.

9.21.2 OTHER PROPERTIES

Each component of this paint based on the specified ingredients should be uniform, stable in storage, and free of grit and coarse particles.

Table 9.62 Analysis of Coal-Tar Pitch

Characteristics	Requirements Type 1 Minimum	Type 1 Maximum	Type 2 Minimum	Type 2 Maximum	Type 3 Minimum	Type 3 Maximum	ASTM Method
Float test at 50°C seconds	N/A	N/A	N/A	N/A	150	220	D139
Softening point, in water, °C	70	75	54	62	N/A	N/A	D36
Insolubles in carbon disulfide, % by weight	N/A	20	N/A	20	N/A	20	D4
Ash, % by weight distillation dry basis, % by weight:	N/A	0.5	N/A	0.5	N/A	0.5	D2415
0–170°C	N/A	N/A	N/A	N/A	N/A	0	N/A
0–250°C	N/A	0	N/A	0	N/A	N/A	N/A
0–270°C	N/A	N/A	N/A	N/A	N/A	6	N/A
0–300°C	N/A	5	N/A	5	N/A	15	N/A
Softening point of residue							
At 300°C; in water, °C	N/A	N/A	N/A	N/A	45	60	D 36

9.21.3 ANALYSIS

The paint should conform to the following composition (analysis) requirements:

- The coal-tar pitch used should meet the requirements for one of the types listed in Table 9.62.
- The epoxy resin should be clear, free of turbidity, crystals, and particulate matter, and should meet the requirements given in Table 9.63.
- The polyamide resin should be clear, free of turbidity and particulate matter, and should meet the requirements given in Table 9.64.

9.21.4 PROPERTIES

The paint should meet the following specifications:

- Odor—The odor should be normal for the materials permitted (ASTM Standard D 1296).
- Color—The color should be black or dark red, as specified in the procurement documents.

The viscosity of component A should not exceed 160 poises when tested as follows:

Table 9.63 Analysis of Epoxy Resin

| Characteristics | Requirements | | ASTM Method |
	Minimum	Maximum	
Epoxide equivalent	180	200	D 1652
Nonvolatile content (1–2 grams after 1 h at 105 ± 2°C), % by weight	99	N/A	N/A
Color, gardner	N/A	5	D 1544
Specific gravity	1.15	1.18	D 1475
Viscosity, brookfield, at 25 °C, poises	100	160	N/A

Table 9.64 Analysis of Polyamide Resin

| Characteristics | Requirements | | ASTM Method |
	Minimum	Maximum	
Amine value[1]	330	360	N/A
Nonvolatile content (1–2 grams) after 1 h at 105 ± 2 C % by weight	97	N/A	N/A
Color, gardner	N/A	9	D 1544
Specific gravity	0.96	0.98	D 1475
Viscosity, brookfield, at 75°C, poises	7	9	N/A

[1]*The amine value is defined as the milligrams of potassium hydroxide equivalent to the amine alkalinity present in a 1-g sample. It is determined by a potentiometric titration with standard perchloric acid according to the following method:*
a) Weigh the approximate amount of well-mixed resin to give a titration in the range of 12–18 ml into a 200-ml, berzelius, tall form beaker on an analytical balance. Cover the beaker with aluminum foil to minimize contact with air.
b) From a graduated cylinder, carefully add 90 ml of solvent (such as nitrobenzene, propylene carbonate, or acetonitrile), insert a stirring bar, cover the beaker with aluminum foil, and stir on a magnetic stirrer to dissolve the sample. Add the solvent immediately after weighing the sample. A fume hood should be used for all operations.
c) From a graduated cylinder, add 20 ml of glacial acetic acid to the sample solution and stir for several minutes.
d) Immerse the electrodes into the sample solution, stir for 2 min and titrate potentiometrically with 0.1 N perchloric acid using the millivolt scale. Record the millivolt readings every 0.1 ml.
e) Plot a graph showing the millivolts against the titration. The endpoint is the midpoint of the inflection on the titration curve.
f) Conduct a blank determination on 90 ml of the solvent and 20 ml of acetic acid. This needs to be done only once for each lot of solvent used. On the majority of lots used, the blank has been found to be zero.
g) Calculate the amine value using the following formula:

$$\text{AMINE VALUE} = \frac{(\text{Sample Titration} - \text{Solvent Blank})\,\text{Normality} \times 56.1}{\text{Weight of Sample}}$$

1. Fill a container having a diameter and a height of not less than 7.6 cm and 9.5 cm, respectively, to a depth of not less than 7.6 cm with a representative sample of component A.
2. Set up a Model RVT or RVF-100 Brookfield Synchro-Electric Viscometer with a No. 7 spindle and with the guard removed.

3. Bring the sample to a temperature of 25°C and stir vigorously for 2 min with a stiff spatula, maintaining this temperature.
4. Immediately after stirring, lower the viscometer until half of the "neck" mark on the spindle is covered.
5. Run the viscometer at 100 rpm for 1 min and take a reading of the position of the point on the dial. If the reading is 40 or less, the viscosity should be considered to be 160 poises or less. If the reading is over 40, immediately start the motor and take additional readings at 1-min intervals. If 1 or more readings of 40 or less are obtained out of 10 readings, taken at 1-min intervals, the viscosity of the material should be considered to be within the specification limits.

The nonvolatile content of component A should not be less than 77% (by weight) when tested as follows:

1. Place a stirrer made of stiff wire into a small, disposable aluminum dish of about 5 cm in diameter and weigh to the nearest 0.1 mg.
2. As rapidly as possible, place between 2 and 3 g of component A into the dish and weigh it immediately to the nearest 0.1 mg.
3. After weighing, spread the material over the bottom of the dish.
4. Heat the dish, wire, and contents in a well-ventilated convection oven maintained at 103–107°C for 3 h.
5. After the material has been in the oven for a few minutes, and periodically thereafter, stir the material.
6. Cool in a desiccator, weigh to the nearest 0.1 mg, and calculate the percentage of nonvolatile on a weight basis.

The mixed paint should meet the following requirements:

- Odor—The odor should be normal for the materials permitted (ASTM Standard D1296).
- Color—The color should be black or dark red as specified in the procurement documents.

The paint should not sag when tested as follows:

1. Prepare approximately 500 ml of the material by thoroughly mixing 100 ml of component B into 400 ml of component A.
2. Determine its viscosity immediately after mixing, by employing a No. 5 spindle.
3. If all of the five readings taken at 1-min intervals are above 50, reduce the viscosity by adding the thinner in small increments until a reading that is less than 50 is obtained.
4. Press a strip of 2.5-cm masking tape across the full width of a solvent-cleaned, 7.6 cm × 15.2 cm, cold-rolled steel panel. The tape should be parallel to and centered on the shorter axis of the panel.
5. Within 5 min after making the final check of viscosity, apply the material to the panel at a wet film thickness of at least 350 μm. The application may be made with a doctor blade with a gap of approximately 630 μm or by brushing.
6. Immediately after applying the material, carefully remove the masking tape and stand the panel in a vertical position (with the bare strip horizontal) in a draft-free, 24–27°C location.
7. Examine the panel after 4 h. Sagging or running the paint into the base area should constitute failure of the material to pass the sag test.
8. Save the mixed paint for the penetration and adhesion tests described in sections 6.11 and 6.12, respectively.

The paint should pass the following penetration test:

1. Clean two 7. 6 cm × 15. 2 cm, cold-rolled steel panels in accordance with ASTM Standard D 609 by spraying solvent.
2. Draw down in accordance with U.S. Federal Standard No. 141, Method 2161, a coat of the paint mixed (including any thinning) for the sag test.
3. Allow the film to dry 18–24 h in a horizontal position at 24–27 c and at a relative humidity of not over 60%.
4. Apply a second coat over and at right angles to the first, using freshly mixed paint prepared identically to that used for the first coat. (Save both paints for preparing adhesion test panels.) The drawdown applicators should provide a total dry film thickness of 510–635 μm, and the coats should be of approximately equal thickness.
5. Allow the second coat to dry in a horizontal position at 24–27°C. After 120 h of curing, clamp the panel onto the table of the penetrometer (ASTM Standard D5) so that the needle is over an area that is within the prescribed thickness range (as measured by SSPCPA 2, "Measurement of Dry Paint Thickness with Magnetic Gages"), and determine the penetration, using a total load of 200 g applied for 5 s at 25°C. The average of the three lowest out of five penetration readings, all taken within 1-cm^2, should not exceed 0.03 cm after 120 h of curing.

The paint should pass the following adhesion test:

1. Sand-blast two steel panels (similar to those used in the penetration test) with a clean, nonmetallic abrasive until a uniform, gray-white surface with a well-developed anchor pattern is achieved.
2. Blow off dust with a blast of clean air.
3. Apply one coat each of the two test batches of paint used for the penetration test panels by brushing, allowing the first coat to dry 18–24 h at 24–27°C before applying the second. Each coat should be applied at a wet film thickness of 250–350 μm.
4. After the final coat has cured for 120 h at 24–27°C, test the adhesion of the paint to the metal with a sharp knife. It should strongly resist being removed from the metal.
5. Also, test the intercoat adhesion by attempting to separate the coats with the knife. Any delamination of the two coats should constitute failure.

The paint should pass the following test for pot life:

1. Mix 100 ml of component B into 400 ml of component A, both of which have been brought to a temperature of 24–27°C before mixing.
2. Pour the material at once into a 0.5-l tin can, seal the can tightly, and store it at 24–27°C.
3. Examine the material 4 h after it was mixed. For its pot life to be considered satisfactory, the mixed material should remain in a fluid condition, and, when thinned with no more than 100 ml of xylene (or where required, the recommended thinner), should be free of lumps and brushable.

9.21.5 STORAGE LIFE AND PACKAGING

The paint (both base component and curing agent) should show no thickening, curdling, gelling, or hard caking when tested as specified in U.S. Federal Standard No. 141, Method 3011, after storage for 12 months from the date of delivery in a full, tightly covered container at a temperature of 10-43°C. The packaging should meet the relevant requirements of ASTM D 3951 (88).

9.21.6 **DIRECTIONS FOR USE**

Coal-tar epoxy-polyamide black or dark red paint is intended for use as a primer, intermediate and finish coat over steel that has been blast cleaned or blast cleaned and primed with a suitable inhibitive primer.

To prepare the paint for application, add the entire contents of the epoxy resin (component B) container to the previously stirred contents of the related container of base (component A) and mix vigorously for at least 2 min with a power agitator equipped with a 7.6-cm-or-longer blade. Some thinning may be desirable for spray application. Use xylene or, where required, the recommended thinner, and 2–20 l batches should be added. Apply the paint as soon after mixing as practicable since the material will thicken substantially over a 2-h period and may set up in the paint tank within 2–4 h during very warm weather unless cooled prior to or after mixing.

This paint is usually applied by spray in two coats to a dry film thickness of 400 μm at its thinnest spots. This requires a spreading rate of 1.5 m^2/l of unthinned paint. In practice, upwards of 300 μm (wet) paint will probably be required for each 200-μm coat to obtain the desired minimum thickness. The drying time between coats under normal coating conditions should not exceed 72 h.

Long drying times between coats may cause poor intercoat adhesion, and it is advisable in warm weather to reduce the maximum interval between coats. Under conditions of hot weather or direct sunlight, it may be necessary to limit the intercoat drying period to 24 h or less. Abusive handling of precoated steel may cause damage to the coating. This is more noticeable at low temperatures or after extended periods of cure.

This paint may be applied to large surfaces by high-pressure airless spraying. For application to complex surfaces, use heavy-duty conventional air atomization spray equipment.

If the method used is brushing, apply with a stiff brush heavily loaded with paint. Apply the paint quickly and smoothly, and avoid excessive brushing.

Do not apply this coating when the receiving surfaces or the ambient temperatures are below 10°C unless it can reasonably be anticipated that the average ambient temperature will be 10°C or higher for the five-day period subsequent to the application of any coat. At temperatures between 10° and 15°C, allow the mixed paint to stand at least 30 min prior to application.

Clean all equipment immediately after use with a suitable solvent. Such cleaning solvents as high-flash aromatic naphtha, xylene, or toluene are satisfactory for cleanup, but can be improved by adding 10–20% of MIBK and 10% isopropyl or normal butyl alcohol.

9.21.7 **SAFETY PRECAUTIONS**

Coal-tar epoxy-polyamide black or dark red paints are hazardous because of their flammability and potential toxicity, so proper safety precautions should be observed. Safe handling practices are required and should include, but not be limited to, the provisions of SSPC-PA Guide 3, "A Guide to Safety in Paint Application," and to the following specifications:

- Keep paint away from heat, sparks, and open flame during storage, mixing, and application. Provide sufficient ventilation to maintain vapor concentration at less than 25% of the lower explosive limit.
- Avoid prolonged or repeated breathing of vapors or spray mist, and prevent contact of the paint with the eyes or skin.

- Clean hands thoroughly after handling paint and before eating or smoking.
- Provide sufficient ventilation to ensure that vapor concentrations do not exceed the published permissible exposure limits. When necessary, supply appropriate personal protective equipment and enforce its use.

This paint may not comply with some air pollution regulations because of its hydrocarbon solvent content. Ingredients in this paint that may pose a hazard include epoxy resin, polyamide resins, hydrocarbon solvent, and coal tars. This paint may contain low concentrations (less than 1% by weight) of materials that are suspected carcinogens. Applicable regulations governing safe handling practices should apply to the use of this paint.

During surface preparation that involves the removal of an old film of this paint, care should be taken to minimize dusting, to protect workers from the dust, and to dispose of coating residues properly.

TECHNICAL GUIDELINES FOR MATERIAL SELECTION— PART 2

10

The first part of the technical guidelines for material selection was presented in Chapter 9. This chapter covers the requirements for the composition, analysis, properties, storage life and packaging, inspection, and labeling of the second part of this discussion, which includes various materials to be used as primers, intermediate coats, and topcoats (finishes).

10.1 VINYL PAINT, (WHITE OR COLORED) AS PRIMER, INTERMEDIATE COAT, AND TOPCOAT (FINISH)

This section covers the minimum requirements for the composition, analysis, properties, storage life and packaging, inspection, and labeling of vinyl paint to be used as primer, intermediate coat, and topcoat (finish).

10.1.1 COMPOSITION

The ingredients and proportions of vinyl paint (white or colored) should be as specified in Table 10.1. The paint based on the specified ingredients should be uniform, stable in storage, and free of grit and coarse particles. Beneficial additives such as antiskinning agents, suspending agents, or wetting aids may be added.

This paint contains approximately 17% by volume of nonvolatile, film-forming solids (pigment and binder).

10.1.2 ANALYSIS

The paint should conform to the composition (analysis) requirements listed in Table 10.2.

10.1.3 PROPERTIES

The paint should meet the requirements given in Table 10.3 and the following criteria:

- Odor—The odor should be normal for the materials permitted (ASTM D1296)
- Color—The color should be white or colored as specified in the procurement documents with reference to Table 10.4.

Essentials of Coating, Painting, and Lining. http://dx.doi.org/10.1016/B978-0-12-801407-3.00010-9

Table 10.1 Composition of Vinyl Paint (White or Colored)

Ingredients	Typical[6] Composition		Ingredient Standards ASTM
	Wt. %	Vol. %	
Pigment (12 ±1%)			
Titanium dioxide	12.0	2.9	D476, Type II Class II
Tinting pigments[1]	N/A	N/A	N/A
Vehicle (88 ±1%)			
Vinyl resin A[2]	8.0	5.9	N/A
Vinyl resin B[3]	8.0	5.9	N/A
Dioctyl phthalate[4]	3 0	2.5	N/A
Methyl isobutyl ketone (MIBK)[5]	34.5	43.0	D1153
Toluene	34.5	39.8	D362
Totals	100.0	100.0	

[1]*Stable, durable tinting pigments should be substituted for a portion of the pigment when a tint is specified in the procurement documents.*
[2]*Vinyl resin A should be a hydroxyl containing a vinyl chloride-acetate copolymer. It should contain 89.5% and 91.5% vinyl chloride, 5.3–7.0% vinyl alcohol, and 2–5.5% vinyl acetate. The inherent viscosity of the resin (ASTM Standard D 1243, Method A) at 20°C should not be less than 0.5.*
[3]*Vinyl resin B should be a carboxyl containing a vinyl chloride acetate copolymer. It should contain 85–87% vinyl chloride, 12–14% vinyl acetate, and 0.5–1% maleic acid. The inherent viscosity of the resin (ASTM Standard D 1243, Method A) at 20°C should not be less than 0.48.*
[4]*Dioctyl phthalate (di-2 ethylhexyl phthalate) should be commercial material that conforms to the following requirements:*

- *Specific gravity at 25°C: 0.980–0.9861*
- *Refractive index at 25°C: 1.4830–1.6859*

[5]*When specified in the procurement documents, suitable high-boiling vinyl solvent may be substituted for a portion of the methyl isobutyl ketone (MIBK) to make the paint amenable to application in hot weather or by brushing.*
[6]*The most favorable composition.*

- Compatibility—There should be no evidence of incompatibility of any of the ingredients of the paint when two volumes of the paint are slowly mixed with one volume of thinner consisting of 85% toluene and 15% MIBK by volume (U.S. Federal Standard No. 141, Method 4203).

The paint should show good adhesion when tested as follows:

Apply one coat (25 μm dry film thickness) of the mixed paint to a clean steel panel free of rust or scale, and also to a similar panel pretreated with wash primer MIL-P 15328, "Primer (Wash) Pretreatment," or SSPC paint 27, "Basic Zinc Chromate-Vinyl Butyral Wash Primer" (12.5 μm dry film thickness). After a 24-h dry, the film being tested on each panel should be subjected to a knife test to determine whether the paint exhibits good adhesion to the undercoats and to the steel.

The paint should be easily applied when tested in accordance with U.S. Federal Standard No. 141, Methods 4331 and 4541. The paint should show no streaking, running, or sagging after drying.

Table 10.2 Analysis of Vinyl Paint (White or Colored)

	Requirements			
Characteristics	Minimum Wt. %	Maximum Wt. %	ASTM Method	U.S. Federal Standard No. 141
Pigment	11.0	13.0	D1208	4021*
Volatiles	67.0	71.0	D2369	N/A
Nonvolatile vehicle Calculated by Difference	18.0	20.0	N/A	4053
Uncombined water	N/A	0.5	D1208	4081
Coarse particles and Skins, as retained on Standard 0.045-mm Sieve Opening (325 Mesh Screen)	N/A	0.25	D185	4092

Using extraction mixture "C" (1:1 toluene and acetone).

Table 10.3 Properties of Vinyl Paint (White or Colored)

	Requirements			
Characteristics	Minimum Wt. %	Maximum Wt. %	ASTM Method	U.S. Federal Standard No. 141
Paint consistency viscosity* Shear Rate 200 rpm				
Grams	100	150	D562	N/A
Krebs	61	72	D562	N/A
Density (kg/l)	0.97	1.05	D1475	N/A
Fineness of grind (μm)	25	N/A	D1210	N/A
Drying time (min)				
Tack free	N/A	15	D1640	4061
Dry hard	N/A	30	D1640	4061

Viscosity 48 hours or more after manufacture.

Table 10.4 Color of Vinyl Paint (White or Colored)	
Paint Color	**Color No. to BS 381 C**
Arctic blue	112
Sea green	217
Brilliant green	221
Canary yellow	309
Light straw	384
Middle brown	411
Signal red	537
Light orange	567
Light gray	631

10.1.4 STORAGE LIFE AND PACKAGING

When in a full, tightly covered container, vinyl paint (white or colored) should show no thickening, curdling, gelling, or hard caking when tested as specified in U.S. Federal Standard No. 141, Method 3011, after being stored for 12 months from date of delivery.

10.1.5 DIRECTIONS FOR USE

Vinyl paint (white or colored) is intended for use as a primer over blast-cleaned steel or over vinyl butyral wash primer. It may also be used as an intermediate or finish coat over itself as primer and over vinyl chloride acetate copolymer paint. Mix thoroughly before use.

The paint should be thinned as necessary with solvent containing not more than 85% toluene and 15% MIBK or methyl ethyl ketone (MEK). The amount of thinning will depend upon the application methods and conditions, and may be as high as 25–33% by volume of paint.

Apply by conventional air spraying. Brushing may be used in small areas. The surface to be painted should be dry and above 2°C, and not less than 3°C above the dew point. Apply so as to obtain the specified film thickness. The minimum dry film thickness should be 25 μm.

A wet film of paint should be deposited on the surface when spraying; the spray gun should be adjusted so that proper atomization is obtained such that dry paint (similar to overspray) is not deposited on the surface. The nozzle should be held about 15 cm from the surface during application.

If application is to be made by brushing, the brush should be heavily loaded with paint; apply it quickly and smoothly. Avoid excessive brushing, and do not go back over the surface until it is thoroughly dry.

At temperatures between 16–27°C, dry for at least 1 h between coats and 72 h before immersion. Varying atmospheric conditions and degrees of ventilation in confined spaces may allow shorter or require longer drying times.

10.1.6 SAFETY PRECAUTIONS

Paints are hazardous because of their flammability and potential toxicity, so proper safety precautions should be observed to protect against these recognized hazards. Safe handling practices are required and should include, but not be limited to, the provisions of SSPC-PA Guide 3, "A Guide to Safety in Paint Application," and to the following specifications:

- Keep paint away from heat, sparks, and open flame during storage, mixing, and application. Provide sufficient ventilation to maintain vapor concentration at less than 25% of the lower explosive limit.
- Avoid prolonged or repeated breathing of vapors or spray mist, and prevent contact of the paint with the eyes or skin.
- Clean hands thoroughly after handling paint and before eating or smoking.
- Provide sufficient ventilation to ensure that vapor concentrations do not exceed the published permissible exposure limits. When necessary, supply appropriate personal protective equipment and enforce its use.

This paint may not comply with some air pollution regulations because of its hydrocarbon solvent content. Ingredients in this paint that may pose a hazard include hydrocarbon solvents. Applicable regulations governing safe handling practices should apply to the use of this paint.

10.2 TWO-PACK, AMINE-ADDUCT, CURED EPOXY PAINT AS PRIMER, INTERMEDIATE, AND TOPCOAT

This section covers the minimum requirements for the composition, properties, storage life, packaging, inspection, and labeling of a chemical-resistant, two-pack, amine-adduct, cured epoxy paint.

10.2.1 COMPOSITION

The material of two-pack, amine-adduct, cured epoxy paint should be furnished in two components. Component A should consist of epoxy resin combined with prime and extender pigments and volatile solvents; component B should consist of a suitable polyamine resin properly combined with volatile solvents and act as the curing agent for component A. Components A and B should be packaged separately and furnished in kit form, and when mixed in accordance with the manufacturer's instructions, a product meeting this specification should result.

The total nonvolatile solids for the admixed components of gloss colors should be a minimum of 60% by volume when tested in accordance with ASTM D 2697.

10.2.2 PROPERTIES

- Odor—The odor of the paint material should not be obnoxious when tested in accordance with U.S. Federal Standard No. 141, Method 4401.
- Color—The color should be as specified by the purchaser with reference to Table 10.5.
- Surface appearance—The paint film should be smooth, uniform, and free of bubbles, pinholes, holidays, and other film irregularities. The spray-applied films, dried under the standard conditions

Table 10.5 Color Specifications

Paint Color	Color No. to BS 381 C	Color No. Rall (Approx.)
Sea green	217	6017
Middle brown	411	8007
Light gray	631	7033
White	595	1000

(25°C ±2°C and relative humidity of 50% ±5%) should provide a hard surface, free of grit, seeds, streaks, orange peel, blisters, or other surface defects when tested in accordance with U.S. Federal Standard No. 141, Method 4541. The paint should be easily applied by brush and spray when tested in accordance with U.S. Federal Standard No. 141, Methods 4321 and 4331. The paint should show no streaking, running, or sagging after drying.

- Setting—When tested in accordance with U.S. Federal Standard No. 141 after standing undisturbed for 6 h, the ready-mixed and thinned paint material should be free of setting precipitation and separation which can not be easily re-dispersed by shaking on a mechanical paint mixer.
- Pot Life—Ready-mixed paint should have a minimum pot life of 6 h, at standard conditions (25°C ±2°C and relative humidity of 50% ±5%).
- Fineness of grind—The fineness of grind of the mixed paint should not be less than 7 (Hegman units) for gloss paints. The tests should be made 1 h after mixing, in accordance with U.S. Federal Standard No. 141, Method 4411.1.
- Flexibility—The mixed paint should show no evidence of cracking, chipping, or flaking when tested in accordance with U.S. Federal Standard No. 141, Method 6221.
- Toxicity—The suitability of paint for contact with food and potable water should be certified by the local health service department when required by the purchaser.

10.2.3 STORAGE LIFE AND PACKAGING

When stored in a full, tightly covered container, the product should meet the requirements of the standards after storage of at least 12 months from the date of delivery in normal conditions.

The paint (both components A and B) as received should show no evidence of levering, skinning, or hard settling of pigment, and the container should not be affected. The material should be easily dispersed in liquid form by hand stirring to form a smooth, homogeneous paint free of persistent foam when tested in accordance with Method 3011.1 of U.S. Federal Standard No. 141, after storing for 12 months from the date of delivery.

The epoxy paint should be packaged as a unit consisting of pigmented compound marked "Component A" and the unpigmented (or clear) hardener marked "Component B." The packaging should meet the relevant requirements of ASTM D 3951(88) unless otherwise specified by the purchaser.

Packing should be accomplished in a manner that will ensure acceptance by a common carrier, at the lowest rate, and will afford protection against physical or mechanical damage during shipment.

In addition to the manufacturer's instructions for use, consisting of complete instructions covering use, surface cleanliness, mixing, thinning, application method, application condition, pot life, wet and dry film thickness per coat, temperature and humidity limitations, and drying time, the following directions should be supplied with each container of paint: "This paint is intended for use on primed substrates. The surface of substrates should be prepared in accordance with standard before applying the primer."

Apply by brush or spray to the specified film thickness or, if none is specified, to at least 125 μm (dry). When application is by spraying, the equipment and operator technique should be adjusted to prevent dry spraying and to deposit a wet film of paint on the substrate. Clean the equipment with suitable thinner both before and after use.

The surface to be painted should be dry, and the surface temperature should be at least 3°C above the dew point. Painting should be in accordance with standards.

In addition to the manufacturer's instructions for safety, the following directions should be supplied with each container of paint:

- This paint is hazardous because of its flammability and potential toxicity, so proper safety precautions should be observed to protect against these recognized hazards. Safe handling practices are required and should include but not be limited to, the provisions of SSPC-PA Guide 3, "A Guide to Safety in Paint Application," and to the following specifications:

  - Keep paint away from heat, sparks, and open flame during storage, mixing, and application. Provide sufficient ventilation to maintain vapor concentration at less than 25% of the lower explosive limit.
  - Avoid prolonged or repeated breathing of vapors or spray mist, and prevent contact of the paint with the eyes or skin.
  - Clean hands thoroughly after handling paint and before eating & smoking.
  - Provide sufficient ventilation to ensure that vapor concentration do not exceed the published permissible exposure limits. When necessary, supply appropriate personal protective equipment and enforce its use.

This paint may not comply with some air pollution regulations because of its hydrocarbon solvent content.

10.3 ORGANIC ZINC-RICH EPOXY PAINT AS PRIMER AND INTERMEDIATE COAT

This section covers the minimum requirements for the composition, analysis, properties, storage life and packaging, inspection, and labeling of organic zinc-rich epoxy paint to be used as primer, intermediate and topcoat.

10.3.1 COMPOSITION

The ingredients and proportions of organic zinc-rich paint should be the following descriptions.

The paint consists of 70% by weight of nonvolatile, film-forming solids (pigments and binder). The zinc-rich paint described in this specification consists of zinc dust, an organic vehicle, and selected additives as required.

Table 10.6A Composition of Organic Zinc-Rich Paint

Ingredients	Composition Wt. %		Ingredient Standards
Pigment (83 ± 2 wt. %)	58	ASTM D520	
Vehicle (17 ± 2 wt. %)	12		N/A
Solvent	30		N/A
Selected additives	As required	N/A	

The major pigment component in paint is zinc dust of the types described in Tables 10.6A and 10.6B. Other pigment components may include curing aids, tinting colors, suspension, and pot life control agents, but they should constitute only a minor part of the total pigment portion so as not to detract from the ability of this paint to protect galvanically.

Table 10.6B Composition of Zinc Dust Pigment

	Type I	Type II
Total zinc calculated as Zn, minimum %	97.5	98.0
Metallic zinc, minimum %	94.0	94.0
Material other than metallic zinc		
ZnO, and admixed CaO, where applicable maximum %	0.75	N/A
Calcium, calculated as CaO, maximum %	0.7	0.7
Lead, calculated as Pb, maximum %	N/A	0.01
Iron, calculated as Fe, maximum %	N/A	0.02
Cadmium, calculated as Cd, maximum %	N/A	0.01
Chlorine, calculated as Cl, maximum %	N/A	0.01
Sulfur, calculated as SO_2, maximum %	N/A	0 01
Moisture and other volatile matters, maximum %	0.10	0.10
Oily or fatty matter, or both, maximum %	N/A	0.05
Zinc oxide (ZnO), maximum %	6.0	remainder
Coarse particles maximum %:		
Total residue retained on a 150-μm Standard sieve opening (Sieve No. 100)	None	0.1
Total residue retained on a 75-μm Standard sieve opening (Sieve No. 200)	N/A	0.8
Total residue retained on a 45-μm Standard sieve opening (Sieve No. 325)	4.0	3.0
Type I general grade *Type II high-purity grade*		

Table 10.7 Epoxy Resin Analysis

Characteristics	Requirements		ASTM Method
	Minimum	**Maximum**	
Epoxide equivalent	450	550	D1652
Color, gardner (40% in butyl carbitol)	N/A	4	D1544

The organic vehicle consists of catalyzed epoxy. The epoxy resin should be a di-epoxide condensation product of biphenol A and epichlorohydrin with a terminal epoxide group that meets the requirements of Table 10.7.

The curing agent consists of a liquid polyamide resin and volatile solvent. The undiluted polyamide resin should meet the requirements of Table 10.8.

10.3.2 **ANALYSIS**

The paint should conform to the composition (analysis) requirements given in Table 10.9.

10.3.3 **PROPERTIES**

Organic zinc-rich paint should meet the qualitative requirements given in the next sections.

The pot life of this paint, when mixed and ready for application in accordance with the manufacturer's instructions, should be a minimum of 4 h at 21°C and 50% relative humidity. Although physical properties (viscosity, etc.) may not change, loss of pot life is indicated by lack of adhesion.

The mixed paint should spray easily and show no streaking, running, sagging, or other objectionable features when tested in accordance with U.S. Federal Standard No. 141, Methods 4331 and 4541.

Steel test panels (ASTM-A 36 hot-rolled steel or equivalent) measuring 10 cm × 15 cm × 1.5 mm or greater should be blast-cleaned to Sa3 (white metal) with a nominal anchor profile from 40–90 μm and coated with zinc-rich paint. The panels should be blast-cleaned and coated on both sides and all edges. The paint should be spray-applied and hardened in accordance with the manufacturer's

Table 10.8 Curing Agent Analysis

Characteristics	Requirements		ASTM Method
	Minimum	**Maximum**	
Amine value[1]	230	250	N/A
Color gardner	N/A	8	D1544
Specific gravity	0.96	0.98	D1475
Viscosity, at 75° C poises	31	37	N/A
[1]*Perchloric acid titration*			

Table 10.9 Composition Analysis of Organic Zinc-Rich Paint		
Characteristics[1]	**Minimum Requirements**	**ASTM Standards**
Total solids, % by weight of paint	70	D2369
Pigment, % by weight of total solids	83	D2371
Total zinc dust[2], % by weight of pigment	93	D521
Total zinc dust[2], % by weight of solids	77	N/A

[1]*The minimum composition requirements of zinc-rich paint are controversial. It is recognized that zinc-rich primer that contains extenders, even though it has less total zinc dust than specified, may be able to pass all the other requirements. However, these compositional requirements are necessary, as certain non-zinc-containing coatings may be able to pass all other requirements of this specification.*
[2]*Zinc dust should meet the requirements for composition of pigment (Table 10.6B).*

recommendations. The dry film thickness should be 60–90 μm unless otherwise specified. Prior to any exposure testing, all panels should be aged for 14 days at 24–26°C and 45–55% relative humidity.

The paint when applied to a 125–150 μm (dry) film thickness, should show no mud cracking when viewed at a 10X magnification. When applied and hardened, the paint should adhere to the steel substrate when subjected to the "Cross-Cut/Tape Test" (ASTM-D3359, Method B).

There should be no separation of the paint film or delamination of an entire square. Spalling loss of adhesion around the perimeter due to cutting of each square is acceptable.

The adhesion rating should be no less than 4B when evaluated according to the procedure of ASTM-D 3359, Method B. When applied and hardened in accordance with Section 6.4 and scribed as described next, the paint should withstand at least 1,000 h of exposure to salt fog (ASTM-B 117) without any blistering or rusting of the coated portion, with no undercutting from the scribe (slight rusting in the scribe mark is permissible, and resulting staining should be ignored).

Any 6-mm-wide strips along the edges of the panel may be ignored. Testing should be done in triplicate.

The scribe mark should be centrally positioned in the lower half of the panel and should consist of an X comprising the diagonals of a 5 × 5 cm square. To ensure the proper positioning, cleanliness, and depth of the scribe mark, a template and scriber or cutting tool with a cutting edge at least 0.8 mm wide should be used. The operator should bear down hard and go over each arm of the cut twice to ensure a clean scribe of sufficient depth to remove any zinc particles from the scribe and to expose clean steel.

Because of the diversity of the potential service environments, this specification may require the zinc-rich paint be further exposed and qualified by at least one additional test relating to the intended exposure. For example, if the intended environment is a petroleum tanker cargo hold that is ballasted with seawater, appropriate test requirements other than those already specified might be:

"Salt Water Immersion (1,000 Hours)," ASTM-D 1308

"Oil Immersion (1,000 Hours)," ASTM-D 1308

A cycling combination of both standards

Comparative testing of all candidate zinc-rich paints will be more meaningful than individual testing of each primer.

Standard tests that may be useful for further qualifications are available from a number of organizations, including the American Society for Testing and Materials (ASTM), U.S. government federal specifications (TT-P, MIL-P, etc.), U.S. Federal Standard No. 141, and the Canadian Government Specifications Board.

However, it should be emphasized that a well-designed nonstandard test may often provide more meaningful information for a given service condition than one or more standard tests.

The minimum flash point, as determined by ASTM D56, should be 10°C. Specific applications, such as the interior of tanks, holds, and other confined spaces normally require a minimum flash point of 38°C.

This paint can be used as a topcoat only when the finished color is not specific.

10.3.4 STORAGE LIFE AND PACKAGING

When stored in a tightly covered, unopened container, the vehicle of the paint should show no thickening, curdling, gelling, gassing, or hard caking after being stored for 12 months from the date of delivery.

The pigment portion of multicomponent, zinc-rich paints should be packaged separately to be mixed with the liquid portion shortly before use. Each container of liquid should be packaged and labeled in accordance with the requirements of Federal Specification PPP-P-1892; it should include directions showing correct proportions of liquid to pigment along with necessary mixing instructions.

All materials supplied under this specification should be subject to timely inspection by the purchaser or the purchaser's authorized representative. The purchaser should have the right to reject any materials supplied that are found to be defective under this specification. In the case of disputes, the arbitration or settlement procedure established in the procurement documents should be followed.

Samples of any or all ingredients used in the manufacture of this paint should be supplied upon request by the purchaser, along with the supplier's name and identification for the materials. Unless otherwise specified, the methods of sampling and testing should be in accordance with U.S. Federal Standard No. 141 and the applicable ASTM method.

10.3.5 DIRECTIONS FOR USE

The manufacturer should supply complete instructions covering the uses, surface preparation, mixing, thinning, application method, application conditions, pot life, wet and dry film thicknesses, temperature and humidity limitations, and drying times with each container of paint.

10.3.6 SAFETY PRECAUTIONS

Paints are hazardous because of their flammability and potential toxicity, so proper safety precautions should be observed to protect against these recognized hazards. Safe handling practices are required and should include, but not be limited to, the provisions of SSPC-PA Guide 3, "A Guide to Safety in Paint Application," and to the following specifications:

- Keep paint away from heat, sparks, and open flame during storage, mixing, and application. Provide sufficient ventilation to maintain vapor concentration at less than 25% of the lower explosive limit.
- Avoid prolonged or repeated breathing of vapors or spray mist, and prevent contact of the paint with the eyes or skin.

- Clean hands thoroughly after handling paint and before eating or smoking.
- Provide sufficient ventilation to ensure that vapor concentrations do not exceed the published permissible exposure limits. When necessary, supply appropriate personal protective equipment and enforce its use.

This paint may not comply with some air pollution regulations because of its hydrocarbon solvent content. Ingredients in this paint, if so formulated, and which may pose a hazard, include lead- and chromate-containing pigments and hydrocarbon solvents. Applicable regulations governing safe handling practices should apply to the use of this paint.

10.4 ZINC SILICATE (INORGANIC ZINC-RICH) PAINT AS PRIMER, INTERMEDIATE COAT, AND TOPCOAT

This section covers the minimum requirements for the composition, analysis, properties, storage life, packaging, inspection, and labeling of zinc silicate (inorganic zinc-rich) paint.

10.4.1 COMPOSITION

The ingredients and proportions of zinc silicate paint should be as specified in Table 10.10 and the next sections.

This paint consists of 78% by weight of nonvolatile, film-forming solids (pigment and binder). The zinc-rich paint described in this specification consists of zinc dust, an inorganic vehicle, and selected additives as required.

The major pigment component in this paint is zinc dust of either of the types described in Table 10.11. Other pigment components may include curing aids, tinting colors, suspension and pot life control agents, but they should constitute only a minor part of the total pigment portion so as not to detract from the ability of this paint to protect galvanically.

Inorganic self-curing vehicles, include soluble alkali metal silicates, quarternary ammonium silicates, phosphates, and modifications thereof.

10.4.2 ANALYSIS

The paint should conform to the composition (analysis) requirements given in Table 10.12. This table defines the minimum compositional requirements of the zinc-rich paint without specifying the vehicle.

Table 10.10 Composition of Zinc Silicate Paint		
Ingredients	**Composition Wt. %**	**Ingredient Standards**
Pigment	66	ASTM D520
Vehicle	12	N/A
Solvent	22	N/A
Selected additives	As required	N/A

Table 10.11 Requirements for Composition of Zinc Dust

	Type I	Type II
Total zinc, calculated as Zn, minimum%	97.5	98.0
Metallic zinc, minimum%	94.0	94.0
Material other than metallic Zinc, ZnO, and admixed CaO, where applicable maximum%	0.75	N/A
Calcium, calculated as CaO, maximum%	0.7	0.7
Lead, calculated as Pb, maximum%	N/A	0.01
Iron, calculated as Fe, maximum%	N/A	0.02
Cadmium, calculated as Cd, maximum%	N/A	0.01
Chlorine, calculated as Cl, maximum%	N/A	0.01
Sulfur, calculated as SO_2, maximum%	N/A	0.01
Moisture and other volatile matter, maximum%	0.10	0.10
Oily or fatty matter, or both, maximum%	N/A	0.05
Zinc oxide (ZnO), maximum%	6.0	Remainder
Coarse particles, maximum%		
Total residue retained on 0.150-mm standard sieve opening (No. 100)	None	0.1
Total residue retained on 0.075-mm standard sieve opening (No. 200)	N/A	0.8
Total residue retained on 0.075-mm standard sieve opening (No. 325)	4.0	3.0

Table 10.12 Requirements

Characteristics	Minimum Requirements[1]	Standard ASTM
Total solid, % by weight of paint	78	D2369
Pigment, % by weight of total solids	85	D2371
Total zinc dust[2], % by weight of pigment	87	D521
Total zinc dust[2], % by weight of total solids	74	N/A

[1]*The minimum composition requirements of a zinc-rich paint are controversial. It is recognized that zinc-rich paints containing extenders, although they have less total zinc dust than specified, may be able to pass all other requirements. However, these compositional requirements are necessary, as certain non-zinc containing coatings may also be able to pass all other requirements of this specification.*
[2]*Zinc dust should meet the requirements for the composition of pigment.*

10.4.3 **PROPERTIES**

The ready-mixed paint should be capable of being broken up with a paddle to smooth, uniform consistency and should not liver, thicken, curdle, gel, or hard-settle, nor show any other objectionable properties in a mixed, freshly opened container.

The mixed paint should spray easily and show no streaking, running, sagging, or other objectionable features when tested in accordance with U.S. Federal Standard No. 141, Methods 4331 and 4541.

Steel test panels (ASTM-A 36 hot-rolled steel or equivalent) measuring 10 cm × 15 cm × 1.5 mm or greater, should be white metal that has been blast-cleaned (Sa3) with a nominal anchor profile from 40.90 μm and coated with zinc-rich paint. The panels should be blast-cleaned and coated on both sides and all edges. The paint should be spray-applied and hardened in accordance with the manufacturer's recommendation. The dry film thickness should be 60–90 μm unless otherwise designated. Prior to any exposure testing, all panels should be aged for 14 days at 24–26°C and 45–55% relative humidity.

When applied in accordance with Section 6.4 to a 125–150 μm (dry) film thickness, the paint should show no mud cracking when viewed at a 10X magnification. When applied and hardened in accordance with Section 6.4, the paint should adhere to the steel substrate when subjected to the cross-cut/tape test given in ASTM-D 3359, Method B.

There should be no separation of the paint film or delamination of an entire square. Spilling loss of adhesion around the perimeter due to cutting of each square is acceptable.

Adhesion rating should be no less than 4B when evaluated according to the procedure of ASTM-D 3359, Method B.

The coating, when applied and hardened in accordance with Section 6.4 and scribed as described next, should withstand at least 1,000 h exposure to salt fog (ASTM-B-117) without any blistering or rusting of the coated portion, with no undercutting from the scribe. (Slight rusting in the scribe mark will be permissible and resulting staining should be ignored.) Strips that are 6 mm wide along the edges of the panel may be ignored. Testing should be done in triplicate.

The scribe mark should be centrally positioned in the lower half of the panel and should consist of an X comprising the diagonals of a 5 × 5 cm square. To ensure proper positioning, cleanliness, and depth of the scribe mark, a template and scriber or cutting tool having a cutting edge at least 0.8 mm wide should be used. The operator should bear down hard and go over each arm of the cut twice to ensure a clean scribe of sufficient depth to remove any zinc particles from the scribe and to expose clean steel.

The minimum flash point, as determined by the Tag Closed Cup (ASTM D56) should be known.

Because of the diversity of potential service environments, this specification may require the zinc-rich paint be further exposed and qualified by at least one additional test relating to the intended exposure. For example, if the intended application is a petroleum tanker cargo hold that is ballasted with seawater, appropriate test requirements other than those already specified might be:

- "Salt Water Immersion (1,000 h)," ASTM-D 1308
- "Oil Immersion (1,000 h)," ASTM-D 1308
- A cycling combination of both standards

Comparative testing of all candidate zinc-rich paints will be more meaningful than individual testing of each paint.

Standard tests that may be useful for further qualification are available from a number of organizations, including ASTM, U.S. Government Federal Specifications (TT-P, MIL-P, etc.) U.S. Federal Standard No. 141, and Canadian Government Specifications Board.

The pot life of the zinc-rich paint, when mixed and ready for application in accordance with the manufacturer's instructions, should be a minimum of 4 h at 21°C and 50% relative humidity. Although physical properties (viscosity, etc.) may not change, loss of pot life is indicated by lack of adhesion when tested in accordance with relevant standards.

The ready-mixed paint should show no thickening, curdling, gelling, gassing, or hard caking after being stored for 24 months from date of delivery in a tightly covered unopened container.

The manufacturer should supply complete instructions covering the uses, surface preparation, mixing, thinning, application method, pot life, wet and dry film thicknesses, temperature and humidity limitations, and drying times with each container of paint.

10.4.4 SAFETY PRECAUTIONS

Paints are hazardous because of their flammability and potential toxicity, so proper safety precautions should be observed to protect against these recognized hazards. Safe handling practices are required and should include, but not be limited to the provisions of SSPC-PA Guide 3, "A Guide to Safety in Paint Application," and to the following specifications:

- Keep paint away from heat, sparks and open flame during storage, mixing, and application. Provide sufficient ventilation to maintain vapor concentration at less than 25% of the lower explosive limit. Avoid prolonged or repeated breathing of vapors or spray mist, and prevent contact of the paint with the eyes and skin.
- Clean hands thoroughly after handling paint and before eating or smoking.
- Provide sufficient ventilation to ensure that vapor concentrations do not exceed the published permissible exposure limits. When necessary, supply appropriate personal protective equipment and enforce its use.

This paint may not comply with some air pollution regulations because of its hydrocarbon solvent content. Ingredients in this paint, if so formulated, and which may pose a hazard, include lead- and chromate-containing pigments and hydrocarbon solvents. Applicable regulations governing safe handling practices should apply to the use of this paint.

10.5 EPOXY-POLYAMIDE PRIMER

This section covers the minimum requirements for the composition, analysis, properties, storage life and packaging, inspection, and labeling of epoxy polyamide primer.

10.5.1 COMPOSITION

The ingredients and proportions of the reference formulations for epoxy polyamide primer should be as specified in Table 10.13. The primer contains approximately 65% by volume of nonvolatile, film-forming solids (pigment and binder).

Table 10.13 Composition of Reference Formulations for Epoxy Polyamide Primers

Ingredients	kg	l	ASTM Standards
Base components:			
Basic lead silico chromate	195	47.8	D1648
Red iron oxide	7.7	1.7	D3722
Magnesium silicate	38.6	13.6	D605
Mica	12.7	4.5	D607
Organo montmorillonite	3.6	2.1	N/A
95/5 Methanol/Water	1.4	1.5	N/A
Epoxy resin	90.3	76.1	N/A
Leveling agent	4.5	4.5	N/A
MIBK	19.5	24.4	D1153
Xylene	57.2	65.7	D331
2-ethoxy ethanol	30.4	32.8	N/A
Totals (base components)	460.7	274.5	N/A
Curing agent components:			
Polyamide resin	48.6	50.0	D364
Xylene	47.2	54.2	N/A
Totals (curing agent comp.)	95.8	104.2	N/A
Totals (formulation)	556.5	378.7	N/A

The curing agent components of the primer should contain a liquid polyamide resin and volatile solvent. The polyamide resin should be a condensation product of dimerized fatty acids and polyamines.

10.5.2 ANALYSIS

The primer should conform to the composition (analysis) requirements laid out in Table 10.14.

10.5.3 PROPERTIES

The epoxy resin should meet the requirements laid out in Table 10.15. The undiluted polyamide resin should meet the requirements given in Table 10.16.

Table 10.14 Analysis of Epoxy Polyamide Primers

Characteristics	Minimum	Maximum	ASTM Standards
Nonvolatiles, % by weight	65	N/A	D2369

Table 10.15 Epoxy Resin Analysis

| Characteristics | Requirements | | ASTM Method |
	Minimum	Maximum	
Epoxide equivalent	450	550	D1652
Color, gardner (40% in butyl carbitol)	N/A	4	D1544

The primer supplied under this specification should be comparable in performance to the reference formulations of Table 10.13. It needs not be composed of the quantities and types of ingredients given in Table 10.13. However, if substitutions of other ingredients are made, the primer should meet the performance requirements of this specification.

After combining the base and curing agent components, the primer should conform to the requirements of Table 10.17. Each component of this primer based on the specified ingredients should be uniform, stable in storage, and free of grit and coarse particles.

The development of solvent (MEK) resistance is required as an indication of satisfactory curing and subsequent chemical resistance. Apply the primer by spray or brush to a clean test panel so that a dry film thickness of 50–75 μm per coat is obtained. Air-dry the panel for five days at 25°C ±2°C and relative humidity of 40–50%. Following the curing period, saturate a small cotton ball with MEK and place on the test panel under a watch glass for 30 min. After a 10-min recovery period, determine the pencil hardness of the coating. The minimum allowable rating is 7B.

Determine pencil hardness as follows:

1. Using a series of drawing leads (either wood clinched or secured in a mechanical holder), expose approximately 6 mm of lead.
2. Using a rotary motion, square the point of the lead against No. 400 grit paper.
3. Hold the lead at approximately 45° and push forward against the film using a pressure just short of breaking the lead.
4. If penetration is not made, repeat using the next harder lead until penetration occurs. Rate the film by indicating the hardest lead that does not penetrate.

Table 10.16 Polyamide Resin Analysis

| Characteristics | Requirements | | ASTM Method |
	Minimum	Maximum	
Amine value[1]	230	250	N/A
Color, gardner	N/A	8	D1544
Specific gravity	0.96	0.98	D1475
Viscosity, brookfield, at 75°C, poises	31	37	N/A

[1]*Perchloric acid titration*

Table 10.17 Properties of Epoxy Polyamide Primers

Characteristics	Minimum	Maximum	ASTM Standards
Paint consistency			
Viscosity shear rate 200 rpm			
Grams	120	220	
Kreb units	65	85	D562
Density (kg/l)	1.4	1.5	D1475
Fineness of grind (μm)	65		D1210
Drying time (24°C 45% R. H.)			
Tack-free (h)	¾	2	D1640
Dry hard (h)	¾	5	
Dry through (h)	¾	8	
Flash point (°C)	27.2	N/A	D1310

Test panels should be made of carbon steel with a minimum size of 10 cm × 20 cm × 0.31 cm unless otherwise specified. They should be blast-cleaned in accordance with SSPC-SP 10. White blast cleaning, air-drying and test conditions should be done at 25°C ± 2°C and 40–50% relative humidity.

To test for salt spray resistance, prepare at least two test panels and apply one prime coat at 63–75 μm (dry) film thickness. Air-dry for 5 days. Protect the backs and edges. Scribe the panels as per ASTM-D-1654 to base metal and expose for 500 h at 5% salt spray in accordance with ASTM-B117. During the test, the panels should be inclined at an angle of 15° off the vertical. At the end of the test period, the primer should have a minimum rust grade rating of 8. Blistering should be no more than Blister Size No. 4, few Photographic standards SSPC-Vis2, "Standard Method of Evaluating Degree of Rusting on Painted Steel Surfaces", or ASTM D 610 may be used for rusting, and ASTM D 714 may be used for blistering.

For the Elcometer adhesion test, prepare test panels using 6-mm-thick steel plate. Apply coatings at 50–75 μm (dry) film thickness per coat in accordance with the following schedule.

Coating	Substrate	Drying times
Primer	Steel	5 days
Intermediate	Primer	72 h for prime
		72 h for intermediate
		72 h for primer
Topcoat	Primer and intermediate	72 h for intermediate
		5 days for topcoat

After combining the base and curing agent components, determine the pot life of the primer as follows: Throughly will half a litre of sample and let stand at (25°C + 2°C) for 8 h.

Note: When mixing larger volumes, more heat will develop with a resultant shortening of the pot life.

The adhesion of the prime coat to the substrate, intercoat adhesion, or cohesion of any coat of the painting system should be determined by the adhesion tester. Prepare the test panels as described above. Lightly sand the coating surface and aluminum dolly, and apply a quick set adhesive containing alpha cyanoacrylate. Allow the adhesive to cure overnight. Scribe the coating and adhesive around the dolly prior to testing. Make a minimum of three trials and report the average. An average of 280,000 kg/m^2 is considered acceptable.

10.5.4 STORAGE LIFE AND PACKAGING

Paint components should show no thickening, curdling, gelling, gassing, or hard caking after being stored for 12 months from the date of delivery in a tightly covered, unopened container.

Note: When mixing larger volumes, more heat will be generated, with a resultant shortening of the pot life.

10.5.5 DIRECTIONS FOR USE

The manufacturer should supply complete instructions covering the uses, surface preparation, mixing, thinning, application method, application conditions, pot life, wet and dry film thicknesses, temperature and humidity limitations, drying times, and other requirements with each container of paint. The following sections describe guidelines for these instructions.

10.5.6 MIXING AND THINNING

Each component should be stirred to a smooth, homogenous mixture. Then the proper amount of base and curing agent components, as recommended by the manufacturer, should be put together and mixed thoroughly. After standing for 30 min at 25°C ± 2°C, the primer may be thinned up to 12% by volume of the total primer for spraying. The primer should be applied within the manufacturer's pot life limitations.

10.5.7 COATING THICKNESS

The primers are usually applied by spray to a dry film thickness of 50–75 μm per coat.

10.5.8 CURE TIME BETWEEN COATS

Under normal conditions, each coat should be air-dried for at least 4 h, but for no more than 72 h, between coats. In very hot weather with surfaces exposed to direct sunlight, it may be necessary to limit the intercoat drying period to 24 h or less. Long drying times between coats may cause poor intercoat adhesion. These coatings should not be applied at temperatures below 10°C.

10.5.9 SAFETY PRECAUTIONS

Paints are hazardous because of their flammabilty and potential toxicity, so proper safety precautions should be observed to protect against these recognized hazards. Safe handling practices are required and should include, but not be limited to, the provisions of SSPC-PA Guide 3, "A Guide to Safety in Paint Application," and to the following specifications:

- Keep paint away from heat, sparks, and open flame during storage, mixing, and application. Provide sufficient ventilation to maintain vapor concentration at less than 25% of the lower explosive limit.
- Avoid prolonged or repeated breathing of vapors or spray mist, and prevent contact of the paint with the eyes or skin.
- Clean hands thoroughly after handling paint and before eating or smoking.
- Provide sufficient ventilation to ensure that vapor concentrations do not exceed the published permissible exposure limits. When necessary, supply appropriate personal protective equipment and enforce its use.

This paint may not comply with some air pollution regulations because of its hydrocarbon solvent content. Ingredients in this paint that may pose a hazard include lead and chromate-containing pigments, hydrocarbon solvents, and plasticizers. Applicable regulations governing safe handling practices should apply to the use of this paint.

10.6 EPOXY POLYAMIDE PAINT AS INTERMEDIATE PAINT

This section covers the minimum requirements for the composition analysis, properties, storage life and packaging, inspection, and labeling of epoxy polyamide intermediate paint.

10.6.1 COMPOSITION

The ingredients and proportions of the reference formulations for epoxy polyamide intermediate paint should be as specified in Table 10.18.

The intermediate contains approximately 65% by volume of nonvolatile, film-forming solids (pigment and binder). The curing agent component of each coating should contain a liquid polyamide resin and volatile solvent. The polyamide resin should be a condensation product of dimerized fatty acids and polyamines.

10.6.2 ANALYSIS

The paint should conform to the composition (analysis) requirements listed in Table 10.19.

10.6.3 PROPERTIES

The undiluted polyamide resin should meet the requirements listed in Table 10.20. Coatings supplied under this specification should be comparable in performance to the reference formulations of Table 10.18. They need not be composed of the quantities and types of ingredients given in Table 10.18. However, if substitutions of other ingredients are made, the coatings should meet the performance requirements of this specification. Table 10.21 shows epoxy resin analysis.

Table 10.18 Composition of Reference Formulations of Epoxy Polyamide Intermediate Paint

Ingredients	Intermediate		Standard ASTM
	kg	l	
Base component:			
Basic lead silico chromate	134.2	32.9	D1648
Red iron oxide	30.8	6.9	D3722
Magnesium silicate	30.8	10.8	D605
Mica	10.5	3.7	D607
Organo montmorillonite	3.6	2.1	N/A
95/5 Methanol/Water	1.4	1.5	N/A
Epoxy resin	91	76.8	N/A
Leveling agent	4.6	4.5	N/A
MIBK	20	25	D1153
Xylene	68	78	D364
2-ethoxy ethanol	29.5	31.8	D331
Total (base component)	424.4	274	
Curing agent component:			
Polyamide resin	81.6	50.5	N/A
Xylene	49.4	56.9	N/A
Totals (coring agent component)	131	107.4	N/A
Totals (formulation)	555.4	381.4	N/A

Table 10.19 Analysis of Epoxy Polyamide Intermediate Paint

Characteristics	Requirements		Standard ASTM
	Minimum	Maximum	
Nonvolatiles, % by weight	60	N/A	D2369

After combining the base and curing agent components, the primer, intermediate, and topcoat should conform to the requirements of Table 10.22. Each component of these paints based on the specified ingredients should be uniform, stable in storage, and free of grit and coarse particles.

10.6.4 SOLVENT RESISTANCE

The development of solvent (MEK) resistance is required as an indication of satisfactory cure and subsequent chemical resistance. Apply the individual coating (primer, intermediate, topcoat) by spray

Table 10.20 Polyamide Resin Analysis

| Characteristics | Requirements | | ASTM |
	Minimum	Maximum	
Amine value[1]	230	250	N/A
Color, gardner	N/A	8	D1544
Specific gravity	0.96	0.98	D1475
Viscosity, brookfield, at 75°C, poises	31	37	N/A

[1]*Perchloric acid titration*

Table 10.21 Epoxy Resin Analysis

| Characteristics | Requirements | | ASTM Method |
	Minimum	Maximum	
Epoxide equivalent	450	550	D1652
Color, gardner (40% in butyl carbitol)	N/A	4	D1544

Table 10.22 Properties of Epoxy Polyamide Intermediate Paint

| Characteristics | Intermediate | | Standard ASTM |
	Minimum	Maximum	
Paint consistency viscosity			D562
Shear rate 200 (rpm)			
Grams	95	190	
Kreb unit	60	89	
Density (kg/l)	1.3	1.4	D1475
Fineness of grind (μm)	65	N/A	D1210
Drying time (24°C, 45% R.H.)			D1640
Tack free (h)	N/A	2	
Dry head (h)	N/A	5	
Dry through (h)	8		
Flash point (°C)	27.2	N/A	D1310

or brush to a clean test panel so that a dry film thickness of 50–70 μm per coat is obtained. Air-dry the panel for five days at 25°C ± 2°C and relative humidity of 40–50%. Following the curing period, saturate a small cotton ball with MEK and place on the test panel under a watch glass for 30 min.

After a 10-min recovery period, determine the pencil hardness of the coating. The minimum allowable rating is 7B.

Determine pencil hardness as follows:

1. Using a series of drawing leads (either wood clinched or secured in a mechanical holder), expose approximately 6 mm, of lead.
2. With a rotary motion, square the point of the lead against No. 400 grit paper.
3. Hold the lead at approximately 45° and push forward against the film, using a pressure just short of breaking the lead.
4. If penetration is not made, repeat using the next harder lead until penetration occurs. Rate the film by indicating the hardest lead that does not penetrate.

10.6.5 TEST PANELS

Test panels should be made of carbon steel with a minimum size of 10 cm × 20 cm × 0.3 cm unless otherwise specified. They should be blast-cleaned in accordance with SSPC-SP10. Air-drying and test conditions should be at 25°C ± 2°C and 40–50% relative humidity.

10.6.6 ELCOMETER ADHESION TEST

For the Elcometer adhesion test, prepare test panels using 6-mm-thick steel plate. Apply coatings at 50–75 μm (dry) film thickness per coat in accordance with the specifications listed in Table 10.23.

The adhesion of the prime coat to the substrate, intercoat adhesion, or cohesion of any coat of the painting system should be determined by the adhesion tester. Prepare test panels as described above. Lightly sand the coating surface and aluminum dolly and apply a quick set adhesive containing alpha cyanoacrylate. Allow the adhesive to cure overnight.

Scribe the coating and adhesive around the dolly prior to testing. Make a minimum of three trials and report the average. An average of 28 kg/cm² is considered acceptable. Table 10.24 shows properties of combined paint.

Table 10.23 Schedule for Elcometer Adhesion Test

Coating	Substrate	Drying Times
Primer	Steel	5 days
Intermediate	Primer	72 h for primer
		72 h for intermediate
Topcoat	Primer and intermediate	72 h for primer
		72 h for intermediate
		5 days for topcoat

Table 10.24 Properties of Combined Paint

Characteristics	Topcoat		Standard ASTM
	Minimum	**Maximum**	
Paint consistency viscosity			
Shear rate 200 (rpm)			
Grams	95	190	D562
Kreb unit	60	80	D1475
Density (kg/l)	1.2	1.3	D1210
Fineness of grind (μm)	65	N/A	D1475
Drying time (24°C, 45% R.H.)			D1640
Tack free (h)	N/A	2	
Dry head (h)	N/A	5	
Dry through (h)		8	
Flash point (°C)	27	N/A	D1310

10.6.7 POT LIFE

Determine the pot life of the individual coatings as follows: Thoroughly mix a 0.5-kg sample of the finished coating and let stand at 25 ± 2°C for 8 h. At the end of this time, there should be no evidence of gelation. The coatings should be in a free-flowing condition and be brushable without needing to be thinned.

10.6.8 DIRECTIONS FOR USE

The manufacturer should supply complete instructions covering the uses, surface preparation, mixing, thinning, application method, application conditions, pot life, wet and dry film thicknesses, temperature and humidity limitations, and drying time with each container of paint. The following text describe the guidelines for the instructions required.

The paint component should be stirred to a smooth, homogeneous mixture. Then the proper amount of base and curing agent components, as recommended by the manufacturer, should be put together and mixed thoroughly. After standing for 30 min at 25°C ± 2°C, coating the paint may be thinned up to 12% by volume of the total coating for spraying. The paint should be applied within the manufacturer's pot life limitations.

The paint is usually applied by spraying to a dry film thickness of 50–75 μm per coat and a total dry thickness of 175–225 μm.

Under normal conditions, each coat should be air-dried for a minimum of 4 h, but for no more than 72 h between coats. In very hot weather with surfaces exposed to direct sunlight, it may be necessary to limit the intercoat drying period to 24 h or less. Long drying times between coats may cause poor intercoat adhesion. These coatings should not be applied at temperatures below 10°C.

10.6.9 SAFETY PRECAUTIONS

Paints are hazardous because of their flammability and potential toxicity, so proper safety precautions should be observed to protect against these recognized hazards. Safe handling practices are required and should include, but not be limited to, the provisions of SSPC-PA Guide 3, "A Guide to Safety in Paint Application," and to the following specifications:

- Keep paint away from heat, sparks, and open flame during storage, mixing, and application. Provide sufficient ventilation to maintain vapor concentration at less than 25% of the lower explosive limit.
- Avoid prolonged or repeated breathing of vapors or spray mist, and prevent contact of the paint with the eyes or skin.
- Clean hands thoroughly after handling paint and before eating or smoking.
- Provide sufficient ventilation to ensure that vapor concentrations do not exceed the published permissible exposure limits.
- When necessary, supply appropriate personal protective equipment and enforce its use.

This paint may not comply with some air pollution regulations because of its hydrocarbon solvent content. Ingredients in this paint that may pose a hazard include lead- and chromate-containing pigments, hydrocarbon solvents, and plasticizers. Applicable regulations governing safe handling practices should apply to the use of this paint.

During surface preparation that involves the removal of an old film of this paint, care should be taken to minimize dusting to protect workers from the dust and to dispose of coating residues properly.

10.7 EPOXY POLYAMIDE PAINT AS TOPCOAT (FINISH)

This section covers the minimum requirements for the composition, analysis, properties, storage life and packaging, inspection, and labeling of epoxy polyamide paint as topcoat (finish).

10.7.1 COMPOSITION

The ingredients and proportions of the reference formulations of epoxy polyamide paint should be as specified in Table 10.25.

This paint contains approximately 60% by volume of nonvolatile, film-forming solids (pigment and binder). The curing agent component of the paint should contain a liquid polyamide resin and volatile solvent. The polyamide resin should be a condensation product of dimerized fatty acids and polyamines.

The base component of the paint should contain an epoxy resin, together with anticorrosion pigments, color pigments, mineral fillers, gellant, leveling agent, and volatile solvents. The epoxy resin should be a di-epoxide condensation product of bisphenol and epichorohydrin with the terminal epoxide group.

10.7.2 ANALYSIS

The paint should conform to the composition (analysis) requirements given in Table 10.26.

Table 10.25 Composition of Reference Formulations of Epoxy Polyamide Paint

Ingredient	Required kg	Required l	ASTM Standards
Base component:			
Basic lead silico chromate	59	14.5	D1648
Rutile titanium dioxide	52	12.5	D476
Chromium oxide	29	5.5	D263
Magnesium silicate	25	9	D605
Mica	8	3	D607
Lampblack	1	0.5	D209
Organo montmorillonite	4	2	N/A
95/5 Methanol / Water	1.5	1.5	N/A
Epoxy resin	96	81	N/A
Leveling agent	5	5	N/A
MIBK	21	26	D1153
Xylene	50	58	D364
2-ethoxy ethanol	31	34	D331
Totals (base component)	382.5	252.5	
Curing agent component:			
Polyamide resin	52	53	N/A
Xylene	62	74	N/A
Totals (coring agent component)	114	127	N/A
Totals (formulation)	496.5	379.5	N/A

Table 10.26 Analysis of Epoxy Polyamide Paint

Characteristics	Requirements Minimum	Requirements Maximum	ASTM Standard Method
Nonvolatiles, % by weight	60	N/A	D2369

10.7.3 PROPERTIES

The epoxy resin should meet the requirements of Table 10.27. The undiluted polyamide resin should meet the requirements of Table 10.28.

The paint supplied under this specification should be comparable in performance to the reference formulation given in Table 10.25. It need not be composed of the quantities and type of ingredients

Table 10.27 Properties of Epoxy Resin

| Characteristics | Requirements | | ASTM Method |
	Minimum	Maximum	
Epoxide equivalent	450	550	D1652
Color, gardner (40% in butyl carbitol)	N/A	4	D1544

Table 10.28 Properties of Polyamide Resin

| Characteristics | Requirements | | ASTM |
	Minimum	Maximum	
Amine value[1]	230	250	N/A
Color, gardner	N/A	8	D1544
Specific gravity	0.96	0.98	D1475
Viscosity, brookfield; at 75°C, poises	31	37	N/A

[1]*Perchloric acid titration*

in the table. However, if substitutions of other ingredients are made, the paint should meet the performance requirements of this specification when incorporated into a painting system.

Each component of this paint based on the specified ingredients should be uniform, stable in storage, and free of grit and coarse particles. After combining the base and curing agent components, the paint should conform to the requirements of Table 10.27 and the following criteria:

- Color—The color should be as specified in procurement documents with reference to Table 10.29.
- Solvent resistance—The development of solvent (MEK) resistance is required as an indication of satisfactory cure and subsequent chemical resistance. Apply the paint by spray or brush to a clean test panel so that a dry film thickness of 50–75 μm per coat is obtained. Air-dry the panel for 5 days at 25°C ± 2°C and relative humidity of 40–50%. Following the curing period, saturate a small cotton ball with MEK and place on the test panel under a watch glass for 30 min. After a 10-min recovery period, determine the pencil hardness of the coating. The minimum allowable rating is "7B."

Determine pencil hardness as follows:

1. Using a series of drawing leads (either wood clinched or secured in a mechanical holder), expose approximately 6 mm of lead.
2. With a rotary motion, square the point of the lead against No. 400 grit paper.
3. Hold the lead at approximately 45° and push forward against the film using a pressure just short of breaking the lead.
4. If penetration is not made, repeat using the next harder lead until penetration occurs. Rate the film by indicating the hardest lead that does not penetrate.

Table 10.29 Reference Colors

Paint Color	Color No. To BS 381 C
Arctic blue	112
Sea green	217
Brilliant green	221
Canary yellow	309
Light straw	384
Middle brown	411
Signal red	537
Light orange	557
Light gray	631
White	595

Table 10.30 Elcometer Adhesion Test Schedule

Coating	Substrate	Drying Times
Primer	Steel	5 days
Intermediate	Primer	72 h for primer
		72 h for intermediate
Topcoat	Primer and Intermediate	72 h for primer
		72 h for intermediate
		5 days for topcoat

Test panels should be made of carbon steel with a minimum size of 10.2 cm × 20.3 cm × 0.3 cm unless otherwise specified. They should be blast-cleaned to SIS 05-5900 a2½. Air-drying and test conditions should be at 25°C ± 2°C and 40–50% relative humidity.

For the Elcometer adhesion test, prepare test panels using 6-mm-thick steel plate. Apply the paint at 50–75 μm (dry) film thickness per coat in accordance with Table 10.30.

The adhesion of the prime coat to the substrate, intercoat adhesion, or cohesion of any coat of the painting system should be determined by the adhesion tester.

Prepare the test panels as described above. Lightly sand the coating surface and aluminum dolly, and apply a quick set adhesive containing alpha cyanoacrylate. Allow the adhesive to cure overnight. Scribe the coating and adhesive around the dolly prior to testing. Make a minimum of three trials and report the average. An average of 28 kg/cm² is considered acceptable.

10.7.4 POT LIFE

Determine the pot life of the paint as follows. Thoroughly mix an 0.5-kg sample of the finished coating and let stand at 25°C ± 2°C for 8 h. At the end of this time, there should be no evidence of gelation. The paint should be in a free-flowing condition and brushable without thinning.

10.7.5 **STORAGE LIFE AND PACKAGING**

The paint (both base component and curing agent) should show no thickening, curding, gelling, or hard caking when tested as specified in U.S. Federal Standard No. 141, method 3011, after being stored for 12 months from date of delivery in a full, tightly covered container, at a temperature of 10–43°C. The packaging should meet the relevant requirement of ASTM D3951-88.

10.7.6 **DIRECTIONS FOR USE**

The manufacturer should supply complete instructions covering the uses, surface preparation, mixing, thinning, application method, application conditions, pot life, wet and dry film thicknesses, temperature and humidity limitations, and drying time with each container of paint. The following text describes the guidelines for the instructions required.

Each coating component should be stirred to a smooth, homogenous mixture. Then the proper amount of base and curing agent components, as recommended by the manufacturer, should be put together and mixed thoroughly. After standing for 30 min at 25°C ± 2°C, the paint may be thinned up to 12% by volume of the total paint for spraying. The paint should be applied within the manufacturer's pot life limitations.

The paint is usually applied by spray to a dry film thickness of 50–75 μm per coat.

Under normal conditions, each coat should be air-dried for a minimum of 4 h, but for no more than 72 h between coats. In very hot weather with surfaces exposed to direct sunlight, it may be necessary to limit the intercoat drying period to 24 h or less. Long drying times between coats may cause poor intercoat adhesion. These coatings should not be applied at temperatures below 10°C.

10.7.7 **SAFETY PRECAUTIONS**

Paints are hazardous because of their flammability and potential toxicity, so proper safety precautions should be observed to protect against these recognized hazards. Safe handling practices are required and should include, but not be limited to, the provisions of SSPC-PA Guide 3, "A Guide to Safety in Paint Application," and to the following specifications:

- Keep paint away from heat, sparks, and open flame during storage, mixing, and application. Provide sufficient ventilation to maintain vapor concentration at less than 25% of the lower explosive limit.
- Avoid prolonged or repeated breathing of vapors or spray mist, and prevent contact of the paint with the eyes or skin.
- Clean hands thoroughly after handling paint and before eating or smoking.
- Provide sufficient ventilation to ensure that vapor concentrations do not exceed the published permissible exposure limits. When necessary, supply appropriate personal protective equipment and enforce its use.

This paint may not comply with some air pollution regulations because of its hydrocarbon solvent content. Ingredients in this paint that may pose a hazard include lead- and chromate-containing pigments, hydrocarbon solvents, and plasticizers. Applicable regulations governing safe handling practices should apply to the use of this paint.

During surface preparation that involves the removal of an old film of this paint, care should be taken to minimize dusting, to protect workers from the dust, and to dispose of coating residues properly.

Table 10.31 Composition of Coal-Tar Mastic Paint

Ingredients	Typical Composition		Standard ASTM
	Minimum	**Maximum**	
Distillation (235°C Wt%)	15	30	D20 (235°C)
Ash content of mastic (Wt%)	15	30	D128 Routine method
Tar acids of mastic (ml/100 g)	N/A	0.6	D453 Distillation to 300°C

10.8 COAL-TAR MASTIC PAINT (COLD-APPLIED)

This section covers the minimum requirements for the composition, properties, storage life and packaging, inspection, and labeling of coal-tar mastic paint (cold-applied).

10.8.1 COMPOSITION

The ingredients and proportions of coal-tar mastic paint should be as specified in Table 10.31 and the following sections. The paint should contain 15–30% by weight of distilled materials at 235°C when tested by ASTM D20.

Coal-tar mastic paint should contain only a homogeneous mixture compound of a coke oven coal-tar mastic pitch, solvents, and inert non-water-absorbent mineral filler. The compound should not contain asphalt or asbestos.

10.8.2 ANALYSIS

The mastic should conform to the composition (analysis) requirements listed in Table 10.32.

10.8.3 PROPERTIES

Test the cold-applied mastic for firm adherence to bare steel, primer (if used), and itself. At 25°C ± 2°C, the mastic should permit easy application by brush or spray, in two successive coats, to a minimum dry film thickness of 760 μm ± 50 μm. At 7°C ± 2°C, the mastic should permit easy brush application.

Table 10.32 Analysis of Coal-Tar Mastic Paint

Characteristics	Minimum Wt%	Maximum Wt%	ASTM Method
Volatile content (235°C)	15	30	D20
Nonvolatile materials (calculated by difference)	85	70	N/A

At 23°C ±1°C and 50% ± 4% relative humidity, apply the mastic to a clean, smooth 305-by-305-by-3-mm thick steel plate, to a uniform wet film thickness of 760 μm ± 50 μm. Immediately after application, suspend the panel in a vertical position at application conditions for 24 h. Then examine for evidence of sag or flow while wet. Prepare a second plate as described, except to a uniform wet film thickness of 380 μm ± 25 μm. Immediately after application, suspend the panel in a vertical position at 71°C ± 2°C for 1 h. Then examine again for evidence of sag or flow while wet.

Expose the coated panel, tested at 23°C to 60 ± 2°C for 16 h. Allow the panel to cool at room temperature for 1 h. Then expose to −23°C ± 2°C for 4 h.

Apply the mastic to two wire-brushed, solvent-cleaned, mild steel plates, each 152-by-152-by-3-mm thick, to a uniform dry film thickness of 760 μm ± 50 μm. Dry for 72 h at room temperature, then test each plate separately, while holding it firmly, coated side up, on a solid horizontal base. Drop a 900-g steel ball from a height of 244 cm so that the impact will be at the center of the plate. Examine the coating for conformance to Table 10.33.

Apply the mastic to a clean glass panel at a uniform dry film thickness of 760 μm ± 50 μm. Dry the panel at 23°C ± 1°C, for 24 h, then suspend vertically in 5% sodium hydroxide, maintained at

Table 10.33 Properties of Coal-Tar Mastic Paint

| Characteristics | Requirements | | ASTM Method | U.S. Federal Standard No. 141 |
	Minimum	Maximum		
Flash point (°C)	35	N/A	D92	N/A
Penetration of distillation residue, mm	5	25	D5, 235 °C Residue sample	N/A
Softening point of distillation residue (°C)	96	115.5	D36, 235 °C Residue sample	N/A
Workability	Satisfactory working and spreading			N/A
Sag	No sag or flow while wet			N/A
Drying set - to - touch dry-to touch	6 h 24 h			4061, Doctor blade application at 11 (m²/l)
Adhesion and protection	Shall not loosen, chalk, crack, peel, run, sag, or otherwise lose protection value			N/A
Resistance to impact	No visible chipping cracking, or detachment from plate, and firm adhesion outside radius of 6 mm from center of impact			N/A
Resistance to alkali	No evidence of disintegration			N/A

23°C ± 1°C, so that half of the coating is immersed. After 30 h, lightly rub the film with a well-rounded glass rod and examine for evidence of disintegration.

The color should be black or dark red.

10.8.4 STORAGE LIFE AND PACKAGING

The mastic should meet all the requirements specified herein after storage of 12 months (minimum) from date of delivery in a full, tightly covered container. The packaging should meet the relevant requirements of ASTM D 3951 unless otherwise specified by the purchaser.

10.8.5 INTENDED USE

The mastic is intended for use on steel structures as necessary to substitute for a hot-applied coal-tar enamel coating. The mastic is used also as a protective coating for dissimilar metals that are in contact, as an electrical insulating coating, and on fiberglass lagging for waterproofing.

It is appropriate for contact with alkaline soils and may be used on underwater marine structures. Generally two coats are used, to a dried film thickness of 510–1,020 μm. Material that is to be exposed to sunlight or weather should be top-coated with MIL-C-15203 coal-tar emulsion.

10.8.6 DIRECTIONS FOR USE

This mastic paint is used to protect steel from corrosion in severe surroundings. It will perform best if applied over blast-cleaned or pickled steel; however, in special cases, it may be used over well-cleaned steel if all rust scale, loose rust, loose mill scale, and loose or nonadherent paint are removed. Oil and grease should be removed. For severe exposure, apply over a thoroughly dry rust-inhibitive primer.

Mix paint thoroughly before use. Under normal conditions, no thinning should be necessary. If thinning should be required, do so using only mineral spirits.

Apply using high-pressure spray equipment. If the paint must be applied by brush, do so with a daubing action.

Apply to the specified film thickness or, if none is specified, to at least 1,600 μm (dry) or approximately 3,200 μm (wet). The surface to be painted should be dry, the surface temperature should be at least 3°C above the dew point, and the temperature of the air should be over 4°C. Do not paint outdoors in rainy weather or if freezing temperatures are expected before the paint dries. Under normal conditions, this mastic will dry for recoating in 240 h, but it will remain soft for long periods.

10.8.7 SAFETY PRECAUTIONS

Coal-tar mastic paints are hazardous because of their flammability and potential toxicity, Proper safety precautions should be observed to protect against these recognized hazards. Safe handling practices are required and should include, but not be limited to, the provision of SSPC-PA Guide 3, "A Guide to Safety in Paint Application," and to the following specifications:

- Keep mastic away from heat, sparks, and open flame during storage, mixing, and application. Provide sufficient ventilation to maintain vapor concentration at less than 25% of the lower explosive limit.

- Avoid prolonged or repeated breathing of vapors or spray mist, and prevent contact of the paint with the eyes or skin.
- Clean hands thoroughly after handling mastic and before eating or smoking.

The paints specified herein may not comply with some air pollution regulations because of their hydrocarbon solvent content. Ingredients in this paint that may pose a hazard include hydrocarbon solvent and coal tars. This paint may contain low concentrations (less than 1% by weight) of materials that are suspected carcinogens. Applicable regulations governing safe handling practices should apply to the use of this paint.

10.9 TWO-PACK ALIPHATIC POLYURETHANE PAINT AS TOPCOAT (FINISH)

This section covers the minimum requirements for the composition analysis, properties, storage life, packaging, inspection, and labeling of two-pack aliphatic polyurethane paint.

10.9.1 COMPOSITION

Aliphatic polyurethane paint should consist of two components. Component A should be pigmented and contain polyester resins. Component B should contain the clear prepolymer aliphatic isocyanate resin and act as the hardener or curing agent for Component A. It should contain no toluene diisocyanate and no aromatic substituted isocyanate. Components A and B should be packaged separately and furnished in kit form. The ingredients used in the manufacture of these products should conform to applicable U.S. federal and military specifications.

The total nonvolatile solids for the admixed components of gloss colors should be a minimum of 52% except black, insignia red, insignia blue, and clear, which should be a minimum of 35%. The total nonvolatile solid content for the admixed components of camouflage and semigloss colors should be 45% except black, insignia red, insignia blue, and clear, which should be a minimum of 35%.

10.9.2 ANALYSIS

The paint should conform to the composition (analysis) requirements given in Table 10.34.

10.9.3 PROPERTIES

Components A and B should meet the requirements of Table 10.35, and mixed paint should meet the following requirements:

- Odor—The odor of the paint material should not be obnoxious when tested in accordance with U.S. Federal Standard No. 141, Test Method 4401.
- Color—The color should be as specified in the procurement documents, with reference to Table 10.36 if required.

Spray several coats of the mixed coating, comprising the two components mixed in the ratio of one part by volume of component A, to one part by volume of component B, and thinned, if required, to meet the viscosity. The thinner should be smooth, uniform, and free of bubbles, pinholes, holidays,

Table 10.34 Analysis of Two-Pack Aliphatic Polyurethane Paint

| Characteristics | Requirements | | U.S. Federal Standard No. 141 | ASTM Standard |
	Minimum	Maximum		
Component A: total solids (vol%) (resin and pigment) Water content (vol%)	70-35-45* N/A	0.5 (Gloss)	4021.1-4042	
Hydroxyl number (based on 100% Resin solids)	N/A	0.75 (Semigloss and camouflage)		D1364
Component B: Total solids (vol%) available isocyanate-content (vol%) total free isocyanate-	35.0 7.0	233 N/A N/A	7381 4042	D2572
Toluene diisocyanate (spot test) admixed paint: volatiles	48-55-65	0.S4 Negative	(Mil-C-83286 B (USAF) N/A 4041.1	D2369
Nonvolatiles (calculated by difference)	52-45-35			

Note:
The solvents contained in the packaged paint should be urethane grade and contain a minimum of alcohol in accordance with the best commercial practices.
*70% for gloss colors; 35% for gloss and semigloss black, insignia red, insignia blue, and clear; 45% for camouflage and semigloss colors.

Table 10.35 Requirement for Ingredients and Proportions of Two-Pack Aliphatic Polyurethane Paint

| Characteristics | Requirements | | U.S. Federal Standard No. 141 | ASTM Standard |
	Minimum	Maximum		
Component A:				
Viscosity (stormer)			4281	D562
Krebs units (for gloss colors)	85	95		
Krebs units (for camouflage colors)	95	110		
Density (kg/l)*	Report		4184	D1475
Fineness of grind (Hegman units)	3	5	4411.1	D1210
Component B: Density (kg/l)* Color (gardener)	Report 3		4184 N/A	D1475 D1544

*The density of the quality conformance sample should not deviate by more than 10% from that of the qualification sample.

Table 10.36 Reference Colors

Paint Color	Color No. To BS 381 C
Arctic blue	112
Sea green	217
Brilliant green	221
Cannary yellow	309
Light straw	384
Middle brown	411
Single red	537
Light orange	567
Light gray	631

and other film irregularities. Spray-applied films, dried under the standard conditions (25°C ±2°C and relative humidity of 50% ± 5%) should provide a hard surface that is free of grit, seeds, streaks, orange peel, blisters, or other surface defects when tested in accordance with U.S. Federal Standard No. 141, Method 4541.

The drying time of the paint should not exceed 2 h for the set-to-touch condition, 1 h for the dry-to-recoat condition, or 6 h for the dry-hard condition when tested in accordance with U.S. Federal Standard No. 141, Method 4061.1. The viscosity of the freshly mixed material should be 17–23 s in a No. 2 Zhan cup. After standing for 6 h, the viscosity should not increase more than 20%.

When tested in accordance with U.S. Federal Standard No. 141 after standing undisturbed for 6 h, the mixed and reduced paint material should be free of curdling, precipitation, and separation that cannot be easily redispersed by shaking on a mechanical paint mixer. The free diisocyanate in the mixed paint should not exceed 1% when tested in accordance with Appendix I of Mil- C-83286 B (USAF).

The paint material after mixing and reducing for spray application should have a minimum pot life of 6 h. After standing for 6 h in a full, closed container at standard conditions (25°C ±2°C and relative humidity of 25% ± 5%).

The fineness of grind of the mixed paint should not be less than 7 for the gloss paints and not less than 5 for the camouflage paint. The tests should be made 1 h after mixing in accordance with U.S. Federal Standard No. 141, Method 4411.1.

The applied film of the paint should meet the requirements of Mil-C-83286 B (USAF).

10.9.4 STORAGE LIFE AND PACKAGING

The paint (both component A and B) should show no thickening, curdling, gelling or hard caking when tested as specified in U.S. Federal Standard No. 141, Method 3011 after being stored for 12 months from date of delivery in a full, tightly covered container. The container should not show evidence of excessive pressure or be deformed by gassing.

The polyurethane paint should be packaged as a unit consisting of a pigmented compound marked "Component A" and a unpigmented (or clear) hardener marked "Component B." The quantity of each component in the kit should be in the ratio of 1 to 1 by volume, respectively. Component B should be packed in full containers that should be thoroughly dry and filled in a dry atmosphere. The packaging should also meet the relevant requirements of ASTM D3951.

10.9.5 DIRECTIONS FOR USE

The manufacturer should supply complete instructions covering the uses, surface preparation, mixing, thinning, application method, application conditions, pot life, wet and dry film thicknesses, temperature and humidity limitations, and drying time with each container of paint. The next sections give guidelines for the instructions required.

Each coating component should be stirred to a smooth, homogenous mixture. Then the proper amount of components A and B, as recommended by the manufacturer, should be put together and mixed thoroughly. The paint film thickness per coat should be specified by the manufacturer.

10.9.6 SAFETY PRECAUTIONS

Paints are hazardous because of their flammability and potential toxicity, so proper safety precautions should be observed to protect against these recognized hazards. Safe handling practices are required and should include, but not be limited to, the provisions of SSPCPA Guide 3, "A Guide to Safety in Paint Application."

The paints specified herein may not comply with some air pollution regulations because of their hydrocarbon solvent content. Ingredients in urethane paints that may pose a hazard include isocyanates and solvents. Applicable regulations governing safe handling practices should apply to the use of urethane coatings.

The main items to keep in mind when working with urethane paint systems are as follows:

- Become informed and aware of the hazards and appropriate control procedures. This can be done by reading the label and the material safety data sheet (if available) or by contacting the supplier of the paint system for other literature and information.
- Follow the recommendations prescribed for use during handling and application as set forth by the supplier.
- Follow all applicable local regulations.

Toxicological research and practical experience have shown that diisocyanates can cause irritation of the skin, respiratory tract, eyes, nose, and throat. In addition, sensitization resulting in allergic dermatitis or asthmatic symptoms can occur following overexposure to diisocyanates. Toxicological research has shown that polyisocyanates have a reduced potential to cause irritation and sensitization relative to their monomeric precursors.

Symptoms of irritation usually include watering of the eyes and a burning sensation in the nose and throat. The amount of irritation is dependent upon the dose, tissue exposed, and individual susceptibility, but it is generally independent of the individual's exposure history.

These acute symptoms are generally reversible soon after the individual is removed from the contaminated area or the material is removed from the skin.

Sensitization is a systemic response that is not limited to the area of contact. It usually occurs as a result of numerous overexposures or one exposure to very high concentrations. Both respiratory and dermal sensitization can occur depending upon the toxicological properties of the diisocyanate, route of exposure, and individual susceptibilities. Exposures subsequent to the exposures that actually resulted in sensitization may cause a very strong allergic reaction.

In the case of respiratory sensitization, the reaction is similar to asthma (i.e., coughing, wheezing, tightness in the chest and shortness of breath). The skin sensitization reaction is allergic dermatitis, which may include symptoms such as rash, itching, hives, and swelling of the arms and legs.

If an individual experiences an irritation response while handling an isocyanate, it should be determined whether the isocyanate was the cause of the irritation. If it was the cause, that indicates that the operation, as performed, allows an overexposure to isocyanates that can result in later sensitization of that worker or others.

A careful evaluation of the controls, protective equipment, and work practices should be done with the goal of reducing the exposure. If irritation persists in spite of proper ventilation and protective measures, the individual must be removed from the areas where isocyanates are being processed or used.

If an individual is sensitized to isocyanates, complete removal from areas of potential exposure is mandated. This is true regardless of the form that the isocyanate is present in. Also, exposure to an isocyanate, other than the one suspected of causing the sensitization, must be avoided.

Solvents are also present in paints. Prolonged or repeated exposure or overexposure to these solvents by either inhalation or direct skin contact may also cause injurious health effects. The effects are dependent upon the solvent, the extent of exposure, and the route of exposure.

Since isocyanates have the potential to irritate and sensitize those working with or around them, it is important that proper steps be taken to protect people who could be exposed to vapor, mist, or overspray. This includes those actually handling the isocyanate as well as those in the immediate vicinity.

Even during brush, roller, and curtain paint applications, it is possible to be exposed to airborne concentrations of solvents and isocyanate vapors. During spray application, not only will vapors be present, but also spray mist or aerosolized droplets.

These droplets contain pigments, solvents, resins, additives, and polymeric materials, as well as isocyanate and unreacted polyisocyanate. Each of these will have its own physiological effect on the exposed organism.

Ideally, control of health hazards posed by vapors and spray mist is performed by engineering methods. Effective engineering should be used whenever possible to eliminate or reduce worker exposure. There are several engineering methods available to reduce exposure to isocyanate vapors and mist.

The most common precaution taken is a properly designed and ventilated enclosure. General ventilation, local ventilation, or isolation may prove adequate under certain conditions. Use of alternative application equipment (e.g., airless or electrostatic spray equipment) may help reduce mist generation during spray painting.

Brush and roller application of the paint may be feasible in some cases. To reduce environmental contamination, exhausted air may need to be cleaned by means of filters or scrubbers. The final design and combination of these control measures is dependent upon the specific application.

Whenever a paint system is spray-applied, it is essential that the applicator be protected from inhalation of both vapors and spray mist by the best possible respiratory protection. Under certain conditions, a fresh air-supplied respirator will be required. In other cases, an air purifier with a particulate filter may be employed. Applicators are urged to consult with their suppliers concerning the type of respiratory protection that is appropriate in a given application.

The appropriate selection and use of a respirator is an important part of protection from work-related chemical hazards. In addition, remember the following:

- Users of respirators must be properly trained.
- Always be sure that the respirator is in good working order.
- Know the equipment's limitations.
- Be sure that the respirator fits properly.
- Clean the equipment after each use.

Respirator manufacturers may be helpful in developing a good respirator program. In addition to respirators, other forms of recommended personal protective equipment include safety glasses or goggles.

Should spray mist get into the eye, rinse immediately and sufficiently with lukewarm water and consult an eye doctor should irritation persist.

Regarding skin contact, it is suggested that as much of the exposed skin area as possible be covered with clothing or skin creams. Cured coating cannot be removed easily.

Application of a protective skin cream to the hands prior to start of work will facilitate the removal of paint splashes or overspray with soap and water. Skin areas covered only by protective creams should be kept to an absolute minimum. Aggressive solvents are unsuitable for cleaning the skin, as they wash away oils and can cause secondary reactions.

10.10 TWO-PACK ALIPHATIC POLYURETHANE PAINT AS TOPCOAT (CAMOUFLAGE COLORS)

This section covers the minimum requirements for the composition, properties, storage life, and packaging of a low-reflective, two-pack aliphatic polyurethane paint for use as topcoat.

10.10.1 COMPOSITION

The paint material should be furnished in two components. Component A should consist of orthophthalic-trimethylopropane polyesters combined with prime and extended pigments and volatile solvents. Component B should consist of an aliphatic isocyanate prepolymer combined with volatile solvents. When mixed according to the manufacturer's recommendations, the product should meet the requirements of this specification.

The hiding pigments should be yellow iron oxide, red iron oxide, and phthalocyanine blue. Organic blacks and antimony sulfide should not be used. Hiding pigments should be chemically pure and free of extenders. Small amounts of other shading pigments and titanium dioxide may be used when necessary to match the color, provided that these pigments have good color permanence. Extender pigments should be a combination of crystalline and diatomaceous silica, be free of talc, and not exceed the amount specified in relevant standards when tested by extracting the pigment as in U.S. Federal Standard No. 141, Method 4021, using only acetone as the extraction solvent, until the supernant is colorless.

Nonvolatile vehicles should preferably be composed as follows:

- Component A—should contain the following:
- Dicarboxylic acids—When tested as per the requirements of ASTM D 2455, the dicarboxylic acids should be orthophthalic, with only trace amounts of other acids.

- Polyols—When tested as per ASTM D 2998, the polyols should be trimethylopropane, with only trace amounts of other polyols.
- Component B—When tested, the nonvolatile vehicle in Component B should be an aliphatic isocyanate prepolymer. It should contain no toluene diisocyanate and no aromatic or aromatic substituted isocyanates.

The volatile content of the mixed coating should consist of a nonphotochemically reactive solvent blend and should conform to the following requirements by volume when tested by U.S. Federal Standard 141, Methods 7355 and 7356:

- Aromatic compounds with eight or more carbon atoms except ethyl benzene: 8% maximum
- Ethyl benzene and toluene: 20% maximum
- Solvents with an olefinic or cyclo-olefinic type of unsaturation: negative
- Ethylene glycol monoethyl ether acetate: 25% minimum

10.10.2 QUANTITATIVE REQUIREMENTS

Component A (polyester) should conform to the quantitative requirements given in Table 10.37.

Component B (isocyanate) should conform to the quantitative requirements of Table 10.38 when tested as described in the next sections.

Table 10.37 Component A (Polyester) Requirements

Characteristics	Minimum	Maximum	U.S. Federal Standard No. 141	ASTM Standard
Nonvolatile, % by weight of component A	61	N/A	4041	
Pigment, % by weight of component A	42	46	4022	
Extender pigment, % by weight of pigment	65	70	5271	
Pigment volume	47	N/A	4312	
Nonvolatile vehicle, % by weight of Component A	18	N/A	4052	
Phthalic anhydride, % by weight of nonvolatile vehicle	44	N/A		D1307
Acid number, based on nonvolatile vehicle	N/A	4	5073	
Hydroxyl number, based on nonvolatile vehicle	165	195	7381	
Alcohols, hydroxyl number units	N/A	10	N/A	
Water, % by weight of Component A	N/A	0.5	4082	
Coarse particles and skins (retained on 0.044-mm No. sieve 325 mesh sieve), % by weight of pigment	N/A	1.0	4092	
Viscosity, krebs stormer shearing rate - 200 rpm g	75	125	4281	
Fineness of grind (Hegman unit)	2	N/A	4411	D1210

Table 10.38 Component B (Isocyanate) Requirements

Characteristics	Minimum	Maximum	U.S. Federal Standard No. 141	ASTM Standard
Nonvolatile, % by weight of component B	73	77	4041	N/A
Viscosity, gardner tubes	F	M	4271	D803
Isocyanate content, % by weight of nonvolatile	18	21	N/A	D2572

To test the reagent, dissolve 32 g of anhydrous dibutylamine in anhydrous chlorobenzene and dilute to 250 ml volume with chlorobenzene. Store in a brown bottle. If high-purity reagent materials are not available, the dibutylamine should be freshly distilled and the chlorobenzene dried over calcium chloride and redistilled.

Weigh a sample of Component B with a mass of no more than 2 g accurately from a dropping bottle into a 250-ml Erlenmeyer flask. Add from a pipet 10 ml of the dibutylamine reagent solution and swirl until clear, but not for less than 2 min. Pipet another 10 ml of a reagent into a separate flask to be titrated as a blank. Add 2–3 drops of a 1% alcoholic solution of bromophenol blue indicator. Add 100 ml of absolute methanol slowly with swirling of the sample. Titrate the excess dibutylamine with aqueous 1 N hydrochloric acid using a 10 ml buret with 0.05-ml divisions, until the color changes from blue to yellow.

Perform the calculation using the following equation:

$$\%NCO = \frac{(ml.\,HCl\,for\,blank - ml.\,HCl\,for\,sample) \times normality \times 4.2}{Weight\,of\,sample\,in\,g}$$

Check for compliance with Table 10.38.

Vacuum-dry a film of the vehicle on a salt plate and scan the infrared spectrum from 2–15 μm. The spectrum will closely resemble the spectrum; that is, it will show the presence of an aliphatic isocyanate and the absence of aromatic bands.

When mixed, according to the manufacturer recommendation, paint should conform to the quantitative requirements listed in Table 10.39.

10.10.3 **REQUIREMENTS**

The color of the paint should match color No. 34087 of U.S. Fed Standard No. 595.

Thoroughly mix 4 parts by volume of Component A with 1 part by volume of Component B. A smooth, homogenous mixture should result, and the paint should be free of grit, seed, skins, or lumps.

Place 3 oz. of the material in a 4-oz. glass jar and do not agitate or disturb for 8 h. At the end of this period, the paint should show no signs of gelatin.

Reduce 4 parts by volume of the mixed paint with 1 part by volume of thinner conforming to Mil-T-81772. Spray the paint on a steel panel to a dry film thickness of between 23 μm and 28 μm and observe for spraying properties in accordance with U.S. Federal Standard No. 141, Method 4331. The paint should spray satisfactorily in all respects and should show no running, sagging, or streaking. The dried film should show no dusting, mottling, or color separation and should present a smooth, lusterless finish, free of seeds.

Table 10.39 Mixed Coating Requirements

Characteristics	Minimum	Maximum	U.S. Federal Standard No. 141
Hiding power (contrast ratio)	0.97	N/A	4122
Drying time			
Dry to touch (min)	5	30	4061
Dry hard (h)	N/A	4	4061
Secular gloss			
60°	N/A	0	6101
65°	N/A	0	6103
Infrared reflectance (%)	25	35	6242

Determine the flexibility in accordance with U.S. Federal Standard No. 141, Method 6221. Draw down a film of the mixed paint with a 50-μm (100-μm gap clearance) film applicator on a No. 31 gauge (272 μm), cold-rolled, luster finish steel panel prepared as in procedure B, "Phosphoric Acid Etch," of U.S. Federal Standard No. 141, Method 2011. Age the film in a horizontal position for 168 h, and then bake for 96 h at 105°C ± 2°C. Condition the panel for a half-hour under referee conditions, and then bend it over a 6.3-mm mandrel. The film of the paint should withstand bending without cracking or flaking.

To test for water resistance, prepare a test panel of the paint and air-dry for 168 h. Coat all exposed unpainted metal surfaces with wax or suitable protective paint and immerse it in water at 23°C ± 1°C for 168 h as in U.S. Federal Standard No. 141, Method 6011. At the end of the test period, remove and examine the sample. The film of the paint should show no blistering or wrinkling and no more than a slight whitening or softening immediately upon removal from the water. After 2 h of air-drying, the portion of the panel that was immersed should be almost indistinguishable with regard to adhesion, hardness, color, and gloss from the portion that was not immersed.

Perform a knife test as in U.S. Federal Standard No. 141, Method 6304, using the flat portion of the panels for the flexibility test. The knife cut should have beveled edges. The film of the paint should be tough and hard and adhere tightly to the substrate. It should be difficult to furrow off with the knife and should not flake, chip, or powder.

Prepare a film of paint as in 5.4, air-dry the specimen for 168 h, and then immerse it for 168 h (at 23°C ± 1°C) in a hydrocarbon fluid conforming to TT-S-735 type III in U.S. Federal Standard No. 141, Method 6011. At the end of the test period, remove and examine the sample. The film of paint should show no blistering or wrinkling and no more than a slight whitening or softening upon removal from the fluid. The test item should be indistinguishable with regard to adhesion, hardness, color, and gloss from a panel prepared at the same time but not immersed.

Draw down the mixed coating using a 50-μm (100-μm gap clearance) film applicator and determine the drying under referee 0.1 conditions in accordance with U.S. Federal Standard No. 141, Method 4061.

10.10.4 STORAGE LIFE AND PACKAGING

The paint (components A and B) should show no thickening, curdling, gelling, or hard caking when tested as specified in U.S. Federal Standard No. 141, Methods 3011 and 4261, after being stored for

12 months from the date of delivery in a full, tightly covered container. The container should not show evidence of excessive pressure or be deformed by gassing.

The polyurethane paint should be packaged as a unit consisting of a pigmented compound marked "Component A" and an unpigmented (or clear) hardener marked "Component B." The quantity of each component in the kit should be in the proportions as required by relevant standards. Component B should be packed in a full container. The containers should be thoroughly dry and filled in a dry atmosphere. The packaging should also meet the relevant requirements of ASTM D3951.

10.10.5 DIRECTIONS FOR USE

The manufacturer should supply complete instructions covering the uses, surface preparation, mixing, thinning, application method, application conditions, pot life, wet and dry film thicknesses, temperature and humidity limitations, and drying time with each container of paint. The following text gives guidelines for the required instructions.

Each coating component should be stirred to a smooth, homogenous mixture. Then the proper amount of Components A and B, as recommended by the manufacturer, should be put together and mixed thoroughly. After standing for ... min at $25°C \pm 2°C$, the paint may be thinned by volume of the total paint for spraying. The paint should be applied within the manufacturer's pot life limitations. The paint is usually applied by spray to a dry film thickness of ... microns per coat.

Under normal conditions, each coat should be air-dried for a minimum of ... h, but no more than ... h between coats. In very hot weather, with surfaces exposed to direct sunlight, if may be necessary to limit the intercoat drying period to ... h or less. Long drying times between coats may cause poor intercoat adhesion.

10.10.6 SAFETY PRECAUTIONS

Paints are hazardous because of their flammability and potential toxicity, so proper safety precautions should be observed to protect against these recognized hazards. Safe handling practices are required and should include, but not be limited to, the provisions of SSPC-PA Guide 3, "A Guide to Safety in Paint Application."

The paints specified herein may not comply with some air pollution regulations because of their hydrocarbon solvent content. Ingredients in urethane paints that may pose a hazard include isocyanates and solvents. Applicable regulations governing safe handling practices should apply to the use of urethane coatings. The main items to consider and keep in mind when working with urethane paint systems are as follows:

- Become informed and aware of the hazards and appropriate control procedures. This can be done by reading the label and the material safety data sheet (if available), or by contacting the supplier of the paint system for other literature and information.
- Follow the recommendations prescribed for use during handling and application as set forth by the supplier.
- Follow all applicable local regulations.

Toxicological research and practical experience have shown that diisocyanates can cause irritation of the skin, respiratory tract, eyes, nose and throat. In addition, sensitization resulting in allergic dermatitis or asthmatic symptoms can occur following overexposure to diisocyanates. Toxicological research has shown that polyisocyanates have a reduced potential to cause irritation and sensitization relative to their monomeric precursors.

Symptoms of irritation usually include watering of the eyes and a burning sensation in the nose and throat. The amount of irritation is dependent upon the dose, tissue exposed, and individual susceptibility, but it is generally independent of the individual's exposure history. These acute symptoms are generally reversible soon after the individual is removed from the contaminated area or the material is removed from the skin.

Sensitization is a systemic response and is not limited to the area of contact. Sensitization usually occurs as a result of numerous overexposures or one exposure to very high concentrations. Both respiratory and dermal sensitization can occur depending upon the toxicologic properties of the diisocyanate, route of exposure, and individual susceptibilities.

Exposures subsequent to the exposures that actually resulted in sensitization may cause a very strong allergic reaction. In the case of respiratory sensitization, the reaction is similar to asthma (i.e., coughing, wheezing, tightness in the chest, and shortness of breath). The skin sensitization reaction is allergic dermatitis, which may include symptoms such as rash, itching, hives, and swelling of the arms and legs.

If an individual experiences an irritation while handling an isocyanate, it should be determined whether the isocyanate was the cause of the irritation. If it was the cause, that indicates that the operation, as performed, allows an overexposure to isocyanates that can result in later sensitization of that worker or others.

A careful evaluation of the controls, protective equipment, and work practices should be made to reduce the exposure. If irritation persists in spite of proper ventilation and protective measures, the individual must be removed from areas where isocyanates are being processed or used.

If an individual is sensitized to isocyanates, complete removal from areas of potential exposure is mandated. This is true regardless of whether the isocyanate is in vapor or mist form. Also, exposure to an isocyanate other than the one suspected of causing the sensitization must be avoided.

Solvents are also present in paint. Prolonged or repeated exposure or overexposure to these solvents by either inhalation or direct skin contact may also cause injurious health effects. The effects are dependent upon the solvent, the extent of exposure, and the route of exposure.

Since isocyanates have the potential to irritate and sensitize those working with or around them, it is important that proper steps be taken to protect people potentially exposed to vapor, mist, or overspray. This includes those actually handling the isocyanate and those in the immediate vicinity.

Even during brush, roller, and painting curtains applications, it is possible to be exposed to airborne concentrations of solvents and isocyanate vapors. During spray application, not only will vapors be present, but also spray mist or aerosolized droplets that contain pigments, solvents, resins, additives, and polymeric materials, as well as isocyanate and unreacted polyisocyanate. Each of these will have its own physiological effect on the exposed organism.

Ideally, control of health hazards posed by vapors and spray mist is performed by engineering methods. Effective engineering should be used whenever possible to eliminate or reduce worker exposure. There are several engineering methods available to reduce exposure to isocyanate vapors and mist.

The most common precaution is a properly designed and ventilated enclosure. General ventilation, local ventilation, or isolation may prove adequate under certain conditions. Use of alternative application equipment (e.g., airless or electrostatic spray equipment) may help reduce spray mist generation during spray painting.

Brush and roller application of the paint may be feasible in some cases. To reduce environmental contamination, exhausted air may need to be cleaned by means of filters or scrubbers. The final design and combination of these control measures is dependent upon the specific application.

Whenever a paint system is spray-applied, it is essential that the applicator be protected from inhalation of both vapors and spray mist by the best possible respiratory protection. Under certain conditions, a fresh air-supplied respirator will be required. In other cases, an air purifier with a particulate filter may be employed. Applicators are urged to consult with their suppliers concerning the type of respiratory protection appropriate in a given application.

The appropriate selection and use of a respirator is an important part of protection from work-related chemical hazards. In addition, the following things should be followed:

• Users of respirators must be properly trained.
• Always be sure the respirator is in good working order.
• Know the limitations of the equipment.
• Be sure that the respirator fits properly.
• Clean the equipment after each use.

Respirator manufacturers may be helpful in developing a good respirator program. In addition to respirators, other forms of recommended personal protective equipment include safety glasses or goggles.

Should spray mist get into the eyes, rinse immediately and sufficiently with lukewarm water and consult an eye doctor should irritation persist. Regarding skin contact, as much of the exposed skin area as possible should be covered with clothing or skin cream. Cured coating cannot be removed easily. Application of a protective skin cream to the hands prior to start of work will facilitate the soap and water removal of paint splashes or overspray.

Skin areas covered only by protective creams should be kept to an absolute minimum. Aggressive solvents are unsuitable for cleaning the skin, as they wash oils away and can cause secondary reactions.

10.11 TWO-PACK AMINE CURED EPOXY RESIN PAINT AS PRIMER, INTERMEDIATE COAT, AND TOPCOAT FOR ATMOSPHERIC ENVIRONMENTS

This section covers the minimum requirements for the composition, properties, and storage life of a two-pack amine cured epoxy resin paint for atmospheric environments.

10.11.1 COMPOSITION

The paint should be furnished in two components. Component A should consist of bisphenol-A epichlorohydrine resin combined with pigments and suitable solvent. Component B should consist of suitable polyamine properly combined with volatile solvents and act as the curing agent for Component A.

Components A and B should be packaged separately and furnished in kit form. When mixed and applied in accordance with the manufacturer's instructions, a product meeting the requirements of the standard specifications should result. The ingredients used in the manufacture of these products should be best commercially available. The total solids for the admixed components of this paint should not be less than 35% by volume when tested in accordance with ASTM D 2697.

Table 10.40 Color Code

Paint Color	Color No. to BS 381°C	Color No. Rall (Approx.)
Arctic blue	112	5024
Sea green	217	6017
Brilliant green	221	6002
Canary yellow	309	1018
Light straw	N/A	1000
Middle brown	411	8007
Signal red	537	3020
Light orange	557	2000
Light gray	631	7033
White	N/A	1000

10.11.2 PROPERTIES

The mixed paint in the ratio recommended by the manufacturer should meet the requirements of the standards.

The odor of the paint material should not be obnoxious when tested in accordance with U.S. Federal Standard No. 141 Test Method 4401.

The color should be as specified by the purchaser, with reference to Table 10.40.

The paint should spray satisfactorily in all respects and should show no running, sagging, or streaking, and the dried film should show no dusting, mottling, or color separation and should present a smooth, lusterless finish free of seeds when sprayed on a steel panel to a dry film thickness between 25 and 28 μm and tested in accordance with U.S. Federal Standard No. 141, Method 4331.

Determine flexibility in accordance with U.S. Federal Standard No. 141, Method 6221. A film of the mixed paint should not be cracking or flaking when subjected to the flexibility test.

The drying time of the paint should meet the purchaser requirements when tested in accordance with U.S. Federal Standard No. 141, Method 4061.1.

The fineness of grind of the mixed paint should not be more than 15 μm for colored paint. The tests should be made 1 h after mixing, in accordance with U.S. Federal Standard No. 141, Method 4411.1, or ASTM D 1210.

10.11.3 STORAGE LIFE AND PACKAGING

The product should meet the requirements of the standards after storage of at least 12 months from the date of delivery in a full tightly covered container, in normal conditions.

The paint (Components A and B) as received should show no evidence of livering, skinning, or hard settling of pigments. The material should be easily mixed in liquid form by hand-stirring to form a smooth, homogeneous paint free of persistent foam when tested in accordance with U.S. Federal Standard No. 141, Method 3011.1, after storing for 12 months from the date of delivery.

The epoxy paint should be packaged as a unit consisting of pigmented compound marked "Component A" and the unpigmented (or clear) hardner marked "Component B." The packaging should meet the relevant requirements of ASTM D 3951 (88) unless otherwise specified by the purchaser.

Packing should be accomplished in a manner that will ensure acceptance by common carrier, at the lowest rate, and will afford protection against physical or mechanical damage during shipment.

10.11.4 DIRECTIONS FOR USE

In addition to the manufacturer's instructions for use, consisting of complete instructions covering the uses, surface cleanliness, mixing, thinning, application method, application condition, pot life, wet and dry film thickness per coat, temperature and humidity limitations, and drying time with each kit, the following directions should be supplied with each container of paint: "The surface of substrates should be prepared in accordance with standards before applying the primer." Apply by spray to the specified film thickness or, if none is specified, to at least 125 μm dry film thickness. The surface to be painted should be dry, and the surface temperature should be at least 3°C above the dew point.

10.11.5 SAFETY PRECAUTIONS

In addition to the manufacturer's instructions for safety, the following directions should be supplied with each container of paint:

This paint is hazardous because of its flammability and potential toxicity, so proper safety precautions should be observed to protect against these recognized hazards. Safe handling practices are required and should include but not be limited to, the provisions of SSPC-PA Guide 3, "A Guide to Safety in Paint Application," and to the following specifications:

- Keep paint away from heat, sparks, and open flame during storage, mixing, and application. Provide sufficient ventilation to maintain vapor concentration at less than 25% of the lower explosive limit.
- Avoid prolonged or repeated breathing of vapors or spray mist, and prevent contact of the paint with the eyes or skin.
- Clean hands thoroughly after handling paint and before eating and smoking.
- Provide sufficient ventilation to ensure that vapor concentration do not exceed the published permissible exposure limits. When necessary, supply appropriate personal protective equipment and enforce its use.

This paint may not comply with some air pollution regulation because of its hydrocarbon solvent content.

10.12 FAST-DRYING SYNTHETIC PRIMER TO BE USED WITH HOT-APPLIED COAL-TAR OR BITUMEN (ASPHALT) ENAMEL

This section covers the minimum requirements for the composition, properties, storage life and packaging, inspection, and labeling of fast-drying synthetic primer for use with hot-applied coal-tar and bitumen (asphalt) enamels.

10.12.1 COMPOSITION

The synthetic primer for cold application should consist of chlorinated rubber and plasticizer and coloring matter, together with solvents needed to give a consistency suitable for application by spray, brush,

Table 10.41 Properties of Fast-Drying Synthetic Primer

Characteristics	Requirements		Test Method
	Minimum	Maximum	
Flow time (4-mm flow cup) at 23°C seconds	35	60	BS 3900 Part A6
Flash point (abel closed cup)°C	23	N/A	BS 2000 Part 170
Volatile matter (105°C to 110°C) loss by mass	N/A	75%	BS 4164 Appendix A

or other approved method. The chlorinated rubber should contain approximately 66% by weight of chlorine. The viscosity (based on a solution of 20% by weight concentration in toluene at 20°C) should fall in the range of 9–14 centipoises (cP), when measured according to ASTM Standard D 445. Up to 50% by weight of the amount of chlorinated rubber could be in the viscosity range of 17 to 25 cP (see also SSPC Paint 17). The plasticizer should be a chlorinated plasticizer according to MIL-C-429 type 1.

Alternative plasticizers can be used, provided they are compatible, high-quality, and chemically resistant.

10.12.2 PROPERTIES

The primer should comply with the requirements in Table 10.41. When dry, it should provide an effective bond between the metal and the subsequent coating in accordance with the appropriate performance requirements given in Tables 10.42 and 10.43. The primer should also meet the following requirements:

- Odor—The odor should be normal for the materials permitted (ASTM Standard D 1296).
- Color—The color should be black.
- Compatibility—There should be no evidence of incompatibility of any of the ingredients of the primer when one volume of primer is slowly mixed with one volume of xylene (U.S. Federal Standard No. 141, Method 4203).

The primer should be easily applied by all three methods (brush, spray, and roller) when tested in accordance with U.S. Federal Standard No. 141, Methods 4321, 4331, and 4541. The primer should show no streaking, running, or sagging after drying.

10.12.3 STORAGE LIFE

The product should show no thickening, curdling, gelling, or hard caking when tested as specified in U.S. Federal Standard No. 141, Method 3011, and should meet the requirements of the standards after storage of at least 24 months from the date of delivery in a full, tightly covered container at normal temperatures.

10.12.4 DIRECTIONS FOR USE

In addition to the manufacturer's instructions for use, the following directions should be supplied with each container of primer:

Table 10.42 Performance Requirements of Fast-Drying Synthetic Primer in Conjunction with Coal-Tar Enamel

Test	Grade 105/15	Grade 105/8	Grade 120/5	Method AWWA	BS 4164
SAG. maximum (mm)					
70°C 24 h	1.5	1.5	N/A	AWWA	Appendix G
80°C 24 h	N/A	N/A	1.5	C203	
Low-temperature cracking and disbonding		10000			
−30°C	None	N/A	N/A	AWWA	Appendix H
−25°C	N/A	None	N/A	C203	
−20°C	N/A	N/A	N/A		
Bend at 0°C first crack minimum (mm)					Appendix J
initial	20	15	N/A		
After heating	15	10	N/A	N/A	
Disbonded area. maximum, mm²					
initial	2000	3000	N/A		
After heating	3000	5000	N/A		
Impact					
disbonded area. maximum mm²					
0°C	15000	N/A	N/A	AWWA	Appendix K
25°C	N/A	1000	N/A	C203	
Peel, initial and delayed, maximum (mm)					
30°C	3.0	N/A	N/A		Appendix L
40°C	3.0	3.0	N/A		
50°C	3.0	3.0	N/A		
60°C	N/A	3.0	3.0		
70°C	N/A	N/A	3.0		
Cathodic disbonding in 28 days maximum (mm)	5	5	5	N/A	Appendix M

Table 10.43 Performance Requirements of Fast-Drying Synthetic Primer in Conjunction with Bitumen Enamel

Test	Grade A	Grade B	Grade C	Method BS 4147
SAG. maximum (mm)				
60°C, 24 h	1.5	N/A	N/A	Appendix E
75°C, 24 h	N/A	1.5	1.5	
Bend at 0°C minimum (mm)	20	15	10	Appendix F
Impact disbonded area, maximum (mm^2)				
0°C	10,000	15,000	N/A	Appendix G
25°C	N/A	N/A	6,500	
Peel, initial and delayed, maximum (mm)				
30°C	3.0	3.0	N/A	
40°C	3.0	3.0	3.0	Appendix H
50°C	3.0	3.0	3.0	
60°C	3.0	3.0	3.0	

This primer is intended for use as a primer on structural steel. The surface of steel should be prepared in accordance with standards before applying the primer. This primer is intended to be followed by hot-applied coal-tar or bitumen enamel conforming to standards. Mix primer thoroughly before use.

Apply by brush or spray to the specified film thickness or, if none is specified, to at least 40 μm (dry) or approximately 125 μm (wet). When application is by spraying, the equipment and operating technique should be properly adjusted to prevent dry spraying and to deposit a wet film of primer on the substrate. Clean the equipment with xylene or the reducing thinner both before and after use.

The surface to be painted should be dry, and the surface temperature should be at least 3°C above the dew point.

In addition to the manufacturer's instructions for safety, the following directions should be supplied with each container of primer:

This primer is hazardous because of its flammability and potential toxicity, so proper safety precautions should be observed to protect against these recognized hazards. Safe handling practices are required and should include, but not be limited to, the provisions of SSPC-PA Guide 3, "A Guide to Safety in Paint Application," and to the following specifications:

- Keep primer away from heat, sparks, and open flame during storage, mixing, and application. Provide sufficient ventilation to maintain vapor concentration at less than 25% of the lower explosive limit.
- Avoid prolonged or repeated breathing of vapors or spray mist, and prevent contact of the paint with the eyes or skin.
- Clean hands thoroughly after handling primer and before eating or smoking.

- Provide sufficient ventilation to ensure that vapor concentrations do not exceed the published permissible exposure limits. When necessary, supply appropriate personal protective equipment and enforce its use.

This primer may not comply with some air pollution regulations because of its hydrocarbon solvent content.

10.13 COAL-TAR PRIMER (COLD-APPLIED) FOR USE WITH HOT-APPLIED COAL-TAR ENAMEL

This section covers the minimum requirements for the composition, properties, storage life and packaging, inspection, and labeling of coal-tar primers of grade A and B (cold-applied) for use with hot-applied coal-tar enamel. Primer grade A is intended for use with enamels grade 105/15 and 105/8, and primer grade B is intended for use with enamel grade 120/5. Therefore, the primer should be selected according to the grade of enamel.

10.13.1 COMPOSITION

Coal-tar primers should consist only of processed coal-tar pitch and refined coal-tar oils (coal-tar oils up to 200°C consist of benzene, toluene, xylene, cumene, etc.), suitably blended to produce a liquid that may be applied cold by brushing or spraying and that will produce an effective bond between the metal and a subsequent coating of coal-tar enamel. The primer should contain no benzene or other toxic or highly volatile solvents. The primer should show no tendency to settle in the container.

10.13.2 PROPERTIES

The coal-tar primer for coating material should meet the requirements of Table 10.44, when tested by the methods specified. When dry, it should provide an effective bond between the metal and the

Table 10.44 Properties of Coal-Tar Primers

Characteristics	Unit	Requirements		Test Method	
		Grade A	Grade B	BSI	ISO 5256 (E) 1985
Ash content (mass on dry extract)	W%	0.5	0.5	N/A	A→M
Softening point (ring and ball)	°C	80–100	100–120	BS 4164 Appendix D	A→D
Viscosity (no. 4 flow cup at 23°C)	Second	20–200	20–200	BS 3900 part A 6 (break point procedure)	N/A
Flash point (ABEL closed cup) min °C	°C	23	23	BS 2000 part 170	N/A
Penetration (at 25°C to 10^{-1} mm)	10^{-1} mm	20–30	10–20	BS 4164 appendix E	A→E

subsequent coating in accordance with the performance requirements given in Table 10.45 and the following specifications:

- Odor—The odor should be normal for the materials permitted (ASTM Standard D.1296).
- Color—The color should be black.

Table 10.45 Performance Requirements of Primer in Conjunction with Coal-Tar Enamel

Test	Grade 105/15	Grade 105/8	Grade 120/5	Method AWWA	BS 4164
SAG. Maximum (mm)					
70°C 24 h	1.5	1.5	N/A	AWWA C 203	Appendix G
80°C 24 h	N/A	N/A	1.5		
Low-temperature cracking and disbonding					
−30°C	None	N/A	N/A	AWWA C 203	Appendix H
−25°C	N/A	None	N/A		
−20°C	N/A	N/A	N/A		
Bend at 0°C first crack, minimum (mm)					
initial	20	15	N/A		
after heating	15	10	N/A		
Disbonded area, maximum, (mm²)					
initial	2000	3000	N/A	N/A	Appendix J
after heating	3000	5000	N/A		
Impact disbonded area, maximum (mm²)					
0°C*	15000	N/A	N/A	AWWA C203	Appendix K
25°C	N/A	1000	N/A		
Peel, initial and delayed, maximum (mm)					Appendix L
30°C	3.0	N/A	N/A		
40°C	3.0	3.0	N/A		
50°C	3.0	3.0	N/A		
60°C	N/A	3.0	3.0		
70°C	N/A	N/A	3.0		
Cathodic disbonding in 28 days maximum (mm)	5	5	5	N/A	Appendix M

Note:
*Apply the enamel not less than 16 h and not more than 72 h after the primer has been applied.

The primer should be capable of application by brush and spray when tested in accordance with U.S. Federal Standard No 141, Methods 4321, 4331, and 4541. The primer should show no streaking, running, or sagging after drying.

10.13.3 STORAGE LIFE, PACKAGING, AND SAMPLING

The product should meet the requirements of the standards after storage for at least 12 months from date of delivery, in a full tightly covered container. The primer should be packaged in uncontaminated steel drums with a capacity of not more than 210 l.

When samples of coating material are required for testing, the purchaser should specify the number of packages to be sampled and the procedure to be adopted. The samples so taken should be identified by the supplier, and half of them should be retained by the purchaser for the purpose of making any tests that may be required.

In coating materials, the filler will settle during storage. To ensure that test samples of these materials are representative, they should be made of equal increments taken from the top, middle, and bottom of the package. The preparation of samples for testing should be in accordance with Method C of ISO 5256.

10.13.4 INSPECTION AND TESTING

All materials supplied under the standard specifications should be subject to timely inspection by the purchaser or the purchaser's authorized representative. The purchaser should have the right to reject any materials supplied that are found to be defective under the standard specifications. In the case of disputes, the arbitration or settlement procedure established in the procurement documents should be followed.

The supplier should be responsible for the performance and costs of all laboratory test requirements as specified in the standards. It should set up and maintain such quality assurance and inspection systems as are necessary to ensure that the materials comply in all respects with the requirements of the standard specifications.

The purchaser's inspectors should have free access to the supplier's work to follow up the progress of the materials covered by the standards and to check the quality of the materials. The supplier should place at the disposal of the purchaser's inspectors all means necessary for carrying out their inspection, free of charge: namely, the results of tests, checking of conformity of materials to standard requirements, checking of markings, and packing and temporary acceptance of materials. Samples submitted to the purchaser will be tested in the purchaser's laboratory or in a responsible commercial laboratory designated by the purchaser.

The supplier should furnish the purchaser with a certified copy of results of tests made by the manufacturer covering physical and performance characteristics of each batch of product to be supplied under the standard specifications. The supplier should furnish to the purchaser or allow the purchaser to collect samples of the material that is representative of each batch of product.

Certified test reports and samples furnished by the supplier should be properly identified with each batch of product. Prior to the acceptance of the supplier's material, samples of material will be tested by the purchaser.

If any sample is found not to conform to the standards, the material represented by that sample will be rejected. If samples of the supplier's material that have been previously accepted are found not to conform to the standards, all such material will be rejected.

After the supplier has obtained approval from the purchaser for the bitumen primer that is proposed to be furnished, the supplier should submit the coating manufacturer's detailed specifications for the bitumen primer supplied with instructions for the handling and application of the material.

Unless otherwise specified in the standard specifications, the methods of sampling and testing should be in accordance with applicable methods of the ASTM, BS 4147, or both.

10.13.5 DIRECTIONS FOR USE

In addition to the manufacturer's instructions for use, the following directions should be supplied with each container of primer:

- This primer is intended for use as a prime coat on steel (mainly steel pipe).
- The surface of steel should be prepared in accordance with the standards (surface preparation before applying the primer).
- This primer is intended to be followed by hot-applied coal-tar enamel conforming to the standards. Mix the primer thoroughly before use.
- Apply by brush to the specified film thickness, or if none is specified, to at least 100 μm (dry).
- The surface to be coated should be dry, and the surface temperature should be at least 3°C above the dew point.

10.13.6 SAFETY PRECAUTIONS

In addition to the manufacturer's instructions for safety, the following directions should be supplied with each container of primer:

This primer is hazardous because of its flammability and potential toxicity, so proper safety precautions should be observed to protect against these recognized hazards. Safe handling practices are required and should include, but not be limited to, the provisions of SSPC-PA Guide 3, "A Guide to Safety in Paint Application," and to the following specifications:

- Keep primer away from heat, sparks, and open flame during storage, mixing, and application. Provide sufficient ventilation to maintain vapor concentration at less than 25% of the lower explosive limit.
- Avoid prolonged or repeated breathing of vapors or spray mist, and prevent contact of the paint with the eyes or skin.
- Clean hands thoroughly after handling primer and before eating or smoking.
- Provide sufficient ventilation to ensure that vapor concentrations do not exceed the published permissible exposure limits. When necessary, supply appropriate personal protective equipment and enforce its use.

This primer may not comply with some air pollution regulations because of its hydrocarbon solvent content.

10.14 COLD-APPLIED BITUMEN PRIMER FOR USE WITH HOT-APPLIED BITUMEN ENAMEL

10.14.1 COMPOSITION

Bitumen primer for cold application should consist of a homogeneous solution of bitumen in hydrocarbon (usually paraffinic hydrocarbon or other suitable solvent) with a consistency suitable for application by brushing or other approved method.

10.14.2 PROPERTIES

The bitumen primer should comply with the requirements given in Table 10.46 when tested by the methods specified. In addition, when dry, it should provide an effective bond between the metal and the subsequent coating, in accordance with the appropriate performance requirements given in Table 10.47. It also should meet the following criteria:

- Odor—The odor should be normal for the materials permitted (ASTM Standard D 1296).
- Color—The color should be black.

10.14.3 WORKING PROPERTIES

The primer should be capable of application by brush and spray when tested in accordance with U.S. Federal Standard No. 141, Methods 4321, 4331, and 4541. The primer should show no streaking, running, or sagging after drying.

Table 10.46 Properties of Cold-Applied Bitumen Primer

Characteristics	Unit	Requirements		Test Method	
		Grade A	Grade B	BSI	ISO 5256 (E) 1985
Ash content (mass on dry extract)	W%	0.5	0.5		A→M
Softening point of bitumen used (ring and ball)	°C	80–100	100–120	BS 2000 Part 58	A→D
Viscosity (no. 4 flow cup at 23°C)	Second	30–200	30–200	N/A	N/A
Flash point (abel closed cup) min°C	°C	23	23	BS 2000 Part 170	BS EN 535
Penetration (of bitumen used) at 25°C to 10^{-1} mm	10^{-1} mm	20–30	10–20	BS 2000 Part 49	A→E
Solubility (of bitumen used) in carbon disulfide or trichloroethylene	minimum % by mass	99	99	BS 2000 Part 58	N/A

Table 10.47 Performance Requirements of Primer in Conjunction with Bitumen Enamel of Suitable Grade

| Test | Unit | Grade A | Grade B | Grade C | Methods | | |
					ISO 5256	BS 4147	BS 4164
Cold bending[1]	mm	≥ 20	≥ 15	≥ 10	Method F	N/A	
Flow[2] (70°C 45°C: 20 h)	mm	≤ 6	≤ 2	≤ 2	Method H	N/A	
Peel, Initial and delayed (maximum)	mm	3.0	3.0	N/A	N/A	Appendix H	
30°C		3.0	3.0	3.0			
40°C		3.0	3.0	3.0			
50°C		3.0	3.0	3.0			
60°C							
Impact, disbonded area (maximum)	mm^2	15000	6500	6500		Appendix G (Revision A)	
0°C[3]		N/A					
25°C							
Cathodic disbonding in 28 days maximum	mm	5	5	5	N/A	N/A	Appendix M
Sag, maximum	mm	1.5	N/A	N/A	N/A	Appendix E	Appendix G
60°C 24 h		N/A	1.5	1.5			
75°C 24 h							

Note: Apply the enamel between 16 and 72 h after the primer has been applied.
[1]*The cold bending test consists of verifying the flexibility at low temperatures of bitumen used as a coating on steel pipes and under conditions simulating the bending of coated pipes.*
[2]*The flow test consists of measuring the displacement of the surface of a coating of a bitumen by its own weight under specified conditions of temperature and time.*
[3]*If the test specimen fails the impact test at 0°C, two additional test specimens should be prepared from the same sample, and the failed test specimen and these other two samples should be tested at 0°C. The material should be deemed to comply with the requirements of the impact test, provided that both of the specimens pass the test.*

10.14.4 STORAGE LIFE, PACKAGING, AND SAMPLING

The product should meet the requirements of the standards after being stored for at least 12 months from the date of delivery in a full tightly covered container. The primer should be packaged in uncontaminated steel drums with a capacity of not more than 210 l.

When samples of coating material are required for testing, the purchaser should specify the number of packages to be sampled and the procedure to be adopted. The samples so taken should be identified by the supplier, and half of them should be retained by the purchaser for the purpose of making such tests as may be required.

In coating materials, the filler will settle during storage. To ensure that test samples of these materials are representative, they should be made of equal increments taken from the top, middle, and bottom of the package. The preparation of samples for testing should be in accordance with Method C of ISO 5256.

10.14.5 INSPECTION AND TESTING

All materials supplied under standard specifications should be subject to timely inspection by the purchaser or the purchaser's authorized representative. The purchaser should have the right to reject any materials supplied that are found to be defective under standard specifications. In the case of disputes, the arbitration or settlement procedure established in the procurement documents should be followed.

The supplier should be responsible for the performance and costs for all laboratory test requirements as specified in the standards. The supplier should set up and maintain such quality assurance and inspection systems as are necessary to ensure that the materials comply in all respects with the requirements of the standard specifications.

The purchaser's inspectors should have free access to the supplier's work to follow up the progress of the materials covered by the standards and to check the quality of materials. The supplier should place at the disposal of the purchaser's inspectors all means necessary for carrying out their inspection, free of charge: namely, the results of tests, checking of the conformity of materials with standard requirements, checking of marking and packing, and temporary acceptance of materials.

Samples submitted to the purchaser will be tested in the purchaser's laboratory or in a responsible commercial laboratory designated by the purchaser. The supplier should furnish the purchaser with a certified copy of results of tests made by the manufacturer covering physical and performance characteristics of each batch of product to be supplied under the standard specifications. The supplier should furnish to the purchaser, or allow the purchaser to collect, samples of the material representative of each batch of product.

Certified test reports and samples furnished by the supplier should be properly identified with each batch of product. Prior to acceptance of the supplier's material, samples of material submitted by the supplier will be tested by the purchaser.

If any sample is found not to conform to the standards, material represented by that sample will be rejected. If samples of the supplier's material that have been previously accepted are found not to conform to the standards, all such material will be rejected.

After the supplier has obtained approval from the purchaser for the bitumen primer that is proposed to be furnished, the supplier should submit the coating manufacturer's detailed specifications for the bitumen primer, along with instructions for the handling and application of the material.

Unless otherwise specified in the standard specifications the methods of sampling and testing should be in accordance with applicable methods of the ASTM, BS 4147, or both.

10.14.6 DIRECTIONS FOR USE

In addition to the manufacturer's instructions for use, the following directions should be supplied with each container of primer.

- This primer is intended for use as a prime coat on structural steel (mainly steel pipes). The surface of steel should be prepared in accordance with standard for surface preparation before applying the primer.

- This primer is intended to be followed by hot-applied bitumen enamel conforming to the standards. Mix primer thoroughly before use.
- Apply by brush to the specified film thickness or, if none is specified, to at least 100 μm (dry).
- The surface to be coated should be dry, and the surface temperature should be at least 3°C above the dew point.

10.14.7 SAFETY PRECAUTIONS

In addition to the manufacturer's instructions for safety, the following directions should be supplied with each container of primer:

This primer is hazardous because of its flammability and potential toxicity, so proper safety precautions should be observed to protect against these recognized hazards. Safe handling practices are required and should include, but not be limited to, the provisions of SSPC-PA Guide 3, "A Guide to Safety in Paint Application," and to the following specifications:

- Keep primer away from heat, sparks, and open flame during storage, mixing, and application. Provide sufficient ventilation to maintain vapor concentration at less than 25% of the lower explosive limit.
- Avoid prolonged or repeated breathing of vapors or spray mist, and prevent contact of the paint with the eyes or skin.
- Clean hands thoroughly after handling primer and before eating or smoking,
- Provide sufficient ventilation to ensure that vapor concentrations do not exceed the published permissible exposure limits. When necessary, supply appropriate personal protective equipment and enforce its use.

This primer may not comply with some air pollution regulations because of its hydrocarbon solvent content.

10.15 COAL-TAR ENAMEL (HOT-APPLIED)

This section covers the minimum requirements for the composition, quality of enamel, properties, storage life and packaging, inspection, and labeling of coal-tar enamel (hot-applied).

10.15.1 COMPOSITION

The coal-tar enamel of either grade (see Table 10.48), as specified by the purchaser, should consist of a uniform mixture of modified coal-tar and inert, nonfibrous filler. The fineness of the inert, nonfibrous filler should be such to pass 90% by weight (minimum) through a 75-μm (200-mesh) standard sieve opening when tested by ASTM D546. The enamel should contain 100% by weight of nonvolatile, film-forming solids (coal tar and filler).

10.15.2 QUALITY OF ENAMEL

The quality of coal-tar enamel is affected by the quality of the coal that is carbonized and by the temperature of carbonization, as well as by the subsequent methods of pitch processing and by the

Table 10.48 Properties of Enamel

Characteristics	Requirements			ASTM and AWWA Method	BS4164 Method
	Grade 105/15	Grade 105/8	Grade 120/5		
Filler content by ignition, % by weight	25–35	25–35	25–35	D 2415	Appendix B
Fineness filler. through 75 μm (200 MESH). % by weight	90 Min	90 Min	90 Min	D 546	BS 179S
Density at 25(°C) g/cm³	1.4–16	1.4–1.6	1.4–1.6	D 71	Appendix C
Softening point (ring and ball)°C	105–116	105–116	120–130*	D 36	Appendix D
Penetration**	10–20	5–12	1–9	AWWA	Appendix E
25°C, 100 (g)	15–55	8–30	3–16	C203	
45°C, 50 (g)					
Flow time (s)					
230°C	9–16	9–16	N/A	N/A	Appendix F
240°C	N/A	N/A	9–24	N/A	

*The softening point range for this grade may be exceeded by agreement between the manufacturer and the purchaser.
**For static conditions above −15°C (5°F), use enamel with 5–10 penetration at 25°C; below −15°C and above −23°C, use 10–15 penetration; and below −23°C and above −29°C, use 15–20 penetration enamel. (Static conditions are those conditions under which the article is not being handled.)

particulars of formulation. To meet the basic quality requirements, coal tar should be produced from coal that has a minimum heating value of 30,000 J/g on a moisture-and mineral-mater-free basis (ASTM D388) and that has been carbonized in a slot-type coke oven at a minimum temperature of 900°C.

10.15.3 PROPERTIES

The coal-tar enamel should comply with the requirements for the appropriate grade given in Table 10.48. Also, in conjunction with the appropriate primer [coal-tar primer (cold-applied), fast-drying synthetic primer for use with hot-applied coal-tar and asphalt enamels, or both], it should comply with the requirements for the appropriate grade given in Table 10.49 and the following criteria:

- Odor—The odor should be normal for the materials permitted (ASTM Standard D1296).
- Color—The color should be black.
- Compatibility—There should be no evidence of incompatibility of any of the ingredients of enamel when two parts of the enamel are melted together.
- Pot life—The pot life of the enamel in molten state should be at least 12 h.
- Application temperature—The application temperature of enamel should be from 220–260°C.

Table 10.49 Properties of Enamel on Primed Metal

Test	Grade 105/15	Grade 105/8	Grade 120/5	Method AWWA	Method BS 4164
Sag, maximum (mm)					
70°C 24 h	1.5	1.5	N/A	AWWA	Appendix G
80°C 24 h		N/A	1.5	C203	
Low-temperature					
Cracking and disbonding					
−30°C	None	N/A	None		
−25°C	N/A	None	N/A	AWWA	Appendix H
−20°C	N/A	N/A	N/A	C203	
Bend at 0°C					
First crack., minimum (mm)					
Initial	20	15	N/A		
after heating	15	10	N/A	N/A	Appendix J
Disbonded area., maximum, (mm^2)					
Initial	2,000	3,000	N/A		
after heating	3,000	5,000	N/A		
Impact disbonded area					
Maximum (mm^2)					
0°C*	15,000	N/A	N/A	AWWA 203	Appendix K
25°C	N/A	10,000	N/A		
Peel initial and delayed., maximum (mm)					
30°C	3.0	N/A	N/A		
40°C	3.0	3.0	N/A		
50°C	3.0	3.0	N/A		
60°C	N/A	3.0	3.0		Appendix L
70°C	N/A	3.0	3.0		
Cathodic disbonding in 28 days maximum (mm)	5	5	5	N/A	Appendix M

Having detailed lines means no tests are to be done.
*If the test specimen fails the impact test at 0°C, two additional test specimens should be prepared from the same sample, and the failed test specimen and these other two samples should be tested at 0°C. The material should be deemed to comply with the requirements of the impact test, provided that both test specimens pass the test.

In addition, the enamel and primer should meet all the requirements specified in the latest editions of AWWA C203 and BS 4164.

10.15.4 STORAGE LIFE AND PACKAGING

The properties of enamel should not change after being stored for at least 12 months from the date of delivery in a full tightly covered container. The coal-tar enamel should be supplied in noncontaminated steel drums with a capacity of not more than 200 l.

10.15.5 DIRECTIONS FOR USE

In addition to the manufacturer's instructions for use, the following directions should be supplied with each container of coal-tar enamel:

- These materials are heavy-duty products to be applied at a minimum thickness of 2.4 mm to provide long-term protection underground and in submarine installations.
- They are applied to iron and steel used over a wide range of service temperatures. They are particularly suitable for flood-coating previously primed products. Agitation of enamel materials in the molten state is necessary to prevent settling of the filler. When these materials are applied externally to pipes, it is usual to incorporate one or more reinforcing layers of inert fabric.
- All materials should be applied in accordance with the manufacturer's instructions. The enamel used should provide a bond between the metal and the coating material that will enable the requirements given in Table 10.49 to be complied with if all the qualities of the coating material are adequate. Care should be exercised to ensure that there is no mixing of materials from different sources or different types. In particular, it should be recognized that the chemical and physical characteristics of bitumen-based coatings differ from those of coal-tar based coatings, and that the two kinds of coating should not be blended to make protective coatings.
- It is also essential to clean out the plant thoroughly when the use of coal-tar coating materials follows that of bitumen coating materials, or vice versa.

10.15.6 SAFETY PRECAUTIONS

These coatings are hazardous because of their flammability and potential toxicity, so proper safety precautions should be observed to protect against these recognized hazards. Safe handling practices are required and should include, but not be limited to, the provisions of SSPC-PA Guide 3, "A Guide to Safety in Paint Application," and the following specifications:

- The coatings specified herein may not comply with some air pollution regulations because of their hydrocarbon solvent content.
- Ingredients in this coating that may pose a hazard include hydrocarbon solvent and coal tars. This coating may contain low concentrations (less than 1% by weight) of materials that are suspected carcinogens. Applicable regulations governing safe handling practices should apply to the use of this coating.
- Hot enamels are applied at temperatures ranging from 220–260°C and extreme care must be used when melting and handling.

10.16 BITUMEN ENAMEL (HOT-APPLIED)

This section covers the minimum requirements for the composition, properties, storage life, packaging, sampling, inspection and testing, and labeling of bitumen (asphalt) enamels grades A, B, and C (hot-applied) that are suitable mainly for protecting the external surfaces of steel pipes and fittings in buried and submerged pipelines. The enamel is to be selected according to the maximum service temperature, local conditions (particularly of climates, storage, handling and exposure), or both. Asphalt enamels are not intended for use on pipelines carrying crude oil, petroleum products, or liquid hydrocarbon, which may dissolve the enamel.

10.16.1 COMPOSITION

The bitumen enamel of either grade (see Table 10.50), as specified by the purchaser, should consist of a uniform mixture of petroleum bitumen and inert, nonfibrous filler, plus suitable additives. The filler should be nonhygroscopic and unreactive with the other constituents of the protection, and resistant to attack by the medium to which it will normally be exposed. It should be stable at the maximum application temperature of the coating material.

The particle size of the filler should meet the following requirements when tested by Method N of ISO 5256:

- 500 μm: 0%
- 90 μm: < 10%

It should be black in color and possess waterproofing and adhesive properties. Suitable additives should be utilized in the enamel formulation in order to resist fungi and to pass the test regarding root penetration.

Table 10.50 Properties of Bitumen (Asphalt) Enamel

Characteristics	Unit	Requirements			Test Method ISO 5256
		Grade A	Grade B	Grade C	
Ash and filler content (mass)	%	≤ 40	≤ 40	≤ 55	Method M
Softening point (ring and ball)	°C	≥ 95	≥ 110	≥ 120	Method D
Penetration* (25°C; 100 g; 5 s.)	10^{-1} mm	< 25	< 20	< 20	Method E
Indentation (25°C; 25 N; 24 h)	mm	≤ 17	≤ 10	≤ 8	Method G
Water absorption (40°C; 5 h)	g/m²	≤ 1.5	≤ 15	≤ 1.5	Method L
Change on heating - Difference in softening point - Difference in penetration	 °C %	 ≤ 10 ≤ 40	 ≤ 10 ≤ 40	 ≤ 10 ≤ 40	Method K
Maximum service temperature	°C	35	60	60	N/A

*The penetration, expressed in millimeters, of a standard rod, placed vertically on the surface of a bitumen and loaded under specified conditions of temperature, load, and time.

Table 10.51 Properties of Enamel on Primed Steel

Test	Unit	Grade A	Grade B	Grade C	Method ISO 5256	BS 4147
Cold bending[1]	mm	≥ 20	≥ 15	≥ 10	Method F	N/A
Flow[2] (70°C; 45° ;20 h)	mm	≤ 6	≤ 2	≤ 2	Method H	N/A
Peel, Initial and delayed (maximum)						
30°C	mm	3.0	3.0	N/A	N/A	⎫
40°C		3.0	3.0	3.0		⎬ Appendix H
50°C		3.0	3.0	3.0		⎭
60°C		3.0	3.0	3.0		
Impact, Disbanded area (maximum)						
0° C[3]	mm²	15,000	N/A	N/A	N/A	⎱ Appendix
25°C		N/A	6,500	6,500		⎰ G(Revision A)

[1]*The test consists of verifying the flexibility at low temperatures of bitumen used as a coating on steel pipes and under conditions simulating the bending of coated pipes.*
[2]*The test consists of measuring the displacement of the surface of a coating of a bitumen by its own weight under specified conditions of temperature and time.*
[3]*If the test specimen fails the impact test at 0°C, two additional test specimens should be prepared from the same sample, and the failed test specimen and these other two samples should be tested at 0°C. The material should be deemed to comply with the requirements of the impact test, provided that both test specimens pass the test.*

The enamel should contain 100% by weight of nonvolatile, film-forming solids (bitumen and filler).

10.16.2 PROPERTIES

The enamel should comply with the requirements for the appropriate grade, as specified by the purchaser, given in Table 1 when tested by the methods specified, and also with the requirements of standards. There should be no evidence of incompatibility of any of the enamel ingredients when two parts of the enamel are melted together.

The application temperature of the enamel should be between 220°C and 240°C. The pot life of the enamel in molten state should be at least 4 h (without agitation).

The bitumen enamel in conjunction with an appropriate primer (bitumen primer for cold application or synthetic primer for coal-tar and bitumen enamel), as specified by the purchaser, should comply with the requirements for the appropriate grade given in Table 10.51 when tested by the methods specified.

Unless otherwise specified, the bitumen enamel and primer should be supplied by the same manufacturer. When the bitumen enamel and primer are from different suppliers, the coating materials should also pass Tests C–J of ISO 5256 (for adhesion and compatibility).

10.16.3 STORAGE LIFE, PACKAGING, AND SAMPLING

The product should meet the requirements of the standards after storage of at least 24 months from date of delivery in a full, tightly covered container. The bitumen enamel should be supplied in one of the following forms:

- In noncontaminated steel drums with a capacity of not more than 200 l
- In craft paper bags, with easily strippable lining containing approximately 50 kg (only in cases of local purchase)

Unless otherwise specified by the purchaser, the number of samples for testing should consist of 10% of the lot, but in no case should it be more than 10 containers of material. The results of the tests on four specimens made from each sample should be averaged for each test to determine conformity to the specified requirements.

In coating materials, the filler will settle during storage. In order to ensure that test samples of these materials are representative, they should consist of equal increments taken from the top, middle, and bottom of the package. Preparation of the samples for testing should be in accordance with Method C of ISO 5256.

10.16.4 DIRECTIONS FOR USE

In addition to the manufacturer's instructions for use, the following directions should be supplied with each container of bitumen enamel:

- These materials are heavy-duty products for application at a minimum thickness of 2.4 mm to provide long-term protection underground and in submarine installations. They are applied to iron and steel.
- The enamel should be maintained moisture- and dirt-free at all times both prior to and during heating and application.
- Agitation of enamel in the molten state is necessary to prevent settling of the filler.
- The hot enamel should be thoroughly and continuously stirred for a minimum of 5 min, and intervals between stirring should not exceed 15 min, regardless of whether the enamel is being dispensed from kettles or is being held ready for use. Iron paddles should be used for stirring. Wooden paddles are not permitted.
- Enamel that has been heated in excess of the maximum allowable temperature or that has been held at the application temperature for a period greater than that specified should be condemned and rejected. Fluxing the enamel will not be permitted.
- All materials should be applied in compliance with the manufacturer's instructions. Care should be exercised to ensure there is no mixing of material from different sources or of different types unless experience has shown that the final product has satisfactory properties. In particular, it should be recognized that the chemical and physical characteristics of coal-tar-based coatings differ from those of bitumen-based coatings, and that the two kinds of coating should not be blended in protective coatings. It is also essential to clean out the plant thoroughly when the use of bitumen coating materials follows that of coal-tar coating materials, or vice versa.

10.16.5 TESTING FOR RESISTANCE TO ROOT PENETRATION

In the procedure to test for resistance to root penetration, three dry flowerpots (unglazed earthenware) approximately 22 cm high are required. A circular strip that is approximately 4 cm wide is applied to the inside of each pot about 10 cm from the bottom using the coating material supplied by the manufacturer. When the strip is completely dry, the pot is filled with soil (containing very little calcium and no compost) of pH 5–6, into which some peat has been mixed. The soil should just cover the bottom edge of the strip.

Three plates, each 2 cm thick, are cast from the material to be tested. Their diameter should be approximately that of the internal diameter (ID) halfway up the flowerpots. Once cool, the cast plates are put on top of the soil layer and the gap between the plate and the pot is carefully sealed with more of the test material. Approximately 9 cm of soil is put on top of the plate, and 35–40 lupine seeds (Lupines albus) are planted in the soil and covered with about 1 cm of soil.

During winter, the seeds should be grown in a heated greenhouse under artificial light. At other times of the year, they can be grown outside. The soil above the plate should be moistened with rainwater fed through a pipe into the pot so that the surface of the test material is kept damp. The soil beneath the plate is kept moist by placing the pots in rain-filled troughs. After 6 weeks (or 8 weeks in winter), the pots are cracked open and the plates are examined for root growth. The bottom side is inspected to see whether any roots have grown through the plate, and the upper side is examined to assess the number of roots penetrating the material and the depth of penetration.

As a control for this test, another pot must be prepared as described above using an 85/40 bitumen plate. At the end of the test period, a large number of lupine-roots must have grown through the plate.

The material cannot be described as root proof if any roots (of any thickness) have grown through the plate or between the plate and the pot, or if roots have penetrated the plate to a depth greater than 5 mm. On the other hand, it can be described as root proof if no roots penetrated further than 5 mm into the plates in two out of the three pots.

The test is valid only if the roots have reached the upper surface of the plates. It should be double-checked by carrying out a long-term test to confirm the depth of penetration and to ascertain whether the prescribed test period is adequate.

10.17 GLASS-FIBER MAT (INNER WRAP)

This specification covers the minimum requirements for machine-applied glass-fiber mat (inner wrap) for reinforcement of hot bitumen and coal-tar enamel coatings.

The glass-fiber mat should be a thin, flexible, nonwoven uniform mat that is longitudinally reinforced across the full sheet width by continuous uniform glass filaments spaced at a maximum of 30 mm apart. The whole should be bonded with a thermosetting resin. The glass used should be chemically resistant borosilicate. The mat should be designed to give the degree of porosity required to ensure good bleed-through of the enamel during the coating process.

At the time of unrolling at ambient temperatures, the successive layers of the mat should not stick to one another. It should be suitable for application in line-travel equipment or in fixed head coating machines in pipe mills and coating yards.

10.17.1 PROPERTIES

The finished glass-fiber mat should meet the requirements of Table 10.52. It should have a smooth surface free of visible defects such as holes, slits, folds, thin areas, leafing, frayed or uneven edges, presence of foreign bodies (oily matter, mud, etc.).

Table 10.52 Physical Characteristics of Glass-Fiber Mat

Characteristics	Units	Values
Weight (minimum)	g/m²	41
Thickness (minimum)	mm	0.3
Tensile strength:		
Longitudinal (minimum)	N/50 mm Width	118
Transversal (minimum)	N/50 mm Width	36
Elmendorf tear strength:		
Longitudinal (minimum)	g.	100
Transversal (minimum)	g.	100
Pliability	N/A	Shall pass test
Porosity	mm/H₂O	0.6–1.9

The binder should be a thermosetting resin that can resist the action of microorganisms and should have a temperature resistance of 280°C (i.e., unaffected under loading in hot enamel at 280°C for 1 min). The binder content should be such that complete impregnation with the coating material is obtained during normal application.

The glass used should be of hydrolytic class III (HGB3) at most. The classification of glass as specified by ISO 719 is shown in Table 10.53, which is related to the consumption of acid and its equivalent of alkali [expressed as sodium oxide (Na_2O)] when tested by the method specified in that standard.

The mat should have an Elmendorf tear strength in the longitudinal direction of not less than 100 g and in the transverse direction of not less than 100 g, as evaluated by Method ASTM D689 or Method TAPPI T414. To test for this, use an Elmendorf tear tester, Thwing-Albert Model 60-16 or equivalent, with a capacity of 0–1,600 g. The instrument should be securely anchored to a table and leveled. The pendulum should be tested for zero by moving it to the left or the pendulum stop and then released. Adjustments should be made if the pointer does not read zero after the pendulum stop has been released. The friction of the apparatus and the condition of the knife should be verified and adjusted, if required, as per ASTM D689. Perform the tests at a temperature of 23°C ± 2°C and 50% ± 10% relative

Table 10.53 Classification of Glass

Class [1]	Consumption Of Hydrochloric Acid Solution [c(HCl)=0.01 mol/l] Per Gram Of Glass Grains (μg/g)	Equivalent Of Alkali Expressed As Mass Of Sodium Oxide (Na_2O) Per Gram Of Glass Grains (μg/g)
HGB 1	Up to and including 0.10	Up to and including 31
HGB 2	From 0,10 Up To and including 0,20	From 31 Up to and including 62
HGB 3	From 0.20 Up To and including 0,85	From 62 Up To and including 264
HGB 4	From 0.85 Up To and including 2,0	From 264 Up To and including 620
HGB 5	From 2,0 Up To and including 3,5	From 620 Up To and including 1,085

[1]HGB stands for the hydrolytic resistance of glass grains according to the boiling water method.

humidity. Cut 10 specimens from the mat using either a sample cutter or template notching machine direction (MD) and machine direction plus cross direction (MD + CD). The sample size should be 76 mm ± 2 mm wide × 63 mm ± 2 mm high.

The test should initially be performed using two plies. The number of plies may subsequently have to be varied as indicated in ASTM D689. The pendulum is moved to the left until the stop is engaged. After the specimen has been clamped into the apparatus with the 63-mm slit in the vertical direction, the knife is depressed to make the slit and the stop depressed to release the pendulum. Record the number of piles, the reading of the pointer, and whether MD or MD + CD tears. The entire procedure is repeated to obtain 10 individual readings, and then the readings are averaged. This average, divided by the number of piles tested, gives the Elmendorf tear strength.

There should be no cracking of a glass-fiber mat when bent over a 6.4-mm-diameter mandrel at 23°C ± 1°C for approximately 2 s, as determined by ASTM D146. Cut five specimens with the dimensions 25 mm × 200 mm, with the long side parallel to the length of the roll, and immerse them in water at 23°C ± 1°C for 10–15 min. Remove each specimen individually and bend it over a 6.4-mm-diameter mandrel through a 90° arc at uniform speed for approximately 2 s. Examine each specimen for cracks or breaks.

When related to pressure differences across the sample, the glass-fiber mat should have a porosity of not less than 0.60 mm and not more than 1.9 mm of water at an average air velocity of 60 m/min, as determined by ASTM D737, modified. Provide five specimens with the dimensions of at least 250 mm × 250 mm, representative of the unsaturated glass-fiber mat to be tested; or test five places on the unsaturated glass-fiber mat as far apart as possible without cutting. The apparatus should consist essentially of a suction fan for drawing air through a known area of unsaturated glass-fiber mat, a circular orifice over which the mat to be tested can be clamped, a means of measuring the pressure drop across the mat, and a means of measuring the volume of air flowing through the mat.

The clamp should effectively eliminate edge leakage. The apparatus should be capable of testing unsaturated glass-fiber mats of different thicknesses and of testing large pieces of glass-fiber mat without cutting. The instrument should be calibrated directly with a precision instrument. Perform all the tests at 23°C ± 1°C and 50% ± 2% relative humidity. Mount the test specimen between the clamp and the circular orifice with sufficient tension to draw the unsaturated glass-fiber mat smooth. It should not be distorted in its own plane. Draw conditioned air through the known area of the mat and through the calibrated flow meter at the rate of 60 m/min and record the pressure reduction across the mat in terms of millimeters of water. Report the average of the test results for the five test specimens or the five different locations on the glass-fiber mat as the porosity.

The mat should have a thickness of not less than 0.3 mm, as determined by ASTM D146, modified. At 10 equally spaced areas selected by random sampling, measure the thickness with an Ames dial reading in 0.25-mm increments. Use a circular foot and anvil, both 645 mm² in area, exerting a pressure of 13.8 kPa. Make all measurements in an atmosphere of 50% relative humidity and at 23°C ± 1°C.

In the case of reinforced glass wrap, the breaking strength in the longitudinal direction should be no less than 118 N/50 mm width. In either case, the breaking strength in the transverse direction should be no less than 36 N/50 mm width, as determined by ASTM D146, modified. Cut 10 specimens with dimensions of 75 mm × 560 mm with the longer dimension along the roll and 10 specimens with the longer dimension across the roll. In those instances where the wrapper width is less than 560 mm, the specimen length should be that of the wrapper length. Impregnate both ends of each specimen with a protective coating of shellac or methacrylate for a distance of 64 mm and allow it to dry.

Test all specimens at 23°C ± 1°C using a tension-testing machine of adequate capacity in which the clamps are attached to swivels that are free to move in any direction. The clamps should be 25 mm × 75 mm and should be covered with masking tape. Grip the specimen 50 mm from each end, leaving 457 mm between the clamps. This gauge length may have to be adjusted, depending on wrapper width. However, the length should be the maximum amount that is consistent with good clamping. Increase the breaking of the load by causing the lower clamps of the machine to travel at a uniform speed of 300 mm/min. Disregard the reading on any specimen that breaks less than 13 mm from either clamp, and test an additional specimen in its place. Report the average of the results of 10 individual tests on specimens cut along the roll as the longitudinal breaking strength and the average of the results of 10 individual tests on specimens cut across the roll as the transverse breaking strength.

The color of the mat should be light, preferably white. The roll sizes, as specified by the purchaser, should be as follows:

- Roll length: 122 m, 244 m
- Roll width: 102 mm, 152 mm, 229 mm, 304 mm, 457 mm

10.17.2 STORAGE LIFE AND PACKAGING

The product should meet the requirements of the standards, after being stored for 24 months from date of delivery in a full, tightly covered container. The mats purchased according to the standards should be rolled onto cardboard tubes with ID of 76 mm (nominal) and packaged in suitable and approved containers so that, during stocking and transport, the full quality of performance is retained. Packing should be weatherproof and strapped on pallets suitable for long-distance shipment.

10.17.3 INSPECTION AND TESTING

All materials supplied under standard specification should be subject to timely inspection by the purchaser or the purchaser's authorized representative. The purchaser should have the right to reject any materials supplied that are found to be defective under standard specification. In the case of disputes, the arbitration or settlement procedure established in the procurement documents should be followed.

The purchaser's inspectors should have free access to the supplier's works to follow up the progress of the materials covered by standard and to check the quality of materials. The supplier should place at the disposal of the purchaser's inspectors all means necessary for carrying out their inspection, free of charge: the results of tests, checking of conformity of materials to standard requirements, checking of markings, and packing and temporary acceptance of materials.

The supplier should set up and maintain such quality control and inspection systems as are necessary to ensure that the materials comply in all respects with the requirements of standard specification. Samples submitted to the purchaser will be tested in the purchaser's laboratory or in a responsible commercial laboratory designated by the purchaser.

The supplier should furnish the purchaser with a certified copy of the results of tests made by the manufacturer covering the physical and performance characteristics of each batch of material to be supplied under standard specification. The supplier should furnish to the purchaser, or allow the purchaser to collect, samples of the material that are representative of each batch of mat. Certified test reports and samples furnished by the supplier should be properly identified with each batch of material.

Prior to acceptance of the supplier's material, samples of material submitted by the supplier will be tested by the purchaser. If any sample is found not to conform to the standards, the material represented by that sample will be rejected. If samples of the supplier's materials that have been previously accepted are found not to conform to the standards, all such material will be rejected.

Unless otherwise specified in the standard specifications, the methods of sampling and testing should be in accordance with applicable methods of the ASTM and ISO 5256.

10.18 COAL-TAR-IMPREGNATED GLASS-FIBER MAT FOR OUTER WRAP

This section covers the minimum requirements for machine-applied, coal-tar-impregnated glass-fiber mat to be used as outerwrap for mechanical protection of hot-applied coal-tar enamel coatings.

The outerwrap should be a nonwoven, thick, glass-fiber mat uniformly impregnated and coated with plasticized coal-tar enamel that is compatible with AWWA C203-86. The glass-fiber mat used should be a uniformly porous mat of chemically resistant borosilicate glass, hydrolytic class III (HGB 3) at most, as specified by ISO 719, which is longitudinally reinforced across the full sheet width by continuous filament glass yarn to provide longitudinal reinforcement. The whole should be bonded with a thermosetting resin binder. The binder content should be such that complete impregnation with the coating material is obtained during normal application. In addition, the binder should be able to resist the action of microorganisms. The weight of the base glass-fiber mat, before coating, should not be less than 50 g/m^2.

The outerwrap should have a controlled bleed-through and must be provided with pin holes to ensure sufficient bleed-through of enamel so that gases, vapors, and air can escape, and also facilitate satisfactory adhesion. It should be rot-, moisture-, and bacteria-proof. The outerwrap surface should be dusted with fine mineral matter (surfacing material) before rolling to prevent sticking between layers under normal conditions of shipment and storage. In addition, at the time of unrolling at ambient temperatures (0–38°C), the successive layers of the outerwrap should not stick to one another.

This outerwrap should be suitable for application by line-travel equipment or in pipe mills and coating yards on a fixed-head coating machine.

10.18.1 PROPERTIES

The finished material should meet the requirements given in Table 10.54. The outerwrap should have a smooth, uniform, surface, free of visible faults such as holes, slits, folds, breaks, badly impregnated areas, delamination, uneven or frayed edges, and presence of foreign bodies (oily matter, mud, etc.). Loose or unbonded surfacing material should be removed from the surface of the wrap by brushing or other suitable means before packaging. The method should be in accordance with ASTM D146.

After test samples from inside the roll have been aged in free air for at least 2 h at 25°C ± 1°C, the average strength in the longitudinal direction should be a minimum of 6,000 N/m of width. The average breaking strength in the transverse direction should be at least 4,000 N/m of width. The test method should be in accordance with ASTM D882, modified. Cut 10 specimens with dimensions of 25 mm × 150 mm, with the longer dimensions along the roll, and 10 additional specimens, with the longer dimensions across the roll. The size of the clamps of the tensile testing machine should be wider than 25 mm, and the clamps should be attached to swivels that are free to move

Table 10.54 Physical Characteristics of Coal-Tar-Impregnated Glass-Fiber Mat

Characteristics	Unit	Requirements
Weight (minimum) (maximum)	g/m²	535 730
Thickness (minimum)	mm	0.75
Breaking strength (Average) Longitudinal (minimum) Transversal (minimum)	N/m Width N/m Width	6,000 4,000
Pliability	N/A	Shall pass test
Weight loss upon heating (maximum)	Wt. %	2

in any direction. Grip the specimen from each end, leaving a distance of at least 76 mm between the jaws.

Initiate the breaking of the load by causing the lower clamp of the machine to travel at a uniform speed of 300 mm/min. Disregard the reading of any specimen that breaks less than 13 mm from either clamp, and test an additional specimen in its place. Report the average of the results of 10 individual tests on specimens cut along the roll as the longitudinal breaking strength and the average of the result of 10 individual tests on specimens cut across the roll as the transverse breaking strength. Readings should be to the nearest pound of force (in newtons).

After test samples from inside the roll have been aged for at least 2 h at 25°C ± 1°C, there should be no cracking of the wrapper when bent over a 25-mm mandrel, as determined by the following test method:

Cut five 150-mm strips with the fiber grain as shown in ASTM D146, and immerse them in water at 25°C for 10–15 min. Bend these strips through 180° at a uniform speed for exactly 2 s around the mandrel.

The thickness should not be less than 0.75 mm. The test method should be in accordance with TAPPI T-411, modified. At 10 equally spaced areas selected by random sampling, measure the thickness with an Ames dial reading in increments of 0.25 mm. Use a circular foot and anvil, both 645 mm² in area, exerting a pressure of 13.8 kPa. Make all measurements in an atmosphere of 50% ± 2% relative humidity and at 23°C ± 1°C.

The loss on heating should be not more than 2%, as determined by the following test method:

1. Cut two samples into 150 mm × 300 mm strips.
2. Remove all loose surfacing material from both sides of each sample to preclude any loose particles falling off in the oven during heating and testing.
3. Weigh each strip.
4. Suspend the strips by wire hooks for 2 h in an oven maintained at 82°C ± 3°C. Care should be taken that the samples do not touch each other or the sides of the oven and that localized overheating of the samples does not take place.
5. Remove the strips from the oven, cool them in a dessicator, and weigh them again.
6. Compute the percentage of loss in weight based on the original weight of the sample. The average of the result on the two samples should be reported as the weight loss upon heating.

The roll sizes, as specified by the purchaser, should be as follows:

- Roll length: 122 m, 244 m
- Roll width: 102 mm, 152 mm, 229 mm, 304 mm, 457 mm

10.18.2 STORAGE LIFE AND PACKAGING

The product should meet the requirements of the standards after being stored for 24 months from the date of delivery in a full, tightly covered container. The outerwrap purchased according to standard should be rolled onto cardboard tubes with ID of 76 mm (nominal) and packaged in suitable and approved containers so that during stocking and transport, the full quality of performance is retained. Packing should be weatherproof and strapped on pallets suitable for long-distance shipment.

10.19 BITUMEN-IMPREGNATED GLASS-FIBER MAT FOR OUTERWRAP

This section covers the minimum requirements for machine-applied bitumen-impregnated glass-fiber mat to be used as outer wrap for the mechanical protection of hot-applied bitumen (asphalt) coating system. The outerwrap should be a nonwoven, thick, glass-fiber mat uniformly impregnated and coated with bitumen (asphalt) enamel that is compatible with ISO 5256. The glass-fiber mat used should be a uniformly porous mat of chemically resistant borosilicate glass hydrolytic class III (HGB 3) at most (as specified by ISO 719), which is longitudinally reinforced across the full sheet width by continuous filament glass yarn to provide longitudinal reinforcement. The entire mat should be bonded with a thermo-setting resin binder. The binder content should be such that complete impregnation with the coating material is obtained during normal application. In addition, the binder should be able to resist the action of microorganisms. The weight of the base glass-fiber mat, before coating, should not be less than 50 g/m^2.

The outerwrap should have a controlled bleed-through and must be provided with pin holes to ensure sufficient bleed-through of enamel so that gases, vapors and air can escape, and also facilitate satisfactory adhesion. It should be rot-, moisture-, and bacteria-proof.

The outerwrap surface should be dusted with fine mineral matter (surfacing material) before rolling to prevent sticking between layers in normal conditions of shipment and storage. In addition, at the time of unrolling at ambient temperatures (0–38°C), the successive layers of the outerwrap should not stick to one another.

This outerwrap should be suitable for application with line-travel equipment or in pipe mills and coating yards on a fixed-head coating machine.

10.19.1 PROPERTIES

The finished material should meet the requirements of Table 10.55. The finished outerwrap should have a smooth, uniform surface, free of visible faults such as holes, slits, folds, breaks, badly impregnated areas, delamination, uneven or frayed edges, and presence of foreign bodies (oily matter, mud, etc.), and other flaws.

Loose or unbonded surfacing material should be removed from the surface of the wrap by brushing or other suitable means before packaging. The test method should be in accordance with ASTM D146.

Table 10.55 Physical Characteristics of Bitumen-Impregnated Glass-Fiber Mat

Characteristic	Unit	Requirements
Weight (minimum) (maximum)	g/m²	535 730
Thickness (minimum)	mm	0.75
Breaking strength (average) Longitudinal (minimum) Transversal (minimum)	N/m width N/m width	6,000 4,000
Pliability	N/A	Shall pass test
Weight loss upon heating (maximum)	Wt. %	2

After test samples from inside the roll have been aged in free air for at least 2 h at 25°C ± 1°C, the average strength in the longitudinal direction should be at least 6,000 N/m of width. The average breaking strength in the transverse direction should be no less than 4,000 N/m of width. The test method should be in accordance with ASTM D882, modified. Cut 10 specimens with dimensions of 25 mm × 150 mm, with the longer dimensions along the roll, and 10 additional specimens, with the longer dimensions across the roll.

The size of the clamps of the tensile testing machine should be wider than 25 mm, and the clamps should be attached to swivels that are free to move in any direction. Grip the specimen from each end, leaving a distance of at least 76 mm between the jaws.

Initiate the breaking of the load by causing the lower clamp of the machine to travel at a uniform speed of 300 mm/min. Disregard the reading of any specimen that breaks less than 6.5 mm away from either clamp, and test an additional specimen in its place. Report the average of the results of 10 individual tests on specimens cut along the roll as the longitudinal breaking strength and the average of the results of 10 individual tests on specimens cut across the roll as the transverse breaking strength. Readings should be to the nearest digit.

After test samples from inside the roll have been aged for at least 2 h at 25°C ± 1°C, there should be no cracking of the wrapper when bent at a uniform speed over a 25-mm mandrel, as determined by the following test method:

Cut five 150-mm strips with the fiber grain as shown in ASTM D146, Figure 1, D-1 to D-5, and immerse them in water at 25°C for 10-15 min. Bend these strips through 180° at a uniform speed for exactly 2 s around the mandrel.

The thickness should not be less than 0.75 mm. The test method should be in accordance with TAPPI T-411, modified. At 10 equally spaced areas selected by random sampling, measure the thickness with an Ames dial reading in increments of 0.25-mm. Use a circular foot and anvil, both 645 mm² in area, exerting a pressure of 13.8 kPa. Make all measurements in an atmosphere of 50% ± 2% relative humidity and at 23°C ± 1°C.

The loss upon heating should be not more than 2%, as determined by the following test method:

1. Cut two samples measuring 150 mm × 300 mm.
2. Remove all loose surfacing material from both sides of the sample to preclude any loose particles falling off in the oven during heating and/or testing.

3. Weigh each strip.
4. Suspend the strips by wire hooks for 2 h in an oven maintained at 82°C ± 3°C. Care should be taken that the samples do not touch each other or the sides of the oven and that localized overheating of the samples does not take place.
5. Remove the strips from the oven, cool them in a dessicator, and weigh them again.
6. Compute the percentage of loss in weight based on the original weight of the sample. The average of the result on the two samples should be reported as the weight loss on heating.

The nominal roll sizes, as specified by the purchaser, may be one of the following:

- Roll length: 120 m, 240 m, 100 mm, 150 mm
- Roll width: 230 mm, 300 mm, 460 mm

10.19.2 STORAGE LIFE AND PACKAGING

The product should meet the requirements of the standards after being stored for 24 months from the date of delivery in a full tightly covered container. The outerwrap purchased according to the standards should be rolled onto cardboard tubes with ID of 80 mm (nominal) and packaged in suitable and approved containers so that during stocking and transport, the full quality of performance is retained.

Packing should be weatherproof and strapped on pallets suitable for long-distance shipment.

10.20 COLD-APPLIED LAMINATED PLASTIC TAPE AS INNER-LAYER TAPE FOR TAPE-COATING SYSTEM OF BURIED STEEL PIPES

This section covers the minimum requirements for cold-applied laminated plastic tape to be used as inner-layer tape (innerwrap) in a tape-coating system to the exterior of all diameters of buried steel pipes through mechanical methods. The main function of the inner-layer tape is to serve as the corrosion-protective coating.

The inner layer should be a prefabricated tape consisting of a plastic backing of a polyethylene film and an adhesive layer of homogeneous elastomeric-sealant component laminated to the polyethylene film. The finished material should provide high electrical resistivity, pliability and conformability, resistance to corrosive environments, low moisture absorption and permeability, ability to remain in place on the pipe during its normal service life, and an effective bond to the primed steel surface. It should be able to resist fungi, bacteria, plant root growth, and excessive mechanical damage during normal operations. The inner-layer tape may be made by any process that produces a product meeting the properties described in the following sections.

Note: The inner-layer tape and primers should be from the same manufacturer.

10.20.1 PROPERTIES

The inner-layer tape should comply with the requirements of Tables 10.56 and 10.57. The outer-layer tape should be compatible with the inner-layer tape.

The plastic backing should be made of high-molecular-weight, film-grade, virgin polyethylene resins with densities in the range of 900–960 kg/m³, as determined by ASTM D 1505, and a nominal

Table 10.56 Standard Dimensions for Inner-Layer Tape (Nominal)

Pipe Diameter	Tape Width	Tape Length
100 mm and under	100 mm	60 m, 120 m
150–300 mm	230 mm	60 m, 120 m
355–610 mm	300 mm	60 m, 120 m, 240 m
660 mm and over	300 mm or 460 mm	60 m, 120 m, 240 m

Table 10.57 Physical Properties of Inner-Layer Tape

Property	Unit	Requirement	Test Method ASTM
Thickness: Total Backing (minimum) Adhesive (minimum)	mm	0.550 ±10% 0.300 0.200	D1000
Tensile strength (minimum)	kg/cm width	5	D1000
Elongation at break (minimum)	%	100	D1000
Adhesion to primed steel (minimum)	kg/cm width	2.2	D1000 (method A)
Adhesion to self (at overlaps) (minimum)	kg/cm width	0.5	D1000
Dielectric strength (minimum)	V/mm (kV/mm)	40	D1000
Insulation resistance (minimum)	megaohms	10^6	D257
Water vapor transmission rate (maximum)	g/m²/ 24 h	3	E96 (method B)
Water absorption (maximum)	%wt	0.1	D570
Cathodic disbonding (maximum)	mm. Diameter	50	G8 (method A)
Heat aging in 30 days at 60°C: Reduction of elongation and Tensile strength (maximum)	%	20	See 6.7
Temperature range: Application Operation	°C	−20–60 −20–60	N/A

melt flow rate (melt index) range of 0.5–2.5 g/10 min, as determined by ASTM D 1238 (condition E), plus suitable additives and carbon black. The minimum thickness should be 0.3 mm, as determined by ASTM D 1000.

The adhesive layer should consist of an elastomer compound based on a stable synthetic rubber and suitable additives. The elastomer content typically should not be less than 20% by weight. The minimum thickness should be 0.2 mm, as determined by ASTM D 1000.

The inner-layer tape should be sufficiently pliable for normal operations and should withstand, without tearing, the tensile force necessary to obtain a tightly wrapped inner coating free of voids. It should be suitable for line-travel application and also shop coating with a wrapping machine, and no significant wrinkles or blisters should develop during application even with exposure to the sun.

The backing should be smooth and uniform, free of visible faults such as slits, folds, breaks, and uneven or frayed edges. The adhesive layer should be smooth and uniform and as free of lumps and bare spots as the best commercial practices will permit. There should be no adhesive transfer when the tape is unwound from the roll.

The color of the plastic backing should be black. The inner-layer tape should be supplied in roll form, wound onto hollow cores with a nominal ID of 80 mm.

After test samples from inside the roll have been aged for 30 days in an air-circulating oven at a constant temperature of 60°C, the tensile strength and the elongation should be determined at 22°C by ASTM D 1000. An average value for tensile strength and elongation should be no less than 80% of the original unaged value.

The inner-layer tape should be furnished in standard widths and lengths consistent with the pipe diameter as specified by the purchaser (see Table 10.55)

10.20.2 STORAGE LIFE AND PACKAGING

The product should meet the requirements of the standards after being stored for 24 months from date of delivery, in a full tightly covered container at temperatures between −20 to 60°C. The inner-layer tapes purchased according to the standard specifications should be packaged in suitable containers to ensure acceptance and safe delivery to their destination. Rolls of inner-layer tape should be packaged in quantities not to exceed the weight limitation of the container specification. Each roll of inner-layer tape should be protected from adhering to other rolls, the container, or to the packaging material itself by the use of separators.

Each container of inner-layer tape should contain application instructions.

10.21 COLD-APPLIED LAMINATED PLASTIC TAPE AS OUTER-LAYER TAPE FOR TAPE-COATING SYSTEM OF BURIED STEEL PIPES

This section covers the minimum requirements for cold-applied laminated plastic tape to be used as outer-layer tape (outerwrap) in a tape-coating system to the exterior of all the diameters of buried steel pipes through mechanical methods. The main function of the outer-layer tape is to provide mechanical protection to the inner-layer tape and to protect the system from environmental hazards.

The outer-layer tape should be a prefabricated tape consisting of a plastic backing of polyethylene film and an adhesive layer of homogeneous elastomeric-sealant component laminated to the polyethylene film. Although the materials used in the outer-layer tape will provide electrical resistivity, low moisture absorption and permeability, and resistance to corrosive environments, the primary function is to provide mechanical protection to the inner-layer tape and to protect the system from the elements. The outer-layer tape should be compounded so that it will be resistant to outdoor weathering.

10.21.1 PROPERTIES

The outer-layer tape should comply with the requirements of Table 10.59, and when applied over the inner-layer tape, it should provide an effective bond to the inner-layer tape in accordance with the appropriate performance requirements given in Table 10.60. The outer-layer tape should also meet the requirements.

The polyethylene should consist of high-molecular-weight film-grade resins with densities in the range of 0.90–0.96 g/cm^3, when determined by ASTM D1505, and suitable additives.

The adhesive layer should be an elastomeric compound composed of a stable synthetic rubber and suitable additives. Typically, the elastomer content should not be less than 20% by weight.

The plastic backing should be smooth and uniform, free of visible faults such as fish eyes, slits, folds, breaks, uneven or frayed edges, and other defects that could affect the appearance or serviceability.

The adhesive layer should be smooth and uniform and as free of lumps and bare spots as the best commercial practices will permit. There should be no adhesive transfer when the tape is unwound from the roll.

The outer-layer tape should be sufficiently pliable for normal operations and should form an effective bond to the inner-layer tape. It should be suitable for line-travel application and shop coating with a wrapping machine, and no significant wrinkles or blisters should develop even with exposure to the sun.

The color of the plastic backing should be black or white, and ultraviolet (UV) resistant. The outer-layer tape should be supplied in roll form, wound onto hollow cores with a nominal ID of 80 mm.

After test samples from inside the roll have been aged for 30 days in an air-circulating oven at a constant temperature of 60°C, the tensile strength and the elongation should be determined at 22°C by ASTM D1000. An average value for tensile strength and elongation should be no less than 80% of the original unaged value.

The outer-layer tape should be furnished in standard widths and lengths consistent with the pipe diameters shown in Table 10.58. The purchaser will specify the roll size of the tape.

10.21.2 STORAGE LIFE AND PACKAGING

The product should meet the requirements of the standards after being stored for 24 months from the date of delivery in the original container at temperatures between −20 to 60°C. The tapes purchased according to the standard specifications should be rolled onto cardboard tubes with ID of 80 mm (nominal) and packaged in suitable and approved containers so that during stocking and transport, the full quality of performance is retained. Each roll of tape should be protected from adhering to other rolls, to the container, or to the packaging material itself by the use of separators. Packing should be weatherproof and strapped on pallets suitable for long-distance shipment.

Table 10.58 Standard Dimensions for Tape (Nominal)		
Pipe Diameter	**Tape Width**	**Tape Length**
100 mm and under	100 mm	60 m, 120 m
150–300 mm	230 mm	60 m, 120 m
355–610 mm	300 mm	60 m, 120 m, 240 m
660 mm and over	300 mm or 460 mm	60 m, 120 m, 240 m

Table 10.59 Physical Properties of Outer-Layer Tape

Property	Unit	Requirement	Test Method ASTM
Thickness	mm	0.760 ± 10%	D1000
Tensile strength (minimum)	kg/cm width	7	D1000
Elongation at break (minimum)	%	100	D1000
Adhesion to inner-layer tape (minimum)	kg/cm width	0.5	D1000 (method A)
Heat aging in 30 days at 60°C: reduction of elongation & tensile strength (maximum)	%	20	See 6.7
Temperature range: application operation	°C	−20 – 60 −20 – 60	N/A

Table 10.60 Performance Requirements of Outer-Layer Tape in Conjunction with Inner-Layer Tape (Total Coating System)

Property	Unit	Requirement	Test Method ASTM
Dielectric strength (minimum)	V/mM (kV/mm)	40	D1000
Cathodic disbondment (maximum)	mm diameter	50	G8 (method A)
Impact resistance (minimum)	N	2.8	G14

10.22 HAND-APPLIED LAMINATED TAPE SUITABLE FOR COLD-APPLIED TAPE-COATING SYSTEM

This section covers the minimum requirements for hand-applied laminated tape to be used for coating special sections, connections, fittings, cable-to-pipe connections, and field repairs of buried steel pipes protected with a cold-applied tape-coating system.

The hand-applied tape should consist of a laminate comprising a stabilized polyethylene backing and an activated adhesive layer of homogeneous elastomer-base compound. The product should provide high electrical resistivity, resistance to corrosive environments, low moisture absorption and permeability, and an effective bond to the primed steel surface. In addition, the tape must be compatible with, and provide an effective bond to, a previously applied coating if present. The tape should also be of such a nature that it resists fungi, bacteria, plant root growth, and excessive mechanical damage during normal operations, and be sufficiently pliable so that it conforms to the surface that is to be coated. It should also withstand, without tearing, the tensile force necessary to obtain a tightly wrapped coating that fills the helix at the overlap and is free of voids.

The tape should be highly conformable to easy hand wrapping even at low temperatures. It should be designed for use with its own primer, and both tape and primer should be from the same manufacturer.

Table 10.61   Physical Properties of Tape

Property	Unit	Requirement	Test Method ASTM

Thickness:
  Total (minimum)
  Backing (minimum)
  Adhesive (minimum)
	mm	0.900
		0.150
		0.650
			D1000

Tensile strength (minimum)	kg/cm width	2.5	D1000
Elongation at break (minimum)	%	150	D1000
Adhesion to primed steel (minimum)	kg/cm width	1.5	D1000 (method A)
Adhesion to self (at overlaps) (minimum)	kg/cm (width)	0.5	D1000
Dielectric strength (minimum)	$V/\mu\,m$	40	D1000
Insulation resistance (minimum)	megaohms	10^6	D257
Water vapor transmission rate (maximum)	$g/m^2/24\,h$	3	E96 (method B)
Water absorption (maximum)	%wt.	0.1	D570
Cathodic disbonding (maximum)	mm (diameter)	50	G8 (method A)
Heat aging in 30 days at 60°C: reduction of elongation & tensile strength (maximum)	%	20	See 6.4

Temperature range:
  Application
  Operation
	°C	$-20-60$
		$-20-60$
			N/A

10.22.1 PROPERTIES

The finished material should meet the requirements of Table 10.61. The backing should be smooth and uniform, free of visible faults such as fish eyes, slits, folds, breaks, and uneven or frayed edges.

The adhesive layer should be smooth and uniform and as free of lumps and bare spots as the best commercial practices will permit. There should be no adhesive transfer when the tape is unwound from the roll.

The color of the plastic backing should be black. The tape should be supplied in rolls wound onto hollow cores with a nominal ID of 38 mm. A removable interleaf (release paper) should be incorporated against the adhesive compound, preferably extending a minimum of 5 mm beyond the width of the tape on each side.

After test samples from inside the roll have been aged for 30 days in an air-circulating oven at a constant temperature of 60°C, the tensile strength and the elongation should be determined at 22°C by ASTM D 1000, an average value for tensile strength and elongation should be not less than 80% of the original unaged value. The roll sizes, as specified by the purchaser, should be as follows:

• Roll length: 10 m, 20 m
• Roll width: 50 mm, 100 mm, 150 mm, 225 mm

10.22.2 STORAGE LIFE, PACKAGING, AND SAMPLING

The product should meet the requirements of the standards after being stored for 24 months from the date of delivery in an original covered container at temperatures between −20 to 60°C. The tapes purchased according to the standard specifications should be packaged in suitable containers to ensure acceptance and safe delivery to their destination. Rolls of tape should be packaged in quantities not to exceed the weight limitations of the container specifications. Each roll of tape should be protected from adhering to other rolls, the container, or to the packaging material itself by the use of release paper.

Unless otherwise specified by the purchaser, the number of samples for testing should consist of 10% of the lot, but in no case should it be more than 10 rolls. The results of the tests on four specimens cut from each sample roll should be averaged for each test specified in the standards to determine conformity to the specified requirements. The numbers and types of test specimens should be in accordance with the ASTM test method for the specific properties to be determined.

10.23 HAND-APPLIED LAMINATED TAPE (SUITABLE FOR HOT-APPLIED COATING SYSTEMS)

This section covers the minimum requirements for hand-applied laminated tape to be used for coating special sections, connections, fittings, cable-to-pipe connections, and field repairs of underground and underwater steel pipelines protected with hot-applied coal-tar or bitumen (asphalt) coating systems.

The hand-applied tape should consist of a laminate comprising a stabilized polyethylene or polyvinylchloride (PVC) backing and a primer-activated adhesive layer of bituminous compound. The product should provide high electrical resistivity, resistance to corrosive environments, low moisture absorption and permeability, and an effective bond to the primed steel surface. In addition, the tape must be compatible with, and provide an effective bond to, a previously applied coating if present. The tape should also be of such a nature that it resists fungi, bacteria, plant root growth, and excessive mechanical damage during normal operations and be sufficiently pliable so that it conforms to the surface that is to be coated. It should also withstand, without tearing, the tensile force necessary to obtain a tightly wrapped coating that fills the helix at the overlap and is free of voids.

The tape should be highly conformable for easy hand wrapping even at low temperatures. It should be designed for use with its own primer, and both tape and primer should be from the same supplier.

10.23.1 PROPERTIES

The finished material should meet the requirements of Table 10.62. The backing should be smooth and uniform, free of visible faults such as slits, folds, breaks, and uneven or frayed edges.

The adhesive layer should be smooth and uniform and as free of lumps and bare spots as the best commercial practices will permit. There should be no adhesive transfer when the tape is unwound from the roll.

The color of the plastic backing should be black. The tape should be supplied in rolls wound onto hollow cores with a typical ID of 38 mm.

A removable interleaf (release paper) should be incorporated against the adhesive compound, preferably extending a minimum of 5 mm beyond the width of the tape on each side.

Table 10.62 Physical Properties of Tape

Property	Unit	Requirement	Test Method ASTM

Thickness:
  Total
  Backing (minimum)
  Adhesive (minimum)
	mm	$0.900 \pm 10\%$
		0.150
		0.650
			D1000

Tensile strength (minimum)	kg/cm (width)	2.5	D1000
Adhesion to primed steel (minimum)	kg/cm (width)	1.5	D1000 (method A)
Adhesion to self (at overlaps) (minimum)	kg/cm (width)	0.5	D1000
Elongation at break (minimum)	%	150	D1000
Dielectric strength (minimum)	V/μm	15	D1000
Insulation resistance (minimum)	megaohms	10^6	D257
Water vapor transmission rate (maximum)	g/m²/24 h	3	E96 (method B)
Water absorption (maximum)	%wt.	0.1	D570
Cathodic disbonding (maximum)	mm (diameter)	40	G8 (method A)
Heat aging in 30 days at 60 °C: reduction of elongation and tensile strength (maximum)	%	20	See 6.4

Temperature range of:
  Application
  Operation
	°C	$-20 - 60$
		$-20 - 60$
			N/A

After test samples from inside the roll have been aged for 30 days in an air-circulating oven at a constant temperature of 60°C, the tensile strength and the elongation should be determined at 22°C by ASTM D 1000, an average value for tensile strength and elongation should be not less than 80% of the original unaged value. The roll sizes, as specified by the purchaser, should be as follows:

- Roll length: 10 m, 20 m
- Roll width: 50 mm, 100 mm, 150 mm, 225 mm

10.23.2 STORAGE LIFE, PACKAGING, AND SAMPLING

The product should meet the requirements of the standards after being stored for 24 months from the date of delivery in a tightly covered container at temperatures between −20 to 60°C.

The tapes purchased according to the standard specifications should be packaged in suitable containers to ensure acceptance and safe delivery to their destination. Rolls of tape should be packaged in quantities not to exceed the weight limitations of the container specifications. Each roll of tape should be protected from adhering to other rolls, the container, or to the packaging material itself by the use of release paper. Each container of tape should contain application instructions.

Unless otherwise specified by purchaser, the number of samples for testing should consist of 10% of the lot, but in no case should be more than 10 rolls. The results of the tests on four specimens cut from each sample roll should be averaged for each test to determine conformity to the specified requirements. The numbers and types of test specimens should be in accordance with the ASTM test method for the specific properties to be determined.

10.24 PERFORATED PLASTIC TAPE (AS ROCKSHIELD) FOR PIPE COATING

This section covers the minimum requirements for perforated plastic tape intended to be used as an outer wrap and rockshield over tape-coated pipes to be buried. It is intended to protect the coating of buried pipes that are subject to aggressive and rocky terrain, soil consolidation or shrinkage stresses, etc.

The plastic tape should consist of a polyethylene plastic film, formed by extrusion molding and roll polishing, perforated on a 15–20 mm^2 pattern with no larger than 1.5-mm-diameter holes to provide a tough flexible protective sheeting with high mechanical properties and good resistance to chemical agents. The hole pattern should be designed so as to encourage the passage of cathodic protection current to the coating surface.

The tape should be resistant to shocks, bacteria when tested with ASTM G22, fungi when tested with ASTM G21, and natural and artificial agents contained in the soil or in the surrounding medium. It should be sufficiently pliable for normal operations, and suitable for line-travel application, as well as shop coating with a wrapping machine.

10.24.1 MATERIALS AND MANUFACTURE

The base material from which the plastic tape is produced should be virgin polyethylene, to which should be added only those antioxidants, UV stabilizers, and pigments necessary for the manufacture of plastic tape that meets the specifications and suit its purposes. The base material should be as uniform in composition and size and as free of contamination as can be achieved by good manufacturing practices. Impurities that are occasionally contained in polymers should not exceed 0.1% by mass.

The nominal density of the base material, when determined in accordance with ASTM D1505, should be greater than 940 kg/m^3. The melt flow rate (i.e., melt index) of the polyethylene resin used should not be less than 0.4 g/10 min when determined by ASTM D1238 (condition E).

The compound should be Class W, as defined in BS 3412:1976 (as amended by amendments 1 and 2), and the antioxidants used should comply with the standards. The carbon black characterization, dispersion, and content should be in accordance with BS 3412:1976.

10.24.2 PROPERTIES

The finished material should comply with the requirements of Table 10.63. The plastic tape should be supplied in roll form, wound onto hollow cores in the dimensions specified. The cores should have a nominal ID of 80 mm.

Table 10.63 Physical Properties of Tape

Property	Unit	Requirement	Test Method ASTM
Density of base polymer (minimum)	kg/m^3	940	D1505
Tensile strength (minimum)	mPa	15	D638*
Elongation (minimum)	%	100	D638*
Brittleness temperature (maximum)	°C	−70	D746
Vicat softening point (minimum)	°C	120	D1525
Hardness	N/A	50–70	D2240 (Shore D)
Thickness (nominal)	mm	1.0	D4801 (Subclause 11.5)
Weight (minimum)**	g/m^2	800	N/A

*Determine the tensile strength at yield and elongation at break, but the speed of grip separation should be 50 mm/min. It is important that these properties be measured in both the transverse and longitudinal directions.
**The actual net weight of each roll should be determined to the nearest 50 g on suitably calibrated equipment.

The plastic tape should have appearance qualities conforming with those produced by good commercial practices. It should be as free as commercially possible of cracks, blisters, bubbles, discolorations, craze, particles of foreign matter, undispersed raw material, and other defects that could affect appearance or serviceability.

The color of the plastic tape should be black. The plastic tape should be furnished in standard widths and lengths consistent with the pipe diameter.

The nominal roll sizes, as will be specified by the purchaser, may be one of the following:

- Roll length: 60 m, 120 m, 240 m
- Roll width: 100 mm, 150 mm, 230 mm, 300 mm, 460 mm

A disk of approximately 50 mm diameter cut from inside the roll should not change in diameter by more than 5% when immersed in boiling water for 30 min and cooled to the ambient temperature.

10.24.3 STORAGE LIFE, PACKAGING, AND SAMPLING

The product should meet the requirements of the standards after being stored for 24 months from date of delivery in a full, tightly covered container. The plastic tapes purchased according to the standard specifications should be packaged in suitable containers to ensure acceptance for storage and safe delivery to their destination. Each roll should be wrapped with at least one layer of polyethylene film or craft paper and be tightly sealed with tape to prevent contamination.

Unless otherwise specified by the purchaser, the number of samples for testing should consist of 10% of the lot, but in no case should it be more than 10 rolls. The results of the tests on four specimens cut from each sample roll should be averaged for each test to determine conformity to the specified requirements. The types of test specimens should be in accordance with the ASTM test method for the specific properties to be determined.

10.25 PLASTIC GRID (AS ROCKSHIELD) FOR PIPE COATING

This section covers the minimum requirements for polyethylene plastic grid sheets, including fastening accessories intended to be used as a rockshield for the mechanical protection of the coating of buried or submarine coated pipes that are subject to aggressive and rocky terrain, soil consolidation or shrinkage stresses, etc., and in conjunction with concrete saddles or set-on weights. The plastic grid should be made of a virgin polyethylene thermoplastic material extruded into a grid mesh pattern to provide a tough and flexible protective sheeting with high mechanical properties and good resistance to chemical agent. The grid pattern should be designed to ensure that the maximum area of exposure is achieved, while maintaining strength, to encourage the passage of cathodic protection current to the coating surface. It should be resistant to shocks, fungi when tested with ASTM G 21, bacteria when tested with ASTM G 22, and artificial agents contained in the soil or in the surrounding medium. The plastic grid may also be the air-blown type.

10.25.1 MATERIALS AND MANUFACTURE

The base material from which the sheet is produced should be polyethylene, to which should be added only those antioxidants, UV stabilizers, and pigments necessary for the manufacture of sheets that meet the specifications and suit the purpose. The base material should be as uniform in composition and size and as free of contamination as can be achieved by good manufacturing practices. Impurities that are occasionally contained in polymers should not exceed 0.1% by mass. The nominal density of the base material, when determined in accordance with ASTM D1505, should be greater than 940 kg/m^3. The nominal melt flow rate (i.e., melt index) of the material should not be less than 0.4 g/10 min when determined in accordance with ASTM D1238 (condition E). The compound should be Class W, as defined in BS 3412:1976 (as amended by amendments 1 and 2), and the antioxidants used should comply with sections 8.1 and 8.2 of that standard. The carbon black characterization, dispersion, and content should be in accordance with BS 3412: 1976.

10.25.2 PROPERTIES

The physical properties of the finished material should comply with the requirements of Table 10.64. The sheets should be supplied flat or in rolls in the dimensions specified by relevant standards, with clean-cut edges.

The sheets should have appearance qualities conforming with those produced by good commercial practices. It should be as free as commercially possible of cracks, discolorations, particles of foreign matter, undispersed raw material, and other defects that could affect appearance or serviceability.

The color of plastic sheets should be black. The sheets should be supplied with the dimensions (width and length) as specified by the purchaser.

A disk of approximately 50 mm diameter cut from the sheet should not change in diameter by more than 3% when immersed in boiling water for 30 min and then allowed to cool.

10.25.3 FASTENING ACCESSORIES

The fastening accessories intended to secure individual sheets placed around the pipe may be either nonslip, quick-fit, self-locking, plastic ties or circumferential strapping with 15-mm-wide plastic bands and associated buckles as specified by the purchaser. These accessories should be made from

Table 10.64 Physical Properties of Plastic Grid Sheets

Property	Unit	Requirement	Test Method ASTM
Density of base polymer (minimum)	kg/m^3	940	D1505
Tensile strength (minimum)	MPa	19	D638[1]
Thickness (nominal)	mm	3.2	D4801 (subclause 11.5)
Weight (minimum)[2]	g/m^2	700	N/A
Brittleness temperature (maximum)	°C	−70	D746
Vicat softening point (minimum)	°C	120	D1525
Hardness	N/A	55–75	D2240 (shore D)
Water absorption	%wt.	0.25	D570

[1]*Determine the tensile strength at break, but the speed of grip separation should be 50 mm/min.*
[2]*The actual net weight of each sheet should be determined to the nearest 50 g on suitably calibrated equipment.*

virgin plastic materials conforming to those produced by good manufacturing practices. Additional requirements specific to the property and application should be identified by the manufacturer upon request.

10.25.4 STORAGE LIFE AND SAMPLING

The product should meet the requirements of the standards after being stored for 24 months from date of delivery in a full tightly covered container. The materials purchased according to the standard specifications should be packaged in suitable containers to ensure acceptance and safe delivery to their destination. Individual fabricated sheets should be packaged in such a manner as to protect the material against physical and mechanical damage and contamination during shipment, handling, and storage. Each container should have application instructions.

Unless otherwise specified by the purchaser, the number of samples for testing should consist of 10% of the lot, but in no case should it be more than 10 sheets. The results of the tests on four specimens cut from each sample sheet should be averaged for each test to determine conformity to the specified requirements. The numbers and types of test specimens should be in accordance with the ASTM test method for the specific properties to be determined.

10.26 HAND-APPLIED PETROLATUM TAPE AND PRIMER

This section sets out the minimum requirements for hand-applied petrolatum tape and primer to be used for the corrosion protection of underground and underwater valves and irregular shape fittings. The standard that applies to this material consists of the following two parts:

- Part 1: Petrolatum Tape
- Part 2: Petrolatum Primer (Priming Solution)

Petrolatum anticorrosive materials should be used neither for parts that contact an organic solvent directly, nor for the external coating of buried pipelines carrying these materials (including crude oil and petroleum products), nor for substances that are buried in soil or water containing crude oil or petroleum products. The reason for this is the danger of petrolatum being dissolved by such organic materials.

10.26.1 PART 1: PETROLATUM TAPE

Part 1 of this specification covers the minimum requirements for composition, properties, storage life, packaging, sampling, inspection and testing, and labeling of petrolatum tape for corrosion protection.

10.26.2 COMPOSITION

The tape should consist of thin, flexible fabric of synthetic fiber cloth uniformly impregnated and coated on both sides with a mixture of petroleum jelly, natural mineral filler, and suitable additives. The base material should be as uniform in composition and size and as free of contamination as can be achieved by good manufacturing practice.

The tape should be designed for use with its own primer, and both tape and primer should be from the same manufacturer.

10.26.3 CLASSIFICATION

The tape should be Type 1 or Type 2, as specified by the purchaser, in terms of its thermal resistance (see Table 10.65).

10.26.4 PROPERTIES

The tape should provide great insulation, corrosion-preventing properties, low moisture absorption and permeability, and noncracking and nonhardening properties. It should be resistant to salts, mineral acids, and alkaline substances. The tape should be simply unrolled in situ onto the primed steel surface.

The finished material should meet the requirements of Table 10.66.

The tape should be wound uniformly and have a smooth, uniform surface free of remarkable deformation, folds, cuts, badly impregnated areas, uneven or frayed edges, foreign bodies, and other defects that affect practical use. The appearance of tapes should be observed visually.

The tape should be highly conformable to easy hand wrapping even at low temperatures (-5–$0°C$) with no occurrence of breakage nor cracking, and no loss of viscosity consistency.

The saponification number of the petrolatum tape should not be greater than 10 mg KOH/g weight of sample. The test method should be in accordance with ASTM D94.

Table 10.65 Classification of Petrolatum Tape	
Type	**Thermal Resistance (°C)**
Type 1	40
Type 2	60

Table 10.66 Physical Properties of Petrolatum Tape

Property	Unit	Requirement	Test Method
Thickness (average)	mm	1.1 (permissible deviation from average = 0.2)	ASTM D1000
Weight (minimum)	kg/m^2	1.3	N/A
Tensile strength (minimum)	kg/cm width	3.5	ASTM D1000
Dielectric strength (minimum) (double layer)	Kv	16	ASTM D149
Water Absorption (maximum)	%	1	ASTM D570
Thermal resistance, 24 h Type 1 Type 2	N/A	No dripping at 40°C ± 2°C No dripping at 60°C ± 2°C	N/A
Tacky adhesion strength (minimum)	kg/cm width	0.5	N/A
Change in pH	N/A	±1	N/A

The tape should be supplied in rolls densely wound onto hollow cores with a typical ID of 38 mm, and preferably covered on one side with a plastic foil. Unless otherwise specified, the material should be of neutral color.

The nominal roll sizes, as will specified by the purchaser, should be as follows:

- Roll length: 5–10 m
- Roll width: 50 mm, 100 mm, 150 mm, and 225 mm

10.26.5 STORAGE LIFE, PACKAGING, AND SAMPLING

The product should meet the requirements of the standards after being stored for 24 months from the date of delivery, in an original covered container in normal conditions. The tapes purchased according to the standard specifications should be packaged in suitable containers to ensure acceptance and safe delivery to their destination.

Each roll should be individually packaged with moisture-proof material, and rolls of tape should be securely packed in parcels or boxes in such a way as to protect them from damage. Packing should be weatherproof and strapped on pallets suitable for long-distance shipment.

Unless otherwise specified, the number of samples for testing should consist of 10% of the lot, but in no case should it be more than 10 rolls. The results of the tests on four specimens cut from each sample roll should be averaged for each test to determine conformity to the specified requirement.

10.26.6 SCOPE

This part of the standard specifications covers the minimum requirements for petrolatum primer (priming solution) to be used in conjunction with petrolatum tape (Part 1) for corrosion protection.

The primer should be based on petroleum-jelly compound and of constituents compatible with the brand of petrolatum tape material blended with the proper type and amount of volatile organic solvent to produce a liquid coating that can be applied cold by brushing or spraying.

The base material should be as uniform in composition and size and as free of contamination as can be achieved by good manufacturing practices. The product should be capable of producing a satisfactory bond between the metal surface and the petrolatum tape.

It should be homogeneous, water-free, stable in storage, free of grit and coarse particles, and resistant to salts, mineral acids and alkaline substances. The primer should contain additives that inhibit corrosion and microbiological attack. The primer should be designed for use with a particular tape, and both tape and primer should be from the same manufacturer.

10.26.7 PROPERTIES

The primer should comply with the requirements of Table 10.67 and the following criteria:

- Odor—The odor should be normal for the materials permitted (ASTM Standard D 1296).
- Color—Unless otherwise specified, the material should be of neutral color.
- Compatibility—There should be no evidence of incompatibility of any of the ingredients of the primer when one volume of the primer is slowly mixed with one volume of its own thinner (U.S. Federal Standard No. 141, Method 4203). The thinner should be defined by the manufacturer.

The primer should have satisfactory brushing properties, with a minimum tendency to produce bubbles during application (U.S. Federal Standard No. 141, Method 4321). No heat should be required to produce an effective bond between the surface to be protected and the subsequently applied tape.

The primer should not settle in the container such that it forms a cake or sludge that cannot be stirred easily by hand (U.S. Federal Standard No. 141, Method 3011).

The covering capacity of primer for surfaces that are clean and free of rust, mill scale, weld spatter, water, grease, dirt, and other loose or deleterious matter should not be less than 8 m^2 per liter of primer with regard to specified adhesion strength of coating system.

Table 10.67 Requirements for Primer			
Property	**Unit**	**Requirement**	**Test Method ASTM**
Specific gravity at 25°C	N/A	0.9±1.1	D70
Viscosity (flow time; ford cup no. 4) at 25°C	S	50–70	D1200
Saponification number (maximum)	mg	10	D94
Temperature range of:	°C	−5–40 −5–60	N/A
• Application • Operation			

The primer should contain no benzene (benzol), chlorinated solvents, hydrolyzable chlorine derivatives, or other materials of a highly toxic nature. The solvent portion of the primer should be certified by the manufacturer to comply with the air pollution control rules and regulations and all safety rules and regulations in effect where the coating is used.

10.26.8 STORAGE LIFE, PACKAGING, AND SAMPLING

The primer should show no thickening, curdling, skinning, gelling, or hard caking after being stored for 24 months from date of delivery in a full, tightly covered container when tested in accordance with U.S. Federal Standard No. 141, Method 3011. The primer should be packaged in perfectly tight containers to prevent solvent from evaporating and being polluted with dust, water, and foreign materials. All containers should be of a suitable shape, with a sufficiently large aperture to allow adequate stirring and mixing. The primer should be furnished in 1-l steel cans, in 3.8-l steel cans, or other suitable containers as specified by the purchaser.

Unless otherwise specified, the number of samples for testing should consist of 10% of the lot, but in no case should it be more than 10 samples. The results of the tests on at least two specimens made from each sample should be averaged for each test to determine conformity to the specified requirements.

10.26.9 DIRECTIONS FOR USE

In addition to the manufacturer's instructions for use, the following directions should be supplied with each container of primer:

- This primer is intended for use as a prime coat on prepared steel surfaces. The surface of steel should be prepared to ST 3 in accordance with the standard before applying primer.
- This primer is intended to be followed by prefabricated hand-applied petrolatum tape conforming to the standard given in Part 1.

10.26.10 SAFETY PRECAUTIONS

In addition to the manufacturer's instructions for safety, the following directions should be supplied with each container of primer:

This primer is hazardous because of its flammability and potential toxicity, so proper safety precautions should be observed to protect against these recognized hazards. Safe handling practices are required and should include, but not be limited to, the provisions of SSPC-PA Guide 3, "A Guide to Safety in Paint Application," and to the following specifications:

- Keep primer away from heat, sparks, and open flame during storage, mixing, and application. Provide sufficient ventilation to maintain vapor concentration at less than 25% of the lower explosive limit.
- Avoid prolonged or repeated breathing of vapors or spray mist, and prevent contact of the primer with the eyes or skin.
- Clean hands thoroughly after handling primer and before eating or smoking.
- Provide sufficient ventilation (if working in closed area) to ensure that the vapor concentrations do not exceed the published permissible exposure limits. When necessary, supply appropriate personal protective equipment and enforce its use.

10.26.11 GENERAL CONDITION OF TESTING

Unless otherwise specified, tapes (and their primers) should be tested on a testing site under the standard atmospheric conditions (i.e., a temperature of 23°C ± 5°C and a relative humidity of 65% ± 10%). The sample rolls of tape (and the primer) should preliminarily be left standing on a testing site under the standard atmospheric conditions for 2 h or longer before removing for the test. The material should be prepared as follows:

1. Place the rolls of tape to be tested on a freely revolving mandrel. Discard the first three layers of tape from the roll.
2. Remove the required length of specimen in a lengthwise direction.
3. Place the specimen on a smooth clean surface or suspend it from one end in free air for the conditioning period, unless otherwise specified.
4. The test specimens should then be prepared as specified in the individual test methods.

10.26.12 DETERMINATION OF THERMAL RESISTANCE

The apparatus to be used in the testing is as follows:

- Thermostatic bath—A hot wind circulating bath capable of adjusting the temperature at 40°C ± 2°C for Type 1 tapes, and at 60°C ± 2°C for Type 2 tapes, constructed to allow the test body to be installed as shown in Figure 10.1.
- Steel pipe—The steel pipe of nominal diameter 50 mm and length of 300 mm, free of scale and other extraneous matters and finished to expose the steel pipe surface, should be used.

The procedure is performed as follows:

1. Cut the sample into a piece 50 mm in width and about 500 mm in length.
2. Stick one of its ends on the upper half of the circumference of the steel pipe.
3. Suspend a weight of 3 kg at the other end.
4. Wind the sample in two layers around the steel pipe by rotating the pipe (refer to Figure 10.2).

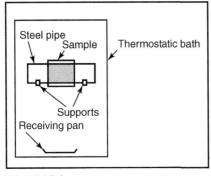

FIGURE 10.1

A hot wind circulating bath capable of adjusting the temperature.

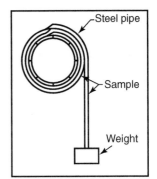

FIGURE 10.2

Wind the sample in two layers around the steel pipe by rotating the pipe.

5. Remove the load, cut off the portion of the sample that remains unwound, and smooth the surface by hand in the winding direction to obtain a smooth test body with no appreciable difference in level.
6. After leaving the test body standing for 30 min to 1 h, place it in the thermostatic bath adjusted at 40°C ±2°C, for Type 1 sample, and at 60°C ± 2°C for Type 2 sample, and keep the test body horizontal (refer to Figure 10.1).
7. After 24 h, observe whether the petrolatum drips.

10.26.13 DETERMINATION OF COLD WORKABILITY

The procedure to determine cold workability employs a thermostatic bath capable of adjusting the temperature to between −5 and 0°C. Leave the sample of 50 mm in width standing in the thermostatic bath for at least 2 h. After immediately removing the outer three layers of the tape under the standard atmospheric conditions, quietly unwind a length of about 1 m in about 3 to 5 s, and examine for the occurrence of breakage, cracking, and a change in viscosity consistency by feeling it with the hands.

10.26.14 DETERMINATION OF CHANGE IN pH

The apparatus used to determine changes in pH are as follows:

- pH meter—The pH meter as specified in ASTM E70.
- Beaker—The beaker of hard quality with nominal capacity of 500 ml.

The procedure is as follows:

1. Introduce 5 ml water (purified by ion exchange) or distilled water into the beaker and completely immerse a test piece 25 mm in width and 110 mm in length in the water.
2. After 24 h has elapsed with the sample kept at room temperature, measure the pH according to ASTM E70.
3. Measure the pH of water without the test piece immersed to obtain the pH value for the blank test.
4. Take the value obtained by subtracting the pH value for the main test from that of the blank test as the change in pH.

10.26.15 DETERMINATION OF TACKY ADHESION STRENGTH

After test samples from inside the roll have been aged for at least 2 h under the standard atmospheric conditions, the average adhesion strength should not be less than 0.5 kg/cm of width. The test method should be in accordance with ASTM D 1000 (Method A), except that the test specimen should be 150 mm in length and 25 mm in width.

Stick the test specimen on one end of the steel plate so as to make the contact surface area 25 mm × 50 mm (see Figure 10.3).

Stick a polyester film of 25 m in thickness and 25 mm × 150 mm in size on the test specimen and pressure-bond it by reciprocating the roller of the pressure-bonding apparatus at a speed of about 50 mm/S.

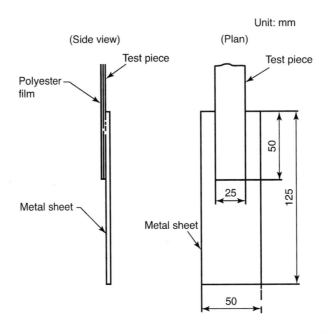

FIGURE 10.3

Method of laminating test piece.

After 30 min of pressure bonding, grip the test piece laminated with the polyester film by the upper chuck of the tensile-testing machine and the metal sheet by the lower chuck, and pull at a speed of 300 mm/min ± 30 mm/min. Thus, read the indicated load value when the test piece is peeled off; that is the tacky adhesion strength of the tape.

10.26.16 DETERMINATION OF DIMENSIONS OF ROLLS

The apparatus used to determine the dimension of rolls is:

- A balance capable of weighing to the nearest 1 g
- A steel ruler capable of measuring to the nearest 1 mm

The test specimen should consist of a single thickness of tape approximately 1,000 mm long removed from a full roll of tape. The specimen should be conditioned for not less than 2 h.

Remove the core from the roll and weigh the tape to the nearest 10 g. If additional tests on the tape are desired, it is permissible to weigh the roll with the core in place first and then subtract the weight of the core after all test specimens have been removed.

Remove a test specimen of tape approximately 1,000 mm long from the roll in accordance with the procedure described in Appendix A (with standard procedure). After conditioning, measure the relaxed length of the specimen to the nearest 10 mm and its width to the nearest 1 mm.

From the net mass and the dimension, calculate the average net mass per unit area for the rolls in the representative sample and record it as the average for the lot.

10.27 HEAT-SHRINKABLE, CROSS-LINKED, TWO-LAYERED POLYETHYLENE COATING

This section covers the minimum requirements for heat-shrinkable, cross-linked, two-layered polyethylene coatings used for external protection of special sections, connections, joints, and fittings of underground and underwater steel pipelines that are coated with extruded polyethylene, fusion-bonded epoxy, coal-tar, and bituminous coatings. It discusses only heat-shrinkable coatings that consist of a polyethylene backing that has been cross-linked by radiation (electron beam) and coated with a non-cross-linked adhesive. For convenience, such coatings are referred to as heat-shrinkable coatings throughout the remainder of this section.

10.27.1 CLASSIFICATION

The prefabricated heat-shrinkable polyethylene coating, as will be specified by the purchaser, should be one of the following types:

- Type 1: Tubular—These coatings are installed before joining the pipe ends by sliding the coating from a free end of the pipe onto the area to be coated.
- Type 2: Wraparound—These coatings are wrapped circumferentially around the pipe area to be coated. Each wraparound coating is provided with either a separate or a built-in closure to secure the overlap during shrinking. The closure should meet all the technical requirements of a Type 3 heat-shrinkable coating as given in Table 10.68.

Table 10.68 Physical Properties of Heat-Shrinkable Coatings (as Supplied)

Property	Unit	Requirement
Width deviation		±10 % width or 6 mm whichever is smaller
Thickness (minimum) Type 1 Type 2 Type 3	mm	1.5 1.5 1.1
Dielectric strength (minimum)	kV/mm	15
Volume resistivity (minimum)	Ω cm	10^{14}
Adhesion to steel (minimum)	kg/cm	1
Adhesion to PE (minimum)	kg/cm	1
Heat shock (test for crosslinking of backing)		No visual cracking, flowing, or dripping
Tensile strength (minimum)	MPa	14
Elongation (minimum)	%	300
Water vapor transmission (maximum)	g/24 h/m^2	0.01
Low-temperature flexibility, −20°C, 25.4 mm mandrel		No cracking

- Type 3: Tape-Type coatings-these are helically wrapped around the pipe area to be coated, with an overlap per the manufacturer's recommendation.

10.27.2 MATERIALS AND MANUFACTURE

Heat-shrinkable coatings consist of materials fabricated from cross-linked polyethylene sheets or tubings precoated with an adhesive. After installation, they should conform to all surface contours of the pipe.

The heat-shrinkable coating should be a laminate that consists of a cross-linked polyethylene backing and a homogeneous adhesive layer. The base polymer of this backing should consist substantially of polyethylene or a polyethylene copolymer. The adhesive may be either a tacky, pressure-sensitive amorphous adhesive (mastic) or a nontacky, semicrystalline adhesive (hot melt).

The polymers used in the manufacture of heat-shrinkable coatings should be virgin material—that is, material that has been through the manufacturing process only once. The finished compound should be free of all foreign matter other than antioxidants, flame retardants, processing aids, cross-linking agents, and pigments as appropriate. The heat-shrinkable polyethylene coating should be extruded, cross-linked, and then expanded to the required dimensions.

10.27.3 REQUIREMENTS

The material should conform to the chemical requirements specified in Tables 10.68 and 10.69.

The product should be homogeneous and essentially free of flaws, defects, pinholes, bubbles, seams, cracks, or inclusions.

Type 1 coatings should be supplied in sleeves of a predetermined diameter to fit the steel pipe. Type 2 coatings may be supplied in individually precut sizes or in roll form. Type 3 coatings should be supplied in roll form.

A roll-form product is wound onto hollow cores with a nominal ID of 76 mm.

The color of the coatings should be black.

Precut coatings should be provided by the manufacturer in standard widths. The manufacturer's recommendations should be followed with regard to appropriate overlap and coverage considerations unless specified otherwise by the purchaser.

Table 10.69 Chemical Requirements	
Property	**Requirement**
Solvent resistance. 24 h at 24°C ±3°C JP-4 Fuel, MIL-T-5624 Lubricating oil, MIL -L-7808 Lubricating oil, MIL -L-23699	
Followed by tests for:	12
Dielectric strength, minimum, kV/mm Tensile strength, minimum, MPa	11
Water absorption, 24 h at 23°C ±2°C, maximum, %	0.2

The width deviation should not exceed 10% of the coating width or 6 mm, whichever is smaller, as determined by one of the following test methods:

- For Type 1 and Type 2 coatings, three discrete product sleeves should be selected at random and placed on a smooth, flat surface.
- For Type 3 coatings, a specimen that is at least 1 m long should be removed from each of three randomly selected rolls of material. The width of the specimen should be measured at several points along the length of the sample to the nearest 1.6 mm.

Type 1 and Type 2 heat-shrinkable coatings should have a minimum thickness of 1.5 mm, and Type 3 coatings should have a minimum thickness of 1.1 mm. The thickness of the coating should be measured at not less than 10 locations on each of the three specimens. The measurement should be made with a micrometer calibrated to read in increments of 0.025 mm and having contact feet of at least 6 mm in diameter. The average of 10 measurements should be as set forth in Table 2. No single measurement should deviate more than 20% from the average value. The thickness of the adhesive should be at least equal to, but no more than three times, the thickness of the cross-linked polyethylene backing.

The heat-shrinkable coatings should be tested for dielectric breakdown in accordance with ASTM D149 using a 25-mm-diameter electrode and a 500-V/s voltage rise. A value below the limit specified in Table 10.68 should constitute a failure of the coating to meet the dielectric strength requirement.

The heat-shrinkable coatings should be tested for heat shock (to determine if the backing of the heat-shrinkable coating is cross-linked) by removing the adhesive from the coating before performing the test and hanging a strip of the prepared backing in a 200°C oven for 4 h. The backing will shrink, but it should show no dripping, flowing, or cracking as specified in Table 10.68. The manufacturer should provide details for removing the adhesive and for performing this test.

The polyethylene backing of the heat-shrinkable coatings should be laminated with appropriate temperature-sensitive paint to be able to change the color when heated.

10.27.4 STORAGE LIFE, PACKAGING, AND SAMPLING

The product should meet the requirements of the standards after being stored for 24 months from the date of delivery in a tightly covered container at temperatures between −20°C and 60°C.

All heat-shrinkable coatings purchased according to the standard specifications should be packaged in suitable containers that ensure acceptance and safe delivery to their destination. Each coated item should be packaged to prevent adherence to the packaging material or the container. Individual types and sizes should be neatly bundled or boxed. Coatings should be packaged in quantities not to exceed the weight limitation of the container specifications. Each container of coating material should contain application instructions.

A lot should consist of all materials that are processed at the same time and under the same conditions and submitted for inspection at one time. Unless otherwise specified by the purchaser, the number of samples for testing should consist of 10% of the lot, but in no case should it be more than 10 samples. The results of the tests on four specimens cut from each sample should be averaged for each test to determine conformity to the specified requirements. The numbers and types of test samples should be in accordance with the ASTM test method for the specific properties to be determined.

10.28 PRIMERS (DITCH AND YARD) TO BE USED WITH COLD-APPLIED, INNER LAYER TAPE (TAPE-COATING SYSTEM OF BURIED STEEL PIPE)

This part of the standard specifications covers the minimum requirements for ditch primer to be used with cold-applied, inner-layer tape in a tape-coating system. The primer is intended for use in line-travel coating applications and the function of this primer is to provide a highly effective bonding medium between yard primer (Part 2 of this specification) and inner-layer tape, as well as between steel surfaces and inner-layer tape.

10.28.1 COMPOSITION

The primer for cold application should compose of synthetic elastomers and resins, anticorrosion inhibitors, stabilizers, and other ingredients blended with the proper type and amount of volatile organic solvent to give a consistency suitable for spray-type, rug-type, and brush-type application. The primer should be uniform, stable in storage, and free of grit and coarse particles. The primer should contain additives that resist the effects of humidity, water condensation, radiation, winds, salty environments, and fungus and bacterial growth.

When required by the purchaser, suitable additives should be utilized in the primer formulation to reduce the incidence of stress-corrosion cracking. The primer should be compatible with the inner-layer tape.

10.28.2 PROPERTIES

The primer should comply with the requirements of Table 10.70. When dry, they should provide a highly effective bonding medium between yard primer and the surface to be protected, and the adhesive layer of the subsequently applied inner-layer tape, to meet the requirements in the table.

The primer should also meet the following criteria:

- Odor—The odor should be normal for the materials permitted (ASTM Standard D 1296).
- Color—The color of the primer should be black.
- Compatibility—There should be no evidence of incompatibility of any of the ingredients of the primer when one volume of the primer is slowly mixed with one volume of its own thinner (U.S. Federal Standard No. 141, method 4203). The thinner should be defined by the manufacturer.

Table 10.70 Properties of Primer in Conjunction with Cold-Applied, Inner-Layer Tape			
Property	**Unit**	**Requirement**	**Test Method ASTM**
Total solid content (minimum)	Wt. %	20	D2369
Density at 25°C	g/cm^3	0.8 ± 0.05	D1475
Viscosity (flow time; ford cup no. 4) at 25°C	s	30-60	D1200
Temperature range of: Application Operation	°C	$-20-60$ $-20-60$	N/A

The primer should have good machine-application properties, with a minimum tendency to produce bubbles during application. No heat should be required to produce an effective bond between the pipe surface to be protected and the subsequently applied inner-layer tape. In addition, the primer should be so designed not to cause shrinkage of inner-layer tape before and after application of outer-layer tape in field coating conditions.

The primer should not settle in the container such that it forms a cake or sludge that cannot be mixed easily by hand or mechanical agitation. It should be quick-drying (within 5–15 min) even at low temperatures.

The covering capacity of the primer for surfaces with a roughness of 50 μm (arithmetical average) and the cleanliness of Sa 2½ should not be less than 5 m^2 per 1 l of primer with regard to the specified adhesion strength of the coating system (see Table 10.71).

A film of the primer tested as described next should withstand bending without cracking or flaking. The test should be run in accordance with U.S. Federal Standard No. 141, Method 6221. The panels should be 8 cm × 16 cm, 20 gauge cold-rolled steel, which has been cleaned in accordance with Federal Standard No. 141, Method 2011, Procedures D and A.

The primer should be applied by spraying to a 75-μm dry film thickness (for wet film, the thickness is approximately 225 μm). After 1 week of drying at 21–27°C, the panel should be bent over a 4-cm mandrel and should show no cracking or loss of adhesion.

A film of the primer tested as described next should show no appreciable film deterioration or excessive change in general appearance. The panels should be 8 cm × 16 cm × 1/3 cm cold-rolled steel prepared according to SIS 055900 "SA 2½" using 0.850–0.425 mm. (20–40 mesh) silica sand. The primer should be applied by spray to a 75-μm dry film thickness (with wet film, the thickness is approximately 225 μm). After 1 week of drying at 21–27°C, the panels should be immersed halfway in distilled water at 38°C.

The panels should be examined every 24 h for blistering, leaching, rusting, or loss of adhesion. Films of the primer prepared and exposed as described above should withstand 500 h immersion with no other defect than a slight discoloration (examine the sample after it has dried for 1 h).

A film of primer tested as described below and examined immediately after removal from the salt spray testing environment should show no more than a trace of rusting and no more than five scattered blisters no larger than 1 mm in diameter. On removal of the primer, there should be no more than a trace of rusting, pitting, or corrosion of the steel.

This test method should be performed in accordance with ASTM Standard B117. The panels should be 8 cm × 16 cm × 1/3 cm cold-rolled steel, sand-blasted in accordance with SIS 055900 "Sa 2½,"

Table 10.71 Performance Requirements of Primer in Conjunction with Cold-Applied, Inner-Layer Tape

Property	Unit	Requirement	Test Method ASTM
Adhesion strength (minimum)	kg/cm width	2.2	D1000 (method A)
Dielectric strength breakdown (minimum)	V/mm (kV/mm)	40	D1000
Cathodic disbonding (maximum)	mm diameter	50	G8 (method A)

using 0.850–0.425 mm (20–40 mesh) silica sand. The primer should be applied by spray to a 75-μm dry film thickness (for wet film, the thickness should be approximately 225 μm). After 1 week of drying at 21–27°C, the panels should be scored as indicated in ASTM Standard B 117. The panels should be exposed in the cabinet at a 30° angle from the vertical, and they should be tested twice and examined at 96-h intervals for a period of 500 h.

Films of the primer prepared and tested as described below should show no rusting, blistering, flaking, loss of adhesion, or excessive change in general appearance. The panels should be 16-cm channel steel in a 30-cm-long section and exposed at a 45° angle southern exposure with the flanges horizontal, each panel should have a weld bead approximately 8 cm long placed at the left end of the panel about 8 cm from the end.

The primer should be applied by spray to a 75-μm dry film thickness (with wet film, the thickness should be approximately 225 μm). After 1 week of drying at 21–27°C, the coated steel should be scored in an X pattern on the right 8 cm from the end. Each stroke of the X must be 10 cm long. Then the panels should be placed on exposure. After 60 days of exposure, they should show no blistering, rusting, or corrosion on their faces and only minor blistering or rusting at the weld and score.

The primer should contain no benzene (benzol), chlorinated solvents, hydrolyzable chlorine derivatives or other materials of highly toxic nature. The solvent portion of the primer should be certified by the manufacturer to comply with the air pollution control rules and regulations and all safety rules and regulations in effect where the coating is used. The primer should be supplied by the manufacturer that supplies the inner-layer tape.

10.28.3 STORAGE LIFE AND PACKAGING

The primer should show no thickening, curdling, skinning, gelling, or hard caking after being stored for 24 months, in normal conditions, from date of delivery in a full, tightly covered container when tested in accordance with U.S. Federal Standard No. 141, Method 3011. The primer should be packaged in perfectly tight containers to prevent solvent from evaporating and being polluted with dust, water, and foreign materials. The primer should be furnished in a new, heavy-gauge steel drum with a capacity specified by the purchaser.

In addition to the manufacturer's instructions for use, the following directions should be supplied with each container of primer:

- This material is intended for use as a ditch primer on primed steel pipes and on prepared steel surfaces. The surface of the steel should be prepared in accordance with the standards before applying the primer.
- This primer is intended to be followed by cold-applied, laminated plastic tape conforming to the standards. Mix the primer thoroughly before use.

10.28.4 SAFETY PRECAUTIONS

In addition to the manufacturer's instructions for safety, the following directions should be supplied with each container of primer:

This primer is hazardous because of its flammability and potential toxicity, so proper safety precautions should be observed to protect against these recognized hazards. Safe handling practices are required and should include, but not be limited to, the provisions of SSPC-PA Guide 3, "A Guide to Safety in Paint Application," and to the following specifications:

- Keep primer away from heat, sparks, and open flame during storage, mixing, and application. Provide sufficient ventilation to maintain vapor concentration at less than 25% of the lower explosive limit.
- Avoid prolonged or repeated breathing of vapors or spray mist, and prevent contact of primer with the eyes or skin.
- Clean hands thoroughly after handling primer and before eating or smoking.
- Provide sufficient ventilation to ensure that vapor concentrations do not exceed the published permissible exposure limits. When necessary, supply appropriate personal protective equipment and enforce its use.

10.29 PRIMER TO BE USED WITH HAND-APPLIED, LAMINATED TAPE SUITABLE FOR COLD-APPLIED TAPE-COATING SYSTEM

This section covers the minimum requirements for primer to be used in conjunction with hand-applied laminated tape for coating special sections, connections, fittings, cable-to-pipe connections, and field repairs of buried steel pipes protected with a cold-applied tape coating system.

10.29.1 COMPOSITION

The primer should compose of synthetic resin and rubber, anticorrosion inhibitor, stabilizer, etc., blended with proper type and amount of volatile organic solvent to produce a free-flowing liquid coating that can be readily applied without heat by brushing. The primer should be uniform, stable in storage, and free of grit and coarse particles. It must contain additives that resist fungus and bacterial growth.

10.29.2 PROPERTIES

The primer should comply with the requirements of Table 10.72. When dry, it should provide a highly effective bonding medium between the surface to be protected and the adhesive layer of the

Table 10.72 Properties of Primer in Conjunction with Hand-Applied, Laminated Tape

Property	Unit	Requirement	Test Method ASTM
Total solid content (minimum)	% by weight	27	D2369
Density (at 25°C)	g/cm^3	0.8 ± 0.03	D1475
Viscosity (flow time; ford cup no. 4) at 25°C	s	35 – 60	D1200

Temperature range of:
  Application
  Operation
	°C	−20 – 60
		−20 – 60	

Table 10.73 Performance Requirements of Primer in Conjunction with Hand-Applied, Laminated Tape

Property	Unit	Requirement	Test Method ASTM
Adhesion strength (minimum)	kg/cm	1.5	D1000 (method A)
Dielectric strength breakdown (minimum)	V/mm (kV/mm)	40	D1000
Cathodic disbonding (maximum)	mm diameter	50	G8 (method A)

subsequently applied tape, complying with the relevant standards, to satisfy the requirements given in Table 10.73 and the following criteria:

- Odor—The odor should be normal for the materials permitted (ASTM Standard D 1296).
- Color—The color of the primer should be black.
- Compatibility—There should be no evidence of incompatibility of any of the ingredients of the primer when one volume of the primer is slowly mixed with one volume of its own thinner (U.S. Federal Standard No. 141, Method 4203). The thinner should be defined by the manufacturer.

The primer should have satisfactory brushing properties, with a minimum tendency to produce bubbles during application. The primer should not pull or have a quick set under the brush. The test method should be in accordance with U.S. Federal Standard No. 141, Method 4321.

The brushed film of the primer should dry to a smooth film of uniform appearance free of grit, seeds, streaks, blisters, or other surface defects when tested in accordance with U.S. Federal Standard No. 141, Method 4541. The primer should not settle in the container so that it forms a cake that cannot be stirred easily by hand (U.S. Federal Standard No. 141, Method 3011).

The primer should be quick-drying (3–10 min) even at low temperatures. The test method should be in accordance with U.S. Federal Standard No. 141, Method 4061.

The covering capacity of the primer for surfaces with roughness of 50 μm (arithmetical average) and the cleanliness of Sa 2½ should not be less than 8 m² per liter of primer with regard to the specified adhesion strength of the coating system (see Table 10.73).

The primer should contain no benzene (benzol), chlorinated solvents, hydrolyzable chlorine derivatives, or other materials of highly toxic nature. The solvent portion of the primer should be certified by the manufacturer to comply with the air pollution control rules and regulations and all safety rules and regulations in effect where the coating is used.

10.29.3 STORAGE LIFE, PACKAGING, AND SAMPLING

The primer should show no thickening, curdling, skinning, gelling, or hard caking after being stored for 24 months, in normal conditions, from date of delivery in a full, tightly covered container when tested in accordance with U.S. Federal Standard No. 141, Method 3011. The primer should be packaged in perfectly tight containers to prevent solvent from evaporating and being polluted with dust, water, and foreign materials.

All containers should be of a suitable shape, with a sufficiently large aperture to allow adequate stirring and mixing. The primer should be furnished in 3.8-l, new steel cans, in 20-l, new steel pails, or other suitable containers as specified by the purchaser.

Unless otherwise specified by the purchaser, the number of samples for testing should consist of 10% of the lot, but in no case should it be more than 10 samples. The result of the tests on at least two specimens made from each sample should be averaged for each test to determine conformity to the specified requirements. The numbers and types of test specimens should be in accordance with the ASTM test method for the specific properties to be determined.

10.29.4 DIRECTIONS FOR USE

In addition to the manufacturer's instructions for use, the following directions should be supplied with each container of primer:

- This primer is intended for use as a prime coat on prepared steel surfaces. The surface of steel should be prepared in accordance with relevant standards before applying the primer.
- This primer is intended to be followed by hand-applied, laminated tape conforming to the standards. Mix primer thoroughly before use.

10.29.5 SAFETY PRECAUTIONS

In addition to the manufacturer's instructions for safety, the following directions should be supplied with each container of primer:

This primer is hazardous because of its flammability and potential toxicity, so proper safety precautions should be observed to protect against these recognized hazards. Safe handling practices are required and should include, but not be limited to, the provisions of SSPC-PA Guide 3, "A Guide to Safety in Paint Application," and to the following specifications:

- Keep primer away from heat, sparks, and open flame during storage, mixing, and application. Provide sufficient ventilation to maintain vapor concentration at less than 25% of the lower explosive limit.
- Avoid prolonged or repeated breathing of vapors or spray mist, and prevent contact of the primer with the eyes or skin.
- Clean hands thoroughly after handling primer and before eating or smoking.
- Provide sufficient ventilation to ensure that vapor concentrations do not exceed the published permissible exposure limits. When necessary, supply appropriate personal protective equipment and enforce its use.

10.30 PRIMER TO BE USED WITH HAND-APPLIED, LAMINATED TAPE SUITABLE FOR HOT-APPLIED TAPE-COATING SYSTEMS

This section covers the minimum requirements for primer to be used in conjunction with hand-applied, laminated tape for coating special sections, connections, fittings, cable-to-pipe connections, and field repairs of underground and underwater steel pipes protected with hot-applied bitumen or coal-tar coating systems.

10.30.1 COMPOSITION

The primer should be based on a bituminous compound and of constituents compatible with the brand of hand-applied, laminated tape blended with the proper type and amount of volatile organic solvent to produce a free-flowing liquid coating that can be readily applied without heat by brushing. The product should be uniform, stable in storage, and free of grit and coarse particles. It should contain additives that inhibit corrosion and microbiological attack.

Note: Any petroleum bitumen used should be a straight-run, steam-, or vacuum-refined residue and should be air-blown in accordance with good commercial practice.

10.30.2 PROPERTIES

The primer should comply with the requirements of Table 10.74. When dry, it should provide a highly effective bonding medium between the surface to be protected and the adhesive layer of the subsequently applied tape, complying with the relevant standards, to perform the requirements given in Table 10.75.

The primer should also meet the requirements of the standards and the following criteria:

- Odor—The odor should be normal for the materials permitted (ASTM Standard D 1296).
- Color—The color of the primer should be black.
- Compatibility—There should be no evidence of incompatibility of any of the ingredients of the primer when one volume of the primer is slowly mixed with one volume of its own thinner (U.S. Federal Standard No. 141, Method 4203). The thinner should be defined by the manufacturer.

The primer should have satisfactory brushing properties, with a minimum tendency to produce bubbles during application. It should not pull or have a quick set under the brush. The test method should be in accordance with U.S. Federal Standard No. 141, Method 4321.

The brushed film of primer should dry to a smooth film of uniform appearance, free of grit, seeds, streaks, blisters, or other surface defects when tested in accordance with U.S. Federal Standard No. 141, Method 4541.

The primer should not settle in the container so it forms a cake that cannot be stirred easily by hand (U.S. Federal Standard No. 141, Method 3011).

The primer should be quick-drying (5–15 min), even at low temperatures.

Table 10.74 Properties of Primer in Conjunction with Hand-Applied, Laminated Tape			
Property	**Unit**	**Requirement**	**Test Method ASTM**
Total solid content (minimum)	% (By weight)	30	D2369
Density (at 25°C)	g/cm³	1 ± 0.1	D1475
Viscosity (flow time: ford cup no. 4) at 25°C	Second	35 – 60	D1200
Temperature range of: Application Operation	°C	−20 – 60 −20 – 60	

Table 10.75 Performance Requirements of Primer in Conjunction with Hand-Applied, Laminated Tape

Property	Unit	Requirement	Test Method ASTM
Adhesion strength (minimum)	kg/cm	15	D1000 (method A)
Dielectric strength breakdown (minimum)	Vμm (kV/mm)	15	D1000
Cathodic disbonding (maximum)	mm (diameter)	40	G8 (method A)

The test method should be in accordance with U.S. Federal Standard No. 141, Method 4061.

The covering capacity of the primer for surfaces with a roughness of 50 μm (arithmetical average) and the cleanliness of Sa 2½ should not be less than 8 m^2 per liter of primer with regard to the specified adhesion strength of the coating system (see Table 10.75).

The primer should contain no benzene (benzol), chlorinated solvents, hydrolyzable chlorine derivatives, or other materials of highly toxic nature. The solvent portion of the primer should be certified by the manufacturer to comply with the air pollution control rules and regulations and all safety rules and regulations in effect where the coating is used.

10.30.3 STORAGE LIFE, PACKAGING, AND SAMPLING

The primer should show no thickening, curdling, skinning, gelling, or hard caking after being stored for 24 months, in normal conditions, from date of delivery in a full, tightly covered container when tested in accordance with U.S. Federal Standard No. 141, Method 3011. The primer should be packaged in perfectly tight containers to prevent solvent from evaporating and being polluted with dust, water, and foreign materials.

All containers should be of a suitable shape, with a sufficiently large aperture to allow adequate stirring and mixing. The primer should be furnished in 3.8-l new steel cans, in 20-l new steel pails or other suitable containers as specified by the purchaser.

Unless otherwise specified by the purchaser, the number of samples for testing should consist of 10% of the lot, but in no case should it be more than 10 samples. The result of the tests on at least two specimens made from each sample should be averaged for each test specified to determine conformity to the specified requirements. The numbers and types of test specimens should be in accordance with the ASTM test method for the specific properties to be determined.

10.30.4 DIRECTIONS FOR USE

In addition to the manufacturer's instructions for use, the following directions should be supplied with each container of primer:

- This material is intended for use as a prime coat on prepared steel surfaces. The surface of steel should be prepared in accordance with the relevant standards before applying the primer.
- This primer is intended to be followed by hand-applied laminated tape conforming to standards. Mix primer thoroughly before use.

10.30.5 SAFETY PRECAUTIONS

In addition to the manufacturer's instructions for safety, the following directions should be supplied with each container of primer:

This primer is hazardous because of its flammability and potential toxicity, so proper safety precautions should be observed to protect against these recognized hazards. Safe handling practices are required and should include, but not be limited to, the provisions of SSPC-PA Guide 3, "A Guide to Safety in Paint Application," and to the following specifications:

- Keep primer away from heat, sparks, and open flame during storage, mixing, and application. Provide sufficient ventilation to maintain vapor concentration at less than 25% of the lower explosive limit.
- Avoid prolonged or repeated breathing of vapors or spray mist, and prevent contact of the primer with the eyes or skin.
- Clean hands thoroughly after handling primer and before eating or smoking.
- Provide sufficient ventilation to ensure that vapor concentrations do not exceed the published permissible exposure limits. When necessary, supply appropriate personal protective equipment and enforce its use.

Glossary of Terms

acid pickling A treatment for the removal of rust and mill scale from steel by immersing it in an acid solution containing an inhibitor. Pickling should be followed by thorough washing and drying before painting.

acrylic latex Aqueous dispersion (thermoplastic or thermosetting) of polymers or copolymers of acrylic acid, methacrylic acid, esters of these acids, or acrylonitrile.

acrylic resin A synthetic resin made from derivatives of acrylic acid.

additive Anything added in small quantities to another substance, usually to improve that substance's properties.

adhesion A state in which two surfaces are held together by interfacial forces that may consist of valence forces, interlocking actions, or both.

adhesion strength The force necessary to remove tape from a prescribed surface when measured in accordance with the specific conditions of a test.

adhesive A substance capable of holding materials together by surface attachment.

aging Storage of paints, varnishes, and other coverings (under defined conditions of temperature, relative humidity, and other environmental elements) in suitable containers, or as dry films of these materials, for the purpose of subsequent testing.

airless spraying A process of atomizing paint by forcing it through an opening at high pressure. This effect is often aided by flashing (vaporization) of the solvents, especially if the paint has been previously heated.

aliphatic solvent A hydrocarbon solvent comprised primarily of paraffinic and cycloparaffinic (naphthenic) hydrocarbon compounds. Aromatic hydrocarbon content may range from less than 1% to about 35%.

alkyd resins Synthetic resins formed by the condensation of polyhydric alcohols with polybasic acids. They may be regarded as complex esters. The most common polyhydric alcohol used is glycerol, and the most common polybasic acid is phthalic anhydride. Modified alkyds are those in which polybasic acid is substituted in part with monobasic acid, of which the fatty acids of vegetable oils are typical.

aluminum paint A coating consisting of a mixture of metallic aluminum pigment in powder or paste form dispersed in a suitable vehicle.

aluminum paste A metallic aluminum flake pigment in paste form, consisting of aluminum, solvent, and various additives. The metallic aluminum pigment can be in the form of very small, coated leaves or amorphous powder, known as *leafing* and *nonleafing,* respectively.

anticorrosion paint or composition A coating used for preventing the corrosion of metals which has been specially formulated to prevent the rusting of iron and steel.

antifouling paint Paint used to prevent the growth of barnacles and other organisms on ships' bottoms, usually containing substances poisonous to organisms.

aromatic solvent A hydrocarbon solvent comprised wholly or primarily of aromatic hydrocarbon compounds. Aromatic solvents containing less than 80% aromatic compounds are frequently designated *partial aromatic solvents.*

acceleration corrosion test A method designed to approximate the deteriorating effects under normal long-term service conditions over a short period of time.

acicular ferrite A highly substructured, nonequiaxed ferrite formed upon continuous cooling by a mixed diffusion and shear mode of transformation that begins at a temperature slightly higher than the transformation temperature range for upper bainite. It is distinguished from bainite in that it has a limited amount of carbon available; thus, there is only a small amount of carbide present.

acid embrittlement A form of hydrogen embrittlement that may be induced in some metals by acid.

active metal A metal that is ready to corrode or being corroded.

active potential The potential of a corroding material.

activity A measure of the chemical potential of a substance, where chemical potential is not equal to concentration, that allows mathematical relations equivalent to those for ideal systems to be used to correlate changes in an experimentally measured quantity with changes in chemical potential.

activity coefficient A characteristic of a quantity expressing the deviation of a solution from ideal thermodynamic behavior; often used in connection with electrolytes.

addition agent A substance added to a solution for the purpose of altering or controlling a process. Examples include wetting agents in acid pickles, brighteners and antipitting agents in plating solutions, and inhibitors.

aeration (1) Exposing to the action of air. (2) Causing air to bubble through. (3) Introducing air into a solution by spraying, stirring, or a similar method. (4) Supplying or infusing with air, as in sand or soil.

aerobic An environment containing oxygen; for instance, normal seawater.

aging A change in the properties of certain metals and alloys that occurs at ambient or moderately elevated temperatures after hot working operations or heat treatments (e.g., quench aging in ferrous alloys, and natural or artificial aging in ferrous and nonferrous alloys) or after a cold working operation (e.g., strain aging). The change in properties is often, but not always, due to a phase change (precipitation), but it never involves a change in chemical composition of the metal or alloy.

alclad A composite wrought product comprised of an aluminum alloy core having on one or both surfaces a metallurgically bonded aluminum or aluminum alloy coating that is anodic to the core, and thus electrochemically protects the core against corrosion.

alkali metal A metal in group 1A of the periodic table—namely, lithium, sodium, potassium, rubidium, cesium, and francium. They form strongly alkaline hydroxides (hence the name).

alkaline (1) Having properties of an alkali. (2) Having a pH greater than 7.

alkyd Reaction products of polyhydric alcohols and polybasic acids.

alkylation (1) A chemical process in which an alkyl radical is introduced into an organic compound by substitution or addition. (2) A refinery process for chemically combining isoparaffin with olefin hydrocarbons.

alligatoring (1) Pronounced wide cracking over the entire surface of a coating with the appearance of alligator hide. (2) The longitudinal splitting of flat slabs in a plane parallel to the rolled surface.

alloy plating The code position of two or more metallic elements.

alloy steel Steel that contains either silicon or manganese in amounts in excess of those in plain carbon steel or that contains any other element or elements as the result of deliberately made alloying additions.

alternate-immersion test A corrosion test in which the specimens are intermittently exposed to a liquid medium at defined time intervals.

aluminizing A process for impregnating the surface of a metal with aluminum in order to obtain protection from oxidation and corrosion.

ammeter An instrument for measuring the magnitude of electric current flow.

amorphous solid A rigid material whose structure lacks crystalline periodicity; that is, the pattern of its constituent atoms or molecules does not repeat periodically in three dimensions. See also *metallic glass*.

anaerobic Lack of oxygen in the electrolyte adjacent to metallic structure.

anaerobic In the absence of air or unreacted or free oxygen.

anchorite A zinc-iron phosphate coating for iron and steel.

anion A negatively charged ion, which migrates to the anode of a galvanic or voltaic cell.

annealing A generic term denoting a treatment consisting of heating to and maintaining a suitable temperature, followed by cooling at a suitable rate. The process is used primarily to soften metallic materials but also simultaneously produce microstructure. The purpose of such changes may be, but is not confined to, the improvement of machinability, facilitation of cold work, improvement of mechanical or electrical properties, and increasing the stability of dimensions. When the term is used by itself, full annealing is implied. When applied only for the relief of stress, the process is properly called *stress relieving* or *stress-relief annealing.*

anode An electrode at which oxidation of the surface or some component of the solution is occurring.

anode corrosion The dissolution of a metal acting as an anode.

anode corrosion efficiency The ratio of actual to theoretical corrosion based on the total current flow calculated by Faraday's law from the quantity of electricity that has passed.

anode effect The effect produced by polarization of the anode in electrolysis. It is characterized by a sudden increase in voltage and a corresponding decrease in amperage due to the anode becoming virtually separated from the electrolyte by a gas film.

anode film (1) The portion of solution in immediate contact with the anode, especially if the concentration gradient is steep. (2) The outer layer of the anode itself.

anode polarization The difference between the potential of an anode passing current and equilibrium potential (or steady-state potential) of the electrode having the same electrode reaction.

anodic cleaning Electrolytic cleaning in which the work is the anode. Also called *reverse-current cleaning.*

anodic coating A film on a metal surface resulting from an electrolytic treatment at the anode.

anodic inhibitor A chemical substance or combination of substances that prevent or reduce, by physical, physio-chemical, or chemical action, the rate of the anodic or oxidation reaction.

anodic metallic coating A coating, composed wholly or partially of an anodic metal (in sufficient quantity to set off electrochemical reaction), which is electrically positive to the substrate to which it is applied.

anodic polarization The change in the initial anode potential resulting from current flow effects at or near the anode surface. Potentially becomes more noble (more positive) because of anodic polarization.

anodic potential An appreciable reduction in corrosion by making a metal an anode and maintaining this highly polarized condition with very little current flow.

anodic protection A technique to reduce corrosion of a metal surface under some conditions by passing sufficient anodic current to it to cause its electrode potential to enter and remain in the passive region.

anodic reaction An electrode reaction equivalent to a transfer of positive charge from the electronic to the ionic conductor. An anodic reaction is an oxidation process. An example common in corrosion is: Me ~ Me n+ + ne.

anodizing Forming a conversion coating on a metal surface by anodic oxidation; most frequently applied to aluminum.

antifouling Intended to prevent the fouling of underwater structures, such as the bottoms of ships; refers to the prevention of marine organism's attachment or growth on a submerged metal surface, generally through chemical toxicity caused by the composition of the metal or coating layer.

antipitting agent An addition agent for electroplating solutions to prevent the formation of pits or large pores in the electrodeposit.

aqueous Pertaining to water; an aqueous solution is a water solution.

artificial aging Aging above room temperature.

atmospheric corrosion The gradual degradation or alteration of a material by contact with substances present in the atmosphere, such as oxygen, carbon dioxide, water vapor, and sulfur and chlorine compounds.

attenuation The progressive decrease in potential and current density along a buried or immersed pipeline in relation to the distance from the point of injection.

batch All materials that are processed in the same conditions and submitted for inspection at one time.

binder The nonvolatile portion of the liquid vehicle of a coating. It binds or cements the pigment particles together and the paint film as a whole to the material to which it is applied. The amount of binder needed to completely wet a pigment is determined primarily by the particle size, shape, chemical composition, and density of the pigment; and the particle size, degree of polymerization, and wetting properties of the binder.

bitumen A very viscous liquid or solid consisting of hydrocarbons and their derivatives, which is soluble in carbon disulfide or trichloroethylene. It is substantially nonvolatile and softens gradually when heated. It is black or brown in color and possesses waterproofing and adhesive properties. It is obtained by refinery processes from petroleum.

bituminous coating Asphalt or tar compound used to provide a protective finish.

black liquor The liquid material remaining from pulpwood cooking in the soda or sulfate paper-making process.

black oxide A black finish on a metal produced by immersing it in hot oxidizing salts or salt solutions.

blast-cleaning Cleaning and roughening of a surface (particularly steel) by the use of a metallic or nonmetallic abrasive that is propelled against a surface by compressed air, centrifugal force, or water.

blast peening Treatment for relieving tensile stress by inducing beneficial compressive stress in the surface by kinetic energy of rounded abrasive particles.

blister A raised area, often dome shaped, resulting from (1) loss of adhesion between a coating or deposit and the base metal or (2) delamination under the pressure of expanding gas trapped in a metal in a near-subsurface zone. Very small blisters may be called pinhead blisters or pepper blisters.

blistering The formation of swellings on the surface of an unbroken paint film by moisture, gases or the development of corrosion products between the metal and the paint film.

blow down (1) Injection of air or water under high pressure through a tube to the anode area for the purpose of purging the annular space and possibly correcting high resistance caused by gas blocking. (2) In connection with boilers or cooling towers, the process of discharging a significant portion of aqueous solution in order to remove accumulated salts, deposits, and other impurities.

blue brittleness Brittleness exhibited by some steels after being heated to a temperature within the range of about 200–370°C, particularly if the steel is worked at the elevated temperature.

blushing Whitening and loss of gloss of a usually organic coating caused by moisture. Also called *blooming*.

bond A piece of metal conductor, either rigid or flexible and usually made of copper, connecting two points on the same structure or different structures to prevent any appreciable change in the potential of the one point with respect to the other.

bond resistance The ohmic resistance of a bond, including the contact resistance at the points of attachment of its extremities

bonderizing A proprietary custom process for phosphatizing.

borosilicate glass Any silicate glass having at least 5% boron oxide (B_2O_3).

braze welding The joining of metals using a technique similar to fusion welding and a filler metal with a lower melting point than the parent metal, but neither using capillary action, as in brazing, nor intentionally melting the parent metal.

brazing A process of joining metals in which, during or after heating, molten filler metal is drawn by capillary action into the space between closed adjacent surfaces of the parts to be joined. In general, the melting point of the filler metal is above 500°C, but it is always below the melting temperature of the parent metal.

brazing alloy A filler metal used in brazing.

breakdown potential The least noble potential where pitting, crevice corrosion, or both will initiate and propagate.

breakaway corrosion A sudden increase in the corrosion rate, especially in high-temperature, "dry" oxidation.

breaking strength The maximum resistance of material to deformation in a tensile test carried to rupture; that is, the breaking load, or force per unit cross-sectional area, of the unstrained specimen.

brightener An agent or combination of agents added to an electroplating bath to produce a smooth, lustrous deposit.

brine Seawater containing a higher concentration of dissolved salt than that of the ocean.

brittle fracture Separation of a solid accompanied by little or no macroscopic plastic deformation. Typically, this type of fracture occurs by rapid crack propagation, with less expenditure of energy than for ductile fracture.

brittleness temperature That temperature, estimated statistically, at which 50% of the specimens would fail the specified test.

brushing Application of a coating by means of a brush.

butt joint A connection between the ends or edges of two parts making an angle to one another of 135–180° inclusive in the region of the joint.

butyl rubber A series of rubberlike products made by the polymerization of a high percentage of a monoolefin like isobutylene and a small amount of a diolefin like butadiene. The resulting product has only a fraction of the unsaturation present in natural rubber, and after vulcanization, it is essentially a cross-linked saturated hydrocarbon. Butyl rubber is essentially a paraffinic hydrocarbon.

calcareous deposits Deposits containing calcium or calcium compounds.

cathodic disbonding The failure of adhesion between a coating and a metallic surface that is directly attributable to cathodic protection conditions and that is often initiated by a defect in the coating system, such as accidental damage, imperfect application, or excessive permeability of the coating.

cathodic protection A technique to reduce the corrosion rate of metal surface by making it a cathode of an electrochemical cell.

cement paint Paint supplied in dry powder form, based essentially on Portland cement, to which pigments are sometimes added for decorative purposes. This dry, powdered paint is mixed with water immediately before use.

centrifugal blast-cleaning The use of motor-driven, bladed wheels to hurl abrasive at a surface with centrifugal force.

ceramic A glazed or unglazed body of crystalline or partly crystalline structure or of glass, produced from essentially inorganic, nonmetallic substances and formed either from a molten mass solidified on cooling, or simultaneously or subsequently matured by the action of the heat.

chalking Formation of a friable powder on the surface of a paint film caused by the disintegration of the binding medium due to disruptive factors during weathering. The chalking of a paint film can be considerably affected by the choice and concentration of the pigment. It can also be affected by the choice of binding medium.

chemical conversion coating A treatment, either chemical or electrochemical, of the metal surface to convert it to another chemical form that provides an insulating barrier of exceedingly low solubility between the metal and its environment, but which is an integral part of the metallic substrate. It provides greater corrosion resistance to the metal and increased adhesion of coatings applied to the metal. Examples of this type of coating include phosphate coatings on steel.

chemical environment An exposure in which a strong concentration of highly corrosive gases, fumes, or chemicals, either in solution or as solids or liquids, contact the surface. The severity may vary tremendously from mild concentration in yard areas to direct immersion in the chemical.

chlorinated hydrocarbon A powerful type of solvent that includes such members as chloroform, carbon tetrachloride, ethylene dichloride, methylene chloride, tetrachlorethane, and trichlorethylene. Generally, they are toxic, and their use is now restricted in some countries. Their main applications include as nonflammable paint removers, cleaning solutions, and special finishes where the presence of residual solvent in the film is a disadvantage.

chemical-resistant resin mortar An intimate mixture of liquid resinous material, selected filler material, and setting agent forming a trowelable mortar that hardens by chemical reaction.

chlorinated rubber Resin formed by the reaction of rubber with chlorine. Unlike rubber, the resulting product is readily soluble and yields solutions of low viscosity. It is sold as white powder, fibers, or as blocks. Commercial products generally contain about 65% chlorine. It has good chemical resistance properties. Mostly chlorinated polymers are now used, i.e., l-butane, polyethylene, etc.

cladding A composite metal containing two or more layers that have been bonded together corralling coextention, welding diffusion, bonding, heavy chemical deposition, or heavy electroplating.

coal tar A mixture of high-molecular-mass hydrocarbons, obtained by distillation of high-temperature coal tar, processed so that to produce base materials for primer grades A and B.

coal-tar epoxy coating A coating in which the binder or vehicle is a combination of coal tar and epoxy resin.

coal-tar urethane coating A coating in which the binder or vehicle is a combination of coal tar with a polyurethane resin.

Coating (noun) Generic term for paints, lacquer, enamels, etc. A liquid, liquefiable, or mastic composition that has been converted to a solid protective, decorative, or functional adherent film after application in a thin layer. A coating is an electrical insulating covering applied to a metal surface as passive protection against external corrosion.

coating (process) The procedure of applying a thin layer of a material in the form of a fluid or powder upon a substance.

coating (product) A thin layer of a material applied by a coating process.

coating applicator qualification Coating application shall be done only by an applicator qualified by field experience in installing the type of coating proposed. To satisfy this requirement, applicators shall submit a list of applications and installations they have worked on and, where such data is available, the service conditions and reported record of performance.

coating system A number of coats separately applied in a predetermined order at suitable intervals to allow for drying or curing

consolidate soil When a soil is subjected to an increase in pressure due to loading at the ground surface, a readjustment in the soil structure occurs. The volume of space between the soil particles decreases, and the soil tends to settle or consolidate over time.

contractor The party that carries out all or part of the design, engineering, procurement, construction, and commissioning for the project. The company may sometimes undertake all or part of the duties of the contractor.

company The party that initiates the project and ultimately pays for its design and construction. The company will generally specify the technical requirements and may also include an agent or consultant authorized to act for the principal.

compatible Two or more paints or varnishes that can be mixed without producing any undesirable effects such as precipitation, coagulation, or gelling are said to be compatible. Different coats of paint that can be associated in a painting system or other coating systems without producing undesirable effects are also compatible.

corrosion protection The separation of metallic material from an attacking medium by paint or coating.

cure To change the properties of a polymer system into a final, more stable, and usable condition by the use of heat, radiation, or reaction with chemical additives.

cementation coating A coating developed on a metal surface by a high-temperature diffusion process (as carburization, calorizing, or chromizing).

cementite A compound of iron and carbon, known chemically as iron carbide and having the approximate chemical formula Fe_3C. It is characterized by an orthorhombic crystal structure. When it occurs as a phase in steel, the chemical composition will be altered by the presence of manganese and other carbide-forming elements.

characteristic resistance In pipes, the electrical resistance between the pipe and remote earth in one direction only from the drain point; measured in ohms.

chelate (1) A molecular structure in which a heterocyclic ring can he formed by the unshared electrons of neighboring atoms. (2) A coordination compound in which a heterocyclic ring is formed by a metal bound to two atoms of the associated ligand.

chelating agent (1) An organic compound in which atoms form more than one coordinate bond with metals in solution. (2) A substance used in metal finishing to control or eliminate certain metallic ions present in undesirable quantities.

chelation A chemical process involving the formation of a heterocyclic ring compound that contains at least one metal cation or hydrogen ion in the ring.

chemical cleaning A method of surface preparation or cleaning involving the use of chemicals (either with or without electrical force) for the removal of mill scale, rust, sediments, and paint. These chemicals can also be introduced into some systems onstream while the system is operating.

chemical conversion coating A protective or decorative coating that is produced deliberately on a metal surface by reaction of the surface with a chosen chemical environment. The thin layer formed by this reaction may perform several or all of the following functions: protecting against corrosion; providing a base for organic coatings; improving the retention of lubricants or compounds; improving abrasion resistance; and providing an absorbent layer for rust-preventive oils and waxes.

chemical potential In a thermodynamic system of several constituents, the rate of change of the Gibbs function of the system with respect to the change in the number of moles of a particular constituent.

chemical vapor deposition A coating process, similar to gas carburizing and carbonitriding, in which a reactant atmosphere gas is fed into a processing chamber, where it decomposes at the surface of the workpiece, liberating one material for either absorption by, or accumulation on, the workpiece. A second material is liberated in gas form and is removed from the processing chamber, along with excess atmosphere gas.

chemisorption The binding of an adsorbate to the surface of a solid by forces whose energy levels approximate those of a chemical bond. See *physisorption.*

chevron pattern A fractographic pattern of radial marks (shear ledges) that look like nested Vs; sometimes called a *herringbone pattern.* Chevron patterns are typically found on brittle fracture surfaces in parts whose widths are considerably greater than their thicknesses. The points of the chevrons can be traced back to the fracture origin.

chromadizing Improving paint adhesion on aluminum or aluminum alloys, mainly aircraft skins, by treatment with a solution of chromic acid. Also called *chromodizing* or *chromatizing,* and not to be confused with *chromating* or *chromizing.*

chromate treatment A treatment of metal in a solution of a hexavalent chromium compound to produce a conversion coating consisting of trivalent and hexavalent chromium compounds.

chromating Performing a chromate treatment.

chrome pickle (1) Producing a chromate conversion coating on magnesium for temporary protection or for a paint base. (2) The solution that produces the conversion coating.

chromizing A surface treatment at elevated temperature, generally carried out in pack, vapor, or salt bath, in which an alloy is formed by the inward diffusion of chromium into the base metal.

clad metal A composite metal containing, two or more layers that have been bonded together. The bonding may have been accomplished by co-rolling, co-extrusion, welding, diffusion bonding, casting, heavy chemical deposition, or heavy electroplating.

cleavage The splitting (fracture) of a crystal on a crystallographic plane of low index.

cleavage fracture A fracture, usually of polycrystalline metal, in which most of the grains have failed by cleavage, resulting in bright reflecting facets. It is associated with low-energy brittle fracture.

cold working Deforming metal plastically under conditions of temperature and strain rate that induce strain hardening; usually, but not necessarily, conducted at room temperature. See *hot working.*

combined carbon The part of the total carbon in steel or cast iron that is present as other than free carbon.

compatibility The ability of a given material to exist unchanged in certain conditions and environments in the presence of some other material.

compressive strength The maximum compressive stress that a material is capable of developing. With a brittle material that fails in compression by fracturing, the compressive strength has a definite value. In the case of ductile, malleable, or semiviscous materials (which do not fail in compression by a shattering fracture), the value obtained for compressive strength is an arbitrary value dependent on the degree of distortion, which is regarded as effective failure of the material.

concentration cell A cell involving an electrolyte and two identical electrodes, with a potential resulting from differences in the chemistry of the environments adjacent to the two electrodes.

concentration polarization That portion of the polarization of a cell produced by concentration changes resulting from the passage of current through the electrolyte.

conductivity The ratio of the electric current density to the electric field in a material. Also called *electrical conductivity* or *specific conductance.*

contact corrosion A term primarily used in Europe to describe galvanic corrosion between dissimilar metals.

contact potential The potential difference at the junction of two dissimilar substances.

continuity bond A metallic connection that provides electrical continuity between metal structures.

controlled galvanic system A cathodic protection system using sacrificial anodes controlled by means of fixed or variable resistors.

conversion coating A coating consisting of a compound of the surface metal, produced by chemical or electrochemical treatments of the metal. Examples include chromate coatings on zinc, cadmium, magnesium, and aluminum and oxide and phosphate coatings on steel. See also *chromate treatment* and *phosphating.*

copper-accelerated salt-spray (CASS) test An accelerated corrosion test for some electrodeposits for anodic coatings on aluminum

copper/copper sulfate reference electrode A reference electrode consisting of copper in a saturated copper sulfate solution.

copper ferrule A ring or cap of copper put around a slender shaft (as a cane or tool handle) to strengthen it or prevent splitting.

corrosion The destruction of a substance; usually a metal, or its properties because of a reaction with its surroundings (environment); i.e., a physiochemical interaction between a metal and its environment that results in changes in the properties of the metal and that may often lead to impairment of the function of the metal, the environment, or the technical system, of which these form a part.

corrosion damage An effect of corrosion that is considered detrimental to the function of the metal, the environment, or the technical system of which these form a part.

corrosion effect A change in any part of the corrosion system caused by corrosion.

corrosion embrittlement The severe loss of ductility of a metal resulting from corrosive attack, usually intergranular and often not visually apparent.

corrosion engineer The person responsible for carrying out corrosion monitoring and interpretation of any relevant data produced.

corrosion fatigue The process in which a metal fractures prematurely under conditions of simultaneous corrosion and repeated cyclic loading at lower stress levels or fewer cycles than would be required in the absence of the corrosive environment.

corrosion fatigue limit The maximum cyclic stress value that a metal can withstand for a specified number of cycles or length of time in a given corrosive environment. See *corrosion fatigue strength.*

corrosion fatigue strength The maximum repeated stress that can be endured by a metal without failure under definite conditions of corrosion and fatigue and for a specific number of stress cycles and a specified period of time.

corrosion inhibitor An inhibitor is a substance that retards a chemical reaction. Thus, a corrosion inhibitor is a substance which, when added to an environment, decreases the rate of attack by the environment on a metal. Corrosion inhibitors are commonly added in small amounts to acids, cooling water, oilwells, and other environments, either continuously or intermittently, to prevent serious corrosion.

corrosion potential The potential of a corroding surface in an electrolyte, relative to a reference electrode.

corrosion product The chemical compound or compounds produced by the reaction of a corroding metal with its environment.

corrosion protection Modification of a corrosion system so that corrosion damage is mitigated.

corrosion rate The rate at which corrosion proceeds, expressed by inches of penetration per year (ipy); mils penetration per year (mpy); milligrams weight loss per square decimeter per day (mdd); microns per year (μm/year), or millimeters per year (mmpy). 1 μm = 0.0395 mils.

corrosion resistance The ability of a metal to withstand corrosion in a given corrosion system.

corrosion system A system consisting of one or more metals and all parts of the environment that influence corrosion.

corrosive agent A substance that, when in contact with a given metal, will react to it.

corrosive environment An environment that contains one or more corrosive agents.

corrosivity The tendency of an environment to cause corrosion in a given corrosion system.

couple A cell developed in an electrolyte resulting from electrical contact between two dissimilar metals. See *galvanic corrosion.*

Coupon A specimen of material exposed to tests or real environments to assess the effect of degradation on the material.

Covered filler rod A filler rod having a covering of flux.

covering power The ability of a solution to give satisfactory plating at very low current densities. a condition that exists in recesses and pits. This term suggests an ability to cover, but not necessarily to build up, a uniform coating, whereas *throwing power* suggests the ability to obtain a coating of uniform thickness of an irregularly shaped object.

Cracking When a metal part fails by cracking, that fact is generally obvious, but the exact type of cracking and the cause are less so. To determine the type of cracking, microscopic examination is necessary. In some instances, the environment plays a minor role, while in others, its role is major.

cracking (of coating) Breaks in a coating that extend through to the underlying surface.

crazing A network of checks or cracks appearing on the surface.

creep Time-dependent strain that occurs under stress. The creep strain occurring at a diminishing rate is called *primary creep;* that occurring at a minimum and almost constant rate, *secondary creep;* and that occurring at an accelerating rate, *tertiary creep.*

creep-rupture embrittlement Embrittlement under creep conditions (for example, of aluminum alloys and steels) that results in abnormally low rupture ductility. In aluminum alloys, iron in amounts above the solubility limit is known to cause such embrittlement; in steels, the phenomenon is related to the amount of impurities (for example, phosphorus, sulfur, copper, arsenic, antimony, and tin) present. In either case, failure occurs by intergranular cracking of the embrittled material.

creep-rupture strength The stress that will cause fracture in a creep test at a given time in a specified constant environment. Also called *stress-rupture strength.*

damp Either moderate absorption or mode rate covering of moisture; implies less wetness than does the word *wet,* and slightly more wetness than that connoted by *moist.*

density The weight per unit volume of material at 23°C, given as the expression $D^{23°C}$, kg/m^3

Three density ranges of polyethylene are generally recognized:

a) low-density polyethylene, from 910–925 kg/m^3;

b) medium-density polyethylene, from 926–940 kg/m^3;

c) high-density polyethylene, greater than 940 kg/m^3.

Note:

These densities refer to the base polymer, without pigment or carbon black, before extrusion. For compounds containing a nominal 2.5% carbon black, a correction factor of 10 kg/m^3 can be used.

descaling Removal of mill scale or caked rust from steel by chemical or mechanical means.

designer The person or party responsible for all or part of the design and engineering of a project.

dew point The temperature at which moisture will condense.

dielectric breakdown (dielectric strength) The voltage at which a single layer of tape will show electrical failure under specific conditions of a test. It is an indication of the ability of a tape to withstand electrical stress.

dielectric strength The voltage at which a single layer of tape will show electrical failure under specific conditions of a test.

disbondment The loss of the bond between a coating and the surface coated.

drying time The time required for an applied film of coating to reach the desired stage of cure, hardness, or nontackiness.

dry-to-handle time The time interval between the application of a coating and its ability to receive the next coat satisfactorily.

dry-to-touch time The interval between application and achieving a tack-free condition.

dichromate treatment A chromate conversion coating produced on magnesium alloys in a boiling solution of sodium dichromate.

dielectric strength The degree of electrical nonconductance of a material; the maximum electric field that a material can withstand without breakdown.

dielectric union Similar to insulated flange, but typically threaded to a pipeline and used on pipe diameters that are 2 inches or less and pressures less than ....

diffusion coating The application of metallic coating, the chemical composition of which was modified by diffusing this at the melting temperature into the substrate. Any process whereby a base metal or alloy is either

(1) coated with another metal or alloy and heated to a sufficient temperature in a suitable environment or (2) exposed to a gaseous or liquid medium containing the other metal or alloy, thus causing diffusion of the coating or of the other metal or alloy into the base metal, with resultant changes in the composition and properties of its surface.

diffusion coefficient A factor of proportionality representing the amount of substance diffusing across a unit area through a unit concentration gradient in unit time.

diffusion-limited current density The current density that corresponds to the maximum transfer rate that a particular species can sustain because of the limitation of diffusion. Often referred to as *limiting current density*.

disbandment The destruction of adhesion between a coating and the surface coated.

discontinuity Any interruption in the normal physical structure or configuration of a part, such as cracks, laps, seams, inclusions, or porosity. A discontinuity may or may not affect the usefulness of the part.

dislocation A linear imperfection in a crystalline array of atoms. Two basic types are recognized: (1) an edge dislocation corresponds to the row of mismatched atoms along the edge formed by an extra, partial plane of atoms within the body of a crystal; (2) a screw dislocation corresponds to the axis of a spiral structure in a crystal, characterized by a distortion that joins normally parallel planes to form a continuous helical ramp (with a pitch of one interplanar distance) winding about the dislocation. Most prevalent is the so-called *mixed dislocation,* which is any combination of an edge dislocation and a screw dislocation.

double layer The interface between an electrode or a suspended particle and an electrolyte created by charge-charge interaction (charge separation), leading to an alignment of oppositely charged ions at the surface of an electrode or particle. The simplest model is represented by a parallel plate condenser of 2×10^{-8} cm in thickness. In general, the electrode will be positively charged with respect to the solution.

drainage Conduction of electric current from an underground metallic structure by means of a metallic conductor. Forced drainage is applied to underground metallic structures by means of an applied electromotive force or sacrificial anode. Natural drainage goes from an underground structure to a more negative (more anodic) structure, such as the negative bus of a trolley substation.

drainage (current requirement) tests Tests with current applied for a short period, usually with temporary anodes and power sources in order to determine the current needed to achieve cathodic protection.

driving EMF (galvanic anode system) The difference between the structure/electrolyte potential and the anode/electrolyte potential.

dry corrosion Gaseous corrosion.

ductility The ability of a material to deform plastically without fracturing, measured by the elongation or reduction of area in a tensile test, by height of cupping in an Erichsen test, or by other means.

edge preparation Squaring, grooving, chamfering, or beveling an edge in preparation for welding.

efflorescence A deposit of salts (usually white) formed on a surface, the substance having emerged in solution from within concrete or masonry and deposited by evaporation.

elastic deformation A change in dimensions directly proportional to and in phase with an increase or decrease in applied force.

elasticity The property of a material, by virtue of which deformation caused by stress disappears upon removal of the stress. A perfectly elastic body completely recovers its original shape and dimensions after the release of stress.

elastic limit The maximum stress that a material is capable of sustaining without any permanent strain (deformation) remaining upon complete release of the stress.

elastomer A natural or synthetic polymer, which at room temperature can be stretched repeatedly to at least twice its original length, and which after removal of the tensile loud will immediately and forcibly return to approximately its original length.

electrical conductivity See *conductivity.*

electrical grounding Provides a low-resistance path to the ground for fault currents in electrical equipment and distribution networks. Since bare copper is commonly used, current requirement calculations must include the copper as a substantial sink for cathodic protection current in order to adequately size it for a cathodic protection system.

electrical isolation The condition of being electrically separated from other metallic structures or the environment.

electrical resistivity The electrical resistance offered by a material to the flow of current multiplied by the cross-sectional area of current flow and per-unit length of the current path; the reciprocal of the conductivity.

electrochemical cell An electrochemical system consisting of an anode and a cathode in metallic contact and immersed in an electrolyte. (The anode and cathode may be different metals or dissimilar areas on the same metal surface.)

electrochemical corrosion Corrosion that is accompanied by a flow of electrons between the cathodic and anodic areas on metallic surfaces.

electrochemical equivalent The weight of an element or group of elements oxidized or reduced at 100% efficiency by the passage of a unit quantity of electricity. Usually expressed as grams per coulomb (1 amp/s).

electrochemical impedance The frequency-dependent, complex-valued proportionality factor (SE/6i) between the applied potential or current and the response signal. This factor is the total opposition (11 or Ill cm-) of an electrochemical system to the passage of a charge. The value is related to the corrosion rate under certain circumstances.

electrochemical impedance spectroscopy (AC impedance) A method to study the impedance of a metal/fluid interface by electrochemical properties by applying a sinusoidal polarization potential to the interface through a range of frequencies.

electrochemical potential The partial derivative of the total electrochemical-free energy at a constituent with respect to the number of moles of this constituent where all factors are kept constant. It is analogous to the chemical potential of a constituent, except that it includes electric as well as chemical contributions to free energy. The potential of an electrode in an electrolyte relative to a reference electrode measured under open circuit conditions.

electrode (1) An electronic conductor used to establish electrical contact with an electrolytic part of a circuit. (2) An electronic conductor in contact with an ionic conductor.

electrode polarization A change of electrode potential with respect to a reference value. Often the free corrosion potential is used as the reference value. The change may be caused, for example, by the application of an external electrical current or by the addition of on oxidant or reductant.

electrodeposition The deposition of a substance on an electrode by passing electric current through an electrolyte.

electrode potential The potential of an electrode in an electrolyte as measured against a reference electrode. The electrode potential does not include any resistance losses in potential in either the solution or external circuit. It represents the reversible work needed to move a unit charge from the electrode surface through the solution to the reference electrode. The potential of an electrode as measured against a reference electrode. The electrode potential does not include any resistance loss in potential in solution due to the current passing to or from the electrode.

electrode reaction An interfacial reaction equivalent to a transfer of charge between electronic and ionic conductors. See also *anodic reaction* and *cathodic reaction*.

electrogalvanizing A process galvanized by electroplating.

electrokinetic potential A potential difference in the solution caused by a residual, unbalanced charge distribution in the adjoining solution, producing a double layer. The electrokinetic potential is different from the electrode potential in that it occurs exclusively in the solution phase; that is, it represents the reversible work necessary to bring a unit charge from infinity in the solution up to the interface in question, but not through the interface. Sometimes called *zeta potential*.

electroless plating A process in which metal ions in a dilute aqueous solution are plated onto a substrate by means of an autocatalytic chemical reduction.

electrolysis The production of chemical changes of the electrolyte by the passage of current through an electrochemical cell.

electrolyte A chemical substance or mixture, usually liquid, containing ions that migrate in an electric field. An ionic conductor (usually in aqueous solution). (1) A chemical substance or mixture, usually liquid, containing ions that migrate in an electric field. (2) A chemical compound or mixture of compounds which when molten or in solution will conduct an electric current.

electrolytic cell An assembly, consisting of a vessel, electrodes, and an electrolyte, in which electrolysis can be carried out.

electrolytic cleaning (1) A process of removing soil, scale, or corrosion from a metal surface by subjecting it as an electrode to an electric current in an electrolytic bath. (2) The process of cleaning, degreasing, of a metal by making it an electrode in a suitable bath.

electromotive force Electrical potential; voltage.

electromotive force series (EMF series) The potential of an electrode in an electrolytic solution when the forward rate of a given reaction is exactly equal to the reverse rate. (The equilibrium potential can be defined only with respect to a specific electrochemical reaction.)

A list of elements arranged according to their standard electrode potentials (hydrogen electrode is a reference point given the value zero), with noble metals, such as gold, being positive and active metals, such as zinc, being negative.

electron flow A movement of electrons in an external circuit connecting an anode and cathode in a corrosion cell; the current flow is arbitrarily considered to be in the opposite direction to the electron flow.

electronegative A qualification applied to a metallic electrode to indicate that its potential is negative with respect to another metallic electrode in the system.

electroplating Electrodeposition of a thin adherent layer of a metal or alloy of desirable chemical, physical, and mechanical properties on a metallic or nonmetallic substrate.

electropolishing A technique commonly used to prepare metallographic specimens, in which a high polish is produced by making the specimen the anode in an electrolytic cell, where preferential dissolution at high points smooths the surface.

electropositive A qualification applied to a metallic electrode to indicate that its potential is positive with respect to another metallic electrode in the system.

electro-slag welding A welding process in which consumable electrodes are fed into a joint containing flux; the current melts the flux, and the flux in turn melts the faces of the joint and the electrodes, allowing the weld metal to form a continuous cast ingot between the joint faces.

electron-beam welding Fusion welding in which the joint is made by fusing the parent metal via the impact of a focused beam of electrons.

electrosmosis The passage of a liquid through a porous medium under the influence of a potential difference.

elongation The increase in length at a break when the tape is tested under specific conditions. The elongation of tape is important as an indicator of its uniformity and quality.

embrittlement Severe loss of ductility of a metal (or alloy). Loss of load-carrying capacity of a metal or alloy; the severe loss of ductility or toughness or both, of a material, usually a metal or alloy. Many forms of embrittlement can lead to brittle fracture. Many forms can occur during thermal treatment or elevated-temperature service (thermally induced embrittlement). Some of these forms of embrittlement, which affect steels, include blue brittleness, 475°C embrittlement, quench-age embrittlement, sigma-phase embrittlement, strain-age embrittlement, temper embrittlement, tempered martensite embrittlement, and thermal embrittlement. In addition, steels and other metals and alloys can be embrittled by environmental conditions (environmentally assisted embrittlement). The forms of environmental embrittlement include acid embrittlement, caustic embrittlement, corrosion embrittlement, creep-rupture embrittlement, hydrogen embrittlement, liquid metal embrittlement, neutron embrittlement, solder embrittlement, solid metal embrittlement, and stress-corrosion cracking.

emulsifier A substance that intimately mixes, modifies the surface tension of colloidal droplets, and disperses dissimilar materials that are ordinarily immiscible, such as oil and water, to produce a stable emulsion. The emulsifier has the double task of promoting emulsification and of stabilizing the finished product.

emulsion A two-phase liquid system in which small droplets of one liquid (the internal phase) are immiscible in and dispersed uniformly throughout a second continuous liquid phase (the external phase).

emulsion paint A paint that constitutes an emulsion of a binder in water. The binder may be oil, oleoresinous varnish, resin, or another emulsifiable binder. Not to be confused with a latex paint, in which the vehicle is latex.

enamel A substance composed of a specially processed coal-tar pitch or bitumen, combined with an inert mineral filler.

encapsulation To protect the assembly by inhibited organic sealant, plastic caps, or cast-potting compound.

endurance limit The maximum stress that a material can withstand for an infinitely large number of fatigue cycles; the maximum cyclic stress level that a metal can withstand without fatigue failure. See also *fatigue strength.*

engineer The person, firm, or employee representing the purchaser tasked with ensuring adequacy of design and quality control.

environment The surroundings or conditions (physical, chemical, and mechanical) in which a material exists.

environment The circumstances, acts, or conditions to which a steel pipeline is subjected.

environmental cracking Brittle fracture of a normally ductile material in which the corrosive effect of the environment is a causative factor. Environmental cracking is a general term that includes corrosion fatigue, high-temperature hydrogen attack, hydrogen blistering, hydrogen embrittlement, liquid metal embrittlement, solid metal embrittlement, stress-corrosion cracking, and sulfide stress cracking. The following terms have been used in the past in connection with environmental cracking, but are becoming obsolete: caustic embrittlement, delayed fracture, season cracking, static fatigue, stepwise cracking, sulfide corrosion cracking, and sulfide stress-corrosion cracking.

environmentally assisted embrittlement See *embrittlement.*

epoxy resin A cross-linking resin based on the reactivity of the epoxide group. One common type is resin made from epichlorhydrin and bisphenol A. Aliphatic polyols such as glycerol may be used instead of the aromatic bisphenol A or bisphenol F.

equilibrium (reversible) potential The potential of an electrode in an electrolytic solution when the forward rate of a given reaction is exactly equal to the reverse rate. The equilibrium potential can be defined only with respect to a specific electrochemical reaction.

erosion Destruction of metals by the abrasive action of moving fluids accelerated by the presence of solid particles in suspension. When corrosion occurs simultaneously, the term "erosion-corrosion" is often used.

erosion A phenomenon manifest in paint films by the wearing away of the finish to expose the substrate or undercoat. The degree of failure depends on the amount of substrate or undercoat visible. Erosion occurs as the result of chalking or by the abrasive action of wind-borne particles of grit.

erosion-corrosion Corrosion that is increased because of the abrasive action of a moving stream; the presence of suspended particles greatly accelerates abrasive action. Erosion-corrosion is characterized by grooves, gullies, waves, rounded holes, and valleys, and usually exhibits a directional pattern. In copper alloy heat exchanger tubes, the attack frequently results in the formation of horseshoe-shaped depressions. Erosion-corrosion is the acceleration of metal loss because of the relative movement between a fluid and a metal surface. Generally, the movement is rapid, and the effects of mechanical wear are involved. Metal is removed as dissolved ions or solid corrosion products that are swept from the surfaces.

etch To roughen a surface by a chemical agent prior to painting in order to increase adhesion.

etching primer A priming paint usually supplied as two separate components that need to be mixed immediately prior to use; thereafter, it is usable for only a limited period. On clean light alloys or ferrous surfaces and on many nonferrous surfaces, such paints give excellent adhesion, partly due to their chemical reaction with the substrate (hence the term *etching primer*), and give a corrosion-inhibiting film that is a very good base for the application of subsequent coats of paint. Also known as *pretreatment primer, wash primer* and *self-etch primer.*

external rendering The application of a coat of mortar over the outside of a framework.

extrusion A process in which heated or unheated plastic is forced through a shaping orifice (i.e., a die) in one continuously formed shape, as in sheet, film, or tubing.

eutectoid (1) An isothermal reversible reaction in which a solid solution is converted to two or more intimately mixed solids upon cooling, the number of solids formed being the same as the number of components in the system. (2) An alloy having the composition indicated by the eutectoid point on an equilibrium diagram. (3) An alloy structure of intermixed solid constituents formed by a eutectoid reaction.

exchange current The rate at which either positive or negative charges are entering or leaving the surface at the point when an electrode reaches dynamic equilibrium in a solution and the rate of anodic dissolution balances the rate of cathodic plating.

exchange current density The rate of charge transfer per unit area when an electrode reaches dynamic equilibrium (at its reversible potential) in a solution; that is, the rate of anodic charge transfer (oxidation) balances the rate of cathodic charge transfer (reduction).

exfoliation A type of subsurface corrosion that occurs and propagates as cracks approximately parallel to the surface. It leaves the metal in a laminated, flaky, or blistered condition and appears most frequently in aluminum alloys or cupronickel.

fabricator The party that manufactures components to perform the duties specified by the company. It is generally considered to be synonymous with the term *manufacturer.*

field painting Surface preparation and painting operations of structural steel or other materials conducted at the project site.

filament A single textile element with a small diameter and a very long length, considered to be continuous.

filler An inert powder that can be incorporated into a bitumen enamel to improve one or more of its useful properties without changing its quality.

film A form of plastic in which the thickness is very low in proportion to its length and width and in which the plastic is present as a continuous phase throughout.

finish coat (topcoat) The final coat in a painting system.

flame cleaning The impingement of an intensely hot flame to the surface of structural steel, resulting in the removal of mill scale and the dehydration of any remaining rust, leaving the surface in a condition suitable for wire brushing, followed by the immediate application of paint.

flammable liquid Any liquid having a flash point below 37.8°C, except any liquid mixture having one or more components with a flash point at or above the upper limit that make up 99% or more of the total volume of the mixture

flash point The minimum temperature (corrected to a barometric pressure of 760 mm Hg) at which a liquid gives off a vapor in sufficient concentration to ignite under specified test conditions.

flint (chert) A very fine-grained, siliceous rock.

fungicide A paint additive that discourages the growth of fungi.

fatigue Subjecting a material to repeated stresses ultimately results in cracking. The environment may have an effect on the fatigue limit of a metal, though this is usually a minor factor. Generally, a fatigue failure is a single fracture, which is transgranular in most common metals. There is normally only a single fracture because stresses on other regions are relieved when the fracture occurs. Characteristic chevron patterns or beach marks can appear on the fracture face.

fatigue crack growth rate The rate of crack extension caused by constant-amplitude fatigue loading, expressed in terms of crack extension per cycle of load application.

fatigue life The number of cycles of stress that can be sustained prior to failure under a stated test condition.

fatigue limit The maximum stress that presumably leads to fatigue fracture in a specified number of stress cycles. If the stress is not completely reversed, the value of the mean stress, the minimum stress, or the stress ratio should also be stated. Compare with *endurance limit.*

fatigue strength The maximum stress that can be sustained for a specified number of cycles without failure, the stress being completely reversed within each cycle unless otherwise stated.

ferrite A solid solution of one or more elements in body-centered cubic iron. Unless otherwise designated (for instance, as chromium ferrite), the solute is generally assumed to be carbon. On some equilibrium diagrams, there are two ferrite regions separated by an austenitic area. The lower area is alpha ferrite; the upper area is delta ferrite. If there is no designation, alpha ferrite is assumed. In magnetic fields, substances having the general formula $M_2 + O_2-$, M_2 3 + $O_{32}-$, the trivalent metal often being iron.

filiform corrosion Corrosion that occurs under film in the form of randomly distributed hairlines.

filler metal Metal added during welding, braze welding, brazing, or surfacing.

filler rod Filler metal in the form of a rod or filler wire.

film A thin layer of material that is not necessarily visible.

finite line If a line terminates in an insulated flange or a dead end, it is called a *finite line*.

flakes Short, discontinuous internal fissures in wrought metals attributed to stresses produced by localized transformation and decreased solubility of hydrogen during cooling after hot working. In a fracture surface, flakes appear as bright silvery areas; on an etched surface, they appear as short, discontinuous cracks. Also called *shatter cracks* or *snowflakes*.

flame hardening The hardening of a metal surface by heating with an oxyacetylene torch, followed by rapid cooling with water or an air jet.

flame spraying Thermal spraying in which coating material is fed into an oxyfuel gas flame, where it is melted. Compressed gas may or may not be used to atomize the coating material and propel it onto the substrate.

flow line A pipeline carrying product from a wellhead to a gas-oil-separator-plant (GOSP), typically 100–250 DN, coated or uncoated, which is usually aboveground on pipe supports with periodic road crossings.

flux Material used during welding, brazing, or braze welding to clean the surfaces of joints, prevent atmospheric oxidation, and to reduce impurities.

fogged metal A metal whose luster has been reduced because of a surface film, usually a corrosion product layer.

foreign structure Any metallic structure that is not intended as part of a cathodic protection system of interest.

fouling An accumulation of deposits; this term includes the accumulation and growth of marine organisms on a submerged metal surface and also includes the accumulation of deposits (usually inorganic) on heat exchanger tubing.

fouling organism Any aquatic organism with a sessile adult stage that attaches to and fouls the underwater structures of ships.

fractography A descriptive treatment of fracture, especially in metals, with specific reference to photographs of the fracture surface. Macrofractography involves photographs at low magnification ($< 25x$); microfractography involves photographs at high magnification ($> 25x$).

fracture mechanics A quantitative analysis for evaluating structural behavior in terms of applied stress, crack length, and specimen or machine component geometry.

fracture toughness A generic term for measuring resistance to extension of a crack. The term is sometimes restricted to results of fracture mechanics tests, which are directly applicable in fracture control. However, the term commonly includes results from simple tests of notched or precracked specimens not based on fracture mechanics analysis. Results from test of the latter type are often useful for fracture control, based on either service experience or empirical correlations with fracture mechanics tests.

free carbon The part of the total carbon in steel or cast iron that is present in elemental form as graphite or temper carbon. See also *combined carbon*.

free corrosion potential Corrosion potential in the absence of net electrical current flowing to or from the metal surface.

free ferrite Ferrite that is formed directly from the decomposition of hypoeutectoid austenite during cooling, without the simultaneous formation of cementite. Also called *proeutectoid ferrite*.

free machining Pertains to the machining characteristics of an alloy to which one or more ingredients have been introduced to give small broken chips, lower power consumption, better surface finish, and longer tool life; among such additions are sulfur or lead to steel, lead to brass, lead and bismuth to aluminum, and sulfur or selenium to stainless steel.

fretting A type of wear that occurs between tight-fitting surfaces subjected to cyclic relative motion of extremely small amplitude. Usually, fretting is accompanied by corrosion, especially of very fine wear debris. The term also refers to metal deterioration caused by repetitive slipping at the interface between two surfaces. See also *fretting corrosion*.

fretting corrosion Another special case of erosion-corrosion that occurs when two heavily loaded metals rub together rapidly, causing damage to one or both metals. Vibration is usually responsible for the damage, but corrosion is also a factor because the frictional heat increases oxidation. In addition, mechanical removal of protective corrosion products continually exposes fresh metal. Fretting corrosion occurs more frequently in air than in water.

furan Resin formed from reactions involving furfuryl alcohol, whether alone or in combination with other constituents.

fusion welding A type of welding in which the weld is made between metals in a molten state, without the application of pressure.

fusion zone The part of the parent metal that is melted into the weld metal.

galvanizing The application of a coating of zinc to steel by a variety of methods.

glass An inorganic fusion product that has cooled to a rigid condition without crystallizing.

glass-fiber mat (mat) A uniform porous mat that is reinforced by strands of glass yarn to give longitudinal tensile strength, with the whole bonded with a thermosetting resin.

gypsum A mineral having a calcium sulfate dehydrate ($CaSO_4$. $2H_2O$) composition.

galvanic Pertaining to the current resulting from the coupling of dissimilar electrodes in an electrolyte.

galvanic anode or sacrificial A metal that, because of its relative position in the electromotive force (EMF) series, provides sacrificial protection to metals that are less negative (lower) in the series when the two are electrically coupled in an electrolyte. The voltage difference between the anode and the structure causes a current flow in the structure that opposes the corrosion current. The common types of galvanic anodes are rod, bracelet, and ribbon.

galvanic cell A cell in which chemical change is the source of electrical energy. It usually consists of two dissimilar conductors in contact with each other and with an electrolyte, or of two similar conductors in contact with each other and with dissimilar electrolytes

galvanic corrosion When two dissimilar metals are in contact with each other and exposed to a conductive environment, a potential exists between them and a current flows. The less-resistant metal becomes anodic, and the more resistant becomes cathodic. Attacks on the less-resistant metal increase, while on the more-resistant one, they decrease.

galvanic couple A pair of dissimilar conductors, commonly metals, in electrical contact.

galvanic couple potential. *mixed potential.*

galvanic current The electric current that flows between metals or conductive nonmetal in a galvanic couple.

galvanic series A list of metals arranged according to their relative corrosion potential in a specific environment; seawater is often used.

galvanize To coat a metal surface with zinc using any of several processes.

galvanizing The accepted term for the coating of iron or steel with zinc by the immersion of the metal in a bath of molten zinc; the word comes from galvano.

galvanneal To produce a zinc-iron alloy coating on iron or steel by keeping the coating molten after hot-dip galvanizing until the zinc alloys completely with the base metal.

galvanometer An instrument for indicating or measuring a small electric current by means of a mechanical motion derived from electromagnetic or electrodynamic forces produced by the current.

galvanostatic An experimental technique whereby an electrode is maintained at a constant current in an electrolyte.

gaseous corrosion Corrosion with gas as the only corrosive agent, without any aqueous phase on the surface of the metal. Also called *dry corrosion*.

gamma iron The face-centered cubic form of pure iron, which is stable from 910 to 1400°C.

general corrosion A form of deterioration that is distributed more or less uniformly over a surface. See also *uniform corrosion.*

Gibbs free energy The thermodynamic function $3G = 5H - TSS$, where H is enthalpy, T is absolute temperature, and S is entropy. Also called *free energy, free enthalpy,* and *Gibbs function.*

glass electrode A glass membrane electrode used to measure pH or hydrogen-ion activity.

grain An individual crystal in a polycrystalline metal or alloy; it may or may not contain twinned regions and subgrains; a portion of a solid metal (usually a fraction of an inch in size), in which the atoms are arranged in an orderly pattern.

grain boundary A narrow zone in a metal corresponding to the transition from one crystallographic orientation to another, thus separating one grain from another; the atoms in each grain are arranged in an orderly pattern; the irregular junction of two adjacent grains.

graphitic corrosion Deterioration of gray cast iron in which the metallic constituents are selectively leached or converted to corrosion products leaving the graphite intact. The term *graphic quotation* is commonly used for this form of corrosion, but is not recommended because it is also used in metallurgy for the decomposition of carbide to graphite; deterioration of gray cast iron in which the metallic constituents are selectively leached or converted to corrosion products leaving the graphite intact.

graphitization A metallurgical term describing the formation of graphite in iron or steel, usually from decomposition of iron carbide at elevated temperatures.

green rot A form of high-temperature corrosion of chromium-bearing alloys in which green chromium oxide (Cr_2O_3) forms, but certain other alloy constituents remain metallic; some simultaneous carburization is sometimes observed.

ground (anode) bed Commonly a group of manufactured electrodes or scrap steel that serves as the anode for the cathodic protection of pipelines, tanks, or other buried metallic structures. Types of ground beds are surface and deep anode configurations.

hand cleaning Surface preparation using hand tools such as wire brushes, scrapers, and chipping hammers.

hazard The likelihood that injury will result when a substance or object is used in a particular quantity or manner. Note that there are really no hazardous substances or objects, only hazardous ways of using them.

heat-shrinkable coatings Coatings that will reduce in dimension from an expanded size to a predetermined size by the application of heat.

high-solid coatings Generally, a coating that contains at least 70% solids by volume. The term *higher solids* is more appropriate for coatings that have a higher percentage of solids than previous (conventional) formulations but still contain less than 70% solids by volume.

homopolymer A polymer consisting of identical monomer units; polyethylene is one example of this.

hot-applied Having a consistency at ambient temperature such that heating is required before application.

hydraulic cement Cement that sets and hardens by chemical interaction with water and capable of doing so underwater.

halogen Any of the elements of a family consisting of fluorine, chlorine, bromine, iodine, and astatine.

hard chromium Chromium that is plated for engineering rather than for a decorative application.

hardenability The relative ability of a ferrous alloy to form martensite when quenched from a temperature above the upper critical temperature. This property is commonly measured as the distance below a quenched surface at which the metal exhibits a specific hardness (such as 50 HRC) or a specific percentage of martensite in the microstructure.

hardfacing Depositing filler metal on a surface by welding, spraying, or braze welding to increase resistance to abrasion, erosion, wear, galling, impact, or cavitation damage.

hard water Water that contains certain salts, such as those of calcium or magnesium, that form insoluble deposits in boilers and form precipitates with soap.

heat-affected zone That portion of the base metal that was not melted during brazing, cutting, or welding, but whose microstructure and mechanical properties were altered by the heat; refers to area adjacent to a weld where the thermal cycle has caused microstructural changes that generally affect corrosion behavior.

heat check A pattern of parallel surface cracks that are formed by alternating rapid heating and cooling of the extreme surface metal; sometimes found on forging dies and piercing punches. There may be two sets of parallel cracks, one perpendicular to the other.

hematite (1) An iron mineral crystallizing in a therhombohedral system; the most important are made of iron. (2) An iron oxide, corresponding to an iron content of approximately 70%.

hermetic seal An impervious seal made by the fusion of metals or ceramic materials (as by brazing, soldering, welding, or fusing glass or ceramic), which prevents the passage of gas or moisture.

high-temperature hydrogen attack A loss of strength and ductility of steel by high-temperature reaction of absorbed hydrogen with carbides in the steel, resulting in decarburization and internal fissuring.

holiday A discontinuity in a coating (such as porosity, cracks, etc.) that allows areas of base metal to be exposed to any corrosive environment that contacts the coated surface.

hot corrosion An accelerated corrosion of metal surfaces that results from the combined effect of oxidation and reactions with sulfur compounds and other contaminants, such as chlorides, to form a molten salt on a metal surface that fluxes, destroys, or disrupts the normal protective oxide.

hot cracking Hot cracking of weldments is caused by the segregation at grain boundaries of low-melting constituents in the weld metal. This can result in grain-boundary tearing under thermal contraction stresses. It can be minimized by the use of low-impurity welding materials and proper joint design. Also called *solidification cracking.*

hot working Deforming metal plastically at such a temperature and strain rate that ecrystallization takes place simultaneously with the deformation, thus avoiding any strain hardening.

hot-dip coating A metallic coating obtained by dipping a base metal into a molten metal.

hot shortness A tendency for some alloys to separate along grain boundaries when stressed or deformed at temperatures near the melting point. Hot shortness is caused by a low-melting constituent, often present only in minute amounts, that is segregated at grain boundaries.

huey test Corrosion testing in a boiling solution of nitric acid. This test is mainly used to detect the susceptibility of stainless steel to intergranular corrosion.

humidity test A corrosion test involving exposure of specimens to controlled levels of humidity and temperature.

hydrogen-assisted cracking (HAC) See *hydrogen embrittlement.*

hydrogen-assisted stress-corrosion cracking (HSCC) See *hydrogen embrittlement.*

hydrogen blistering The formation of blisters on or below a metal surface from excessive internal hydrogen pressure; formation of blisterlike bulges on a ductile metal surface caused by internal hydrogen pressures. Hydrogen may be formed during cleaning, plating, corrosion, and so forth.

hydrogen-controlled electrode A covered electrode which, when used correctly, produces less than a specified amount of diffusible hydrogen in the weld deposit.

hydrogen damage At moderate temperatures, hydrogen damage can occur as a result of a corrosion reaction on a surface or cathodic protection. Atomic hydrogen diffuses into the metal and collects at internal voids or laminations, where it combines to form more voluminous molecular hydrogen. In steels, blisters sometimes occur. At higher temperatures and pressures, atomic hydrogen can diffuse into steel and collect at grain boundaries. Either molecular hydrogen is then formed, or the hydrogen reacts with iron carbides to form methane, resulting in cracking and decarburization. Hydrogen cracking is intergranular and highly branched, but not continuous.

inhibitor A material used, normally in small amounts, to arrest or retard a chemical reaction, especially corrosion.

inert filler Finely divided mineral powder or inorganic fiber that is not substantially hygroscopic, not electrically conducting, and does not react with other ingredients of the coating material or with the environment in which it will be used.

inhibitive pigment Pigment that assists in the prevention of corrosion or some other undesirable effect.

inhibitor The general term for compounds or materials that slow down or stop an undesired chemical change such as corrosion, oxidation, polymerization, drying, skinning, and mildew growth.

inorganic coating A coating based on silicates or phosphates and usually pigmented with metallic zinc. See also *cement paint* and *zinc-rich primer.*

insulation resistance The insulation resistance between two electrodes that are in contact with, or embedded in, a specimen, is the ratio of the direct voltage applied to the electrodes to the total current between them.

intermediate coat (undercoat) The paint intended to be used between the primer and topcoat in a paint system.

ionic transport Corrosion of a metal is an electrochemical reaction between the metal and its environment, which results in a waste of metal. Thus, corrosion is a combination of the chemical effect of transported ions in a corrosive environment to the metal surface with an associated electrical energy (corrosion current).

intergranular corrosion Metals are composed of grains or crystals that form as solidification occurs. A crystal grows until it meets another advancing crystal. The regions of disarray between crystals are called *grain boundaries,* which differ in composition from the crystal center. Intergranular corrosion is the selective attack of the grain boundary or an adjacent zone. The most common example of intergranular corrosion is that of sensitized austenitic stainless steels in heat-affected zones at welds. Intergranular corrosion usually leaves the surface roughened, but definite diagnosis must be made by microscopic examination. Corrosion which occurs preferentially at grain boundaries.

intergranular cracking Cracking or fracturing that occurs between the grains or crystals in a polycrystalline aggregate. Also called *intercrystalline cracking.* See also *transgranular cracking.*

intergranular fracture The brittle breaking of a metal between the grains, or crystals, that form it. Also called *intercrystalline fracture.* See also *transgranular fracture.*

intergranular stress-corrosion cracking (IGSCC) Stress-corrosion cracking which occurs along grain boundaries.

internal oxidation The formation of isolated particles of corrosion products beneath a metal surface, which occurs as the result of preferential oxidation of certain alloy constituents by inward diffusion of oxygen, nitrogen, sulfur, or other substances.

intumescence The swelling or bubbling of a coating, usually because of heating; this term is currently used in space and fire-protection applications.

ion An atom or group of atoms that has gained or lost one or more outer electrons and thus carries an electric charge. Positive ions, or *cations,* are deficient in outer electrons. Negative ions, or *anions,* have an excess of outer electrons.

ion erosion Deterioration of material caused by ion impact.

ion exchange The reversible interchange of ions between a liquid and solid, with no substantial structural changes in the solid.

iron rot Deterioration of wood in contact with iron-based alloys.

isocorrosion diagram A graph or chart that shows constant corrosion behavior with changing solution (environment) composition and temperature.

lacquer A coating composition that is based on synthetic thermoplastic film-forming material dissolved in organic solvent and that dries primarily by solvent evaporation. Typical lacquers include coatings based on vinyl resins, acrylic resins, chlorinated rubber resins, and other resins.

laitance A milky-white deposit on new concrete.

laminate A product made by bonding together two or more layers of material or materials.

lime Calcium oxide (CaO); also, a general term for various chemical and physical forms of quick lime, hydrated lime, and hydraulic hydrated lime.

lining Any sheet or layer of material attached directly to the inside face of formwork to improve or otherwise alter its quality and surface texture.

lot An indefinite number of containers of materials manufactured by a single plant run through the same processing equipment, with no change in ingredient materials. Also known as a *batch.*

low-temperature flexibility The resistance to cracking of heat-shrinkable coatings when wrapped around prescribed mandrels at specified temperatures.

lump In porcelain enamels, a rounded projection in the enamel surface, usually considered a defect.

laminar scale Rust formation in heavy layers.

lamellar tearing Occurs in the base metal adjacent to weldments due to high through-thickness strains introduced by weld metal shrinkage in highly restrained joints. Tearing occurs by decohesion and linking along the working direction of the base metal; cracks usually run roughly parallel to the fusion line and are steplike in appearance. This phenomenon can be minimized by designing joints to minimize weld shrinkage stresses and joint restraint.

Langelier saturation index An index calculated from total dissolved solids, calcium concentration, total alkalinity, pH, and solution temperature that shows the tendency of a water solution to precipitate or dissolve calcium carbonate.

leakage or coating resistance Leakage resistance of pipe or resistance of pipe radially to remote earth. This includes the resistance of the coating (if any) and is affected by the resistivity of the environment; measured in ohm-kilometers.

ledeburite The eutectic of the iron-carbon system, the constituents of which are austenite and cementite. The austenite decomposes into ferrite and cementite upon cooling below the temperature at which transformation of austenite to ferrite or ferrite plus cementite is completed.

ligand The molecule, ion, or group bound to the central atom in a chelate or a coordination compound.

limiting current density The maximum current density that can be used to obtain a desired electrode reaction without undue interference, such as from polarization.

linear elastic fracture mechanics A method of fracture analysis that can determine the stress (or load) required to induce fracture instability in a structure containing a crack like flaw of known size and shape.

linear pipe resistance The pipe resistance measured in ohms per unit length, which can be calculated from the specific resistivity of steel or iron and regarding the pipeline as an annular cylinder. The specific resistivity of steel pipe will normally vary from 15 to 23 microhm-cm, depending on its chemistry. Absent specific test results, it is normal to use a value of 18 microhm-cm.

linear polarization resistance (LPR) At small applied polarization potentials, the relationship between the applied potentials approximates the polarization resistance. See also *polarization resistance*.

lipophilic Having an amenity for oil.

liquid metal embrittlement Catastrophic brittle failure of a normally ductile metal when in contact with a liquid metal and subsequently stressed in tension.

local action Corrosion due to the action of local cells; that is, galvanic cells resulting from inhomogeneities between adjacent areas on a metal surface exposed to an electrolyte.

local cell A galvanic cell resulting from inhomogeneities between areas on a metal surface in an electrolyte. The inhomogeneities may be of physical or chemical nature in either the metal or its environment.

long-line current Current that flows through the Earth from an anodic to a cathodic area of a continuous metallic structure usually used only where the areas are separated by considerable distance and where the current results from concentration-cell action.

luggin probe (Luggin-Haber capillary) A small tube or capillary filled with electrolyte, terminating close to the metal surface being studied, and used to provide an ionically conducting path without diffusion between an electrode being studied and a reference electrode.

manual cleaning Includes hand cleaning and power tool cleaning.

manufacturer The person, firm, or corporation that manufactures and provides the coatings under the provisions of relevant standard.

marine atmosphere Frequent and relatively high concentration of salt mist, but it does not imply direct contact with salt spray or splashing waves; it contains a high concentration of chloride in contrast to the high concentration of sulfur dioxide in industrial environments.

masonry Construction composed of shaped or molded units, usually small enough to be handled by one person and composed of stone, ceramic, brick or tile, concrete, glass, or the like.

masonry cement A hydraulic cement for use in mortar for masonry construction, containing a type of cement and one or more other materials such as hydrated lime, limestone chalk, talc, slag, or clay.

maximum amplitude A term used in relevant standard is defined as the greatest vertical distance between the summit of any peak on a blast-cleaned surface and the bottom of an immediately adjacent trough. This does not take into account any exceptionally high, "rogue" peaks, which are liable to occur on a blast-cleaned surface as a result of embedded particles of abrasive. Such peaks are very undesirable, and their size and number may be a subject of special agreement between the parties to a contract.

metallic coating One or more layer of metal on a steel base (base material).

metal spraying Application of a spray coat of metal (usually zinc or aluminum) onto a prepared surface. The metal to be sprayed is rendered molten by passing it, in wire or powder form, through a flame pistol that projects the semimolten metal onto the surface by means of a jet of compressed air.

mill scale Mill scale is the term used for the surface oxides produced during hot rolling of steel. It breaks and flakes when the steel is flexed and paint applied over it may fail prematurely. The extent of such failures is unpredictable but they frequently occur within a few weeks of painting, particularly in aggressive environments. No protective coating can give lifelong protection unless both the scale and rust are removed.

macroscopic Visible at magnifications up to 25x.

macrostructure The structure of metals as revealed by macroscopic examination of the etched surface of a polished specimen.

magnetite Naturally occurring magnetic iron oxide (Fe_3O_4).

manual welding Welding in which the means of making it are held in the hand.

martensite A generic term for microstructures formed by diffusionless phase transformation in which the parent and product phases have a specific crystallographic relationship. Martensite is characterized by an acicular pattern in the microstructure in both ferrous and nonferrous alloys. In alloys where the solute atoms occupy interstitial positions in the martensitic lattice (such as carbon in iron), the structure is hard and highly strained; but where the solute atoms occupy substitutional positions (such as nickel in iron), the martensite is soft and ductile. The amount of high-temperature phase that transforms to martensite on cooling depends to a large extent on the lowest temperature attained, there being a rather distinct beginning temperature (Ms) and a temperature at which the transformation is essentially complete (Mf).

mechanical plating Plating wherein fine metal powders are peened onto the work by tumbling or other means.

metal-arc welding Arc welding using a consumable electrode.

metal-insert-gas (MIG) welding Welding using a consumable electrode.

metal dusting Accelerated deterioration of metals in carbonaceous gases at elevated temperatures to form a dustlike corrosion product; a unique form of high-temperature corrosion that forms a dustlike corrosion product and sometimes develops hemispherical pits on a susceptible metal surface; simultaneous carburization is generally observed.

metal ion concentration cell A galvanic cell caused by a difference in metal ion concentration at two locations on the same metal surface.

metallic glass An alloy having an amorphous or glassy structure. See also *amorphous solid*.

metallizing (1) The application of an electrically conductive metallic layer to the surface of nonconductors. (2) The application of metallic coatings by nonelectrolytic procedures such as spraying of molten metal and deposition from the vapor phase.

meteor perforation Perforation of material in outer space resulting from meteor strikes.

microbiologically influenced corrosion (MIC) Corrosion that substantially increases as the result of the presence of bacteria, such as sulfate reducing bacteria (SRB) or acid-producing bacteria (APB).

microscopic Visible at magnifications above 25x.

microstructure The structure of a prepared surface of a metal as revealed by a microscope at a magnification exceeding 25x.

mill scale An oxide layer on metals or alloys produced by metal rolling, hot-forming, welding, or heat treatment; especially applicable to iron and steel.

mixed potential The potential of a specimen (or specimens in a galvanic couple) when two or more electrochemical reactions are occurring. Also called *galvanic couple potential.*

molal solution The concentration of a solution expressed in moles of solute divided by 1000 g of solvent.

molar solution Aqueous solution that contains 1 mole (gram-molecular weight) of solute in 1 l of the solution.

mole The mass that is numerically equal (in grams) to the relative molecular mass of a substance. It is the amount of substance of a system that contains as many elementary units (6.023 exp23) as there are atoms of carbon in 0.012 kg of the pure nuclide C12; the elementary unit must be specified and may be an atom, molecule, ion, electron, photon, or even a specified group of such units.

monomer A molecule, usually an organic compound, with the ability to join with a number of identical molecules to form a polymer.

nominal parameter One of several parameters (e.g., weight, thickness, or density) specified on product labels, invoices, sales literature, and the like. The actual parameters shall not be less than 95% of the nominal parameters.

nonsaline water Potable and nonpotable water applicable to river installations, sewage treatment tanks, water tanks, and domestic water systems.

noble metal A metal that is not very reactive (e.g., silver, gold, or copper) and may be found naturally on Earth in metallic form.

normalizing Heating a ferrous alloy to a suitable temperature above the transformation range and then cooling in air to a temperature substantially below the transformation range.

orange peel A surface condition characterized by an irregular waviness of the porcelain enamel resembling an orange skin in texture; sometimes considered a defect.

organic Being or composed of hydrocarbons or their derivatives, or matter of plant or animal origin.

organic acid A chemical compound with one or more carboxyl radicals (COOH) in its structure; examples are butyric acid [$CH_3(CH_2)_2COOH$], maleic acid (HOOCCH-CHCOOH), and benzoic acid (C_6H_5COOH).

organic zinc coating A paint containing zinc powder pigment and an organic (i.e., containing carbon) resin.

organic zinc-rich paint A coating containing zinc powder pigment and an organic resin.

overaging Aging under conditions of time and temperature greater than those required to obtain maximum change in a certain property, so that the property is altered in the direction of the initial value.

overheating Heating a metal or alloy to such a high temperature that its properties are impaired. When the original properties cannot be restored by further heat treating, by mechanical working, or by a combination of working and heat treating, it is known as *burning.*

overload When a metal part has been subjected to a single stress beyond its tensile strength, it can fail by overloading. The fracture can be either ductile or brittle, depending on factors such as the metal's hardness and operating temperature. In most cases, a single fracture results.

oxyacetylene welding Gas welding in which fuel gas is acetylene and which is burned in an oxygen atmosphere.

oxidation (1) A corrosion reaction in which the corroded metal forms an oxide; usually applied to reaction with a gas containing elemental oxygen, such as air. (2) A reaction in which there is an increase in valence resulting from a loss of electrons. See also *reduction.*

oxidized surface With steel, a surface having a thin, tightly adhering, oxidized skin (from straw to blue in color), extending in from the edge of a coil or sheet.

oxidizing agent A compound that causes oxidation, thereby itself being reduced.

oxygen concentration cell A galvanic cell resulting from difference in oxygen concentrations between two locations

ozone A powerfully oxidizing allotropic form of the element oxygen. The ozone molecule contains three atoms (O_3). Ozone gas is decidedly blue, and both liquid and solid ozone are an opaque blue-black color, similar to that of ink.

paint Any pigmented liquid, liquifiable, or mastic composition designed for application to a substrate in a thin layer that is converted to an opaque solid film after application. Used for protection, decoration, or identification, or to serve some other function.

paint (or coating) One or more separate, coherent layers consisting of nonperformed materials and a binder normally of organic nature.

passivation The act of making a substance inert or unreactive.

penetration The depth, expressed in units of 0.1 mm, to which a standard needle placed vertically on the surface of the sample of bitumen enamel, and loaded with a 100-g weight under the specified conditions of temperature (25°C) and time (5 s), will enter.

petrolatum Petroleum jelly used for impregnation.

petroleum bitumen A mixture of high-molecular-mass hydrocarbons derived from petroleum by the oxidation of suitable selected bases to a varying extent, possibly by adding fillers, in order to produce a base material conforming to either grade A or B.

petroleum jelly A purified mixture of semisolid hydrocarbons obtained from petroleum.

phosphating Pretreatment of steel and certain other metal surfaces by chemical solutions containing metal phosphates and phosphoric acid as the main ingredients, to form a thin, inert, adherent, corrosion-inhibiting phosphate layer that serves as a good base for subsequent paint coats.

pigment Finely ground, natural or synthetic, inorganic or organic, and insoluble dispersed particles (powder) that, when dispersed in a liquid vehicle to make paint, may provide in addition to color many of the essential properties of paint: opacity, hardness, durability, and corrosion resistance.

pin hole A film defect characterized by small, porelike flaws in a coating that extend entirely through the applied film and have the general appearance of pinpricks when viewed in reflecting light. The term is generally applied to holes caused by solvent bubbling, moisture, other volatile products, or the presence of extraneous particles in the applied film.

plaster A cementitious material that, when mixed with a suitable amount of water, forms a plastic mass or paste that when applied to a surface, adheres to it and subsequently hardens.

plastic A material that contains as an essential ingredient a high polymer, and which, at some stage in its processing into finished products, can be shaped by flow.

pliability The quality or state of being flexible in bending or creasing.

polyethylene plastic A plastic or resin prepared by the polymerization of no less than 85 wt.% ethylene and no less than 95 wt.% of total olefins.

porcelain Glazed or unglazed vitreous ceramic whiteware that is matured like ceramic and glazed together in the same firing operation.

porosity The rate of air flow through a mat under a differential pressure between the two fabric surfaces; related to air permeability.

powder organic coating A product containing pigments, resins, and other additives that is applied in the form of a powder to a metallic substrate and then fused to form a coherent, continuous finish.

power tool cleaning The use of pneumatic and electric portable power tools to prepare substrate for painting.

precast concrete A concrete member that is cast and cured in other than its final position.

prefabrication primer Quick-drying material applied as a thin film to a metal surface after cleaning (e.g., by a blast-cleaning process) to give protection for the period before and during fabrication. These primers shall not interfere seriously with conventional welding or cutting operations or give off toxic fumes during such operations.

pretreatment Usually refers to the chemical treatment of unpainted metal surface before painting.

primer The first complete coat of paint of a painting system applied to a surface. Such paints are designed to provide adequate adhesion to new surfaces and are formulated to meet the special requirements of those surfaces. The type of primer varies with the surface, its condition, and the total painting system to be used. Primers for steel work contain special anticorrosive pigments, such as red lead, zinc chromate, and zinc powder.

parent metal The metal to be joined; also known as the *base metal.*

parkerizing The trade name for the production of phosphate coating on steel articles by immersion in an aqueous solution of manganese or zinc acid with phosphate.

partial annealing An imprecise term used to denote a treatment given cold-worked material to reduce its strength to a controlled level or to effect stress relief. To be meaningful, the type of material, the degree of cold work, and the time-temperature schedule must be stated.

parting The selective attack of one or more components of a solid solution alloy; e.g., dezincification and dealumination.

parts per billion A measure of proportion by weight, equivalent to one unit weight of a material per billion (10^9) unit weights of a compound. One part per billion is equivalent to 1 mg/kg.

parts per million A measure of proportion by weight, equivalent to one unit weight of a material per million (10^6) unit weights of a compound. One part per million is equivalent to 1 mg/g.

passivation A reduction of the anodic reaction rate of an electrode involved in electrochemical action such as corrosion.

passivator A type of inhibitor that appreciably changes the potential of a metal to a more noble (positive) value.

passive The state of a metal when it is much more resistant to corrosion than its position in the electromotive force (EMF) series would predict. Passivation is a surface phenomenon.

passive-active cell (1) A cell, the electromotive force (EMF) of which is due to the potential difference between a metal in an active state and the same metal in a passive state. (2) A corrosion cell in which the anode is a metal in the active state and the cathode is the same metal in the passive state.

passivity A metal or alloy that is thermodynamically unstable in a given electrolytic solution is said to be passive when it remains visibly unchanged for a prolonged period. The following should be noted:

- During passivation, the appearance may change if the passivating film is sufficiently thick (e.g., interference films);
- The electrode potential of a passive metal is always appreciably more noble than its potential in the active state;
- Passivity is an anodic phenomenon and thus control of corrosion by decreasing cathodic reactivity (e.g. amalgamated zinc in sulfuric acid) or by cathodic protection is not passivity.

patina A coating (usually green) that forms on the surface of metals, such as copper and copper alloys, exposed to the atmosphere. Also used to describe the appearance of a weathered surface of any metal.

pearlite A metastable lamellar aggregate of ferrite and cementite resulting from the transformation of austenite at temperatures above the bainite range.

peen plating Deposition of the coating metal, in powder form, on the substrate by a tumbling action in the presence of a peening shot.

phosphating Forming an adherent phosphate coating on a metal by immersion in a suitable aqueous phosphate solution. See also *phosphatizing*.

phosphatizing The forming of a thin inert phosphate coating on a surface, usually accomplished by treating with H_3PO_4 (phosphoric acid).

pH A measure of the acidity or alkalinity of a solution; the negative logarithm of the hydrogen-ion activity. Denotes the degree of acidity or basicity of a solution. At 25°C, 7.0 is the neutral value. Values below 7.0 indicate increasing acidity; values above it, increasing basicity.

physical vapor deposition A coating process whereby the cleaned and masked component to be coated is heated and rotated on a spindle above the streaming vapor generated by melting and evaporating a coating material source bar with a focused electron beam in an evacuated chamber.

physisorption The binding of an adsorbate to the surface of a solid by forces whose energy levels approximate those of condensation. See also *chemisorption*.

pickle A solution, usually acid, used to remove mill scale or other corrosion products from a metal.

pickle / pickling A form of chemical and electrolytic removal or loosening of mill scale and corrosion products from the surface of a metal in a chemical solution (usually acidic). Electrolytic pickling can be anodic or cathodic, depending on the polarization of metal in the solution.

pitting Highly localized corrosion resulting in deep penetration at only a few spots.

pitting factor The ratio of the depth of the deepest pit resulting from corrosion divided by the average penetration as calculated from weight loss.

plane strain In linear elastic fracture mechanics, the stress condition in which there is zero strain in a direction normal to both the axis of applied tensile stress and the direction of crack growth (that is, parallel to the crack front); most nearly achieved in loading thick plates along a direction parallel to the plate surface. Under plane-strain conditions, the plane of fracture instability is normal to the axis of the principal tensile stress.

plane stress In linear elastic fracture mechanics, the condition in which the stress in the thickness direction is zero; most nearly achieved in loading very thin sheet along a direction parallel to the surface of the sheet. Under plane-stress conditions, the plane of fracture instability is inclined 45° to the axis of the principal tensile stress.

plasma plating Deposition on critical areas of metal coatings resistant to wear and abrasion, by means of a high-velocity, high-temperature ionized inert gas jet.

plasma spraying A thermal spraying process in which the coating material is melted with heat from a plasma torch that generates a nontransferred arc: i.e., molten coating material is propelled against the base metal by hot, ionized gas issuing from the torch.

plastic deformation The permanent (inelastic) distortion of metals under applied stresses that strain the material beyond its elastic limit.

plasticity The property that enables a material to undergo permanent deformation without rupture.

polarization The deviation from the open circuit potential of an electrode resulting from the passage of current.

polarization curve A plot of current density versus electrode potential for a specific electrode-electrolyte combination.

polarization resistance The slope (dE/di) at the corrosion potential of a potential (E) versus current density (i) curve. (It is inversely proportional to the corrosion current density when the polarization resistance technique is applicable.)

polyester A resin formed by the condensation of polybasic and monobasic acids with polyhydric alcohols.

polymer A chain of organic molecules produced by the joining of primary units called *monomers.*

potential Any of various functions from which intensity or velocity at any point in a field may be calculated. The driving influence of an electrochemical reaction.

potential survey The measurement of potential of a structure or pipeline relative to a reference electrode potential pitting.

potentiodynamic (potentiokinetic) The technique for varying the potential of an electrode in a continuous manner at a preset rate.

potentiostat An electronic device that maintains an electrode at a constant potential; used in anodic protection devices.

poultice corrosion A term used in the automotive industry to describe the corrosion of vehicle body parts due to the collection of road salts and debris on ledges and in pockets that are kept moist by weather and washing.

pourbaix (potential-pH) diagram A plot of the redox potential of a corroding system versus the pH of the system, compiled using thermodynamic data and the Nernst equation. The diagram shows regions within which the metal itself or some of its compounds are stable.

powder metallurgy The art of producing metal powders and utilizing metal powders for the production of massive materials and shaped objects.

precious metal One of the relatively scarce and valuable metals: gold, silver, and the platinum-group metals. Also called *noble metal.*

precipitation hardening Hardening caused by the precipitation of a constituent from a supersaturated solid solution. See also *age hardening* and *aging.*

precipitation heat treatment Artificial aging in which a constituent precipitated from a supersaturated solid solution.

precracked specimen A specimen that is notched and subjected to alternating stresses until a crack has developed at the root of the notch.

pressure welding A welding process in which a weld is made by a sufficient pressure to cause plastic flow of the surfaces, which may or may not be heated.

primary current distribution The current distribution in an electrolytic cell that is free of polarization.

primary passive potential (passivation potential) The potential corresponding to the maximum active current density (critical anodic current density) of an electrode that exhibits active-passive corrosion behavior.

principal stress (normal) The maximum or minimum value at the normal stress at a point in a plane considered with respect to all possible orientations of the considered plane. On such principal planes, the shear stress is zero. There are three principal stresses on three mutually perpendicular planes. The state of stress at a point may be uniaxial, a state of stress in which two of the three principal stresses are zero; biaxial, a state of stress in which only one of the three principal stresses is zero; and triaxial, a state of stress in which none of the principal stresses is zero. The term *multiaxial stress* refers to either biaxial or triaxial stress.

profile An anchor pattern on a surface produced by abrasive blasting or acid treatment.

protection current The current made to flow into a metallic structure from its ectrolytic environment to effect cathodic protection of the structure.

protective potential The threshold value of the corrosion potential that has to be reached to enter a protective potential range. This term is used in cathodic protection to refer to the minimum potential required to suppress corrosion.

protective potential range A range of corrosion potential values in which unacceptable corrosion resistance is achieved for a particular purpose.

quartz/silica Glass made either by flame hydrolysis of silicon tetrachloride or by melting silica, usually in the form of granular quartz (i.e., fused silica).

quench-age embrittlement Embrittlement of low-carbon steel resulting from the precipitation of solute carbon of existing dislocations and from precipitation hardening of the steel caused by differences in the solid solubility of carbon in ferrite at different temperatures. This type of embrittlement usually is caused by rapid cooling of the steel from temperature slightly below the temperature at which austenite begins to form, and it can be minimized by quenching from lower temperature.

quench aging Aging induced by rapid cooling after solution heat treatment.

quench cracking Fracture of a metal during quenching from elevated temperature. Most frequently observed in hardened carbon steel, alloy steel, or tool steel, parts of high hardness and low toughness. These cracks often emanate from filets, holes, corners, or other stress raisers and result from high stresses due to the volume changes accompanying transformation to martensite.

quench hardening (1) In ferrous alloys, hardening by austenitizing and then cooling at a rate such that a substantial amount of austenite transforms to martensite. (2) In copper and titanium alloys, hardening by solution treating and quenching to develop a martensite-like structure.

quenching The rapid cooling of metals (often steel) from a suitable elevated temperature. This generally is accomplished by immersion in water, oil, polymer solution, or salt, although forced air is sometimes used.

refractories Nonmetallic materials with chemical and physical properties applicable for structures and system components exposed to environments above 538°C (1000°F).

release paper A sheet serving as a protectant or carrier, or both, for an adhesive film or mass, which is easily removed from the film or mass prior to use.

resin A solid, semisolid, or pseudosolid organic material that has an indefinite and often high relative molecular mass, exhibits a tendency to flow when subjected to stress, usually has a softening or melting range, and usually fractures conchoidally. In a broad sense, the term is used to designate any polymer that is a basic material for plastics.

rubber A material capable of quickly and forcibly recovering from all deformations. It can be modified to be essentially insoluble, but it can swell in boiling solvents (e.g., benzene, methyl ethyl ketone (MEK), and ethanol/toluene azeotrope. Rubber in its modified state, free of diluents, stretched at 18–29°C and held for 1 min. before release, retracts within 1 min. to less than 1.5 or 2 times its original length.

rural environment An atmospheric exposure that is virtually unpolluted by smoke and sulfur gases, and which is sufficiently inland to be unaffected by salt contaminations or the high humidity of coastal areas.

rust A reddish, brittle coating formed on iron or ferrous metals resulting from exposure to humid atmosphere or chemical attack.

redox potential The potential of a reversible oxidation-reduction electrode measured with respect to a reference electrode, corrected to the hydrogen electrode, in a given electrolyte.

reducing agent A compound that causes reduction, thereby itself becoming oxidized.

reduction A reaction in which there is a decrease in valence resulting from a gain in electrons. See also *oxidation.*

reference electrode A nonpolarizable electrode with a known and highly reproducible potential used for potentiometric and voltammetric analyses.

refractory metal A metal having an extremely high melting point; for example, tungsten, molybdenum, tantalum, niobium, chromium, vanadium, and rhenium. In the broad sense, this term refers to a metal having a melting point above the range for iron, cobalt, and nickel.

relative humidity The ratio, expressed as a percentage, of the amount of water vapor present in a given volume of air at a given temperature to the amount required to saturate the air at that temperature.

remedial bond A bond installed between a primary and a secondary structure in order to eliminate or reduce corrosion interaction.

remote earth An area in which the structure-to-electrolyte potential change is negligible with changes in reference electrode position away from the structure.

residual stress A stress that remains within a body as a result of plastic deformation.

resistance The opposition that a device or material offers to the flow of direct current, equal to the voltage drop across the element divided by the current through the element. Also called *electrical resistance.*

resistance bond A bond either incorporating resistors or of adequate resistance in itself to limit the flow of current.

resistance welding A type of welding in which force is applied to surfaces in contact and in which the heat for welding is produced by the passage of electric current through the electrical resistance at, and adjacent to, these surfaces.

resolution In ultrasonics, the ability of a system to give simultaneous separate indications from discontinuities that are close together, both in depth and lateral position.

reverse current switch A switch that is installed in series with interference bonds where stray current is known to reverse direction. Prevents serious corrosion caused by reversed current discharging to the electrolyte, by interrupting the reversed current. A failed switch becomes an open circuit or a solid bond.

ringworm corrosion Localized corrosion frequently observed in oil-well tubing in which a circumferential attack is observed near a region of metal "upset."

riser (1) That section of pipeline extending from the ocean floor up the platform. Also, the vertical tube in a steam generator convection bank that circulates water and steam upward. (2) A reservoir of molten metal connected to a casting to provide additional metal to the casting, required as the result of shrinkage before and during solidification.

run The metal method or deposited during one passage of an electrode, torch, or blow-pipe.

rust A visible corrosion product consisting of hydrated oxides of iron; applied only to ferrous alloys. See also *white rust.*

rusting Corrosion of iron or an iron-based alloy to form a reddish-brown product that is primarily hydrated ferric oxide.

sagging (1) A defect characterized by a wavy line or lines appearing on those surfaces of porcelain enamel that have been fired in a vertical position. (2) A defect characterized by irreversible downward bending in an article insufficiently supported during the firing cycle.

salt fog test An accelerated corrosion test in which specimens are exposed to a fine mist of a solution usually containing sodium chloride, but sometimes modified with other chemicals.

salt spray test A test applied to metal finishes to determine their anticorrosive properties, involving the spraying of common salt (sodium chloride) solution on the surface of a coating steel panel.

sand blast To use sand, flint, or a similar nonmetallic abrasive propelled by an air blast on metal, masonry, concrete, or other hard substance to remove dirt, rust, or paint.

sanding An abrasive process used to level a coated surface prior to the application of a further coat.

saponification number The number of milligrams of potassium hydroxide that is consumed by 1 g of oil under the conditions of the test.

saturated calomel electrode A reference electrode composed of mercury, mercurous chloride (known as *calomel*), and a saturated aqueous chloride solution.

scaling (1) The formation at high temperatures of thick corrosion layers on a metal surface. (2) The deposition of water-insoluble constituents on a metal surface.

scarifying A method of preparing concrete surfaces for coating. Scarifiers are sharp rotating knives in a self-contained unit resembling a plant sweeper.

season cracking An obsolete historical term usually applied to stress-corrosion crackling of brass. A term usually applied to stress corrosion cracking of brass.

seawater Ocean and other saline water and estuary water.

selective leaching A corrosion process also called *parting* or *de-alloying,* or more specifically, it can be called *de-zincification* in the case of brasses, *de-nickelification* in cupronickels, etc. Selective leaching may occur in a plug form or in a more evenly distributed layer type. Stagnant conditions and regions under deposits are conductive to selective leaching. In brasses, it can occur at pH extremes in water; high dissolved solids and high temperature also promote selective leaching. The overall dimensions of a part do not change drastically, but appreciable weakening can occur.

semiautomatic welding A type of welding in which some of the variables are automatically controlled, but manual guidance is necessary.

sensitization heat treatment A heat treatment, whether accidental, intentional, or incidental (as during welding), that causes the precipitation of constituents at grain boundaries, often causing the alloy to become susceptible to intergranullar corrosion, cracking or stress corrosion cracking (SCC).

sensitization In austenitic stainless steel, the precipitation of chromium carbide, usually at grain boundaries, upon exposure to temperatures of 550–850°C.

shade A color produced by a pigment or dye mixture with some black in it.

shear A type of force that causes or tends to cause two contiguous parts of the same body to slide relative to each other in a direction parallel to their plane of contact.

shear strength The stress required to produce fracture in the plane of cross section, the conditions of loading being such that the directions of force and of resistance are parallel and opposite, although their paths are offset by a specified minimum amount.

sherodizing The coating of iron or steel with zinc by heating it in zinc powder at a temperature below the melting point of zinc.

sieve A metallic plate or sheet, a woven wire cloth, or other similar device, with regularly spaced apertures of uniform size, mounted in a suitable frame or holder for use in separating material according to size.

sigma phase A hard, brittle, nonmagnetic intermediate phase with a tetragonal crystal structure, containing 30 atoms per unit cell, space group $P4_2$mnm, occurring in many binary and ternary alloys of the transition elements. The composition of this phase in the various systems is not the same, and the phase usually exhibits a wide range of homogeneity. Alloying with a third transition element usually enlarges the field of homogeneity and extends it deep into the ternary section.

sigma-phase embrittlement A type of embrittlement of iron-chromium alloys (most notably austenitic stainless steel) caused by precipitation at grain boundaries of the hard, brittle intermetallic sigma phase during long periods of exposure to temperatures between approximately 560 and 980°C. Sigma-phase embrittlement results in severe loss in toughness and ductility and can make the embrittled material susceptible to intergranular corrosion. See also *sensitization.*

silver/silver chloride electrode (Ag/AgCl) A reference electrode consisting of silver coated with silver chloride in an electrolyte containing chloride ions.

slip Plastic deformation by the irreversible shear displacement (translation) of one part of a crystal relative to another in a definite crystallographic direction and usually on a specific crystallographic plane. Sometimes called *glide.*

slow strain rate technique An experimental method for evaluating a material's susceptibility to stress-corrosion cracking. It involves pulling the specimen to failure in uniaxial tension at a controlled slow strain rate while it is in the test environment and examining it for evidence of stress-corrosion cracking.

slushing compound An obsolete term describing oil or grease coatings used to provide temporary protection against atmospheric corrosion.

smelt Molten slag; in the pulp and paper industry, the cooking chemicals tapped from the recovery boiler as molten material and dissolved in the smelt tank as green liquor.

S-N diagram A plot showing the relationship of stress (S) and the number of cycles (N) before fracture in fatigue testing.

softening point (ring and ball) The temperature at which a disk of material contained in a ring undergoes a standard deformation caused by the weight of a ball under standardized test conditions.

soft water Water that is free of magnesium or calcium salts.

soil resistivity The electrical resistivity of the soil. This characteristic is important in cathodic protection systems, as it affects current distribution through the soil and potentials on the protected structure.

solder embrittlement The reduction in mechanical properties of a metal as a result of local penetration of solder along grain boundaries.

solid-metal embrittlement The occurrence of embrittlement in a material below the melting point of the embrittling species.

solid solution A single, solid, homogeneous crystalline phase containing two or more chemical species.

solute The component of either a liquid or solid solution that is present to a lesser or minor extent; the component that is dissolved in the solution.

solution In chemistry, a homogeneous dispersion of two or more kinds of molecular or ionic species. A solution may be composed of any combination of liquids, solids, or gases, but it always consists of a single phase.

solution heat treatment Heating an alloy to a suitable temperature, maintaining that temperature long enough to cause one or more constituents to become a solid solution, and then cooling rapidly enough to hold these constituents in solution.

solution potential A type of electrode potential where the half-cell reaction involves only the metal electrode and its ion.

solvent A volatile liquid, which is used in the manufacture of primer to dissolve or disperse the film-forming constituents, and which evaporates during drying and therefore does not become a part of the dried film. Solvents are used to control the consistency and character of the primer and to regulate application properties. Aliphatic solvents are mild solvents derived from petroleum, such as mineral spirit. Aromatic solvents are strong solvents derived from coal tar and certain petroleum types, such as toluene, xylene, and solvent naphta.

solvent The component of either a liquid or solid solution that is present to a greater or major extent; the component that dissolves the solute.

sour gas A gaseous environment containing hydrogen sulfide and carbon dioxide in hydrocarbon reservoirs. Prolonged exposure to sour gas can lead to hydrogen damage, sulfide-stress cracking, stress-corrosion cracking in ferrous alloys, or any combination.

sour water Wastewater containing fetid materials, usually sulfur compounds.

spalling The spontaneous chipping, fragmentation, or separation of a surface or surface coating.

splash zone A wind and water area of floating and tidal structures, such as wharfs, piers, seawalls, and platforms, or frequent salt spray.

spraying A method of application in which the coating material is broken up into fine mist that is directed onto the surface to be coated. This atomization process is usually, but not necessarily, effected by a compressed air jet.

stabilizer A substance added, usually in small proportions, to retard undesirable chemical or physical changes.

stain Matter that colors wood, plaster, or other masonry by penetration without hiding it and without leaving any perceptible surface film.

stoneware Vitreous or semivitreous ceramic of a fine texture made primarily from nonrefractory fire clay.

stucco A cement plaster used for coating exterior walls and other exterior surfaces of building.

surface preparation Any method of treating a surface to get it ready for painting. Swedish standards include photographic depictions of the surface appearance of hand and power tool cleaning and various grades of blast-cleaning over four initial mill scale and rust conditions of steel.

surface profile Surface profile is a measurement of the roughness of the surface which results from abrasive blast-cleaning. The height of the profile produced on the surface is measured from the bottoms of the lowest valleys to the tops of the highest peaks.

synthetic primer A primer containing solvent and whose base consists of resins and synthetic plasticizers.

tearing A defect in the surface of porcelain enamel, characterized by short breaks or cracks that have been healed.

tear strength The force required to either start or continue to propagate a tear in a fabric under specified conditions.

tensile strength The maximum resistance of material to deformation in a tensile test carried to rupture; that is, the breaking load, or force per unit cross-sectional area, of the unstrained specimen.

terne An alloy of lead containing 3–15% Sn, which is used as a hot-dip coating for steel sheets or plates. Terne coatings, which are smooth and dull in appearance, give steel better corrosion resistance and enhance its ability to be formed, soldered, or painted.

terne plate Deposition of lead-tin alloy on iron or steel sheets by the hot-dip process.

test access hole Provides a means of contacting soil through concrete or asphalt for measuring structure-to-soil potentials. Contains no wires and is usually capped but easily accessible.

test station Permanent wires attached to the structure and led to a convenient location for electrical measurements. Used at points where the structure or soil is otherwise inaccessible for electrical testing (underground or underwater).

thermal electromotive force The electromotive force generated in a circuit containing two dissimilar metals when one junction is at a temperature different from that of the other. See also *thermocouple*.

thermal embrittlement An intergranular fracture of maraging steel with decreased toughness resulting from improper processing after hot working. This type of embrittlement occurs upon heating above 1095°C and then slow cooling through the temperature range of 815–980°C, and has been attributed to precipitation of titanium carbides and titanium carbonitrides at austenite grain boundaries during cooling through the critical temperature range.

thermal cutting The parting or shaping of materials by the application of heat with or without a stream of cutting oxygen.

thermal spraying A group of coating or welding processes in which finely divided metallic or nonmetallic materials are deposited in a molten or semimolten condition to form a coating. The coating material may be in the form of powder, ceramic rod, wire, or molten materials. See also *flame spraying* and *plasma spraying*.

thermocouple A device for measuring temperatures consisting of lengths of two dissimilar metals or alloys that are electrically joined at one end and connected to a voltage-measuring instrument at the other end.

thermogalvanic corrosion Corrosion resulting from an electrochemical cell caused by a thermal gradient.

thermoplastic A plastic that repeatedly can be softened by heating and hardened by cooling through a temperature range characteristic of the plastic, and that in softened state can be shaped by flow into articles by molding or extrusion.

thermoplastic material A plastic that repeatedly will soften by heating and harden by cooling within a temperature range characteristic for the plastic. In the softened state. it can be shaped by flow into articles; e.g. by molding/extrusion.

A plastic being substantially infusible and insoluble after curing by heat or other means

thermosetting resin A plastic that, after having been cured by heat or other means, is substantially infusible and insoluble.

thinner A volatile liquid added to a primer to facilitate application and aid penetration by lowering the viscosity.

threshold limit value (TLV) A concentration of airborne material that experts agree can be inhaled for a working lifetime by almost all workers without any injury. The few workers who will be affected will develop their symptoms so slowly that periodic medical examination can be expected to detect the issue while the effects are still reversible.

threshold stress The critical gross section stress at the onset of stress-corrosion cracking under specified conditions.

throwing power (1) The relationship between the current density at a point on a surface and its distance from the counter-electrode. The greater the ratio of the surface resistivity shown by the electrode reaction to the volume resistivity of the electrolyte, the better is the throwing power of the process. (2) The ability of a plating solution to produce a uniform metal distribution on an irregularly shaped cathode.

TIG-welding Tungsten inert-gas arc welding using a nonconsumable electrode of pure or activated tungsten.

tinning The process of coating metal with a very thin layer of molten solder or brazing filler metal.

tie coat An intermediate coat used to bond different types of paint coats; used to improve the adhesion of succeeding coatings.

torsion A twisting deformation of a solid body about an axis in which lines that were initially parallel to the axis become helixes.

torsional stress The shear stress on a transverse cross section resulting from a u-twisting action.

total carbon The sum of free carbon and combined carbon (including carbon in solution) in a ferrous alloy.

toughness The ability of a metal to absorb energy and deform plastically before fracturing.

total solids The nonvolatile matter in a coating composition; i.e., the ingredients of a coating composition which, after drying, are left behind and constitute dry film.

transcrystalline See *transgranular.*

transcrystalline cracking See *transgranular cracking.*

transference The movement of ions through the electrolyte associated with the passage of the electric current. Also called *transport* or *migration.*

transgranular Through or across crystals or grains. Also called *intracrystalline* or *transcrystalline.*

transgranular cracking Cracking or fracturing that occurs through or across a crystal or grain. Also called *transcrystalline cracking.* See also *intergranular cracking.*

transgranular fracture A break through or across the crystals or grains of a metal. Also called *transcrystalline fracture* or *intracrystalline fracture.* See also *intergranular fracture.*

transition metal A metal in which the available electron energy levels are occupied in such a way that the d-band contains less than its maximum number of 10 electrons per atom; for example, iron, cobalt, nickel, and tungsten. The distinctive properties of the transition metals result from incompletely filled d-levels.

transition temperature (1) An arbitrarily defined temperature that lies within the range in which metal fracture characteristics (as usually determined by tests of notched specimens) change rapidly, such as from primarily fibrous (shear) to primarily crystalline (cleavage) fracture. (2) A term sometimes used to denote an arbitrarily defined temperature within a range in which the ductility changes rapidly with temperature.

transpassive region The region of an anodic polarization curve, noble to and above the passive potential range, in which there is a significant increase in current density (increased metal dissolution) as the potential becomes more positive (noble).

transpassive state (1) A state of anodically passivated metal characterized by a considerable increase of the corrosion current, in the absence of pitting, when the potential is increased. (2) The noble region of potential where an electrode exhibits at higher than passive current density.

triaxial stress See *principal stress (normal)*.

tuberculation The formation of localized corrosion products scattered over the surface in the form of knoblike mounds called *tubercles*.

typical value A value exhibiting the essential characteristic of a special type of material.

U-bend specimen A horseshoe-shaped test piece used to detect the susceptibility of a material to stress corrosion cracking.

ultimate strength The maximum stress (tensile, compressive, or shear) that a material can sustain without fracture, determined by dividing the maximum load by the original cross-sectional area of the specimen. Also called *nominal strength* or *maximum strength*.

ultrasonic measurement The timing of the transmission of ultrasonic sound waves through a material to determine the material's thickness.

underfilm corrosion Corrosion that occurs under organic films in the form of randomly distributed, threadlike filaments or spots. In many cases, this is identical to filiform corrosion.

uniaxial stress See *principal stress (normal)*.

uniform corrosion (1) A type of corrosion attack (deterioration) uniformly distributed over metal surface. (2) Corrosion that proceeds at approximately the same rate over a metal surface. Also called *general corrosion*. General corrosion is characterized by a chemical or electrochemical reaction that occurs uniformly over the exposed surface. Anodic and cathodic sites shift constantly so that corrosion spreads over the entire metal surface. Identifying general corrosion is usually simple, but determining its cause is often difficult. Chemical dissolution by acids, bases or chelants frequently results in general corrosion.

vicat softening point The temperature at which a flat-ended needle of a 1-mm^2 circular cross section will penetrate a thermoplastic specimen to a depth of 1 mm under a specified load using a selected uniform rate of temperature increase.

viscosity The property of a liquid to resist shear deformation increasingly with increasing rate of deformation. Dynamic viscosity is the ratio of the applied shear stress to the velocity gradient; kinematic viscosity is the ratio of the dynamic viscosity to the density of the liquid, both measured at the same temperature.

The SI unit for dynamic viscosity is the pascal second (Pa.s). The traditional unit is the centipoise (cp); 1cp = 1 mili Pa.s.

The SI unit for kinematic viscosity is the square-meter per second (m^2/s). The traditional unit is the centistokes (cst); 1 cst = 1 mm^2/s.

vacuum deposition/vapor deposition/gas plating Deposition of a metal coating by the precipitation (sometimes in vacuum) of metal vapor on the treated surface. Vapor may be produced by thermal decomposition, cathode sputtering, or evaporation of molten metal in air or inert gas.

valence A positive number that characterizes the combining power of an element for other elements, as measured by the number of bonds to other atoms that one atom of the given element forms upon chemical combination: hydrogen is assigned valence 1, and the valence is the number of hydrogen atoms with which an atom of the given element combines, or the equivalent.

vapor plating The deposition of a metal or compound on a heated surface by the reduction or decomposition of a volatile compound at a temperature below the melting points of the deposit and the base material. The reduction is usually accomplished by a gaseous reducing agent such as hydrogen. The decomposition process may involve thermal dissociation or reaction with the base material. Occasionally used to designate deposition on cold surfaces by vacuum evaporation. See also *vacuum deposition*.

void A term generally applied to describe holidays, holes, and skips in a film of paint. Also used to describe shrinkage in castings and weld.

water-blasting Blast-cleaning of metal using high-velocity water with or without the addition of an abrasive.

water immersion An exposure in which the surface is in direct contact with fresh water or saltwater.

water vapor transmission rate The steady water vapor flow in unit time through unit area of a body, normal to specific parallel surfaces, under specific conditions of temperature and humidity at each surface.

weather resistance The ability of a material to resist all ambient weather conditions, including changes of temperature, precipitation, effect of wind and humidity, sunlight, oxygen and other gases and impurities in the atmosphere, ultraviolet (UV) rays, radiation, and ozone.

weight-coat A steel-mesh-reinforced concrete layer applied over a primary coating system; provides negative buoyancy for submarine pipelines.

weld A union between pieces of metal at faces rendered plastic or liquid by heat or by pressure, or by both. A filler metal whose melting temperature is of the same order as that of the parent metal may or may not be used.

welding The making of a weld.

weld decay Intergranular corrosion, usually of stainless steel or certain nickel-based alloys, that occurs as the result of sensitization in the heat-affected zone during a welding operation.

weld metal All metal melted during the making of a weld and retained in the weld.

weld zone The zone containing the weld metal and the heat-affected zone.

wetting A condition in which the interfacial tension between a liquid and a solid is such that the contact angle is 0 to 90°.

wetting agent A substance that reduces the surface tension of a liquid, thereby causing it to spread more readily on a solid surface.

white liquor Cooking liquor from the kraft pulping process produced by recausticizing green liquor with lime.

white rust Zinc oxide; the powdery product of corrosion of zinc or zinc-coated surfaces.

wood preservation Treatment of wood with chemical substances that reduce its susceptibility to deterioration by fungi, insects, and marine borers.

working electrode The test or specimen electrode in an electrochemical cell.

yield Evidence of plastic deformation in structural materials. Also called *plastic flow* or *creep*.

yield point The first stress in a material, usually less than the maximum attainable stress, at which an increase in strain occurs without an increase in stress. Only certain metals—those that exhibit a localized, heterogeneous type of transition from elastic deformation to plastic deformation—produce a yield point. If there is a decrease in stress after yielding, a distinction may be made between the upper and lower yield points. The load at which a sudden drop in the flow curve occurs is called the *upper yield point*. The constant load shown on the flow curve is known as the *lower yield point*.

yield strength The stress at which a material exhibits a specified deviation from proportionality of stress and strain. An onset of 0.2% is used for many metals.

yield stress The stress level in a material at or above the yield strength but below the ultimate strength; i.e., a stress in the plastic range.

zinc reference electrode A 99.9% metallic zinc rod, with an iron content not exceeding 0.0014%, which makes contact directly with the electrolyte around the structure for measurement purposes. Some of the zinc alloys used for galvanic anodes are also suitable. The metallic electrode can be made in any convenient form.

zinc-rich primer An anticorrosive primer for iron and steel that incorporates zinc dust in a concentration sufficient to give electrical conductivity in the dried film, thus enabling the zinc metal to corrode preferentially to the substrate (i.e., to give galvanic protection).

References

American Concrete Institute (ACI)
"Manual of Concrete Practice 1989"

American National Standard Institute (ANSI)
8701, "Protection Tools"
ANSI Z129.1, "Precautionary Labeling of Hazardous Industrial Chemicals"

American Petroleum Institute (API)
API PR5L5, "Recommended Practice for Marine Transportation of Line Pipe"
API-RP-1631, "Interior Lining of Underground Storage Tanks"
API-PR-10E, "Recommended Practice for Application of Cement Lining to Steel Tubular Goods, Handling Installation, and Joining"

American Society of Mechanical Engineering (ASME)
ASME-SA-263, "Corrosion-Resisting Chromium Steel Clad Plate, Sheet, and Strip"
ASME-SA-264, "Stainless Chromium-Nickel Steel Clad Plate, Sheet, and Strip"
ASME-SA-265, "Nickel-Base Alloy Clad Steel Plate"
ASME-SA-578, "Straight-Beam Ultrasonic Examination of Plain and Clad Steel Plates for Special Application"

American Society for Testing and Materials (ASTM)
ASTM-A-106, "Specification for Seamless Carbon Steel Pipe for High-Temperature Service"
ASTM-A-185, "Standard Specification for Steel Welded Wire Reinforcement, Plain, for Concrete"
ASTM-A-240, "Specification for Stainless and Heat-Resisting Chromium and Chromium-Nickel Steel Plate, Sheet, and Strip for Fusion-Welded Unfired Pressure Vessel"
ASTM-A-262, "Practices for Detecting Susceptibility to Intergranular Attack in Austenitic Stainless Steel"
ASTM-A-283, "Specification for Low- and Intermediate-Tensile-Strength Carbon Steel Plates"
ASTM-A-285, "Specification for Pressure Vessel Plates Carbon Steel, Low- and Intermediate-Tensile Strength"
ASTM-A-314, "Specification for Stainless and Heat-Resisting Steel Billets and Bars for Forging"
ASTM-A-390, "Specification for Zinc-Coated (Galvanized) Steel Poultry Fabric"
ASTM-A-424, "Specification for Steel Sheet for Porcelain Enameling"
ASTM-A-497, "Specification for Welded Deformed Steel Wire Fabric for Concrete Reinforcement"
ASTM-A-516, "Specification for Pressure Vessel Plates, Carbon Steel, for Moderate- and Lower-Temperature Service"
ASTM-A-575, "Specification for Steel Bars, Carbon, Merchant Quality"
ASTM-A 615 M, "Specification for Deformed and Plain Billet-Steel Bars for Concrete Reinforcement (Metric)"
ASTM-B-432, "Copper and Copper Alloy Clad Steel Plate"
ASTM-C-27, "Classification of Fire Clay and High-Alumina Refractory Brick"
ASTM-C-33, "Specification for Concrete Aggregates"
ASTM-C-35, "Specification for Inorganic Aggregate for Use in Gypsum Plaster"
ASTM-C-150, "Specification for Portland Cement"
ASTM-C-155, "Classification of Insulating Fire Brick"
ASTM-C-171, "Specification for Sheet Materials for Curing Concrete"

Essentials of Coating, Painting, and Lining. http://dx.doi.org/10.1016/B978-0-12-801407-3.00012-2

ASTM-C-309, "Specification for Liquid Membrane-Forming Compounds for Curing Concrete"

ASTM-C-494, "Specification for Chemical Admixture for Concrete"

ASTM-C-581, "Code of Practice for Determining Chemical Resistance of Thermosetting Resin Used in Glass Fiber Reinforced Structure"

ASTM-C-616, "Specification for Fly Ash and Raw or Calcined Natural Pozzolan for Use as a Mineral Admixture in Portland Cement Concrete"

ASTM-C-618, "Specification for Flyash and Raw or Calcined Natural Pozzolan for Use as a Mineral Admixture in Portland Cement Concrete"

ASTM-C-660, "Production and Preparation of Gray-Iron Casting for Porcelain Enameling"

ASTM-D-531, "Test Method for Rubber Property"

ASTM-D-543, "Test Method for Resistance of Plastics to Chemical Reagents"

ASTM-D-785, "Test Method for Rockwell Hardness of Plastic and Electrical Insulating Material"

ASTM-D-1415, "Test Method for Rubber Property-International Hardness"

ASTM-D-2240, "Test Method for Rubber Property-Durometer Hardness"

ASTM-G-14, "Test Method for Impact Resistance of Pipeline Coating (Falling Weight Tests)"

ASTM-G-28, "Method for Detecting Susceptibility to Intergranular Attack in Wrought Nickel-Rich, Chromium-Bearing Alloys"

ASTM-G-48, "Test Method for Pitting and Crevice Corrosion Resistance of Stainless Steel and Related Alloys by The Use of Ferric Chloride Solution"

B117, "Salt Spray (Fog) Testing"

D5, "Penetration of Bituminous Materials"

D83, "Red Lead Pigment"

D185, "Coarse Particles in Pigments, Pastes, and Paints"

D212, "Pure Chrome Green"

D234, "Raw Linseed Oil"

D235, "Petroleum Spirits (Mineral Spirits)"

D362, "Industrial Grade Toluene"

D476, "Titanium Dioxide Pigments"

D 427, "Test Method for Shrinkage Factors of Soils"

D 522, "Test Method for Elongation of Attached Organic Coating with Conical Mandrel Apparatus"

D523, "Specular Gloss"

D562, "Consistency of Paints Using the Stormer Viscometer"

D 570, "Test Method for Water Absorbtion of Plastic"

D600, "Liquid Paint Driers"

D562, "Consistency of Paints Using the Stormer Viscometer"

D602, "Barium Sulfate Pigments"

D605, "Magnesium Silicate Pigment"

D 610-85, "Method of Evaluation Degree of Rusting on Painted Steel"

D 696, "Test Method for Coefficient of Linear Thermal Expansion of Plastic"

D 714-87, "Method of Evaluation Degree of Blistering of Paints"

D 746, "Test Method for Brittleness Temperature of Plastic and Elastomers by Impact"

D 785, "Test Method for Rockwell Hardness of Plastic and Electrical Insulating"

D962, "Aluminum Pigments, Powder and Paste, for Paints"

D 1002, "Test Method for Strength Properties of Adhesive In shear by Tension Loading (Metal-to-Metal)"

D1010, "Asphalt Emulsions for Use as Protective Coatings for Metal"

D1153, "Methyl Isobutyl Ketone"

D1208, "Common Properties of Certain Pigments"
D1210, "Fineness of Dispersion of Pigment Vehicle Systems"
D1243, "Dilute Solution Viscosity of Vinyl Chloride Polymers"
D1296, "Odors of Volatile Solvents and Diluents"
D1475, "Density of Paint, Varnish, Lacquer and Related Products"
D1542, "Quantitative Test for Rosin in Varnishes"
D2369, "Volatile Content of Paints"
D2371, "Pigment Content of Solvent-Type Paints"
D2801, "Leveling Characteristics of Paints by Draw-Down Method"
D3278, "Flash Point of Liquids by Setaflash Closed Tester"
D3722, "Natural Red and Brown Iron Oxides"
D3951, "Standard Practice for Commercial Packaging"
D1475, "Density of Paint, Varnish, Lacquer, and Related Products"
D 1525, "Test Method for Vicat Softening Temperature of Plastics"
D1542, "Quantitative Test for Rosin in Varnishes"
D 1603, "Test Method for Carbon Black in Olefin Plastic"
D1640, "Drying, Curing, or Film Formation of Organic Coatings at Room Temperature"
D 1673, "Test Method for Relative Permitivity and Displation Factor of Expanded Cellular Plastics Used for Electrical Insulation"
D 1693, "Test Method for Environment Stress-Cracking of Ethylene Plastics"
D1856, "Recovery of Asphalt from Solution by Abson Method"
D 2240, "Test Method for Rubber Property-Durometer Hardness"
D2369, "Volatile Content of Paints"
D2371, "Pigment Content of Solvent Type Paints"
D2801, "Leveling Characteristics of Paints by Draw Down Method"
D 3176, "Method for Ultimate Analysis of Coal and Cock"
D 3180, "Method for Calculation Coal and Cock Analysis"
D3278, "Flash Point of Liquids by Setaflash Closed Tester"
D 3640-80, "Standard Guide Lines of Emission Control in Vapor Process"
D3951-88, "Standard Practice for Commercial Packaging"
E 161-87, "Precision Electroformed Sieves"
E 323-80, "Perforated-Plate Sieves for Testing Purposes"
F 941-85, "Practice for Inspection of Marine Surface Preparation and Coating Application"
G 8, "Test Method for Cathodic Disbonding of Pipeline Coatings"
G 13, "Test Method for Impact Resistance of Pipeline Coating (Limestone Prob Test)"
G 14, "Test Method for Impact Resistance of Pipeline Coating (Falling Weight Test)"
G 19, "Test Method for Disbonding Characteristics of Pipeline Coating by Direct Soil Burial"

American Water Works Association (AWWA)

AWWA-210-84, "Liquid Epoxy Coating Systems for Interior and Exterior of Steel Water Pipelines"
AWWA C 203, "Coal-Tar Protective Coatings and Lining for Pipelines—Enamel and Tape Hot-Applied"
AWWA C 205, "Cement-Mortar Protective Lining and Coating"
AWWA C 209, "Cold-Applied Tape Coating for the Exterior at Special Sections"
AWWA C 210, "Liquid Epoxy Coating Systems for Interior and Exterior of Steel Water Pipelines"
AWWA C 213, "Fusion-Bonded Epoxy Coating for the Interior and Exterior of Steel Water Pipelines"
AWWA C 214, "Tape Coating Systems for the Exterior of Steel Water Pipeline"

AWWA C 215, "Extruded Polyolefin Coatings for the Exterior of Steel Water Pipelines"
AWWA C 216, "Heat-Shrinkable Cross-Linked Polyolefin Coatings for the Exterior of Special Sections, Connections, and Fittings for Steel Water Pipelines"

Australian Standard (AS)
AS 1518, "Extruden High-Density Polyethylene Protective Coating for Pipes"
AS 2043, "Coal-Tar and Synthetic (Fast-Dry) Primers for Steel Pipes"
AS 2046, "Code of Practice for the Coating and Lining of Steel Pipes with Coal-Tar Primer/Enamel Systems"
AS 2518, "Fusion-Bonded Low-Density Polyethylene Coating for Pipes and Fittings"

British Standard Institution (BSI)
187, "Specification for Calcium Silicate Bricks"
427, "Vickers Hardness Test"
524, "Specification for Refined Cresylic Acid"
580, "Specification for Trichloroethylene"
805, "Specification for Toluenes"
1191, Parts 1 and 2, "Specification for Gypsum Building Plasters"
2451, "Chilled Iron Shot and Grit"
2713, "Specification for 2-Ethoxy Ethanol"
3591, "Specification for Industrial Methylated Spirits"
4072, "Wood Preservatives by Means of Copper/Chromium/Arsenic Composition"
4764, "Specification for Powder Cement Paint"
5056, "Specification for Copper Naphthenate Wood Preservatives"
5262, "External Rendered Finishes"
5589, "Code of Practice for Preservation of Timber"
5707, Part 2, "Specification for Pentachlorophenol Wood Preservative"
6073, Part 1, "Specification for Precast Concrete Masonry Units"
BS 381 C, "Colors for Identification, Coding, and Special Purposes"
BS 470, "Specification for Inspection, Access, and Entry Opening for Pressure Vessels"
BS 490: Part 10, "Testing for Physical Properties of Rubber and Plastics"
BS 1133, "Temporary Protection of Metal Surfaces Against Corrosion"
BS 2015, "Glossary of Paint Terms"
BS 2035, "Specification for Cast-Iron Flanged Pipes and Flanged Fitting"
BS 2562, "Protection of Iron and Steel Against Corrosion and Oxidation at Elevated Temperature"
BS 3412, "Polyethylene Materials for Moulding and Extrusion"
BS 3416, "Specification for Bitumen-Based Coatings for Cold Application, Suitable for Use in Contact with Potable Water"
BS 4232, "Specification for Surface Finish of Blast-Cleaned Steel for Painting,"
2nd Quality
BS 5493, "Code of Practice for Protective Coating of Iron and Steel Structures Against Corrosion"
BS 6374, "Lining of Equipment with Polymeric Materials for The Process Industries"
Part 1: "Specification for Lining with Sheet Applied Thermoplastics"
Part 2: "Specification for Lining with Non-Sheet Applied Thermoplastics"
Part 3: "Specification for Lining with Stoved Thermosetting Resins"
Part 4: "Specification for Lining with Cold Curing Thermosetting Resins"
Part 5: "Specification for Lining with Rubbers"
BS 8007, "Code of Practice for Design of Concrete Structures for Retaining Aqueous Liquids"

BS 8110, "Structural Use of Concrete"
BS EN 1089, "Transportable Gas Cylinder"
BS 381C, "Colors for Identification, Coding, and Special Purposes"
BS 534, "Specification for Steel Pipes, Joints, and Specials for Water and Sewage"
BS 729, "Hot-Dip Galvanized Coatings on Iron and Steel Articles"
BS 1133, "Packaging Code"
BS 1319, "Specification for Medical Gas Cylinders, Valves, and Yoke Connections"
BS 1706, "Electroplated Coatings of Cadmium and Zinc on Iron and Steel"
BS 1710, "Preparation of Steel Sub Surface Before Application of Paint and Related Products"
BS 7079, "Color Identification"
BS 2015, "Glossary of Paint Terms"
BS 2569, "Sprayed Metal Coatings"
BS 3294, "The Use of High-Strength Friction Grip Bolts in Structural Steel Work"
BS 3382, "Electroplated Coatings on Threaded Components"
BS 3698, "Calcium Plumbate Priming Paints"
BS 4147, "Hot-Applied Bitumen Based Coating for Ferrous Products"
BS 4164, "Coal-Tar-Based Hot-Applied Coating Materials for Protecting Iron and Steel, Including Suitable Primers Where Required"
BS 4604, "The Use of High-Strength Friction Grip Bolts in Structural Steel Work, Metric Series"
BS 4652, "Metallic Zinc-Rich Priming Paint (Organic Media)"
BS 4800, "Specification for Paint Colors for Building Purposes"
BS 4921, "Sherardized Coatings on Iron and Steel Articles"
BS 5493, "Code of Practice for Protective Coating of Iron and Steel Structures Against Corrosion"
CP 3012, "Cleaning and Preparation of Metal Surface"

British Standards (BS)
BS 3900, "Paint and Varnishes Tests"
BS 4482, "Specification for Cold Reduced Steel Wire for the Reinforcement of Concrete"

Codes of Practice (CP)
CP 110, "Structural Use of Concrete"
CP 114, "Structural Use of Reinforced Concrete in Buildings"
CP 117, "Composite Construction in Structural Steel and Concrete"
CP 1021, "Cathodic Protection"

Deutsches Institiute für Normung (DIN)
18364, "Contact Procedure for Building Works"
53209, "Designation of Degree of Blistering of Paint Coats"
53210, "Designation of Degree of Rusting of Paint Coats"
55928, Part 28, "Corrosion Protection of Steel Structure by Organic, Inorganic and Metallic Coatings"
DIN 18364, "Works for Protection Against Corrosion of Steel and Aluminum"
DIN 30670, "Polyethylene Sheathing of Steel Tubes and Steel Shapes and Fittings"
DIN 50049, "Materials Testing Certificates"
(Parts 1, 2 & 3) 1986
DIN 53122, "Testing of Plastics Films; Elastomer Films; Paper, Board and Other Sheet Materials; Determination of Water Vapor Transmission Rate; Gravimetric Method"
DIN 53152, "Testing of Paints, Varnishes and Similar Coating Materials, Bend Test on Paint Coating and Similar Coatings"

DIN 53155, "Testing of Paints, Varnishes and Similar Coating Materials, Chip-Test of Coatings According to Peters"

DIN 53380, "Testing of Plastic Films; Determination of the Gas Transmission Rate"

DIN 53455, "Testing of Plastic; Tensile Test"

DIN-53456, "Testing of Plastics"

DIN-53505, "Testing of Rubber, Elastomer, and Plastic, Shore Hardness Testing A & D"

DIN-53519, "Testing of Elastomers, Determination of Indentation Hardness of Soft Rubber"

DIN 53460, "Determination of the Vicat Softening Temperature of Thermoplastic"

DIN 53479, "Testing of Plastics and Elastomers; Determination of Density"

DIN 53495, "Testing of Plastics; Determination of Water Absorption"

DIN 53505, "Testing of Rubber, Elastomers, and Plastics; Shore Hardness Testing A and D"

DIN 53735, "Testing of Plastics, Determination of the Melt Flow Index of Thermoplastics"

DIN 55990 T2, "Testing of Paints, Varnishes and Similar Coating Materials-Powder Coatings, Determination of Particle Size Distribution"

DIN 55990 T3, "Testing of Paints Varnishes and Similar Coating Materials-Powder Coatings, Determination of Density"

DIN 55990 T8, "Testing of Paints-Varnishes and Similar Coating Materials, Powder Coatings Assessment of the Chemical Storage Stability"

DIN 67530, "Reflectometer as a Means for Glass Assessment of Plane Surfaces of Paint Coatings and Plastic"

International Electrotechnical Commission (IEC)

IEC 243, "Methods of Test for Electric Strength of Solid Insulating Materials"

International Standardization for Organization (ISO)

ISO 868, "Testing of Rubber, Elastomer, and Plastics Shore Hardness Testing A and D"

ISO 1133D, "Plastics-Determination of the Metal Mass-Flow Rate (MFR) and the Metal Volume-Flow Rate (MVR) of Thermoplastic"

ISO 3233/BS 3900, "Paint and Varnishes—Determination of Volume of Dry Coating, Part A10 (Nonvolatile Matter) Obtained from a Given Volume of Liquid Coating"

ISO 3274, "Instruments for the Measurement of Surface Roughness by the Profile Method-Contact (Stylus) Instruments of Consecutive Profile Transformation-Contact Profile Meters, System M"

ISO 4288, "Rules and Procedures for the Measurement of Surface Roughness Using Stylus Instruments"

ISO 4599, "Test Method for Environment Stress Cracking of Ethylene Plastics"

ISO 8501, "Preparation of Steel Substrate Before Application of Paint and Related Products"

ISO 9002, "Quality Systems-Model for Quality Assurance in Production, Installation and Servicing"

ISO 9004, "Quality Management and Quality System Elements Guidelines"

ISO R527, "Testing of Plastic, Tensile Test"

National Association of Corrosion Engineers (NACE)

NACE-6H160, "Glass Lining and Vitreous Enamels"

NACE-6K157, "Acid Proof Vessel Construction with Membrane and Brick Lining"

RP-01-72, "Surface Preparation of Steel by Water Blasting"

RP-01-84, "Recommendation Practice, and Repair of Lining Systems"

NACE-Coating and Lining Handbook

Steel Structure Painting Council (SSPC)

Vol. 1, "Good Painting Practice"

Vol. 2, "System and Specification"

SSPC-Paint No. 8, "Aluminum Vinyl Paint"
SSPC-PA Guide 3, "A Guide to Safety in Paint Application"
SSPC-Vol. 1, "Good Painting Practice"
SSPE-Vol. 2, "Systems and Specification"
SSPC 104, "White or Tinted Alkyd Paint"

Standardiserings-Kommissionen I Sverige (SIS)
SIS 05 5900, "Swedish Standards Institution Practice, Surface Preparation Standard for Painting Steel Surface"

Steel Structures Painting Council (SSPC)
SSPC No. 12, "Cold-Applied Asphalt Mastic"
SSPC-PA Guide 3, "A Guide to Safety in Paint Application"

Swedish Standard (SIS)
05 5900, "Rust Levels of Steel Structure and Quality Levels for Preparation of Steel Surface for Rust-Protecting Paints".

U.S. Federal Standards (UFS)
Method 3011, "Condition in Container"
Method 3021, "Skinning (Partially Filled Container)"
Method 4021, "Pigment Content (Centrifuge)"
Method 4053, "Nonvolatile Vehicle Content"
Method 4061, "Drying Time"
Method 4081, "Water Content (Reflex Method)"
Method 4092, "Coarse Particles and Skins"
Method 4203, "Reducibility and Dilution Stability"
Method 4311, "Volume Percentage of Pigment in Total Nonvolatile Matter (calculation Method)"
Method 4321, "Brushing Properties"
Method 4331, "Spraying Properties"
Method 4494, "Sag Test (Multinotch Blade)"
Method 4541, "Working Properties and Appearance of Dried Film"
Method 6221, "Flexibility"
MIL-A-15206, "Aluminum Stearate, Technical"
MIL-P-15328, "Primer (Wash) Pretreatment"
MIL-P-15929, "Primer Coating, Shipboard, Vinyl-Red Lead (for Hot Spray)" (Federal Test Method Standard No. 141)
TT-C-520, "Coating Compounds, Bituminous, Solvent Type, Underbody for Motor Vehicles"
TT-E-489, "Enamel, Alkyd, Gloss (for Exterior and Interior Surfaces)"
TT-L-215, "Linseed Oil, Raw (for Use in Organic Coatings)"
TT-R-266, "Resin, Alkyd Solutions"
TT-T-291, "Thinner, Paint, Mineral Spirits, Regular and Odorless" (Federal Test Method Standard No. 141)
Aalund, L.R., 1992. Polypropylene system scores high as pipeline anti-corrosion coating. Oil and Gas Journal 90 (50), 42–45.
Agarwal, A.K., Garg, A., Srivastava, D.K., Shukla, M.K., 2007. Comparative wear performance of titanium-based coatings for automotive applications using exhaust gas recirculation. Surface and Coatings Technology 201 (13), 6182–6188.
Ahmad, Z., 1996. Corrosion phenomena in coastal area of Gulf. British Corrosion Journal 31 (3), 191–197.
Almeida, E., Morcillo, M., Rosales, B., 2000. Atmospheric corrosion of zinc. Part 2: Marine atmospheres. British Corrosion Journal 35 (4), 289–296.

Almeida, M.E.M., 2005. Minimisation of steel atmospheric corrosion: Updated structure of intervention. Progress in Organic Coatings 54 (2), 81–90.

Al-Mithin, A.W., Al-Ahmad, A.W., Vinayak Sardesai, G., Kumar, Santhosh, 2010. Integrity management of aged equipment in sour oil and gas service Kuwait oil company approach, American Society of Mechanical Engineers. Pressure Vessels and Piping Division (Publication) PVP7, 141–147.

Al-Sulaiman, S., Al-Mithin, A.W., Murray, G., Biedermann, A.J., & Islam, M., 2008. Advantages and limitations of using field test kits for determining bacterial proliferation in oil field waters. NACE—International Corrosion Conference Series, pp. 086551–0865518.

Amaya, H., Ueda, M., Sagara, M., Noda, K., 2007. Effect of variation of local PH on stress corrosion cracking of stainless steels in high concentration brines, NACE—International Corrosion Conference Series, pp. 070961–0709615.

Asami, K., Kikuchi, M., 2003. In-depth distribution of rusts on a plain carbon steel and weathering steels exposed to coastal-industrial atmosphere for 17 years. Corrosion Science 45 (11), 2671–2688.

Ault, J.P., 2006. The use of coatings for corrosion on offshore oil. Journal of Protective Coatings and Linings 23 (4), 42–46.

Babiarz, M., 2009. Tips for maintenance coating work on offshore oil and gas rigs. Journal of Protective Coatings and Linings 26 (8), 33–35.

Bagherzadeh, M.R., Mahdavi, F., 2007. Preparation of epoxy-clay nanocomposite and investigation on its anti-corrosive behavior in epoxy coating. Progress in Organic Coatings 60 (2), 117–120.

Benmoussa, A., Hadjel, M., Traisnel, M., 2006. Corrosion behavior of API 5L X-60 pipeline steel exposed to near-neutral pH soil simulating solution. Materials and Corrosion 57 (10), 771–777, doi: 10.1002/maco.200503964.

Benson, R.C., 2002. A review of soil resistivity measurements for grounding, corrosion assessment, and cathodic protection. Materials Performance 41 (1), 28–34.

Boardman, W., Casserly, T., Ramamurti, R., 2010. Advanced high-temperature corrosion and erosion-resistant internal coating for oil industry applications, Materials Science and Technology Conference and Exhibition 2010, MS and T'10 3, 1515–1524.

Bradshaw, A., Simms, N.J., Nicholls, J.R., 2013. Hot corrosion tests on corrosion-resistant coatings developed for gas turbines burning biomass and waste-derived fuel gases. Surface and Coatings Technology 228, 248–257.

Bull, S.J., Bhat, D.G., Staia, M.H., 2003. Properties and performance of commercial TiCN coatings. Part 2: Tribological performance. Surface and Coatings Technology 163–164, 507–514.

Byeli, A.V., Minevich, A.A., Stepanenko, A.V., Gick, L.A., Kholodilov, O.V., 1992. Wear resistance and structure of (Ti, Al)N coatings. Journal of Physics D: Applied Physics 25 (1 A), A292–A296.

Cairney, J.M., Harris, S.G., Munroe, P.R., Doyle, E.D., 2004. Transmission electron microscopy of TiN and TiAlN thin films using specimens prepared by focused ion beam milling. Surface and Coatings Technology 183 (2–3), 239–246.

Cerniti, S., Condanni, D., Marangoni, M., Barteri, M., & Podrini, A, (1995). "Carbon Steel Tubing Internally Coated with Resins: Mechanical and Corrosion Testing for Application in Corrosive Environments," Offshore Mediterranean Conference (OMC 95), March 15–17, Ravenna, Italy.

Chen, C., Khobaib, M., Curliss, D., 2003. Epoxy layered-silicate nanocomposites. Progress in Organic Coatings 47 (3–4), 376–383.

Chen, X., Li, X.G., Du, C.W., Cheng, Y.F., 2009. Effect of cathodic protection on corrosion of pipeline steel under disbonded coating. Corrosion Science 51 (9), 2242–2245.

Cheng, Y.F., 2005. Corrosion of X-65 pipeline steel in carbon dioxide-containing solutions. Bulletin of Electrochemistry 21 (11), 503–511.

Chu, R., Chen, W., Wang, S.-H., King, F., Jack, T.R., Fessler, R.R., 2004. Microstructure dependence of stress corrosion cracking initiation in X-65 pipeline steel exposed to a near-neutral pH soil environment. Corrosion 60 (3), 275–283.

Ciaraldi, S., Abdallah, A., Attia, A.M., El Leil, H.A., Bedair, S., Konecki, M., Lain, A., 1999. Rehabilitation of GUPCO's massive and aged pipeline infrastructure in the Gulf of Suez. Corrosion Prevention and Control 46 (5), 122–132.

Colombitta, C., 1993. "The Subsea Pipeline Network Operated by Agip S.p.A. in the Italian Seas," Offshore Mediterranean Conference (OMC 93), March 11-13, Italy.

Condanni, D., 1997. "Corrosion Cost in Offshore Developments," Black Sea '97, October 2-3, Ankara, Turkey.

Cornago, M. ENI S.p.A.-Agip Division and R. Malfanti, IMC engineering S.r.l. (1998). The paint and coatings costs for the corrosion prevention in the construction and maintenance phases the Agip experience. Symposium on Corrosion Control By Coatings, Cathodic Protection and Inhibitors in Seawater, Dubrovnik (Croatia) October.

Cornago, M., 1999. Rev2 protective coating and hot dip galvanizing. ENI E&P Division, ENI Functional Specification 20000 VAR.PAI.FUN, September.

Cornago, M. ENI E&P Division, 2004. ENI Functional Specification 20550 PIP.COR.FUN. RevO External Coatings for Corrosion Protection of Steel Pipes and Components, February.

Cornago, M. ENI E&P Division, 2004. ENI Functional Specification 20551 PIP.COR.FUN. RevO Internal Coatings for Corrosion Protection of Steel Pipes and Components, February.

Cornago, M., 2004. ENI E&P Division, ENI Functional Specification 20552 VAR.PAI.FUN, February.

Cornago, M., & P. Cavassi, ENI S.p.A.-Agip Division and R. Malfanti, IMC engineering S.r.l. (1998). The corrosion cost impact on the oil and gas production industry—PCE '98 Conference, The Hague (NL), April 1–3.

Corvo, F., Perez, T., Dzib, L.R., Martin, Y., Castañeda, A., Gonzalez, E., Perez, J., 2008. Outdoor-indoor corrosion of metals in tropical coastal atmospheres. Corrosion Science 50 (1), 220–230.

Cribb, W.R., Grensing, F.C., 2011. Spinodal copper alloy C72900—New high-strength antifriction alloy system. Canadian Metallurgical Quarterly 50 (3), 232–239.

Crook, P., 1994. Cobalt-based alloys resist wear, corrosion, and heat. Advanced Materials and Processes 145 (4), 27–30.

Crouch, A.E., Goyen, T., Porter, P., 2004. New method uses conformable array to map external pipeline corrosion. Oil and Gas Journal 102 (41), 55–59.

Ebina, T., Mizukami, F., 2007. Flexible transparent clay films with heat-resistant and high gas-barrier properties. Advanced Materials 19 (18), 2450–2453.

Endo, S., Fujita, S., Nagae, M., 1997. New line pipe resists preferential corrosion at welds. Oil and Gas Journal 95 (11), 92–102.

Esaklul, K.A., Ahmed, T.M., 2009. Prevention of failures of high strength fasteners in use in offshore and subsea applications. Engineering Failure Analysis 16 (4), 1195–1202.

Eslami, A., Fang, B., Kania, R., Worthingham, B., Been, J., Eadie, R., Chen, W., 2010. Stress corrosion cracking initiation under the disbonded coating of pipeline steel in near-neutral pH environment. Corrosion Science 52 (11), 3750–3756.

Eslami, A., Kania, R., Worthingham, B., Boven, G.V., Eadie, R., Chen, W., 2011. Effect of CO_2 and R-ratio on near-neutral pH stress corrosion cracking initiation under a disbonded coating of pipeline steel. Corrosion Science 53 (6), 2318–2327.

Evan, J.G., Moyle, P.R., 2006. Bulletin 2209-L, U.S. Department of the Interior. U.S. Geological Survey, Reston, Virginia.

Evans, N., 1993. Case for corrosion surveillance. Corrosion Prevention and Control 40 (2), 32–35.

Fatah, M.C., Ismail, M.C., 2011. Empirical equation of CO_2 corrosion with presence of low concentrations of acetic acid under turbulent flow conditions. Corrosion Engineering Science and Technology 46 (1), 49–55.

Feliu, S., Morcillo, M., Feliu, Jr., S., 1993. The prediction of atmospheric corrosion from meteorological and pollution parameters–II. Long-term forecasts. Corrosion Science 34 (3), 415–422.

Foss, M., Gulbrandsen, E., Sjöblom, J., 2008. Alteration of wettability of corroding carbon steel surface by carbon dioxide corrosion inhibitors—Effect on carbon dioxide corrosion rate and contact angle. Corrosion 64 (12), 905–919.

Funahashi, M., Bushman, J.B., 1991. Technical review of 100 mV polarization shift criterion for reinforcing steel in concrete. Corrosion 47 (5), 376–386.

Gafarov, N.A., Goncharov, A.A., Kushnarenko, V.M., Shchepinov, D.N., Chirkov, Yu.A., 2003. An analysis of the failures of equipment and pipelines at the Orenburg oil-gas condensate deposit. Protection of Metals 39 (3), 294–296.

Glazov, N.P., 2001. Enhancement of corrosion protection of steel pipelines. Zashchita Metallov 37 (5), 464–470.

Gray, L.G.S., Appleman, B.R., 2003. EIS electrochemical impedance spectroscopy—A tool to predict remaining coating life? Journal of Protective Coatings and Linings 20 (2), 66–74.

Greday, Y., 2006. Corrosion behaviour of different hot rolled steels. Corrosion Science 48 (2), 472–480.

Grimes, K., 1995. Stress corrosion crack in-line pig shows promise in tests. Pipe Line and Gas Industry 78 (3), 37–40.

Groysman, A., 2005. Anticorrosion techniques for aboveground storage tanks. Materials Performance 44 (11), 40–43.

Gu, X.-P., Long, D., Qiu, Y.-B., Zhao, Z.-G., 2009. Review on anti-corrosion joint coating technology of oil and gas pipelines. Corrosion and Protection 30 (2), 261–264.

Hall, S.C., 1998. Cathodic protection criteria for prestressed concrete pipe—An update. Materials Performance 37 (11), 14–21.

Han, W., Yu, G., Wang, Z., Wang, J., 2007. Characterisation of initial atmospheric corrosion carbon steels by field exposure and laboratory simulation. Corrosion Science 49 (7), 2920–2935.

He, H., Wang, X.-F., Jin, H., 2004. Some problems existing in the standard for oil and gas industry SY/T 4078-1995. Corrosion and Protection 25 (9), 402–405.

Hendrix, D.E., 1997. Hydrogen embrittlement of high-strength fasteners in atmospheric service. Materials Performance 36 (12), 54–56.

Holtsbaum, W. Brian, 2005. Electrical safety and cathodic protection rectifiers. Materials Performance 44 (6), 26–28.

Islam, M.S., Tong, L., Falzon, P.J., 2014. Influence of metal surface preparation on its surface profile, contact angle, surface energy, and adhesion with glass fibre prepreg. International Journal of Adhesion and Adhesives 51, 32–41.

Ivanov, E.S., Zavadskaya, V.P., Getmanskii, M.D., 1983. Influence of inhibitors on the corrosion fatigue of steel in hydrogen-sulfide-containing media. Soviet Materials Science 18 (5), 443–445.

Jack, T.R., Wilmott, M.J., Sutherby, R.L., Worthingham, R.G., 1996. External corrosion of line pipe—A summary of research activities. Materials Performance 35 (3), 18–24.

Jayaraman, A., Saxena, R.C., 1995. Corrosion and its control in petroleum refineries—A review. Corrosion Prevention and Control 42 (6), 123–131.

Jouen, S., Hannoyer, B., Barbier, A., Kasperek, J., Jean, M., 2004. A comparison of runoff rates between Cu, Ni, Sn, and Zn in the first steps of exposition in a French industrial atmosphere. Materials Chemistry and Physics 85 (1), 73–80.

Johansen, B., Kegel, T., 1996. The effects of oil coating on the measurement of gas flow using sharp-edged orifice flowmeters. American Society of Mechanical Engineers. Fluids Engineering Division (Publication) FED, 236, 639–646.

Jong, C.K., Do, J.S., Kang, O.Y., Jae, B.C., Yoon, S.C., Young, J.K., Song, C.C., Woo, S.K., 2005. Development of corroded gas pipeline assessment program based on limit load solution. Key Engineering Materials, 47–52, 297–300 I.

Karlsson, L., Hultman, L., Sundgren, J.-E., 2000. Influence of residual stresses on the mechanical properties of TiCxN1-x (x = 0, 0.15, 0. 45) thin films deposited by arc evaporation. Thin Solid Films 371 (1), 167–177.

Kim, K.H., Lee, S.H., 1996. Comparative studies of TiN and Ti1-xAlxN by plasma-assisted chemical vapor deposition using a TiCl4/AlCl3/N2/H2/Ar gas mixture. Thin Solid Films 283 (1–2), 165–170.

Konstantinov, I.O., Burushkin, O.S., Ryabchenko, V.N., Sheshukov, A.I., 1996. Radiometric corrosion monitoring of the industrial oil pipelines. Zashchita Metallov 1, 104–110.

Kunisada, N., Choi, K.-H., Korai, Y., Mochida, I., Nakano, K., 2004. Optimum coating of USY as a support component of NiMoS on alumina for deep HDS of gas oil. Applied Catalysis A: General 276 (1–2), 51–59.

Kuskov, V.N., Kolenchin, N.F., 2014. Prospects for the replacement of steel parts for oil and gas equipment by aluminium parts. WIT Transactions on Ecology and the Environment 190 (2), 801–808.

Lagad, V.V., Srinivasan, S., & Kane, R.D. (2008). Facilitating internal corrosion direct assessment using advanced flow and corrosion prediction models NACE—International Corrosion Conference Series, 081311-0813110.

Larsen, K.R., 2013. Failure modes of coatings on offshore oil and gas. Materials Performance 52 (8), 26–29.

Lebaron, P.C., Wang, Z., Pinnavaia, T.J., 1999. Polymer-layered silicate nanocomposites: An overview. Applied Clay Science 15 (1–2), 11–29.

Lee, S.-M., Kho, Y.-T., Na, K.-H., 2002. Effect of passivity of the oxide film on low-pH stress corrosion cracking of API 5L X-65 pipeline steel in bicarbonate solution. Corrosion 58 (4), 329–336.

Leth-Olsen, H., 2005. CO_2 corrosion in bromide and formate well-completion brines, Second SPE International Symposium on Oilfield Corrosion 2005 Corrosion Control in Oil and Gas Production—"Fluids, Chemicals, Materials, and More", Proceedings, art. no. SPE 95072, 67–75.

Li, Q.X., Wang, Z.Y., Han, W., Han, E.H., 2008. Characterization of the rust formed on weathering steel exposed to Qinghai salt lake atmosphere. Corrosion Science 50 (2), 365–371.

Li, S.-L., Yu, C.-Y., 2005. Corrosion survey and analysis of equipment in an oil refining complex. Corrosion and Protection 26 (7), 310–312, 314.

Lin, C.-Y., Pan, J.-Y., 2000. Corrosion characteristics of furnaces burning with emulsified diesel oil containing sodium sulphate. Corrosion Prevention and Control 47 (3), 83–92.

Liu, H., Wang, Y., 2001. A discussion on the method of evaluating residual corrosion strength of in-service pressure pipeline. Tianranqi Gongye/Natural Gas Industry 21 (6), 90–92.

Luo, B., Huo, J.-W., Li, X., Zhao, Z.-G., 2009. Comparison and selection of external anti-corrosive joint coating materials of oil and gas pipelines. Corrosion and Protection 30 (2), 258–260.

Lutkenhaus, J.L., Olivetti, E.A., Verploegen, E.A., Cord, B.M., Sadoway, D.R., Hammond, P.T., 2007. Anisotropic structure and transport in self-assembled layered polymer-clay nanocomposites. Langmuir 23 (16), 8515–8521.

Macdonald, K.A., Cosham, A., 2005. Best practice for the assessment of defects in pipelines—Gouges and dents. Engineering Failure Analysis 12 (5 SPEC. ISS.), 720–745.

Mahajanam, S.P.V., Joosten, M.W., 2011. Guidelines for filler-material selection to minimize preferential weld corrosion in pipeline steels. SPE Projects, Facilities and Construction 6 (1), 5–12.

McKennis, J., Bae, N., Kimura, M., Shimamoto, K., Sato, H., 2008. A new chemical mechanistic postulate regarding annular environmentally assisted cracking (AEAC)—Importance of cations and key contaminants in packer fluids in cracking of martensitic stainless steel. NACE—International Corrosion Conference Series, 084831-0848316.

Merlo, A.M., 2003. The contribution of surface engineering to the product performance in the automotive industry. Surface and Coatings Technology, 174–175, 21–26.

Morales, J., Martín-Krijer, S., Díaz, F., Hernández-Borges, J., González, S., 2005. Atmospheric corrosion in subtropical areas: Influences of time of wetness and deficiency of the ISO 9223 norm. Corrosion Science 47 (8), 2005–2019.

Morcillo, M., Almeida, E., Marrocos, M., Rosales, B., 2001. Atmospheric corrosion of copper in Ibero-America. Corrosion 57 (11), 967–980.

Morshed, A., 2008. Improving asset corrosion management using KPIs. Materials Performance 47 (5), 50–54.

Natesan, M., Selvaraj, S., Manickam, T., Venkatachari, G., 2008. Corrosion behavior of metals and alloys in marine-industrial environment. Science and Technology of Advanced Materials 9 (4), art. no. 045002.

Natesan, M., Venkatachari, G., Palaniswamy, N., 2006. Kinetics of atmospheric corrosion of mild steel, zinc, galvanized iron and aluminium at 10 exposure stations in India. Corrosion Science 48 (11), 3584–3608.

Niu, L., Cheng, Y.F., 2007. Corrosion behavior of X-70 pipe steel in near-neutral pH solution. Applied Surface Science 253 (21), 8626–8631.

Noveiri, E., Goodarzi, E.M., 2012. The study of corrosion in oil and gas transfer pipeline and its removal by nano-composite coating. WSEAS Transactions on Heat and Mass Transfer 6 (4), 101–111.

Odnevall Wallinder, I., Leygraf, C., 2001. Seasonal variations in corrosion rate and runoff rate of copper roofs in an urban and a rural atmospheric environment. Corrosion Science 43 (12), 2379–2396.

Oesch, S., Heimgartner, P., 1996. Materials and Corrosion 47, 425–438.

Olsen, D.W., 2008. Mineral commodities summary garnet (industrial), January, U.S. Geological Survey, Reston, VA.

Pan, A., He, L., Yang, S., Niu, M., 2014. The effect of side chains on the reactive rate and surface wettability of pentablock copolymers by ATRP. Journal of Applied Polymer Science 131 (9), 40209.

Papavinasam, S., Arsenault, B., Attard, M., Revie, R.W., 2012. Metallic underlayer coating as third line of protection of underground oil and gas pipelines from external corrosion. Corrosion 68 (12), 1146–1153.

Perdomo, J.J., Song, I., 2000. Chemical and electrochemical conditions on steel under disbonded coatings: The effect of applied potential, solution resistivity, crevice thickness, and holiday size. Corrosion Science 42 (8), 1389–1415.

Pérez, F.J., Martínez, L., Hierro, M.P., Gómez, C., Portela, A.L., Pucci, G.N., Duday, D., Song, F., Bichon, B., Fassett, R., Boss, T., Lu, A., 2009. Cased pipe segments could be less safe than uncased segments. Oil and Gas Journal 107 (15), 50–56.

Rabartdinov, Z.R., Denislamov, I.Z., 2012. Exploitation of oil fields with H_2S in the North-Western part of Bashkortostan. Neftyanoe Khozyaistvo–Oil Industry (8), 96–98.

Priest, M., Taylor, C.M., 2000. Automobile engine tribology—Approaching the surface. Wear 241 (2), 193–203.

Romano, M., Dabiri, M., Kehr, A., 2005. The ins and outs of pipeline coatings: Coatings used to protect oil and gas pipelines. Journal of Protective Coatings and Linings 22 (7), 40–47.

Roy, S.K., Ho, K.H., 1994. Corrosion of steel in tropical marine atmospheres. British Corrosion Journal 29 (4), 287–291.

Samie, F., Tidblad, J., Kucera, V., Leygraf, C., 2007. Atmospheric corrosion effects of HNO_3-Comparison of laboratory-exposed copper, zinc, and carbon steel. Atmospheric Environment 41 (23), 4888–4896.

Selvaraj, M., Sathianarayanan, S., Muthukrishnan, S., Venkatachari, G., 2009. Corrosion protective performance studies of coating systems in soda ash chemical industry. Progress in Organic Coatings 66 (3), 206–212.

Soares, C.G., Garbatov, Y., Zayed, A., Wang, G., 2009. Influence of environmental factors on corrosion of ship structures in marine atmosphere. Corrosion Science 51 (9), 2014–2026.

Sarathi, R., Sahu, R.K., Rajeshkumar, P., 2007. Understanding the thermal, mechanical, and electrical properties of epoxy nanocomposites. Materials Science and Engineering A, 445-446, 567–578.

Serednyts'kyi, Ya.A., 2000. Polyurethane materials as anticorrosive coatings of pipelines. Materials Science 36 (3), 415–421.

Shao, X.-D., 2008. Application of repair technology of internal protection coatings of pipelines for oilfild and relevant economical analysis. Corrosion Science and Protection Technology 20 (3), 229–231.

Schneider, J.M., Voevodin, A., Rebholz, C., Matthews, A., Hogg, J.H.C., Lewis, D.B., Ives, M., Smith, L., 1999. Control of corrosion in oil and gas production tubing. British Corrosion Journal 34 (4), 247–253.

Song, F.M., Sridhar, N., 2008. Modeling pipeline crevice corrosion under a disbonded coating with or without cathodic protection under transient and steady state conditions. Corrosion Science 50 (1), 70–83.

Stalder, F., 1997. Pipeline failures. Materials Science Forum 247, 139–146.

Stanley, L., Jones, B.L., 2004. Microbe-assisted external corrosion in oil and pipelines. Pipeline World (Aug. 4), 5–10.

Su, Y.L., Kao, W.H., 1999. Effect of thickness and carbon content on tribological behaviour and mechanical properties of Ti-C:H coatings. Wear 236 (1–2), 221–234.

Totlani, M.K., Athavale, S.N., 2000. Electroless nickel for corrosion control in chemical, oil, and gas industries. Corrosion Reviews 18 (2–3), 155–179.

Triantafyllidis, K.S., LeBaron, P.C., Park, I., Pinnavaia, T.J., 2006. Epoxy-clay fabric film composites with unprecedented oxygen-barrier properties. Chemistry of Materials 18 (18), 4393–4398.

Tsaprailis, H., Garfias-Mesias, L.F., 2011. Rapid evaluation of metallic coatings on large cylinders exposed to marine environments. Corrosion 67 (12), art. no. 126002.

Velazquez, J.C., Valor, A., Caleyo, F., Venegas, V., Espina-Hernandez, J.H., Hallen, J.M., Lopez, M.R., 2009. Pitting corrosion models improve integrity management, reliability. Oil and Gas Journal 107 (28), 56–62.

Walsh, T., 2006. Ageing assets and the fatigue challenge. Journal of Offshore Technology 14 (1), 22–24.

Wang, C., Neville, A., 2008. Inhibitor performance on corrosion and erosion/corrosion under turbulent flow with sand and CO_2—An AC impedance study. SPE Production and Operations 23 (2), 215–220.

Wang, K., Wang, L., Wu, J., Chen, L., He, C., 2005. Preparation of highly exfoliated epoxy/clay nanocomposites by, "slurry compounding": Process and mechanisms. Langmuir 21 (8), 3613–3618.

Wang, W., Free, M.L., Horsup, D., 2005. Prediction and measurement of corrosion inhibition of mild steel by imidazolines in brine solutions. Metallurgical and Materials Transactions B: Process Metallurgy and Materials Processing Science 36 (3), 335–341.

Wang, X.-F., Hu, Y.-C., Li, Y.-P., Zhou, Q., 2011. Preparation of epoxy/montmorillonite nanocomposite coating and its application in the high-temperature oil-gas environment with H2S. Advanced Materials Research, 154–155, 508–514.

Yamamoto, T., 2000. Filed test on the corrosion behavior of internally coated oil pipelines located in Daqing and Sichuan oil field in China. Zairyo to Kankyo. Corrosion Engineering 49 (4), 209–214.

Yan, M.C., Wang, J.Q., Han, E.H., Ke, W., 2007. Electrochemical measurements using combination microelectrode in crevice simulating disbonded of pipeline coatings under catholic protection. Corrosion Engineering Science and Technology 42 (1), 42–49.

Yang, H., Geng, W.P., Hu, J., 2014. An experimental study on anti-corrosion coating of ground oil and gas equipment. Advanced Materials Research, 850–851, 204–207.

Yang, Q., Seo, D.Y., Zhao, L.R., Zeng, X.T., 2004. Erosion resistance performance of magnetron sputtering deposited TiAlN coatings. Surface and Coatings Technology, 188-189 (1-3 SPEC.ISS.), 168–173.

Yuan, G., Wang, Q., Shen, R., Yuan, J., Lu, L., Wang, C., 2008. String corrosion and string protection during constructing and operating gas storage facility in bedded salt deposit. SPE Production and Operations 23 (1), 63–67.

Zhang, J.-P., 2003. Application of glass-flake-doped coatings to long-term anti-corrosion. Corrosion and Protection 24 (9), 401–402.

Zhang, L., Li, X.G., Du, C.W., Cheng, Y.F., 2009. Corrosion and stress corrosion cracking behavior of X70 pipeline steel in a CO 2-containing solution. Journal of Materials Engineering and Performance 18 (3), 319–323.

Zhao, X.H., Bai, Z.Q., Lin, K., Han, Y., 2011. The corrosion behavior about two Ni-based alloys in CO_2 /H2S environments. Advanced Materials Research, 152–153, 1624–1631.

Zhu, X.Y., Modi, H., Ayala, A., Kilbane, J.J., 2006. Technical note: Rapid detection and quantification of microbes related to microbiologically influenced corrosion using quantitative polymerase chain reaction. Corrosion 62 (11), 950–955.

(2007) Paints and varnishes—Corrosion protection of steel structures by protective paint systems International Standard ISO 12944—Part 1s to 8 International Organization for Standardization, Geneva.

(1992) Corrosion of metals and alloys—Corrosivity of atmospheres—Classification.

International Standard ISO 9223 International Organization for the Standardization, Geneva.

(2007) API Bulletin 91, First Edition American Petroleum Institute (API), June. API Publishing Services, Washington, DC.

(2005) *Drill Bits Newsletter, 15* (5). International Association of Drilling Contractors. May.

(2008) National Emphasis Program—Crystalline silica. Occupational Safety and Health Administration, Directive Number CPL 03-00-007, January 24, Washington DC.

(2008) Coatings inspection guidelines specific to inspectors employed by ProTec. Progressive Technical Services Inc. (ProTec). Information provided by David Miles, ProTec, Belle Chase, Louisiana.

(2007) QuikDeck suspended access system: The Access Advantage ThyssenKrupp Safway, Inc. Brochure ORN 1801, Rev A 6/07, Waukesha, WI.

Index

Printed and bound by CPI Group (UK) Ltd, Croydon, CR0 4YY

08/05/2025

01864800-0004